"*Knowledge is of two kinds. We know a subject ourselves, or we know where we can find information upon it.*"

JAMES BOSWELL
THE LIFE OF SAMUEL JOHNSON, 1791

The Complete Car Cost Guide
1993 Edition

Credits

Publisher
Peter Levy

Editor
Steve Gross

Text Editor
Peter Bohr

Editorial Staff
Vibha Bansal
Cheryl Ford Smith
Jamison J. Redding

Information Director
Charles A. Donaldson

Database Manager
Donald J. Oates

Database Supervisor
Deborah L. Eldridge

Database/Analysis
Jennifer Chu
Joannie M. Donaldson
Gary Grunwald
Dean Mello
Larry Montalvo
Anh Nguyen
Timothy Nugent
Timothy William Oates
Kristina A. Olson
Luiz Payne

Production
William Analla
Anna-Marie S. Herreria
Victoria Liou
Martha Flint Topor

Programming Manager
Timothy H. Denney

Programming
Stephen Smith

Art Director
Vincent Paul Mendolia

Photo Scanning
Gary Eldridge

Cover Photography
Jamie Krueger

Public Relations
Benton McMillian-Gordon

Thanks

A special thank you to:

Mr. Bryce Benjamin
Mr. Jonathan Bulkeley, Money Magazine
Mr. Walter Burns
Mr. Louis R. Cooper, Automotive Alliance LTD.
Mr. John Dinkel
Mr. Christopher L. Ford, Western National Warranty Corporation
Ms. Carol Jaech, Santa Clara County Library System
Mr. Tom Kempner
Mr. Wayne Oler
Mr. Ron Raymond
Mr. Steve Wood, GE Capital Fleet Services

Thanks to the public relations and marketing departments of all the manufacturers whose vehicles are represented in this book for their willingness to provide information and answer questions.

Thanks to the people of the United States Government, Department of Commerce, Department of Energy, and Department of Transportation for their information and support.

Thanks to Linotext.

Thanks to the technical support staff at ACI US for their technical information.

Finally, a special thanks to all of our customers for making this book a success.

Ordering Information

Are you in the market for a pickup truck, station wagon, mini or full-size van, or utility vehicle? If so, pick up a copy of *The Complete Small Truck Cost Guide*.

Copies of *The Complete Small Truck Cost Guide* and *The Complete Car Cost Guide* may be ordered directly from the publisher.

Orders should be sent to IntelliChoice, Inc., 1135 So. Saratoga/Sunnyvale Rd., San Jose, CA 95129.

Phone orders may be placed toll-free at 1-800-CAR-BOOK (1-800-227-2665). Call (408) 554-8711 for current information on pricing and availability of these books.

The Complete Small Truck Cost Guide
(ISBN 0-941443-16-7)

The Complete Car Cost Guide
(ISBN 0-941443-15-9)

Are you interested in a custom report comparing any two vehicles of your choice?
See the advertisement in the center of this book for details.

Notice

This book was compiled solely by IntelliChoice, Inc. IntelliChoice is an independent information research firm not sponsored by or associated with any automobile manufacturer. IntelliChoice has conducted research to obtain the information contained in *The Complete Car Cost Guide* in the manner IntelliChoice considers to be commercially reasonable from sources IntelliChoice considers to be reliable. IntelliChoice has made no independent verification of the accuracy of the information it has received. *The Complete Car Cost Guide* contains and is based upon the information actually known by IntelliChoice at the time the book was compiled. IntelliChoice makes no guaranty, warranty or representation with respect to the completeness, adequacy or accuracy of the information contained in *The Complete Car Cost Guide* other than as expressly set forth herein. All prices and specifications are subject to change without notice.

Restrictions

This publication may not be transmitted, reproduced, or quoted in whole or in part by mimeograph or any other printed means, by any electronic means, or for presentation on radio, television, videotape, or film without express written permission from the publisher.

Copyrights

Copyright © 1993 by IntelliChoice, Inc., 1135 So. Saratoga/Sunnyvale Road, San Jose, CA 95129. *The Complete Car Cost Guide* (ISBN 0-941443-15-9). Fourth-class postage paid at San Jose, CA. IntelliChoice and ArmChair Compare are registered trademarks of IntelliChoice, Inc. Just-the-Facts and the IntelliChoice logo (line drawing of a face in a circle) are trademarks of IntelliChoice, Inc.

The insurance information used herein includes copyrighted material of Insurance Services Office, Inc. with its permission. Copyright © 1993 by Insurance Services Office, Inc.

All manufacturer and product names and logos are trademarks or registered trademarks of their respective holders.

Production

This book was derived from IntelliCar™, a comprehensive automotive database developed with 4th Dimension® version 3.0 from ACI US Inc. 4th Dimension 3.0 is the most advanced relational database available for Macintosh computers. For information about 4th Dimension, contact ACI US at 408-252-4444.

This book was designed and produced using Macintosh® Quadra computer systems. Every page was printed directly onto film negatives. Photos were scanned using Adobe Photoshop® on a Microtek® scanning system.

Library of Congress Catalog Card Number: 88-83221

IntelliChoice, Inc.
1135 South Saratoga/Sunnyvale Road
San Jose, California 95129-3660
Phone (408) 554-8711
Fax (408) 253-4822

Printed and bound in the United States of America

To Our Continuing Readers

We have made a few changes to our cost calculations that may interest our continuing readers. The changes are as follows:

Engines

>Engine size is now also considered when determining which engine is used on a particular model for ownership cost calculations. In previous editions, horsepower was the only criterion.
>
>See page 53A for a more complete explanation.

Repairs

>The figures used for the "Repair" cost category are now based on service contract pricing. This allows more vehicles to have complete repair profiles, but results in slightly higher repair costs than in prior editions of this book.
>
>For further explanation of this approach, please see page 61A.

From the Publisher

Welcome to the seventh annual edition of *The Complete Car Cost Guide*, the only publication that helps you evaluate how much an automobile really "costs."

The fundamental principle of this book—that it's not the purchase price, but the amount it takes to own and operate a new vehicle that determines its true value—is more relevant than ever in 1993.

From a price perspective, 1992 was a buyer's dream. The economy was weak and fewer people were buying cars. Automakers, anxious to move inventory, responded by making great deals easily available to the fortunate few who were ready to buy.

This year, the economy is showing sparks of life, and more people are buying cars. Consequently, manufacturers are making more profitable deals. They can now afford to promote features other than price, and are equipping more models with anti-lock brakes, air bags, and other safety features. While it's definitely a step in the right direction, it makes selecting a car more difficult. With more to choose from, it's harder to choose.

What this means is, car buyers who want to strike a great bargain in 1993 must be more than just lucky. They must be prepared.

That's where *The Complete Car Cost Guide* can help.

This book is designed to prepare shoppers like you to choose and buy a car. To get you started, we've divided the book into three distinct areas. The first, AutoExplorer™, will help you select vehicles that fit your taste and budget. The second area guides you through selecting and purchasing a new car. Finally, the third section presents a comprehensive analysis of each car, helping you evaluate and compare all your potential choices.

While it may not be easy to find the car that's exactly right for you, I'm confident this book will help inform and prepare you to make an intelligent decision.

Good luck with your search!

And thank you for your ongoing and enthusiastic support of this book. We will always work hard to earn your respect.

Sincerely,

Peter S. Levy
Publisher

Table of contents

Short Cuts

11A	Insurance Comparison Worksheet
14A	Leasing
22A	The Ten-Minute Technical Inspection
38A	A Handy Guide to Options and Technical Features
43A	Six Steps To a Good Deal
69A	Best Overall Values
99A	Lease Interest Rates
108A	Convert a loan amount into a monthly payment
110A	Cash rebates vs. discount finance rates

xi 1993 AutoExplorer™

1A Introduction

5A Section 1: Ownership Costs

5A	**Standing Costs**
6A	**Depreciation**
8A	**Insurance**
9A	Liability Coverage
9A	Collision and Comprehensive Coverage
10A	Other Coverages
10A	Insurance Symbols
10A	Shop Around
12A	**Finance Costs**
12A	Pay Cash or Take Out a Loan?
13A	Striking a Balance: Monthly Payments vs. Interest Expense
13A	Finding Yourself "Upside Down"
14A	**Leasing**
14A	The Allure of Leasing
14A	The Catch
15A	A New Wrinkle
16A	The Fine Print of Leasing
17A	The Best Lease Deal
18A	Leasing and Taxes
18A	**State Fees**
19A	**Running Costs**
19A	**Fuel Costs**
19A	Is All Gas Alike?
20A	**Maintenance**
20A	Some Thoughts on Extended Service Intervals
21A	Keep on Shining
21A	Maintenance Warranties
22A	The Ten-Minute Technical Inspection
23A	**Repairs**
23A	Warranties
23A	Service Bulletins
24A	Service Contracts (Mechanical Breakdown Insurance)
24A	Numbing the Pain of Car Repair
24A	Reading your Garage Floor

29A Section 2: Choosing and Buying a Car

29A	**Checking Out a New Car**
30A	**Options and Special Features**
30A	Option packages
30A	**Choosing Safety**
30A	Anatomy of a Car Crash
31A	Choosing a safe Car
32A	Seatbelts
32A	Air Bags
33A	Car Size
33A	Structural Crashworthiness
34A	Side Impact Protection

Table of contents

34A	Head Restraints
34A	Crash Avoidance Features
36A	**What Price Should You Pay?**
37A	**Rebates and Sales Incentives**
38A	**A Handy Guide to Options and Technical Features**
40A	**When is the Best Time to Buy a New Car?**
40A	Best Time of Year
40A	Best Time of Month
40A	**Buying, Financing, and Selling Your Old Car**
41A	**Finance & Insurance Manager**
43A	**Six Steps to a Good Deal**
44A	**No Dicker Stickers**
45A	**Brokers**
45A	Broker Varieties
46A	**Keeping Your Old Car vs. Buying a New One**
47A	To Fix or Not to Fix
48A	**Unloading your Old Car**

51A Section 3: Annotated Vehicle Charts

52A	**Vehicle Description**
53A	**Purchase Price**
54A	**Warranty and Maintenance**
55A	**Ownership Costs**
56A	**Ownership Costs (Assumptions)**
57A	**Ownership Cost Derivation**
57A	Depreciation
58A	Insurance
59A	Finance Costs
59A	State Fees
59A	Fuel
60A	Maintenance
61A	Repairs
62A	**Insurance Adjustment Table**
63A	**Mileage Adjustment Table**
64A	**State Fees, Regulations, and Insurance Information**

1 Section 4: Vehicle Charts

67A Section 5: Appendices

69A	**Appendix A:**	**Best Overall Value**
83A	**Appendix B:**	**Highest/Lowest Cost**
87A	**Appendix C:**	**Financing Sources**
88A	**Appendix D:**	**Insurance Companies**
89A	**Appendix E:**	**Service Contract Locator**
90A	**Appendix F:**	**Better Business Bureau Locations**
96A	**Appendix G:**	**Automobile Manufacturers**
99A	**Appendix H:**	**Lease Interest Rates**
108A	**Appendix I:**	**Monthly Payments**
110A	**Appendix J:**	**Discount Financing**

119A Index

Figures

1A	1. Ownership Cost Comparison
7A	2. Depreciation Comparison
8A	3. Insurance Coverages
10A	4. Insurance Symbols
11A	5. Insurance Worksheet
13A	6. Interest Expense
15A	7. Leasing Comparison
38A	8. A Handy Guide To Options and Technical Features
46A	9. Projected Five-year Comparison
56A	10. Ownership Cost (Assumptions)
62A	11. Insurance Adjustment Table
63A	12. Mileage Adjustment Table
64A	13. State Fees, Regulations, and Insurance Information

The Complete Car Cost Guide

1993 AutoExplorer™

All prices listed below are based on the manufacturer's suggested retail price and do not include factory or dealer installed options. Actual prices may vary due to rebates and local dealer markups.

Subcompact	Under $8,500	$8,501–$10,500	$10,501–$15,000	Over $15,000
Dodge	Colt (C)	Colt (S) Colt GL (C,S) Shadow (H) Shadow ES (H)	Daytona (H) Daytona ES (H) Shadow ES (CN) Shadow Highline (CN)	
Eagle	Summit DL (C)	Summit DL (S) Summit ES (C,S)	Talon DL (C) Talon ES (C)	Talon TSi Turbo (C) Talon TSi Turbo AWD (C,4)
Ford	Escort Pony (H) Festiva GL (H) Festiva L (H)	Escort LX (H,S)	Escort GT (H) Escort LX-E (S)	
Geo	Metro (H) Metro XFi (H)	Metro LSi (CN,H) Prizm (S)	Prizm LSi (S) Storm 2+2 (C) Storm 2+2 GSi (C)	
Honda			Civic del Sol S (C) Civic del Sol Si (C)	
Hyundai	Excel (H,S) Excel GS (H)	Elantra (S) Elantra GLS (S) Excel GL (S) Scoupe (C) Scoupe LS (C)	Scoupe Turbo (C)	
Mazda	323 (H)	323 SE (H)	MX-3 (C) MX-3 GS (C) Protege DX (S) Protege LX (S)	
Mercury		Tracer (S)	Capri (CN) Tracer LTS (S)	Capri XR2 (CN)
Mitsubishi	Mirage S (C)	Mirage ES (C,S) Mirage LS (C) Mirage S (S)	Eclipse (C) Eclipse GS (C) Eclipse GS 16V (C) Mirage LS (S)	Eclipse GS 16V Turbo (C) Eclipse GSX 16V Turbo AWD (C,4)
Nissan		Sentra E (S)	240SX (C) NX 1600 (C) NX 2000 (C) Sentra GXE (S) Sentra SE (S 2 Door) Sentra SE-R (S 2 Door) Sentra XE (S)	240SX (CN,H) 240SX SE (C,H)
Plymouth	Colt (C)	Colt (S) Colt GL (C,S) Sundance (H)	Laser (H) Laser RS (H) Sundance Duster (H)	Laser RS Turbo (H) Laser RS Turbo AWD (H)
Subaru	Justy (H)	Justy GL (H) Justy GL 4WD (H,4), 4 Door Justy GL 4WD (H,4) Loyale (S)	Loyale 4WD (S,4)	
Suzuki	Swift GA (H,S)	Swift GS (S) Swift GT (H)		

C—Coupe   S—Sedan   H—Hatchback   CN—Convertible   4—4-Wheel Drive

The Complete Car Cost Guide

All prices listed below are based on the manufacturer's suggested retail price and do not include factory or dealer installed options. Actual prices may vary due to rebates and local dealer markups.

Subcompact (Cont'd)	Under $8,500	$8,501-$10,500	$10,501-$15,000	Over $15,000
Toyota	Tercel (S 2 Door)	Tercel DX (S)	Corolla (S) Corolla DX (S) Paseo (C) Tercel LE (S)	Corolla LE (S)
Volkswagen	Fox (S 2 Door)	Fox GL (S)		Cabriolet (CN) Cabriolet Classic (CN)

Compact	Under $12,000	$12,001-$15,000	$15,001-$18,000	Over $18,000
Acura		Integra LS (H) Integra RS (H,S)	Integra GS (H,S) Integra LS (S) Integra LS Special (H)	Integra GS-R (H)
Buick		Skylark Custom (C,S) Skylark Limited (C,S)	Skylark Gran Sport (C,S)	
Chevrolet	Beretta (C) Cavalier RS (C,S) Cavalier VL (C,S) Corsica LT (S)	Beretta GT (C) Cavalier Z24 (C)	Beretta GTZ (C) Cavalier RS (CN)	Cavalier Z24 (CN)
Chrysler		LeBaron (C) LeBaron LE (S)	LeBaron (CN) LeBaron GTC (C) LeBaron Landau (S) LeBaron LX (C)	LeBaron GTC (CN) LeBaron LX (CN) New Yorker Salon (S)
Dodge	Spirit Highline (S)	Dynasty (S) Spirit ES (S)	Dynasty LE (S)	
Ford	Mustang LX (C,H) Tempo GL (S)	Probe (H) Tempo LX (S)	Mustang LX (CN) Probe GT (H)	
Honda	Civic CX (H) Civic DX (C,H,S) Civic LX (S) Civic VX (H)	Civic EX (C) Civic Si (H)	Civic EX (S) Prelude S (C)	Prelude Si (C)
Hyundai		Sonata (S) Sonata GLS (S)		
Infiniti				G20 (S)
Mazda		626 DX (S)	626 LX (S) MX-6 (C)	626 ES (S) MX-6 LS (C)
Mercury	Topaz GS (S)			
Mitsubishi		Galant S (S)	Galant ES (S) Galant LS (S)	
Nissan		Altima GXE (S) Altima XE (S)	Altima SE (S)	Altima GLE (S) Maxima GXE (S) Maxima SE (S)

C—Coupe S—Sedan H—Hatchback CN—Convertible 4—4-Wheel Drive

1993 AutoExplorer™

The Complete Car Cost Guide

All prices listed below are based on the manufacturer's suggested retail price and do not include factory or dealer installed options. Actual prices may vary due to rebates and local dealer markups.

Compact (Cont'd)

	Under $12,000	$12,001-$15,000	$15,001-$18,000	Over $18,000
Oldsmobile		Achieva S (C,S) Achieva SL (C,S)		
Plymouth	Acclaim (S)			
Pontiac	LeMans SE (S) LeMans SE Aerocoupe (H) LeMans Value Leader Aerocpe (H) Sunbird LE (C,S) Sunbird SE (C,S)	Grand Am GT (C,S) Grand Am SE (C,S) Sunbird GT (C)	Sunbird SE (CN)	
Saab				900 S (CN,H,S) 900 Turbo (CN,H)
Saturn	SC1 (C) SL (S) SL1 (S) SL2 (S)	SC2 (C)		
Subaru			Legacy L (S) Legacy L AWD (S,4)	Legacy LS (S) Legacy LS AWD (S,4) Legacy LSi AWD (S,4) Legacy Sport AWD (S,4)
Toyota		Celica ST (C)	Camry DX (S) Camry DX V6 (S) Camry LE (S) Celica GT (C,H)	Camry LE V6 (S) Camry SE V6 (S) Camry XLE (S) Camry XLE V6 (S) Celica All-Trac Turbo (H,4) Celica GT (CN) Celica GT-S (H)
Volkswagen			Passat GL (S)	Passat GLX (S)
Volvo				240 (S)

Luxury

	Under $35,000	$35,001-$40,000	$40,001-$55,000	Over $55,000
Acura	Legend (S) Legend L (C,S)	Legend LS (C,S)		
Alfa Romeo	164 L (S) 164 S (S)			
Audi	100 (S) 100 S (S) 90 CS (S) 90 CS Quattro Sport (S,4) 90 S (S)	100 CS (S)	100 CS Quattro (S,4) 100 CS Quattro (W,4) S4 (S,4)	V8 Quattro (S,4)
BMW	325 i (S) 325 iS (C)	325 iC (CN) 525 i (S) 525 i Touring (W)	535 i (S) 740 i (S)	740 iL (S) 750 iL (S)
Buick	Park Avenue (S) Park Avenue Ultra (S)			

C—Coupe   S—Sedan   H—Hatchback   CN—Convertible   4—4-Wheel Drive

1993 AutoExplorer™

The Complete Car Cost Guide

All prices listed below are based on the manufacturer's suggested retail price and do not include factory or dealer installed options. Actual prices may vary due to rebates and local dealer markups.

Luxury (Cont'd)

	Under $35,000	$35,001-$40,000	$40,001-$55,000	Over $55,000
Cadillac	DeVille Coupe (C) DeVille Sedan (S) Eldorado (C) Fleetwood (S)	DeVille Touring Sedan (S) Seville (S) Sixty Special Sedan (S)	Seville Touring Sedan (S)	Allante (CN) Allante HT (CN)
Chrysler	Imperial (S)			
Infiniti	J30 (S)		Q45 (S)	
Jaguar			XJ6 (S) XJS (C)	XJ6 Vanden Plas (S) XJS (CN)
Lexus	ES 300 (S) SC 300 (C)		LS 400 (S) SC 400 (C)	
Lincoln	Continental Executive (S) Town Car Executive (S)	Continental Signature (S) Mark VIII (C) Town Car Cartier Designer (S) Town Car Signature (S)		
Mercedes Benz	190E 2.3 (S) 190E 2.6 (S)		300 D 2.5 Turbo (S) 300 E (S) 300 E 2.8 (S) 300 TE (W)	300 CE (C) 300 CE (CN) 300 E 4Matic (S,4) 300 SD (S) 300 SE (S) 300 SL (CN) 300 TE 4Matic (W,4) 400 E (S) 400 SEL (S) 500 E (S) 500 SEL (S) 500 SL (CN) 600 SEL (S)
Oldsmobile	Ninety-Eight Regency (S) Ninety-Eight Regency Elite (S) Ninety-Eight Touring Sedan (S)			
Saab	9000 CDE (S) 9000 CDE Turbo (S) 9000 CSE (H)	9000 CSE Turbo (H)		
Volvo		960 (S) 960 (W)		

Midsize/Large

	Under $16,000	$16,001-$19,000	$19,001-$23,000	Over $23,000
Acura				Vigor GS (S) Vigor LS (S)
BMW				318 i (S) 318 iS (C)
Buick	Century Custom (C,S) Century Special (S)	Century Limited (S) Regal Custom (C,S) Regal Limited (C,S)	LeSabre Custom (S) LeSabre Limited (S) Regal Gran Sport (C,S) Roadmaster (S)	Riviera (C) Roadmaster Limited (S)

C—Coupe   S—Sedan   H—Hatchback   CN—Convertible   4—4-Wheel Drive

The Complete Car Cost Guide

All prices listed below are based on the manufacturer's suggested retail price and do not include factory or dealer installed options. Actual prices may vary due to rebates and local dealer markups.

Midsize/Large (Cont'd)	Under $16,000	$16,001-$19,000	$19,001-$23,000	Over $23,000
Chevrolet	Lumina (C,S) Lumina Euro (C,S)	Caprice Classic (S) Lumina Z34 (C)	Caprice Classic LS (S)	
Chrysler		Concorde (S)	New Yorker Fifth Avenue (S)	
Dodge	Intrepid (S)	Intrepid ES (S)		
Eagle		Vision ESi (S)	Vision TSi (S)	
Ford	Taurus GL (S) Thunderbird LX (C)	Taurus LX (S)	Crown Victoria (S) Crown Victoria LX (S) Thunderbird Super Coupe (C)	Taurus SHO (S)
Honda	Accord DX (C,S)	Accord EX (C,S) Accord LX (C,S)	Accord SE (C,S)	
Mazda				929 (S)
Mercury	Cougar XR7 (C)	Sable GS (S) Sable LS (S)	Grand Marquis GS (S) Grand Marquis LS (S)	
Mitsubishi			Diamante ES (S)	Diamante LS (S)
Oldsmobile	Cutlass Ciera S (S) Cutlass Supreme S (C,S)	Cutlass Ciera SL (S)	Cutlass Supreme Conv. (CN) Cutlass Supreme Int'l (C) Cutlass Supreme Int'l (S) Eighty-Eight Royale (S) Eighty-Eight Royale LS (S)	
Pontiac	Grand Prix LE (S) Grand Prix SE (C)	Grand Prix SE (S)	Bonneville SE (S) Grand Prix GT (C) Grand Prix STE (S)	Bonneville SSE (S) Bonneville SSEi (S)
Saab				9000 CD (S) 9000 CD Turbo (S) 9000 CS (H) 9000 CS Turbo (H)
Volvo				850 GLT (S) 940 (S) 940 Turbo (S)

Sport	Under $20,000	$20,001-$35,000	$35,001-$60,000	Over $60,000
Acura				NSX (C)
Alfa Romeo		Spider (CN) Spider Veloce (CN)		
BMW				850 Ci (C) M5 (S)
Chevrolet		Corvette (C)	Corvette (CN)	Corvette ZR-1 (C)
Dodge	Daytona IROC (H) Daytona IROC R/T (H) Stealth (C)	Stealth ES (C) Stealth R/T (C) Stealth R/T Turbo AWD (C,4)	Viper RT/10 (CN)	

C—Coupe   S—Sedan   H—Hatchback   CN—Convertible   4—4-Wheel Drive

The Complete Car Cost Guide

All prices listed below are based on the manufacturer's suggested retail price and do not include factory or dealer installed options. Actual prices may vary due to rebates and local dealer markups.

Sport (Cont'd)	*Under $20,000*	*$20,001-$35,000*	*$35,001-$60,000*	*Over $60,000*
Ford	Mustang GT (H) Mustang LX 5.0L (C,H)	Mustang GT (CN) Mustang LX 5.0L (CN)		
Mazda	MX-5 Miata (CN)	RX-7 (C)		
Mitsubishi		3000GT (C) 3000GT SL (C)	3000GT VR-4 (C,4)	
Nissan		300ZX (C) 300ZX 2+2 (C)	300ZX (CN) 300ZX Turbo (C)	
Porsche			911 RS America (C) 968 (C) 968 Cabriolet (CN)	911 America (CN) 911 Carrera 2 (C) 911 Carrera 2 Cabriolet (CN) 911 Carrera 2 Targa (C) 911 Carrera 4 (C,4) 911 Carrera 4 Cabriolet (CN,4) 911 Carrera 4 Targa (C,4) 928 GTS (C)
Toyota	MR2 (C)	MR2 Turbo (C)		
Volkswagen		Corrado SLC (C)		

Wagons	*Under $12,000*	*$12,001-$15,000*	*$15,001-$19,000*	*Over $19,000*
Audi				100 CS Quattro (W,4)
BMW				525 i Touring (W)
Buick		Century Special (W)	Century Custom (W)	Roadmaster Estate Wagon (W)
Chevrolet	Cavalier RS (W) Cavalier VL (W)			Caprice Classic (W)
Eagle	Summit DL (W)	Summit AWD (W,4) Summit LX (W)		
Ford	Escort LX (W)		Taurus GL (W)	Taurus LX (W)
Honda			Accord LX (W)	Accord EX (W)
Mercedes Benz				300 TE (W) 300 TE 4Matic (W,4)
Mercury	Tracer (W)		Sable GS (W)	Sable LS (W)
Mitsubishi	Expo LRV (W)	Expo (W) Expo AWD (W,4) Expo LRV AWD (W,4) Expo LRV Sport (W)	Expo SP (W) Expo SP AWD (W,4)	
Oldsmobile		Cutlass Ciera Cruiser S (W)	Cutlass Ciera Cruiser SL (W)	
Plymouth	Colt Vista (W)	Colt Vista AWD (W,4) Colt Vista SE (W)		

C—Coupe   S—Sedan   H—Hatchback   CN—Convertible   4—4-Wheel Drive

The Complete Car Cost Guide

All prices listed below are based on the manufacturer's suggested retail price and do not include factory or dealer installed options. Actual prices may vary due to rebates and local dealer markups.

Wagons (Cont'd)	Under $12,000	$12,001-$15,000	$15,001-$19,000	Over $19,000
Saturn	SW1 (W)	SW2 (W)		
Subaru	Loyale (W)	Loyale 4WD (W,4)	Legacy L (W) Legacy L AWD (W,4)	Legacy LS (W) Legacy LS AWD (W,4) Legacy LSi AWD (W,4) Legacy Touring Wagon AWD (W,4)
Toyota		Corolla DX (W)	Camry DX (W)	Camry LE (W) Camry LE V6 (W)
Volkswagen				Passat GLX (W)
Volvo				240 (W) 940 (W) 940 Turbo (W) 960 (W)

C—Coupe   S—Sedan   H—Hatchback   CN—Convertible   4—4-Wheel Drive

1993 AutoExplorer™

The Complete Car Cost Guide

Introduction

An inexpensive car to buy could be expensive to own.

Do you want to get a good value the next time you buy a new car? Of course. Why else would you be reading this book?

What does "good value" mean anyway? Most people think it means choosing a nice-looking, reliable car, and negotiating the lowest possible price for it. So they trek through dealerships comparing car features and prices.

This may be a reasonable sounding approach, but it doesn't have much to do with getting a good value. That's because ultimately a car's value has less to do with its initial price than it does with its cost to own and operate.

You don't believe it? Let's take a closer look. The price you pay to buy a car is not a cost. "Price" and "cost" are related, but there's a distinct difference between them. A car's price is simply an amount of your hard-earned cash that you must exchange to own the car. There's no expense involved at the time of the purchase because, at this point, the car is still worth exactly what you just paid for it. If you paid the dealer $10,000, he or she will have the money, but you will have something of equal value — the car.

Ah, but as soon as you slip behind the wheel, turn on the stereo, drive home, and park your new car in the driveway hoping to attract envious glances from the neighbors, you'll start to run up a tab. The car will begin to depreciate. It will use fuel. You'll have to pay insurance premiums, state fees and taxes, finance charges, and at some point down the road, repair bills. These are the car's ownership costs. When you want the best value, it's more prudent to consider these costs than just the purchase price.

By the time the typical car is five years old, the cost to own and operate it will exceed its original purchase price. Even more important, similarly priced cars can have very different ownership costs over a five-year period *(See Figure 1)*. An inexpensive car to buy could be an expensive car to own.

That's why it's important to understand ownership costs when you search for the car that will truly be the best value.

The Complete Car Cost Guide is unlike any other reference source. It's the only one that gives you a complete financial profile of today's new cars and detailed cost projections for five years into the future.

Figure 1

Ownership Cost Comparison

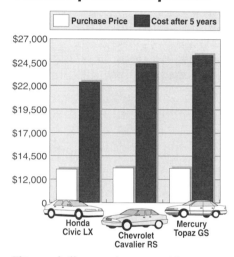

This graph illustrates how cars with a similar purchase price can have different ownership costs over time. In this example, the Topaz owner will spend about $1,000 more than the Cavalier owner and over $4,000 more than the Civic owner in the first five years. (Prices are based on "Total Target Price" listed in the vehicle charts, beginning on page 1.)

Sources:

IntelliChoice, Inc. is not associated with any automobile manufacturer or supplier. We strive to be completely impartial in our analyses and conclusions. Our primary responsibility is to assist the new car shopper.

The research used to compile the vehicle data and ratings in this book has been done in entirety by IntelliChoice, Inc. Where appropriate, we make use of information supplied by the United States Government, by automobile manufacturers, and by a variety of industry sources. If you have technical questions about any area of this book, please write to the Director of Research, IntelliChoice, Inc., 1135 S. Saratoga/Sunnyvale Road, San Jose, CA 95129.

Specifically, this book provides you with:

Detailed ownership costs. Allows you to compare the costs of owning different makes and models so you can choose the best value.

Expected annual costs for five years. Helps you budget your annual expenses.

Expected resale value for five years. Allows you to negotiate a favorable agreement if you plan to lease a car.

Dealer cost and list price. Helps you to negotiate the best possible deal for your new car.

Warranty information. Lets you see how well your car is protected by the manufacturer.

Advice. Gives you tips on selecting options, on choosing a dealer, on maintaining your car's resale value, on financing or leasing your new car, and on selling your old car.

Ratings. Lets you see at a glance which cars deliver the best overall value.

The Complete Car Cost Guide is an economic guide. It doesn't offer subjective reviews of a car's comfort, handling, or style. It doesn't list braking or zero-to-sixty miles-per-hour performance figures. Nor does it provide the results of government crash tests.

Of course performance, comfort, handling, style, and safety are important factors in buying a new car. But then, if you can't afford to buy and operate the car you want, nothing else really matters, does it?

Section One
Ownership Costs

Ownership Costs

Understanding ownership costs can save you money. Regardless of what you may read in new car reviews or hear in new car ads, it's not possible to know whether a vehicle is a good or a poor value until you learn about all the major ownership costs: depreciation, insurance, financing costs, maintenance, repairs, and more.

For example, a higher-priced car that holds its value over the years may be a bargain compared to a lower-priced car that plunges in value like Christmas tree ornaments on December 26th. In just one year, depreciation could eliminate the apparent initial advantage of the car with the lower purchase price. However, even if the lower-priced vehicle holds its value well, it may still cost a fortune to insure. So which is the better value?

Here's a more specific example. You may be told that a particular vehicle, such as a Ford Festiva, is very economical because it gets such great gas mileage. Beware! A Festiva may get good mileage but, since it rates average or below average in depreciation and insurance, its *overall* economy rates only average compared to other cars in its class. You cannot determine the overall value of a car, or any savings you may realize in selecting one car over another, without considering all the ownership costs. Information on overall economy is what this book provides.

All the major ownership costs can be divided into two categories: standing costs and running costs. Standing costs, also called fixed costs, are what you pay to own the car whether you ever take it out of the driveway or not. Running costs, also called variable costs, are the expenses you incur as you use the car.

Standing Costs

Standing costs include depreciation, insurance, financing, and various government fees. (We do not factor into our analysis the cost of garaging your car, but if you live in Manhattan or downtown San Francisco, you should keep in mind that this expense can add substantially to your car's standing cost.)

The good news about standing costs is that they usually decrease with time. Depreciation is less each year, insurance premiums may decline as the value of the car decreases, state fees usually decrease on older cars, and finance charges will be eliminated once you pay off the loan.

The seven major ownership categories:

- *Depreciation*
- *Insurance*
- *Financing*
- *State fees*
- *Fuel*
- *Repairs*
- *Maintenance*

The Complete Car Cost Guide

Depreciation Tips

Before You Buy

* *Choose a car with low expected depreciation. Let the vehicle charts beginning on page 1 in this book be your guide.*

* *Select a popular, highly regarded model, or a car from a prestigious automaker. Avoid offbeat or less well-known makes and models.*

* *Choose a car from an automaker that doesn't frequently change body styles. If you are set on buying a vehicle that does have frequent style changes, make sure to buy soon after a change.*

* *Select a larger engine if it's an option offered on the car you want. A larger engine generally adds to a car's resale value.*

* *Select appropriate features and options for your geographic area. Cars in southern California or Florida are expected to have air conditioning, for instance. High-quality stereo sound systems almost always increase a car's value. And keep in mind that pink cars don't sell well anywhere.*

* *If a manufacturer is guaranteeing the resale value of a vehicle you are considering, make sure you understand exactly what's being guaranteed. For instance, how is the "guaranteed" resale value determined? How long is the guarantee? Do you have to buy the same vehicle in order to receive your guaranteed resale value?*

* *Consider buying a convertible. As they say in the car-selling trade, when the top goes down, the price — even the resale price — goes up (Caveat: Keep in mind that other ownership costs may be very high on a convertible.)*

* *Find out if the car you're considering has a "sister" model, a near-identical car sold as a different brand or model. If possible, choose the one that holds its value better. For instance, the 1991 Plymouth Laser RS sold for more than its sister, the Mitsubishi Eclipse GS, but the Eclipse GS is worth more today.*

Depreciation

This is the big one — at least for most cars during the first few years they're on the road. In fact, some cars depreciate as much as 20 percent or more during the first year alone.

Depreciation is the amount something decreases in value over time. Although it's a very real cost, depreciation is tricky because you don't get a monthly bill for it. You don't continually dole out money from your pocket to pay depreciation as you do, for example, every time you visit your mechanic.

But, make no mistake about it; you will pay for depreciation when the time comes to sell your car and buy another one. For instance, let's assume you purchase a car in 1993 for $15,000. If you unload it five years later for $5,000, you'll have spent $10,000 in depreciation to own the car.

The cost of depreciation should be a primary factor in your choice of which car to buy. If you expect to keep the car five years or less, it may be the most important factor.

Not all cars depreciate as much or at the same rate. Often there's even a difference between two- and four-door versions of the same model. And why is this? It's one of life's little mysteries, but here are some likely reasons:

Cars that capture the public's fancy tend to hold their value. Many people would like to own a new Mercedes-Benz or Porsche, but few can afford to. So there's strong demand for used cars of these brands, which keeps resale values up.

Cars with an obscure or unfavorable public image, or cars commonly thought trouble-prone or costly to maintain generally have lower resale prices. Audi has not fully recovered from unfavorable publicity caused by sudden acceleration problems.

When the new-car price of a popular model increases, the price of used versions of the same car may also rise. The price of a used car is based, in part, upon the price of its new model. Therefore, when the price of a new model has a significant increase relative to other new vehicles, used versions of the same model tend to hold their value. However, this is not always true when all prices are rising. According to the *Wall Street Journal*, while new car prices have risen an average of $3,000 since 1987, used car prices rose from 1987 to 1991 an average of just $400.

Some manufacturers actively intervene in the used car market to support the resale value of their cars. For example, some manufacturers have programs that guarantee the resale value of selected models. Also, most manufacturers have made changes to the ways they market their late-model fleet cars in an attempt to support resale values.

Quotas often result in higher prices for popular new cars, and eventually for used cars as well. Due to export quotas, some Japanese car models are in short supply. Basic supply and demand suggests that this will increase the prices of these vehicles, both in the new and used markets.

A model change can lower resale value. When a model is dropped or completely restyled by a carmaker, cars of that model already on the road are likely to decrease in value more than they would otherwise. For example, Toyota redesigned the Celica for the 1990 model year. According to *Kelley Blue Book*, a 1990 Celica ST Coupe was worth 84% of its original price in January of 1992. However, a 1989 Celica ST Coupe was worth just 78% of its original price in January of 1991. The redesigned 1990 model held its value much better than the 1989.

A car's depreciation will vary depending on whether the car accumulates either very high or very low mileage for its age. The "normal" annual mileage range is generally considered to be between 10,000 and 15,000 miles.

The "base" model in a line-up usually holds its value better than the higher priced models. A "GL" edition may cost several thousand dollars more than the base edition when new, but may have a resale value of just a few hundred dollars more after two or more years.

A vehicle's class is often an indicator of its resale value. Subcompacts and luxury vehicles typically hold their values better than compact, midsize, or larger cars. However, if getting the highest resale value is your goal, you're probably better off with a van, utility vehicle, or pickup truck. (You may want to consult *The Complete Small Truck Cost Guide* for details on depreciation and other ownership costs of these vehicles.)

Depreciation Tips

Before You Buy continued

• Consider buying a one or two-year-old model, rather than brand new. Odd advice coming from a new car book, but it could be worth the price of this book a hundred times over. Since depreciation expense is greatest in a car's first year, you can save thousands of dollars by waiting a year. The best cars to purchase after a year are those that depreciate most in that first year. For instance, you may only save about $500 or $600 by buying a one-year-old Honda Accord, whereas you may save more than $3,000 by buying a one-year-old Chrysler Lebaron. The charts in this book show which models are best suited for this strategy.

• Maybe you shouldn't buy a new car at all! (See section "Keeping Your Old Car vs. Buying a New One" on page 46A.)

After You've Bought

• Keep your car longer. Cars typically depreciate the most during the first three years; after five years, depreciation will be minimal.

• When it comes time to sell your car, don't trade it in to a dealer. You'll almost always get a higher price — and minimize depreciation — by selling it yourself to a private party.

• Perhaps most importantly, keep up the mechanical condition and appearance of your car and save all receipts for maintenance work that you've done on your car. Low mileage "cream puffs," no matter what make or model, will always have plenty of buyers and will command the highest resale value. (See maintenance tips on page 21A.)

Figure 2
Depreciation Comparison

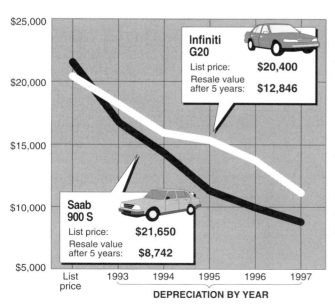

This graph illustrates how two cars with a similar purchase price have vastly different depreciation rates and, therefore, have substantially different resale values after 5 years. However, even though one car appears to be a better value than the other, you should not draw that conclusion until you have examined all of the major cost areas.

Insurance Tips

Before You Buy

• *Choose your car carefully. Sports and specialty cars generally have the most costly claims records and are also favorites of thieves. Insurers frown on cars that offer less protection to drivers and passengers during an accident—usually small or midsize imported cars. Insurers are also wary of cars with powerful engines that could get unskilled drivers into trouble; companies often tag as a high-performance vehicle any car that has a turbocharger or a 0-60 mph acceleration time of under 7 or 8 seconds. Finally, many insurers would rather avoid cars with price tags of $35,000 or more because of their high replacement cost.*

• *Make a package deal. When you insure more than one car with the same company, you will usually qualify for a 10-25 percent discount off the total premium for each car. And if you insure your home with the same company, you may get a discount of up to 15 percent off the total.*

• *Ask your state insurance regulator for consumer information on auto insurance companies, particularly if you're considering purchasing insurance from an unfamiliar company. You can check with your local Better Business Bureau to see if any complaints have been filed against companies you are considering. (Appendix F on page 90A lists Better Business Bureau locations.) About 30 states provide price or complaint information on auto insurers, according to Consumer Insurance Interest Group. You can find the phone number of your insurance regulator by calling your state's Department of Motor Vehicles, listed in the white pages.*

• *Buy a car with an airbag and an anti-theft device. The former could save you 10-40 percent on the bodily injury portion of your premium, while the latter could save you 5-25 percent on the comprehensive and collision portion. Anti-lock brakes also qualify for discounts with some insurers.*

Insurance

Ugh — insurance! The only thing worse than paying for it is not having it when you need it.

Insurance is the price you pay to protect yourself, your passengers, and your car against unexpected losses. It's required by law, and if you've ever insured a car, you know it makes up a formidable portion of your total ownership expense. In fact, the typical automobile insurance premium increased at almost twice the rate of other consumer prices during the last decade. Unlike depreciation, insurance is a very visible expense — your insurance bill is a regular visitor to your mailbox.

Although you pay a single auto insurance premium, you're actually protecting several entities: your car (that's the collision and comprehensive part of your bill), and yourself and others (the bodily injury and property damage parts).

In most cases, where you live and the type of car you own will determine the price of your collision and comprehensive coverage. The price of your bodily injury and property damage coverage (also known as "liability" coverage) depends on your driving record, as well as on who you are, where you live, and how frequently you drive your car.

Figure 3

Insurance Coverages

Type of coverage	What it covers
Collision	Pays for damages to your car that result from a collision with another vehicle or object.
Comprehensive	Pays for damages to your vehicle other than those caused in a collision (e.g. flood, fire, hail, theft, or vandalism).
Personal Liability	Pays claims against you, and covers the cost of legal defense if your car damages property or injures or kills someone in an accident.
Property Damage	Pays for legal defense and claims if your car damages someone else's property.
Medical Payments	Pays for medical expenses of the driver and passengers of your car in an accident.
Un/Underinsured Motorist	Pays for injuries caused by an uninsured or hit-and-run driver.

Ownership Costs

Though the price of auto insurance is high, by shopping around, by eliminating coverage you don't need, and by taking no more than necessary on the rest, you may be able to trim your bill by hundreds of dollars.

Liability Coverage

(about 60% of your premium)

When it comes to insuring yourself against damage or harm to others, you certainly don't want to skimp, leaving yourself vulnerable to a lawsuit that could take away your assets. Most states require you to buy bodily injury and property damage liability coverage. While you may be able to satisfy the law with bodily injury coverage that pays $20,000 for each person you injure, up to $40,000 per accident, and property damage coverage that pays $5,000, these are minimum figures.

Most insurance advisors recommend bodily-injury coverage of at least $100,000 a person up to $300,000 an accident, and property damage coverage of $50,000. If your net worth is more than $300,000, you should carry coverage of $200,000 a person and $500,000 an accident, and perhaps an "umbrella" policy that will bolster coverage even further. A $1 million umbrella policy will typically add $120-$150 to your yearly bill.

You won't want to skimp on uninsured or underinsured motorist coverage. This pays for injuries to your passengers and for expenses health plans don't cover if you are involved in an accident with an uninsured driver. Keep in mind that one in ten drivers across the country doesn't have any auto insurance.

On the other hand, don't buy more coverage than your net worth requires. If you don't own a home and you've wiped out your bank account to put a down payment on a new Hyundai, you may not need that $1 million umbrella policy.

Collision and Comprehensive Coverage

(about 40% of your premium)

Many people can also save on other parts of their policy. Collision coverage pays for damages to your car caused by an accident, while comprehensive coverage pays for damages to your car caused by other risks, such as theft or fire. Unlike liability coverage, collision and comprehensive coverage is often subject to a deductible — an amount you must pay before you can collect.

Having a high deductible can save you substantial sums on the collision and comprehensive portion of your total insurance bill. For example, one insurer charges $99 a year more for a policy with a $300 deductible than for one with a $500 deductible. This means you'd pay almost $100 per year to save, at most, $200 if you have an accident. But if you instead take that $100 and put it in the bank each year, you'd soon have enough to pay the full deductible in the event of a collision.

Because you don't want to collect on minor mishaps and risk raising your premiums, you'll want a high deductible anyway. Collecting $50 beyond a $250 deductible to replace a $300 cracked windshield could raise your total premium by several hundred dollars.

Insurance Tips

After You've Bought

• *Stay clean. No single factor is more important than your motor vehicle report (MVR) in determining the premiums for your liability coverage. Each time you receive a traffic ticket, you risk increasing your future premiums. In general, you will pay substantially higher premiums if you receive more than one or two tickets during a three-year period. If your state allows you to attend traffic school in exchange for removing a violation from your MVR, take advantage of the option.*

• *Stay a stranger to your insurance company. Claims, as well as tickets, upset insurers. Don't involve your insurance company in repairing door dings or annoyances like cracked windshields even if you could collect a little something beyond your deductible. Pay for these repairs out of your own pocket. If someone damages your car and you don't live in a no-fault insurance state, try to collect from the other person's insurance company before you involve your own insurer. Obtain a police report that indicates you were not at fault in the accident.*

• *If you own a business, put your car in your company's name. You may be able to persuade the insurer who writes your business policy to cover the car as a company vehicle at a lower rate than you would receive on a personal policy.*

• *Get married. Insurers prefer drivers between the ages of 30 and 65. Some insurers give discounts to drivers aged 55-65. But once you're over 70, you are considered risky, as you are if you're under 30. However, insurers generally treat a married guy or gal under 30 years old as they would someone in the lower risk 30-65 age group. Marriage, they figure, puts an end to your partying days and keeps you off the streets at night.*

Insurance Tips

After You've Bought continued

• *Every time you receive your insurance bill, review your coverage. Make sure your bill is accurate and that you're receiving all discounts to which you're entitled. If there's been a change in your insurance profile (e.g., you've just turned 30), inform your insurer. If your car is over five or six years old, consider eliminating your collision coverage.*

Figure 4

Insurance Symbols

Approximate Original Sticker Price	Insurance Symbol*
0 — $6,500	1
$6,501 — $8,000	2
$8,001 — $9,000	3
$9,001 — $10,000	4
$10,001 — $11,250	5
$11,251 — $12,500	6
$12,501 — $13,750	7
$13,751 — $15,000	8
$15,001 — $16,250	10
$16,251 — $17,500	11
$17,501 — $18,750	12
$18,751 — $20,000	13
$20,001 — $22,000	14
$22,001 — $24,000	15
$24,001 — $26,000	16
$26,001 — $28,000	17
$28,001 — $30,000	18
$30,001 — $33,000	19
$33,001 — $36,000	20
$36,001 — $40,000	21
$40,001 — $45,000	22
$45,001 — $50,000	23
$50,001 — $60,000	24
$60,001 — $70,000	25
$70,001 — $80,000	26
$80,001 and above	27

*The higher the symbol, the more costly to insure.

Other Coverages

Medical coverage pays hospital and doctor fees for the driver and passengers. But if your health insurance already covers these things, you may not need medical coverage included with your auto insurance.

However, if you live in a state with no-fault insurance, you may be required to buy personal-injury protection that covers your medical bills. But again, you may be able to cut some of the costs for this coverage if your heath plan covers you in an auto accident.

There are a number of other minor types of auto insurance, like towing coverage. If you're a member of an auto club, you may not need towing coverage, for instance.

Insurance Symbols

Insurers use a car rating system to determine the premium price for collision and comprehensive coverage. For many insurers, the symbols are compiled by Insurance Services Office in New York, NY.

Every car is given a rating symbol between 1 and 27; the higher the rating number, the more costly the premium. All things being equal, two identically priced cars will have the same insurance rating.

However, insurance companies often loathe the cars most of us love — luxury and exciting high-performance cars — and these cars may be subject to surcharges. In addition, some cars are often more expensive to repair, incur more extensive damage in collisions, and are more likely to be stolen than others. So ratings are adjusted to account for these risk factors.

Insurance rating figures used in this book are based on "Insurance Symbols," *(See Figure 4)* with the rating for a particular vehicle adjusted for any special risk factors.

Shop Around

The insurance business is essentially a gigantic book-making operation that bets premiums against the probabilities of having to pay out on an accident or injury claim. Except in a handful of states where bureaucrats set the rates, it's up to each insurer to develop its own probabilities and set its rates accordingly. Rates and underwriting guidelines will reflect what market segments an insurer wishes to target and how efficiently the company is managed.

All of which is to say, rates for the same driver and car can vary dramatically from company to company. If you live in a state with competitive insurance rates, start shopping by obtaining a quote from one of the major firms in the industry — State Farm, Allstate, Farmers, Nationwide, USAA, AAA, and Aetna are among the largest. (See Appendix D for the addresses and phone numbers of the ten largest insurance companies.) Then contact an independent agent who can query other companies for you. Compare them all. Use the worksheet on the following page to compare rates. You'll find that premium prices can differ by as much as 50 percent for the same coverage.

The Complete Car Cost Guide

Figure 5   Insurance Worksheet

Insurance Comparison Worksheet

	Write desired coverage here	Write premium quotes here

Company name 1:_____ 2:_____

Level of coverage you desire for:
  Bodily injury liability:
  Property-damage liability:
  Medical payments:
  Personal-injury protection (no-fault states):
  Collision
    a. $100 deductible:
    b. $250 deductible:
    c. $500 deductible:
  Comprehensive:
    a. $50 deductible:
    b. $250 deductible:
    c. $500 deductible:
  Uninsured motorist:
  Underinsured motorist:

SUBTOTAL A: ☐   ☐

Other coverages you might consider
  Towing and labor:
  Rental-car reimbursement:

+ SUBTOTAL B: ☐   ☐

Do any other charges apply?
  Membership fee:
  Surcharges:

+ SUBTOTAL C: ☐   ☐

Do you qualify for any discounts?
  Theft Deterrent System:
  Passive Restraint System:
  Accident free driving record:
  Multi-car (if adding a vehicle):

− SUBTOTAL D: ☐   ☐

= TOTAL PREMIUM: ☐   ☐
(Subtotals A, plus B, plus C, minus subtotal D)

Ownership Costs

Finance Tips

• *Consider buying the car with a home equity loan. First, interest rates on home equity loans are often a little lower than on a car loan. Second, payment schedules are often more flexible; that is, some home equity loans allow you to pay interest only. Third, and perhaps most significant, interest on home equity loans may be tax deductible — a big savings if you are able to itemize your deductions on your tax return. But be aware that securing a home equity loan just to buy a car might cost you more. That's because there are often significant up-front fees to set up a home equity loan.*

• *You might also consider borrowing against a cash-value life insurance policy. The interest rate will often be lower than on a car loan, and you never have to pay back the loan if you don't want to. Of course if you die or cash-in the policy, the proceeds will be reduced by the outstanding balance.*

• *Consider all discount financing offers on your dream car. Dealers and manufacturers are making some terrific offers — 1.9% through 7.8% 48-month loan offers aren't unusual.*

• *Compare discount financing to a cash rebate if both are offered. "Buy now and we'll give you $500 cash, or provide a 4.8% loan for four years." Which should you choose? It depends on a lot of factors. Appendix J lets you easily determine which option is more advantageous for you.*

• *If you can afford to buy the car out-right, do so — it's usually the least expensive way to buy a car.*

• *If you take out a loan, put as large a down payment on the car as you can manage. And pay as much per month as you can possibly budget to keep the length of the loan to a minimum.*

• *Be aware that there are several ways to lower monthly payments, but only a lower interest rate or a lower amount borrowed will lower your total interest expense.*

Finance Costs

It is now time for a *very* brief accounting lesson. When we talk about finance costs, it is important to distinguish between what is actually a cost and what is not. Let's say you bought a new car for $12,000. You no longer have the money, but you do have a car of equal value — an asset. If a bank loaned you $12,000 to buy the car, you owe the bank that money but it still hasn't cost you anything. However, when you borrow money from any lending institution, not only will the lender ask you to pay back the money they loaned you, but they will also charge you interest. The money that they loan you is not a cost — the interest charge is. End of lesson.

Pay Cash or Take Out A Loan?

For most car buyers, the answer to that question is easy: Let me sign those loan docs! Most people don't have $16,000 (the average price of a new car these days) lying around in their bank accounts. So if they want a new car, they'll have to borrow to the hilt to get it.

But, what if a buyer has a choice: to pay cash or to take out a loan. Let's say that Debbie Debtor and Chris Cash each buy a $16,000 car, and each has exactly $16,000 in the bank. Chris pays cash for her car. Debbie puts 10% down, finances the $14,400 difference with a 48-month, 11% loan, and puts her $14,400 in a bank earning 6% interest. She withdraws $338.18 from the bank each month to pay her car payment, which leaves her with a bank balance of $0 after 48 months. Who comes out ahead?

Chris doesn't earn any interest on her money, but didn't pay any either, so she has a net cost of zero. Debbie must pay $372.18 each month in car payments, but she can only take out $338.18 each month from her bank account (or else she would deplete her bank account before 48 months). She has to pay an additional $34 per month for 48 months, or a total of $1,632 more than Chris.

Another way to look at it is that Debbie will pay $3,464 in interest over the 48 months, but will earn only $1,832 on her money in the bank, a difference of $1,632. To make matters worse, Debbie will probably pay income tax on the $1,832 that her money earned in the bank.

It used to be that Debbie could deduct her interest payments from her income taxes, which could have made her come out ahead of Chris, but interest deductions are no longer allowed.

The bottom line: as long as the after-tax interest rate on the car loan is more than the after-tax interest rate that you could earn on your money, paying cash is less expensive than taking a loan.

Striking a Balance: Monthly Payments vs. Interest Expense

Determining the cost of a loan is inherently complex. There are a myriad of terms that affect the final cost of the loan. Through it all, a few simple truths exist:

- The higher the interest rate, the higher the monthly payment.
- The more you borrow, the higher the monthly payment.
- The longer the period of your loan, the lower the monthly payment.

In each of these cases, the cost of the loan will be higher.

While the first two points are fairly intuitive, the third point catches a lot of people off guard. A car salesperson will frequently attempt to lower your monthly payment, sometimes quite substantially, by stretching out the loan period. But, buyer beware, this will ultimately cost you more. For example, the monthly payment on a $10,000 loan at 12% for five years is $40 less than the same loan over a four-year period. However, the total interest expense is $700 more for the five-year loan versus the four-year loan. This is the cost you pay for the privilege of stretching out your payments for one more year *(See Figure 6)*.

Finding Yourself "Upside Down"

Nowadays, it seems you have to pay as much for a new car as you did a few years ago for a house. But as new-car prices rose dramatically in recent years, lenders found a clever way to allow people to continue to buy new cars — they simply extended the length of loans, thus keeping monthly payments affordable.

In the past, 24-month or 36-month car loans were the norm. But today, 60- and even 72-month loans are common. With the longer loans, however, it takes longer to reach a positive equity position in a car and owe less on it than it's worth. As soon as you drive that shiny new car off the dealer's lot, the car plunges in value — thanks to that ol' devil depreciation. But with a shorter length loan, after a year or so of making payments, your car's value will begin to be worth more than you owe on it. Until then, you're "upside-down," as they say in the auto business.

With longer loans however, you could be upside down for two, three, or four years. And therein lies the rub: If during that time you want to trade in your car on another one, you'll be in the frustrating situation of owing more on your old car than it's worth, thus making it all the more expensive for you to buy the newer car.

So if you must take a longer loan in order to lower the monthly payments on the car you really want, plan on keeping the car for nearly the life of the loan or more. But if you can't keep it that long, be sure you choose a car that holds its resale value well *(See vehicle charts)* because that will reduce the time it takes you to reach a positive equity position.

Figure 6

Interest Expense

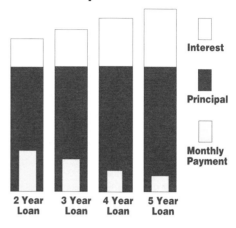

As the length of the loan increases, the interest cost also increases, yet the monthly payment decreases. The principal, however, remains the same regardless of the length of the payment period. The total interest that you pay can be computed by adding up the monthly payments and subtracting the amount that you initially borrowed (loan amount). You can convert the loan amount into a monthly payment using Appendix I.

Finance Tips

• *If you take out a lengthy loan on a car, refer to the vehicle charts in this book and buy one that will retain a high proportion of its value. This will shorten the time you are "upside down."*

• *When is an interest rate not an interest rate? When it doesn't include the "hidden" costs of a loan. All interest rates are not comparable. One rate may include a loan origination fee, and expensive "simple" interest, while another may have no loan fee and cheaper "compound" interest. APR (Annual Percentage Rate) is a very specific term that factors in any hidden fees and tells you the rate you will actually pay when all fees are taken into account. It is the "apples-to-apples" comparable rate. Lenders are required by law to tell you the APR of your loan. You should use this rate, and only this rate, in your comparisons.*

• *Shop around at different banks, savings and loans, thrifts, and credit unions to compare loan interest rates. Then compare those to ones offered by the car dealer and his financing partners, such as GMAC.*

• *Your auto insurance company may also offer financing, sometimes on very favorable terms. Ask your agent.*

Leasing

In the past few years, leasing has exploded in popularity. Automakers from General Motors to Mercedes-Benz are offering attractive lease programs and inundating their dealers with sales brochures touting all the charms of leasing. As a result, leasing now accounts for about one quarter of all new-car sales to individuals. Some predict that by the middle of the decade half of new-car sales will be leases.

The Allure of Leasing

No question about it, leasing has its advantages over buying:

• You don't have to come up with a big down payment. With the price of the typical new car these days hovering around $16,000, the usual 20 percent down payment can amount to a hefty sum. And if you're in the market for really expensive machinery in the $50,000-100,000 range, you're looking at tying up $10,000-20,000 in cash just for the down payment.

Besides the difficulty for some people in just coming up with such substantial sums of cash, others don't want to pull their cash resources out of particularly lucrative investments to buy a car.

• Your monthly payments on the car you want will often be lower if you lease. Here's why: When you buy a $30,000 car for instance, you make payments based on that price, minus the down payment. But when you lease, the payments are lower because the car isn't yours when the lease is up; that means you don't have to pay for the whole car. The car that cost $30,000 might still be worth $15,000 after a 36-month lease, so the leasing company would base your payments on $15,000, not $30,000.

• Or, for a given monthly payment you can lease a more expensive car than you could afford if you bought it with a loan. Again, the same reasoning holds — you're not paying for the entire car because you won't own it at the end of the lease.

• You can walk away from the leased car at the end of the contract and let the leasing company have the headache of reselling it.

• You can drive a more expensive car and trade it in more frequently. With luxury cars carrying price tags of $40,000, $50,000 or more these days, many people can only afford to buy them with long-term loans lasting five or even six years in order to get the monthly payment down to a manageable level. But because lease payments are often lower than loan payments, you may be able to take a shorter-term lease and still have an affordable monthly payment.

The Catch

No down payment, lower monthly payments, no resale hassles — these are nice advantages, all right. But unfortunately, they don't come free.

Make no mistake, in the end leasing is almost always more expensive than buying a car outright with cash. (However, if you have some surefire investment that will pay a very high return for your cash, leasing may be less costly. But in these days of four- and five-percent money-market rates, few of us have such high-paying, risk-free investment opportunities).

Moreover, leasing a car is often more expensive than buying it with a loan. At the end of the lease, you have nothing to show for all those monthly payments except memories and the wad of gasoline-credit receipts jammed in the glove compartment.

A New Wrinkle

In general, leasing may be the most expensive way to put a new car in your garage. But now there's an added wrinkle in the lease game — factory subsidized leasing programs. These programs not only provide all the usual advantages of leasing, but can actually save you money compared to buying the car with a loan — and possibly even compared to paying cash.

To Lease or Not to Lease

Leasing may be right for you if:

• *The manufacturer is subsidizing the lease.*

• *You prefer to use your cash in investments other than a car.*

• *You would like to drive a more expensive car for a lower initial cash outlay.*

• *You like to trade in your car every two, three, or four years on a new model.*

• *You want to avoid the hassle of disposing of a used car.*

Leasing may not be right for you if:

• *Saving money is a major consideration.*

• *You usually drive your car more than 15,000 miles a year.*

• *You are not inclined to take care of your car.*

• *You object to paying penalties that will be imposed if you terminate a lease early for any reason.*

• *You want to significantly modify your car.*

Figure 7

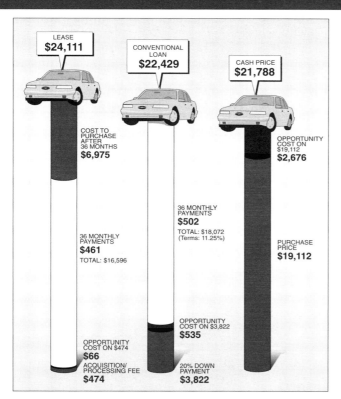

Leasing Comparison

The leasing example above describes a typical lease arrangement for a mid-priced 1993 Sedan. The example shows if you pay cash for the car, you will end up paying less than if you finance the car or lease it for 36 months. This is true even after taking into account any investment income— or "opportunity cost"— that you may earn if you instead lease the car and invest the cash (or finance the car and invest the down payment) in a bank certificate of deposit.

In this example we assume the bank certificate of deposit has an after-tax annual yield of 4.5%. We also assume that the lessee will purchase the vehicle at the end of the lease for $6,975—the vehicle's resale value. To determine resale value, we use a residual rate of 37%.

Comparing Lease Interest Rates

The interest rate used to calculate a lease is the one true measure of the cost of a lease. Unfortunately, this rate is very difficult to obtain from a lessor. To determine the lessor's interest rate, so you can compare leases on an apples-to-apples basis, use Appendix H on page 99A.

Ownership Costs

The Fine Print Of Leasing

Unfortunately, the details of leasing make new-car shopping all the more painfully complex. And the savvy lessee must learn a new vocabulary.

The Residual. This is the predetermined/projected value of the car at the end of the lease — and it's the key to the monthly payments, as well as the ultimate total cost of the deal.

The higher the residual value of the car, the lower your monthly payments. Residual value is essentially a measure of the car's expected depreciation. The lessee pays for the depreciation while he or she drives the car. So if you select a car that depreciates very little, your monthly payment will be lower.

Furthermore, the residual will determine the total cost of the lease deal and whether or not you should purchase the car when the lease is up. In Figure 7, each comparison assumed the car would actually have a market value at the end of the lease equal to the predetermined sales price or residual.

But if a car were to actually have a market value less than the residual at the end of the lease, the lessee armed with this knowledge would be prudent to walk away from the car. On the other hand, if the actual market value were greater than the residual, then the lessee should buy the car at the predetermined residual, sell it, and pocket the difference.

How does the leasing company determine a residual? Like anything that involves projecting future values — be it stocks or used-car prices — there's a certain amount of guesswork required. Leasing companies may offer widely varying residuals for the same make and model; a factory/dealer lease plan may offer a 50 percent residual, while a bank may be willing to go only 40 percent.

Regardless of whether you may want to buy the car at the end of the lease, it's in your best interest to negotiate the highest possible residual.

The vehicle charts in this book are an excellent guide to the expected depreciation/resale value of any particular car you may be considering leasing.

Open and Closed End Leases. Most leases today are closed-end leases, meaning that the residual value is fixed at the beginning of the lease.

With an open-end lease there is still a residual value set at the beginning of the lease. But if the car is worth less than the residual value at the lease's end, the lessee must pay the difference. In other words, the lessee is assuming the risk for depreciation with an open-end lease, which negates one of the big advantages of leasing.

Purchase Option. In our examples, we've assumed that the lessee has the right to purchase the car at the end of the lease for the car's residual value set at the beginning of the lease. But unfortunately, not all leases allow this.

The Money Factor. This is the phrase used for the finance rate upon which the lease payments are based. The lower the rate, the lower the monthly payments. Automakers are subsidizing their lease plans by offering special low money factors. Obviously, you'll want the lowest rate you can find.

Capital Cost Reduction. This is a fancy name for a cash down payment. A large down payment will, of course, reduce the monthly payments, but it will also negate one of the big advantages of leasing. However, if you own your present car, you may be able to use it as the down payment to start the lease.

Early Termination. Because a car's depreciation is the highest in the first few months after it leaves the dealer's lot, and because the depreciation is spread out over the life of the lease in equal amounts, lessees who break a lease early have almost always used up more of a car's value than they've paid for.

So lease plans generally carry steep penalties for early termination. Be aware of what those penalties are before you sign the lease.

Deficiency Liability Waiver or Gap Protection. Even if you had no intention of terminating the lease early, it will automatically happen if the car is "totaled" in an accident or stolen. Your own auto insurance will cover the actual cash value of the car at the time of

continued on next page

The Fine Print Of Leasing continued

its loss. But that may not be enough to cover the lease payoff balance and early-termination penalties.

Some manufacturers, General Motors and Mercedes, for instance, have lease plans that include valuable gap insurance to cover any difference owed in case of an accidental loss.

Mileage Allowance. Here's another potential trap for the lessee. The typical lease requires that the lessee drive no more than an average of 15,000 miles per year over the term of the lease. If the total mileage is greater, there is a charge of anywhere between 8 cents and 25 cents per mile over the limit, depending upon the lease.

Some lease plans allow you to pay the penalty for excess mileage ahead of time at a lower rate. With GMAC's plans, if you know you'll be driving more than an average of 15,000 miles per year, the prepayment is 8 cents a mile. Otherwise, GMAC charges 10 cents a mile over the 15,000-mile limit if at the end of the lease you discover you've driven more than the limit.

Wear and Tear. It's your responsibility to keep the car in good condition. Return the car with a dented fender, bald tires, or a ruined engine because of lack of routine maintenance, and you'll be charged for the repairs. Some wear and tear is allowed, of course. But if you aren't inclined to take reasonable care of your cars, then a leased car may not be for you.

Miscellaneous Fees. Expect a number of assorted fees with any lease. A security deposit, equal to one month's payment and refundable at the end of the lease, is usually required. There is often an "acquisition" fee — a sort of processing fee. And there may be a "disposition" fee of $150-$250 at the end of the lease for getting the car ready to be sold.

A New Wrinkle continued

The automakers subsidize their plans by boosting the predetermined sales price at the end of the agreement (thus reducing the cost on which the monthly lease payments are based) and perhaps more significantly, offering reduced lease financing costs. These lease financing costs work out to as little as half — or even less — than the going rate for conventional auto loans.

The advantages of subsidized lease programs to the automakers are several when compared to other marketing schemes. For one, turned-in lease cars mean a steady supply of good used cars for their dealers. For another, the programs give the dealers a further opportunity to pitch the customer's business when the lease expires; dealers may wave a security deposit or month's payment for a repeat customer.

At the beginning of the 1993 model year, GMAC was offering a "Smartlease" on a 1993 Oldsmobile Achieva for $199 a month. With GMAC's subsidized lease, the customer may save money compared to either taking out a conventional loan or paying cash.

The Best Lease Deal

Always keep in mind that most everything is negotiable when leasing a car. One of the biggest

Ownership Costs

State Fees Tips

• If your county allows you to bring your personal vehicle into the county tax-free, and if you live near a county with a lower sales tax rate, consider buying your car in the other county. A 1% reduction on your tax rate, on a $20,000 car, will save you $200.

• Pay your state taxes and fees on time to avoid costly penalties.

• If you have a choice, register your vehicle as a passenger vehicle. State fees are usually lower on passenger vehicles than on commercial vehicles.

mistakes you can make is walking into a dealership and announcing that you can pay so much a month to lease a car.

Instead, start by negotiating the price of the car, just as you would if you were buying it outright. Then negotiate the residual value and any other provisions in the lease, such as excess mileage fees and the purchase-option price.

Choosing a car with a history of low depreciation and then negotiating the most favorable sales price for the car will equal the lowest monthly lease payments.

Leasing and Taxes

Some folks are under the impression that there are special income tax advantages to leasing a car, especially now that you can no longer write off interest on consumer loans.

The fact is, whether you buy or lease a car, you can write off associated expenses if the car is a legitimate business expense.

But, for those who can't write off a car as a business expense, tax reform did not necessarily make leasing more attractive. Instead, it made buying a car with a loan less attractive.

You may save on sales taxes by leasing a car if you turn it in at the end of the lease. When you buy a car, you pay sales tax on the full purchase price. But when you lease, you don't pay tax on the car's residual value, unless of course, you buy the car at the end of the lease.

State Fees

The government will have its paw out as soon as you buy your new car. You'll have to pay tax and license, and registration fees. Furthermore, you'll be required to pay taxes and/or fees every year for as long as you own the car.

Some states base their fees on the original price of the car and/or on its age. Others base the fees on the car's weight. Still others have fixed fees.

The Complete Car Cost Guide

Running Costs

Running costs, or variable costs, are the costs you bear every time you take your car out of your garage. They include fuel, maintenance, and repairs. (Bridge tolls, occasional parking fees, and when you're naughty, traffic tickets are running costs, too. However, they vary too greatly to be considered in this book.)

Running costs tend to increase as a car gets older. But remember that standing costs usually decrease over time. Usually the increase in running costs is outweighed by the decrease in standing costs, so overall, the car is less costly to own and operate when it becomes an old crock — a good reason to hang on to your car for several years after you buy it. For example, after age five or so, the car may require more repairs, but depreciation costs fall to practically nothing.

Fuel Costs

Fuel is the one expense car owners face almost on a daily basis. Of course you can reduce your fuel costs by driving less, though that's often easier said than done.

You can also buy a fuel-miserly car instead of a gas-hog; the vehicle charts in this book will show you which is which. Keep in mind that if you buy a fuel-inefficient car, your costs rise in two respects. First, there's the cost of the extra fuel. Second, a gas-guzzler tax imposed on the automaker by the federal government (the tax doubled in 1991) may also be included in the car's purchase price — so you'll end up paying it.

But beyond the type of car you buy, there are many other considerations in reducing fuel costs.

Is all gas alike?

Advertising hype about "Lead-Free Super," "Ultra-High Test Unleaded," "Extra-Mile Regular," "Irregular Regular," etc., can make buying gasoline a little confusing. You may begin to feel like a recent cartoon character; faced with all the choices, he tells the gas station attendant, "Just surprise me!"

It's an amusing approach, but not wise. It can even be illegal. Since the mid-1970s, most new cars must use unleaded gas or their emission-controlling catalytic converters will be damaged.

But you still must make the choice between regular unleaded or the more expensive premium "high-octane" unleaded. A car's engine develops power because a mixture of gasoline and air is burned in the engine's combustion chambers. If this mixture burns too rapidly, there's an explosion in the combustion chambers. The engine won't fly apart, but the explosion will set up a vibration that you will hear as a ringing or "knocking" sound, especially when you accelerate briskly. Inside the engine, the effect is not unlike a hammer blow to the top of the piston. If the blows are severe enough, they can damage the engine.

By matching the engine's octane requirements to a gasoline's octane rating, you'll get a nice,

Fuel Tips

Before You Buy

• Buy a fuel-efficient car. Refer to the vehicle charts in this book for fuel-economy ratings

• Be aware that buying a car that requires premium fuel will cost you considerably more at the pump. If you drive a car that gets 20 mpg 15,000 miles per year, you will pay an additional $112 per year if premium fuel costs 15 cents per gallon more than regular fuel. The charts in this book indicate which vehicles require premium fuel.

After You've Bought

• Buy gas with the proper octane rating for your car.

• If you own a car with fuel injection, buy a name-brand gasoline with a detergent additive that helps prevent clogged fuel injectors.

• Keep your car's engine tuned, and its tires properly inflated and aligned.

• Avoid putting the pedal to the metal; accelerate smoothly instead.

• At highway speeds, it's more fuel-efficient to close the windows and turn on the air conditioner. At speeds over 40 mph, open windows create wind resistance that increases fuel consumption.

• The most fuel-efficient speed range is 35 to 55 mph.

• After starting a cold engine, don't let it idle for more than 30 seconds; instead drive slowly for several miles until the engine temperature warms up.

• If you think you'll be waiting for more than a minute — at a railroad crossing, for instance — turn off the engine.

• If you have a car with a manual transmission, learn the proper shift points.

Ownership Costs

The Complete Car Cost Guide

Consider these facts:

- *Misaligned front wheels can increase fuel consumption by 2 percent.*

- *Underinflated tires can increase fuel consumption by 5 percent.*

- *A malfunctioning thermostat in the cooling system can increase fuel consumption by 7 percent.*

- *Worn spark plugs and other ignition components, as well as clogged air filters, can increase fuel consumption by 11 percent*

- *Altogether, these maladies could increase your fuel costs by a whopping 25 percent. The point here is obvious — a well-tuned car uses less fuel.*

even burn of the gas/air mixture. You can get an idea of your car's octane requirements by looking in your car's owner's manual. Unfortunately, finding the optimum choice of fuel for your car is not quite that simple. An engine's octane requirements can vary according to its age, the outside air temperature, humidity, and altitude. All this means you have to experiment with different grades and brands until you find one that eliminates knocking in your car.

Several years ago there really wasn't much difference between brands of gasoline, despite the big advertising bucks the oil companies spent trying to make motorists believe there was. But now the situation has changed. Today, with the near elimination of lead that was used to boost gasoline octane ratings, and with the popularity of fuel-injected cars, oil companies are paying more attention to the way they refine their gas and to the additives they use.

You'll not only find that octane ratings — even for premium grades of gas — will differ between brands, but that cars with fuel-injected engines will run better on certain brands. Most new cars have fuel injection, and the injectors are very sensitive to fuel contamination. The annoying symptoms of clogged fuel injectors include lack of power, stumbling, and even stalling.

The major oil companies are now promoting gasoline with additives that clean fuel injectors. The stuff — like Chevron's patented Techroline additive — really works.

Maintenance

Automakers do sometimes build cars that should simply have "headache" written on the side. But happily, most modern cars, no matter the make or model, are darned impressive machines — especially if they're given periodic care.

Regular maintenance includes all services required by the manufacturer to maintain the car's warranty, services suggested by the manufacturer to ensure trouble-free operation, and other regular service, such as tune-ups and replacing brake pads, exhaust systems, tires, belts, fluids, and filters.

Some Thoughts on Extended Service Intervals

In the old days — the really old days — cars required routine service every few hundred miles. During the 1960s and 1970s, 3,000-mile oil and lubrication service intervals were the norm. Today, automakers have stretched recommended oil changes to 7,500 miles or more, while chassis components are sealed so they never need lubrication.

A program of routine maintenance, including oil changes, tune-ups, radiator flushes, tire pressure checks, and so on, can save you big money in several ways:

It will prevent premature break-downs and budget-busting repairs.

It will make your car last longer, so you won't need to buy a new car as often. That means you avoid the costs of depreciation, taxes, and finance charges that go with a new car.

When the time comes to sell your car, you'll get a higher price for it.

Routine maintenance is required in order to keep your new-car warranty in effect.

Routine maintenance may even improve your car's fuel economy.

Ownership Costs

The Complete Car Cost Guide

There's no question that technology has eliminated or reduced the need for several routine service chores. Electronic ignitions, for instance, have greatly prolonged the life of spark plugs. But when it comes to oil, many mechanics question the whole business of extended service intervals. They still prefer to see the oil changed every 3,000 miles or so; after all, oil is an engine's life blood.

Automaker recommendations can be tricky. In the fine print of the owner's manual, some automakers specify extended intervals for cars used in "normal driving," which they define as high-speed, highway driving. They classify puttering around town or getting stuck in commuter traffic as "severe driving," to which the extended changes don't apply.

Extended service intervals have one drawback: If your mechanic sees your car less often, he has fewer opportunities to spot potential problems, which makes it all the more important for you to inspect your car frequently. But if you're serious about giving your car a long, trouble-free life, change the oil and filter every 3,000 to 5,000 miles. It's worth a few extra dollars in the long run.

Keep On Shining

Many a love affair with a car is based on appearances. Today, thanks to automobile "detailers," the affair doesn't necessarily have to dim with age. With a combination of skill, elbow grease, and the right products, a professional detailer can keep a car looking like new almost indefinitely. Treating your car to the attentions of a detailer isn't merely an extravagance; it's protection for your investment. An old car that looks new will have the highest resale value.

A complete detail job is far more than just a car wash. Every area where grease clings and dirt collects receives the detailer's attention — including vent louvers, door jambs, the trunk, and the engine. The car's paint is washed, rubbed out with polish, and then waxed. The carpets and upholstery are vacuumed and shampooed. Emblems, badges, chrome and plastic trim, and the wheels are all cleaned — with toothbrushes if necessary.

Having your car detailed once or twice a year should be part of your routine maintenance program. It's far cheaper than buying a new car, or even a new paint job.

Maintenance Warranties

As of this writing, five manufactures offered maintenance warranties on their 1993 models: Audi and Hyundai offer warranties on all of their models, while BMW, Alfa Romeo, and Mazda only cover some of their vehicles.

Maintenance warranties cover the services required to maintain a new vehicle's overall warranty — typically tune-ups, oil changes, and minor adjustments — and are offered for between one year and four years.

The actual dollar value of a maintenance warranty ranges from nothing (if you don't take advantage of it) to several hundred or a thousand dollars for three years of maintenance. Most of the really costly maintenance services occur after three years, e.g., tire replacement, exhaust system work, etc. That's one of the reasons you'll rarely see maintenance warranties extend much beyond three years.

Maintenance Tips

Before You Buy

- *Many new cars are designed and built to minimize maintenance costs. For instance, many GM cars now feature long-lasting stainless steel mufflers, and distributor-less ignition systems. Use the charts to compare vehicles' overall maintenance costs, as well as costs for tune-ups and brake service.*

- *Take advantage of any maintenance warranty offered on your car, but don't choose a car just because it comes with a maintenance warranty. Use the charts to see which vehicles feature this type of warranty.*

After You've Bought

- *The most important tip of all: read your car's owner's manual.*

- *Don't skimp on routine maintenance. A regular maintenance program will protect your warranty, save you from costly, unexpected repairs, and will extend the life of your car.*

- *Make a quick inspection of your car every two weeks for problems in the making. (See "The Ten-Minute Technical Inspection" on the next page.)*

- *Be wary of extended service intervals; at minimum, have your car's oil and filter changed every 3,000 to 5,000 miles.*

- *Keep all your service receipts; a complete set can increase your car's resale value if it proves your car was well maintained.*

The Complete Car Cost Guide

The Ten-Minute Technical Inspection

Your mechanic may deftly tune your car's engine, but that won't matter a bit if the engine fries itself because you failed to notice that the radiator ran out of coolant. So roll up your sleeves, because now you're going to learn your way around your car. And in the process, you may thwart a costly, irritating, and possibly dangerous breakdown in the making. Here goes:

✓ First, pop open the hood. Do you see the engine? Good. Actually that's quite an accomplishment, what with engines mounted sideways and all the complicated paraphernalia automakers hang on them these days.

✓ Now find the engine dipstick; it's sticking up somewhere along the side of the engine. (If you can't find it, look in your car's owner's manual for the location.) Pull it out, wipe it off, stick it back in, and pull it out again. Does the oil level reach somewhere between the hatch marks on the dipstick? If not, add some. If it needs more than a quart, take the car to your mechanic and find out why it is losing oil.

✓ Now find the radiator or the radiator overflow tank. Take off the cap and look inside. Do you see some greenish liquid that looks like lime Kool-Aid? If not, add water. Again, if it takes more than a quart, you'd better have your mechanic take a look. And if it's not green, you'll need to have the radiator drained and refilled with a 50/50 mixture of coolant and water.

✓ While you're at the radiator, examine all the rubber hoses running from the radiator to the engine. Look at any other hoses in the engine compartment for that matter. Are they cracked, or do they look so brittle that they should be? Are there any bulges, or are they squishy soft? If so, have them replaced. Chances are you'll not find the right size in a provincial gas station when a hose bursts on your next trip out of town.

✓ If your car isn't quite as up to date as today's news, look for a distributor. It's a round thing with thick wires running out of the top. Follow each of the wires to the engine. Are they firmly attached to the distributor at one end, and to the spark plugs at the other? Are the little rubber booties that cover the ends in place? But if your car is a new or nearly new model, the distributor and plug wires probably won't be visible, so you can skip this step. Indeed, some of the newest cars don't even have traditional distributors or plug wires.

✓ Next, find the belt (or belts) that run the alternator, air conditioning compressor, and perhaps the radiator cooling fan or other accessories. Make sure it isn't frayed or cracked. Find a spot where it's suspended between two pulleys and push down with your finger; if it gives more than a half inch, it needs tightening.

✓ Now find the battery. If it's a refillable type, take off the caps and check all six cells to see that there's water. If the terminals look like they've grown moss, scrub them with a little baking soda and water. And if you see signs of battery acid on the pan where the battery resides, wash the pan thoroughly and check the battery case for leaks. Acid will eventually eat right through metal.

✓ Now start the engine. Look back under the hood. Do you see the fan whirring? Does the engine settle into a smooth idle, or does it jerk around while it runs? Do you see any leaks from any hoses? Do you hear any ominous sounds? Once the novelty of all this has worn off, stop the engine and shut the hood.

✓ Glance under the car. Your car can hold up to 11 different fluids of one kind or another, and all of them should be in the car, not on the driveway. Examine the exhaust system for rusty holes. If your car has front-wheel drive, glance in back of the front wheels. Do you see a greasy mess behind there? The rubber boots that cover the constant velocity joints may be torn or missing. CV joints are relatively expensive to replace, and the boots protect them from dirt and sand that can ruin them.

✓ Get up, turn on all the lights and walk around the car to see that they work. Don't forget to check the turn signals. Look at the windshield wipers; make sure they're not shredding or you'll have a scratched windshield after the next rain.

✓ Now look at all four tires. Is there plenty of tread? Are the sidewalls cracking? Look especially closely at the front tires for signs of uneven wear. If they're scalloped or worn excessively on one side, either the suspension is tired or there's an alignment problem. Buy a tire gauge for a couple of bucks, and check the pressures. Don't forget the spare tire, too. Correct pressures are listed in your owner's manual.

That's it. You've just learned where all the more important fallible things are. Make this inspection a ritual every other week, and you'll forestall many expensive repairs.

Ownership Costs — page 22A

Repairs

One of the blessings of a new car is a warranty. If something goes kerplunk in the night, at least you won't have to pay to fix it.

But as your car gets older you won't have the assurance of a warranty. However, follow the maintenance advice in the preceding section, and you will keep repair costs — even on an old car — to a minimum. Preventive maintenance will prevent costly breakdowns.

Warranties

That old one-year/12,000 miles, whichever came first, warranty seems to have gone the way of necker knobs and eight-track stereos.

Now the standard fare with most automakers, American or foreign, is a three-year/36,000-mile plan. They usually provide "complete" or "bumper-to-bumper" protection for all parts of the car during the warranty period — except for normal wear 'n tear on things like tires or brake linings.

The two years or 24,000 miles of added coverage is certainly good news for car buyers. And there's even more good news: some automakers — mostly those who build luxury cars — are offering even longer coverage for 1993. At Mercedes, you'll be covered for four years and 50,000 miles, for instance. The same goes for BMW and Cadillac.

Some automakers distinguish between complete, bumper-to-bumper coverage and powertrain-only warranties (the powertrain includes the engine, transmission and other parts of the drivetrain). Chrysler offers buyers of its 1993 products a choice of a one-year/12,000-mile basic warranty PLUS a seven-year/70,000-mile powertrain warranty, or a three-year/36,000-mile bumper-to-bumper warranty. Toyota gives its customers a three-year/36,000-mile warranty PLUS an additional two years of coverage, on the powertrain.

Almost all automakers are also throwing in five-, six-, or seven-year warranties that cover the car against rusting.

Audi adds yet another twist to its bumper-to-bumper three-year/50,000-mile plan. Not only does it cover unexpected failures of components, but it includes routine maintenance, such as oil changes and tune-ups. The Audi buyer pays only for gasoline and new tires, if needed.

Keep in mind too, that no matter what the vehicle, federal law says that emissions-control related parts must be covered by its maker for five years or 50,000 miles. And that other items, including tires, batteries or sound systems, have their own warranties.

Service Bulletins

Manufacturers sometimes issue "service bulletins" for certain problems. A service bulletin can be an authorization to dealers to fix a particular problem on a car for free — even if it is no longer covered by the warranty. Manufacturers do this when a large number of vehicles experience the same problem. Service bulletins are generally not publicized, and are often overlooked. Before you pay for any repairs, you should first check with your dealer's

Warranties

Basic warranty: *Comes with the car. If it is a three-year/36,000-mile warranty, the coverage ends at three years or 36,000 miles, whichever comes first.*

Powertrain warranty: *Covers the engine, transmission, and other parts of the drivetrain.*

Bumper-to-bumper warranty: *More extensive than a base warranty. Covers everything except the tires and items that normally wear out, such as windshield wipers and light bulbs.*

Safety Restraint Warranty: *Covers seat belts and air bags, although the warranty terms for each of these may differ. This coverage is usually equal to or longer than the powertrain warranty.*

Corrosion warranty: *Covers rust and deterioration caused by natural elements. Terms vary widely but often are six to seven years.*

Optional extended warranty (a.k.a. Mechanical Breakdown Insurance, Service Contracts): *Longer and more extensive coverage than the basic warranty, available for an extra charge.*

Emissions warranty: *Covers exhaust-related parts such as catalytic converters. By law, the coverage applies to all cars for five years or 50,000 miles.*

Service Contract Caution

Many independent service contract companies have gone out of business over the past several years, making it difficult or impossible for claims to be paid. If you are considering an independent service contract, make sure to ask who the underwriter is. If you aren't certain of the underwriter's financial strength, it may be wiser to choose a factory-backed plan.

service manager to see if the repair costs are covered by a service bulletin.

Service Contracts (Mechanical Breakdown Insurance)

Dealers offer extended service contracts as a protection plan to car buyers when the basic warranty expires. Even though the dealers sell contracts to their customers, they are often backed up by either the auto manufacturer, an insurance company, or some other independent contractor. If you are buying a service contract for the extra coverage, make sure you're not duplicating coverage on the basic warranty. Many factory warranties are much longer today than they were just a few years ago.

Just like the price of the car itself, the price of the service contract is negotiable. In fact, the dealer markup on a service contract is higher than the markup on the car and other options. You should be able to negotiate a significant discount on the price of the service contract.

With certain brands, you may have a choice between a manufacturer's service contract and an independent service contract. If this is the case, compare each contract on the components covered, the term of the contract, the deductibles on each repair, the overall cost of the contract, and the convenience it offers. Independent contracts will usually allow repairs at your choice of repair facilities, whereas manufacturer programs require that you return your car to a franchised dealer (except for emergencies).

One thing to consider when you evaluate a service contract is that the peace of mind it offers may well exceed its price. A service contract offers comfort and security — the knowledge that you are covered in the event of a major problem. To some people, this is well worth the price.

Numbing the Pain of Car Repairs

For many car owners, a visit to a repair shop is accompanied by all the joy and excitement of going to the dentist's chair. There are ways to make the visit less painful, however.

If your car is under warranty, you're pretty much married to an authorized dealer for repairs. But if you receive lousy service from say, the Subaru dealer who sold you your Loyale, there's no reason why you can't high-tail it across town to another authorized Subaru dealer for service. All dealers must honor a manufacturer's warranty whether or not they originally sold the car.

Once your car is out of warranty *(See the previous section on warranties),* you have a couple of other choices. You can take your car to an independent mechanic, or to a specialist who only works on specific components: radiators, brakes, mufflers, and so on.

Independent mechanics and specialists will often charge less than a dealer. But make sure they have a working knowledge of your kind of car, can obtain the correct parts, and have the diagnostic equipment necessary to repair complex electronic gadgets found in so many cars today.

How do you find a competent, trustworthy mechanic? There's a certain amount of trial and error involved. But recommendations from friends or owners of cars like yours are usually your best bet. In addition, mechanics usually know the scoop on other mechanics in town; if you buy a Chevy and know a good Toyota mechanic, ask him or her to recommend someone who works on Chevys.

Whether you take your car to a dealer, independent, or specialist, be polite. It's a great American pastime to bad-mouth auto mechanics — and unfortunately, some deserve it. But in the long run, a car is only as good as the mechanic who takes care of it. Here are some pointers to help the two of you have a long and happy relationship:

- A good mechanic may have grease on his overalls, but don't treat him as if he's not intelligent. You can't be a dummy and properly repair complex, modern cars.

- Follow a trusted mechanic's repair recommendations. A good mechanic will spot potential problems before they occur.

- Look for ASE (Automotive Service Excellence) certification. ASE certified mechanics have passed rigorous industry tests and continually update their technical know-how.

- If you're not sure whether to trust a mechanic, ask to watch while he works on your car. If that's not possible, you should at least expect him to explain the problem in simple language and to show you the defective part while it's still on the car. If he's a good mechanic, he'll appreciate your interest.

- It's proper business practice for you to receive a written estimate before any work is performed. In some states it's the law. And if you authorize a repair and the mechanic later finds more is involved, he should get your permission to raise the estimate before he proceeds with the repair.

- If you don't trust your mechanic and if a quote seems too high, be sure you understand what's involved with the repair. Then before you agree to the repair, telephone one or two other mechanics for quotes so you can compare.

- When talking to your mechanic, be as specific as possible. "There's a high-pitched squeal that seems to come from the right rear wheel between 50 and 60 mph," is a lot more helpful than, "The wheel makes funny noises."

- If you're unhappy about warranty work, don't hesitate to complain to the manufacturer's regional service representative. Your car's owner's manual will usually give an address. If you're unhappy about work you paid for, contact your state's bureau of automotive repair. As a last resort, small claims courts are generally sympathetic to car owners who can present a well-documented case of an auto repair rip-off.

Repair Tips

Before You Buy

- *Don't reject a car just because the charts show a high expected repair cost. The cost of repair is often the smallest of the seven major ownership costs, since most cars today are very reliable.*

After You've Bought

- *Keep receipts and document all the services performed on your car. If you can't prove that you've followed the manufacturer's recommended service program, you could invalidate your warranty.*

- *Before you pay for a repair, see if the automaker has issued a service bulletin for the problem. Even if your car is no longer under warranty, the repairs may be covered.*

- *Cultivate a good relationship with a mechanic. Interview mechanics in your area before you desperately need one.*

- *Don't be afraid to switch dealers if your car is under warranty and you're dissatisfied with the service you are getting.*

- *Don't be afraid to try an independent mechanic for routine service or, if your car is out of warranty, for repair. They're often less expensive than a dealer's service department.*

- *Expect full explanations and estimates of cost before you authorize your mechanic to perform any work.*

Reading Your Garage Floor

Have you ever looked at the garage floor underneath your car and noticed a veritable palette of pretty colors? There may be rich browns, pale yellows, electric greens, or deep reds.

Unfortunately, they shouldn't be there. Your car contains more varieties of fluids than the local tavern — well, nearly a dozen anyway. And sooner or later, one or more of them is going to leak, leaving a tell-tale puddle.

Once every week or two, bend down and take a peek. Some leaks are benign. But others could indicate your car is about to give you grief.

In addition to the color, the smell and feel of the fluid, as well as its location on the floor, will give you clues as to which fuel is which.

Brown or black. Probably engine oil, especially if you find it underneath the front half of the car where the engine resides (in most vehicles, anyway). If your car has a manual transmission, it could be coming from there as well. In either instance, it will of course feel oily when you touch it. Engine oil leaks are especially noticeable after the car has run hard at highway speeds.

Your engine contains several quarts of oil and won't miss the amount in a drip the size of a quarter. Older engines and transmissions are especially prone to dribble a little oil.

But if the leaks are substantial, you might find the engine running dangerously low on oil. Have the leak checked by a mechanic. It could be coming from a faulty seal or gasket that's costly to repair. Or it may be something easily fixed, like a loose oil filter.

Brownish or blackish oil dripping from your car's aft end could indicate a leaking differential. It may just be a loose drain plug. Again, the danger is running out of oil, in which case the differential gears will grind themselves into oblivion.

Red or pink. You'll probably find this hue under the automatic transmission, beneath the center of a rear-wheel-drive car, and towards the front of a front-wheel-drive car. A small leak is no cause for immediate worry; once again, the danger is running out completely if the leak worsens, or if you fail to check the transmission fluid level regularly.

Red, amber, or clear fluid, oily to the touch. This could be coming from the power steering, if your car has such a gadget. The power steering unit is usually located near the front of the car, to one side of the engine. If the fluid level is too low, the power steering pump may be damaged.

Green, with a distinctive sweet odor. This is radiator coolant, and it will leak from the front half of the car where the radiator, overflow tank, water pump, and various hoses reside.

Coolant should never leak. A coolant leak may either indicate something quite serious like a bad water pump, or something simple to repair like a loose hose clamp. But a coolant leak will usually lead to an overheated — and perhaps ruined — engine.

Reddish brown fluid that evaporates quickly means gasoline. A usual clue for a gas leak is a vague smell, perhaps in the trunk or engine compartment.

Gas leaks are dangerous. If you see a puddle, and a dab of the stuff on your finger smells of gas, don't drive the car. Call the auto-club truck — immediately. Otherwise, your car may become a Molotov cocktail.

Clear fluid with a medicinal smell. This may come from the hydraulic clutch system (if your car has a manual transmission). Or it may come from the braking system; both the clutch and brakes use the same kind of fluid.

The reservoirs for both are located in the engine compartment; if one reservoir is low, you'll know which system is leaking. If the clutch system becomes too low on fluid, you won't be able to engage the clutch. Even worse, if the brake system is too low, you won't be able to stop the car. So in either case, have a mechanic check out the leak right away.

Clear water on the floor? No problem with this one. It means the air conditioner is doing its job of removing excess humidity from the car's interior.

Section Two
Choosing and Buying a Car

The Complete Car Cost Guide

Choosing and Buying a Car

It's inevitable. It happens to every car driver. Sooner or later, you'll have fantasies. Fantasies of buying a brand-new car. Fantasies of gazing into the deep, lustrous, unblemished paint of a brand-new car. Fantasies of how you'll look — a sneer on your aristocratic lips, perhaps — as you sit behind the wheel of a brand-new car. Fantasies of not having to worry about repairs and breakdowns of a brand-new car. Freedom, power, sex, and elan — a brand-new car can provide it all.

...pop! Hold on. Back to reality for a moment. All this fantasizing can make you forget mundane matters like budget-busting monthly payments or how you're going to get rid of the old clunker you're currently driving.

Checking Out A New Car

Chances are, when you walk in to the showroom door of your local new-car dealer, a beaming salesperson will pounce on you. Hold him or her at bay for a few minutes while you contemplate the car alone. Stand back and take a long look. Does it grab you? View it from all angles. Even if you're just buying transportation, it's nice to own a car you like to look at every day.

Now get a little closer and look for indications of quality construction. Are the gaps between the fenders and hood even? Do the doors hang straight? Is the paint truly smooth, or is it thick and crinkly like orange peel? Open the hood and the trunk. Run your hand around the inside edges; you won't find spurs or jagged edges on a well-made car. Is the engine a rat's maze, or does the wiring seem neatly bundled, and are the components accessible for service? Is the trunk easy to load and large enough to meet your needs?

Now slide behind the wheel, and make sure the seat gives you support at the small of your back and under your thighs. After several hours of driving, firm-feeling seats, by the way, can actually be less fatiguing than soft seats. Make sure you can comfortably reach all the controls. Look out all the windows and notice if there are any large blind spots that might make lane changes exciting events.

The salesperson will probably want to ride with you on the test drive. Once again, ask him or her to can the sales chatter. Concentrate on the car. Try a variety of roads and speeds; some cars ride fine on smooth roads but lose their aplomb on rough ones. Note the car's noise level with the windows and sunroof open, or if it's a convertible, with the top down. A noisy car can be tiresome on the highway. If the car has a manual transmission, be sure you're comfortable with both the position

Before you completely surrender yourself to your dream car, resist temptation for a few moments and ask yourself some prudent questions:

• *How much can I afford to spend each month? (Appendix I will show you how to convert a monthly payment into a car price.) What's really more important, the kids' college tuition or my dream car?*

• *How long do I expect to own the car? If these buying-a-new-car fantasies come infrequently, say once every eight or ten years, then depreciation is less of a factor and you can consider vehicles with a higher rate of depreciation.*

• *How well do I tend to maintain a car? If you're the kind to drive 'em and leave 'em, you'd better steer clear of fussy, high-maintenance cars even if you do find them attractive.*

• *How many miles a year will I drive? If you pile on the miles, a car with poor fuel economy and high maintenance costs could bankrupt you.*

• *How do I drive? If you live in a big city, a fancy car could mean astronomical insurance costs. If you live in the mountains, an under-powered econobox might not get you where you need to go. If you live in sunny Florida, a car with four-wheel drive might be a silly extravagance.*

Choosing and Buying a Car

Test Drive Tip

If you fall in love with a car during the test drive, don't let the salesperson know it. It will be easier to negotiate later if the salesperson does not sense that you have become emotionally attached to the car.

Consider renting the kind of car you think you might want to buy for a day or two before you make a final decision. Major rent-a-car agencies offer a wide variety of new cars in their fleets; sometimes they get the hottest new models before dealers. Drive the car as you expect to use it — over your regular route to work, for instance, to see if it suits your needs.

Don't let the dealer salesperson talk you into options you don't need or want. Rustproofing is rarely needed on today's cars, which already come with lengthy rust-through warranties. Nor are dealer fabric treatments. If the dealer only has fully optioned cars in his stock, try another dealer.

---

and the operation of the gearshift lever. Front-wheel drive cars in particular often have balky transmissions that can make shifting a chore.

Take plenty of time on the test drive; really get to know the car. After all, it may become your daily companion for several years. If the car passes muster, you'll probably be eager to sign on the dotted line. But control the urge, because there are several other matters to consider first.

Options and Special Features

Options can be a mother lode of profit for dealers and automakers alike. But do you really want a car "loaded" with options, or even one with advanced technical features that come as standard equipment? Some options and features are truly useful and add to a car's resale value. Others end up being expensive maintenance headaches. *(Refer to Figure 8 on page 38A-39A to help make your choices a little easier.)*

Option Packages

Manufacturers often group options together in packages, sometimes providing a discount on the package over the price of the individual options. Option packages reduce the manufacturer's production cost, since they have to make fewer variations of each car. It can also save you money if you want those options anyway. On the other hand, you can waste money on these packages if they mean you pay for options you don't want.

To properly evaluate an option package, first determine the options you really want, then add the prices of the individual options and compare them to the total price of the option package.

It's also not uncommon for a dealer to put cars on the lot with accessories and options already installed. "Sorry, you have to buy it as is," the salesperson will tell you. However, if you don't want this car, try another dealership, or try to custom order, if you don't mind waiting. On the other hand, if you're willing to purchase a car that is already on the dealer's lot, you may be able to negotiate a better deal than if you have the car custom ordered.

Choosing Safety

Ponder this for a moment: According to the National Highway Traffic Safety Administration (NHTSA), two out of three motorists are involved in an accident at some time that injures someone in the car.

Other statistics show that every motorist can expect to be in an automobile crash once every 10 years. And for about one out of 20, it will cause serious injury.

With numbers like this, it's no wonder that more and more new-car buyers have found the safety religion. Today, safety ranks right up there with price, quality and reliability on most buyers' shopping lists.

Anatomy of a Car Crash

NHTSA research shows that the major causes of injury inside the car during a collision are, in descending order, the steering wheel, the instrument panel, the doors, the windshield, the front roof pillar, the glove box area, the roof edges, and the roof itself. It's not surprising then, that front-seat passengers are more likely to be injured than rear-seat passengers who are protected by the padding of the front seat backs.

NHTSA data also shows that 51 percent of deaths occur in head-on impacts, 27 percent in side impacts and only 4 percent in rear impacts. Rollovers are particularly lethal because they are more likely to eject unrestrained passengers from the car than other types of collisions. And fatality rates are 25 times higher for ejected passengers than for those who remain in the car.

The objectives of safety features are 1) to keep the occupants inside the vehicle; 2) to keep them from banging around inside; 3) to absorb some of the forces of impact rather than transfer it to the occupants; or 4) to help prevent a collision from happening in the first place.

Choosing a Safe Car

NHTSA's book of *Federal Motor Vehicle Safety Standards And Regulations* may be chock-a-block with good intentions. But while some mandated safety features — center-mounted brake lights, for instance — are indeed standard on all 1993 cars, others — like side-impact protection — are being phased in over several years. Still other safety features — anti-lock brakes, for example — aren't required by law at all, and are only available at the manufacturer's discretion.

In other words, there is still plenty of variation in the safety features of 1993 models, despite all the rules and regulations. Here are the important features that should be on your safety wish list:

Seatbelts

A three-point seat belt is a belt with three attachment points: one on the side pillar or on the door, and one on each side of the driver's or passenger's hips.

The three-point belt is superior to the older lap-only seat belt for a couple of reasons. Not only do three-point belts reduce the likelihood of the wearer hitting the steering wheel, windshield or other interior surfaces, but they spread the crash forces over more of the body and reduce the strain of the lap belt on the lower body. All 1993 passenger cars are required to have some form of three-point seat belts for both front- AND rear-seat occupants.

Beginning in 1990, NHTSA began requiring some form of "passive" or automatic restraint system in passenger cars for the driver.

And what is a "passive system?" Either an air bag or some form of automatic seat belt, of which there are three. The first uses a motorized shoulder belt that positions itself around the occupant when the door is closed and the ignition is switched on. The second is also connected to the door, but is not motorized. With either of these, the lap belt must still be manually fastened by the driver or passenger.

A third type of automatic belt has both lap and shoulder belts attached to the door, and forces the driver or passenger to slide under this web as they enter the car.

There are several drawbacks to these automatic belt systems. In some accidents, the door is torn off the car; if the seat belts are attached to the door, the seat belts will obviously go with it. Many people also find the automatic belts an obstacle to getting in and out of the car, and find ways to defeat them. Automatic belts also may not fit as well as manual belts. And finally, some of the automatic belts still require the

Safety Tips

• When you choose a car, make sure that you will be able to see and recognize safety hazards approaching from all directions. Any part of the car that obstructs your view, such as headrests, front or back roof pillars, large areas of reflection, or poorly placed rear-deck brake lights, poses a potential danger. And don't forget to consider the placement of passengers and cargo.

• When you're sitting in the dealer show room checking out your dream car, don't forget to try out the car's seatbelt system. Can you easily reach the belt from your driving position, pull it across your body in one motion, and snap it in place without having to search for the buckle? Does the lap belt fit snugly across your pelvis (not your stomach), and does the shoulder belt fit snugly across your torso and over your collar bone without riding up on your neck? Check to see if the shoulder mounting is adjustable, to provide you with the most comfortable fit. Reach down and see if the seatbelt release is easy to find and operate, and that the belt retracts smoothly to its original position. If the seatbelt system fails any of these tests, you may not use it consistently, and because seatbelts are so important, you're better off looking at another car.

Don't Forget the Kids

If you have a family, there are several additional safety features you will want to look for when choosing a car.

• *First, you want seatbelts designed to properly hold a child restraint in place, with an automatic locking retractor. This means that once the seatbelt is fastened, it will remain firmly in place without loosening so the child restraint stays put.*

• *You also want the driver to be able to lock automatic windows so children can't play with them. For the same reason, you want child-proof door locks, so small children cannot open rear doors from the inside.*

• *Also, small, wide-angle "child-view" mirrors are available that snap on to the regular rear-view mirror, and allow you to keep tabs on the kids in the back seat without taking your eyes far from the road.*

• **Built-in Child Seats** — *Chrysler pioneered the idea of built-in child restraint seat in its vans. The seat folds out from the regular, adult's seat, and comes complete with its own seat belts. It's recommended for kids weighing 20 to 45 pounds.*

Ford and GM both expect to offer the seats in some of their mini-vans later in 1993. Chrysler will also have them in the rear seats of its Eagle Vision, Concorde, and Intrepid.

user to manually fasten the lap belt portion; if they forget to fasten it, they risk "submarining" under the shoulder harness in an accident.

The ideal restraint system is a pair of air bags — one for the driver and one for the passenger — combined with manual, fasten-'em-yourself three-point seat belts. The shoulder belt should be attached to the central pillar of the car, not to the door. And the lap belt should ride across your pelvis, not across your stomach.

Above all, a seat belt must be comfortable when it's snug up against the body. If it's not comfortable, it's tempting not to wear it. And if it's not snug, it may not provide the protection it should. In fact, slackened seat belts may INCREASE the risk of injury.

For these reasons, some 1993 cars have belt adjusters and belt tensioners. The former is a simple slide device that allows the shoulder belt anchor on the car's pillar to be moved up and down to accommodate occupants of various heights and body builds. It's not rocket-science, and it's a wonder all cars don't have them.

The latter device is a little more high-tech and is typically found on upscale cars like all or some models of Acuras, Audis, BMWs, Lexus', Mercedes, Saabs, and Volvos. The tensioners automatically tighten the belts around the occupants in the first milliseconds of a crash so there's absolutely no slack in the belts.

Air bags

Pundits once called them "automotive whoopee cushions." But air bags have proven important safety features since 1990, when automakers first began installing them in relatively large numbers on their cars. Costs have come down, doubts about reliability have been laid to rest, and car buyers are eager to have them.

Yet, despite their acceptance, air bags still aren't in all cars. For 1993, some two-thirds of all new cars will have them — but for the most part, on the driver's side only. Passenger-side air bags are still relatively rare.

It should be common knowledge why it's vital to wear a seat belt even in a car with air bags. That is, because air bags don't protect in side impacts, rear impacts or rollovers — accidents that account for nearly half of all highway fatalities.

But if you're a dedicated seat belt user, you may wonder why you need an air bag. A pamphlet entitled "Shopping For A Safer Car" from the Insurance Institute for Highway Safety (IIHS), an insurance industry research group, explains: "Because (even) the best belts — those with tensioners — allow some occupant movement in a crash as the belt pulls tightly around the reel. Plus, there's stretch designed into safety belts to keep people from stopping as abruptly in a crash as the car does. This combination of looseness and stretch means front-seat occupants wearing belts can still move forward enough in a serious crash to hit the steering wheel, dashboard, or windshield... Serving as a pillow between car occupants and the vehicle interior, airbags cushion people's heads and faces."

Make sure to check the charts in this guide to see which new cars come with air bags.

Car Size

Once upon a time, before NHTSA, air bags, three-point seat belts and all the rest, if you wanted the safest car you went out and bought the biggest car you could find and depended on its mass to protect you.

Well, the laws of physics haven't changed. Modern technology may have made small cars safer, but if you're going to be involved in a crash, your survival odds are still higher in a large car.

In one recent study by the IIHS, insurance-company injury claims showed that among the 17 cars with the highest number of claims, 15 of the cars were small. Two were midsize, and not one was large. Conversely, among the nine cars with the lowest number of injury claims, seven were large cars, two were midsize and not one was small.

The big break in the death rate seems to come with cars that have a wheelbase of just over 100 inches (that's considered a mid-size car). According to IIHS data, cars with wheelbases of 105 inches or more have 1.4 deaths or less per 10,000 registered vehicles, while cars with 104-inch wheelbases or less have 2.1 or more deaths per 10,000 registered vehicles.

Structural Crashworthiness

Large car or small, there are differences in injury rates between cars in the SAME size range.

Besides the choices each automaker has regarding seat belts and automatic restraint systems, automakers can choose various structural features that make their cars more crashworthy.

Automakers build two important structures into a car to protect the occupants. The first is a collapsible energy-absorbing structure designed to crush in a controlled manner, absorbing the energy of

Safety Features

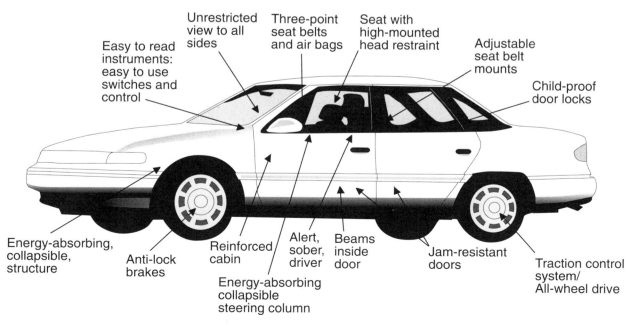

Crash Avoidance Safety Features Every Car Should Have

- Anti-lock brakes
- Legible instruments
- Easily operated controls and switches
- Clear field of vision to all sides
- Day/night rear-view mirror
- Dual side mirrors
- Center high-mounted brake light
- Windshield wiper/washer system
- Adequate heat/defrost system
- Electric rear-window defroster
- Comfortable but firmly-padded seats with driver's seat height adjustment
- Light, bright exterior color

Optional Crash Avoidance Active Safety Features

- Rear window wiper
- Headlight washing system
- All-wheel drive
- Traction control
- Active suspension
- Halogen headlights
- Power mirrors and locks
- Full-size spare tire
- Rear light bulb, tire pressure, brake system sensors

Crash Survival Features

- Deformable crash structure
- Collapsible steering column
- Easy to fasten and release, 3-point seat belts
- Automatic seat-belt tensioners
- Seat belt buckle that moves forward or back along with the seat
- Adjustable seat belt shoulder mount
- Driver's side airbag
- Front passenger's side airbag
- Adjustable or fixed head restraint properly positioned to prevent whiplash
- Jam-resistant door latches
- Safety glass side door anti-intrusion beams
- Roll bar (convertible vehicles)

the crash and increasing the time it takes for the car to come to a stop. The second structure is a reinforced, protective cabin that surrounds the car's occupants and protects them from injury by keeping the exterior impact from reaching them.

Each year NHTSA crash tests popular models into a fixed barrier at 35 mph. The tests measure the crash's impact on the driver and passenger dummies' heads, chests, and thighbones, using all the car's standard safety equipment. And the results, even among different models in the same size class, do indeed vary greatly. You can obtain information on NHTSA's crash tests by calling 1-800-424-9393.

The IIHS also publishes periodic reports, assembled from claims data collected from the institute's sponsoring insurance companies, that show injury frequencies for popular models. For further information, write IIHS, 1005 North Glebe Road, Arlington, VA 22201.

Side Impact Protection

Some 1993 models will meet new federal requirements for beefed-up doors and side structures. Automakers must phase in side-impact protection between 1993 and 1996; as a rule, larger and more expensive cars will get the improvements first.

The standard calls for a car to withstand a broadside hit by another vehicle traveling 30 mph. Most automakers will probably meet the requirement by adding stronger steel beams beneath the body panels and adding interior padding.

Side impacts are second only to head-on collisions as the most serious traffic accidents. If a model meets the standard, you and your passengers will have better protection for your chest and pelvic areas during a collision.

Head Restraints

Of course, these are popularly known as "head rests," a terrible misnomer. Their purpose is not to give your head a place to rest, but to prevent it from snapping back sharply in a rear-end collision.

Though all cars must have front-seat head restraints, not all head restraints are alike. There are two basic kinds, adjustable and fixed. Adjustable head restraints are fine — if they're adjusted high enough (about ear level) and far enough forward to hit the occupant's head, not his or her neck. Problem is, few people take the time to adjust their head restraints.

The fixed type, like those found in Porsches or Volvos, are always high enough, and are probably the best kind for most people. Look too, for head restraints in the rear seats.

Crash Avoidance Features

Automakers actually build safety into their cars in two different ways. There are the crash survival features already discussed — from seat belts to head restraints. But of equal importance are features that help you avoid a crash in the first place.

In this respect, almost anything that gives you better control over your car is a safety feature — from good acceleration that allows you to merge safely with the flow of traffic, to good ventilation that keeps you alert during long spells behind the wheel.

Choosing and Buying a Car

But here are a few particularly important crash-avoidance features you'll want to consider for your 1993 dream car:

Anti-lock Brakes (ABS). Slam on the brakes during an emergency, and chances are your car will become Mr. Toad's wild ride as it skids down the road, especially if it's rain-slicked or icy. Locking up the wheels — or skidding — is dangerous because it not only increases the distance before the cars stops, but because the driver loses the ability to steer the car.

ABS is standard equipment on most luxury cars. But GM is leading the way in equipping popular-priced cars; nearly all GM models will come with ABS by 1995. And ABS will come on more models than ever in 1993.

Check the charts in this guide to see which cars have ABS.

Traction Control/All-Wheel Drive. Traction control does for acceleration what ABS does for deceleration. Traction control and all-wheel-drive are pricey features. But they provide more control, and hence an extra margin of safety on wet or icy roads.

Proper Ergonomics. A car's sound system may reproduce music that's the next best thing to a live performance, but if the system's controls are so tiny that you need your bifocals to find them, they're a detriment to safety.

Convenient placement and logical arrangement around the cockpit of a car's controls is an important, yet often overlooked, safety feature.

Good Visibility. Clear vision from inside the car would seem to be an obvious given in the design of any modern car. Yet all too often cars have poor outward visibility, with window size and placement dictated more by style than by function.

Heavily tinted windows can also dangerously obscure vision, especially for older drivers at night.

When negotiating a price, keep in mind that the following three prices are very different:

1) The dealer's cost or price (a.k.a. factory invoice) — the maximum price the dealer paid the factory for the car.

2) The manufacturer's suggested list price.

3) The car's sticker price, which usually differs from the suggested list price, and includes special equipment, options, preparation charges, and other fees.

What Price Should You Pay?

Wouldn't it be a relief if new-car prices were no more negotiable than a can of tuna at the supermarket? Unfortunately, that's not the nature of the game.

And in many ways it is a game. Unless you've been trained, don't expect to beat the dealer — he or she is a pro, after all. But if you do a little homework, you can arrive at a fair deal for both yourself and the dealer.

Sometimes a dealer can purchase a car for less than the figure shown as dealer cost. This is because manufacturers frequently allow volume discounts, and often provide other dollar incentives, such as rebates, for dealers to "move" cars. Since there are many ways for dealers to hide their true costs, be wary of any numbers they quote you.

Knowing the dealer cost does not automatically guarantee that you will get the deal of the century. But, it is an effective weapon to have as part of your negotiating arsenal. Not only will it give you a good indication of how much profit a dealer is trying to make on each sale, but it will also serve as an excellent point from which to begin negotiating. Realistically, you should set your goal on purchasing a new car at the "target price" listed in the vehicle charts starting on page 1. This figure is calculated using the dealer cost as a base, and factoring in current market conditions to arrive at a price that represents a good deal to you and a fair profit for the dealer.

Once upon a time, the manufacturer's suggested list price was the price at which the dealer was willing to start negotiating. These days the situation can be quite different; this truth becomes apparent when you discover that a particular model is priced hundreds, sometimes thousands, of dollars above the manufacturer's list price. This is simply a function of supply and demand. If a particular car is in great demand, a dealer can sell it for whatever the market will bear. You may be able to bargain some money off the inflated price, but you will find it close to impossible to buy a car at or below list. In this instance, check dealers outside your local market area. They may offer better pricing on your model.

On the other hand, it will be much easier to negotiate a better deal on overstocked cars or models that are not selling well. Dealers who are carrying less popular models that have been sitting on their lots for a long time are stuck with a double-edged liability. Not only are they losing profit on potential sales, but they are paying finance charges and other expenses while they keep the cars. They may be happy to get them off the lot at a bargain price.

Beware...

Frequently, a dealer will try to tack on an extra charge called an "advertising allowance," or, in industry parlance, a "pack." The dealer will argue that this figure is a legitimate business expense that he has to recover in addition to any profit on each car sold. However, this is just accounting hocus-pocus. Like any other business, the dealer will cover his costs through the profit from the sale of his merchandise, and not by adding a special fee. The profit range suggested in the "target price" should reasonably compensate the dealer.

As of 1991, Congress imposed a new tax on expensive automobiles. There is a 10 percent tax on the portion of a car's price that exceeds $30,000. For example, the buyer of a car with a price tag of $40,000 would have to pay a luxury tax of $1,000 — that is, 10 percent of $10,000.

Rebates and Sales Incentives

Cash-back rebates have become as much a part of the new-car sales routine as tire-kicking and test drives.

In recent years, manufacturers have added a number of other items to their sales-incentive menu as well. Automakers may offer a choice of discount financing or a cash rebate, for instance. (See Appendix J on page 110A)

Now there are discount leases, with automakers subsidizing some of the leasing costs the buyer would otherwise incur. (See "Leasing," page 14A.)

In addition to these monetary incentives, there are some new and intriguing non-monetary sales incentives.

Automakers and dealers alike are pushing customer service as a reason to buy their particular products. Virtually every automaker offers free towing in the event of a breakdown that's covered under warranty, and a growing number of automakers are providing roadside assistance. Pontiac, for example, is expanding its roadside assistance program in 1993 to include free help when the owner is out of fuel, locked out, or has a dead battery. If you're 150 miles or more from home and your new car breaks down, Oldsmobile covers food and lodging expense up to $500.

Several automakers are offering their customers courtesy transportation or loaner cars when their cars are taken to the dealer for service or repairs.

Oldsmobile and Saturn permit new-car buyers to return their cars to their dealers within a certain period for a full-credit exchange on another Olds or Saturn if they're unhappy with their car.

Rebate Tips

Don't rely on the dealer to inform you of available rebates and incentives. Each issue of Automotive News, *available at many public libraries, lists current consumer and dealer incentives. If you would like a complete report on a particular vehicle, including current consumer and dealer incentives, you can order the ArmChair Compare® Report, as advertised in this book.*

Don't expect to find rebates from most automakers specializing in luxury or sports cars. Even among companies offering rebates, some give them only to dealers. Sometimes, rebates are offered only in certain sales regions — and sometimes a regional rebate is less than the rebate available in the rest of the nation.

Don't buy a car just because it comes with a large rebate. A rebate tacked on a poor value doesn't automatically make it a good value. Use the charts in this book to determine the good values, then see which of those have a rebate.

Rebate: *Cash given by the auto manufacturer in return for a sale.*

Incentive: *The bait a manufacturer uses to fish for sales. An incentive may be a cash rebate, discount options packages, low-interest financing, and much more.*

Customer incentive: *An automaker's financial or non-financial offer intended to entice the customer to buy.*

Dealer incentive: *A financial offer by the manufacturer to the dealer for making a sale. The dealer may or may not pass on the incentive to the customer. If it is passed on, it may take one of many forms, such as a price discount, a rebate, free options, or free maintenance.*

The Complete Car Cost Guide

Figure 8

A Handy Guide To Options And Technical Features

Any option or feature you add will raise the sticker price of a new vehicle but may not increase the price when it comes time for resale. Just like cars, car options vary in how rapidly they depreciate. For example, $500 spent on air conditioning may add $400 to the vehicle price when it's time to resell, but $500 spent on a leather interior may add little or nothing to the resale price. Some options (e.g., anti-lock brakes) that are clearly beneficial may not be available on a particular model, or might not fit into your budget. When selecting your options, it is important to weigh emotional, safety, and practical aspects along with the economic ones. A car with wisely chosen options will prove to be a better value.

Options	Benefits	Drawbacks
Active Suspension: Electronically controlled hydraulic actuators that respond to changing road surfaces and driver input to keep the car level at all times.	Allows for comfortable ride and excellent emergency handling.	Expensive, with a lot of mechanical complexity. Unnecessary on most cars.
Adjustable Shock Absorbers: Shock absorbers that can be adjusted by either the driver or an on-board computer.	Allows you to adjust the ride according to different driving conditions. Sport handling means a firm, more controlled ride, and Touring means a softer, more comfortable ride.	Expensive to buy; expensive to replace when they wear out.
Air Bags: Hidden in the steering wheel, and on some cars, in the passenger side of the dashboard, they inflate during a collision to protect driver and/or passenger. They are not substitutes for seatbelts.	Have been proven to significantly reduce injuries in cases of head-on collision.	They only provide protection in a frontal collision, and are expensive to replace after use.
Air Conditioning	Reduces fatigue in hot weather. Adds considerably to resale value of car.	Expensive to buy. Somewhat expensive to maintain, and will become more so as ban on Freon takes effect in the near future. Increases fuel consumption.
Anti-Lock Brakes or ABS: Electronic sensors and computer control prevent the brakes from locking up.	Prevents the car from skidding uncontrollably during a panic stop. Shortens stopping distances.	Expensive to buy. May require some additional maintenance.
Automatic Transmission	Makes driving easier in stop-and-go traffic and on mountain or hilly roads. Adds to resale value if appropriate for vehicle.	May decrease fuel economy, though not as much as in years past. May decrease acceleration. May decrease resale value if on inappropriate vehicle.
Continuously Variable Transmission: An automatic transmission that, instead of using three or four separate gears, uses belts to produce continuously changing speed ratios.	Better fuel economy than with a manual transmission, because it lets the engine work in its most efficient range.	It's a complex technology whose reliability hasn't been proven. May be very expensive to maintain.
Cruise Control	Reduces driver fatigue on long trips, and may help increase fuel economy.	May require some maintenance.
Digital Instrumentation	Amusing—if you're into video games.	Can be confusing and distracting; also very expensive to repair.
Diesel Engine	Fuel economy.	Slow acceleration, difficult to start in extremely cold weather, unpleasant fumes, fuel sometimes hard to find.
Four-Wheel Disc Brakes: Cars often have "disc" brakes on the front wheels and "drum" brakes on the rear wheels. However, under hard use, disc brakes remain cooler than drum brakes, which means they retain their stopping power longer. Disc brakes on all four wheels are preferable.	Retain stopping power longer; disc brake pads cheaper to replace.	Disc brake pads wear more quickly than drum brake pads.
Four-Wheel or All-Wheel Drive	Increases traction; particularly useful in snow, rain, and in off-road driving.	Adds complexity and weight, which raises car's price and lowers its fuel economy.
Four-Wheel Steering: At low speeds, the rear wheels turn in the opposite direction to the front wheels; at higher speeds, all four wheels turn in the same direction.	Makes the car generally easier to steer, particularly during parking.	Added initial cost, and possibly added maintenance and repair cost.
Fuel Injection: Replaces carburetor as the device that sends fuel and air to the engine's combustion chambers.	Improves engine response and fuel economy. Reduces exhaust pollutants.	Fuel injectors are sensitive to impurities in gas; may require detergent gas or additives to prevent clogging.

Choosing and Buying a Car

The Complete Car Cost Guide

Options	Benefits	Drawbacks
Head-up Instrument Display: Projects a car's speedometer reading or other instrument displays onto the windshield in front of the driver.	Lets drivers read instrument displays without taking their eyes from the road.	Unnecessary gizmo. Only essential for fighter pilots.
Heavy Duty Cooling System	Necessary for hauling heavy loads in hot weather, especially on a vehicle with air conditioning.	No disadvantages except extra initial cost.
Heavy Duty Suspension	Necessary for hauling heavy loads.	Rougher ride when not hauling a load.
Keyless Entry: A keypad mounted on the front doors that unlocks them when you enter a five digit code.	Entry is possible, even if the keys are locked in the car. Also, allows a third party access to the car without the keys.	You have to remember a five-digit code in order to use the system.
Larger Engine: Engine size is noted by the number of cylinders, and by "displacement." Displacement can be measured in liters, cubic centimeters or cubic inches. The more cylinders and the greater the displacement, the larger the engine. Larger engines generally provide more torque and more horsepower.	Quieter and smoother during highway cruising; added power for passing, for mountain travel, and for hauling a trailer.	Slightly higher maintenance costs; possibly higher costs for fuel, insurance, and state registration.
Multiple Valves: All engines have at least two valves for each cylinder. But some have three or four valves for each cylinder. The additional valves allow a more efficient movement of the air/fuel mixture in the combustion chambers so the engine develops more power.	Better acceleration without sacrificing fuel economy.	Added complication; possibly higher repair costs.
Overhead Cam: Some engines have single or double overhead camshafts instead of pushrods to open and close the valves as the engine operates.	Better engine response at high speeds.	Nothing really, except slightly higher manufacturing costs.
Power Seats	Impresses those who like gadgets.	Something else to go wrong.
Power Steering, and Power Steering with Variable Assist: At higher speeds, simple power steering can make the steering feel too light. Variable-assist or speed-sensitive steering is a type of power steering that automatically boosts the power assist at low speeds and reduces it at high speeds.	Makes it easier to turn the steering wheel, especially when parking. Necessary for large, heavy vehicles; can increase resale value for those vehicles.	Added initial cost; added maintenance cost.
Power Windows/Locks	Convenient; a safety feature when children are in car.	Higher initial cost; increased repair costs.
Rear Window Defogger	Increases outward vision in icy or humid conditions.	Can crack window if left on in hot weather.
Rustproofing/Undercoating	No benefit, unless car isn't rustproofed/undercoated by the factory.	Unnecessary because almost all new cars have rustproofing and undercoating.
Stereo Sound System	Improves resale value of car. (Tip: Aftermarket products are often superior in price and sound to factory or dealer-installed units.)	Makes car a target for theft.
Sunroof/T-Top/Moonroof	Fun. Can increase resale value substantially.	Added wind noise; may reduce interior headroom, and may leak or squeak if moldings deteriorate.
Tinted Windows	Keeps car cooler in sunny weather, thus reducing sun-caused damage to upholstery and dashboard. Reduces load on air conditioner.	Slightly higher initial cost.
Turbocharger/Supercharger: A turbocharger is operated by hot exhaust gases that continually rush out of the engine as it runs. A supercharger is driven by a belt or a shaft from the engine. Both devices force extra air and fuel into the engine's combustion chambers for more power.	Provides the acceleration of a larger engine, but without higher fuel consumption.	History of repair problems; requires extra maintenance; often more costly to insure.
Traction Control: Uses sensors to determine when wheels begin to slip and reduces engine speed/applies brakes to maintain optimum traction.	Allows for smooth acceleration on slippery roads. Worth having if you regularly drive wet, muddy, or icy roads.	Expensive, with a lot of mechanical complexity. No benefit under normal driving conditions.

Choosing and Buying a Car

When is the Best Time to Buy a New Car?

Best Time of Year

The winter months usually present good bargain opportunities. With increased cash outlays for clothing, utility bills, and holiday expenses, auto sales are slow and salespeople are eager to make a sale. Additionally, manufacturers try to clear their previous inventory to make room for the new models. And sometimes heated competition between manufacturers create great bargains for shoppers. For example, at the end of 1992, Ford practically gave away its Taurus in an effort to claim that it was the #1 car sold in the United States.

The summer months can also be a good time to shop. This is when sales of the current-year model have slowed down and consumers are waiting for the new model year cars to be introduced. But be cautious during springtime! That's when consumers come out of their winter doldrums and car sales increase. Dealers are less inclined to make price concessions while sales are up.

Best Time of Month

The end of the month is the best time to buy a new car. A salesperson prospers by consistently meeting or exceeding his or her monthly quota. To increase a salesperson's incentive, dealers often have contests and offer bonuses for exceeding monthly quotas. As a result, salespeople are more likely to sacrifice a small part of their commission at the end of the month if it will help them meet or exceed a quota.

Buying, Financing, and Selling Your Old Car

There is one golden rule when it comes to negotiating a new car: Buying a new car, financing it, and trading in your old car may all occur at the same time, but these are three separate transactions and should be negotiated separately.

Imagine this scenario. A salesman is negotiating with a customer about the price of a new car. Negotiations have come to a standstill because the customer says he can not afford a penny more than $13,000 for a particular car that is priced at $13,200. The dialogue might sound like this:

"Well, what are your current monthly payments?" asks the salesman.

"$275 a month," the customer replies, "and I only have two payments left."

"How about this," says the salesman, "if I can get you into that new car for $275 a month, have we got a deal?"

"That sounds reasonable," answers the customer.

The salesman smiles, "Do you want to trade your car in? I could probably get your monthly payments even lower."

"Well, sure. I'll trade my car in if you can give me a good deal," says the customer.

The salesman checks out the customer's car and says, "How about this, I will buy out the balance of your loan, give you a $2,000 check on top of that to use as a down payment for a new car, and then I'll finance the rest for 60 months at a monthly payment of only $250 — $25 lower than you're currently paying. Now, doesn't that sound terrific?"

"It certainly does!"

But, it isn't so terrific. In fact, the customer is going to pay more for his new car than he initially wanted to. The salesman has confused him. $250 a month has the illusion of a good deal because it is less than the customer's current payments. However, upon closer examination we can see what is really happening. First of all, the salesman offered the customer $2,000 in addition to "buying out" his current loan on the old car. That means the customer receives a total value of $2,550 for his car. $550 will be paid to the title holder of the old car (for the last two payments) and the $2,000 will be applied as a down payment for the new car. If we then calculate the present value of $250 dollars per month over a 60-month term at 11% interest, it works out to be about $11,600. Combining all this together, we find that the customer has agreed to pay the equivalent of $13,600 for his car: $2,000 from the trade-in and $11,600 from the loan. This is $600 more than the customer originally said he'd pay. To add insult to injury, the customer is without a doubt going to get less on his trade-in than if he sells the car himself.

Even though this example is contrived, situations like this happen every day. Salespeople are experts at this tactic and employ it every chance they get. Even if you know all your facts, it is still easy to get confused. It can be difficult enough trying to negotiate a single transaction with a salesman, but when you try to do all three at once, you are stacking the deck in his favor. Don't even try; keep these three transactions separate.

Finance & Insurance Manager

Whew, you did it! It took hours of haggling. But you and the dealer have finally come to terms on your new car.

You bargained hard to get what you consider a fair price for the car and its optional equipment. You even managed to get the dealer to give you a reasonable price in trade for that old jalopy you're now driving. And the salesman has quoted you a competitive interest rate on dealer-arranged financing.

Why, you can practically feel the keys to your new dream car dropping into the palm of your hand. All that's left to do is a quick trip to the dealer's finance and insurance (F&I) office to wrap up a little paperwork. Time to relax, right?

Wrong, not unless you want to risk spending more than you had planned.

The dealer's F&I person is often one of his most adept salespersons. In a few strokes of the pen, he or she can turn a good deal for the customer into a bonanza for the dealer.

For instance, the F&I person can write the contract with a higher interest rate than the car salesman had quoted. He or she can add on an overpriced extended warranty. ("It'll only cost you a few dollars a month," the F&I person will say.)

F&I folks also love to tack on life and accident insurance designed

Evaluating the Dealer

Other than price, service is what really sets dealers apart. A reasonable approach to selecting a dealer is to talk to the service department before you talk to a salesperson.

to pay off the car loan should you die or become disabled. It's often expensive, and most people already have disability and life insurance, and don't need the additional coverage anyway.

Then there's the "Rule of 78," a clever accounting method that can only benefit the dealer. Few car buyers understand the Rule of 78. Yet those who take dealer-arranged financing are often asked to initial a little box on the loan agreement that states interest on the loan will be calculated using the Rule of 78.

Compared to a simple-interest loan, a Rule-of-78 loan increases the portion of each monthly payment that goes to the lender as interest in the early part of the loan. It decreases the portion of each monthly payment that goes to the lender as interest in the last part of the loan.

In other words, the lender gets his interest sooner while the amount you owe gets paid off later.

For example, on a $10,000 loan for 24 months at 12 percent interest, the monthly payment will be $470.80. If it is a simple-interest loan, $100 of the first month's payment will go to the lender as interest. The balance, $370.80, will be used to pay off the principal, $10,000. Under the Rule of 78, nearly $104 of the first month's payment will go as interest, leaving a smaller amount to payoff the principal. By the 24th month, $4.64 will go as interest, if the lender uses simple-interest calculations. Under the Rule of 78, $4.33 will go to the lender as interest.

Now these may seem like insignificant differences. But on larger loans for longer terms, the differences are more profound. On a 60-month, $16,500 loan at 10.5 percent interest, a little more than half the total interest will be paid just 18 months into the loan, but only 24 percent of the principal.

Using the Rule of 78 does not change the amount of the loan's monthly payment. And it doesn't change the total amount of interest you'll pay on the loan — IF you keep the loan (and the car) for the entire length of the loan.

And that's the rub. What if you want to trade in the car long before the end of the loan's term? It means that you'll owe more on the car than you'd otherwise owe if it had been a simple-interest loan.

Or what if the car is stolen or wrecked early in the loan's term? Your insurance company will reimburse you for the value of the lost car, not for what you owe on your loan. If you owe more than the car was worth — and the possibility of that is greater under the Rule of 78 — you'll be stuck with paying the difference out of your pocket.

While many dealers and finance companies use the Rule of 78, banks and credit unions are more likely to offer simple-interest loans.

The point should be clear: Don't sign any contract unless you completely understand it. And until those keys actually drop into your hand, pay close attention to every transaction between you and the dealer.

The Complete Car Cost Guide

Six Steps To A Good Deal

Haggling, horse-trading, bartering. Call it what you will. Though some buyers may consider it great sport to try and beat the last penny of profit out of an auto dealer, most consumers hate the hassle.

Consider the sad saga of Mrs. B. Innocently enough, she went to a showroom and announced that she had $10,000 to spend on a modest new car for her college-student daughter. A nice salesman showed her a nice car, and eventually they came to a nice agreement. Or so Mrs. B thought.

After telephoning her daughter with the good news, Mrs. B returned to the dealership the next day, ready to buy. "I'm sorry," said the salesman, "but my sales manager won't let me sell the car for less than $12,000." Yep, it was the old bait and switch — and Mrs. B was furious. "How can they get away with this?" she demanded.

They can and they do. But in today's highly competitive auto market, dealers and automakers are beginning to realize that many car buyers would just as soon face a firing squad as go through the tribulations of negotiating for a new car. Thank goodness the system is beginning to change.

But until then, here are six steps to help you deal with the dealer:

1) Know the car you want and how much you can afford to pay. Experts say your annual income should be at least twice the price of the new car you buy. Check the "AutoExplorer™" section in the front pages of this book. This will give you an idea of all the models offered by each manufacturer in several price ranges. Don't waste everyone's time test-driving and negotiating for a car you can't possibly afford.

2) Be prepared. Review the charts in this book for cars you're interested in. Study the suggested retail prices and dealer's cost for the model and options you want. Knowing these gives you a vital negotiating advantage.

Find out about rebates *(See page 37A)*, and if a choice between a cash rebate and discount financing is given, use Appendix J to determine which is best for you.

3) Qualify the dealer. Call several dealers within a reasonable driving distance and ask each if they have the car on their lot with the options you want. You'll nearly always get a better deal if the dealer doesn't have to trade with another dealer for the car you want. Ask for a price quote on the phone, but understand that it may be a teaser price just to get you into the dealership.

4) Remember the golden rule of car-buying: First negotiate the price of the car. Then discuss financing. Then discuss the value of your trade-in. Always keep the three transactions separate; salespeople are adept at giving away with one hand and taking a lot more with the other.

For instance, they may talk about monthly payments instead of the car's price; by stretching out the length of a loan — with correspondingly lower monthly payments — they can make an overpriced car seem affordable.

5) Keep cool. The salesperson may try to pressure you — "Are you ready to buy today?" The salesperson may play on your sympathy — "I'll lose my job if I agree to ..." The salesperson may intimidate you — "The dealer cost figures you have are all wrong." Or the salesperson may indicate you have a deal and then turn you over to the sales manager who nixes it; together, they'll work on you to sweeten the deal in their favor.

Through it all, stay confident of your research, and remember the Scout's motto: Be prepared. Be prepared to put down a deposit as soon as you've arrived a fair price, and if you feel mistreated, be prepared to walk out of the dealership.

6) Let the dealer make a buck. Bear in mind that "profit" is not a dirty word, and that no dealer is going to sell you a $10 bill for a buck and remain in business.

Choosing and Buying a Car

No Dicker Stickers

Take-it-or-leave-it price tags have suddenly become all the rage in the auto industry. A handful of enlightened individual dealers across the country, as well as all the dealers of one automaker, have decided to face head-on the traditional notion that the new-car showroom should take on the atmosphere of a third-world bazaar.

General Motors' newest division, Saturn Corporation, took the lead in eliminating the haggling over a car's price at its dealerships. Instead of a commission, Saturn salespeople are paid salaries, which helps remove some of the incentive to use high-pressure sales tactics new-car shoppers detest. And most important, the sticker prices on new Saturns aren't negotiable.

Individual dealers for other automakers, after noting Saturn's success, are trying no-haggle policies in their own showrooms. And surveys have shown that most customers love it.

Instead of boosting the window sticker prices by thousands with the expectation of knocking them down during the grueling negotiating process, the dealers tack on a reasonable profit to the manufacturer's invoice price and refuse to bargain down the price with the customer.

Though no-dicker-stickers eliminate part of the new-car-shopping hassle, they are not however, a panacea for haggle-phobic customers. Even no-dicker-sticker dealers have other ways to line their pockets with your money.

Buyers who wish to trade in their old car to the dealer must still negotiate a price for their car.

Dealer "F&I" (finance and insurance) managers can still pad dealer-provided loan or lease agreements with high or unnecessary extra charges if the customer lets his or her guard down.

And because dealer overhead can vary depending upon a dealer's location, the no-dicker prices will vary. Car buyers still need to shop around for the best price, even among one-price dealerships.

Moreover, salespersons may still pressure you to buy a car the dealer has in stock rather than trade with another dealer so you can have the car outfitted exactly the way you want.

Finally, if you really think you can beat the dealer at his own game (few people actually can), if you enjoy the challenge of a good haggle, and if you have the time and patience to try, there's still no substitute for hard, astute bargaining at a traditional dealer.

For dealers other than Saturn, no-dicker sticker prices tend to be well below the manufacturer's suggested retail prices, but they still tend to be $500 to $1,000 over the dealer's cost. A hard-bargaining customer at a traditional dealership should be able to pay as little as $200 to $500 over the dealer's cost.

Brokers

It's an odd thing, this business of buying a new car.

Next to a home, an automobile is the second most expensive purchase most of us ever make. Yet when we buy a home, we can count on the counsel of our real estate agent, escrow officers, title company personnel, loan officers and perhaps even our attorney. Moreover, there's usually a lengthy waiting period until the close of escrow, during which time we can find a way to back out of the deal.

Yet when we face a new car dealer and his agents — salespersons, sales managers, finance and insurance managers — we face them alone. And we're expected to close the deal in less than a day.

No wonder then, auto brokers around the country have stepped forward to help the beleaguered new-car shopper. Like a real estate agent, an auto broker's mission is to negotiate the purchase for the buyer. An auto broker can be the answer for car buyers who fear they can't beat the dealer at his own game — or don't wish to take the time to try.

Most brokers bypass dealer salespersons. Instead, they deal only with dealer sales managers or fleet managers. Successful brokers have good working relationships with their local dealers, and can avoid much of the time-consuming sales games the individual buyer faces in the showroom.

Brokers are also familiar with their local dealers' inventories. They keep on top of which models are hot and which are not. When a broker can locate a dealer with three of the exact cars you want, chances are good the broker can negotiate a more favorable deal on one of the cars than if the dealer has to order or trade with another dealer to obtain your dream car.

Of course, a competent auto broker knows the tricks of the trade and isn't likely to fall for them.

Auto brokers usually charge an up-front fee of $200 to $400 for their service. But a good broker may be able to arrange a better deal on the car you want than you could by yourself, even including his fee.

Broker Varieties

Not all brokers are the same, however. Some brokers are like a shopping friend; they'll accompany you to dealerships and do the negotiating for you in your presence once you've found the car you want. At the end of the process, you'll pay the dealer and take delivery of the car from him.

Other brokers take your order, and then operating over the phone, find the car you want at a dealer that may be close by or several hundred miles away. The broker will take delivery of the car, and you will then purchase it from him.

Finally, other brokers act as a referral service by putting you in touch with a dealer who promises to sell you the car at a favorable price. Auto clubs and credit unions often work with these referral brokers. It's up to you, however, to do the final negotiating.

Regardless of which type you use, you're relying on the broker to make you a truly good deal — one better than you could cut yourself.

How Much Can a Broker Save You?

If you're a good negotiator, you can almost always negotiate a better deal than what a broker offers. A broker has to make money, which means the price you pay the broker will always be more than the price the broker paid the dealer — a price you could probably get on your own. However, if you're not comfortable negotiating, a broker can get you a real bargain.

The Complete Car Cost Guide

Figure 9

Projected Five-year Comparison:

1993 Corsica LT vs.1988 Corsica

Cost areas	1993 Model	1988 Model
Depreciation	$8,494	$2,500
Financing	$2,378	$0
Insurance	$7,035	$5,187
State Fees	$560	$350
Fuel	$3,612	$3,695
Maintenance	$4,438	??
Repair	$651	??
Total	$27,168	$11,732
Difference	$15,436	

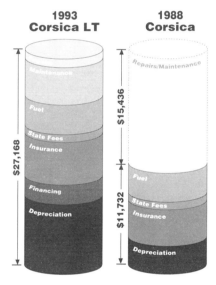

The 1988 Corsica could have 5 year maintenance/repair costs up to $15,436 - and still have lower overall costs than a new Corsica.

Keeping Your Old Car vs. Buying a New One

Ok, you've done your homework. You've studied the charts in this book, compared your choices, and now you are ready to buy a new car. All you need to do is get rid of your old clunker and purchase the car of your dreams. Hold on for just a minute.

You probably have a list of reasons to buy a new car. Reasons that might include the newest safety features, modern styling, or perhaps you need a larger vehicle to fit your growing family. These are just a few; the list goes on and on. Still, with all the good reasons to buy a new car, there is a very compelling one to consider holding on to the old car. Cost!

Taking a purely left-brain, economic approach, keeping your old car could save you thousands of dollars.

It is true that older cars generally cost more to maintain, and are more likely to break down, resulting in an expensive and aggravating repair. But before you push the panic button, look at the economics.

For example: Let's assume you own a 1988 Chevrolet Corsica and you are faced with the decision of buying a new one or keeping the old one. How do the numbers stack up? Figure 9 shows a comparison of the projected five-year ownership costs for your 1988 Chevrolet Corsica to a new 1993 model. (The example is based on the assumption that you drive the national average of 14,000 miles per year.)

Why is there such a large difference in expected ownership costs? For one thing, your used car will depreciate at a rate much slower than the new one. Cars depreciate the fastest in the first few years. Therefore, your 1988 Corsica has already been through its highest depreciation. You can see from the example that the 1993 car will depreciate $5,994 more than the 1988 edition over the next five years.

More good news: Your insurance rates are likely to be lower on a used model. The example shows a $1,848 cost advantage for the used car. You may even consider raising the deductible on the collision portion of your premium, or dropping collision entirely, lowering your insurance payments even more.

And then there are finance charges. Chances are you have paid back your loan or are close to making that last payment. The prospect of four or five years of paying loan interest may put a dent in your budget planning.

As for various annual taxes and license fees, again the used car will cost approximately $210 less over five years, depending on the state you live in.

Fuel costs are likely to be slightly higher for the 1988 model. It is common for vehicles to drop off in fuel performance as they get older. Additionally, technological improvements for newer models often result in higher mileage per gallon.

After tallying up these costs, if you decide to keep your old car for another five years you will save $15,436 ($27,168 minus $11,732), right?

Well, not quite. What about maintenance and repair costs for the used car? There is no reliable data that will show how much you will have to spend on maintenance and repairs over the next five years, but we can assume that it will be more than you'll spend on the new car — particularly when you consider that the old car is out of warranty. However, this example does show how much you can budget each year for maintenance and repairs. If you spread the $15,436 over five years, you can pay about $3,100 per year before it would be more cost effective to buy a new Corsica.

The old heap doesn't look so bad now, does it?

To Fix or Not to Fix

There may come a time when you're faced with the decision of whether to repair your car or sell it and buy another one. Barring any sentimental reasons you may have for holding on to that loveable old heap, you'll want to make your decision based on objective financial criteria. As illustrated in Figure 9, it probably makes sense to repair your car and keep it.

However, even if you decide to get rid of the car and buy a new one, you still have to decide whether to repair it.

Unless you've got money to burn, if it costs as much to repair the car as it would to buy another one, it makes sense to skip the repairs and sell it for its scrap value. (And that's if your lucky. More often, you have to pay someone to tow away the old work horse!)

It's more likely, however, that the cost of repairs will be substantially less than the cost of replacing your current car. In this case, the decision of whether or not to fix your car is not so straightforward. It all depends on what the car would be worth with the repairs vs. what you could sell it for without the repairs.

So, should you fix it before selling it, or sell it as it is? In most cases, you will be better off fixing it. By spending a few hundred dollars to repair your car before selling it, you can increase its value by thousands of dollars. A car in good condition will be much easier to sell and it will command a much higher price.

For example, let's say you have a 1986 Nissan Sentra, which, if it were in good condition, would be worth $3,500. But, let's assume this car requires $500 in repairs.

If you decide not to repair it, you'll have your choice of selling it to a buyer who is looking for a fixer-upper or trading it to a dealer. In either case, because the car is not in good condition, you will be selling at a substantial discount, usually the wholesale price. In our example, the Sentra, without the repairs, would be worth about $2000. In addition, the buyer will probably take into consideration the cost of the needed repairs and deduct the $500 repair from the trade-in figure and pay you a net $1,500.

If, however, you opt to fix the car before selling it, the car can command its $3,500 full retail value. So, although you're out the $500 in repairs, you'll still net $3,000--$1,500 more than if you sold it without repairing it.

Here's an example of a good ad:

BMW '81 320i, Ascot-Gray, A/C, stereo, sunroof. Perfect cond. $6,500. (219) 689-0000

Unloading Your Old Car

Finally, you've considered all of the advantages and disadvantages of keeping your old car, and you have concluded that your best option is to buy a new one. So what should you do about that old lump you're now driving? The easiest solution would be to trade it in on the new car — no ads to write, no waiting by the phone for replies, and no worries about smog or safety certificates (in many states, the seller's responsibility). Of course, you pay for the convenience. The dealer has to make a fair profit on any transaction, and if he takes your old car in trade, he's got to be able to turn around and sell it for more than he or she paid you. So the dealer will probably offer you the "wholesale price," or even less if the car needs fixing. But, if you sold the car yourself, you could probably get a figure between the dealer wholesale price and the dealer's full retail price.

Selling it yourself needn't be an ordeal. It just takes a little marketing. As anyone on Madison Avenue will tell you, a successful sale involves the right packaging, the right price, and the right advertising. First, make an honest assessment of the old heap. If relatively inexpensive items aren't working, fix them — things like mufflers, the radio, or the speedometer.

It's amazing, but the vast majority of buyers will overlook mechanical ills, sometimes major ones, if the paint shines and the interior seems fresh. And conversely, a car in good mechanical shape but with a ratty appearance will turn off most buyers. So by all means, have your car detailed before you try to sell it yourself, or even before you trade it in to the dealer. It will be worth the extra cost because a good-looking car will not only fetch a higher price, it'll sell faster.

Once you've transformed the old heap into a "meticulously maintained, fine motorcar," attach a price to it. Look in your local newspaper's classifieds, in tabloids like *The Recycler* or *Auto Trader* (usually available at convenience stores), or in any used vehicle pricing guide such as *Kelley Blue Book* or *NADA Used Car Guide,* and see how similar cars are priced. Then write your own ad: include the brand name, the year, the model, and any important features such as automatic transmission or air conditioning. Mention the condition of the car. And always list a price even if you're willing to bargain. (However, never use phrases like "asking $7,000" with the price. You don't want to broadcast your willingness to negotiate.) If you have space in the ad, mention the car's color.

Now just sit back and let the calls come in. When someone makes you an acceptable offer, always ask for cash or a bank check. Never take a personal check.

Section Three
Annotated Vehicle Charts

The Complete Car Cost Guide

Annotated Vehicle Charts

This section will give you an in-depth explanation of every aspect of the vehicle charts in this book. On pages 52A-55A you will find a sample vehicle chart divided into distinct sections: vehicle description; purchase price; warranty and maintenance; and ownership costs.

The text on pages 57A-61A explains the formulas we used and the assumptions made to calculate the ownership costs for every vehicle included in this book.

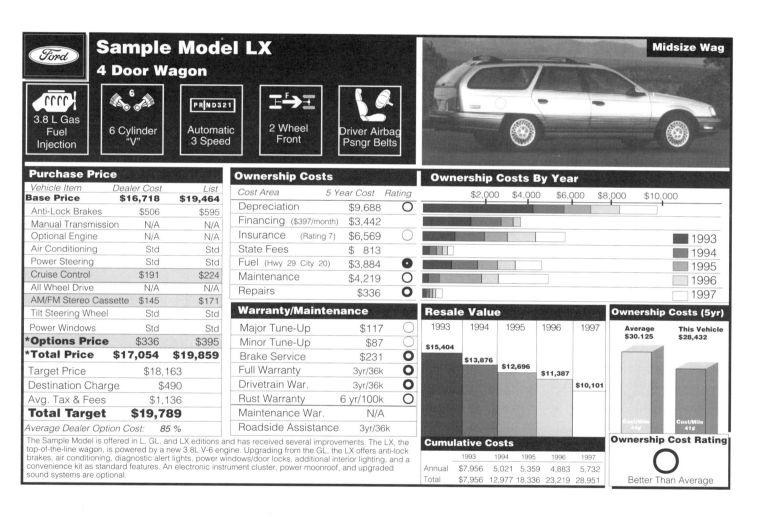

It is very important for a new car buyer to distinguish between *price* and *cost*. The consumer who compares vehicles based only on purchase price is using a set of criteria that can be misleading. The true cost of owning a car is not the money you exchange for the "pink slip," the true cost is what you pay after you drive your new car off the dealer's lot. This cost includes depreciation, interest payments, insurance, state fees, fuel, maintenance, and repairs. Therefore, when you shop and compare new cars, use the purchase price to find the cars that fit your budget, and use the expected ownership costs to determine the car that is the best value.

Annotated Vehicle Charts

Vehicle Description

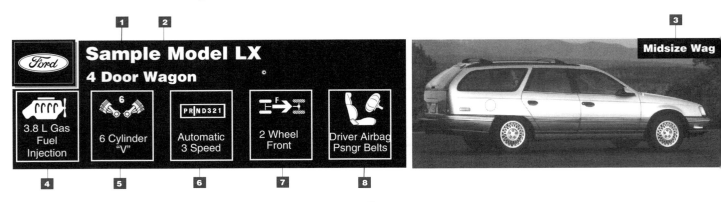

1 The number of doors and body style of the vehicle, defined as follows:

Convertible	An open-topped vehicle.
Coupe	Usually 2-door and bucket seat design, including a traditional hood and trunk.
Hatchback	(Also known as fastback or liftback) A sloping design from the rear window backward (often a very steep slope). The rear window and rear trunk lid are often one piece.
Sedan	(Also known as hardtop or notchback) Usually a 4-door design, including a traditional hood and trunk.
Wagon	Usually square-looking rear end with rear cargo door that provides for expanded cargo area behind the seats.

2 The vehicle class, defined as follows:

Compact	An automobile with a wheelbase over 98.9 and under 104.9 inches.
Compact Wagon	A wagon with a wheelbase over 98.9 and under 104.9 inches.
Large	An automobile with a wheelbase over 110 inches.
Large Wagon	A wagon with a wheelbase over 110 inches.
Luxury	Priced over 30K with the exception of "Sport" vehicles or model lines that straddle $30K.
Midsize	An automobile with a wheelbase over 104.9 and under 109.9 inches.
Midsize Wagon	A wagon with a wheelbase over 104.9 and under 109.9 inches.
Sport	A performance oriented vehicle.
Subcompact	An automobile with a wheelbase under 98.9 inches.
Subcompact Wagon	A wagon with a wheelbase under 98.9 inches.

3 The brand and model name of the vehicle.

4 The engine type/engine aspiration

Gas	Standard piston powered engine.
Turbo	Turbo charged piston engine.
Diesel	Diesel design piston engine requiring diesel fuel.
Turbodiesel	Turbo charged diesel design piston engine.
Rotary	Wankel engine (no pistons).
Carburetor	Mechanical air/fuel mixing device.
Fuel Injection	A pump/injector method of metering fuel.

5 Cylinders

In a rotary engine, this number refers to the number of chambers.

In-line	Cylinders are in a straight line.
Opposing	Cylinders are in two rows facing each other.
V-shaped	Cylinders are in two rows that form a V.

6 Transmission

Manual 3-speed, Manual 4-speed, Manual 5-speed, Manual 6-speed, Automatic 3-speed, Automatic 4-speed, Automatic 5-speed.

7 Drive

Front Wheel Drive	Front wheels deliver power.
Rear Wheel Drive	Rear wheels deliver power.
4 WD On-Demand (a.k.a. Part-time)	All-wheel drive that the driver manually engages and disengages.
4 WD Full-Time (a.k.a. AWD)	All-wheel drive that automatically delivers power to wheels as needed.

8 Restraint system

Dual Airbags	Driver and passenger airbags are standard.
Driver Airbag Psngr Opt	Driver airbag is standard—passenger airbag is optional.
Driver Airbag Psngr Belts	Driver airbag is standard—passenger airbag is not available.
Belts Standard Airbag Opt	Automatic seat belts are standard for driver and passenger—airbags are optional for driver and passenger.
Belts Standard Driver Air Opt	Automatic seat belts are standard for driver and passenger—airbags are optional for driver only.
Automatic Seatbelts	Automatic seatbelts are standard for driver—airbags are not available.
Manual Seatbelts	Manual seatbelts are standard for driver—automatic seatbelts are optional.
Manual Seatbelts Only	Manual belts are standard for driver—automatic belts are not available.

Annotated Vehicle Charts

The Complete Car Cost Guide

Purchase Price

	1	**2**	**3**
Vehicle Item		Dealer Cost	List
Base Price		**$9,216**	**$10,320**
ABS Brakes		Std	Std
Auto 3 Speed Trans.		$417	$490
2.3 L Eng.		$561 **	$660
Air Conditioning		$574	$675
Power Steering		Std	Std
Cruise Control		$149	$175
All Wheel Drive		N/A	N/A
AM/FM w/Cass.		Dlr	Dlr
Tilt Wheel		$106	$125
Power Windows		$242	$285
***Options Price**		*$417*	*$490*
***Total Price**		**$9,633**	**$10,810**
Target Price			$10,869
Destination Charge			$400
Avg. Tax & Fees			$724
Total Target $			**$11,593**
Average Dealer Option Cost:			91 %

(Labels **4** point to the feature list column; **5** to *Total Price; **6** to Target Price; **7** to Destination Charge; **8** to Avg. Tax & Fees; **9** to Total Target $; **10** to Average Dealer Option Cost.)

1 The left-hand column identifies specific car items. Each line represents a different item, such as the Base Price and various options.

2 Dealer Cost is the dealer's factory invoiced cost for each car item. It is the price paid by the dealer, not including any manufacturer's incentive or holdback.

3 List price. This is the manufacturer's suggested retail price for the particular car item.

4 The feature list. The list includes the most popular major features/options in use today. If a feature does not have a price listed next to it, it will have one of the following designations:

Dlr	Dealer Installed Equipment
N/A	Not available on this model
N/C	No Charge
Pkg	Available as part of a package of options
Std	Standard equipment for this model
N/R	Not reported at time of printing
Grp	Available only with another option

Some features have a double asterisk ("**") next to the price. This indicates that the particular feature can only be purchased at that price if certain other requirements are met. If there is a choice, we show the option price that can be purchased without restrictions. We only resort to list options with restrictions if there is no unrestricted way to purchase the item.

5 Total Price is the sum of the base car price plus the options price.

6 The Target Price is the price that you can reasonably expect to negotiate for the vehicle as configured; however, it is not necessarily the lowest price you may find. It includes an average acceptable mark-up for the dealer, which varies from model to model based on market conditions. Recent market conditions have been included in the Target Prices shown.

7 The destination charge is usually the same anywhere in the country, but may vary in some locations.

8 Forty seven out of fifty states have a sales tax on vehicles, which ranges from 1.5% to 8%. We have used an average tax rate of 5%, based on the price of the specific vehicle shown. Furthermore, most states have annual registration fees, which are often paid to the dealer at the time of new car delivery. An average figure is used here based on the price of the specific vehicle. You can refer to Figure 13 on page 64A to determine the applicable fees for your state.

9 The Total Target is the grand total price that you can reasonably expect to pay for the vehicle as configured. It is the sum of the Target Price, plus all taxes, destination fees, and registration fees.

10 The Average Dealer Option Cost is the average percent of list price that the dealer pays for options. You can use this to estimate the dealer's cost for any option not specifically shown in the chart. For example, if the Average Dealer Option Cost shown is 91%, and the suggested retail price of leather seats is $400, then the dealer's cost for the leather seats is about .91 x $400, or $364. This is shown as "N/A" on vehicles that have no available manufacturer-installed options.

* Only the shaded options are included in the Options Price and Total Price figures. They are also included in ownership cost calculations. Other options are listed for your convenience.

Shading— All shaded option are included in the total price of the vehicle, as well as in all ownership cost calculations. In an effort to make true "apples-to-apples" comparisons, we attempt to include the same equipment on similar vehicles. (It would not be fair for one compact car to be evaluated with a manual transmission while another is evaluated with an automatic transmission.)

The options that we include/shade differ by class, as shown in the chart below. Generally, we include/shade options that are equipped in two-thirds or more of the vehicles sold in that class. For instance, in recent model years, over 80% of all midsize vehicles were equipped with cruise control, yet less than 33% of subcompact models had this feature. Therefore, we include/shade cruise control for the midsize class, but not for subcompacts.

Additionally, the only options that are included/shaded are those with a price, or those marked "N/C" (No Charge). Items that are standard equipment, items marked "**" (restrictions), items marked "Dlr" (Dealer Installed), and items marked "Pkg" or "Grp" (available only as part of a package of options) are not included/shaded.

Optional engines are shaded and included in the analysis based on horsepower and engine size. For each class of vehicle, we establish a "target" horsepower and engine size based on the average horsepower and the average size of all standard engines on all models in the class. If the standard engine of a particular model meets the size and horsepower targets, it is used in the analysis. If the standard engine is too small or has insufficient horsepower, then we choose the least expensive engine available without purchase restrictions that does meet the size and horsepower targets. If no engine meets the requirements, the largest engine is used.

This table shows features that are either shaded or not shaded for each vehicle class.
Yes-Shaded
No-Not Shaded

Shaded Features By Vehicle Class

	Cruise Control	Air Cond.	Auto Trans.	Power Steer	4 Wheel Drive	AM/FM Cassette	Tilt Wheel	Power Windows	ABS Brakes	Gas Guzzler Tax
Subcompact	No	No	No	Yes	No	No	No	No	No	Yes
Compact	No	Yes	Yes	Yes	No	Yes	No	No	No	Yes
Midsize	Yes	Yes	Yes	Yes	No	Yes	No	No	No	Yes
Large	Yes	Yes	Yes	Yes	No	Yes	No	Yes	No	Yes
Luxury	Yes	Yes	Yes	Yes	No	Yes	Yes	Yes	Yes	Yes
Sport	No	Yes	No	Yes	No	Yes	No	No	No	Yes
Subcmpct Wagon	No	No	No	Yes	No	No	No	No	No	Yes
Compact Wagon	No	Yes	Yes	Yes	No	Yes	No	No	No	Yes
Midsize Wagon	Yes	Yes	Yes	Yes	No	Yes	No	No	No	Yes
Large Wagon	Yes	Yes	Yes	Yes	No	Yes	No	Yes	No	Yes

Annotated Vehicle Charts

Warranty and Maintenance

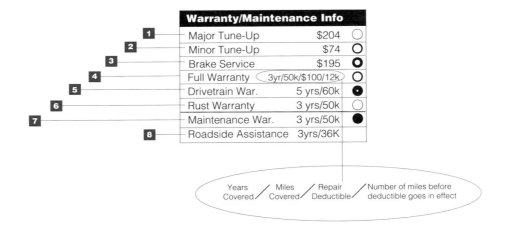

NOTE: For all items in this section, the rating symbol compares the expected dollar expense for the specific vehicle against the average expected dollar expense for vehicles of the same category. Parts prices are manufacturer labeled, while labor rates are the national average for the brand.

1 The average cost of a major tune-up (minor tune-up, distributor cap and rotor replacement, fuel filter and PCV valve replacement, and ignition system inspection).

2 The average cost of a minor tune-up (replacing the air cleaner and spark plugs, inspecting the distributor cap, rotor, and ignition wires, checking compression, and adjusting the ignition timing and idle speed).

3 The average cost of brake service (replacing the pads on disk brakes, or shoes on drum brakes, on all four wheels, and checking, bleeding, and adjusting the brake system).

4 The full warranty on the vehicle. The full warranty typically covers parts and labor for any factory-installed part that is defective in material or workmanship under normal use. This typically excludes tires, emission system parts (covered by a separate warranty), expendable maintenance items, glass breakage, and air conditioning lubricant. Some manufacturers, Saab for example, include a deductible on each repair after a specific number of miles.

5 Coverage on the engine, transaxle, transmission, and axle and drive components. For some manufacturers, this warranty applies to the original purchaser only. In other cases, the coverage changes if the car is sold. Also, some manufacturers require a fixed payment for any repair covered under this warranty.

6 Rust-through coverage. This applies to perforation only, generally meaning complete rust-through in a sheet metal panel. Surface corrosion resulting from stone chips or paint scratches are typically not covered.

7 Maintenance Warranty. With this coverage, all routine services and oil changes will be paid for when the work is performed at an authorized dealer (includes wiper blades, brake pads, light bulbs, wheel alignments, and other wear items).

8 Roadside Assistance. The length of time and mileage for which the manufacturer will provide a toll-free 800 number and free towing in the event of a breakdown. Many manufacturers provide additional services as part of their roadside assistance plan. To determine all the components of a roadside assistance plan for a particular vehicle, ask your dealer.

The Complete Car Cost Guide

Ownership Costs

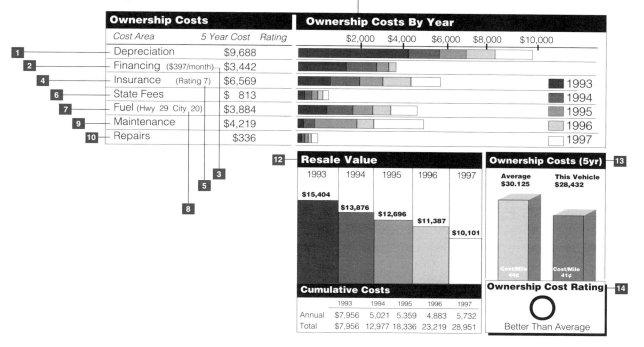

NOTE: For all rating symbols other than "overall rating," the symbol compares the expected figure for the specific vehicle against the average expected figure for vehicles of the same category and price range. The overall rating symbol considers all vehicles in addition to vehicles of the same category and price range.

1 The total cost of depreciation over five years. (See page 57A)

2 The total cost of financing over five full years. Financing cost includes the interest on a loan, but not the principal. (See Page 59A)

3 The monthly payment. (See page 59A)

4 The total cost of insurance over five years. (See page 58A)

5 The insurance rating for this vehicle. If a surcharge applies, the rating is followed by "+," "++," "sport," or "sport+," depending on the degree of the surcharge. The charge is also reflected in the premium. If a rating is estimated, it is followed by "[Est]."

6 The total cost of state fees over five years. (See page 59A)

7 The total cost of fuel over five years. (See page 59A)

8 The EPA reported miles per gallon figures for the specific vehicle. "Hwy" refers to highway driving. If a figure is estimated, it is indicated by "[Est]." If premium gasoline is required, it is indicated by "Prem."

9 The total cost of maintenance over five years. (See page 60A)

10 The total cost of repair insurance over five years. (See page 61A)

11 **Ownership Costs By Year.** In this graph, each cost area is represented by a horizontal bar that shows the total of that cost over a five-year period. The bar is broken up into different shaded sections, with each section representing one year between 1993 and 1997. For example, look at the depreciation bar in the sample chart. The bar consists of five shaded segments that extend to $9,688. This is the total amount of depreciation for this vehicle over the next five years However, if you want to be more specific, you can measure the distance of a shaded segment to find out the depreciation for a single year. For instance, note that the shade representing 1994 extends from about $4,200 to $5,800. The difference between these two figures, $1,600, is the amount of depreciation for 1994. (In some cost areas, most often "Repairs," it can be difficult to distinguish where one shade ends and the next begins. However, since shorter bars always mean less cost compared to longer ones, it is a positive sign when shaded segments are extremely close together.)

12 **Resale value.** The bar graph illustrates the vehicle's expected resale value over time. This graph allows you to determine what this vehicle is likely to be worth at annual intervals during the five-year period.

13 **Ownership Costs (5yr).** This graph illustrates the vehicle's total expected ownership costs after a five-year period and how it compares to the average cost of vehicles in its class and price range. Additionally, the cost per mile is shown on the bottom of each bar.

14 **Ownership Cost Rating.** The ownership cost rating is a summary of a vehicle's economic value, considering its purchase price and overall ownership costs. This rating compares a vehicle's economic value to that of other vehicles of similar price and style.

Why are there no rating symbols for "Financing" and "State Fees?" Because these two costs are *mathematically derived* from the purchase price of the car, using our economic assumptions. If two cars have the exact same purchase price, they will have the exact same finance charge and state fees, so there is no point in comparing these areas. On the other hand, these two cars could have very different depreciation, insurance, fuel, repair, and maintenance costs.

Annotated Vehicle Charts

The Complete Car Cost Guide

Figure 10

Ownership Costs (Assumptions)

Ownership Period	5 years
Annual Inflation	4.0%
Annual Mileage	14,000 miles
Dealer Cost/Manufacturer Suggested List Price/Destination Fees	As reported by manufacturers. Actual Dealer cost may be somewhat lower than listed dealer cost due to manufacturer allowances.
Target Purchase Price	Based on dealer cost of an individual model with options shown and shaded. Varies by market condition, by class, and by individual model. Target price is always more than dealer cost, usually below list price, but sometimes higher.
Luxury/Gas Guzzler taxes:	Rates that went into effect on January 1, 1991.
State Sales/Use tax/ State Registration fees	5% sales tax is used as a nationwide average; 1% initial state registration fee based on MSRP. Refer to Figure 11 on page 60A to determine the exact figure for your state.
Resale Value	There is a large range in used car prices, ranging from a low "wholesale auction" price, to a high "private party" transaction price. Prices here are assumed to be close to the private party price. Each individual vehicle will follow a pattern set by earlier models, by brand, by country of origin, by vehicle class, by vehicle price range, and by area of the country.
Insurance	Principal operator is under age 65; all drivers have more than six years experience with no chargeable accidents; personal use; lives in a suburban/urban community, with: Collision — $500 Deductible Comprehensive — $500 Deductible Personal liability — $100,000/$300,000 Medical — $25,000 Property — $50,000 Uninsured driver — $25,000/$50,000
Finance Costs	20% down payment on a 48 month loan. Annual interest rate of 9.5%.
State Fees	1% of MSRP in the first year, reduced by .1% in each successive year. Refer to Figure 13 on page 64A to determine the exact figure for your state.
Fuel	U.S. Government EPA mileage figures. Mileage is 60% highway, 40% city. Fuel cost per gallon is $1.15 for unleaded regular, $1.27 for premium, and $1.30 for diesel (subject to inflation).
Repairs	Cost of a $0 deductible extended service contract that will pay for repairs for 5 years or 70,000 miles. Figures used are actual prices averaged from two nationally available service contract providers.
Maintenance	Services performed generally at manufacturer's suggested intervals where stated. Other services done at selected intervals. (See page 60A) Cost per service is based upon industry-standard service times and national labor rate averages by brand. Parts prices are based on manufacturers suggested list price where available.

Annotated Vehicle Charts

Ownership Cost Derivation

The following pages describe the methods used, as well as the basic assumptions made for each ownership cost area shown in the vehicle charts.

Depreciation

The vehicle charts show, for each new vehicle, the projected annual and cumulative amount of depreciation over five years. A rating symbol gives you an easy way to compare one car's depreciation to all other cars of a similar class and price. Additionally, the charts have a resale section that shows the expected resale value of the car each year over the next five years.

Depreciation and resale value are two measures of the same element. A car's purchase price less depreciation equals a car's resale value. By the same token, a car's purchase price less its resale value equals its depreciation.

We project the future resale value of a car on the basis of several factors. We assume that the vehicle will be purchased at the Total Target Price, which is the total "out-the-door" price that the buyer can expect to pay. This price includes state and local taxes, destination fees, and luxury car and/or Gas Guzzler taxes if applicable. We then factor in the historical depreciation for that specific car model. We also consider depreciation trends by brand, vehicle category, price, and country of origin. For example, to forecast the expected resale value of a 1993 Ford Mustang LX Sedan, we consider the resale value history of previous Mustang LX Sedans, of other Ford models, of compact class cars, of American-built cars, and of cars in the $10,000 to $12,000 price range. When actual history is lacking for the exact model, we rely more on these other areas. To this statistical method, we add our own economic, industry, and model-specific expectations to determine the expected resale value for that particular vehicle.

Depreciation is determined by a car's resale value, and can vary somewhat depending on where you're selling the car, who you're selling it to, optional equipment, and the condition and mileage of the car. We assume values for the Western United States (East coast values are a few percentage points higher), and that your car is sold to a private buyer. Furthermore, in our resale value/depreciation calculations, we include the cost of optional equipment (the shaded options in the vehicle charts). Additionally, we assume that the car is in good, but not mint, condition, and that you've driven it approximately 14,000 miles per year.

Depreciation is determined by a car's resale value, and can vary somewhat depending on where you're selling the car, who you're selling it to, optional equipment, and the condition and mileage of the car.

The Complete Car Cost Guide

It's almost always true that an older car will cost less to insure than a newer car of the same model. However, over time, inflation makes the insurance premium increase faster than the age of the car makes it decrease, so the overall cost goes up.

Insurance

The vehicle charts show the insurance industry symbol (see Figure 4 on page 10A) of each car, courtesy of Insurance Services Office. The lower the symbol, the less expensive collision and comprehensive insurance will be. Additionally, the charts show, based on the assumed coverages and deductibles, how much you can expect to spend on insurance over five years. Our rating symbol lets you easily compare the cost of insuring different cars in the same class and price range.

Many personal, geographic, and political factors determine your exact insurance premium. We assume that you're under age 65. You have a good driving record, drive about 14,000 miles per year, and have no inexperienced drivers in your household.

Insurance rates vary tremendously depending on where you live. We assume that you live in a suburban area in a state with average insurance rates. You can refer to Figure 11 on page 62A to adjust the insurance figures to match your state's figures.

You may be wondering why the annual insurance premium shown in the vehicle charts goes up every year, especially since you may know that it generally costs less to insure a used car than a new one. It's almost always true that an older car will cost less to insure than a newer car of the same model. However, over time, inflation makes the premium increase faster than the age of the car makes it decrease, so the overall cost goes up.

Of all car costs, insurance is the most variable. Your actual rates will probably differ somewhat from those found in *The Complete Car Cost Guide*. However, when comparing cars, the difference between our projections and your actual rates will be consistent from car to car. If you need more precise information, we recommend that you call a local insurance agent to determine the actual insurance costs for each automobile that you consider.

The Complete Car Cost Guide

Finance Costs

The vehicle charts show the total interest charges you can expect to pay for the car over five years. They also show your monthly car payment if you purchase the car at the "target purchase price."

There is no rating symbol for finance costs. This is because these costs do not depend on the car you buy, but are related to the purchase price. If two different cars are purchased for the same amount, they will have the same finance cost.

Our financial projections are based on the latest economic information. We assume that you will put down 20% immediately and take a 48 month, 9.5% loan.

If you're considering a loan with terms other than these, you can use Appendix I to determine your monthly payment.

Even if you pay cash for your car instead of borrowing money to pay for it, you should factor a finance cost into the cost of ownership. The reason for this is that if you didn't invest the money in the car, it could be in the bank earning interest.

State Fees

The charts show the amount you can expect to pay in state taxes and registration fees over five years. Our state fee calculations assume initial state fees of 1% of the current MSRP of the vehicle, reduced by .1% in each successive year.

Registration fees and taxes vary by state. You can use Figure 13 (see Page 64A) to determine the exact fee structure for your state.

Fuel

The vehicle charts show EPA (U.S. Environmental Protection Agency) fuel mileage figures for both "highway" and "city" driving. If automatic transmission is standard or is a shaded option, the fuel figures are for automatic transmission; otherwise, the figures are based on manual transmission. The charts also show the amount you can expect to pay in fuel expenses over a five-year period. A rating symbol gives you an easy way to compare this vehicle's fuel expense against all other vehicles of a similar class and price.

To determine actual fuel costs, we assume that vehicles will perform closer to the "highway" mileage figure. We assume that fuel will cost $1.15 per gallon for unleaded regular, $1.27 per gallon for unleaded premium, and $1.30 for diesel. Additionally, we assume that fuel costs will increase by the rate of inflation and that you will drive 14,000 miles per year.

Annotated Vehicle Charts

Maintenance

The vehicle charts show the amount you can expect to pay for routine maintenance expenses over five years. A rating symbol gives you an easy way to compare a particular car's maintenance costs against all other cars of a similar class and price. The charts also show average costs to perform the standard maintenance services of a major tune-up, a minor tune-up, and a standard brake service.

We include the following services in determining our maintenance cost figures:

- Oil changes
- Major tune-ups
- Minor tune-ups
- Basic brake service
- Basic clutch service
- Basic automatic transmission service
- Alignments
- Front bearing service
- Cooling system maintenance
- Shock absorber service
- Muffler replacement
- Hoses and belts (including timing belt)
- Fluids and filters
- Tires
- Batteries
- Headlamp replacement
- Regular inspections

We assume that services take place at an authorized dealer location, using standard flat labor times, an average hourly labor rate for each individual brand, and manufacturer's suggested retail prices for parts, where available. You can often get better prices than these. (We use them for consistency from car to car.)

We generally factor in a manufacturer suggested mileage or time interval to perform a particular service. If the manufacturer does not specify, we apply our own intervals for particular services. We assume that most trips are longer than 5 miles ("normal" driving conditions).

Repairs

The vehicle charts project the amount you can expect to pay for repair expenses over five years or 70,000 miles. A rating symbol helps you easily compare one car's expected repair costs against all other cars in the same class. The charts also show manufacturers' warranties for full coverage, power train coverage, corrosion coverage, maintenance coverage, and roadside assistance coverage. Repair costs in the charts take into account all of these manufacturer warranties.

To determine the cost projection, we take advantage of service contract pricing. Conceptually, a service contract is an insurance policy protecting against repair costs. In exchange for your "premium," the service contract provider agrees to pay for any needed repairs. (See page 24A for a more detailed description of service contracts.) To provide this service profitably, the service contract provider has to make sure that, on average, the actual cost of repairs on the vehicle is less than the premium price. Service contract providers take great pains in pricing their products low enough to stay competitive, yet high enough to make a profit.

Therefore, we can expect that the service contract price is an excellent estimate of likely repair costs (in fact, their price is somewhat higher than the likely repair costs, allowing for a reasonable profit).

Whether or not you purchase a service contract, you can use the service contract price to estimate the cost of repairs.

We utilize pricing from two separate service contract providers to determine the repair costs shown in this book. The plans used from both providers are for five years or 70,000 miles, and are $0 deductible plans, meaning you pay nothing during the plan period other than the initial premium. The price shown is what you can reasonably expect to pay. It allows for a reasonable profit for the service contract provider, as well as for the dealer who sells the contract.

It's important to separate hype from reality when considering how a car's expected repair incidence will influence your decision on which car to buy.

At one time, there were vast differences in relative reliability among vehicles. Today, most any new vehicle you purchase is likely to be highly reliable. Other than minor state fees, the expected cost of repairs in the first five years will be less than any other cost associated with your new car! That's right, and it's true even for the vehicles with the highest expected repair cost. The cost of depreciation, insurance, interest, fuel, and maintenance will all be considerably higher than the cost of repairs.

Therefore, you are likely to jump to the wrong economic conclusion if you purchase a car because you heard it's "very reliable," or avoid a car because you heard it's a "lemon." In fact, the difference in repair cost between the car with the highest expected repair cost and the one with the lowest expected repair cost is only $955.

The Complete Car Cost Guide

Figure 11

Insurance Factors

	1990 Rank	1990 Index
Alabama	31	0.87
Alaska	7	1.29
Arizona	20	1.1
Arkansas	40	0.79
California	4	1.47
Colorado	21	1.05
Connecticut	2	1.51
DC	9	1.28
Delaware	8	1.29
Florida	19	1.11
Georgia	13	1.18
Hawaii	14	1.16
Idaho	45	0.72
Illinois	25	0.95
Indiana	41	0.78
Iowa	51	0.55
Kansas	47	0.71
Kentucky	48	0.7
Louisiana	17	1.11
Maine	29	0.89
Maryland	12	1.24
Massachusetts	11	1.27
Michigan	15	1.15
Minnesota	36	0.84
Mississippi	34	0.85
Missouri	32	0.87
Montana	44	0.73
Nebraska	46	0.71
Nevada	18	1.11
New Hampshire	6	1.3
New Jersey	1	1.71
New Mexico	30	0.89
New York	5	1.31
North Carolina	33	0.86
North Dakota	50	0.64
Ohio	28	0.9
Oklahoma	26	0.91
Oregon	23	0.99
Pennsylvania	10	1.27
Rhode Island	3	1.49
South Carolina	27	0.9
South Dakota	49	0.66
Tennessee	39	0.82
Texas	16	1.13
Utah	38	0.82
Vermont	35	0.84
Virginia	24	0.95
Washington	22	1.03
West Virginia	37	0.84
Wisconsin	43	0.73
Wyoming	42	0.75
NATIONAL AVG		1

Insurance Adjustment Table

To calculate the total five-year insurance cost for each vehicle listed in the "Vehicle Charts," we use a national average. To be more precise, the Insurance Factors table (Figure 11) will allow you to adjust the total insurance cost listed in the vehicle charts for the state in which you live. To make this adjustment, simply find your state on the chart and then multiply its Index by the total insurance cost shown in the vehicle charts. The figures in the left column (Rank) shows the highest insurance rate (1- New Jersey) to the lowest (51-Iowa).

Remember that insurance rates can vary greatly, even within a state. For an exact insurance rate, you should contact an insurance agent. We recommend that you get quotes from several different agents before you actually buy your insurance.

Annotated Vehicle Charts

Mileage Adjustment Table

To calculate the total five-year fuel cost for each vehicle listed in the charts, we assume that the vehicle is driven 14,000 miles per year, and that the price of gasoline is $1.15 per gallon. If your annual mileage differs from our assumptions, or if you expect to pay more (or less) for gas, then this table allows you to adjust the total fuel cost for any vehicle in which you are interested. To make this adjustment, first find your annual mileage in the left-hand column. Then, find the column that corresponds with the price you pay for a gallon of gasoline. Finally, multiply the number (factor) shown in this column by the total fuel cost shown in the vehicle charts.

Figure 12

Annual Mileage	Fuel Price Per Gallon								
	$0.90	$1.00	$1.05	$1.10	$1.15	$1.20	$1.25	$1.35	$1.50
4,000	0.22	0.25	0.26	0.27	0.29	0.3	0.31	0.34	0.37
6,000	0.34	0.37	0.39	0.41	0.43	0.45	0.47	0.5	0.56
8,000	0.45	0.5	0.52	0.55	0.57	0.6	0.62	0.67	0.75
10,000	0.56	0.62	0.65	0.68	0.71	0.75	0.78	0.84	0.93
12,000	0.67	0.75	0.78	0.82	0.86	0.89	0.93	1.01	1.12
14,000	0.78	0.87	0.91	0.96	1	1.04	1.09	1.17	1.3
16,000	0.89	0.99	1.04	1.09	1.14	1.19	1.24	1.34	1.49
18,000	1.01	1.12	1.17	1.23	1.29	1.34	1.4	1.51	1.68
20,000	1.12	1.24	1.3	1.37	1.43	1.49	1.55	1.68	1.86
22,000	1.23	1.37	1.43	1.5	1.57	1.64	1.71	1.84	2.05
24,000	1.34	1.49	1.57	1.64	1.71	1.79	1.86	2.01	2.24
26,000	1.45	1.61	1.7	1.78	1.86	1.94	2.02	2.18	2.42
28,000	1.57	1.74	1.83	1.91	2	2.09	2.17	2.35	2.61
30,000	1.68	1.86	1.96	2.05	2.14	2.24	2.33	2.52	2.8
35,000	1.96	2.17	2.28	2.39	2.5	2.61	2.72	2.93	3.26
40,000	2.24	2.48	2.61	2.73	2.86	2.98	3.11	3.35	3.73
45,000	2.52	2.8	2.93	3.07	3.21	3.35	3.49	3.77	4.19
50,000	2.8	3.11	3.26	3.42	3.57	3.73	3.88	4.19	4.66
55,000	3.07	3.42	3.59	3.76	3.93	4.1	4.27	4.61	5.12
60,000	3.35	3.73	3.91	4.1	4.29	4.47	4.66	5.03	5.59
65,000	3.63	4.04	4.24	4.44	4.64	4.84	5.05	5.45	6.06
70,000	3.91	4.35	4.57	4.78	5	5.22	5.43	5.87	6.52
75,000	4.19	4.66	4.89	5.12	5.36	5.59	5.82	6.29	6.99
80,000	4.47	4.97	5.22	5.47	5.71	5.96	6.21	6.71	7.45
85,000	4.75	5.28	5.54	5.81	6.07	6.34	6.6	7.13	7.92
90,000	5.03	5.59	5.87	6.15	6.43	6.71	6.99	7.55	8.39
95,000	5.31	5.9	6.2	6.49	6.79	7.08	7.38	7.97	8.85
100,000	5.59	6.21	6.52	6.83	7.14	7.45	7.76	8.39	9.32

The Complete Car Cost Guide

Figure 13 **State Fees, Regulations, and Insurance Information**

State	Taxes Sales/Use Tax Rate	Based on	Additional Local Tax	Registration Fees Fixed Fee	By Value[1]	By Weight[2]	Insurance Competitive	No Fault	Minimum Financial Responsibility
Alabama	2.00%	Net of trade	Yes	$15.00	$14.70	$0.00	Somewhat	No	20/40/10
Alaska	0.00%		Yes	$5.00	$35.00	$0.00	Somewhat	No	50/100/25
Arizona	5.00%	Net of trade	Yes	$4.00	$360.00	$0.00	Yes	No	15/30/10
Arkansas	4.50%	Full Value	Yes	$5.00	$0.00	$25.00	Yes	Yes	25/50/15
California	7.25%	Full Value	Yes	$10.00	$308.00	$0.00	Somewhat	No	15/30/5
Colorado	3.00%	Net of trade	Yes	$5.50	$0.00	$15.00	Yes	Yes	25/50/15
Connecticut	6.00%	Net of trade	Yes	$16.00	$0.00	$0.00	Somewhat	Yes	20/40/10
DC	6.00%	Full Value	No	$10.00	$0.00	$45.00	Somewhat	Yes	10/20/05
Delaware	2.00%	Net of trade	No	$15.00	$0.00	$0.00	Somewhat	Yes	15/30/10
Florida	6.00%	Net of trade	No	$31.25	$0.00	$30.60	Yes	Yes	10/20/10
Georgia	3.00%	Net of trade	Yes	$5.00	$0.00	$0.00	Somewhat	No	10/20/05
Hawaii	4.00%	Net of trade	Yes	$0.00	$0.00	$15.00	Somewhat	Yes	25/25/10
Idaho	5.00%	Net of trade	No	$8.00	$0.00	$0.00	Yes	No	25/50/15
Illinois	6.25%	Net of trade	Yes	$5.00	$0.00	$0.00	Yes	No	20/40/15
Indiana	5.00%	Net of trade	No	$5.00	$0.00	$12.75	Somewhat	No	25/50/10
Iowa	4.00%	Net of trade	No	$15.00	$150.00	$12.00	Somewhat	No	20/40/15
Kansas	4.90%	Net of trade	Yes	$3.50	$0.00	$27.25	Somewhat	Yes	25/50/10
Kentucky	6.00%	Full Value	Yes	$6.00	$0.00	$0.00	Yes	Yes	25/50/10
Louisiana	4.00%	Net of trade	Yes	$24.50	$10.00	$0.00	Somewhat	No	10/20/10
Maine	6.00%	Net of trade	Yes	$10.00	$0.00	$20.00	Yes	No	20/40/10
Maryland	5.00%	Full Value	No	$12.00	$0.00	$35.00	Somewhat	Yes	20/40/10
Massachusetts	5.00%	Net of trade	Yes	$50.00	$0.00	$20.00	No	Yes	10/20/10
Michigan	4.00%	Full Value	No	$11.00	$81.00	$0.00	Somewhat	Yes	20/40/10
Minnesota	6.50%	Net of trade	No	$2.00	$192.50	$0.00	Yes	Yes	30/60/10
Mississippi	3.00%	Net of trade	Yes	$4.00	$0.00	$0.00	Somewhat	No	10/20/05
Missouri	4.23%	Net of trade	Yes	$7.50	$0.00	$15.50	Yes	No	25/50/10
Montana	1.50%	Full Value	Yes	$5.00	$300.00	$0.00	Yes	No	25/50/10
Nebraska	5.00%	Net of trade	Yes	$6.00	$0.00	$0.00	Somewhat	No	25/50/25
Nevada	7.00%	Unique trade	Yes	$20.00	$210.00	$0.00	Yes	No	15/30/10
New Hampshire	0.00%		No	$20.00	$27.00	$31.20	Somewhat	Yes	25/50/25
New Jersey	6.00%	Net of trade	No	$0.00	$0.00	$67.90	Somewhat	Yes	15/30/5
New Mexico	3.00%	Net of trade	No	$5.45	$0.00	$42.00	Somewhat	No	25/50/10
New York	4.00%	Full Value	Yes	$5.00	$0.00	$22.50	Somewhat	Yes	10/20/05
North Carolina	3.00%	Net of trade	Yes	$35.00	$0.00	$0.00	Somewhat	No	25/50/10
North Dakota	5.00%	Net of trade	No	$7.00	$0.00	$50.00	Somewhat	Yes	25/50/25
Ohio	5.00%	Net of trade	Yes	$3.00	$21.50	$0.00	Yes	No	12.5/25/7.5
Oklahoma	3.25%	Full Value	No	$11.00	$187.50	$0.00	Somewhat	No	10/20/10
Oregon	0.00%		No	$14.00	$0.00	$30.00	Yes	Yes	25/50/10
Pennsylvania	6.00%	Net of trade	No	$15.00	$0.00	$24.00	Yes	Yes	15/30/5
Rhode Island	7.00%	Net of trade	Yes	$0.00	$0.00	$30.00	Yes	No	25/50/10
South Carolina	5.00%	Net of trade	Yes	$5.00	$0.00	$0.00	Somewhat	Yes	15/30/5
South Dakota	3.00%	Net of trade	No	$5.00	$0.00	$30.00	Yes	Yes	25/50/25
Tennessee	5.50%	Net of trade	Yes	$6.50	$0.00	$20.75	Somewhat	No	15/30/10
Texas	6.25%	Net of trade	Yes	$13.00	$58.50	$0.00	No	Yes	10/20/05
Utah	5.88%	Net of trade	Yes	$6.00	$0.00	$12.00	Yes	Yes	15/30/5
Vermont	5.00%	Net of trade	No	$10.00	$0.00	$42.00	No	No	20/40/10
Virginia	3.00%	Full Value	Yes	$10.00	$0.00	$30.00	Yes	Yes	25/50/20
Washington	7.50%	Full Value	Yes	$4.25	$368.10	$27.85	Somewhat	Yes	25/50/10
West Virginia	5.00%	Net of trade	Yes	$5.00	$0.00	$31.50	Somewhat	No	20/40/10
Wisconsin	5.00%	Net of trade	Yes	$5.00	$0.00	$0.00	Yes	Yes	25/50/10
Wyoming	3.00%	Net of trade	Yes	$5.00	$270.00	$0.00	Yes	No	25/50/20

Annotated Vehicle Charts

[1] Fee based on a $15,000 vehicle.
[2] Fee based on 3,000 pound vehicle weight.

Section Four
Vehicle Charts

Acura Integra RS
2 Door Hatchback

1.8L 140 hp Gas Fuel Inject. | 4 Cylinder In-Line | Manual 5 Speed | 2 Wheel Front | Automatic Seatbelts

Compact — GS Model Shown

Purchase Price

Car Item	Dealer Cost	List
Base Price	$10,861	$12,930
Anti-Lock Brakes	N/A	N/A
Automatic 4 Speed	$630	$750
Optional Engine	N/A	N/A
Air Conditioning	Dlr	Dlr
Power Steering	Std	Std
Cruise Control	N/A	N/A
All Wheel Drive	N/A	N/A
AM/FM Stereo Cassette	Dlr	Dlr
Steering Wheel, Tilt	Std	Std
Power Windows	N/A	N/A
*Options Price	$630	$750
*Total Price	$11,491	$13,680
Target Price	$12,222	
Destination Charge	$365	
Avg. Tax & Fees	$769	
Total Target $	**$13,356**	
Average Dealer Option Cost:	84%	

Ownership Costs

Cost Area	5 Year Cost	Rate
Depreciation	$4,046	○
Financing ($268/month)	$2,200	
Insurance (Rating 13)	$8,061	●
State Fees	$560	
Fuel (Hwy 29 City 23)	$3,321	○
Maintenance	$4,882	◉
Repairs	$571	○

Warranty/Maintenance Info

Major Tune-Up	$259	●
Minor Tune-Up	$128	●
Brake Service	$208	○
Overall Warranty	4 yr/50k	○
Drivetrain Warranty	4 yr/50k	○
Rust Warranty	4 yr/unlim. mi	○
Maintenance Warranty	N/A	
Roadside Assistance	N/A	

Resale Value
1993: $12,124 | 1994: $11,468 | 1995: $10,870 | 1996: $10,110 | 1997: $9,310

Cumulative Costs
	1993	1994	1995	1996	1997
Annual	$4,435	$3,889	$4,812	$3,814	$6,691
Total	$4,435	$8,324	$13,136	$16,950	$23,641

Ownership Costs (5yr)
Average: $26,241 | This Car: $23,641
Cost/Mile 37¢ | Cost/Mile 34¢

Ownership Cost Rating
○ Excellent

The 1993 Integra is available in eight models - RS, LS, LS Special, GS and GS-R hatchbacks; and RS, LS, and GS sedans. New for 1993, the Integra RS hatchback is offered in two new colors (Granada Black Pearl and Isle Green). A new 4-year/50,000-mile limited bumper-to-bumper warranty with a 4-year/unlimited-mile outer body rust-through warranty is standard. Other features include air dam, full wheel covers, wraparound body side moldings, integral fog lamps, and an optional body-color rear spoiler.

Acura Integra LS
2 Door Hatchback

1.8L 140 hp Gas Fuel Inject. | 4 Cylinder In-Line | Manual 5 Speed | 2 Wheel Front | Automatic Seatbelts

Compact

Purchase Price

Car Item	Dealer Cost	List
Base Price	$12,461	$14,835
Anti-Lock Brakes	N/A	N/A
Automatic 4 Speed	$630	$750
Optional Engine	N/A	N/A
Air Conditioning	Dlr	Dlr
Power Steering	Std	Std
Cruise Control	Std	Std
All Wheel Drive	N/A	N/A
AM/FM Stereo Cassette	Std	Std
Steering Wheel, Tilt	Std	Std
Power Windows	Std	Std
*Options Price	$630	$750
*Total Price	$13,091	$15,585
Target Price	$13,966	
Destination Charge	$365	
Avg. Tax & Fees	$877	
Total Target $	**$15,208**	
Average Dealer Option Cost:	84%	

Ownership Costs

Cost Area	5 Year Cost	Rate
Depreciation	$5,099	○
Financing ($306/month)	$2,505	
Insurance (Rating 14)	$8,238	●
State Fees	$640	
Fuel (Hwy 29 City 23)	$3,321	○
Maintenance	$4,882	◉
Repairs	$571	○

Warranty/Maintenance Info

Major Tune-Up	$259	●
Minor Tune-Up	$128	●
Brake Service	$208	○
Overall Warranty	4 yr/50k	○
Drivetrain Warranty	4 yr/50k	○
Rust Warranty	4 yr/unlim. mi	○
Maintenance Warranty	N/A	
Roadside Assistance	N/A	

Resale Value
1993: $13,172 | 1994: $12,425 | 1995: $11,776 | 1996: $10,959 | 1997: $10,109

Cumulative Costs
	1993	1994	1995	1996	1997
Annual	$5,419	$4,127	$4,975	$3,944	$6,791
Total	$5,419	$9,546	$14,521	$18,465	$25,256

Ownership Costs (5yr)
Average: $27,876 | This Car: $25,256
Cost/Mile 40¢ | Cost/Mile 36¢

Ownership Cost Rating
○ Excellent

The 1993 Integra is available in eight models - RS, LS, LS Special, GS and GS-R hatchbacks; and RS, LS, and GS sedans. New for 1993, the Integra LS hatchback is offered in two new colors (Granada Black Pearl and Isle Green). The LS offers a new 4-year/50,000-mile limited bumper-to-bumper warranty with a 4-year/unlimited-mile outer body rust-through warranty. Other features include power antenna, dual power mirrors, power sunroof with sliding sunshade, and cruise control.

---

* Includes shaded options
** Other purchase requirements apply

● Poor | ◐ Worse Than Average | ○ Average | ○ Better Than Average | ○ Excellent | ⊖ Insufficient Information

©1993 by IntelliChoice, Inc. (408) 554-8711 All Rights Reserved. Reproduction Prohibited.
Refer to *Section 3: Annotated Vehicle Charts* for an explanation of these charts.

Acura Integra LS Special
2 Door Hatchback

 1.8L 140 hp Gas Fuel Inject.   4 Cylinder In-Line   Manual 5 Speed   2 Wheel Front Automatic Seatbelts

Compact — Base Model Shown

Purchase Price

Car Item	Dealer Cost	List
Base Price	**$13,721**	**$16,335**
Anti-Lock Brakes	N/A	N/A
Automatic 4 Speed	$630	$750
Optional Engine	N/A	N/A
Air Conditioning	Dlr	Dlr
Power Steering	Std	Std
Cruise Control	Std	Std
All Wheel Drive	N/A	N/A
AM/FM Stereo Cassette	Std	Std
Steering Wheel, Tilt	Std	Std
Power Windows	Std	Std
*Options Price	$630	$750
*Total Price	$14,351	$17,085
Target Price	$15,347	
Destination Charge	$365	
Avg. Tax & Fees	$961	
Total Target $	**$16,673**	
Average Dealer Option Cost:	**84%**	

The 1993 Integra is available in eight models - RS, LS, LS Special, GS and GS-R hatchbacks; and RS, LS, and GS sedans. The Integra Special Edition is new for 1993. Features for the Special include leather-trimmed seats, body-colored side moldings, alloy wheels with high performance Michelin XGT-V4 tires and a color-keyed rear spoiler. Other features include two new exterior colors (Granada Black Pearl and Isle Green Pearl) and a speed-sensitive, variable power-assisted, rack-and-pinion steering system.

Ownership Costs

Cost Area	5 Year Cost	Rate
Depreciation	$5,592	○
Financing ($335/month)	$2,746	
Insurance (Rating 16)	$8,784	●
State Fees	$699	
Fuel (Hwy 29 City 23)	$3,321	○
Maintenance	$4,882	◉
Repairs	$571	○

Warranty/Maintenance Info

Major Tune-Up	$259	●
Minor Tune-Up	$128	●
Brake Service	$208	○
Overall Warranty	4 yr/50k	○
Drivetrain Warranty	4 yr/50k	○
Rust Warranty	4 yr/unlim. mi	○
Maintenance Warranty	N/A	
Roadside Assistance	N/A	

Ownership Cost By Year

Scale: $2,000 – $10,000; years 1993–1997

Resale Value

1993	1994	1995	1996	1997
$14,396	$13,587	$12,877	$11,992	$11,081

Cumulative Costs

	1993	1994	1995	1996	1997
Annual	$5,877	$4,382	$5,205	$4,152	$6,979
Total	$5,877	$10,259	$15,464	$19,616	$26,595

Ownership Costs (5yr)

Average	This Car
$29,164	$26,595
Cost/Mile 42¢	Cost/Mile 38¢

Ownership Cost Rating

○ Excellent

Acura Integra GS
2 Door Hatchback

 1.8L 140 hp Gas Fuel Inject. 4 Cylinder In-Line   Manual 5 Speed   2 Wheel Front Automatic Seatbelts

Compact

Purchase Price

Car Item	Dealer Cost	List
Base Price	**$14,284**	**$17,005**
Anti-Lock Brakes	Std	Std
Automatic 4 Speed	$630	$750
Optional Engine	N/A	N/A
Air Conditioning	Dlr	Dlr
Power Steering	Std	Std
Cruise Control	Std	Std
All Wheel Drive	N/A	N/A
AM/FM Stereo Cassette	Std	Std
Steering Wheel, Tilt	Std	Std
Power Windows	Std	Std
*Options Price	$630	$750
*Total Price	$14,914	$17,755
Target Price	$15,966	
Destination Charge	$365	
Avg. Tax & Fees	$998	
Total Target $	**$17,329**	
Average Dealer Option Cost:	**84%**	

The 1993 Integra is available in eight models - RS, LS, LS Special, GS and GS-R hatchbacks; and RS, LS, and GS sedans. New for 1993, the Integra GS hatchback is offered in two new colors- Granada Black Pearl and Isle Green. A new 4-year/50,000-mile limited bumper-to-bumper warranty with 4-year/unlimited-mile outer body rust-through warranty is standard. Other features include color-keyed bumpers and front air dam, alloy wheels, wraparound bodyside moldings, sport driver's seat, and integral fog lamps.

Ownership Costs

Cost Area	5 Year Cost	Rate
Depreciation	$6,013	○
Financing ($348/month)	$2,854	
Insurance (Rating 17)	$8,719	●
State Fees	$725	
Fuel (Hwy 29 City 23)	$3,321	○
Maintenance	$4,882	◉
Repairs	$571	○

Warranty/Maintenance Info

Major Tune-Up	$259	●
Minor Tune-Up	$128	●
Brake Service	$208	○
Overall Warranty	4 yr/50k	○
Drivetrain Warranty	4 yr/50k	○
Rust Warranty	4 yr/unlim. mi	○
Maintenance Warranty	N/A	
Roadside Assistance	N/A	

Resale Value

1993	1994	1995	1996	1997
$14,653	$13,853	$13,134	$12,237	$11,316

Cumulative Costs

	1993	1994	1995	1996	1997
Annual	$6,315	$4,400	$5,227	$4,164	$6,979
Total	$6,315	$10,715	$15,942	$20,106	$27,085

Ownership Costs (5yr)

Average	This Car
$29,740	$27,085
Cost/Mile 42¢	Cost/Mile 39¢

Ownership Cost Rating

○ Excellent

---

page 2

* Includes shaded options
** Other purchase requirements apply

● Poor ◉ Worse Than Average   Average   Better Than Average   Excellent   Insufficient Information

©1993 by IntelliChoice, Inc. (408) 554-8711 All Rights Reserved. Reproduction Prohibited.
Refer to *Section 3: Annotated Vehicle Charts* for an explanation of these charts.

Acura Integra GS-R
2 Door Hatchback

Compact

- 1.7L 160 hp Gas Fuel Inject.
- 4 Cylinder In-Line
- Manual 5 Speed
- 2 Wheel Front
- Automatic Seatbelts

Purchase Price

Car Item	Dealer Cost	List
Base Price	**$15,338**	**$18,260**
Anti-Lock Brakes	Std	Std
Automatic Transmission	N/A	N/A
Optional Engine	N/A	N/A
Air Conditioning	Dlr	Dlr
Power Steering	Std	Std
Cruise Control	Std	Std
All Wheel Drive	N/A	N/A
AM/FM Stereo Cassette	Std	Std
Steering Wheel, Tilt	Std	Std
Power Windows	Std	Std
*Options Price	$0	$0
*Total Price	$15,338	$18,260
Target Price	$16,452	
Destination Charge	$365	
Avg. Tax & Fees	$1,027	
Total Target $	**$17,844**	
Average Dealer Option Cost:	N/A	

The 1993 Integra is available in eight models - RS, LS, LS Special, GS and GS-R hatchbacks; and RS, LS, and GS sedans. New for 1993, the Integra GS-R has a 4-yr./50,000-mile limited bumper-to-bumper warranty with a 4-yr./unlimited-mile outer body rust-through warranty. Other features include body-color front and rear bumpers and front air dam, alloy wheels and a body-color rear spoiler. All Integra models feature a 4-wheel disc bake system.

Ownership Costs

Cost Area	5 Year Cost	Rate
Depreciation	$6,393	○
Financing ($359/month)	$2,940	
Insurance (Rating 17)	$8,719	●
State Fees	$745	
Fuel (Hwy 29 City 24)	$3,258	◉
Maintenance	$4,737	○
Repairs	$571	○

Warranty/Maintenance Info

Major Tune-Up	$259	●
Minor Tune-Up	$128	●
Brake Service	$208	○
Overall Warranty	4 yr/50k	○
Drivetrain Warranty	4 yr/50k	○
Rust Warranty	4 yr/unlim. mi	◉
Maintenance Warranty	N/A	
Roadside Assistance	N/A	

Ownership Cost By Year

Legend: 1993, 1994, 1995, 1996, 1997

Resale Value

1993	1994	1995	1996	1997
$15,020	$14,163	$13,386	$12,429	$11,451

Cumulative Costs

	1993	1994	1995	1996	1997
Annual	$6,491	$4,476	$5,225	$4,221	$6,950
Total	$6,491	$10,967	$16,192	$20,413	$27,363

Ownership Costs (5yr)

Average	This Car
$30,173	$27,363
Cost/Mile 43¢	Cost/Mile 39¢

Ownership Cost Rating

Excellent

Acura Integra RS
4 Door Sedan

Compact

- 1.8L 140 hp Gas Fuel Inject.
- 4 Cylinder In-Line
- Manual 5 Speed
- 2 Wheel Front
- Automatic Seatbelts

GS Model Shown

Purchase Price

Car Item	Dealer Cost	List
Base Price	**$11,638**	**$13,855**
Anti-Lock Brakes	N/A	N/A
Automatic 4 Speed	$630	$750
Optional Engine	N/A	N/A
Air Conditioning	Dlr	Dlr
Power Steering	Std	Std
Cruise Control	N/A	N/A
All Wheel Drive	N/A	N/A
AM/FM Stereo Cassette	Dlr	Dlr
Steering Wheel, Tilt	Std	Std
Power Windows	N/A	N/A
*Options Price	$630	$750
*Total Price	$12,268	$14,605
Target Price	$13,068	
Destination Charge	$365	
Avg. Tax & Fees	$822	
Total Target $	**$14,255**	
Average Dealer Option Cost:	84%	

The 1993 Integra is available in eight models - RS, LS, LS Special, GS and GS-R hatchbacks; and RS, LS, and GS sedans. New for 1993, the Integra is offered in two fresh colors (Granada Black Pearl and Isle Green Pearl). The RS sedan features dual door mirrors, body-color front and rear bumpers and front air dam, integral fog lamps, full wheel covers, 60/40 split fold-down rear seatbacks, and a 4-year/50,000-mile limited bumper-to-bumper warranty with a 4-year/unlimited-mile outer body rust-through warranty.

Ownership Costs

Cost Area	5 Year Cost	Rate
Depreciation	$5,545	○
Financing ($287/month)	$2,349	
Insurance (Rating 11)	$7,693	◉
State Fees	$600	
Fuel (Hwy 29 City 23)	$3,321	○
Maintenance	$4,882	◉
Repairs	$571	○

Warranty/Maintenance Info

Major Tune-Up	$259	●
Minor Tune-Up	$128	●
Brake Service	$208	○
Overall Warranty	4 yr/50k	○
Drivetrain Warranty	4 yr/50k	○
Rust Warranty	4 yr/unlim. mi	○
Maintenance Warranty	N/A	
Roadside Assistance	N/A	

Resale Value

1993	1994	1995	1996	1997
$12,043	$11,231	$10,488	$9,610	$8,710

Cumulative Costs

	1993	1994	1995	1996	1997
Annual	$5,419	$4,029	$4,921	$3,874	$6,718
Total	$5,419	$9,448	$14,369	$18,243	$24,961

Ownership Costs (5yr)

Average	This Car
$27,035	$24,961
Cost/Mile 39¢	Cost/Mile 36¢

Ownership Cost Rating

Excellent

---

* Includes shaded options
** Other purchase requirements apply

Legend: ● Poor | ◉ Worse Than Average | ○ Average | ○ Better Than Average | ○ Excellent | ⊖ Insufficient Information

©1993 by IntelliChoice, Inc. (408) 554-8711 All Rights Reserved. Reproduction Prohibited.
Refer to *Section 3: Annotated Vehicle Charts* for an explanation of these charts.

Acura Integra LS — 4 Door Sedan

Compact

- 1.8L 140 hp Gas Fuel Inject.
- 4 Cylinder In-Line
- Manual 5 Speed
- 2 Wheel Front
- Automatic Seatbelts

GS Model Shown

Purchase Price

Car Item	Dealer Cost	List
Base Price	**$13,091**	**$15,585**
Anti-Lock Brakes	N/A	N/A
Automatic 4 Speed	$630	$750
Optional Engine	N/A	N/A
Air Conditioning	Dlr	Dlr
Power Steering	Std	Std
Cruise Control	Std	Std
All Wheel Drive	N/A	N/A
AM/FM Stereo Cassette	Std	Std
Steering Wheel, Tilt	Std	Std
Power Windows	Std	Std
*Options Price	$630	$750
*Total Price	$13,721	$16,335
Target Price	$14,656	
Destination Charge	$345	
Avg. Tax & Fees	$917	
Total Target $	**$15,918**	
Average Dealer Option Cost:	**84%**	

Ownership Costs

Cost Area	5 Year Cost	Rate
Depreciation	$6,570	○
Financing ($320/month)	$2,622	
Insurance (Rating 12)	$7,898	◉
State Fees	$667	
Fuel (Hwy 29 City 23)	$3,321	○
Maintenance	$4,882	◉
Repairs	$571	○

Warranty/Maintenance Info

Major Tune-Up	$259	●
Minor Tune-Up	$128	●
Brake Service	$208	○
Overall Warranty	4 yr/50k	○
Drivetrain Warranty	4 yr/50k	○
Rust Warranty	4 yr/unlim. mi	○
Maintenance Warranty	N/A	
Roadside Assistance	N/A	

Resale Value

1993	1994	1995	1996	1997
$12,801	$11,940	$11,195	$10,278	$9,348

Cumulative Costs

	1993	1994	1995	1996	1997
Annual	$6,493	$4,219	$5,031	$3,986	$6,802
Total	$6,493	$10,712	$15,743	$19,729	$26,531

Ownership Costs (5yr)

Average	This Car
$28,520	$26,531
Cost/Mile 41¢	Cost/Mile 38¢

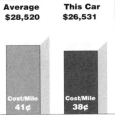

Ownership Cost Rating: ○ Better Than Average

The 1993 Integra is available in eight models - RS, LS, LS Special, GS and GS-R hatchbacks; and RS, LS and GS sedans. New for 1993, the Integra is offered in two fresh colors (Granada Black Pearl and Isle Green Pearl). The LS sedan features power operated dual door mirrors, body-color front and rear bumpers and front air dam, integral fog lamps, full wheel covers, 60/40 split fold-down rear seatbacks and reclining front bucket seats.

Acura Integra GS — 4 Door Sedan

Compact

- 1.8L 140 hp Gas Fuel Inject.
- 4 Cylinder In-Line
- Manual 5 Speed
- 2 Wheel Front
- Automatic Seatbelts

Purchase Price

Car Item	Dealer Cost	List
Base Price	**$14,738**	**$17,545**
Anti-Lock Brakes	Std	Std
Automatic 4 Speed	$630	$750
Optional Engine	N/A	N/A
Air Conditioning	Dlr	Dlr
Power Steering	Std	Std
Cruise Control	Std	Std
All Wheel Drive	N/A	N/A
AM/FM Stereo Cassette	Std	Std
Steering Wheel, Tilt	Std	Std
Power Windows	Std	Std
*Options Price	$630	$750
*Total Price	$15,368	$18,295
Target Price	$16,466	
Destination Charge	$365	
Avg. Tax & Fees	$1,029	
Total Target $	**$17,860**	
Average Dealer Option Cost:	**84%**	

Ownership Costs

Cost Area	5 Year Cost	Rate
Depreciation	$7,469	○
Financing ($359/month)	$2,942	
Insurance (Rating 14)	$7,956	◉
State Fees	$747	
Fuel (Hwy 29 City 23)	$3,321	○
Maintenance	$4,882	◉
Repairs	$571	○

Warranty/Maintenance Info

Major Tune-Up	$259	●
Minor Tune-Up	$128	●
Brake Service	$208	○
Overall Warranty	4 yr/50k	○
Drivetrain Warranty	4 yr/50k	○
Rust Warranty	4 yr/unlim. mi	○
Maintenance Warranty	N/A	
Roadside Assistance	N/A	

Resale Value

1993	1994	1995	1996	1997
$14,320	$13,350	$12,478	$11,427	$10,391

Cumulative Costs

	1993	1994	1995	1996	1997
Annual	$7,080	$4,457	$5,249	$4,170	$6,932
Total	$7,080	$11,537	$16,786	$20,956	$27,888

Ownership Costs (5yr)

Average	This Car
$30,203	$27,888
Cost/Mile 43¢	Cost/Mile 40¢

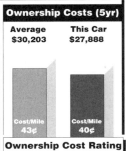

Ownership Cost Rating: ○ Better Than Average

The 1993 Integra is available in eight models - RS, LS, LS Special, GS and GS-R hatchbacks; and RS, LS and GS sedans. New for 1993, the Integra is offered in two new colors (Granada Black Pearl and Isle Green Pearl). The GS sedan features dual power mirrors, body-color front and rear bumpers and front air dam, fog lamps, full wheel covers, 60/40 split fold-down rear seatbacks, child-proof rear door locks, and a new 4-year/50K-mile limited bumper-to-bumper warranty.

---

* Includes shaded options
** Other purchase requirements apply

● Poor | ◉ Worse Than Average | ◐ Average | ○ Better Than Average | ○ Excellent | ⊖ Insufficient Information

©1993 by IntelliChoice, Inc. (408) 554-8711 All Rights Reserved. Reproduction Prohibited.
Refer to *Section 3: Annotated Vehicle Charts* for an explanation of these charts.

Acura Legend L
2 Door Coupe

Luxury

 3.2L 230 hp Gas Fuel Inject.
 6 Cylinder "V"
 Manual 6 Speed
 2 Wheel Front
 Driver/Psngr Airbags Std

LS Model Shown

Purchase Price

Car Item	Dealer Cost	List
Base Price	**$28,249**	**$34,450**
Anti-Lock Brakes	Std	Std
Automatic 4 Speed	$656	$800
Optional Engine	N/A	N/A
Air Conditioning	Std	Std
Power Steering	Std	Std
Cruise Control	Std	Std
All Wheel Drive	N/A	N/A
AM/FM Stereo Cassette	N/A	N/A
Steering Wheel, Scope	Std	Std
Power Windows	Std	Std
***Options Price**	**$656**	**$800**
***Total Price**	**$28,905**	**$35,250**
Target Price		$31,758
Destination Charge		$365
Avg. Tax & Fees		$1,962
Luxury Tax		$212
Total Target $		**$34,297**

The 1993 Legend is available in five models - (Base) Legend sedan; and L and LS coupes and sedans. New for 1993, the Legend L coupe has a passenger's side airbag supplemental restraint system (SRS), NVH (noise, vibration, harshness) improvements - new engine-mount system, thicker window glass and additional sound-absorbing insulation materials throughout the vehicle. Also standard are cast aluminum wheels, heated dual power mirrors and leather trimmed interior.

Ownership Costs

Cost Area	5 Year Cost	Rate
Depreciation	$13,413	○
Financing ($689/month)	$5,649	
Insurance (Rating 18 [Est.])	$9,058	○
State Fees	$1,425	
Fuel (Hwy 23 City 18 -Prem.)	$4,652	○
Maintenance	$5,309	○
Repairs	$600	○

Warranty/Maintenance Info

Major Tune-Up	$200	○
Minor Tune-Up	$137	○
Brake Service	$210	○
Overall Warranty	4 yr/50k	○
Drivetrain Warranty	4 yr/50k	○
Rust Warranty	4 yr/unlim. mi	●
Maintenance Warranty	N/A	
Roadside Assistance	N/A	

Ownership Cost By Year

$2,000 $4,000 $6,000 $8,000 $10,000 $12,000 $14,000

■ 1993
■ 1994
■ 1995
■ 1996
□ 1997

Resale Value

1993	1994	1995	1996	1997
$30,079	$27,739	$25,500	$23,123	$20,884

Cumulative Costs

	1993	1994	1995	1996	1997
Annual	$9,510	$7,315	$7,831	$6,315	$9,135
Total	$9,510	$16,825	$24,656	$30,971	$40,106

Ownership Costs (5yr)

Average	This Car
$45,815	$40,106
Cost/Mile 65¢	Cost/Mile 57¢

Ownership Cost Rating

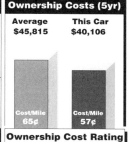
Excellent

Acura Legend LS
2 Door Coupe

Luxury

 3.2L 230 hp Gas Fuel Inject.
 6 Cylinder "V"
 Manual 6 Speed
 2 Wheel Front
 Driver/Psngr Airbags Std

Purchase Price

Car Item	Dealer Cost	List
Base Price	**$31,037**	**$37,850**
Anti-Lock Brakes	Std	Std
Automatic 4 Speed	$656	$800
Optional Engine	N/A	N/A
Auto Climate Control	Std	Std
Power Steering	Std	Std
Cruise Control	Std	Std
All Wheel Drive	N/A	N/A
AM/FM Stereo Cassette	Std	Std
Steering Wheel, Scope	Std	Std
Power Windows	Std	Std
***Options Price**	**$656**	**$800**
***Total Price**	**$31,693**	**$38,650**
Target Price		$34,862
Destination Charge		$365
Avg. Tax & Fees		$2,151
Luxury Tax		$523
Total Target $		**$37,901**

The 1993 Legend is available in five models - (Base) Legend sedan; and L and LS coupes and sedans. New for 1993, the Legend LS coupe has a passenger's side airbag supplemental restraint system (SRS), NVH (noise, vibration, harshness) improvements - new engine-mount system, thicker window glass and additional sound-absorbing insulation materials throughout the vehicle. The LS upgrades the L with standard power passenger seat, illuminated entry system, burled walnut trim and heated front seats.

Ownership Costs

Cost Area	5 Year Cost	Rate
Depreciation	$16,441	○
Financing ($762/month)	$6,244	
Insurance (Rating 16 [Est.])	$8,449	○
State Fees	$1,560	
Fuel (Hwy 23 City 18 -Prem.)	$4,652	○
Maintenance	$5,397	○
Repairs	$600	○

Warranty/Maintenance Info

Major Tune-Up	$200	○
Minor Tune-Up	$137	○
Brake Service	$210	○
Overall Warranty	4 yr/50k	○
Drivetrain Warranty	4 yr/50k	○
Rust Warranty	4 yr/unlim. mi	●
Maintenance Warranty	N/A	
Roadside Assistance	N/A	

Ownership Cost By Year

$5,000 $10,000 $15,000 $20,000

■ 1993
■ 1994
■ 1995
■ 1996
□ 1997

Resale Value

1993	1994	1995	1996	1997
$30,131	$28,009	$25,904	$23,617	$21,460

Cumulative Costs

	1993	1994	1995	1996	1997
Annual	$13,232	$7,195	$7,762	$6,167	$8,987
Total	$13,232	$20,427	$28,189	$34,356	$43,343

Ownership Costs (5yr)

Average	This Car
$48,360	$43,343
Cost/Mile 69¢	Cost/Mile 62¢

Ownership Cost Rating

Excellent

---

* Includes shaded options
** Other purchase requirements apply

● Poor   ◐ Worse Than Average   ○ Average   ◯ Better Than Average   ◌ Excellent   ⊖ Insufficient Information

©1993 by IntelliChoice, Inc. (408) 554-8711   All Rights Reserved. Reproduction Prohibited.
Refer to *Section 3: Annotated Vehicle Charts* for an explanation of these charts.

Acura Legend
4 Door Sedan

 3.2L 200 hp Gas Fuel Inject.
 6 Cylinder "V"
Manual 5 Speed
 2 Wheel Front
 Driver/Psngr Airbags Std

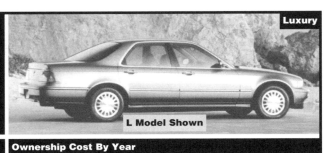
Luxury — L Model Shown

Purchase Price		
Car Item	Dealer Cost	List
Base Price	**$23,944**	**$29,200**
Anti-Lock Brakes	Std	Std
Automatic 4 Speed	$656	$800
Optional Engine	N/A	N/A
Air Conditioning	Std	Std
Power Steering	Std	Std
Cruise Control	Std	Std
All Wheel Drive	N/A	N/A
AM/FM Stereo Cassette	Std	Std
Steering Wheel, Scope	Std	Std
Power Windows	Std	Std
*Options Price	$656	$800
*Total Price	$24,600	$30,000
Target Price		$26,814
Destination Charge		$365
Avg. Tax & Fees		$1,663
Total Target $		**$28,842**
Average Dealer Option Cost:		**82%**

The 1993 Acura Legend is available in five models - L and LS coupes; and (Base) Legend, L and LS sedans. New for 1993, the Base sedan features a passenger's side airbag supplemental restraint system (SRS), NVH (noise, vibration, harshness) improvements - new engine-mount system, thicker window glass and additional sound-absorbing insulation materials throughout the vehicle. Also featured are redesigned alloy wheels, smoother-shifting automatic transmission, and two new exterior colors.

Ownership Costs		
Cost Area	5 Year Cost	Rate
Depreciation	$12,379	○
Financing ($580/month)	$4,751	
Insurance (Rating 13)	$7,797	◐
State Fees	$1,215	
Fuel (Hwy 24 City 19 -Prem.)	$4,436	◐
Maintenance	$5,739	◐
Repairs	$600	○

Warranty/Maintenance Info		
Major Tune-Up	$200	○
Minor Tune-Up	$137	◐
Brake Service	$210	◐
Overall Warranty	4 yr/50k	◐
Drivetrain Warranty	4 yr/50k	◐
Rust Warranty	4 yr/unlim. mi	●
Maintenance Warranty	N/A	
Roadside Assistance	N/A	

Ownership Cost By Year

Resale Value

1993	1994	1995	1996	1997
$24,044	$22,055	$20,209	$18,291	$16,463

Cumulative Costs

	1993	1994	1995	1996	1997
Annual	$9,392	$6,352	$7,129	$5,448	$8,596
Total	$9,392	$15,744	$22,873	$28,321	$36,917

Ownership Costs (5yr)

Average	This Car
$41,885	$36,917
Cost/Mile 60¢	Cost/Mile 53¢

Ownership Cost Rating: ○ Excellent

Acura Legend L
4 Door Sedan

 3.2L 200 hp Gas Fuel Inject.
 6 Cylinder "V"
 Manual 5 Speed
 2 Wheel Front
 Driver/Psngr Airbags Std

Luxury

Purchase Price		
Car Item	Dealer Cost	List
Base Price	**$25,584**	**$31,200**
Anti-Lock Brakes	Std	Std
Automatic 4 Speed	$656	$800
Optional Engine	N/A	N/A
Air Conditioning	Std	Std
Power Steering	Std	Std
Cruise Control	Std	Std
All Wheel Drive	N/A	N/A
AM/FM Stereo Cassette	Std	Std
Steering Wheel, Scope	Std	Std
Power Windows	Std	Std
*Options Price	$656	$800
*Total Price	$26,240	$32,000
Target Price		$28,689
Destination Charge		$365
Avg. Tax & Fees		$1,777
Total Target $		**$30,831**
Average Dealer Option Cost:		**82%**

The 1993 Acura Legend is available in five models - L and LS coupes; and (Base) Legend, L and LS sedans. New for 1993, the L sedan features a passenger's side airbag supplemental restraint system (SRS), an 8-speaker Acura/Bose music system, body-colored side moldings, a dual-mode power door lock system, redesigned alloy wheels, smoother-shifting automatic transmission, and two new exterior colors. Legend L models have an upgraded eight-way power driver's seat with a memory function.

Ownership Costs		
Cost Area	5 Year Cost	Rate
Depreciation	$12,488	○
Financing ($620/month)	$5,078	
Insurance (Rating 14)	$7,956	◐
State Fees	$1,295	
Fuel (Hwy 24 City 19 -Prem.)	$4,436	◐
Maintenance	$5,739	◐
Repairs	$600	○

Warranty/Maintenance Info		
Major Tune-Up	$200	○
Minor Tune-Up	$137	◐
Brake Service	$210	◐
Overall Warranty	4 yr/50k	◐
Drivetrain Warranty	4 yr/50k	◐
Rust Warranty	4 yr/unlim. mi	●
Maintenance Warranty	N/A	
Roadside Assistance	N/A	

Resale Value

1993	1994	1995	1996	1997
$26,284	$24,242	$22,276	$20,240	$18,343

Cumulative Costs

	1993	1994	1995	1996	1997
Annual	$9,326	$6,556	$7,362	$5,637	$8,711
Total	$9,326	$15,882	$23,244	$28,881	$37,592

Ownership Costs (5yr)

Average	This Car
$43,382	$37,592
Cost/Mile 62¢	Cost/Mile 54¢

Ownership Cost Rating: ○ Excellent

---

*Includes shaded options
**Other purchase requirements apply

● Poor ◐ Worse Than Average ○ Average ○ Better Than Average ○ Excellent ⊖ Insufficient Information

©1993 by IntelliChoice, Inc. (408) 554-8711 All Rights Reserved. Reproduction Prohibited.
Refer to *Section 3: Annotated Vehicle Charts* for an explanation of these charts.

Acura Legend LS
4 Door Sedan

Luxury

 3.2L 200 hp Gas Fuel Inject.
 6 Cylinder "V"
 Manual 5 Speed
 2 Wheel Front
 Driver/Psngr Airbags Std

Purchase Price

Car Item	Dealer Cost	List
Base Price	**$29,274**	**$35,700**
Anti-Lock Brakes	Std	Std
Automatic 4 Speed	$656	$800
Optional Engine	N/A	N/A
Auto Climate Control	Std	Std
Power Steering	Std	Std
Cruise Control	Std	Std
All Wheel Drive	N/A	N/A
AM/FM Stereo Cassette	Std	Std
Steering Wheel, Scope	Std	Std
Power Windows	Std	Std
*Options Price	$656	$800
*Total Price	$29,930	$36,500
Target Price	$32,946	
Destination Charge	$365	
Avg. Tax & Fees	$2,035	
Luxury Tax	$331	
Total Target $	**$35,677**	

The 1993 Acura Legend is available in five models - L and LS coupes; and (Base) Legend, L and LS sedans. New for 1993, the LS sedan features a passenger's side airbag supplemental restraint system (SRS), an 8-speaker Acura/Bose premium music system, dual-mode power door lock system, redesigned alloy wheels, smoother-shifting automatic transmission, and two new colors. LS models have an upgraded 8-way power driver's seat with a memory function, a 4-way power passenger seat, and heated front seats.

Ownership Costs

Cost Area	5 Year Cost	Rate
Depreciation	$15,572	○
Financing ($717/month)	$5,877	
Insurance (Rating 15)	$8,199	○
State Fees	$1,475	
Fuel (Hwy 24 City 19 -Prem.)	$4,436	○
Maintenance	$5,739	○
Repairs	$600	○

Warranty/Maintenance Info

Major Tune-Up	$200	○
Minor Tune-Up	$137	○
Brake Service	$210	○
Overall Warranty	4 yr/50k	○
Drivetrain Warranty	4 yr/50k	○
Rust Warranty	4 yr/unlim. mi	○
Maintenance Warranty	N/A	
Roadside Assistance	N/A	

Ownership Cost By Year

Resale Value

1993	1994	1995	1996	1997
$27,688	$25,824	$23,949	$21,971	$20,105

Cumulative Costs

	1993	1994	1995	1996	1997
Annual	$13,191	$6,715	$7,513	$5,719	$8,760
Total	$13,191	$19,906	$27,419	$33,138	$41,898

Ownership Costs (5yr)

Average	This Car
$46,750	$41,898
Cost/Mile 67¢	Cost/Mile 60¢

Ownership Cost Rating
○ Excellent

Acura NSX
2 Door Coupe

Sport

 3.0L 270 hp Gas Fuel Inject.
 6 Cylinder "V"
 Manual 5 Speed
 2 Wheel Rear
 Driver/Psngr Airbags Std

Purchase Price

Car Item	Dealer Cost	List
Base Price	**$53,300**	**$65,000**
Anti-Lock Brakes	Std	Std
Automatic 4 Speed	$3,280	$4,000
Optional Engine	N/A	N/A
Auto Climate Control	Std	Std
Power Steering	Grp	Grp
Cruise Control	Std	Std
All Wheel Drive	N/A	N/A
AM/FM Stereo Cassette	Std	Std
Steering Wheel, Scope	Std	Std
Power Windows	Std	Std
*Options Price	$0	$0
*Total Price	$53,300	$65,000
Target Price	$58,630	
Destination Charge	$600	
Avg. Tax & Fees	$3,618	
Luxury Tax	$2,923	
Total Target $	**$65,771**	

The 1993 NSX is available only as a two-seat sports coupe. New for 1993, the NSX features a passenger's side airbag Supplemental Restraint System (SRS), a redesigned center console, automatic seat belt tensioners, and a 4-year/50,000-mile limited warranty. The NSX is the world's first aluminum production car. Only 25 hand-built vehicles can be produced per working day on the low-volume production line.

Ownership Costs

Cost Area	5 Year Cost	Rate
Depreciation		⊖
Financing ($1,322/month)	$10,834	
Insurance (26 Sport+ [Est.])	$18,179	●
State Fees	$2,624	
Fuel (Hwy 24 City 19 -Prem.)	$4,436	○
Maintenance	$7,189	●
Repairs	$955	○

Warranty/Maintenance Info

Major Tune-Up	$249	○
Minor Tune-Up	$154	◉
Brake Service	$464	●
Overall Warranty	4 yr/50k	○
Drivetrain Warranty	4 yr/50k	○
Rust Warranty	4 yr/unlim. mi	○
Maintenance Warranty	N/A	
Roadside Assistance	N/A	

Ownership Cost By Year

Insufficient Depreciation Information

Resale Value

Insufficient Information

Cumulative Costs

	1993	1994	1995	1996	1997
Annual	Insufficient Information				
Total	Insufficient Information				

Ownership Costs (5yr)

Insufficient Information

Ownership Cost Rating
⊖ Insufficient Information

---

* Includes shaded options
** Other purchase requirements apply

● Poor   ◉ Worse Than Average   ○ Average   ○ Better Than Average   ○ Excellent   ⊖ Insufficient Information

©1993 by IntelliChoice, Inc. (408) 554-8711   All Rights Reserved. Reproduction Prohibited.
Refer to *Section 3: Annotated Vehicle Charts* for an explanation of these charts.

page 7

Acura Vigor LS
4 Door Sedan

 2.5L 176 hp Gas Fuel Inject. | 5 Cylinder In-Line | Manual 5 Speed | 2 Wheel Front | Driver Airbag Psngr Belts

GS Model Shown — Large

Purchase Price

Car Item	Dealer Cost	List
Base Price	**$20,140**	**$24,265**
Anti-Lock Brakes	Std	Std
Automatic 4 Speed	$622	$750
Optional Engine	N/A	N/A
Air Conditioning	Std	Std
Power Steering	Std	Std
Cruise Control	Std	Std
All Wheel Drive	N/A	N/A
AM/FM Stereo Cassette	Std	Std
Steering Wheel, Tilt	Std	Std
Power Windows	Std	Std
*Options Price	$622	$750
*Total Price	**$20,762**	**$25,015**
Target Price	$22,472	
Destination Charge	$365	
Avg. Tax & Fees	$1,396	
Total Target $	**$24,233**	
Average Dealer Option Cost:	83%	

The 1993 Vigor is available in two models - GS and LS sedans. New for 1993, the Vigor LS offers a passenger's side airbag supplemental restraint system, smoother shifting automatic transmission, NVH (noise, vibration, harshness) improvements - additional sound-absorbing insulation materials, increased rear leg room, new front grille design, and two new colors. A new 4-yr.-/50K-mile limited bumper-to-bumper warranty with a 4-yr./unlimited-mile outer body rust-through warranty is standard.

Ownership Costs

Cost Area	5 Year Cost	Rate
Depreciation		⊖
Financing ($487/month)	$3,992	
Insurance (Rating 15)	$8,199	◯
State Fees	$1,015	
Fuel (Hwy 26 City 20 -Prem.)	$4,149	◯
Maintenance	$6,480	●
Repairs	$560	◯

Warranty/Maintenance Info

Major Tune-Up	$287	●
Minor Tune-Up	$138	◉
Brake Service	$250	◯
Overall Warranty	4 yr/50k	◯
Drivetrain Warranty	4 yr/50k	◯
Rust Warranty	4 yr/unlim. mi	◯
Maintenance Warranty	N/A	
Roadside Assistance	N/A	

Ownership Cost By Year

Insufficient Depreciation Information

1993, 1994, 1995, 1996, 1997

Resale Value
Insufficient Information

Ownership Costs (5yr)
Insufficient Information

Cumulative Costs
	1993	1994	1995	1996	1997
Annual	*Insufficient Information*				
Total	*Insufficient Information*				

Ownership Cost Rating
⊖ Insufficient Information

Acura Vigor GS
4 Door Sedan

 2.5L 176 hp Gas Fuel Inject. | 5 Cylinder In-Line | Manual 5 Speed | 2 Wheel Front | Driver/Psngr Airbags Std

Large

Purchase Price

Car Item	Dealer Cost	List
Base Price	**$22,203**	**$26,750**
Anti-Lock Brakes	Std	Std
Automatic 4 Speed	$622	$750
Optional Engine	N/A	N/A
Air Conditioning	Std	Std
Power Steering	Std	Std
Cruise Control	Std	Std
All Wheel Drive	N/A	N/A
AM/FM Stereo Cassette	Std	Std
Steering Wheel, Tilt	Std	Std
Power Windows	Std	Std
*Options Price	$622	$750
*Total Price	**$22,825**	**$27,500**
Target Price	$24,799	
Destination Charge	$365	
Avg. Tax & Fees	$1,537	
Total Target $	**$26,701**	
Average Dealer Option Cost:	83%	

The 1993 Vigor is available in two models - GS and LS sedans. New for 1993, the Vigor GS offers a passenger's side airbag supplemental restraint system, smoother shifting automatic transmission, NVH (noise, vibration, harshness) improvements - additional sound-absorbing insulation materials, increased rear leg room, new front grille design, and two new colors. A new 4-yr.-/50K-mile limited bumper-to-bumper warranty with a 4-yr./unlimited-mile outer body rust-through warranty is standard.

Ownership Costs

Cost Area	5 Year Cost	Rate
Depreciation		⊖
Financing ($537/month)	$4,398	
Insurance (Rating 16)	$8,449	◯
State Fees	$1,115	
Fuel (Hwy 26 City 20 -Prem.)	$4,149	◯
Maintenance	$6,480	●
Repairs	$560	◯

Warranty/Maintenance Info

Major Tune-Up	$287	●
Minor Tune-Up	$138	◉
Brake Service	$250	◯
Overall Warranty	4 yr/50k	◯
Drivetrain Warranty	4 yr/50k	◯
Rust Warranty	4 yr/unlim. mi	◯
Maintenance Warranty	N/A	
Roadside Assistance	N/A	

Ownership Cost By Year
Insufficient Depreciation Information

1993, 1994, 1995, 1996, 1997

Resale Value
Insufficient Information

Ownership Costs (5yr)
Insufficient Information

Cumulative Costs
	1993	1994	1995	1996	1997
Annual	*Insufficient Information*				
Total	*Insufficient Information*				

Ownership Cost Rating
⊖ Insufficient Information

---

page 8

* Includes shaded options
** Other purchase requirements apply

● Poor | ◉ Worse Than Average | ◯ Average | ◯ Better Than Average | ◯ Excellent | ⊖ Insufficient Information

©1993 by IntelliChoice, Inc. (408) 554-8711 All Rights Reserved. Reproduction Prohibited.
Refer to *Section 3: Annotated Vehicle Charts* for an explanation of these charts.

Alfa Romeo 164 L
4 Door Sedan

Luxury

 3.0L 183 hp Gas Fuel Inject.
 6 Cylinder "V"
 Manual 5 Speed
 2 Wheel Rear
 Driver Airbag Psngr Belts

Purchase Price

Car Item	Dealer Cost	List
Base Price	**$24,797**	**$30,240**
Anti-Lock Brakes	Std	Std
Automatic 4 Speed	$607	$740
Optional Engine	N/A	N/A
Auto Climate Control	Std	Std
Power Steering	Std	Std
Cruise Control	Std	Std
All Wheel Drive	N/A	N/A
AM/FM Stereo Cassette	Std	Std
Steering Wheel, Scope	Std	Std
Power Windows	Std	Std
*****Options Price**	$607	$740
*****Total Price**	**$25,404**	**$30,980**
Target Price		$27,385
Destination Charge		$395
Avg. Tax & Fees		$1,703
Total Target $		**$29,483**
Average Dealer Option Cost:		82%

The 1993 Alfa Romeo 164 is available in two models - L and S sedans. For 1993, the L sedan features an upgraded sound system, an optional trunk-mounted compact disc changer and new exterior color choices. The L sedan interior features six-way adjustable sport bucket seats, leather upholstery, front and rear seat heating elements, a leather-wrapped steering wheel and 15-inch cast alloy luxury wheels. Options include an anti-theft security system. Dealer cost prices are estimates.

Ownership Costs

Cost Area	5 Year Cost	Rate
Depreciation	$15,820	◐
Financing ($593/month)	$4,857	
Insurance (Rating 19)	$9,777	◐
State Fees	$1,255	
Fuel (Hwy 25 City 18 -Prem.)	$4,451	◐
Maintenance	$7,688	●
Repairs	$1,196	◐

Warranty/Maintenance Info

Major Tune-Up	$722	●
Minor Tune-Up	$126	○
Brake Service	$431	●
Overall Warranty	3 yr/36k	◐
Drivetrain Warranty	3 yr/36k	◐
Rust Warranty	6 yr/60k	◐
Maintenance Warranty	3 yr/36k	○
Roadside Assistance	3 yr/30k	

Resale Value

1993	1994	1995	1996	1997
$20,659	$17,983	$16,641	$15,040	$13,663

Cumulative Costs

	1993	1994	1995	1996	1997
Annual	$13,788	$7,208	$7,930	$5,841	$10,277
Total	$13,788	$20,996	$28,926	$34,767	$45,044

Ownership Costs (5yr)

Average	This Car
$42,618	$45,044
Cost/Mile 61¢	Cost/Mile 64¢

Ownership Cost Rating

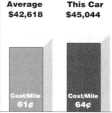

Worse Than Average

Alfa Romeo 164 S
4 Door Sedan

Luxury

 3.0L 200 hp Gas Fuel Inject.
 6 Cylinder "V"
 Manual 5 Speed
 2 Wheel Rear
 Driver Airbag Psngr Belts

Purchase Price

Car Item	Dealer Cost	List
Base Price	**$28,692**	**$34,990**
Anti-Lock Brakes	Std	Std
Automatic Transmission	N/A	N/A
Optional Engine	N/A	N/A
Auto Climate Control	Std	Std
Power Steering	Std	Std
Cruise Control	Std	Std
All Wheel Drive	N/A	N/A
AM/FM Stereo Cassette	Std	Std
Steering Wheel, Scope	Std	Std
Power Windows	Std	Std
*****Options Price**	$0	$0
*****Total Price**	**$28,692**	**$34,990**
Target Price		$31,121
Destination Charge		$395
Avg. Tax & Fees		$1,930
Luxury Tax		$152
Total Target $		**$33,598**

The 1993 Alfa Romeo 164 is available in two models - L and S sedans. For 1993, the S sedan features an upgraded sound system, an optional trunk-mounted compact disc changer and new exterior color choices. The S sedan upgrades the L sedan with exterior ground effects and 15-inch cast alloy sport wheels. Options include eight-way adjustable sport bucket seats. For 1990, the 164 was one of the top-selling Alfa Romeo models. Dealer cost prices are estimates.

Ownership Costs

Cost Area	5 Year Cost	Rate
Depreciation	$18,294	◐
Financing ($675/month)	$5,534	
Insurance (Rating 20)	$9,788	◐
State Fees	$1,415	
Fuel (Hwy 25 City 17 -Prem.)	$4,578	◐
Maintenance	$8,872	●
Repairs	$1,196	◐

Warranty/Maintenance Info

Major Tune-Up	$722	●
Minor Tune-Up	$126	○
Brake Service	$431	●
Overall Warranty	3 yr/36k	◐
Drivetrain Warranty	3 yr/36k	◐
Rust Warranty	6 yr/60k	◐
Maintenance Warranty	3 yr/36k	○
Roadside Assistance	3 yr/30k	

Resale Value

1993	1994	1995	1996	1997
$25,040	$21,013	$19,216	$17,104	$15,304

Cumulative Costs

	1993	1994	1995	1996	1997
Annual	$13,869	$8,833	$8,464	$6,458	$12,053
Total	$13,869	$22,702	$31,166	$37,624	$49,677

Ownership Costs (5yr)

Average	This Car
$45,620	$49,677
Cost/Mile 65¢	Cost/Mile 71¢

Ownership Cost Rating

Worse Than Average

---

* Includes shaded options
** Other purchase requirements apply

● Poor   ◐ Worse Than Average   ◐ Average   ○ Better Than Average   ○ Excellent   ⊖ Insufficient Information

©1993 by IntelliChoice, Inc. (408) 554-8711   All Rights Reserved. Reproduction Prohibited.
Refer to *Section 3: Annotated Vehicle Charts* for an explanation of these charts.

Alfa Romeo Spider
2 Door Convertible

Sport

 2.0L 120 hp Gas Fuel Inject.
 4 Cylinder In-Line
 Manual 5 Speed
 2 Wheel Rear
 Driver Airbag Psngr Belts

Purchase Price

Car Item	Dealer Cost	List
Base Price	**$18,064**	**$21,764**
Anti-Lock Brakes	N/A	N/A
Automatic 3 Speed	$508	$612
Optional Engine	N/A	N/A
Air Conditioning	$842	$1,015
Power Steering	Std	Std
Cruise Control	N/A	N/A
All Wheel Drive	N/A	N/A
AM/FM Stereo Cassette	Std	Std
Steering Wheel, Tilt	N/A	N/A
Power Windows	Std	Std
*Options Price	$842	$1,015
*Total Price	$18,906	$22,779
Target Price	$20,162	
Destination Charge	$395	
Avg. Tax & Fees	$1,260	
Total Target $	**$21,817**	
Average Dealer Option Cost:	83%	

Ownership Costs

Cost Area	5 Year Cost	Rate
Depreciation	$13,440	●
Financing ($438/month)	$3,594	
Insurance (Rating 20 Sport)	$13,829	●
State Fees	$927	
Fuel (Hwy 30 City 22 -Prem.)	$3,677	○
Maintenance	$4,983	○
Repairs	$1,196	◉

Warranty/Maintenance Info

Major Tune-Up	$345	●
Minor Tune-Up	$105	○
Brake Service	$189	○
Overall Warranty	3 yr/36k	◯
Drivetrain Warranty	3 yr/36k	◯
Rust Warranty	6 yr/60k	◯
Maintenance Warranty	N/A	
Roadside Assistance	3 yr/30k	

Resale Value: 1993 $14,153 | 1994 $12,460 | 1995 $11,069 | 1996 $9,735 | 1997 $8,377

Ownership Costs (5yr): Average $40,106 | This Car $41,646 | Cost/Mile 57¢ | Cost/Mile 59¢

Cumulative Costs

	1993	1994	1995	1996	1997
Annual	$12,672	$6,616	$7,283	$6,731	$8,344
Total	$12,672	$19,288	$26,571	$33,302	$41,646

Ownership Cost Rating: Average

The 1993 Spider is available in two models - (Base) Spider and Veloce convertibles. For 1993, the Spider features an all-aluminum 2.0L four cylinder engine, a manually operated folding convertible top with a soft headliner and boot cover, a front air dam and dual remote outside mirrors. The Base Spider offers vinyl seats with suede leather trim, 14-inch stamped steel wheels with full covers and a vinyl-wrapped steering wheel. Options include alloy wheels and metallic paint.

Alfa Romeo Spider Veloce
2 Door Convertible

Sport

 2.0L 120 hp Gas Fuel Inject.
4 Cylinder In-Line
Manual 5 Speed
 2 Wheel Rear
 Driver Airbag Psngr Belts

Purchase Price

Car Item	Dealer Cost	List
Base Price	**$20,642**	**$24,870**
Anti-Lock Brakes	N/A	N/A
Automatic 3 Speed	$508	$612
Optional Engine	N/A	N/A
Air Conditioning	Std	Std
Power Steering	Std	Std
Cruise Control	N/A	N/A
All Wheel Drive	N/A	N/A
AM/FM Stereo Cassette	Std	Std
Steering Wheel, Tilt	N/A	N/A
Power Windows	Std	Std
*Options Price	$0	$0
*Total Price	$20,642	$24,870
Target Price	$22,105	
Destination Charge	$395	
Avg. Tax & Fees	$1,378	
Total Target $	**$23,878**	
Average Dealer Option Cost:	83%	

Ownership Costs

Cost Area	5 Year Cost	Rate
Depreciation	$14,540	●
Financing ($480/month)	$3,933	
Insurance (Rating 21 Sport)	$14,453	●
State Fees	$1,011	
Fuel (Hwy 30 City 22 -Prem.)	$3,677	○
Maintenance	$5,082	○
Repairs	$1,196	◉

Warranty/Maintenance Info

Major Tune-Up	$345	●
Minor Tune-Up	$105	○
Brake Service	$189	○
Overall Warranty	3 yr/36k	◯
Drivetrain Warranty	3 yr/36k	◯
Rust Warranty	6 yr/60k	◯
Maintenance Warranty	N/A	
Roadside Assistance	3 yr/30k	

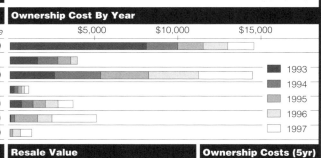

Resale Value: 1993 $15,681 | 1994 $13,802 | 1995 $12,262 | 1996 $10,792 | 1997 $9,338

Ownership Costs (5yr): Average $42,112 | This Car $43,892 | Cost/Mile 60¢ | Cost/Mile 63¢

Cumulative Costs

	1993	1994	1995	1996	1997
Annual	$13,483	$7,047	$7,688	$7,037	$8,637
Total	$13,483	$20,530	$28,218	$35,255	$43,892

Ownership Cost Rating: Average

The 1993 Spider is available in two models - Spider (Base) and Veloce convertibles. For 1993, the Spider features an all-aluminum 2.0L four cylinder engine, a manually operated folding convertible top with a soft headliner and boot cover and dual remote outside mirrors. The Spider Veloce upgrades the Base convertible with all leather seats, a leather-wrapped steering wheel, 15-inch light alloy wheels and an optional removable hardtop with a headliner and an electric rear window defroster.

---

page 10

* Includes shaded options
** Other purchase requirements apply

● Poor | ◉ Worse Than Average | ○ Average | ◯ Better Than Average | ◯ Excellent | ⊖ Insufficient Information

©1993 by IntelliChoice, Inc. (408) 554-8711   All Rights Reserved. Reproduction Prohibited.
Refer to *Section 3: Annotated Vehicle Charts* for an explanation of these charts.

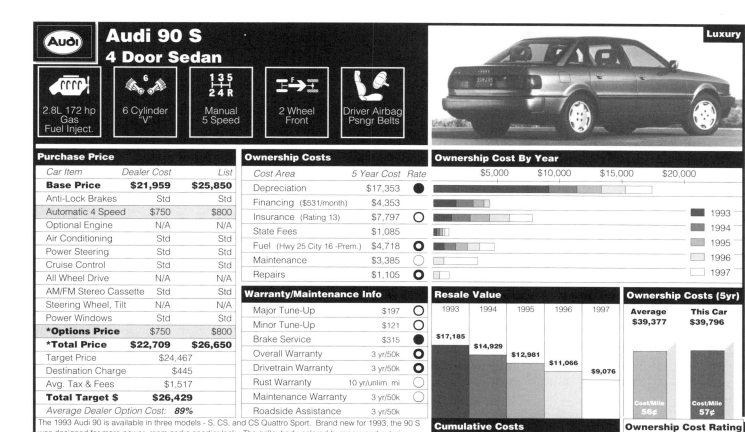

Audi 90 S — 4 Door Sedan — Luxury

- 2.8L 172 hp Gas Fuel Inject.
- 6 Cylinder "V"
- Manual 5 Speed
- 2 Wheel Front
- Driver Airbag Psngr Belts

Purchase Price

Car Item	Dealer Cost	List
Base Price	$21,959	$25,850
Anti-Lock Brakes	Std	Std
Automatic 4 Speed	$750	$800
Optional Engine	N/A	N/A
Air Conditioning	Std	Std
Power Steering	Std	Std
Cruise Control	Std	Std
All Wheel Drive	N/A	N/A
AM/FM Stereo Cassette	Std	Std
Steering Wheel, Tilt	N/A	N/A
Power Windows	Std	Std
*Options Price	$750	$800
*Total Price	$22,709	$26,650
Target Price	$24,467	
Destination Charge	$445	
Avg. Tax & Fees	$1,517	
Total Target $	$26,429	
Average Dealer Option Cost:	89%	

The 1993 Audi 90 is available in three models - S, CS, and CS Quattro Sport. Brand new for 1993, the 90 S was designed for more power, room and a sportier look. The grille, body-colored bumpers and exterior panels are all new. Standard features include an anti-theft alarm system, tinted glass, central locking system, headlight washer system, retained accessory power for windows and front and rear fog lights. A two-way power sunroof and an All Weather Package are available as optional equipment.

Ownership Costs

Cost Area	5 Year Cost	Rate
Depreciation	$17,353	●
Financing ($531/month)	$4,353	
Insurance (Rating 13)	$7,797	○
State Fees	$1,085	
Fuel (Hwy 25 City 16 - Prem.)	$4,718	○
Maintenance	$3,385	○
Repairs	$1,105	○

Warranty/Maintenance Info

Major Tune-Up	$197	○
Minor Tune-Up	$121	○
Brake Service	$315	●
Overall Warranty	3 yr/50k	○
Drivetrain Warranty	3 yr/50k	○
Rust Warranty	10 yr/unlim. mi	
Maintenance Warranty	3 yr/50k	○
Roadside Assistance	3 yr/50k	

Resale Value

1993	1994	1995	1996	1997
$17,185	$14,929	$12,981	$11,066	$9,076

Cumulative Costs

	1993	1994	1995	1996	1997
Annual	$13,639	$6,264	$5,527	$6,354	$8,012
Total	$13,639	$19,903	$25,430	$31,784	$39,796

Ownership Costs (5yr)

- Average: $39,377
- This Car: $39,796
- Cost/Mile Average: 56¢
- Cost/Mile This Car: 57¢

Ownership Cost Rating: Average

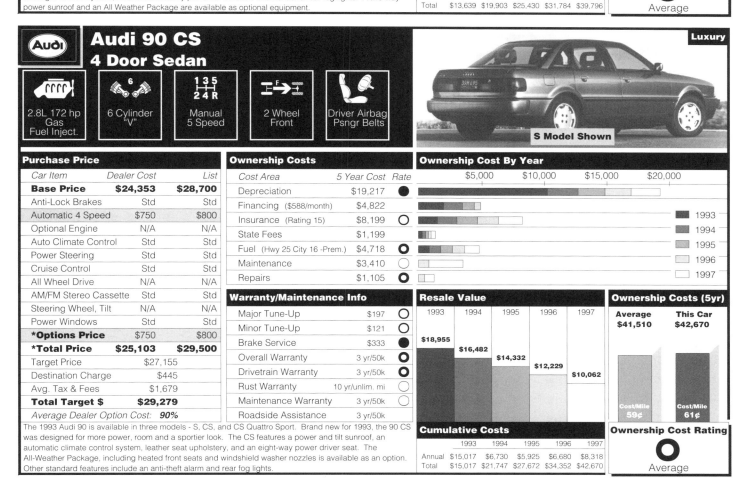

Audi 90 CS — 4 Door Sedan — Luxury

- 2.8L 172 hp Gas Fuel Inject.
- 6 Cylinder "V"
- Manual 5 Speed
- 2 Wheel Front
- Driver Airbag Psngr Belts

S Model Shown

Purchase Price

Car Item	Dealer Cost	List
Base Price	$24,353	$28,700
Anti-Lock Brakes	Std	Std
Automatic 4 Speed	$750	$800
Optional Engine	N/A	N/A
Auto Climate Control	Std	Std
Power Steering	Std	Std
Cruise Control	Std	Std
All Wheel Drive	N/A	N/A
AM/FM Stereo Cassette	Std	Std
Steering Wheel, Tilt	N/A	N/A
Power Windows	Std	Std
*Options Price	$750	$800
*Total Price	$25,103	$29,500
Target Price	$27,155	
Destination Charge	$445	
Avg. Tax & Fees	$1,679	
Total Target $	$29,279	
Average Dealer Option Cost:	90%	

The 1993 Audi 90 is available in three models - S, CS, and CS Quattro Sport. Brand new for 1993, the 90 CS was designed for more power, room and a sportier look. The CS features a power and tilt sunroof, an automatic climate control system, leather seat upholstery, and an eight-way power driver seat. The All-Weather Package, including heated front seats and windshield washer nozzles is available as an option. Other standard features include an anti-theft alarm system and rear fog lights.

Ownership Costs

Cost Area	5 Year Cost	Rate
Depreciation	$19,217	●
Financing ($588/month)	$4,822	
Insurance (Rating 15)	$8,199	○
State Fees	$1,199	
Fuel (Hwy 25 City 16 - Prem.)	$4,718	○
Maintenance	$3,410	○
Repairs	$1,105	○

Warranty/Maintenance Info

Major Tune-Up	$197	○
Minor Tune-Up	$121	○
Brake Service	$333	●
Overall Warranty	3 yr/50k	○
Drivetrain Warranty	3 yr/50k	○
Rust Warranty	10 yr/unlim. mi	
Maintenance Warranty	3 yr/50k	○
Roadside Assistance	3 yr/50k	

Resale Value

1993	1994	1995	1996	1997
$18,955	$16,482	$14,332	$12,229	$10,062

Cumulative Costs

	1993	1994	1995	1996	1997
Annual	$15,017	$6,730	$5,925	$6,680	$8,318
Total	$15,017	$21,747	$27,672	$34,352	$42,670

Ownership Costs (5yr)

- Average: $41,510
- This Car: $42,670
- Cost/Mile Average: 59¢
- Cost/Mile This Car: 61¢

Ownership Cost Rating: Average

---

* Includes shaded options
** Other purchase requirements apply

Legend: ● Poor — ◐ Worse Than Average — ◑ Average — ○ Better Than Average — ○ Excellent — ⊖ Insufficient Information

©1993 by IntelliChoice, Inc. (408) 554-8711 All Rights Reserved. Reproduction Prohibited.
Refer to *Section 3: Annotated Vehicle Charts* for an explanation of these charts.

Audi 90 CS Quattro Sport
4 Door Sedan

 2.8L 172 hp Gas Fuel Inject.
 6 Cylinder "V"
 Manual 5 Speed
 4 Wheel Full-Time
 Driver Airbag Psngr Belts

Luxury

Purchase Price

Car Item	Dealer Cost	List
Base Price	$27,335	$32,250
Anti-Lock Brakes	Std	Std
Automatic Transmission	N/A	N/A
Optional Engine	N/A	N/A
Auto Climate Control	Std	Std
Power Steering	Std	Std
Cruise Control	Std	Std
4 Wheel Full-Time Drive	Std	Std
AM/FM Stereo Cassette	Std	Std
Steering Wheel, Tilt	N/A	N/A
Power Windows	Std	Std
***Options Price**	$0	$0
***Total Price**	$27,335	$32,250
Target Price	$29,718	
Destination Charge	$445	
Avg. Tax & Fees	$1,835	
Luxury Tax	$16	
Total Target $	**$32,014**	

The 1993 Audi 90 is available in three models - S, CS, and CS Quattro Sport. Brand new for 1993, the 90 CS Quattro Sport features a rear wing spoiler, a lowered sport-tuned suspension, power and tilt sunroof, an automatic climate control system, front sport seats, and leather seat upholstery. Other standard features include an anti-theft alarm system and front and rear fog lights. The All-Weather Package, including heated front seats and windshield washer nozzles is available as an option.

Ownership Costs

Cost Area	5 Year Cost	Rate
Depreciation	$20,016	●
Financing ($643/month)	$5,274	
Insurance (Rating 16)	$8,449	○
State Fees	$1,308	
Fuel (Hwy 22 City 17 -Prem.)	$4,892	●
Maintenance	$5,058	○
Repairs	$1,195	◐

Warranty/Maintenance Info

Major Tune-Up	$197	○
Minor Tune-Up	$121	○
Brake Service	$333	●
Overall Warranty	3 yr/50k	○
Drivetrain Warranty	3 yr/50k	○
Rust Warranty	10 yr/unlim. mi	○
Maintenance Warranty	3 yr/50k	○
Roadside Assistance	3 yr/50k	

Ownership Cost By Year
1993, 1994, 1995, 1996, 1997

Resale Value
1993	1994	1995	1996	1997
$20,570	$18,149	$16,065	$14,009	$11,998

Cumulative Costs
	1993	1994	1995	1996	1997
Annual	$16,431	$6,924	$6,056	$7,090	$9,691
Total	$16,431	$23,355	$29,411	$36,501	$46,192

Ownership Costs (5yr)
Average $43,569 — This Car $46,192
Cost/Mile 62¢ — Cost/Mile 66¢

Ownership Cost Rating
● Worse Than Average

Audi 100
4 Door Sedan

 2.8L 172 hp Gas Fuel Inject.
 6 Cylinder "V"
 Manual 5 Speed
 2 Wheel Front
 Driver/Psngr Airbags Std

Luxury

Purchase Price

Car Item	Dealer Cost	List
Base Price	$25,477	$30,400
Anti-Lock Brakes	Std	Std
Automatic 4 Speed	$750	$800
Optional Engine	N/A	N/A
Air Conditioning	Std	Std
Power Steering	Std	Std
Cruise Control	Std	Std
All Wheel Drive	N/A	N/A
AM/FM Stereo Cassette	Std	Std
Steering Wheel, Scope	Std	Std
Power Windows	Std	Std
***Options Price**	$750	$800
***Total Price**	$26,227	$31,200
Target Price	$28,425	
Destination Charge	$445	
Avg. Tax & Fees	$1,760	
Total Target $	**$30,630**	
Average Dealer Option Cost:	94%	

The Audi 100 is available in five models-(Base) 100, S and CS Quattro sedan and wagon. New for 1993, the 100 sedan features a driver's-side airbag to an already array of safety features. Additional features include anti-lock brakes, a front seat belt tensioning system and an automatic transmission with five seperate shift programs. Interior features include walnut wood inlays for dash/door panels, reclining bucket seats with velour upholstery, CFC-free air conditioner and adjustable steering column.

Ownership Costs

Cost Area	5 Year Cost	Rate
Depreciation	$19,473	●
Financing ($616/month)	$5,046	
Insurance (Rating 15)	$8,199	○
State Fees	$1,266	
Fuel (Hwy 24 City 19 -Prem.)	$4,436	◐
Maintenance	$3,212	○
Repairs	$1,105	◐

Warranty/Maintenance Info

Major Tune-Up	$166	○
Minor Tune-Up	$121	○
Brake Service	$319	●
Overall Warranty	3 yr/50k	○
Drivetrain Warranty	3 yr/50k	○
Rust Warranty	10 yr/unlim. mi	○
Maintenance Warranty	3 yr/50k	○
Roadside Assistance	3 yr/50k	

Resale Value
1993	1994	1995	1996	1997
$19,410	$16,902	$14,853	$12,891	$11,157

Cumulative Costs
	1993	1994	1995	1996	1997
Annual	$15,971	$6,796	$5,825	$6,524	$7,621
Total	$15,971	$22,767	$28,592	$35,116	$42,737

Ownership Costs (5yr)
Average $42,783 — This Car $42,737
Cost/Mile 61¢ — Cost/Mile 61¢

Ownership Cost Rating
○ Average

---

page 12

* Includes shaded options
** Other purchase requirements apply

● Poor | ◐ Worse Than Average | ○ Average | ○ Better Than Average | ○ Excellent | ⊖ Insufficient Information

©1993 by IntelliChoice, Inc. (408) 554-8711 All Rights Reserved. Reproduction Prohibited.
Refer to *Section 3: Annotated Vehicle Charts* for an explanation of these charts.

Audi 100 S
4 Door Sedan

Luxury

| 2.8L 172 hp Gas Fuel Inject. | 6 Cylinder "V" | Manual 5 Speed | 2 Wheel Front | Driver/Psngr Airbags Std |

Purchase Price

Car Item	Dealer Cost	List
Base Price	**$27,843**	**$33,250**
Anti-Lock Brakes	Std	Std
Automatic 4 Speed	$750	$800
Optional Engine	N/A	N/A
Air Conditioning	Std	Std
Power Steering	Std	Std
Cruise Control	Std	Std
All Wheel Drive	N/A	N/A
AM/FM Stereo Cassette	Std	Std
Steering Wheel, Scope	Std	Std
Power Windows	Std	Std
***Options Price**	**$750**	**$800**
***Total Price**	**$28,593**	**$34,050**
Target Price		$31,112
Destination Charge		$445
Avg. Tax & Fees		$1,923
Luxury Tax		$156
Total Target $		**$33,636**

The Audi 100 is available in five models-(Base) 100, S and CS Quattro sedan and wagon. New for 1993, the 100 S sedan features driver and passenger airbags. Standard features include 5-speed manual transmission, MacPherson strut front suspension, 4-wheel disc brakes and body-side moldings. Additional features include walnut wood inlays for dash/door panels, velour upholstery and 8-way power front bucket seats with adjustable headrests. The option list includes leather upholstery and cellular phone.

Ownership Costs

Cost Area	5 Year Cost	Rate
Depreciation	$21,419	●
Financing ($676/month)	$5,541	
Insurance (Rating 16)	$8,449	○
State Fees	$1,379	
Fuel (Hwy 24 City 19 -Prem.)	$4,436	◐
Maintenance	$3,212	○
Repairs	$1,105	◐

Warranty/Maintenance Info

Major Tune-Up	$166	○
Minor Tune-Up	$121	○
Brake Service	$319	●
Overall Warranty	3 yr/50k	○
Drivetrain Warranty	3 yr/50k	○
Rust Warranty	10 yr/unlim. mi	○
Maintenance Warranty	3 yr/50k	○
Roadside Assistance	3 yr/50k	

Ownership Cost By Year

1993, 1994, 1995, 1996, 1997

Resale Value

1993	1994	1995	1996	1997
$21,172	$18,451	$16,215	$14,083	$12,217

Cumulative Costs

	1993	1994	1995	1996	1997
Annual	$17,496	$7,237	$6,184	$6,800	$7,824
Total	$17,496	$24,733	$30,917	$37,717	$45,541

Ownership Costs (5yr)

Average	This Car
$44,916	$45,541
Cost/Mile 64¢	Cost/Mile 65¢

Ownership Cost Rating

○ Average

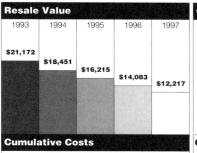

Audi 100 CS
4 Door Sedan

Luxury

| 2.8L 172 hp Gas Fuel Inject. | 6 Cylinder "V" | Manual 5 Speed | 2 Wheel Front | Driver/Psngr Airbags Std |

Purchase Price

Car Item	Dealer Cost	List
Base Price	**$31,578**	**$37,750**
Anti-Lock Brakes	Std	Std
Automatic 4 Speed	$750	$800
Optional Engine	N/A	N/A
Auto Climate Control	Std	Std
Power Steering	Std	Std
Cruise Control	Std	Std
All Wheel Drive	N/A	N/A
AM/FM Stereo Cassette	Std	Std
Steering Wheel, Scope	Std	Std
Power Windows	Std	Std
***Options Price**	**$750**	**$800**
***Total Price**	**$32,328**	**$38,550**
Target Price		$35,238
Destination Charge		$445
Avg. Tax & Fees		$2,174
Luxury Tax		$568
Total Target $		**$38,425**

The Audi 100 is available in five models-(Base) 100, S and CS Quattro sedan and wagon. New for 1993, the 100 CS sedan adds driver and passenger airbags to complement such existing standard equipment as anti-lock brakes and a front seat belt tensioning system. Additional standard features include an automatic climate control system, power two-way moonroof, eight-way power front seats, leather upholstery and walnut wood inlays. A cellular telephone and 10-disc CD player is optional on all CS models.

Ownership Costs

Cost Area	5 Year Cost	Rate
Depreciation	$24,505	●
Financing ($772/month)	$6,330	
Insurance (Rating 17)	$8,719	○
State Fees	$1,560	
Fuel (Hwy 24 City 19 -Prem.)	$4,436	◐
Maintenance	$3,212	○
Repairs	$1,105	◐

Warranty/Maintenance Info

Major Tune-Up	$166	○
Minor Tune-Up	$121	○
Brake Service	$319	●
Overall Warranty	3 yr/50k	○
Drivetrain Warranty	3 yr/50k	○
Rust Warranty	10 yr/unlim. mi	○
Maintenance Warranty	3 yr/50k	○
Roadside Assistance	3 yr/50k	

Resale Value

1993	1994	1995	1996	1997
$24,005	$20,941	$18,403	$15,998	$13,920

Cumulative Costs

	1993	1994	1995	1996	1997
Annual	$19,876	$7,919	$6,732	$7,219	$8,121
Total	$19,876	$27,795	$34,527	$41,746	$49,867

Ownership Costs (5yr)

Average	This Car
$48,285	$49,867
Cost/Mile 69¢	Cost/Mile 71¢

Ownership Cost Rating

○ Average

---

* Includes shaded options
** Other purchase requirements apply

● Poor ◐ Worse Than Average ○ Average ○ Better Than Average ○ Excellent ⊖ Insufficient Information

©1993 by IntelliChoice, Inc. (408) 554-8711 All Rights Reserved. Reproduction Prohibited.
Refer to *Section 3: Annotated Vehicle Charts* for an explanation of these charts.

page 13

Audi 100 CS Quattro
4 Door Sedan

 2.8L 172 hp Gas Fuel Inject.
 6 Cylinder "V"
 Manual 5 Speed
 4 Wheel Full-Time
 Driver/Psngr Airbags Std

Luxury

Purchase Price

Car Item	Dealer Cost	List
Base Price	$34,234	$40,950
Anti-Lock Brakes	Std	Std
Automatic 4 Speed	$750	$800
Optional Engine	N/A	N/A
Auto Climate Control	Std	Std
Power Steering	Std	Std
Cruise Control	Std	Std
4 Wheel Full-Time Drive	Std	Std
AM/FM Stereo Cassette	Std	Std
Steering Wheel, Scope	Std	Std
Power Windows	Std	Std
***Options Price**	$750	$800
***Total Price**	$34,984	$41,750
Target Price	$38,133	
Destination Charge	$445	
Avg. Tax & Fees	$2,351	
Luxury Tax	$858	
Total Target $	$41,787	

The Audi 100 is available in five models-(Base) 100, S and CS Quattro sedan and wagon. New for 1993, the 100 CS Quattro sedan features a driver and passenger airbag. The Quattro model is powered by a 2.8-liter, 172 horsepower V6 engine. It is the first in the 100 series to mate an automatic transmissions with Audi's all-wheel-drive system. Standard features include leather upholstery, eight-way power front seats, front fog lights, a tilt and slide power moonroof and an infrared remote locking system.

Ownership Costs

Cost Area	5 Year Cost	Rate
Depreciation	$26,571	●
Financing ($840/month)	$6,884	
Insurance (Rating 18)	$9,058	○
State Fees	$1,688	
Fuel (Hwy 22 City 18 -Prem.)	$4,766	●
Maintenance	$3,754	◐
Repairs	$1,425	●

Warranty/Maintenance Info

Major Tune-Up	$166	◐
Minor Tune-Up	$121	○
Brake Service	$319	●
Overall Warranty	3 yr/50k	◐
Drivetrain Warranty	3 yr/50k	◐
Rust Warranty	10 yr/unlim. mi	
Maintenance Warranty	3 yr/50k	◐
Roadside Assistance	3 yr/50k	

Ownership Cost By Year
Legend: 1993, 1994, 1995, 1996, 1997

Resale Value
1993	1994	1995	1996	1997
$26,019	$22,710	$20,001	$17,445	$15,216

Cumulative Costs
	1993	1994	1995	1996	1997
Annual	$21,609	$8,495	$7,173	$7,960	$8,909
Total	$21,609	$30,104	$37,277	$45,237	$54,146

Ownership Costs (5yr)
Average	This Car
$50,681	$54,146
Cost/Mile 72¢	Cost/Mile 77¢

Ownership Cost Rating
● Worse Than Average

Audi 100 CS Quattro
4 Door Wagon

 2.8L 172 hp Gas Fuel Inject.
 6 Cylinder "V"
 Automatic 4 Speed
 4 Wheel Full-Time
 Driver/Psngr Airbags Std

Luxury

Purchase Price

Car Item	Dealer Cost	List
Base Price	$36,973	$44,250
Anti-Lock Brakes	Std	Std
Manual Transmission	N/A	N/A
Optional Engine	N/A	N/A
Auto Climate Control	Std	Std
Power Steering	Std	Std
Cruise Control	Std	Std
4 Wheel Full-Time Drive	Std	Std
AM/FM Stereo Cassette	Std	Std
Steering Wheel, Scope	Std	Std
Power Windows	Std	Std
***Options Price**	$0	$0
***Total Price**	$36,973	$44,250
Target Price	$40,301	
Destination Charge	$445	
Avg. Tax & Fees	$2,484	
Luxury Tax	$1,075	
Total Target $	$44,305	

The Audi 100 is available in five models-(Base) 100, S and CS Quattro sedan and wagon. New for 1993, the 100 CS Quattro wagon features a driver and passenger airbag to complement such existing standard equipment as anti-lock brakes, a front seat belt tensioning system and permanent all-wheel-drive. In addition to the automatic transmission, the Wagon features as standard eequipment leather seat upholstery and a Cold Weather Package which includes heated front seats with individual temperature controls.

Ownership Costs

Cost Area	5 Year Cost	Rate
Depreciation	$26,285	●
Financing ($890/month)	$7,298	
Insurance (Rating 18)	$9,058	○
State Fees	$1,788	
Fuel (Hwy 22 City 18 -Prem.)	$4,766	●
Maintenance	$3,754	◐
Repairs	$1,425	●

Warranty/Maintenance Info

Major Tune-Up	$166	◐
Minor Tune-Up	$121	○
Brake Service	$319	●
Overall Warranty	3 yr/50k	◐
Drivetrain Warranty	3 yr/50k	◐
Rust Warranty	10 yr/unlim. mi	
Maintenance Warranty	3 yr/50k	◐
Roadside Assistance	3 yr/50k	

Resale Value
1993	1994	1995	1996	1997
$29,761	$26,442	$23,569	$20,457	$18,020

Cumulative Costs
	1993	1994	1995	1996	1997
Annual	$20,583	$8,656	$7,439	$8,564	$9,132
Total	$20,583	$29,239	$36,678	$45,242	$54,374

Ownership Costs (5yr)
Average	This Car
$52,552	$54,374
Cost/Mile 75¢	Cost/Mile 78¢

Ownership Cost Rating
○ Average

---

page 14 — *Includes shaded options **Other purchase requirements apply

Legend: ● Poor ◐ Worse Than Average ◑ Average ○ Better Than Average ○ Excellent ⊖ Insufficient Information

©1993 by IntelliChoice, Inc. (408) 554-8711 All Rights Reserved. Reproduction Prohibited.
Refer to *Section 3: Annotated Vehicle Charts* for an explanation of these charts.

Audi S4
4 Door Sedan

Luxury

2.2L 227 hp Turbo Gas Fuel Inject. | 5 Cylinder In-Line | Manual 5 Speed | 4 Wheel Full-Time | Driver/Psngr Airbags Std

Purchase Price

Car Item	Dealer Cost	List
Base Price	**$39,131**	**$46,850**
Anti-Lock Brakes	Std	Std
Automatic Transmission	N/A	N/A
Optional Engine	N/A	N/A
Auto Climate Control	Std	Std
Power Steering	Std	Std
Cruise Control	Std	Std
4 Wheel Full-Time Drive	Std	Std
AM/FM Stereo Cassette	Std	Std
Steering Wheel, Scope	Std	Std
Power Windows	Std	Std
***Options Price**	$0	$0
***Total Price**	**$39,131**	**$46,850**
Target Price	$42,457	
Destination Charge	$445	
Avg. Tax & Fees	$2,618	
Luxury Tax	$1,290	
Total Target $	**$46,810**	

The 1993 S4 is available in one model edition - as a four-door all-wheel drive sedan. New for 1993, the S4 includes a passenger side airbag, a power two-way glass moonroof, and smaller Quattro nomenclature on the grill and rear decklid. The cockpit includes sport front bucket seats w/lumbar and thigh adjustments and a gray-faced instrument panel. Standard leather seat upholstery is available in black and ecru. Other features include cruise control, power central locking and alloy wheels.

Ownership Costs

Cost Area	5 Year Cost	Rate
Depreciation		⊖
Financing ($941/month)	$7,711	
Insurance (Rating 19)	$9,358	○
State Fees	$1,892	
Fuel (Hwy 23 City 18 -Prem.)	$4,652	●
Maintenance	$6,446	●
Repairs		⊖

Warranty/Maintenance Info

Major Tune-Up	$179	○
Minor Tune-Up	$120	○
Brake Service	$443	●
Overall Warranty	3 yr/50k	○
Drivetrain Warranty	3 yr/50k	○
Rust Warranty	10 yr/unlim. mi	○
Maintenance Warranty	3 yr/50k	○
Roadside Assistance	3 yr/50k	

Ownership Cost By Year

Insufficient Depreciation Information

■ 1993 ■ 1994 ■ 1995 □ 1996 □ 1997

Resale Value

Insufficient Information

Cumulative Costs

	1993	1994	1995	1996	1997
Annual	Insufficient Information				
Total	Insufficient Information				

Ownership Costs (5yr)

Insufficient Information

Ownership Cost Rating

⊖ Insufficient Information

Audi V8 Quattro
4 Door Sedan

Luxury

4.2L 276 hp Gas Fuel Inject. | 8 Cylinder "V" | Automatic 4 Speed | 4 Wheel Full-Time | Driver/Psngr Airbags Std

Purchase Price

Car Item	Dealer Cost	List
Base Price	**$47,057**	**$56,400**
Anti-Lock Brakes	Std	Std
Manual Transmission	N/A	N/A
Optional Engine	N/A	N/A
Auto Climate Control	Std	Std
Power Steering	Std	Std
Cruise Control	Std	Std
4 Wheel Full-Time Drive	Std	Std
AM/FM Stereo Cassette	Std	Std
Steering Wheel, Scope	Std	Std
Power Windows	Std	Std
***Options Price**	$0	$0
***Total Price**	**$47,057**	**$56,400**
Target Price	$51,292	
Destination Charge	$445	
Avg. Tax & Fees	$3,155	
Luxury/Gas Guzzler Tax	$4,274	
Total Target $	**$59,166**	

The 1993 Audi V-8 Quattro is available in one model edition. For 1993, the V-8 Quattro features aerodynamic styling, flush door handles and integrated front and rear fog lamps. The V-8 Quattro comes equipped with dual airbags, burled walnut inlaid trim, leather upholstery and a six-function trip information computer. Other features include a theft deterrent system, compact disc changer, heated windshield washer nozzles, tinted glass, an automatic climate control system, and child safety door locks.

Ownership Costs

Cost Area	5 Year Cost	Rate
Depreciation	$35,307	○
Financing ($1,189/month)	$9,746	
Insurance (Rating 23)	$11,066	○
State Fees	$2,274	
Fuel (Hwy 20 City 14 -Prem.)	$5,639	●
Maintenance	$5,306	○
Repairs	$1,505	●

Warranty/Maintenance Info

Major Tune-Up	$205	○
Minor Tune-Up	$119	○
Brake Service	$418	●
Overall Warranty	3 yr/50k	○
Drivetrain Warranty	3 yr/50k	○
Rust Warranty	10 yr/unlim. mi	○
Maintenance Warranty	3 yr/50k	○
Roadside Assistance	3 yr/50k	

Resale Value

1993	1994	1995	1996	1997
$40,506	$37,238	$32,638	$27,993	$23,859

Cumulative Costs

	1993	1994	1995	1996	1997
Annual	$26,372	$10,034	$10,323	$11,170	$12,944
Total	$26,372	$36,406	$46,729	$57,899	$70,843

Ownership Costs (5yr)

Average	This Car
$62,163	$70,843
Cost/Mile 89¢	Cost/Mile $1.01

Ownership Cost Rating

● Poor

---

* Includes shaded options
** Other purchase requirements apply

● Poor | ◐ Worse Than Average | ○ Average | ○ Better Than Average | ○ Excellent | ⊖ Insufficient Information

©1993 by IntelliChoice, Inc. (408) 554-8711 All Rights Reserved. Reproduction Prohibited.
Refer to *Section 3: Annotated Vehicle Charts* for an explanation of these charts.

BMW 318 iS
2 Door Coupe — Midsize

- 1.8L 138 hp Gas Fuel Inject.
- 4 Cylinder In-Line
- Manual 5 Speed
- 2 Wheel Rear
- Driver Airbag Psngr Belts

Purchase Price

Car Item	Dealer Cost	List
Base Price	**$20,795**	**$24,810**
Anti-Lock Brakes	Std	Std
Automatic 4 Speed	$700	$850
Optional Engine	N/A	N/A
Auto Climate Control	Std	Std
Power Steering	Std	Std
Cruise Control	$370	$455
All Wheel Drive	N/A	N/A
AM/FM Stereo Cassette	Std	Std
Steering Wheel, Tilt	N/A	N/A
Power Windows	Std	Std
*Options Price	$1,070	$1,305
*Total Price	$21,865	$26,115
Target Price	$23,969	
Destination Charge	$405	
Avg. Tax & Fees	$1,484	
Total Target $	**$25,858**	
Average Dealer Option Cost:	*82%*	

The BMW 3-Series is available in five models–318iS and 325iS coupes, 325i and 318i sedans, and 325i convertible. New for 1993, the 318iS coupe features more attractive interior material textures and increased head room for the driver. Standard features include 15-inch alloy wheels, split fold-down rear seats, power two-way sunroof and CFC-free air conditioning. The option list includes two-stage heated front seats, six-disc CD player/changer, leather upholstery and BMW cellular telephone.

Ownership Costs

Cost Area	5 Year Cost	Rate
Depreciation	$10,523	○
Financing ($520/month)	$4,259	
Insurance (Rating 21 [Est.])	$10,706	●
State Fees	$1,061	
Fuel (Hwy 30 City 22 -Prem.)	$3,677	○
Maintenance	$4,855	◐
Repairs	$831	◐

Warranty/Maintenance Info

Major Tune-Up	$295	●
Minor Tune-Up	$190	●
Brake Service	$255	◐
Overall Warranty	4 yr/50k	◐
Drivetrain Warranty	4 yr/50k	◐
Rust Warranty	6 yr/unlim. mi	◐
Maintenance Warranty	N/A	
Roadside Assistance	4 yr/50k	

Ownership Cost By Year
1993 / 1994 / 1995 / 1996 / 1997

Resale Value
1993	1994	1995	1996	1997
$22,937	$20,873	$19,058	$17,118	$15,335

Cumulative Costs
	1993	1994	1995	1996	1997
Annual	$7,664	$6,725	$7,443	$5,779	$8,301
Total	$7,664	$14,389	$21,832	$27,611	$35,912

Ownership Costs (5yr)
- Average: $37,971 — Cost/Mile 54¢
- This Car: $35,912 — Cost/Mile 51¢

Ownership Cost Rating
○ Better Than Average

BMW 318 i
4 Door Sedan — Midsize

- 1.8L 138 hp Gas Fuel Inject.
- 4 Cylinder In-Line
- Manual 5 Speed
- 2 Wheel Rear
- Driver Airbag Psngr Belts

Purchase Price

Car Item	Dealer Cost	List
Base Price	**$19,875**	**$23,710**
Anti-Lock Brakes	Std	Std
Automatic 4 Speed	$700	$850
Optional Engine	N/A	N/A
Auto Climate Control	Std	Std
Power Steering	Std	Std
Cruise Control	$370	$455
All Wheel Drive	N/A	N/A
AM/FM Stereo Cassette	Std	Std
Steering Wheel, Tilt	N/A	N/A
Power Windows	Std	Std
*Options Price	$1,070	$1,305
*Total Price	$20,945	$25,015
Target Price	$22,916	
Destination Charge	$405	
Avg. Tax & Fees	$1,420	
Total Target $	**$24,741**	
Average Dealer Option Cost:	*82%*	

The BMW 3-Series is available in five models–318iS and 325iS coupes, 325i and 318i sedans, and 325i convertible. New for 1993, the 318i sedan features new material textures on dash, seatbacks, doors and consoles. Standard features include a driver's-side airbag, automatic front seatbelt tensioners, two-way sunroof and CFC-free air conditioning. Available options include two-stage heated front seats, new split fold-down rear seat, CD player/changer and BMW cellular telephone.

Ownership Costs

Cost Area	5 Year Cost	Rate
Depreciation	$9,543	○
Financing ($497/month)	$4,076	
Insurance (Rating 18 [Est.])	$9,453	◉
State Fees	$1,017	
Fuel (Hwy 30 City 22 -Prem.)	$3,677	○
Maintenance	$4,218	○
Repairs	$831	◐

Warranty/Maintenance Info

Major Tune-Up	$295	●
Minor Tune-Up	$190	●
Brake Service	$255	◐
Overall Warranty	4 yr/50k	◐
Drivetrain Warranty	4 yr/50k	◐
Rust Warranty	6 yr/unlim. mi	◐
Maintenance Warranty	N/A	
Roadside Assistance	4 yr/50k	

Resale Value
1993	1994	1995	1996	1997
$23,012	$20,951	$19,059	$17,067	$15,198

Cumulative Costs
	1993	1994	1995	1996	1997
Annual	$6,153	$6,414	$6,918	$5,550	$7,780
Total	$6,153	$12,567	$19,485	$25,035	$32,815

Ownership Costs (5yr)
- Average: $36,846 — Cost/Mile 53¢
- This Car: $32,815 — Cost/Mile 47¢

Ownership Cost Rating
○ Excellent

---

page 16 — * Includes shaded options — ** Other purchase requirements apply

Legend: ● Poor — ◉ Worse Than Average — ○ Average — ○ Better Than Average — ○ Excellent — ⊖ Insufficient Information

©1993 by IntelliChoice, Inc. (408) 554-8711 All Rights Reserved. Reproduction Prohibited.

Refer to *Section 3: Annotated Vehicle Charts* for an explanation of these charts.

BMW 325 iS
2 Door Coupe

Luxury

- 2.5L 189 hp Gas Fuel Inject.
- 6 Cylinder In-Line
- Manual 5 Speed
- 2 Wheel Rear
- Driver Airbag Psngr Belts

Purchase Price

Car Item	Dealer Cost	List
Base Price	**$25,940**	**$30,950**
Anti-Lock Brakes	Std	Std
Automatic 4 Speed	$700	$850
Optional Engine	N/A	N/A
Auto Climate Control	Std	Std
Power Steering	Std	Std
Cruise Control	Std	Std
All Wheel Drive	N/A	N/A
AM/FM Stereo Cassette	Std	Std
Steering Wheel, Tilt	Std	Std
Power Windows	Std	Std
*Options Price	$700	$850
*Total Price	$26,640	$31,800
Target Price	$29,521	
Destination Charge	$405	
Avg. Tax & Fees	$1,818	
Total Target $	**$31,744**	
Average Dealer Option Cost:	82%	

The BMW 3-Series is available in five models–318iS and 325iS coupes, 325i and 318i sedans, and 325i convertible. New for 1993, the 325iS coupe features a variable valve timing and knock control. Interior features new material textures on dash, seatbacks, doors and console. Other features include eight-way power front seats, leather upholstery, fold-down rear seats and 15-inch alloy wheels. Other options includes two packages: the Sports Package and Inclement Weather Package.

Ownership Costs

Cost Area	5 Year Cost	Rate
Depreciation	$13,589	◯
Financing ($638/month)	$5,229	
Insurance (Rating 24 [Est.])	$12,323	●
State Fees	$1,288	
Fuel (Hwy 28 City 20 -Prem.)	$3,991	◯
Maintenance	$5,110	◯
Repairs	$985	◯

Warranty/Maintenance Info

Major Tune-Up	$343 ●
Minor Tune-Up	$212 ●
Brake Service	$255 ◯
Overall Warranty	4 yr/50k ◯
Drivetrain Warranty	4 yr/50k ◯
Rust Warranty	6 yr/unlim. mi ◯
Maintenance Warranty	N/A
Roadside Assistance	4 yr/50k

Ownership Cost By Year

1993, 1994, 1995, 1996, 1997

Resale Value

1993	1994	1995	1996	1997
$25,898	$24,155	$22,102	$20,140	$18,155

Cumulative Costs

	1993	1994	1995	1996	1997
Annual	$11,412	$7,151	$8,591	$6,356	$9,005
Total	$11,412	$18,563	$27,154	$33,510	$42,515

Ownership Costs (5yr)

- Average: $43,232 — Cost/Mile 62¢
- This Car: $42,515 — Cost/Mile 61¢

Ownership Cost Rating

◯ Average

BMW 325 i
4 Door Sedan

Luxury

- 2.5L 189 hp Gas Fuel Inject.
- 6 Cylinder In-Line
- Manual 5 Speed
- 2 Wheel Rear
- Driver Airbag Psngr Belts

Purchase Price

Car Item	Dealer Cost	List
Base Price	**$24,850**	**$29,650**
Anti-Lock Brakes	Std	Std
Automatic 4 Speed	$700	$850
Optional Engine	N/A	N/A
Auto Climate Control	Std	Std
Power Steering	Std	Std
Cruise Control	Std	Std
All Wheel Drive	N/A	N/A
AM/FM Stereo Cassette	Std	Std
Steering Wheel, Tilt	N/A	N/A
Power Windows	Std	Std
*Options Price	$700	$850
*Total Price	$25,550	$30,500
Target Price	$28,249	
Destination Charge	$405	
Avg. Tax & Fees	$1,742	
Total Target $	**$30,396**	
Average Dealer Option Cost:	82%	

The BMW 3-Series is available in five models–318iS and 325iS coupes, 325i and 318i sedans, and 325i convertible. New for 1993, the 325i sedan features split fold-down rear seats. Other new options include sports suspension, metallic paint, leather seating areas, onboard computer, CD player/changer and BMW cellular telephone. Standard features include eight-way power front seats with adjustable head restraints, power two-way sunroof, driver's-side airbag with automatic front seatbelt tensioners.

Ownership Costs

Cost Area	5 Year Cost	Rate
Depreciation	$11,966	◯
Financing ($611/month)	$5,008	
Insurance (Rating 21+ [Est.])	$12,847	●
State Fees	$1,235	
Fuel (Hwy 28 City 20 -Prem.)	$3,991	◯
Maintenance	$5,110	◯
Repairs	$985	◯

Warranty/Maintenance Info

Major Tune-Up	$343 ●
Minor Tune-Up	$212 ●
Brake Service	$255 ◯
Overall Warranty	4 yr/50k ◯
Drivetrain Warranty	4 yr/50k ◯
Rust Warranty	6 yr/unlim. mi ◯
Maintenance Warranty	N/A
Roadside Assistance	4 yr/50k

Resale Value

1993	1994	1995	1996	1997
$23,569	$22,383	$21,246	$19,822	$18,430

Cumulative Costs

	1993	1994	1995	1996	1997
Annual	$12,385	$6,614	$7,725	$5,902	$8,516
Total	$12,385	$18,999	$26,724	$32,626	$41,142

Ownership Costs (5yr)

- Average: $42,259 — Cost/Mile 60¢
- This Car: $41,142 — Cost/Mile 59¢

Ownership Cost Rating

◯ Better Than Average

---

* Includes shaded options
** Other purchase requirements apply

Legend: ● Poor | ◕ Worse Than Average | ◐ Average | ◯ Better Than Average | ○ Excellent | ⊖ Insufficient Information

©1993 by *IntelliChoice, Inc.* (408) 554-8711 All Rights Reserved. Reproduction Prohibited.
Refer to *Section 3: Annotated Vehicle Charts* for an explanation of these charts.

BMW 325 iC
2 Door Convertible

Luxury

- 2.5L 168 hp Gas Fuel Inject.
- 6 Cylinder In-Line
- Manual 5 Speed
- 2 Wheel Rear
- Driver Airbag Psngr Belts

Purchase Price

Car Item	Dealer Cost	List
Base Price	**$30,395**	**$36,320**
Anti-Lock Brakes	Std	Std
Automatic Transmission	N/A	N/A
Optional Engine	N/A	N/A
Auto Climate Control	Std	Std
Power Steering	Std	Std
Cruise Control	Std	Std
All Wheel Drive	N/A	N/A
AM/FM Stereo Cassette	Std	Std
Steering Wheel, Tilt	N/A	N/A
Power Windows	Std	Std
***Options Price**	**$0**	**$0**
***Total Price**	**$30,395**	**$36,320**
Target Price	$33,890	
Destination Charge	$405	
Avg. Tax & Fees	$2,082	
Luxury Tax	$430	
Total Target $	**$36,807**	

The BMW 3-Series is available in five models–318iS and 325iS coupes, 325i and 318i sedans, and 325i convertible. New for 1993, the 325i convertible now comes standard with leather and wood interior trim. Other new features include automatic front seatbelt tensioners, variable valve timing and a redesigned outside mirrors for improved aerodynamics. Standard equipment includes sporty cross-spoke wheels, multi-adjustable sports front seats, central locking and a power-operated convertible top.

Ownership Costs

Cost Area	5 Year Cost	Rate
Depreciation	$13,797	○
Financing ($740/month)	$6,063	
Insurance (Rating 26 [Est.])	$13,714	●
State Fees	$1,469	
Fuel (Hwy 24 City 17 -Prem.)	$4,673	◉
Maintenance	$4,794	○
Repairs	$985	○

Warranty/Maintenance Info

Major Tune-Up	$314	●
Minor Tune-Up	$217	●
Brake Service	$173	○
Overall Warranty	4 yr/50k	○
Drivetrain Warranty	4 yr/50k	○
Rust Warranty	6 yr/unlim. mi	○
Maintenance Warranty	N/A	
Roadside Assistance	4 yr/50k	

Ownership Cost By Year

1993, 1994, 1995, 1996, 1997

Resale Value

1993	1994	1995	1996	1997
$31,094	$28,938	$26,957	$24,983	$23,010

Cumulative Costs

	1993	1994	1995	1996	1997
Annual	$12,051	$8,265	$8,758	$7,525	$8,896
Total	$12,051	$20,316	$29,074	$36,599	$45,495

Ownership Costs (5yr)

Average	This Car
$46,616	$45,495
Cost/Mile 67¢	Cost/Mile 65¢

Ownership Cost Rating
○ Average

BMW 525 i
4 Door Sedan

Luxury

- 2.5L 189 hp Gas Fuel Inject.
- 6 Cylinder In-Line
- Manual 5 Speed
- 2 Wheel Rear
- Driver Airbag Psngr Belts

Purchase Price

Car Item	Dealer Cost	List
Base Price	**$31,095**	**$37,100**
Anti-Lock Brakes	Std	Std
Automatic 4 Speed	$700	$850
Optional Engine	N/A	N/A
Air Conditioning	Std	Std
Power Steering	Std	Std
Cruise Control	Std	Std
All Wheel Drive	N/A	N/A
AM/FM Stereo Cassette	Std	Std
Steering Wheel, Scope	Std	Std
Power Windows	Std	Std
***Options Price**	**$700**	**$850**
***Total Price**	**$31,795**	**$37,950**
Target Price	$35,451	
Destination Charge	$405	
Avg. Tax & Fees	$2,177	
Luxury Tax	$586	
Total Target $	**$38,619**	

The BMW 5-Series is available in four models–525i sedan and Touring wagon, and 535i and M5 sedans. New for 1993, the 525i sedan features walnut interior trim and upgraded leather upholstery as standard equipment. Other new features include outside mirrors that further reduce wind noise, variable valve timing and electronic knock control. Standard features include gathered leather upholstery, heated driver's-door lock and 10-way power front seats including power-adjustable head restraints.

Ownership Costs

Cost Area	5 Year Cost	Rate
Depreciation	$17,159	○
Financing ($776/month)	$6,361	
Insurance (Rating 20 [Est.])	$10,244	○
State Fees	$1,534	
Fuel (Hwy 25 City 18 -Prem.)	$4,451	○
Maintenance	$3,721	○
Repairs	$1,055	○

Warranty/Maintenance Info

Major Tune-Up	$299	●
Minor Tune-Up	$207	●
Brake Service	$209	○
Overall Warranty	4 yr/50k	○
Drivetrain Warranty	4 yr/50k	○
Rust Warranty	6 yr/unlim. mi	○
Maintenance Warranty	N/A	
Roadside Assistance	4 yr/50k	

Resale Value

1993	1994	1995	1996	1997
$30,660	$28,491	$26,264	$23,664	$21,460

Cumulative Costs

	1993	1994	1995	1996	1997
Annual	$13,760	$7,648	$8,059	$6,750	$8,308
Total	$13,760	$21,408	$29,467	$36,217	$44,525

Ownership Costs (5yr)

Average	This Car
$47,836	$44,525
Cost/Mile 68¢	Cost/Mile 64¢

Ownership Cost Rating
○ Excellent

---

* Includes shaded options
** Other purchase requirements apply

● Poor   ◉ Worse Than Average   ○ Average   ○ Better Than Average   ○ Excellent   ⊖ Insufficient Information

©1993 by IntelliChoice, Inc. (408) 554-8711   All Rights Reserved. Reproduction Prohibited.
Refer to *Section 3: Annotated Vehicle Charts* for an explanation of these charts.

BMW 535i — 4 Door Sedan — Luxury

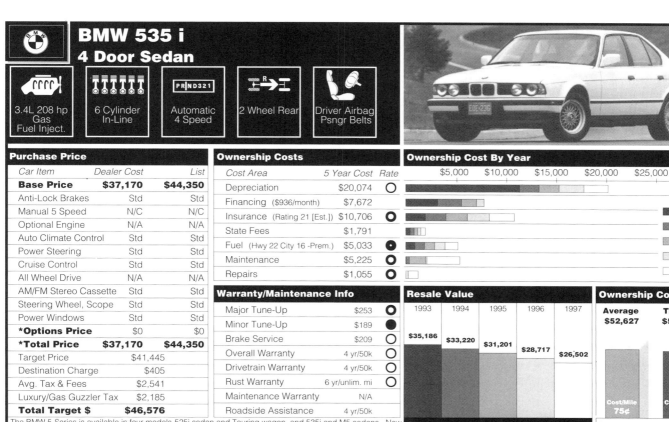

- 3.4L 208 hp Gas Fuel Inject.
- 6 Cylinder In-Line
- Automatic 4 Speed
- 2 Wheel Rear
- Driver Airbag Psngr Belts

Purchase Price

Car Item	Dealer Cost	List
Base Price	**$37,170**	**$44,350**
Anti-Lock Brakes	Std	Std
Manual 5 Speed	N/C	N/C
Optional Engine	N/A	N/A
Auto Climate Control	Std	Std
Power Steering	Std	Std
Cruise Control	Std	Std
All Wheel Drive	N/A	N/A
AM/FM Stereo Cassette	Std	Std
Steering Wheel, Scope	Std	Std
Power Windows	Std	Std
*Options Price	$0	$0
*Total Price	$37,170	$44,350
Target Price		$41,445
Destination Charge		$405
Avg. Tax & Fees		$2,541
Luxury/Gas Guzzler Tax		$2,185
Total Target $		**$46,576**

The BMW 5-Series is available in four models–525i sedan and Touring wagon, and 535i and M5 sedans. New for 1993, the 535i sedan features automatic front seatbelt tensioners and gathered leather interior trim as standard equipment. Standard features include power two-way sunroof, onboard computer and 10-way power front seats including power-adjustable head restraints. Option list includes a six-disc CD player, Inclement Weather Package, Memory Package and BMW cellular telephone.

Ownership Costs

Cost Area	5 Year Cost	Rate
Depreciation	$20,074	○
Financing ($936/month)	$7,672	
Insurance (Rating 21 [Est.])	$10,706	○
State Fees	$1,791	
Fuel (Hwy 22 City 16 -Prem.)	$5,033	◉
Maintenance	$5,225	○
Repairs	$1,055	○

Warranty/Maintenance Info

Major Tune-Up	$253	○
Minor Tune-Up	$189	●
Brake Service	$209	○
Overall Warranty	4 yr/50k	○
Drivetrain Warranty	4 yr/50k	○
Rust Warranty	6 yr/unlim. mi	○
Maintenance Warranty	N/A	
Roadside Assistance	4 yr/50k	○

Ownership Cost By Year

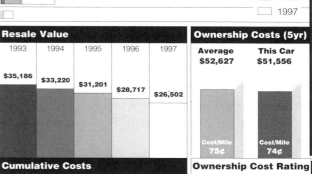

1993, 1994, 1995, 1996, 1997

Resale Value

1993	1994	1995	1996	1997
$35,186	$33,220	$31,201	$28,717	$26,502

Cumulative Costs

	1993	1994	1995	1996	1997
Annual	$17,988	$8,111	$8,678	$7,007	$9,772
Total	$17,988	$26,099	$34,777	$41,784	$51,556

Ownership Costs (5yr)

Average	This Car
$52,627	$51,556
Cost/Mile 75¢	Cost/Mile 74¢

Ownership Cost Rating
○ Average

BMW M5 — 4 Door Sedan — Sport

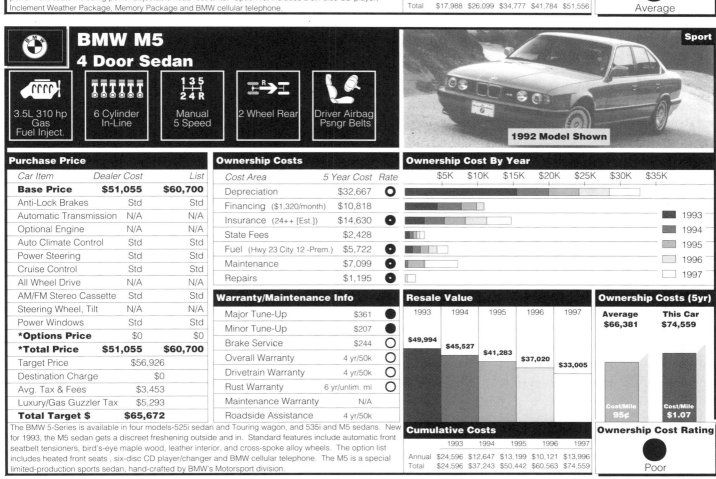

1992 Model Shown

- 3.5L 310 hp Gas Fuel Inject.
- 6 Cylinder In-Line
- Manual 5 Speed
- 2 Wheel Rear
- Driver Airbag Psngr Belts

Purchase Price

Car Item	Dealer Cost	List
Base Price	**$51,055**	**$60,700**
Anti-Lock Brakes	Std	Std
Automatic Transmission	N/A	N/A
Optional Engine	N/A	N/A
Auto Climate Control	Std	Std
Power Steering	Std	Std
Cruise Control	Std	Std
All Wheel Drive	N/A	N/A
AM/FM Stereo Cassette	Std	Std
Steering Wheel, Tilt	N/A	N/A
Power Windows	Std	Std
*Options Price	$0	$0
*Total Price	$51,055	$60,700
Target Price		$56,926
Destination Charge		$0
Avg. Tax & Fees		$3,453
Luxury/Gas Guzzler Tax		$5,293
Total Target $		**$65,672**

The BMW 5-Series is available in four models–525i sedan and Touring wagon, and 535i and M5 sedans. New for 1993, the M5 sedan gets a discreet freshening outside and in. Standard features include automatic front seatbelt tensioners, bird's-eye maple wood, leather interior, and cross-spoke alloy wheels. The option list includes heated front seats, six-disc CD player/changer and BMW cellular telephone. The M5 is a special limited-production sports sedan, hand-crafted by BMW's Motorsport division.

Ownership Costs

Cost Area	5 Year Cost	Rate
Depreciation	$32,667	○
Financing ($1,320/month)	$10,818	
Insurance (24++ [Est.])	$14,630	◉
State Fees	$2,428	
Fuel (Hwy 23 City 12 -Prem.)	$5,722	◉
Maintenance	$7,099	◉
Repairs	$1,195	◉

Warranty/Maintenance Info

Major Tune-Up	$361	●
Minor Tune-Up	$207	●
Brake Service	$244	○
Overall Warranty	4 yr/50k	○
Drivetrain Warranty	4 yr/50k	○
Rust Warranty	6 yr/unlim. mi	○
Maintenance Warranty	N/A	
Roadside Assistance	4 yr/50k	○

Resale Value

1993	1994	1995	1996	1997
$49,994	$45,527	$41,283	$37,020	$33,005

Cumulative Costs

	1993	1994	1995	1996	1997
Annual	$24,596	$12,647	$13,199	$10,121	$13,996
Total	$24,596	$37,243	$50,442	$60,563	$74,559

Ownership Costs (5yr)

Average	This Car
$66,381	$74,559
Cost/Mile 95¢	Cost/Mile $1.07

Ownership Cost Rating
● Poor

---

* Includes shaded options
** Other purchase requirements apply

● Poor   ◉ Worse Than Average   ○ Average   ○ Better Than Average   ○ Excellent   ⊖ Insufficient Information

©1993 by IntelliChoice, Inc. (408) 554-8711 All Rights Reserved. Reproduction Prohibited.
Refer to *Section 3: Annotated Vehicle Charts* for an explanation of these charts.

BMW 525 i Touring
4 Door Wagon

2.5L 189 hp Gas Fuel Inject. | 6 Cylinder In-Line | Automatic 4 Speed | 2 Wheel Rear | Driver Airbag Psngr Belts

 Luxury

Purchase Price

Car Item	Dealer Cost	List
Base Price	**$33,355**	**$39,800**
Anti-Lock Brakes	Std	Std
Manual Transmission	N/A	N/A
Optional Engine	N/A	N/A
Air Conditioning	Std	Std
Power Steering	Std	Std
Cruise Control	Std	Std
All Wheel Drive	N/A	N/A
AM/FM Stereo Cassette	Std	Std
Steering Wheel, Scope	Std	Std
Power Windows	Std	Std
*Options Price	$0	$0
*Total Price	$33,355	$39,800
Target Price	$37,191	
Destination Charge	$405	
Avg. Tax & Fees	$2,282	
Luxury Tax	$760	
Total Target $	**$40,638**	

The BMW 5-Series is available in four models-525i sedan and Touring wagon, and 535i and M5 sedans. New for 1993, the 525i Touring features wood interior trim and automatic front seatbelt tensioners. Standard features include multi-function tailgate, roof-rack system and split fold-down seatbacks. Other features include a twin-panel sunroof, heated outside mirrors and optional Inclement Weather Package. The 525i Touring is the first BMW sports-wagon model to be offered in North America.

Ownership Costs

Cost Area	5 Year Cost	Rate
Depreciation	$19,954	◯
Financing ($817/month)	$6,693	
Insurance (Rating 20 [Est.])	$10,244	◯
State Fees	$1,608	
Fuel (Hwy 25 City 18 -Prem.)	$4,451	◯
Maintenance	$4,308	◯
Repairs	$1,055	◯

Warranty/Maintenance Info

Major Tune-Up	$299	●
Minor Tune-Up	$207	●
Brake Service	$209	◯
Overall Warranty	4 yr/50k	◯
Drivetrain Warranty	4 yr/50k	◯
Rust Warranty	6 yr/unlim. mi	◯
Maintenance Warranty	N/A	
Roadside Assistance	4 yr/50k	

Ownership Cost By Year

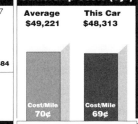

1993, 1994, 1995, 1996, 1997

Resale Value

1993	1994	1995	1996	1997
$31,163	$28,774	$26,072	$23,348	$20,684

Cumulative Costs

	1993	1994	1995	1996	1997
Annual	$15,432	$7,989	$8,897	$6,911	$9,084
Total	$15,432	$23,421	$32,318	$39,229	$48,313

Ownership Costs (5yr)

Average	This Car
$49,221	$48,313
Cost/Mile 70¢	Cost/Mile 69¢

Ownership Cost Rating
◯ Average

BMW 740 i
4 Door Sedan

4.0L 282 hp Gas Fuel Inject. | 8 Cylinder "V" | Automatic 5 Speed | 2 Wheel Rear | Driver/Psngr Airbags Std

 Luxury

Purchase Price

Car Item	Dealer Cost	List
Base Price	**$44,150**	**$54,000**
Anti-Lock Brakes	Std	Std
Manual Transmission	N/A	N/A
Optional Engine	N/A	N/A
Auto Climate Control	Std	Std
Power Steering	Std	Std
Cruise Control	Std	Std
All Wheel Drive	N/A	N/A
AM/FM Stereo Cassette	Std	Std
Steering, Tilt w/Memory	Std	Std
Power Windows	Std	Std
*Options Price	$0	$0
*Total Price	$44,150	$54,000
Target Price	$49,227	
Destination Charge	$405	
Avg. Tax & Fees	$3,026	
Luxury/Gas Guzzler Tax	$3,263	
Total Target $	**$55,921**	

The BMW 7-Series is available in four models-740i, 740iL and 750iL sedans. New for 1993, the 740i sedan features an all-new 282 hp, dual-overhead-cam V-8 engine. Other new features include passenger's-side airbag, automatic front seatbelt tensioners and ventilated rear disc brakes. The 740i also features gathered leather upholstery, wood inlays in walnut trim and an improved audio system. The option list includes the Inclement Weather Package and Automatic Stability Control Plus Traction.

Ownership Costs

Cost Area	5 Year Cost	Rate
Depreciation	$25,480	◯
Financing ($1,124/month)	$9,211	
Insurance (Rating 23 [Est.])	$11,066	◯
State Fees	$2,176	
Fuel (Hwy 22 City 16 -Prem.)	$5,033	◉
Maintenance	$5,216	◯
Repairs	$1,275	◉

Warranty/Maintenance Info

Major Tune-Up	$351	●
Minor Tune-Up	$219	●
Brake Service	$253	◯
Overall Warranty	4 yr/50k	◯
Drivetrain Warranty	4 yr/50k	◯
Rust Warranty	6 yr/unlim. mi	◯
Maintenance Warranty	N/A	
Roadside Assistance	4 yr/50k	

Resale Value

1993	1994	1995	1996	1997
$42,010	$39,043	$36,250	$33,354	$30,441

Cumulative Costs

	1993	1994	1995	1996	1997
Annual	$21,326	$9,779	$10,170	$7,773	$10,409
Total	$21,326	$31,105	$41,275	$49,048	$59,457

Ownership Costs (5yr)

Average	This Car
$60,031	$59,457
Cost/Mile 86¢	Cost/Mile 85¢

Ownership Cost Rating
◯ Average

---

page 20 | * Includes shaded options ** Other purchase requirements apply

● Poor | ◉ Worse Than Average | ◐ Average | ◯ Better Than Average | ◯ Excellent | ⊖ Insufficient Information

©1993 by IntelliChoice, Inc. (408) 554-8711 All Rights Reserved. Reproduction Prohibited.
Refer to *Section 3: Annotated Vehicle Charts* for an explanation of these charts.

BMW 740 iL
4 Door Sedan

4.0L 282 hp Gas Fuel Inject. | 8 Cylinder "V" | Automatic 5 Speed | 2 Wheel Rear | Driver/Psngr Airbags Std

i Model Shown — Luxury

Purchase Price

Car Item	Dealer Cost	List
Base Price	**$47,420**	**$58,000**
Anti-Lock Brakes	Std	Std
Manual Transmission	N/A	N/A
Optional Engine	N/A	N/A
Auto Climate Control	Std	Std
Power Steering	Std	Std
Cruise Control	Std	Std
All Wheel Drive	N/A	N/A
AM/FM Stereo Cassette	Std	Std
Steering, Tilt w/Memory	Std	Std
Power Windows	Std	Std
*Options Price	$0	$0
*Total Price	$47,420	$58,000
Target Price	$52,873	
Destination Charge	$405	
Avg. Tax & Fees	$3,248	
Luxury Tax	$2,328	
Total Target $	**$58,854**	

Ownership Costs

Cost Area	5 Year Cost	Rate
Depreciation	$26,866	○
Financing ($1,183/month)	$9,694	
Insurance (Rating 23 [Est.])	$11,066	○
State Fees	$2,336	
Fuel (Hwy 22 City 15 -Prem.)	$5,194	◉
Maintenance	$5,216	○
Repairs	$1,275	◉

Warranty/Maintenance Info

Major Tune-Up	$351	●
Minor Tune-Up	$219	●
Brake Service	$253	○
Overall Warranty	4 yr/50k	○
Drivetrain Warranty	4 yr/50k	○
Rust Warranty	6 yr/unlim. mi	○
Maintenance Warranty	N/A	
Roadside Assistance	4 yr/50k	

Ownership Cost By Year

1993 / 1994 / 1995 / 1996 / 1997

Resale Value

1993	1994	1995	1996	1997
$45,572	$42,099	$38,799	$35,277	$31,988

Cumulative Costs

	1993	1994	1995	1996	1997
Annual	$20,968	$10,502	$10,837	$8,496	$10,844
Total	$20,968	$31,470	$42,307	$50,803	$61,647

Ownership Costs (5yr)

	Average	This Car
	$63,584	$61,647
Cost/Mile	91¢	88¢

Ownership Cost Rating

○ Better Than Average

The BMW 7-Series is available in four models-740i, 740iL and 750iL sedans. New for 1993, the 740iL sedan features and all-new 4.0-liter dual-overhead-cam V-8 engine. Standard features include ventilated rear disc brakes, passenger's-side airbag and automatic front seatbelt tensioners. New interior features include gathered leather upholstery, improved audio system, contrasting wood lays in walnut trim and a newly designed center console. An available rear sunshade is now power-operated.

BMW 750 iL
4 Door Sedan

5.0L 296 hp Gas Fuel Inject. | 12 Cylinder "V" | Automatic 4 Speed | 2 Wheel Rear | Driver/Psngr Airbags Std

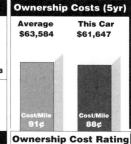

Luxury

Purchase Price

Car Item	Dealer Cost	List
Base Price	**$66,145**	**$80,900**
Anti-Lock Brakes	Std	Std
Manual Transmission	N/A	N/A
Optional Engine	N/A	N/A
Auto Climate Control	Std	Std
Power Steering	Std	Std
Cruise Control	Std	Std
All Wheel Drive	N/A	N/A
AM/FM Stereo CD	Std	Std
Steering, Tilt w/Memory	Std	Std
Power Windows	Std	Std
*Options Price	$0	$0
*Total Price	$66,145	$80,900
Target Price	$73,752	
Destination Charge	$405	
Avg. Tax & Fees	$4,521	
Luxury/Gas Guzzler Tax	$7,416	
Total Target $	**$86,094**	

Ownership Costs

Cost Area	5 Year Cost	Rate
Depreciation	$47,893	○
Financing ($1,730/month)	$14,182	
Insurance (Rating 26+ [Est.])	$15,582	◉
State Fees	$3,252	
Fuel (Hwy 18 City 12 -Prem.)	$6,420	●
Maintenance	$5,086	○
Repairs	$1,505	●

Warranty/Maintenance Info

Major Tune-Up	$402	●
Minor Tune-Up	$208	●
Brake Service	$231	○
Overall Warranty	4 yr/50k	○
Drivetrain Warranty	4 yr/50k	○
Rust Warranty	6 yr/unlim. mi	○
Maintenance Warranty	4 yr/50k	◯
Roadside Assistance	4 yr/50k	

Resale Value

1993	1994	1995	1996	1997
$60,071	$54,445	$49,077	$43,522	$38,201

Cumulative Costs

	1993	1994	1995	1996	1997
Annual	$36,804	$15,016	$14,036	$12,493	$15,571
Total	$36,804	$51,820	$65,856	$78,349	$93,920

Ownership Costs (5yr)

	Average	This Car
	$83,924	$93,920
Cost/Mile	$1.20	$1.34

Ownership Cost Rating

● Worse Than Average

The BMW 7-Series is available in four models-740i, 740iL and 750iL sedans. New for 1993, the 750iL sedan features Xenon low-beam headlamps. Interior features include dual airbags, wood trim, more extensively gathered leather upholstery and new audio system. Other features include front and rear heated seats, a power-operated rear-window sunshade and a pass through with ski bag as standard equipment. Also new: a soft-close trunklid that automatically closes after being lowered.

---

* Includes shaded options
** Other purchase requirements apply

● Poor | ◉ Worse Than Average | ○ Average | ○ Better Than Average | ◯ Excellent | ⊖ Insufficient Information

©1993 by IntelliChoice, Inc. (408) 554-8711   All Rights Reserved. Reproduction Prohibited.
Refer to *Section 3: Annotated Vehicle Charts* **for an explanation of these charts.**

BMW 850 Ci
2 Door Coupe

 5.0L 296 hp Gas Fuel Inject.
 12 Cylinder "V"
 Automatic 4 Speed
 2 Wheel Rear
Driver/Psngr Airbags Std

Sport

Purchase Price

Car Item	Dealer Cost	List
Base Price	$68,190	$83,400
Anti-Lock Brakes	Std	Std
Manual 6 Speed	N/C	N/C
Optional Engine	N/A	N/A
Auto Climate Control	Std	Std
Power Steering	Std	Std
Cruise Control	Std	Std
All Wheel Drive	N/A	N/A
AM/FM Stereo CD	Std	Std
Steering, Tilt w/Memory	Std	Std
Power Windows	Std	Std
*Options Price	$0	$0
*Total Price	$68,190	$83,400
Target Price		$76,032
Destination Charge		$405
Avg. Tax & Fees		$4,660
Luxury/Gas Guzzler Tax		$7,644
Total Target $		**$88,741**

The 850Ci coupe is available in a one model edition. New for 1993, the 850Ci features an upgraded interior and dual airbags. The interior features gathered leather upholstery, bird's-eye maple wood trim, and a split fold-down rear seats. Other features include upgraded material textures for the instrument panel, doors and center console. The 850Ci also features a two-stage heated front seats with timer, power windows, rear-window defroster and a power two-way sunroof with one-touch operation.

Ownership Costs

Cost Area	5 Year Cost	Rate
Depreciation		⊖
Financing ($1,784/month)	$14,618	
Insurance (27++ [Est.])	$17,028	●
State Fees	$3,352	
Fuel (Hwy 18 City 12)	$5,813	●
Maintenance	$5,474	○
Repairs	$1,505	●

Warranty/Maintenance Info

Major Tune-Up	$424	●
Minor Tune-Up	$230	●
Brake Service	$323	◐
Overall Warranty	4 yr/50k	○
Drivetrain Warranty	4 yr/50k	○
Rust Warranty	6 yr/unlim. mi	○
Maintenance Warranty	4 yr/50k	○
Roadside Assistance	4 yr/50k	

Ownership Cost By Year

Insufficient Depreciation Information

1993, 1994, 1995, 1996, 1997

Resale Value

Insufficient Information

Cumulative Costs

	1993	1994	1995	1996	1997
Annual	Insufficient Information				
Total	Insufficient Information				

Ownership Costs (5yr)

Insufficient Information

Ownership Cost Rating

⊖ Insufficient Information

Buick Century Custom
2 Door Coupe

 2.2L 110 hp Gas Fuel Inject.
4 Cylinder In-Line
 Automatic 3 Speed
 2 Wheel Front
Driver Airbag Psngr Belts

Midsize

Purchase Price

Car Item	Dealer Cost	List
Base Price	$13,668	$15,620
Anti-Lock Brakes	N/A	N/A
4 Spd Auto	$172	$200
3.3L 160 hp Gas	$568	$660
Air Conditioning	Std	Std
Power Steering	Std	Std
Cruise Control	Pkg	Pkg
All Wheel Drive	N/A	N/A
AM/FM Stereo Cassette	$120	$140
Steering Wheel, Tilt	Std	Std
Power Windows	Pkg	Pkg
*Options Price	$688	$800
*Total Price	$14,356	$16,420
Target Price		$15,699
Destination Charge		$500
Avg. Tax & Fees		$979
Total Target $		**$17,178**
Average Dealer Option Cost:	**86%**	

The 1993 Century is available in six models - Special, Custom, and Limited sedans; Special and Custom wagons; and the Custom coupe. New for 1993, the Custom coupe features four new colors (Light Driftwood Metallic, Bright White, Ruby Red Metallic, and Dark Cherry Metallic). The Custom upgrades the Special with armrest with front storage bin, driver's airbag, tilt steering wheel and dual visor vanity mirrors.

Ownership Costs

Cost Area	5 Year Cost	Rate
Depreciation	$9,474	○
Financing ($345/month)	$2,830	
Insurance (Rating 11)	$7,693	○
State Fees	$676	
Fuel (Hwy 26 City 19)	$3,847	◐
Maintenance	$4,610	○
Repairs	$709	○

Warranty/Maintenance Info

Major Tune-Up	$196	○
Minor Tune-Up	$138	●
Brake Service	$213	○
Overall Warranty	3 yr/36k	○
Drivetrain Warranty	3 yr/36k	○
Rust Warranty	6 yr/100k	○
Maintenance Warranty	N/A	
Roadside Assistance	3 yr/36k	

Resale Value

1993	1994	1995	1996	1997
$11,995	$10,707	$9,673	$8,640	$7,704

Ownership Costs (5yr)

Average	This Car
$28,058	$29,839
Cost/Mile 40¢	Cost/Mile 43¢

Cumulative Costs

	1993	1994	1995	1996	1997
Annual	$8,713	$4,833	$5,278	$5,117	$5,898
Total	$8,713	$13,546	$18,824	$23,941	$29,839

Ownership Cost Rating

● Worse Than Average

---

page **22**

* Includes shaded options
** Other purchase requirements apply

● Poor ◐ Worse Than Average ○ Average ○ Better Than Average ○ Excellent ⊖ Insufficient Information

©1993 by IntelliChoice, Inc. (408) 554-8711 All Rights Reserved. Reproduction Prohibited.
Refer to *Section 3: Annotated Vehicle Charts* for an explanation of these charts.

Buick Century Special
4 Door Sedan

Midsize

 2.2L 110 hp Gas Fuel Inject.
 4 Cylinder In-Line
 Automatic 3 Speed
 2 Wheel Front
Belts Std, Driv Air Opt

Limited Model Shown

Purchase Price

Car Item	Dealer Cost	List
Base Price	$12,713	$14,205
Anti-Lock Brakes	N/A	N/A
4 Spd Auto	$172	$200
3.3L 160 hp Gas	$568	$660
Air Conditioning	Std	Std
Power Steering	Std	Std
Cruise Control	Pkg	** Pkg
All Wheel Drive	N/A	N/A
AM/FM Stereo Cassette	$120	$140
Steering Wheel, Tilt	Pkg	Pkg
Power Windows	Pkg	Pkg
***Options Price**	$688	$800
***Total Price**	$13,401	$15,005
Target Price		$14,620
Destination Charge		$500
Avg. Tax & Fees		$911
Total Target $		$16,031
Average Dealer Option Cost:		86%

The 1993 Century is available in six models - Special, Custom, and Limited sedans; Special and Custom wagons; and the Custom coupe. New for 1993, the Special sedan features four new colors (Light Driftwood Metallic, Bright White, Ruby Red Metallic, and Dark Cherry Metallic). Other features include deluxe wheel covers, automatic door locks, clearcoat paint and "headlights on" warning chime.

Ownership Costs

Cost Area	5 Year Cost	Rate
Depreciation	$8,619	○
Financing ($322/month)	$2,641	
Insurance (Rating 4)	$6,658	◎
State Fees	$621	
Fuel (Hwy 26 City 19)	$3,847	◉
Maintenance	$4,610	○
Repairs	$709	○

Warranty/Maintenance Info

Major Tune-Up	$196	○
Minor Tune-Up	$138	◉
Brake Service	$213	○
Overall Warranty	3 yr/36k	○
Drivetrain Warranty	3 yr/36k	○
Rust Warranty	6 yr/100k	○
Maintenance Warranty	N/A	
Roadside Assistance	3 yr/36k	

Ownership Cost By Year

(1993-1997 bar chart)

Resale Value

1993	1994	1995	1996	1997
$11,446	$10,206	$9,244	$8,301	$7,412

Cumulative Costs

	1993	1994	1995	1996	1997
Annual	$7,831	$4,515	$4,952	$4,789	$5,618
Total	$7,831	$12,346	$17,298	$22,087	$27,705

Ownership Costs (5yr)

Average	This Car
$26,611	$27,705
Cost/Mile 38¢	Cost/Mile 40¢

Ownership Cost Rating

Worse Than Average

Buick Century Custom
4 Door Sedan

Midsize

 2.2L 110 hp Gas Fuel Inject.
 4 Cylinder In-Line
 Automatic 3 Speed
 2 Wheel Front
 Driver Airbag Psngr Belts

2 Dr Coupe Shown

Purchase Price

Car Item	Dealer Cost	List
Base Price	$13,917	$15,905
Anti-Lock Brakes	N/A	N/A
4 Spd Auto	$172	$200
3.3L 160 hp Gas	$568	$660
Air Conditioning	Std	Std
Power Steering	Std	Std
Cruise Control	Pkg	Pkg
All Wheel Drive	N/A	N/A
AM/FM Stereo Cassette	$120	$140
Steering Wheel, Tilt	Std	Std
Power Windows	Pkg	Pkg
***Options Price**	$688	$800
***Total Price**	$14,605	$16,705
Target Price		$15,981
Destination Charge		$500
Avg. Tax & Fees		$996
Total Target $		$17,477
Average Dealer Option Cost:		86%

The 1993 Century is available in six models - Special, Custom, and Limited sedans; Special and Custom wagons; and the Custom coupe. New for 1993, the Custom sedan features four new colors (Light Driftwood Metallic, Bright White, Ruby Red Metallic, and Dark Cherry Metallic). The Custom upgrades the Special with armrest with front storage bin, driver's airbag, wheel lip moldings, tilt steering wheel and dual visor vanity mirrors.

Ownership Costs

Cost Area	5 Year Cost	Rate
Depreciation	$9,451	○
Financing ($351/month)	$2,879	
Insurance (Rating 6)	$6,919	○
State Fees	$688	
Fuel (Hwy 26 City 19)	$3,847	◉
Maintenance	$4,610	○
Repairs	$709	○

Warranty/Maintenance Info

Major Tune-Up	$196	○
Minor Tune-Up	$138	◉
Brake Service	$213	○
Overall Warranty	3 yr/36k	○
Drivetrain Warranty	3 yr/36k	○
Rust Warranty	6 yr/100k	○
Maintenance Warranty	N/A	
Roadside Assistance	3 yr/36k	

Resale Value

1993	1994	1995	1996	1997
$12,451	$11,092	$10,023	$8,975	$8,026

Cumulative Costs

	1993	1994	1995	1996	1997
Annual	$8,436	$4,775	$5,172	$4,976	$5,744
Total	$8,436	$13,211	$18,383	$23,359	$29,103

Ownership Costs (5yr)

Average	This Car
$28,349	$29,103
Cost/Mile 40¢	Cost/Mile 42¢

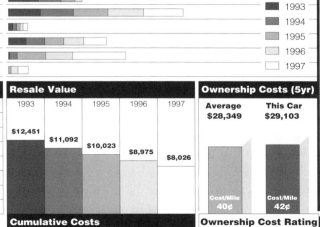

Ownership Cost Rating

○ Average

---

* Includes shaded options
** Other purchase requirements apply

 Poor   Worse Than Average   Average   Better Than Average   Excellent    Insufficient Information

©1993 by IntelliChoice, Inc. (408) 554-8711 All Rights Reserved. Reproduction Prohibited.
Refer to *Section 3: Annotated Vehicle Charts* for an explanation of these charts.

Buick Century Limited
4 Door Sedan

Midsize

 2.2L 110 hp Gas Fuel Inject. | 4 Cylinder In-Line | Automatic 3 Speed | 2 Wheel Front | Driver Airbag Psngr Belts

Purchase Price

Car Item	Dealer Cost	List
Base Price	**$14,757**	**$16,865**
Anti-Lock Brakes	N/A	N/A
4 Spd Auto	$172	$200
3.3L 160 hp Gas	$568	$660
Air Conditioning	Std	Std
Power Steering	Std	Std
Cruise Control	Pkg	Pkg
All Wheel Drive	N/A	N/A
AM/FM Stereo Cassette	$120	$140
Steering Wheel, Tilt	Std	Std
Power Windows	Pkg	Pkg
*****Options Price**	**$688**	**$800**
*****Total Price**	**$15,445**	**$17,665**
Target Price	$16,935	
Destination Charge	$500	
Avg. Tax & Fees	$1,054	
Total Target $	**$18,489**	
Average Dealer Option Cost:	**86%**	

The 1993 Century is available in six models - Special, Custom, and Limited sedans; Special and Custom wagons; and the Custom coupe. New for 1993, the Limited sedan features four new colors (Light Driftwood Metallic, Bright White, Ruby Red Metallic, and Dark Cherry Metallic). The Limited sedan upgrades the Custom with bodyside moldings, temperature and voltmeter guages, trip odometer and reading/map lights.

Ownership Costs

Cost Area	5 Year Cost	Rate
Depreciation	$10,238	◐
Financing ($372/month)	$3,046	
Insurance (Rating 7)	$7,035	○
State Fees	$726	
Fuel (Hwy 26 City 19)	$3,847	●
Maintenance	$4,615	◐
Repairs	$709	◐

Warranty/Maintenance Info

Major Tune-Up	$196	◐
Minor Tune-Up	$138	◉
Brake Service	$213	◐
Overall Warranty	3 yr/36k	◐
Drivetrain Warranty	3 yr/36k	◐
Rust Warranty	6 yr/100k	○
Maintenance Warranty	N/A	
Roadside Assistance	3 yr/36k	

Ownership Cost By Year

(1993-1997 bar chart)

Resale Value

1993	1994	1995	1996	1997
$12,588	$11,180	$10,141	$9,151	$8,251

Cumulative Costs

	1993	1994	1995	1996	1997
Annual	$9,413	$4,906	$5,207	$4,961	$5,729
Total	$9,413	$14,319	$19,526	$24,487	$30,216

Ownership Costs (5yr)

Average	This Car
$29,331	$30,216
Cost/Mile 42¢	Cost/Mile 43¢

Ownership Cost Rating
○ Average

Buick Century Special
4 Door Wagon

Midsize Wagon

 2.2L 110 hp Gas Fuel Inject. | 4 Cylinder In-Line | Automatic 3 Speed | 2 Wheel Front | Belts Std, Driv Air Opt

Purchase Price

Car Item	Dealer Cost	List
Base Price	**$13,389**	**$14,960**
Anti-Lock Brakes	N/A	N/A
4 Spd Auto	$172	$200
3.3L 160 hp Gas	$568	$660
Air Conditioning	Std	Std
Power Steering	Std	Std
Cruise Control	Pkg	** Pkg
All Wheel Drive	N/A	N/A
AM/FM Stereo Cassette	$120	$140
Steering Wheel, Tilt	Pkg	Pkg
Power Windows	Pkg	Pkg
*****Options Price**	**$688**	**$800**
*****Total Price**	**$14,077**	**$15,760**
Target Price	$15,383	
Destination Charge	$500	
Avg. Tax & Fees	$957	
Total Target $	**$16,840**	
Average Dealer Option Cost:	**86%**	

The 1993 Century is available in six models - Special, Custom, and Limited sedans; Special and Custom wagons; and the Custom coupe. New for 1993, the Special wagon features four new colors (Light Driftwood Metallic, Bright White, Ruby Red Metallic, and Dark Cherry Metallic). Other features include deluxe wheel covers, all-season tires, cloth notchback seat covering, 55/45 seats with a front-seat armrest, and a new trim level has been added.

Ownership Costs

Cost Area	5 Year Cost	Rate
Depreciation	$9,462	◉
Financing ($338/month)	$2,774	
Insurance (Rating 5)	$6,786	○
State Fees	$651	
Fuel (Hwy 26 City 19)	$3,847	◐
Maintenance	$4,611	◐
Repairs	$709	◐

Warranty/Maintenance Info

Major Tune-Up	$196	◐
Minor Tune-Up	$138	◉
Brake Service	$213	◐
Overall Warranty	3 yr/36k	◐
Drivetrain Warranty	3 yr/36k	◐
Rust Warranty	6 yr/100k	○
Maintenance Warranty	N/A	
Roadside Assistance	3 yr/36k	

Resale Value

1993	1994	1995	1996	1997
$11,423	$10,130	$9,185	$8,269	$7,378

Cumulative Costs

	1993	1994	1995	1996	1997
Annual	$8,750	$4,641	$4,992	$4,804	$5,653
Total	$8,750	$13,391	$18,383	$23,187	$28,840

Ownership Costs (5yr)

Average	This Car
$27,383	$28,840
Cost/Mile 39¢	Cost/Mile 41¢

Ownership Cost Rating
◐ Worse Than Average

---

page 24

* Includes shaded options
** Other purchase requirements apply

● Poor | ◉ Worse Than Average | ○ Average | ○ Better Than Average | ○ Excellent | ⊖ Insufficient Information

©1993 by IntelliChoice, Inc. (408) 554-8711 All Rights Reserved. Reproduction Prohibited.
Refer to *Section 3: Annotated Vehicle Charts* for an explanation of these charts.

Buick Century Custom
4 Door Wagon

Midsize Wagon

- 2.2L 110 hp Gas Fuel Inject.
- 4 Cylinder In-Line
- Automatic 3 Speed
- 2 Wheel Front
- Driver Airbag Psngr Belts

Purchase Price

Car Item	Dealer Cost	List
Base Price	$15,094	$17,250
Anti-Lock Brakes	N/A	N/A
4 Spd Auto	$172	$200
3.3L 160 hp Gas	$568	$660
Air Conditioning	Std	Std
Power Steering	Std	Std
Cruise Control	Pkg	** Pkg
All Wheel Drive	N/A	N/A
AM/FM Stereo Cassette	$120	$140
Steering Wheel, Tilt	Std	Std
Power Windows	Pkg	Pkg
***Options Price**	$688	$800
***Total Price**	$15,782	$18,050
Target Price		$17,320
Destination Charge		$500
Avg. Tax & Fees		$1,077
Total Target $		**$18,897**
Average Dealer Option Cost:		*86%*

The 1993 Century is available in six models - Special, Custom, and Limited sedans; Special and Custom wagons; and the Custom coupe. New for 1993, the Custom wagon features four new colors (Light Driftwood Metallic, Bright White, Ruby Red Metallic, and Dark Cherry Metallic). Other features include door edge guards, temperature and voltmeter guages, reading/map lights, trip odometer, and split folding rear seat.

Ownership Costs

Cost Area	5 Year Cost	Rate
Depreciation	$10,633	●
Financing ($380/month)	$3,113	
Insurance (Rating 7)	$7,035	○
State Fees	$742	
Fuel (Hwy 26 City 19)	$3,847	○
Maintenance	$4,611	○
Repairs	$709	○

Warranty/Maintenance Info

Major Tune-Up	$196	○
Minor Tune-Up	$138	●
Brake Service	$213	○
Overall Warranty	3 yr/36k	○
Drivetrain Warranty	3 yr/36k	○
Rust Warranty	6 yr/100k	○
Maintenance Warranty	N/A	
Roadside Assistance	3 yr/36k	

Ownership Cost By Year

1993, 1994, 1995, 1996, 1997

Resale Value

1993	1994	1995	1996	1997
$12,961	$11,471	$10,354	$9,290	$8,264

Cumulative Costs

	1993	1994	1995	1996	1997
Annual	$9,480	$5,013	$5,299	$5,044	$5,854
Total	$9,480	$14,493	$19,792	$24,836	$30,690

Ownership Costs (5yr)

Average	This Car
$29,725	$30,690
Cost/Mile 42¢	Cost/Mile 44¢

Ownership Cost Rating

○ Average

Buick LeSabre Custom
4 Door Sedan

Large

- 3.8L 170 hp Gas Fuel Inject.
- 6 Cylinder "V"
- Automatic 4 Speed
- 2 Wheel Front
- Driver Airbag Psngr Belts

Purchase Price

Car Item	Dealer Cost	List
Base Price	$17,443	$19,935
Anti-Lock Brakes	Std	Std
Manual Transmission	N/A	N/A
Optional Engine	N/A	N/A
Air Conditioning	Std	Std
Power Steering	Std	Std
Cruise Control	Pkg	Pkg
All Wheel Drive	N/A	N/A
AM/FM Stereo Cassette	$120	$140
Steering Wheel, Tilt	Std	Std
Power Windows	Std	Std
***Options Price**	$120	$140
***Total Price**	$17,563	$20,075
Target Price		$19,116
Destination Charge		$555
Avg. Tax & Fees		$1,190
Total Target $		**$20,861**
Average Dealer Option Cost:		*86%*

The 1993 LeSabre is available in two models - Custom and Limited sedans. New for 1993, the LeSabre features an enhanced 3.8L V-6 engine, dimmable door control floodlighting, five fresh colors (Bright White, Light Windsor Gray Metallic, Polo Green, Dark Cherry Metallic, and Smokey Amethyst Metallic), power door locks, and optional 15-inch custom wheel covers. Other features include a wraparound instrument panel, flush-mounted glass to reduce wind noise, and an optional Touring Package.

Ownership Costs

Cost Area	5 Year Cost	Rate
Depreciation	$10,633	○
Financing ($419/month)	$3,436	
Insurance (Rating 8)	$7,040	○
State Fees	$825	
Fuel (Hwy 28 City 19)	$3,704	○
Maintenance	$5,368	●
Repairs	$709	○

Warranty/Maintenance Info

Major Tune-Up	$196	○
Minor Tune-Up	$139	●
Brake Service	$249	○
Overall Warranty	3 yr/36k	○
Drivetrain Warranty	3 yr/36k	○
Rust Warranty	6 yr/100k	○
Maintenance Warranty	N/A	
Roadside Assistance	3 yr/36k	

Resale Value

1993	1994	1995	1996	1997
$15,541	$14,034	$12,792	$11,531	$10,228

Cumulative Costs

	1993	1994	1995	1996	1997
Annual	$8,993	$5,117	$5,594	$4,724	$7,287
Total	$8,993	$14,110	$19,704	$24,428	$31,715

Ownership Costs (5yr)

Average	This Car
$32,379	$31,715
Cost/Mile 46¢	Cost/Mile 45¢

Ownership Cost Rating

○ Average

---

* Includes shaded options
** Other purchase requirements apply

● Poor | ◐ Worse Than Average | ○ Average | ◯ Better Than Average | ◯ Excellent | ⊖ Insufficient Information

©1993 by IntelliChoice, Inc. (408) 554-8711 All Rights Reserved. Reproduction Prohibited.
Refer to *Section 3: Annotated Vehicle Charts* for an explanation of these charts.

page **25**

Buick LeSabre Limited
4 Door Sedan

Large

Purchase Price		
Car Item	Dealer Cost	List
Base Price	**$19,018**	**$21,735**
Anti-Lock Brakes	Std	Std
Manual Transmission	N/A	N/A
Optional Engine	N/A	N/A
Air Conditioning	Std	Std
Power Steering	Std	Std
Cruise Control	Pkg	Pkg
All Wheel Drive	N/A	N/A
AM/FM Stereo Cassette	$120	$140
Steering Wheel, Tilt	Std	Std
Power Windows	Std	Std
*Options Price	$120	$140
*Total Price	$19,138	$21,875
Target Price	$20,900	
Destination Charge	$555	
Avg. Tax & Fees	$1,297	
Total Target $	**$22,752**	
Average Dealer Option Cost:	**86%**	

The 1993 LeSabre is available in two models - Custom and Limited sedans. New for 1993, the LeSabre Limited features an enhanced 3.8L V-6 engine, dimmable door control floodlighting, five fresh colors (Bright White, Light Windsor Gray Metallic, Polo Green, Dark Cherry Metallic, and Smokey Amethyst Metallic), power door locks, and variable-effort steering. Other features include a wraparound instrument panel, flush-mounted glass to reduce wind noise, and an optional Touring Package.

Ownership Costs		
Cost Area	5 Year Cost	Rate
Depreciation	$12,187	○
Financing ($457/month)	$3,747	
Insurance (Rating 10)	$7,274	○
State Fees	$897	
Fuel (Hwy 28 City 19)	$3,704	○
Maintenance	$5,368	●
Repairs	$709	○

Warranty/Maintenance Info		
Major Tune-Up	$196	○
Minor Tune-Up	$139	●
Brake Service	$249	○
Overall Warranty	3 yr/36k	○
Drivetrain Warranty	3 yr/36k	○
Rust Warranty	6 yr/100k	○
Maintenance Warranty	N/A	
Roadside Assistance	3 yr/36k	

Ownership Cost By Year

Resale Value
1993: $16,041 | 1994: $14,441 | 1995: $13,183 | 1996: $11,956 | 1997: $10,565

Cumulative Costs
	1993	1994	1995	1996	1997
Annual	$10,575	$5,368	$5,732	$4,774	$7,437
Total	$10,575	$15,943	$21,675	$26,449	$33,886

Ownership Costs (5yr)
Average: $34,164 | This Car: $33,886
Cost/Mile: 49¢ | Cost/Mile: 48¢

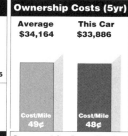

Ownership Cost Rating
○ Average

Buick Park Avenue
4 Door Sedan

Luxury

Purchase Price		
Car Item	Dealer Cost	List
Base Price	**$22,525**	**$26,040**
Anti-Lock Brakes	Std	Std
Manual Transmission	N/A	N/A
Optional Engine	N/A	N/A
Air Conditioning	Std	Std
Power Steering	Std	Std
Cruise Control	Std	Std
All Wheel Drive	N/A	N/A
AM/FM Stereo Cassette	Std	Std
Steering Wheel, Tilt	Std	Std
Power Windows	Std	Std
*Options Price	$0	$0
*Total Price	$22,525	$26,040
Target Price	$24,783	
Destination Charge	$600	
Avg. Tax & Fees	$1,535	
Total Target $	**$26,918**	
Average Dealer Option Cost:	**86%**	

The 1993 Park Avenue is available in two models - (Base) Park Avenue and Ultra sedans. New for 1993, the Base Park Avenue features revised instrument panel gauge graphics and transmission selector pattern, new grilles, tail lamp lens design, lighted visor vanity mirror with cloth cover, a (mid-year) new steering wheel design, and four new colors (Brt. White, Lt. Windsor Gray, Dk. Cherry, and Smokey Amethyst). New options include an automatic ride control system and 15-inch wire wheel covers.

Ownership Costs		
Cost Area	5 Year Cost	Rate
Depreciation	$14,430	○
Financing ($541/month)	$4,434	
Insurance (Rating 8)	$7,040	○
State Fees	$1,065	
Fuel (Hwy 27 City 19)	$3,775	○
Maintenance	$5,461	○
Repairs	$730	○

Warranty/Maintenance Info		
Major Tune-Up	$193	○
Minor Tune-Up	$139	○
Brake Service	$230	○
Overall Warranty	3 yr/36k	○
Drivetrain Warranty	3 yr/36k	○
Rust Warranty	6 yr/100k	○
Maintenance Warranty	N/A	
Roadside Assistance	3 yr/36k	

Ownership Cost By Year
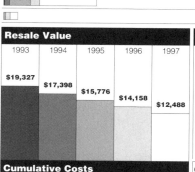

Resale Value
1993: $19,327 | 1994: $17,398 | 1995: $15,776 | 1996: $14,158 | 1997: $12,488

Cumulative Costs
	1993	1994	1995	1996	1997
Annual	$11,753	$5,925	$6,238	$5,206	$7,813
Total	$11,753	$17,678	$23,916	$29,122	$36,935

Ownership Costs (5yr)
Average: $38,920 | This Car: $36,935
Cost/Mile: 56¢ | Cost/Mile: 53¢

Ownership Cost Rating
○ Excellent

---

page 26

* Includes shaded options
** Other purchase requirements apply

● Poor | ◐ Worse Than Average | ○ Average | ○ Better Than Average | ○ Excellent | ⊖ Insufficient Information

©1993 by IntelliChoice, Inc. (408) 554-8711 All Rights Reserved. Reproduction Prohibited.
Refer to *Section 3: Annotated Vehicle Charts* for an explanation of these charts.

Buick Park Avenue Ultra
4 Door Sedan

 3.8L 205 hp Suprchrgd Fuel Inject.
 6 Cylinder "V"
 Automatic 4 Speed
 2 Wheel Front
Driver Airbag Psngr Belts

Luxury

Purchase Price

Car Item	Dealer Cost	List
Base Price	$25,427	$29,395
Anti-Lock Brakes	Std	Std
Manual Transmission	N/A	N/A
Optional Engine	N/A	N/A
Auto Climate Control	Std	Std
Power Steering	Std	Std
Cruise Control	Std	Std
All Wheel Drive	N/A	N/A
AM/FM Stereo Cassette	Std	Std
Steering Wheel, Tilt	Std	Std
Power Windows	Std	Std
***Options Price**	$0	$0
***Total Price**	$25,427	$29,395
Target Price	$28,147	
Destination Charge	$600	
Avg. Tax & Fees	$1,737	
Total Target $	**$30,484**	
Average Dealer Option Cost:	86%	

The 1993 Park Avenue is available in two models - (Base) Park Avenue and Ultra sedans. New for 1993, the Ultra Park Avenue features revised instrument panel gauge graphics and transmission selector pattern, new grilles, tail lamp lens design, lighted visor vanity mirror with cloth cover, 6-way power driver and passenger seats, and four new colors (Brt. White, Lt. Windsor Gray, Dk. Cherry, and Jadestone). New options include an automatic ride control system and 15-inch wire wheel covers.

Ownership Costs

Cost Area	5 Year Cost	Rate
Depreciation	$15,845	○
Financing ($613/month)	$5,023	
Insurance (Rating 10)	$7,274	○
State Fees	$1,200	
Fuel (Hwy 27 City 17)	$3,991	○
Maintenance	$5,504	○
Repairs	$730	○

Warranty/Maintenance Info

Major Tune-Up	$193	○
Minor Tune-Up	$139	○
Brake Service	$230	○
Overall Warranty	3 yr/36k	○
Drivetrain Warranty	3 yr/36k	○
Rust Warranty	6 yr/100k	○
Maintenance Warranty	N/A	
Roadside Assistance	3 yr/36k	

Ownership Cost By Year
1993, 1994, 1995, 1996, 1997

Resale Value
1993	1994	1995	1996	1997
$21,116	$19,239	$17,745	$16,299	$14,639

Cumulative Costs
	1993	1994	1995	1996	1997
Annual	$13,892	$6,173	$6,367	$5,195	$7,940
Total	$13,892	$20,065	$26,432	$31,627	$39,567

Ownership Costs (5yr)
Average	This Car
$41,432	$39,567
Cost/Mile 59¢	Cost/Mile 57¢

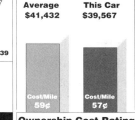

Ownership Cost Rating
○ **Excellent**

Buick Regal Custom
2 Door Coupe

 3.1L 140 hp Gas Fuel Inject.
 6 Cylinder "V"
 Automatic 4 Speed
 2 Wheel Front
 Automatic Seatbelts

Midsize

Purchase Price

Car Item	Dealer Cost	List
Base Price	$14,534	$16,610
Anti-Lock Brakes	$387	$450
Manual Transmission	N/A	N/A
3.8L 170 hp Gas	$340	** $395
Air Conditioning	Std	Std
Power Steering	Std	Std
Cruise Control	Pkg	Pkg
All Wheel Drive	N/A	N/A
AM/FM Stereo Cassette	$249	** $290
Steering Wheel, Tilt	Std	Std
Power Windows	$237	$275
***Options Price**	$237	$275
***Total Price**	$14,771	$16,885
Target Price	$15,820	
Destination Charge	$505	
Avg. Tax & Fees	$990	
Total Target $	**$17,315**	
Average Dealer Option Cost:	86%	

The 1993 Regal is available in six models - Custom, Gran Sport, and Limited coupes and sedans. New for 1993, the Custom has an improved drivetrain. The optional V-6 engine with tuned port injection has been enhanced for quieter, more responsive performance and provides improved fuel economy. Regal has a more sculptured front and rear appearance with the hood integrated with the new grille and headlamps. New front seats are featured that are orthopedically contoured. Analog gauges are standard.

Ownership Costs

Cost Area	5 Year Cost	Rate
Depreciation	$9,948	●
Financing ($348/month)	$2,853	
Insurance (Rating 8)	$7,223	○
State Fees	$696	
Fuel (Hwy 30 City 19)	$3,579	○
Maintenance	$5,432	●
Repairs	$709	○

Warranty/Maintenance Info

Major Tune-Up	$193	○
Minor Tune-Up	$142	●
Brake Service	$201	○
Overall Warranty	3 yr/36k	○
Drivetrain Warranty	3 yr/36k	○
Rust Warranty	6 yr/100k	○
Maintenance Warranty	N/A	
Roadside Assistance	3 yr/36k	

Ownership Cost By Year
1993, 1994, 1995, 1996, 1997

Resale Value
1993	1994	1995	1996	1997
$12,617	$11,019	$9,782	$8,516	$7,367

Cumulative Costs
	1993	1994	1995	1996	1997
Annual	$8,107	$5,011	$5,401	$4,676	$7,245
Total	$8,107	$13,118	$18,519	$23,195	$30,440

Ownership Costs (5yr)
Average	This Car
$28,533	$30,440
Cost/Mile 41¢	Cost/Mile 43¢

Ownership Cost Rating
● **Worse Than Average**

---

* Includes shaded options
** Other purchase requirements apply

● Poor Worse Than Average ○ Average Better Than Average ⊖ Excellent Insufficient Information

©1993 by IntelliChoice, Inc. (408) 554-8711 All Rights Reserved. Reproduction Prohibited.
Refer to *Section 3: Annotated Vehicle Charts* for an explanation of these charts.

Buick Regal Limited
2 Door Coupe

Midsize

 3.1L 140 hp Gas Fuel Inject.
 6 Cylinder "V"
 Automatic 4 Speed
 2 Wheel Front
 Automatic Seatbelts

Purchase Price

Car Item	Dealer Cost	List
Base Price	**$15,978**	**$18,260**
Anti-Lock Brakes	Std	Std
Manual Transmission	N/A	N/A
3.8L 170 hp Gas	$340	** $395
Air Conditioning	Std	Std
Power Steering	Std	Std
Cruise Control	Pkg	Pkg
All Wheel Drive	N/A	N/A
AM/FM Stereo Cassette	$129	** $150
Steering Wheel, Tilt	Std	Std
Power Windows	$237	$275
*Options Price	$237	$275
*Total Price	$16,215	$18,535
Target Price	$17,414	
Destination Charge	$505	
Avg. Tax & Fees	$1,086	
Total Target $	**$19,005**	
Average Dealer Option Cost:	**86%**	

The 1993 Regal is available in six models - Custom, Gran Sport, and Limited coupes and sedans. New for 1993, the Limited has an improved drivetrain. The optional V-6 engine with tuned port injection has been enhanced for a quieter, more responsive performance and provides improved fuel economy. Regal has a more sculptured front and rear appearance with the hood integrated with the new grille and headlamps. New front seats are featured that are orthopedically contoured. Analog gauges are standard.

Ownership Costs

Cost Area	5 Year Cost	Rate
Depreciation	$11,126	●
Financing ($382/month)	$3,131	
Insurance (Rating 10)	$7,479	○
State Fees	$760	
Fuel (Hwy 30 City 19)	$3,579	○
Maintenance	$5,432	●
Repairs	$709	○

Warranty/Maintenance Info

Major Tune-Up	$193	○
Minor Tune-Up	$142	●
Brake Service	$201	
Overall Warranty	3 yr/36k	○
Drivetrain Warranty	3 yr/36k	○
Rust Warranty	6 yr/100k	○
Maintenance Warranty	N/A	
Roadside Assistance	3 yr/36k	

Ownership Cost By Year

1993, 1994, 1995, 1996, 1997

Resale Value

1993	1994	1995	1996	1997
$13,468	$11,790	$10,454	$9,118	$7,879

Cumulative Costs

	1993	1994	1995	1996	1997
Annual	$9,125	$5,241	$5,620	$4,830	$7,400
Total	$9,125	$14,366	$19,986	$24,816	$32,216

Ownership Costs (5yr)

Average	This Car
$30,221	$32,216
Cost/Mile 43¢	Cost/Mile 46¢

Ownership Cost Rating
● Worse Than Average

Buick Regal Gran Sport
2 Door Coupe

Midsize

 3.8L 170 hp Gas Fuel Inject.
 6 Cylinder "V"
 Automatic 4 Speed
 2 Wheel Front
 Automatic Seatbelts

Purchase Price

Car Item	Dealer Cost	List
Base Price	**$16,708**	**$19,095**
Anti-Lock Brakes	Std	Std
Manual Transmission	N/A	N/A
Optional Engine	N/A	N/A
Air Conditioning	Std	Std
Power Steering	Std	Std
Cruise Control	Pkg	Pkg
All Wheel Drive	N/A	N/A
AM/FM Stereo Cassette	Std	Std
Steering Wheel, Tilt	Std	Std
Power Windows	$237	$275
*Options Price	$237	$275
*Total Price	$16,945	$19,370
Target Price	$18,223	
Destination Charge	$505	
Avg. Tax & Fees	$1,135	
Total Target $	**$19,863**	
Average Dealer Option Cost:	**86%**	

The 1993 Regal is available in six models - Custom, Gran Sport, and Limited coupes and sedans. New for 1993, the Gran Sport has an improved drivetrain. Its V-6 engine with tuned port injection has been enhanced for a quieter, more responsive performance and provides improved fuel economy. Regal has a more sculptured front and rear appearance with the hood integrated with the new grille and headlamps. Other features include new intake and exhaust manifolds and standard analog gauges.

Ownership Costs

Cost Area	5 Year Cost	Rate
Depreciation	$11,370	●
Financing ($399/month)	$3,273	
Insurance (Rating 11)	$7,693	○
State Fees	$795	
Fuel (Hwy 28 City 19)	$3,704	○
Maintenance	$5,612	●
Repairs	$709	○

Warranty/Maintenance Info

Major Tune-Up	$190	○
Minor Tune-Up	$138	●
Brake Service	$201	
Overall Warranty	3 yr/36k	○
Drivetrain Warranty	3 yr/36k	○
Rust Warranty	6 yr/100k	○
Maintenance Warranty	N/A	
Roadside Assistance	3 yr/36k	

Resale Value

1993	1994	1995	1996	1997
$14,208	$12,503	$11,147	$9,758	$8,493

Cumulative Costs

	1993	1994	1995	1996	1997
Annual	$9,373	$5,382	$5,812	$5,008	$7,581
Total	$9,373	$14,755	$20,567	$25,575	$33,156

Ownership Costs (5yr)

Average	This Car
$31,074	$33,156
Cost/Mile 44¢	Cost/Mile 47¢

Ownership Cost Rating
● Worse Than Average

---

page **28**

* Includes shaded options
** Other purchase requirements apply

● Poor   ◐ Worse Than Average   ◯ Average   ○ Better Than Average   ○ Excellent   ⊖ Insufficient Information

©1993 by IntelliChoice, Inc. (408) 554-8711   All Rights Reserved. Reproduction Prohibited.
Refer to *Section 3: Annotated Vehicle Charts* for an explanation of these charts.

Buick Regal Custom
4 Door Sedan

Midsize

 3.1L 140 hp Gas Fuel Inject.
 6 Cylinder "V"
 Automatic 4 Speed
 2 Wheel Front
 Automatic Seatbelts

Purchase Price

Car Item	Dealer Cost	List
Base Price	**$14,757**	**$16,865**
Anti-Lock Brakes	$387	$450
Manual Transmission	N/A	N/A
3.8L 170 hp Gas	$340	** $395
Air Conditioning	Std	Std
Power Steering	Std	Std
Cruise Control	Pkg	Pkg
All Wheel Drive	N/A	N/A
AM/FM Stereo Cassette	$249	** $290
Steering Wheel, Tilt	Std	Std
Power Windows	$237	$275
*Options Price	$237	$275
*Total Price	$14,994	$17,140
Target Price	$16,066	
Destination Charge	$505	
Avg. Tax & Fees	$1,005	
Total Target $	**$17,576**	
Average Dealer Option Cost:	**86%**	

Ownership Costs

Cost Area	5 Year Cost	Rate
Depreciation	$9,638	○
Financing ($353/month)	$2,895	
Insurance (Rating 6)	$6,919	○
State Fees	$706	
Fuel (Hwy 30 City 19)	$3,579	○
Maintenance	$5,432	●
Repairs	$709	○

Warranty/Maintenance Info

Major Tune-Up	$193	○
Minor Tune-Up	$142	●
Brake Service	$201	○
Overall Warranty	3 yr/36k	○
Drivetrain Warranty	3 yr/36k	○
Rust Warranty	6 yr/100k	○
Maintenance Warranty	N/A	
Roadside Assistance	3 yr/36k	

Ownership Cost By Year
1993 / 1994 / 1995 / 1996 / 1997

Resale Value
1993	1994	1995	1996	1997
$13,169	$11,585	$10,348	$9,082	$7,938

Cumulative Costs
	1993	1994	1995	1996	1997
Annual	$7,779	$4,954	$5,351	$4,618	$7,176
Total	$7,779	$12,733	$18,084	$22,702	$29,878

Ownership Costs (5yr)
Average	This Car
$28,794	$29,878
Cost/Mile 41¢	Cost/Mile 43¢

Ownership Cost Rating
○ Average

The 1993 Regal is available in six models - Custom, Gran Sport, and Limited coupes and sedans. New for 1993, the Custom sedan has an improved drivetrain. The optional V-6 engine w/tuned port injection has been enhanced for a quieter, more responsive performance and provides improved fuel economy. Regal has a more sculptured front and rear appearance with the hood integrated with the new grille and headlamps. Front seats are featured that are orthopedically contoured. Analog gauges are standard.

Buick Regal Limited
4 Door Sedan

Midsize

 3.1L 140 hp Gas Fuel Inject.
 6 Cylinder "V"
 Automatic 4 Speed
 2 Wheel Front
 Automatic Seatbelts

Purchase Price

Car Item	Dealer Cost	List
Base Price	**$16,153**	**$18,460**
Anti-Lock Brakes	Std	Std
Manual Transmission	N/A	N/A
3.8L 170 hp Gas	$340	** $395
Air Conditioning	Std	Std
Power Steering	Std	Std
Cruise Control	Pkg	Pkg
All Wheel Drive	N/A	N/A
AM/FM Stereo Cassette	$129	** $150
Steering Wheel, Tilt	Std	Std
Power Windows	$237	$275
*Options Price	$237	$275
*Total Price	$16,390	$18,735
Target Price	$17,608	
Destination Charge	$505	
Avg. Tax & Fees	$1,098	
Total Target $	**$19,211**	
Average Dealer Option Cost:	**86%**	

Ownership Costs

Cost Area	5 Year Cost	Rate
Depreciation	$10,843	●
Financing ($386/month)	$3,164	
Insurance (Rating 8)	$7,223	○
State Fees	$769	
Fuel (Hwy 30 City 19)	$3,579	○
Maintenance	$5,432	●
Repairs	$709	○

Warranty/Maintenance Info

Major Tune-Up	$193	○
Minor Tune-Up	$142	●
Brake Service	$201	○
Overall Warranty	3 yr/36k	○
Drivetrain Warranty	3 yr/36k	○
Rust Warranty	6 yr/100k	○
Maintenance Warranty	N/A	
Roadside Assistance	3 yr/36k	

Ownership Cost By Year
1993 / 1994 / 1995 / 1996 / 1997

Resale Value
1993	1994	1995	1996	1997
$14,031	$12,318	$10,967	$9,618	$8,368

Cumulative Costs
	1993	1994	1995	1996	1997
Annual	$8,737	$5,239	$5,591	$4,795	$7,357
Total	$8,737	$13,976	$19,567	$24,362	$31,719

Ownership Costs (5yr)
Average	This Car
$30,425	$31,719
Cost/Mile 43¢	Cost/Mile 45¢

Ownership Cost Rating
● Worse Than Average

The 1993 Regal is available in six models - Custom, Gran Sport, and Limited coupes and sedans. New for 1993, the Limited sedan has an improved drivetrain. The optional V-6 engine w/tuned port injection has been enhanced for a quieter, more responsive performance and provides improved fuel economy. Regal has a sculptured front and rear appearance with the hood integrated with the new grille and headlamps. Front seats are orthopedically contoured. Analog gauges are standard equipment.

---

* Includes shaded options
** Other purchase requirements apply

● Poor   ◐ Worse Than Average   ◑ Average   ○ Better Than Average   ○ Excellent   ⊖ Insufficient Information

©1993 by IntelliChoice, Inc. (408) 554-8711 All Rights Reserved. Reproduction Prohibited.
Refer to *Section 3: Annotated Vehicle Charts* for an explanation of these charts.

Buick Regal Gran Sport
4 Door Sedan

Midsize

 3.8L 170 hp Gas Fuel Inject.
 6 Cylinder "V"
 Automatic 4 Speed
 2 Wheel Front
 Automatic Seatbelts

Purchase Price

Car Item	Dealer Cost	List
Base Price	**$16,896**	**$19,310**
Anti-Lock Brakes	Std	Std
Manual Transmission	N/A	N/A
Optional Engine	N/A	N/A
Air Conditioning	Std	Std
Power Steering	Std	Std
Cruise Control	Pkg	Pkg
All Wheel Drive	N/A	N/A
AM/FM Stereo Cassette	Std	Std
Steering Wheel, Tilt	Std	Std
Power Windows	$237	$275
***Options Price**	**$237**	**$275**
***Total Price**	**$17,133**	**$19,585**
Target Price	$18,432	
Destination Charge	$505	
Avg. Tax & Fees	$1,148	
Total Target $	**$20,085**	
Average Dealer Option Cost:	***86%***	

The 1993 Regal is available in six models - Custom, Gran Sport, and Limited coupes and sedans. New for 1993, the Gran Sport sedan has an improved drivetrain. Its V-6 engine with tuned port injection has been enhanced for a quieter, more responsive performance and provides improved fuel economy. Regal has a more sculptured front and rear appearance with the hood integrated with the new grille and headlamps. Other features include new intake and exhaust manifolds and standard analog gauges.

Ownership Costs

Cost Area	5 Year Cost	Rate
Depreciation	$11,236	○
Financing ($404/month)	$3,308	
Insurance (Rating 8)	$7,223	○
State Fees	$805	
Fuel (Hwy 28 City 19)	$3,704	○
Maintenance	$5,612	●
Repairs	$709	○

Warranty/Maintenance Info

Major Tune-Up	$190	○
Minor Tune-Up	$138	●
Brake Service	$201	○
Overall Warranty	3 yr/36k	○
Drivetrain Warranty	3 yr/36k	○
Rust Warranty	6 yr/100k	○
Maintenance Warranty	N/A	
Roadside Assistance	3 yr/36k	

Ownership Cost By Year

Legend: 1993, 1994, 1995, 1996, 1997

Resale Value

1993	1994	1995	1996	1997
$14,660	$12,918	$11,527	$10,125	$8,849

Cumulative Costs

	1993	1994	1995	1996	1997
Annual	$9,074	$5,342	$5,762	$4,927	$7,492
Total	$9,074	$14,416	$20,178	$25,105	$32,597

Ownership Costs (5yr)

Average	This Car
$31,294	$32,597
Cost/Mile 45¢	Cost/Mile 47¢

Ownership Cost Rating

● Worse Than Average

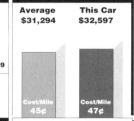

Buick Riviera
2 Door Coupe

Midsize

 3.8L 170 hp Gas Fuel Inject.
 6 Cylinder "V"
 Automatic 4 Speed
 2 Wheel Front
 Driver Airbag Psngr Belts

Purchase Price

Car Item	Dealer Cost	List
Base Price	**$22,767**	**$26,320**
Anti-Lock Brakes	Std	Std
Manual Transmission	N/A	N/A
Optional Engine	N/A	N/A
Auto Climate Control	Std	Std
Power Steering	Std	Std
Cruise Control	Std	Std
All Wheel Drive	N/A	N/A
AM/FM Stereo Cassette	Std	Std
Steering Wheel, Tilt	Std	Std
Power Windows	Std	Std
***Options Price**	**$0**	**$0**
***Total Price**	**$22,767**	**$26,320**
Target Price	$24,563	
Destination Charge	$600	
Avg. Tax & Fees	$1,527	
Total Target $	**$26,690**	
Average Dealer Option Cost:	***86%***	

The 1993 Buick Riviera is available as a two-door coupe. New for 1993, the Riviera features dark cherry metallic as a new color. Riviera's design is enhanced with a new combination of white diamond exterior color and white leather. Interiors are also available in red, blue, and a new beige. Models ordered with Gran Touring suspension have new 16-inch aluminum wheels with GA touring tires, fast-ratio power steering, and Gran Touring suspension. Buick's "dyna-ride" suspension is standard.

Ownership Costs

Cost Area	5 Year Cost	Rate
Depreciation	$14,485	○
Financing ($536/month)	$4,396	
Insurance (Rating 14)	$7,956	○
State Fees	$1,076	
Fuel (Hwy 27 City 19)	$3,775	○
Maintenance	$5,455	●
Repairs	$985	●

Warranty/Maintenance Info

Major Tune-Up	$183	○
Minor Tune-Up	$138	●
Brake Service	$223	○
Overall Warranty	3 yr/36k	○
Drivetrain Warranty	3 yr/36k	○
Rust Warranty	6 yr/100k	○
Maintenance Warranty	N/A	
Roadside Assistance	3 yr/36k	

Resale Value

1993	1994	1995	1996	1997
$18,622	$16,771	$15,252	$13,662	$12,205

Cumulative Costs

	1993	1994	1995	1996	1997
Annual	$12,386	$6,013	$6,315	$5,465	$7,949
Total	$12,386	$18,399	$24,714	$30,179	$38,128

Ownership Costs (5yr)

Average	This Car
$38,181	$38,128
Cost/Mile 55¢	Cost/Mile 54¢

Ownership Cost Rating

○ Average

---

page 30

* Includes shaded options
** Other purchase requirements apply

● Poor | ◐ Worse Than Average | ○ Average | ◯ Better Than Average | ○ Excellent | ⊖ Insufficient Information

©1993 by IntelliChoice, Inc. (408) 554-8711 All Rights Reserved. Reproduction Prohibited.
Refer to *Section 3: Annotated Vehicle Charts* for an explanation of these charts.

Buick Roadmaster
4 Door Sedan

5.7L 180 hp Gas Fuel Inject. | 8 Cylinder "V" | Automatic 4 Speed | 2 Wheel Rear | Driver Airbag Psngr Belts

Large

Purchase Price

Car Item	Dealer Cost	List
Base Price	**$19,736**	**$22,555**
Anti-Lock Brakes	Std	Std
Manual Transmission	N/A	N/A
Optional Engine	N/A	N/A
Air Conditioning	Std	Std
Power Steering	Std	Std
Cruise Control	Std	Std
All Wheel Drive	N/A	N/A
AM/FM Stereo Cassette	$280	** $325
Steering Wheel, Tilt	Std	Std
Power Windows	Std	Std
***Options Price**	**$0**	**$0**
***Total Price**	**$19,736**	**$22,555**
Target Price		$21,345
Destination Charge		$555
Avg. Tax & Fees		$1,326
Total Target $		**$23,226**
Average Dealer Option Cost:		**86%**

The 1993 Roadmaster is available in three models - (Base) Roadmaster and Limited sedans and the Estate wagon. New for 1993, the Roadmaster features an acoustics upgrade, a locking system for power windows which allows the driver to control operation of the windows, and two new exterior colors (Ruby Red Metallic and Dark Cherry Metallic). Available features include leather seats, leather-wrapped steering wheel, towing package, and a cassette, graphic equalizer and compact disc player.

Ownership Costs

Cost Area	5 Year Cost	Rate
Depreciation	$12,955	◯
Financing ($467/month)	$3,827	
Insurance (Rating 11)	$7,693	◯
State Fees	$925	
Fuel (Hwy 25 City 16)	$4,272	●
Maintenance	$4,489	◯
Repairs	$709	◯

Warranty/Maintenance Info

Major Tune-Up	$198	◯
Minor Tune-Up	$133	◉
Brake Service	$230	◯
Overall Warranty	3 yr/36k	◯
Drivetrain Warranty	3 yr/36k	◯
Rust Warranty	6 yr/100k	◯
Maintenance Warranty	N/A	
Roadside Assistance	3 yr/36k	

Ownership Cost By Year

$2,000 $4,000 $6,000 $8,000 $10,000 $12,000 $14,000

■ 1993
■ 1994
■ 1995
■ 1996
□ 1997

Resale Value

1993	1994	1995	1996	1997
$16,634	$14,760	$13,188	$11,764	$10,271

Cumulative Costs

	1993	1994	1995	1996	1997
Annual	$10,682	$5,869	$6,421	$5,539	$6,359
Total	$10,682	$16,551	$22,972	$28,511	$34,870

Ownership Costs (5yr)

Average	This Car
$34,839	$34,870
Cost/Mile 50¢	Cost/Mile 50¢

Ownership Cost Rating

◯ Average

Buick Roadmaster Limited
4 Door Sedan

5.7L 180 hp Gas Fuel Inject. | 8 Cylinder "V" | Automatic 4 Speed | 2 Wheel Rear | Driver Airbag Psngr Belts

Large

Purchase Price

Car Item	Dealer Cost	List
Base Price	**$21,805**	**$24,920**
Anti-Lock Brakes	Std	Std
Manual Transmission	N/A	N/A
Optional Engine	N/A	N/A
Auto Climate Control	Std	Std
Power Steering	Std	Std
Cruise Control	Std	Std
All Wheel Drive	N/A	N/A
AM/FM Stereo Cassette	Std	Std
Steering Wheel, Tilt	Std	Std
Power Windows	Std	Std
***Options Price**	**$0**	**$0**
***Total Price**	**$21,805**	**$24,920**
Target Price		$23,674
Destination Charge		$555
Avg. Tax & Fees		$1,466
Total Target $		**$25,695**
Average Dealer Option Cost:		**86%**

The 1993 Roadmaster is available in three models - (Base) Roadmaster and Limited sedans and the Estate wagon. New for 1993, the Limited features variable-effort steering, an acoustics upgrade, a locking system for power windows which allows the driver to control operation of the windows, and two new exterior colors have been added (Ruby Red Metallic and Dark Cherry Metallic). Available features include leather seats, trailer towing package, and a cassette, graphic equalizer and compact disc player.

Ownership Costs

Cost Area	5 Year Cost	Rate
Depreciation	$14,392	◯
Financing ($516/month)	$4,232	
Insurance (Rating 11)	$7,693	◯
State Fees	$1,019	
Fuel (Hwy 25 City 16)	$4,272	●
Maintenance	$4,489	◯
Repairs	$709	◯

Warranty/Maintenance Info

Major Tune-Up	$198	◯
Minor Tune-Up	$133	◉
Brake Service	$230	◯
Overall Warranty	3 yr/36k	◯
Drivetrain Warranty	3 yr/36k	◯
Rust Warranty	6 yr/100k	◯
Maintenance Warranty	N/A	
Roadside Assistance	3 yr/36k	

Ownership Cost By Year

$5,000 $10,000 $15,000

■ 1993
■ 1994
■ 1995
■ 1996
□ 1997

Resale Value

1993	1994	1995	1996	1997
$18,612	$16,438	$14,605	$12,971	$11,303

Cumulative Costs

	1993	1994	1995	1996	1997
Annual	$11,366	$6,317	$6,781	$5,794	$6,548
Total	$11,366	$17,683	$24,464	$30,258	$36,806

Ownership Costs (5yr)

Average	This Car
$37,185	$36,806
Cost/Mile 53¢	Cost/Mile 53¢

Ownership Cost Rating

◯ Average

---

* Includes shaded options
** Other purchase requirements apply

● Poor   ◉ Worse Than Average   ◯ Average   ◯ Better Than Average   ◯ Excellent   ⊖ Insufficient Information

©1993 by IntelliChoice, Inc. (408) 554-8711   All Rights Reserved. Reproduction Prohibited.
Refer to *Section 3: Annotated Vehicle Charts* for an explanation of these charts.

page 31

Buick Roadmaster Estate Wagon
4 Door Wagon

5.7L 180 hp Gas Fuel Inject. | 8 Cylinder "V" | Automatic 4 Speed | 2 Wheel Rear | Driver Airbag Psngr Belts

Large Wagon

Purchase Price

Car Item	Dealer Cost	List
Base Price	**$20,869**	**$23,850**
Anti-Lock Brakes	Std	Std
Manual Transmission	N/A	N/A
Optional Engine	N/A	N/A
Air Conditioning	Std	Std
Power Steering	Std	Std
Cruise Control	Pkg	Pkg
All Wheel Drive	N/A	N/A
AM/FM Stereo Cassette	$159	** $185
Steering Wheel, Tilt	Std	Std
Power Windows	Std	Std
***Options Price**	**$0**	**$0**
***Total Price**	**$20,869**	**$23,850**
Target Price		$22,618
Destination Charge		$555
Avg. Tax & Fees		$1,403
Total Target $		**$24,576**
Average Dealer Option Cost:		**86%**

Ownership Costs

Cost Area	5 Year Cost	Rate
Depreciation	$14,566	●
Financing ($494/month)	$4,048	
Insurance (Rating 12)	$7,898	○
State Fees	$976	
Fuel (Hwy 25 City 16)	$4,272	◐
Maintenance	$4,307	◉
Repairs	$709	○

Warranty/Maintenance Info

Major Tune-Up	$198	◐
Minor Tune-Up	$133	◐
Brake Service	$230	◐
Overall Warranty	3 yr/36k	◐
Drivetrain Warranty	3 yr/36k	◐
Rust Warranty	6 yr/100k	○
Maintenance Warranty	N/A	
Roadside Assistance	3 yr/36k	

Ownership Cost By Year — 1993, 1994, 1995, 1996, 1997

Resale Value: 1993 $16,451; 1994 $14,727; 1995 $13,068; 1996 $11,571; 1997 $10,010

Cumulative Costs

	1993	1994	1995	1996	1997
Annual	$12,358	$5,840	$6,567	$5,559	$6,452
Total	$12,358	$18,198	$24,765	$30,324	$36,776

Ownership Costs (5yr): Average $36,123 — This Car $36,776
Cost/Mile 52¢ — Cost/Mile 53¢

Ownership Cost Rating: Worse Than Average

The 1993 Roadmaster is available in three models - (Base) Roadmaster and Limited sedans and the Estate wagon. New for 1993, the Estate wagon features a locking system for power windows which allows the driver to control operation of the windows and two new exterior colors. The Estate wagon continues to have the vista roof, a dark-tinted glass panel in the roof. A vista cover for the roof glass is optional. A two-way tailgate provides instant access to the rear cargo or seating area.

Buick Skylark Custom
2 Door Coupe

2.3L 115 hp Gas Fuel Inject. | 4 Cylinder In-Line | Automatic 3 Speed | 2 Wheel Front | Automatic Seatbelts

Compact — 4 Door Shown

Purchase Price

Car Item	Dealer Cost	List
Base Price	**$12,113**	**$12,955**
Anti-Lock Brakes	Std	Std
Manual Transmission	N/A	N/A
3.3L 160 hp Gas	$396	$460
Air Conditioning	Pkg	Pkg
Power Steering	Std	Std
Cruise Control	Pkg	Pkg
All Wheel Drive	N/A	N/A
AM/FM Stereo Cassette	Pkg	Pkg
Steering Wheel, Tilt	Pkg	Pkg
Power Windows	Pkg	Pkg
***Options Price**	**$396**	**$460**
***Total Price**	**$12,509**	**$13,415**
Target Price		$13,419
Destination Charge		$475
Avg. Tax & Fees		$834
Total Target $		**$14,728**
Average Dealer Option Cost:		**86%**

Ownership Costs

Cost Area	5 Year Cost	Rate
Depreciation	$8,899	◉
Financing ($296/month)	$2,426	
Insurance (Rating 11)	$7,693	○
State Fees	$555	
Fuel (Hwy 29 City 20)	$3,548	○
Maintenance	$4,640	○
Repairs	$651	○

Warranty/Maintenance Info

Major Tune-Up	$192	○
Minor Tune-Up	$138	◉
Brake Service	$212	○
Overall Warranty	3 yr/36k	○
Drivetrain Warranty	3 yr/36k	○
Rust Warranty	6 yr/100k	○
Maintenance Warranty	N/A	
Roadside Assistance	3 yr/36k	

Resale Value: 1993 $9,950; 1994 $8,773; 1995 $7,792; 1996 $6,789; 1997 $5,829

Cumulative Costs

	1993	1994	1995	1996	1997
Annual	$8,055	$4,505	$5,068	$4,384	$6,400
Total	$8,055	$12,560	$17,628	$22,012	$28,412

Ownership Costs (5yr): Average $26,013 — This Car $28,412
Cost/Mile 37¢ — Cost/Mile 41¢

Ownership Cost Rating: Worse Than Average

The 1993 Skylark is available in six models - Custom, Gran Sport and Limited coupes or sedans. New for 1993, the Custom coupe features a two-function battery "run-down" protection system, 14-inch deluxe wheel covers, fixed rear seatbacks, 55/45 split bench front seats with a storage armrest, and two new exterior colors (Light Driftwood Metallic and Smokey Amethyst Metallic). Other features include Dyna-ride suspension, power rack and pinion steering and rear-door child locks.

---

page **32**

*Includes shaded options
**Other purchase requirements apply

● Poor ◉ Worse Than Average ◐ Average ○ Better Than Average ○ Excellent ⊖ Insufficient Information

©1993 by IntelliChoice, Inc. (408) 554-8711 All Rights Reserved. Reproduction Prohibited.
Refer to *Section 3: Annotated Vehicle Charts* for an explanation of these charts.

Buick Skylark Limited
2 Door Coupe

Compact

 2.3L 115 hp Gas Fuel Inject. | 4 Cylinder In-Line | Automatic 3 Speed | 2 Wheel Front | Automatic Seatbelts

Purchase Price

Car Item	Dealer Cost	List
Base Price	**$12,557**	**$13,875**
Anti-Lock Brakes	Std	Std
Manual Transmission	N/A	N/A
3.3L 160 hp Gas	$396	$460
Air Conditioning	Pkg	Pkg
Power Steering	Std	Std
Cruise Control	Pkg	Pkg
All Wheel Drive	N/A	N/A
AM/FM Stereo CD	$537	$624
Steering Wheel, Tilt	Pkg	Pkg
Power Windows	Pkg	Pkg
*Options Price	$933	$1,084
*Total Price	$13,490	$14,959
Target Price		$14,485
Destination Charge		$475
Avg. Tax & Fees		$902
Total Target $		**$15,862**
Average Dealer Option Cost:		86%

Ownership Costs

Cost Area	5 Year Cost	Rate
Depreciation	$9,359	●
Financing ($319/month)	$2,613	
Insurance (Rating 11)	$7,693	○
State Fees	$617	
Fuel (Hwy 29 City 19)	$3,640	○
Maintenance	$4,640	○
Repairs	$651	○

Warranty/Maintenance Info

Major Tune-Up	$192	○
Minor Tune-Up	$138	●
Brake Service	$212	○
Overall Warranty	3 yr/36k	○
Drivetrain Warranty	3 yr/36k	○
Rust Warranty	6 yr/100k	○
Maintenance Warranty	N/A	
Roadside Assistance	3 yr/36k	

Ownership Cost By Year

1993, 1994, 1995, 1996, 1997

Resale Value

1993	1994	1995	1996	1997
$11,038	$9,741	$8,635	$7,532	$6,503

Cumulative Costs

	1993	1994	1995	1996	1997
Annual	$8,210	$4,716	$5,260	$4,528	$6,499
Total	$8,210	$12,926	$18,186	$22,714	$29,213

Ownership Costs (5yr)

Average	This Car
$27,339	$29,213
Cost/Mile 39¢	Cost/Mile 42¢

Ownership Cost Rating

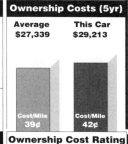

Worse Than Average

The 1993 Skylark is available in six models - Custom, Gran Sport and Limited coupes or sedans. New for 1993, the Limited coupe features a two-function battery "run-down" protection system, 14-inch deluxe wheel covers, fixed rear seatbacks, 55/45 split bench front seats with a storage armrest, and two new exterior colors (Light Driftwood Metallic and Smokey Amethyst Metallic). Also featured is "Dyna-ride" suspension, which is designed specifically to enhance ride without sacrificing handling.

Buick Skylark Gran Sport
2 Door Coupe

Compact

 3.3L 160 hp Gas Fuel Inject. | 6 Cylinder "V" | Automatic 3 Speed | 2 Wheel Front | Automatic Seatbelts

Purchase Price

Car Item	Dealer Cost	List
Base Price	**$14,263**	**$15,760**
Anti-Lock Brakes	Std	Std
Manual Transmission	N/A	N/A
Optional Engine	N/A	N/A
Air Conditioning	Pkg	Pkg
Power Steering	Std	Std
Cruise Control	Pkg	Pkg
All Wheel Drive	N/A	N/A
AM/FM Stereo CD	$537	$624
Steering Wheel, Tilt	Pkg	Pkg
Power Windows	Pkg	Pkg
*Options Price	$537	$624
*Total Price	$14,800	$16,384
Target Price		$15,948
Destination Charge		$475
Avg. Tax & Fees		$990
Total Target $		**$17,413**
Average Dealer Option Cost:		86%

Ownership Costs

Cost Area	5 Year Cost	Rate
Depreciation	$10,508	●
Financing ($350/month)	$2,869	
Insurance (Rating 12)	$7,898	○
State Fees	$675	
Fuel (Hwy 29 City 20)	$3,548	○
Maintenance	$5,858	●
Repairs	$651	○

Warranty/Maintenance Info

Major Tune-Up	$192	○
Minor Tune-Up	$138	●
Brake Service	$212	○
Overall Warranty	3 yr/36k	○
Drivetrain Warranty	3 yr/36k	○
Rust Warranty	6 yr/100k	○
Maintenance Warranty	N/A	
Roadside Assistance	3 yr/36k	

Resale Value

1993	1994	1995	1996	1997
$11,132	$9,943	$8,954	$7,927	$6,905

Cumulative Costs

	1993	1994	1995	1996	1997
Annual	$9,810	$4,723	$5,465	$4,503	$7,506
Total	$9,810	$14,533	$19,998	$24,501	$32,007

Ownership Costs (5yr)

Average	This Car
$28,562	$32,007
Cost/Mile 41¢	Cost/Mile 46¢

Ownership Cost Rating

Poor

The 1993 Skylark is available in six models - Custom, Gran Sport and Limited coupes or sedans. New for 1993, the Gran Sport coupe features a two-function battery "run-down" protection system, 14-inch deluxe wheel covers, 55/45 split bench front seats with a storage armrest, and two new exterior colors (Light Driftwood Metallic and Smokey Amethyst Metallic). Other features include a Gran Touring suspension and optional adjustable ride control system. Lower accent paint is optional.

---

* Includes shaded options
** Other purchase requirements apply

● Poor | Worse Than Average | ○ Average | ○ Better Than Average | ○ Excellent | ⊖ Insufficient Information

©1993 by IntelliChoice, Inc. (408) 554-8711 All Rights Reserved. Reproduction Prohibited.
Refer to *Section 3: Annotated Vehicle Charts* for an explanation of these charts.

Buick Skylark Custom
4 Door Sedan

Compact

2.3L 115 hp Gas Fuel Inject. | 4 Cylinder In-Line | Automatic 3 Speed | 2 Wheel Front | Automatic Seatbelts

Purchase Price

Car Item	Dealer Cost	List
Base Price	**$12,113**	**$12,955**
Anti-Lock Brakes	Std	Std
Manual Transmission	N/A	N/A
3.3L 160 hp Gas	$396	$460
Air Conditioning	Pkg	Pkg
Power Steering	Std	Std
Cruise Control	Pkg	Pkg
All Wheel Drive	N/A	N/A
AM/FM Stereo Cassette	Pkg	Pkg
Steering Wheel, Tilt	Pkg	Pkg
Power Windows	Pkg	Pkg
*Options Price	$396	$460
*Total Price	$12,509	$13,415
Target Price		$13,419
Destination Charge		$475
Avg. Tax & Fees		$834
Total Target $		**$14,728**
Average Dealer Option Cost:		86%

The 1993 Skylark is available in six models - Custom, Gran Sport and Limited coupes or sedans. New for 1993, the Custom sedan features a two-function battery "run-down" protection system, 14-inch deluxe wheel covers, fixed rear seatbacks, 55/45 split bench front seats with a storage armrest, and two new exterior colors (Light Driftwood Metallic and Smokey Amethyst Metallic). Other features include Dyna-ride suspension, power rack and pinion steering, all-season radial tires, and rear-door child locks.

Ownership Costs

Cost Area	5 Year Cost	Rate
Depreciation	$8,663	○
Financing ($296/month)	$2,426	
Insurance (Rating 7)	$7,035	○
State Fees	$555	
Fuel (Hwy 29 City 20)	$3,548	○
Maintenance	$4,640	○
Repairs	$651	○

Warranty/Maintenance Info

Major Tune-Up	$192	○
Minor Tune-Up	$138	●
Brake Service	$212	○
Overall Warranty	3 yr/36k	○
Drivetrain Warranty	3 yr/36k	○
Rust Warranty	6 yr/100k	○
Maintenance Warranty	N/A	
Roadside Assistance	3 yr/36k	

Ownership Cost By Year

$2,000 — $4,000 — $6,000 — $8,000 — $10,000

- 1993
- 1994
- 1995
- 1996
- 1997

Resale Value

1993	1994	1995	1996	1997
$10,200	$9,051	$8,056	$7,053	$6,065

Cumulative Costs

	1993	1994	1995	1996	1997
Annual	$7,684	$4,351	$4,951	$4,247	$6,285
Total	$7,684	$12,035	$16,986	$21,233	$27,518

Ownership Costs (5yr)

Average	This Car
$26,013	$27,518
Cost/Mile 37¢	Cost/Mile 39¢

Ownership Cost Rating

◐ Worse Than Average

Buick Skylark Limited
4 Door Sedan

    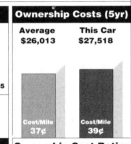

Compact

2.3L 115 hp Gas Fuel Inject. | 4 Cylinder In-Line | Automatic 3 Speed | 2 Wheel Front | Automatic Seatbelts

Purchase Price

Car Item	Dealer Cost	List
Base Price	**$12,557**	**$13,875**
Anti-Lock Brakes	Std	Std
Manual Transmission	N/A	N/A
3.3L 160 hp Gas	$396	$460
Air Conditioning	Pkg	Pkg
Power Steering	Std	Std
Cruise Control	Pkg	Pkg
All Wheel Drive	N/A	N/A
AM/FM Stereo CD	$537	$624
Steering Wheel, Tilt	Pkg	Pkg
Power Windows	Pkg	Pkg
*Options Price	$933	$1,084
*Total Price	$13,490	$14,959
Target Price		$14,485
Destination Charge		$475
Avg. Tax & Fees		$902
Total Target $		**$15,862**
Average Dealer Option Cost:		86%

The 1993 Skylark is available in six models - Custom, Gran Sport and Limited coupes or sedans. New for 1993, the Limited sedan features a two-function battery "run-down" protection system, 14-inch deluxe wheel covers, fixed rear seatbacks, 55/45 split bench front seats with a storage armrest, and two new exterior colors (Light Driftwood Metallic and Smokey Amethyst Metallic). Also featured is Dyna-ride suspension, which is designed specifically to enhance ride without sacrificing handling.

Ownership Costs

Cost Area	5 Year Cost	Rate
Depreciation	$8,588	○
Financing ($319/month)	$2,613	
Insurance (Rating 7)	$7,035	○
State Fees	$617	
Fuel (Hwy 29 City 20)	$3,548	○
Maintenance	$4,640	○
Repairs	$651	○

Warranty/Maintenance Info

Major Tune-Up	$192	○
Minor Tune-Up	$138	●
Brake Service	$212	○
Overall Warranty	3 yr/36k	○
Drivetrain Warranty	3 yr/36k	○
Rust Warranty	6 yr/100k	○
Maintenance Warranty	N/A	
Roadside Assistance	3 yr/36k	

Ownership Cost By Year

$2,000 — $4,000 — $6,000 — $8,000 — $10,000

- 1993
- 1994
- 1995
- 1996
- 1997

Resale Value

1993	1994	1995	1996	1997
$11,177	$10,219	$9,376	$8,319	$7,274

Cumulative Costs

	1993	1994	1995	1996	1997
Annual	$7,933	$4,233	$4,848	$4,326	$6,352
Total	$7,933	$12,166	$17,014	$21,340	$27,692

Ownership Costs (5yr)

Average	This Car
$27,339	$27,692
Cost/Mile 39¢	Cost/Mile 40¢

Ownership Cost Rating

○ Average

---

* Includes shaded options
** Other purchase requirements apply

● Poor   ◐ Worse Than Average   ○ Average   ○ Better Than Average   ○ Excellent   ⊖ Insufficient Information

©1993 by IntelliChoice, Inc. (408) 554-8711   All Rights Reserved. Reproduction Prohibited.
Refer to *Section 3: Annotated Vehicle Charts* for an explanation of these charts.

Buick Skylark Gran Sport
4 Door Sedan

Compact

 3.3L 160 hp Gas Fuel Inject.
 6 Cylinder "V"
 Automatic 3 Speed
 2 Wheel Front
 Automatic Seatbelts

Purchase Price

Car Item	Dealer Cost	List
Base Price	**$14,263**	**$15,760**
Anti-Lock Brakes	Std	Std
Manual Transmission	N/A	N/A
Optional Engine	N/A	N/A
Air Conditioning	Pkg	Pkg
Power Steering	Std	Std
Cruise Control	Pkg	Pkg
All Wheel Drive	N/A	N/A
AM/FM Stereo CD	$537	$624
Steering Wheel, Tilt	Pkg	Pkg
Power Windows	Pkg	Pkg
*Options Price	$537	$624
*Total Price	$14,800	$16,384
Target Price	$15,948	
Destination Charge	$475	
Avg. Tax & Fees	$990	
Total Target $	**$17,413**	
Average Dealer Option Cost:	*86%*	

Ownership Costs

Cost Area	5 Year Cost	Rate
Depreciation	$10,340	●
Financing ($350/month)	$2,869	
Insurance (Rating 8)	$7,223	○
State Fees	$675	
Fuel (Hwy 29 City 20)	$3,548	○
Maintenance	$5,858	●
Repairs	$651	○

Warranty/Maintenance Info

Major Tune-Up	$192	○
Minor Tune-Up	$138	●
Brake Service	$212	○
Overall Warranty	3 yr/36k	○
Drivetrain Warranty	3 yr/36k	○
Rust Warranty	6 yr/100k	○
Maintenance Warranty	N/A	
Roadside Assistance	3 yr/36k	

Ownership Cost By Year — 1993, 1994, 1995, 1996, 1997

Resale Value: 1993 $11,554; 1994 $10,280; 1995 $9,224; 1996 $8,146; 1997 $7,073

Cumulative Costs

	1993	1994	1995	1996	1997
Annual	$9,264	$4,678	$5,397	$4,414	$7,411
Total	$9,264	$13,942	$19,339	$23,753	$31,164

Ownership Costs (5yr) — Average $28,562 / This Car $31,164 — Cost/Mile 41¢ / 45¢

Ownership Cost Rating: Worse Than Average

The 1993 Skylark is available in six models - Custom, Gran Sport and Limited coupes or sedans. New for 1993, the Gran Sport sedan features a two-function battery "run-down" protection system, 14-inch deluxe wheel covers, fixed rear seatbacks, 55/45 split bench front seats with a storage armrest, and two new exterior colors (Light Driftwood Metallic and Smokey Amethyst Metallic). Other features include a Gran Touring suspension and optional adjustable ride control system. Lower accent paint is optional.

Cadillac Allante
2 Door Convertible

Luxury

 4.6L 295 hp Gas Fuel Inject.
 8 Cylinder "V"
 Automatic 4 Speed
 2 Wheel Front
 Driver Airbag Psngr Belts

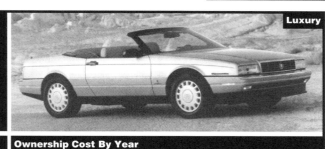

Purchase Price

Car Item	Dealer Cost	List
Base Price	**$52,979**	**$61,675**
Anti-Lock Brakes	Std	Std
Manual Transmission	N/A	N/A
Optional Engine	N/A	N/A
Auto Climate Control	Std	Std
Power Steering	Std	Std
Cruise Control	Std	Std
All Wheel Drive	N/A	N/A
AM/FM Stereo CD	Std	Std
Steering Wheel, Tilt	Std	Std
Power Windows	Std	Std
*Options Price	$0	$0
*Total Price	$52,979	$61,675
Target Price	$54,568	
Destination Charge	$0	
Avg. Tax & Fees	$3,345	
Luxury/Gas Guzzler Tax	$4,157	
Total Target $	**$62,070**	

Ownership Costs

Cost Area	5 Year Cost	Rate
Depreciation	$40,541	●
Financing ($1,248/month)	$10,225	
Insurance (Rating 25 Sport+)	$17,284	●
State Fees	$2,467	
Fuel (Hwy 21 City 14 -Prem.)	$5,504	●
Maintenance	$5,982	●
Repairs	$751	○

Warranty/Maintenance Info

Major Tune-Up	$126	○
Minor Tune-Up	$51	○
Brake Service	$214	○
Overall Warranty	4 yr/50k	○
Drivetrain Warranty	7 yr/100k	○
Rust Warranty	7 yr/100k	○
Maintenance Warranty	N/A	
Roadside Assistance	1 yr/12k	

Resale Value: 1993 $39,234; 1994 $33,941; 1995 $29,263; 1996 $25,173; 1997 $21,529

Cumulative Costs

	1993	1994	1995	1996	1997
Annual	$32,052	$13,643	$13,523	$10,736	$12,801
Total	$32,052	$45,695	$59,218	$69,954	$82,755

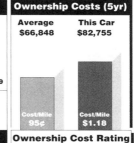

Ownership Costs (5yr) — Average $66,848 / This Car $82,755 — Cost/Mile 95¢ / $1.18

Ownership Cost Rating: Poor

The 1993 Allante is available in two model editions - (Base) Allante and Allante HT convertibles. New for 1993, the Base Allante is the first Cadillac to be equipped with the new 32 valve, 290 bhp Northstar V8 engine and a Road Sensing Suspension. The 1993 Allante also features new lightweight cast aluminum 16" x 7" wheels and all-season Goodyear Eagle tires. Allante's hand-made Italian body features a new front fascia spoiler for better high-speed aerodynamics and stability.

---

* Includes shaded options
** Other purchase requirements apply

● Poor   ◉ Worse Than Average   ○ Average   ○ Better Than Average   ○ Excellent   ⊖ Insufficient Information

©1993 by IntelliChoice, Inc. (408) 554-8711   All Rights Reserved. Reproduction Prohibited.
Refer to *Section 3: Annotated Vehicle Charts* for an explanation of these charts.

Cadillac Allante HT
2 Door Convertible

4.6L 295 hp Gas Fuel Inject. | 8 Cylinder "V" | Automatic 4 Speed | 2 Wheel Front | Driver Airbag Psngr Belts

Convertible Model Shown — Luxury

Purchase Price

Car Item	Dealer Cost	List
Base Price	**$55,877**	**$65,125**
Anti-Lock Brakes	Std	Std
Manual Transmission	N/A	N/A
Optional Engine	N/A	N/A
Auto Climate Control	Std	Std
Power Steering	Std	Std
Cruise Control	Std	Std
All Wheel Drive	N/A	N/A
AM/FM Stereo CD	Std	Std
Steering Wheel, Tilt	Std	Std
Power Windows	Std	Std
*Options Price	$0	$0
*Total Price	$55,877	$65,125
Target Price	$57,553	
Destination Charge	$0	
Avg. Tax & Fees	$3,529	
Luxury Tax	$2,755	
Total Target $	**$63,837**	

Ownership Costs

Cost Area	5 Year Cost	Rate
Depreciation	$41,259	●
Financing ($1,283/month)	$10,516	
Insurance (26 Sport+ [Est.])	$18,179	◉
State Fees	$2,605	
Fuel (Hwy 21 City 14 -Prem.)	$5,504	◉
Maintenance	$5,982	◉
Repairs	$751	○

Warranty/Maintenance Info

Major Tune-Up	$126	○
Minor Tune-Up	$51	○
Brake Service	$214	○
Overall Warranty	4 yr/50k	○
Drivetrain Warranty	7 yr/100k	○
Rust Warranty	7 yr/100k	○
Maintenance Warranty	N/A	
Roadside Assistance	1 yr/12k	

Ownership Cost By Year

Bars for 1993–1997 ($5K–$45K scale)

Resale Value

1993	1994	1995	1996	1997
$41,126	$35,608	$30,729	$26,413	$22,578

Cumulative Costs

	1993	1994	1995	1996	1997
Annual	$32,247	$14,161	$13,989	$11,193	$13,206
Total	$32,247	$46,408	$60,397	$71,590	$84,796

Ownership Costs (5yr)

Average	This Car
$69,913	$84,796
Cost/Mile $1.00	Cost/Mile $1.21

Ownership Cost Rating
● **Poor**

The 1993 Allante is available in two model editions - (Base) Allante and Allante HT convertibles. New for 1993, the Allante HT is the first Cadillac to be equipped with the new 32 valve, 290 bhp Northstar V8 engine and a Road Sensing Suspension. The 1993 Allante HT also features new lightweight cast aluminum 16" x 7" wheels and all-season Goodyear Eagle tires. Allante's hand-made Italian body features a new front fascia spoiler for better high-speed aerodynamics and stability.

Cadillac DeVille Coupe
2 Door Coupe

4.9L 200 hp Gas Fuel Inject. | 8 Cylinder "V" | Automatic 4 Speed | 2 Wheel Front | Driver Airbag Psngr Belts

Sedan Model Shown — Luxury

Purchase Price

Car Item	Dealer Cost	List
Base Price	**$29,336**	**$33,915**
Anti-Lock Brakes	Std	Std
Manual Transmission	N/A	N/A
Optional Engine	N/A	N/A
Auto Climate Control	Std	Std
Power Steering	Std	Std
Cruise Control	Std	Std
All Wheel Drive	N/A	N/A
AM/FM Stereo Cassette	Std	Std
Steering Wheel, Tilt	Std	Std
Power Windows	Std	Std
*Options Price	$0	$0
*Total Price	$29,336	$33,915
Target Price	$33,332	
Destination Charge	$600	
Avg. Tax & Fees	$2,042	
Luxury Tax	$393	
Total Target $	**$36,367**	

Ownership Costs

Cost Area	5 Year Cost	Rate
Depreciation	$21,080	○
Financing ($731/month)	$5,990	
Insurance (Rating 16)	$8,784	○
State Fees	$1,381	
Fuel (Hwy 25 City 16 -Prem.)	$4,718	○
Maintenance	$5,950	◉
Repairs	$751	○

Warranty/Maintenance Info

Major Tune-Up	$181	○
Minor Tune-Up	$118	○
Brake Service	$217	○
Overall Warranty	4 yr/50k	○
Drivetrain Warranty	4 yr/50k	○
Rust Warranty	6 yr/100k	○
Maintenance Warranty	N/A	
Roadside Assistance	1 yr/12k	

Resale Value

1993	1994	1995	1996	1997
$24,310	$21,891	$19,750	$17,506	$15,287

Cumulative Costs

	1993	1994	1995	1996	1997
Annual	$17,442	$7,460	$7,590	$6,205	$9,957
Total	$17,442	$24,902	$32,492	$38,697	$48,654

Ownership Costs (5yr)

Average	This Car
$44,815	$48,654
Cost/Mile 64¢	Cost/Mile 70¢

Ownership Cost Rating
◉ **Worse Than Average**

The Cadillac DeVille is available in three model editions - (Base) Coupe and Sedan DeVille, and DeVille Touring Sedan. New for 1993, the Coupe DeVille offers new interior and exterior colors (Dark Cherry and Bronze), speed sensitive steering, speed sensitive suspension, and redesigned front-end styling. The coupe also offers automatic traction control, Twilight Sentinel headlights, cornering lights, Solar glass, and optional leather seats.

---

* Includes shaded options
** Other purchase requirements apply

● Poor   ◉ Worse Than Average   ◐ Average   ○ Better Than Average   ○ Excellent   ⊖ Insufficient Information

©1993 by *IntelliChoice, Inc.* (408) 554-8711   All Rights Reserved. Reproduction Prohibited.
Refer to *Section 3: Annotated Vehicle Charts* for an explanation of these charts.

Cadillac DeVille Sedan
4 Door Sedan

Luxury

4.9L 200 hp Gas Fuel Inject. | 8 Cylinder "V" | Automatic 4 Speed | 2 Wheel Front | Driver Airbag Psngr Belts

Purchase Price

Car Item	Dealer Cost	List
Base Price	**$28,536**	**$32,990**
Anti-Lock Brakes	Std	Std
Manual Transmission	N/A	N/A
Optional Engine	N/A	N/A
Auto Climate Control	Std	Std
Power Steering	Std	Std
Cruise Control	Std	Std
All Wheel Drive	N/A	N/A
AM/FM Stereo Cassette	Std	Std
Steering Wheel, Tilt	Std	Std
Power Windows	Std	Std
*Options Price	$0	$0
*Total Price	$28,536	$32,990
Target Price		$32,361
Destination Charge		$600
Avg. Tax & Fees		$1,984
Luxury Tax		$296
Total Target $		**$35,241**

The Cadillac DeVille is available in three model editions - (Base) Coupe and Sedan DeVille, and DeVille Touring Sedan. New for 1993, the Sedan DeVille offers new interior and exterior colors (Dark Cherry and Bronze), speed sensitive steering, speed sensitive suspension, and redesigned front-end styling. The Base model also offers automatic traction control, Twilight Sentinel headlights, cornering lights, Solar glass, and optional leather seats.

Ownership Costs

Cost Area	5 Year Cost	Rate
Depreciation	$20,054	●
Financing ($708/month)	$5,805	
Insurance (Rating 12)	$7,898	○
State Fees	$1,344	
Fuel (Hwy 25 City 16 -Prem.)	$4,718	●
Maintenance	$6,012	●
Repairs	$751	●

Warranty/Maintenance Info

Major Tune-Up	$181	○
Minor Tune-Up	$118	○
Brake Service	$217	○
Overall Warranty	4 yr/50k	○
Drivetrain Warranty	4 yr/50k	○
Rust Warranty	6 yr/100k	○
Maintenance Warranty	N/A	
Roadside Assistance	1 yr/12k	

Ownership Cost By Year

1993, 1994, 1995, 1996, 1997

Resale Value

1993	1994	1995	1996	1997
$24,583	$22,024	$19,807	$17,488	$15,187

Cumulative Costs

	1993	1994	1995	1996	1997
Annual	$15,793	$7,363	$7,467	$6,093	$9,866
Total	$15,793	$23,156	$30,623	$36,716	$46,582

Ownership Costs (5yr)

Average	This Car
$44,123	$46,582
Cost/Mile 63¢	Cost/Mile 67¢

Ownership Cost Rating
Worse Than Average

Cadillac DeVille Touring Sedan
4 Door Sedan

Luxury

4.9L 200 hp Gas Fuel Inject. | 8 Cylinder "V" | Automatic 4 Speed | 2 Wheel Front | Driver Airbag Psngr Belts

Purchase Price

Car Item	Dealer Cost	List
Base Price	**$31,408**	**$36,310**
Anti-Lock Brakes	Std	Std
Manual Transmission	N/A	N/A
Optional Engine	N/A	N/A
Auto Climate Control	Std	Std
Power Steering	Std	Std
Cruise Control	Std	Std
All Wheel Drive	N/A	N/A
AM/FM Stereo Cassette	Std	Std
Steering Wheel, Tilt	Std	Std
Power Windows	Std	Std
*Options Price	$0	$0
*Total Price	$31,408	$36,310
Target Price		$35,648
Destination Charge		$600
Avg. Tax & Fees		$2,181
Luxury Tax		$625
Total Target $		**$39,054**

The Cadillac DeVille is available in three model editions - (Base) Coupe and Sedan DeVille, and DeVille Touring Sedan. New for 1993, all DeVille models offer new interior and exterior colors (Dark Cherry and Bronze), speed sensitive steering, speed sensitive suspension, and redesigned front-end styling. The DeVille Touring Sedan features tan leather interior with walnut trim, and an optional engine block heater and heated windshield.

Ownership Costs

Cost Area	5 Year Cost	Rate
Depreciation	$23,905	●
Financing ($785/month)	$6,434	
Insurance (Rating 14)	$8,238	○
State Fees	$1,475	
Fuel (Hwy 25 City 16 -Prem.)	$4,718	●
Maintenance	$6,127	●
Repairs	$751	●

Warranty/Maintenance Info

Major Tune-Up	$181	○
Minor Tune-Up	$118	○
Brake Service	$217	○
Overall Warranty	4 yr/50k	○
Drivetrain Warranty	4 yr/50k	○
Rust Warranty	6 yr/100k	○
Maintenance Warranty	N/A	
Roadside Assistance	1 yr/12k	

Resale Value

1993	1994	1995	1996	1997
$24,442	$22,012	$19,866	$17,543	$15,149

Cumulative Costs

	1993	1994	1995	1996	1997
Annual	$20,105	$7,526	$7,670	$6,236	$10,111
Total	$20,105	$27,631	$35,301	$41,537	$51,648

Ownership Costs (5yr)

Average	This Car
$46,608	$51,648
Cost/Mile 67¢	Cost/Mile 74¢

Ownership Cost Rating
Worse Than Average

---

* Includes shaded options
** Other purchase requirements apply

● Poor   ◐ Worse Than Average   ◔ Average   ○ Better Than Average   ○ Excellent   ⊖ Insufficient Information

©1993 by IntelliChoice, Inc. (408) 554-8711 All Rights Reserved. Reproduction Prohibited.
Refer to *Section 3: Annotated Vehicle Charts* for an explanation of these charts.

Cadillac Eldorado
2 Door Coupe

Luxury

 4.9L 200 hp Gas Fuel Inject.
 8 Cylinder "V"
 Automatic 4 Speed
 2 Wheel Front
 Driver/Psngr Airbags Std

Purchase Price

Car Item	Dealer Cost	List
Base Price	**$29,401**	**$33,990**
Anti-Lock Brakes	Std	Std
Manual Transmission	N/A	N/A
4.6L 295 hp Gas	Pkg	Pkg
Auto Climate Control	Std	Std
Power Steering	Std	Std
Cruise Control	Std	Std
All Wheel Drive	N/A	N/A
AM/FM Stereo Cassette	Std	Std
Steering Wheel, Tilt	Std	Std
Power Windows	Std	Std
*Options Price	$0	$0
***Total Price**	**$29,401**	**$33,990**
Target Price	$32,223	
Destination Charge	$600	
Avg. Tax & Fees	$1,987	
Luxury Tax	$282	
Total Target $	**$35,092**	

Ownership Costs

Cost Area	5 Year Cost	Rate
Depreciation	$21,012	◉
Financing ($705/month)	$5,781	
Insurance (Rating 17)	$9,084	◉
State Fees	$1,384	
Fuel (Hwy 25 City 16 -Prem.)	$4,718	◉
Maintenance	$6,856	●
Repairs	$751	◉

Warranty/Maintenance Info

Major Tune-Up	$165	○
Minor Tune-Up	$118	○
Brake Service	$213	○
Overall Warranty	4 yr/50k	◉
Drivetrain Warranty	4 yr/50k	◉
Rust Warranty	6 yr/100k	◉
Maintenance Warranty	N/A	
Roadside Assistance	1 yr/12k	

Ownership Cost By Year

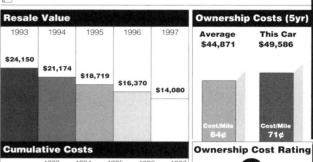

Resale Value

1993: $24,150 — 1994: $21,174 — 1995: $18,719 — 1996: $16,370 — 1997: $14,080

Cumulative Costs

	1993	1994	1995	1996	1997
Annual	$16,296	$8,009	$7,955	$6,386	$10,940
Total	$16,296	$24,305	$32,260	$38,646	$49,586

Ownership Costs (5yr)

Average: $44,871
This Car: $49,586
Cost/Mile: 64¢ / 71¢

Ownership Cost Rating

 Worse Than Average

The Cadillac Eldorado is offered in one model edition with an optional Touring Package. New for 1993, the Eldorado features speed sensitive steering, dual airbags, and front seat and suspension enhancements. Also, two optional sport trims are now available - a Sport Appearance Package (which includes exterior changes) and the Sport Performance Package which includes the Appearance Package as well as a Northstar powertrain, traction control, special suspension, and larger roll resistance tires.

Cadillac Fleetwood
4 Door Sedan

Luxury

 5.7L 185 hp Gas Fuel Inject.
 8 Cylinder "V"
 Automatic 4 Speed
 2 Wheel Rear
 Driver/Psngr Airbags Std

Purchase Price

Car Item	Dealer Cost	List
Base Price	**$29,401**	**$33,990**
Anti-Lock Brakes	Std	Std
Manual Transmission	N/A	N/A
Optional Engine	N/A	N/A
Auto Climate Control	Std	Std
Power Steering	Std	Std
Cruise Control	Std	Std
All Wheel Drive	N/A	N/A
AM/FM Stereo Cassette	Std	Std
Steering Wheel, Tilt	Std	Std
Power Windows	Std	Std
*Options Price	$0	$0
***Total Price**	**$29,401**	**$33,990**
Target Price	$30,886	
Destination Charge	$600	
Avg. Tax & Fees	$1,920	
Luxury Tax	$149	
Total Target $	**$33,555**	

Ownership Costs

Cost Area	5 Year Cost	Rate
Depreciation	$19,260	◉
Financing ($674/month)	$5,528	
Insurance (Rating 16)	$8,784	○
State Fees	$1,384	
Fuel (Hwy 25 City 16)	$4,272	◉
Maintenance	$4,036	○
Repairs	$751	○

Warranty/Maintenance Info

Major Tune-Up	$157	○
Minor Tune-Up	$118	○
Brake Service	$217	○
Overall Warranty	4 yr/50k	○
Drivetrain Warranty	4 yr/50k	○
Rust Warranty	6 yr/100k	○
Maintenance Warranty	N/A	
Roadside Assistance	1 yr/12k	

Resale Value

1993: $22,614 — 1994: $20,406 — 1995: $18,510 — 1996: $16,443 — 1997: $14,295

Cumulative Costs

	1993	1994	1995	1996	1997
Annual	$16,051	$7,018	$7,227	$6,229	$7,490
Total	$16,051	$23,069	$30,296	$36,525	$44,015

Ownership Costs (5yr)

Average: $44,871
This Car: $44,015
Cost/Mile: 64¢ / 63¢

Ownership Cost Rating

○ Average

The Cadillac Fleetwood is available as a four-door sedan. New for 1993, the Fleetwood has been redesigned with such interior and exterior refinements as a longer, more aerodynamic body, front integrated grille, and flush glass design with larger windows. Standard equipment includes power pull-down trunk, dual airbags, PASS-Key II theft deterrent system, power adjustable seats, "turn signal on" warning indicator, and retained accessory power. Also available is a Fleetwood Brougham Package.

---

page 38

* Includes shaded options
** Other purchase requirements apply

● Poor | ◉ Worse Than Average | Average | ○ Better Than Average | ○ Excellent | ⊖ Insufficient Information

©1993 by *IntelliChoice, Inc.* (408) 554-8711 All Rights Reserved. Reproduction Prohibited.

Refer to *Section 3: Annotated Vehicle Charts* for an explanation of these charts.

Cadillac Seville
4 Door Sedan

4.9L 200 hp Gas Fuel Inject. | 8 Cylinder "V" | Automatic 4 Speed | 2 Wheel Front | Driver/Psngr Airbags Std

Luxury

Purchase Price

Car Item	Dealer Cost	List
Base Price	**$31,996**	**$36,990**
Anti-Lock Brakes	Std	Std
Manual Transmission	N/A	N/A
Optional Engine	N/A	N/A
Auto Climate Control	Std	Std
Power Steering	Std	Std
Cruise Control	Std	Std
All Wheel Drive	N/A	N/A
AM/FM Stereo Cassette	Std	Std
Steering Wheel, Tilt	Std	Std
Power Windows	Std	Std
*Options Price	$0	$0
*Total Price	$31,996	$36,990
Target Price	$36,315	
Destination Charge	$600	
Avg. Tax & Fees	$2,222	
Luxury Tax	$692	
Total Target $	**$39,829**	

The Cadillac Seville is available in two model editions - (Base) and STS four-door sedans. New for 1993, the Seville offers speed sensitive steering and suspension design, standard power reclining front seats, and dual airbags. The base sedan comes equipped with heated windshield, vehicle information center, automatic climate control, and retained accessory power. Also new this year is an available express open sunroof which can be partially or fully open with the touch of a button.

Ownership Costs

Cost Area	5 Year Cost	Rate
Depreciation	$24,233	O
Financing ($800/month)	$6,562	
Insurance (Rating 15)	$8,507	O
State Fees	$1,504	
Fuel (Hwy 25 City 16 -Prem.)	$4,718	O
Maintenance	$6,844	●
Repairs	$791	O

Warranty/Maintenance Info

Major Tune-Up	$165	O
Minor Tune-Up	$118	O
Brake Service	$213	O
Overall Warranty	4 yr/50k	
Drivetrain Warranty	4 yr/50k	
Rust Warranty	6 yr/100k	O
Maintenance Warranty	N/A	
Roadside Assistance	1 yr/12k	

Ownership Cost By Year
1993, 1994, 1995, 1996, 1997

Resale Value
1993	1994	1995	1996	1997
$25,928	$22,961	$20,434	$17,993	$15,596

Cumulative Costs
	1993	1994	1995	1996	1997
Annual	$19,504	$8,160	$8,091	$6,446	$10,958
Total	$19,504	$27,664	$35,755	$42,201	$53,159

Ownership Costs (5yr)
Average	This Car
$47,117	$53,159
Cost/Mile 67¢	Cost/Mile 76¢

Ownership Cost Rating
● Poor

Cadillac Seville Touring Sedan
4 Door Sedan

4.6L 295 hp Gas Fuel Inject. | 8 Cylinder "V" | Automatic 4 Speed | 2 Wheel Front | Driver/Psngr Airbags Std

Luxury

Purchase Price

Car Item	Dealer Cost	List
Base Price	**$36,321**	**$41,990**
Anti-Lock Brakes	Std	Std
Manual Transmission	N/A	N/A
Optional Engine	N/A	N/A
Auto Climate Control	Std	Std
Power Steering	Std	Std
Cruise Control	Std	Std
All Wheel Drive	N/A	N/A
AM/FM Stereo Cassette	Std	Std
Steering Wheel, Tilt	Std	Std
Power Windows	Std	Std
*Options Price	$0	$0
*Total Price	$36,321	$41,990
Target Price	$41,224	
Destination Charge	$600	
Avg. Tax & Fees	$2,517	
Luxury Tax	$1,182	
Total Target $	**$45,523**	

The Cadillac Seville is available in two model editions - (Base) and STS four-door sedans. New for 1993, the Seville offers speed sensitive steering and suspension design, standard power reclining front seats, and dual airbags. The Seville Touring Sedan has standard wood trim, leather interior, power lumbar, a new road sensitive suspension, Northstar engine, and automatic traction control. Also new is an available express open sunroof which can be partially or fully open with the touch of a button.

Ownership Costs

Cost Area	5 Year Cost	Rate
Depreciation	$26,995	O
Financing ($915/month)	$7,500	
Insurance (Rating 16++)	$10,561	O
State Fees	$1,704	
Fuel (Hwy 25 City 16 -Prem.)	$4,718	O
Maintenance	$7,737	●
Repairs	$791	O

Warranty/Maintenance Info

Major Tune-Up	$126	O
Minor Tune-Up	$51	O
Brake Service	$213	O
Overall Warranty	4 yr/50k	O
Drivetrain Warranty	4 yr/50k	O
Rust Warranty	6 yr/100k	O
Maintenance Warranty	N/A	
Roadside Assistance	1 yr/12k	

Resale Value
1993	1994	1995	1996	1997
$29,809	$26,546	$23,801	$21,127	$18,528

Cumulative Costs
	1993	1994	1995	1996	1997
Annual	$22,218	$9,152	$9,437	$7,138	$12,062
Total	$22,218	$31,370	$40,807	$47,945	$60,007

Ownership Costs (5yr)
Average	This Car
$50,860	$60,007
Cost/Mile 73¢	Cost/Mile 86¢

Ownership Cost Rating
● Poor

---

* Includes shaded options
** Other purchase requirements apply

● Poor | Worse Than Average | Average | ◯ Better Than Average | ◯ Excellent | ⊖ Insufficient Information

©1993 by IntelliChoice, Inc. (408) 554-8711 All Rights Reserved. Reproduction Prohibited.
Refer to *Section 3: Annotated Vehicle Charts* for an explanation of these charts.

Cadillac Sixty Special Sedan
4 Door Sedan

Luxury

4.9L 200 hp Gas Fuel Inject. | 8 Cylinder "V" | Automatic 4 Speed | 2 Wheel Front | Driver Airbag Psngr Belts

Purchase Price

Car Item	Dealer Cost	List
Base Price	**$32,204**	**$37,230**
Anti-Lock Brakes	Std	Std
Manual Transmission	N/A	N/A
Optional Engine	N/A	N/A
Auto Climate Control	Std	Std
Power Steering	Std	Std
Cruise Control	Std	Std
All Wheel Drive	N/A	N/A
AM/FM Stereo Cassette	Std	Std
Steering Wheel, Tilt	Std	Std
Power Windows	Std	Std
*Options Price	$0	$0
*Total Price	$32,204	$37,230
Target Price		$33,814
Destination Charge		$600
Avg. Tax & Fees		$2,099
Luxury Tax		$441
Total Target $		**$36,954**

The Cadillac Sixty Special is available as a four-door sedan. New for 1993, the Sixty Special features speed sensitive steering, new seat design and front end styling, and new interior and exterior colors. The Sixty Special offers a speed sensitive suspension, white sidewall tires, dual heated power mirrors, Twilight Sentinel headlights, Solar glass, retained accessory power and illuminated entry system. An optional Ultra Seating Package is also available.

Ownership Costs

Cost Area	5 Year Cost	Rate
Depreciation	$22,565	●
Financing ($743/month)	$6,088	
Insurance (Rating 14)	$8,238	○
State Fees	$1,513	
Fuel (Hwy 25 City 16 -Prem.)	$4,718	○
Maintenance	$6,012	●
Repairs	$791	○

Warranty/Maintenance Info

Major Tune-Up	$181	○
Minor Tune-Up	$118	○
Brake Service	$217	○
Overall Warranty	4 yr/50k	○
Drivetrain Warranty	4 yr/50k	○
Rust Warranty	6 yr/100k	○
Maintenance Warranty	N/A	
Roadside Assistance	1 yr/12k	

Ownership Cost By Year

Legend: 1993, 1994, 1995, 1996, 1997

Resale Value

1993	1994	1995	1996	1997
$24,065	$21,470	$19,202	$16,830	$14,389

Cumulative Costs

	1993	1994	1995	1996	1997
Annual	$18,247	$7,591	$7,676	$6,277	$10,134
Total	$18,247	$25,838	$33,514	$39,791	$49,925

Ownership Costs (5yr)

Average	This Car
$47,297	$49,925
Cost/Mile 68¢	Cost/Mile 71¢

Ownership Cost Rating

○ Average

Chevrolet Beretta
2 Door Coupe

Compact

2.2L 110 hp Gas Fuel Inject. | 4 Cylinder In-Line | Manual 5 Speed | 2 Wheel Front | Driver Airbag Psngr Belts

Purchase Price

Car Item	Dealer Cost	List
Base Price	**$10,312**	**$11,395**
Anti-Lock Brakes	Std	Std
Automatic 3 Speed	$477	$555
3.1L 140 hp Gas	$516	$600
Air Conditioning	$692	$805
Power Steering	Std	Std
Cruise Control	$194	$225
All Wheel Drive	N/A	N/A
AM/FM Stereo Cassette	$120	$140
Steering Wheel, Tilt	$125	$145
Power Windows	$237	$275
*Options Price	$1,805	$2,100
*Total Price	$12,117	$13,495
Target Price		$13,140
Destination Charge		$475
Avg. Tax & Fees		$821
Total Target $		**$14,436**
Average Dealer Option Cost:	**86%**	

The 1993 Chevrolet Beretta is available in three models - (Base) Beretta, GT, and GTZ coupes. New for 1993, the (Base) Beretta features improved ride quality and reduced noise level, new exterior colors, low-oil level sensor with warning light, and a concentric slave clutch. Standard features include bodyside moldings, dual remote outside mirrors, a headlamps-on reminder chime and tinted glass. All body panels except the roof are rust resistant. Full gauge instrument cluster is an option.

Ownership Costs

Cost Area	5 Year Cost	Rate
Depreciation	$8,418	●
Financing ($290/month)	$2,378	
Insurance (Rating 14)	$8,238	●
State Fees	$560	
Fuel (Hwy 28 City 20)	$3,612	○
Maintenance	$4,445	○
Repairs	$651	○

Warranty/Maintenance Info

Major Tune-Up	$142	○
Minor Tune-Up	$106	○
Brake Service	$224	○
Overall Warranty	3 yr/36k	○
Drivetrain Warranty	3 yr/36k	○
Rust Warranty	6 yr/100k	○
Maintenance Warranty	N/A	
Roadside Assistance	3 yr/36k	

Ownership Cost By Year

Legend: 1993, 1994, 1995, 1996, 1997

Resale Value

1993	1994	1995	1996	1997
$9,978	$8,809	$7,808	$6,857	$6,018

Cumulative Costs

	1993	1994	1995	1996	1997
Annual	$7,826	$4,562	$5,085	$4,208	$6,621
Total	$7,826	$12,388	$17,473	$21,681	$28,302

Ownership Costs (5yr)

Average	This Car
$26,082	$28,302
Cost/Mile 37¢	Cost/Mile 40¢

Ownership Cost Rating

● Worse Than Average

---

page 40

* Includes shaded options
** Other purchase requirements apply

Legend: ● Poor | ◐ Worse Than Average | ○ Average | ◯ Better Than Average | ◯ Excellent | ⊖ Insufficient Information

©1993 by IntelliChoice, Inc. (408) 554-8711 All Rights Reserved. Reproduction Prohibited.
Refer to *Section 3: Annotated Vehicle Charts* for an explanation of these charts.

Chevrolet Beretta GT
2 Door Coupe

Compact

- 2.2L 110 hp Gas Fuel Inject.
- 4 Cylinder In-Line
- Manual 5 Speed
- 2 Wheel Front
- Driver Airbag Psngr Belts

Purchase Price

Car Item	Dealer Cost	List
Base Price	**$11,760**	**$12,995**
Anti-Lock Brakes	Std	Std
Automatic 3 Speed	$477	$555
3.1L 140 hp Gas	$516	$600
Air Conditioning	$692	$805
Power Steering	Std	Std
Cruise Control	$194	$225
All Wheel Drive	N/A	N/A
AM/FM Stereo Cassette	$120	$140
Steering Wheel, Tilt	$125	$145
Power Windows	$237	$275
*Options Price	$1,805	$2,100
*Total Price	**$13,565**	**$15,095**
Target Price	$14,763	
Destination Charge	$475	
Avg. Tax & Fees	$918	
Total Target $	**$16,156**	
Average Dealer Option Cost:	86%	

Ownership Costs

Cost Area	5 Year Cost	Rate
Depreciation	$9,236	○
Financing ($325/month)	$2,662	
Insurance (Rating 15)	$8,507	●
State Fees	$623	
Fuel (Hwy 28 City 20)	$3,612	○
Maintenance	$5,455	●
Repairs	$651	○

Warranty/Maintenance Info

Major Tune-Up	$142	○
Minor Tune-Up	$106	○
Brake Service	$214	○
Overall Warranty	3 yr/36k	○
Drivetrain Warranty	3 yr/36k	○
Rust Warranty	6 yr/100k	○
Maintenance Warranty	N/A	
Roadside Assistance	3 yr/36k	

Resale Value

1993	1994	1995	1996	1997
$11,196	$9,997	$8,897	$7,817	$6,920

Cumulative Costs

	1993	1994	1995	1996	1997
Annual	$8,512	$4,746	$5,793	$4,424	$7,271
Total	$8,512	$13,258	$19,051	$23,475	$30,746

Ownership Costs (5yr)

Average	This Car
$27,456	$30,746
Cost/Mile 39¢	Cost/Mile 44¢

Ownership Cost Rating: Poor

The 1993 Chevrolet Beretta is available in three models - (Base) Beretta, GT, and GTZ coupes. New for 1993, the GT coupe features a low-oil level sensor with warning light, improved ride quality and reduced noise level, and two new exterior colors. The GT upgrades the (Base) Beretta with a full gauge instrument cluster, and a Level II Sport Suspension. Standard features include dual remote outside mirrors, tinted glass, bodyside moldings and a new brake/transmission shift interlock.

Chevrolet Beretta GTZ
2 Door Coupe

Compact

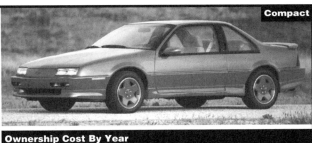

- 2.3L 180 hp Gas Fuel Inject.
- 4 Cylinder In-Line
- Manual 5 Speed
- 2 Wheel Front
- Driver Airbag Psngr Belts

Purchase Price

Car Item	Dealer Cost	List
Base Price	**$14,475**	**$15,995**
Anti-Lock Brakes	Std	Std
Automatic 3 Speed	$477	** $555
3.1L 140 hp Gas	($129)	** ($150)
Air Conditioning	Std	Std
Power Steering	Std	Std
Cruise Control	$194	$225
All Wheel Drive	N/A	N/A
AM/FM Stereo Cassette	$120	$140
Steering Wheel, Tilt	$125	$145
Power Windows	$237	$275
*Options Price	$120	$140
*Total Price	**$14,595**	**$16,135**
Target Price	$15,992	
Destination Charge	$475	
Avg. Tax & Fees	$989	
Total Target $	**$17,456**	
Average Dealer Option Cost:	86%	

Ownership Costs

Cost Area	5 Year Cost	Rate
Depreciation	$10,906	●
Financing ($351/month)	$2,876	
Insurance (Rating 16+)	$10,541	●
State Fees	$664	
Fuel (Hwy 30 City 21 -Prem.)	$3,760	○
Maintenance	$5,272	◐
Repairs	$651	○

Warranty/Maintenance Info

Major Tune-Up	$113	○
Minor Tune-Up	$78	○
Brake Service	$214	○
Overall Warranty	3 yr/36k	○
Drivetrain Warranty	3 yr/36k	○
Rust Warranty	6 yr/100k	○
Maintenance Warranty	N/A	
Roadside Assistance	3 yr/36k	

Resale Value

1993	1994	1995	1996	1997
$11,120	$9,797	$8,640	$7,494	$6,550

Cumulative Costs

	1993	1994	1995	1996	1997
Annual	$10,387	$5,331	$6,287	$4,929	$7,735
Total	$10,387	$15,718	$22,005	$26,934	$34,669

Ownership Costs (5yr)

Average	This Car
$28,349	$34,669
Cost/Mile 40¢	Cost/Mile 50¢

Ownership Cost Rating: Poor

The 1993 Chevrolet Beretta is available in three models - (Base) Beretta, GT, and GTZ coupes. New for 1993, the GTZ coupe features improved ride quality and reduced noise level, two new exterior colors and a low-oil level sensor with warning light. The GTZ upgrades the GT with a high output engine, and Level III Special Sport Performance Suspension. Standard features include dual remote outside mirrors, bodyside moldings, brake/transmission shift interlock, and tinted glass.

---

* Includes shaded options
** Other purchase requirements apply

● Poor | ◐ Worse Than Average | ◑ Average | ○ Better Than Average | ○ Excellent | ⊖ Insufficient Information

©1993 by IntelliChoice, Inc. (408) 554-8711 All Rights Reserved. Reproduction Prohibited.
Refer to *Section 3: Annotated Vehicle Charts* for an explanation of these charts.

Chevrolet Camaro
2 Door Coupe

Compact

 3.4L 160 hp Gas Fuel Inject.
 6 Cylinder "V"
 Manual 5 Speed
 2 Wheel Rear
Driver/Psngr Airbags Std

Purchase Price

Car Item	Dealer Cost	List
Base Price	**N/R**	**N/R**
Anti-Lock Brakes	Std	Std
Automatic 4 Speed	N/R	N/R
Optional Engine	N/A	N/A
Air Conditioning	Pkg	Pkg
Power Steering	Std	Std
Cruise Control	Pkg	Pkg
All Wheel Drive	N/A	N/A
AM/FM Stereo Cassette	Std	Std
Steering Wheel, Tilt	Std	Std
Power Windows	Pkg	Pkg
*Options Price	$0	$0
*Total Price	N/R	N/R
Target Price		N/R
Destination Charge		N/R
Avg. Tax & Fees		N/R
Total Target $		**N/R**
Average Dealer Option Cost:		N/A

Ownership Costs

Cost Area	5 Year Cost	Rate
Depreciation		⊖
Financing ($0/month)		
Insurance (Rating N/R)		⊖
State Fees		
Fuel (Hwy 28 City 19)	$3,704	○
Maintenance		⊖
Repairs	$709	○

Warranty/Maintenance Info

Major Tune-Up	$0	⊖
Minor Tune-Up	$89	○
Brake Service	$221	○
Overall Warranty	3 yr/36k	○
Drivetrain Warranty	3 yr/36k	○
Rust Warranty	6 yr/100k	○
Maintenance Warranty	N/A	
Roadside Assistance	3 yr/36k	

Ownership Cost By Year

$1,000 $2,000 $3,000 $4,000

- Insufficient Depreciation Information
- Insufficient Financing Information
- Insufficient Insurance Information
- Insufficient State Fee Information
- (Fuel bar)
- Insufficient Maintenance Information
- (Repairs bar)

Legend: 1993, 1994, 1995, 1996, 1997

Resale Value
Insufficient Information

Ownership Costs (5yr)
Insufficient Information

Cumulative Costs
	1993	1994	1995	1996	1997
Annual	Insufficient Information				
Total	Insufficient Information				

Ownership Cost Rating
⊖ Insufficient Information

The 1993 Camaro is available in two models - (Base) and Z28 coupe. The Camaro coupe has been redesigned for 1993, and the convertible model is no longer offered. Standard features on the coupe are bolt-on wheel covers, dual air bags, four-wheel anti-lock brake system, intermittent wipers, tinted glass, and stainless steel exhaust system. Options include compact disc player, removable roof panels, and rear-window defogger. Pricing was not available at time of printing.

Chevrolet Camaro Z28
2 Door Coupe

Sport

 5.7L 275 hp Gas Fuel Inject.
 8 Cylinder "V"
 Manual 6 Speed
 2 Wheel Rear
Driver/Psngr Airbags Std

Purchase Price

Car Item	Dealer Cost	List
Base Price	**N/R**	**N/R**
Anti-Lock Brakes	Std	Std
Automatic 4 Speed	N/R	N/R
Optional Engine	N/A	N/A
Air Conditioning	Pkg	Pkg
Power Steering	Std	Std
Cruise Control	Pkg	Pkg
All Wheel Drive	N/A	N/A
AM/FM Stereo Cassette	Std	Std
Steering Wheel, Tilt	Std	Std
Power Windows	Pkg	Pkg
*Options Price	$0	$0
*Total Price	N/R	N/R
Target Price		N/R
Destination Charge		N/R
Avg. Tax & Fees		N/R
Total Target $		**N/R**
Average Dealer Option Cost:		N/A

Ownership Costs

Cost Area	5 Year Cost	Rate
Depreciation		⊖
Financing ($0/month)		
Insurance (Rating N/R)		⊖
State Fees		
Fuel (Hwy 25 City 17)	$4,145	○
Maintenance	$5,236	◐
Repairs	$750	○

Warranty/Maintenance Info

Major Tune-Up	$185	◐
Minor Tune-Up	$132	○
Brake Service	$341	●
Overall Warranty	3 yr/36k	○
Drivetrain Warranty	3 yr/36k	○
Rust Warranty	6 yr/100k	○
Maintenance Warranty	N/A	
Roadside Assistance	3 yr/36k	

Ownership Cost By Year

$1,000 $2,000 $3,000 $4,000 $5,000 $6,000

- Insufficient Depreciation Information
- Insufficient Financing Information
- Insufficient Insurance Information
- Insufficient State Fee Information
- (Fuel bar)
- (Maintenance bar)
- (Repairs bar)

Legend: 1993, 1994, 1995, 1996, 1997

Resale Value
Insufficient Information

Ownership Costs (5yr)
Insufficient Information

Cumulative Costs
	1993	1994	1995	1996	1997
Annual	Insufficient Information				
Total	Insufficient Information				

Ownership Cost Rating
⊖ Insufficient Information

The 1993 Camaro is available in two models - (Base) and Z28 coupe. The Camaro has been redesigned for 1993, and the convertible model is no longer on the line. Standard features on this model include four-wheel anti-lock brakes, 16" aluminum wheels, power four-wheel disc brakes, tinted glass, and stainless steel exhaust system. Options include body-side moldings, fog lamps, removable roof panels, rear-window defogger, and sunshades with T-Top. Pricing was not available at time of printing.

---

page 42

* Includes shaded options
** Other purchase requirements apply

● Poor   ◐ Worse Than Average   ○ Average   ○ Better Than Average   ○ Excellent   ⊖ Insufficient Information

©1993 by IntelliChoice, Inc. (408) 554-8711 All Rights Reserved. Reproduction Prohibited.
Refer to *Section 3: Annotated Vehicle Charts* for an explanation of these charts.

Chevrolet Caprice Classic
4 Door Sedan

 5.0L 170 hp Gas Fuel Inject.
 8 Cylinder "V"
 Automatic 4 Speed
 2 Wheel Rear
 Driver Airbag Psngr Belts

Large

Purchase Price

Car Item	Dealer Cost	List
Base Price	**$15,746**	**$17,995**
Anti-Lock Brakes	Std	Std
Manual Transmission	N/A	N/A
Optional Engine	N/A	N/A
Air Conditioning	Std	Std
Power Steering	Std	Std
Cruise Control	$194	$225
All Wheel Drive	N/A	N/A
AM/FM Stereo Cassette	$151	** $175
Steering Wheel, Tilt	Std	Std
Power Windows	$292	$340
***Options Price**	**$486**	**$565**
***Total Price**	**$16,232**	**$18,560**
Target Price	$17,305	
Destination Charge	$555	
Avg. Tax & Fees	$1,084	
Total Target $	**$18,944**	
Average Dealer Option Cost:	86%	

Ownership Costs

Cost Area	5 Year Cost	Rate
Depreciation	$12,242	●
Financing ($381/month)	$3,121	
Insurance (Rating 7)	$7,035	○
State Fees	$765	
Fuel (Hwy 26 City 17)	$4,064	◎
Maintenance	$3,792	◎
Repairs	$709	◎

Warranty/Maintenance Info

Major Tune-Up	$173	○
Minor Tune-Up	$124	○
Brake Service	$222	○
Overall Warranty	3 yr/36k	◎
Drivetrain Warranty	3 yr/36k	◎
Rust Warranty	6 yr/100k	○
Maintenance Warranty	N/A	
Roadside Assistance	3 yr/36k	

Ownership Cost By Year

Bars for 1993, 1994, 1995, 1996, 1997 ($2,000–$14,000)

Resale Value

1993	1994	1995	1996	1997
$13,140	$11,233	$9,661	$8,102	$6,702

Cumulative Costs

	1993	1994	1995	1996	1997
Annual	$9,396	$5,457	$5,936	$5,029	$5,910
Total	$9,396	$14,853	$20,789	$25,818	$31,728

Ownership Costs (5yr)

Average: $30,876 — This Car: $31,728
Cost/Mile: 44¢ — Cost/Mile: 45¢

Ownership Cost Rating

Worse Than Average

The 1993 Chevrolet Caprice Classic is available in three models - (Base) Caprice Classic and LS sedans, and (Base) Classic wagon. New for 1993, the Base Caprice Classic features new exterior enhancements, wide bodyside moldings, base wheel covers, rear reading lamps and a power window lockout switch added to the driver's door. Standard features include stainless steel exhaust system, tinted glass, child security rear door locks, flush-mounted headlamps, and dual body-color aero sport mirrors.

Chevrolet Caprice Classic LS
4 Door Sedan

 5.0L 170 hp Gas Fuel Inject.
 8 Cylinder "V"
 Automatic 4 Speed
 2 Wheel Rear
 Driver Airbag Psngr Belts

Large

Purchase Price

Car Item	Dealer Cost	List
Base Price	**$17,496**	**$19,995**
Anti-Lock Brakes	Std	Std
Manual Transmission	N/A	N/A
5.7L 180 hp Gas	Pkg	Pkg
Air Conditioning	Std	Std
Power Steering	Std	Std
Cruise Control	$194	$225
All Wheel Drive	N/A	N/A
AM/FM Stereo Cassette	$280	$325
Steering Wheel, Tilt	Std	Std
Power Windows	Std	Std
***Options Price**	**$474**	**$550**
***Total Price**	**$17,970**	**$20,545**
Target Price	$19,215	
Destination Charge	$555	
Avg. Tax & Fees	$1,200	
Total Target $	**$20,970**	
Average Dealer Option Cost:	86%	

Ownership Costs

Cost Area	5 Year Cost	Rate
Depreciation	$12,590	●
Financing ($421/month)	$3,455	
Insurance (Rating 8)	$7,223	○
State Fees	$845	
Fuel (Hwy 26 City 17)	$4,064	◎
Maintenance	$3,792	◎
Repairs	$709	◎

Warranty/Maintenance Info

Major Tune-Up	$173	○
Minor Tune-Up	$124	○
Brake Service	$222	○
Overall Warranty	3 yr/36k	◎
Drivetrain Warranty	3 yr/36k	◎
Rust Warranty	6 yr/100k	○
Maintenance Warranty	N/A	
Roadside Assistance	3 yr/36k	

Resale Value

1993	1994	1995	1996	1997
$14,369	$12,684	$11,204	$9,811	$8,380

Cumulative Costs

	1993	1994	1995	1996	1997
Annual	$10,387	$5,394	$5,963	$4,940	$5,994
Total	$10,387	$15,781	$21,744	$26,684	$32,678

Ownership Costs (5yr)

Average: $32,845 — This Car: $32,678
Cost/Mile: 47¢ — Cost/Mile: 47¢

Ownership Cost Rating

Average

The 1993 Chevrolet Caprice Classic is available in three models - (Base) Caprice Classic and LS sedans, and (Base) Classic wagon. New for 1993, the LS sedan features rear-end styling touches, gold accents, rear reading lamps and door-edge guards. The LS upgrades the Base with cast aluminum wheels with lock and custom cloth seat trim. Standard features include corrosion-resistance two-side-galvanized steel for most body panels, steel-belted radial tires, and dual body-color aero sport mirrors.

---

* Includes shaded options
** Other purchase requirements apply

Legend: ● Poor ◐ Worse Than Average ◎ Average ○ Better Than Average ○ Excellent ⊖ Insufficient Information

©1993 by IntelliChoice, Inc. (408) 554-8711 All Rights Reserved. Reproduction Prohibited.
Refer to *Section 3: Annotated Vehicle Charts* for an explanation of these charts.

Chevrolet Caprice Classic
4 Door Wagon

Large Wagon

 5.0L 170 hp Gas Fuel Inject.
 8 Cylinder "V"
 Automatic 4 Speed
 2 Wheel Rear
 Driver Airbag Psngr Belts

Purchase Price

Car Item	Dealer Cost	List
Base Price	**$17,128**	**$19,575**
Anti-Lock Brakes	Std	Std
Manual Transmission	N/A	N/A
5.7L 180 hp Gas	$215	$250
Air Conditioning	Std	Std
Power Steering	Std	Std
Cruise Control	$194	$225
All Wheel Drive	N/A	N/A
AM/FM Stereo Cassette	$151	** $175
Steering Wheel, Tilt	Std	Std
Power Windows	$292	$340
***Options Price**	**$486**	**$565**
***Total Price**	**$17,614**	**$20,140**
Target Price	$18,823	
Destination Charge	$555	
Avg. Tax & Fees	$1,176	
Total Target $	**$20,554**	
Average Dealer Option Cost:	86%	

The 1993 Chevrolet Caprice Classic is available in three models – (Base) Caprice Classic and LS sedans, and (Base) Classic wagon. New for 1993, the Base Caprice Classic wagon adds several new interior and exterior colors, redesigned wheel covers, and newly optional rear reading lamps and door edge guards. Among the standard equipment is a stainless steel exhaust system, pull-down center arm rest, child security rear door locks, and dual body-color aero sport mirrors. The 5.7L V-8 engine is optional.

Ownership Costs

Cost Area	5 Year Cost	Rate
Depreciation	$11,017	●
Financing ($413/month)	$3,385	
Insurance (Rating 8)	$7,223	○
State Fees	$828	
Fuel (Hwy 26 City 17)	$4,064	○
Maintenance	$3,815	○
Repairs	$709	○

Warranty/Maintenance Info

Major Tune-Up	$173	○
Minor Tune-Up	$124	○
Brake Service	$222	○
Overall Warranty	3 yr/36k	○
Drivetrain Warranty	3 yr/36k	○
Rust Warranty	6 yr/100k	○
Maintenance Warranty	N/A	
Roadside Assistance	3 yr/36k	

Ownership Cost By Year

Resale Value

1993	1994	1995	1996	1997
$14,761	$13,167	$11,984	$10,802	$9,537

Cumulative Costs

	1993	1994	1995	1996	1997
Annual	$9,546	$5,277	$5,660	$4,721	$5,837
Total	$9,546	$14,823	$20,483	$25,204	$31,041

Ownership Costs (5yr)

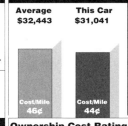

Average: $32,443 — This Car: $31,041
Cost/Mile: 46¢ — Cost/Mile: 44¢

Ownership Cost Rating: ○ Better Than Average

Chevrolet Cavalier VL
2 Door Coupe

Compact

 2.2L 110 hp Gas Fuel Inject.
 4 Cylinder In-Line
Manual 5 Speed
 2 Wheel Front
 Automatic Seatbelts

Purchase Price

Car Item	Dealer Cost	List
Base Price	**$8,051**	**$8,520**
Anti-Lock Brakes	Std	Std
Automatic 3 Speed	$426	$495
Optional Engine	N/A	N/A
Air Conditioning	$641	$745
Power Steering	Std	Std
Cruise Control	Pkg	Pkg
All Wheel Drive	N/A	N/A
AM/FM Stereo Cassette	$406	$472
Steering Wheel, Tilt	$125	$145
Power Windows	$228	$265
***Options Price**	**$1,473**	**$1,712**
***Total Price**	**$9,524**	**$10,232**
Target Price	$10,131	
Destination Charge	$475	
Avg. Tax & Fees	$637	
Total Target $	**$11,243**	
Average Dealer Option Cost:	86%	

The 1993 Chevrolet Cavalier is available in nine models – Z24 coupe, RS and Z24 convertibles, and VL and RS wagons, coupes and sedans. New for 1993, the VL coupe features wind noise improvements, three new exterior colors, and dual visor vanity mirrors as optional equipment. Standard features include a stainless steel exhaust system, self-aligning steering wheel, steel wheels with bolt-on full wheel covers, and cloth reclining front bucket seats with integral head restraints.

Ownership Costs

Cost Area	5 Year Cost	Rate
Depreciation	$6,502	○
Financing ($226/month)	$1,852	
Insurance (Rating 6)	$6,919	●
State Fees	$428	
Fuel (Hwy 32 City 23)	$3,152	○
Maintenance	$4,060	○
Repairs	$651	○

Warranty/Maintenance Info

Major Tune-Up	$128	○
Minor Tune-Up	$84	○
Brake Service	$216	○
Overall Warranty	3 yr/36k	○
Drivetrain Warranty	3 yr/36k	○
Rust Warranty	6 yr/100k	○
Maintenance Warranty	N/A	
Roadside Assistance	3 yr/36k	

Ownership Cost By Year

Resale Value

1993	1994	1995	1996	1997
$8,230	$7,173	$6,386	$5,591	$4,741

Cumulative Costs

	1993	1994	1995	1996	1997
Annual	$5,796	$3,881	$4,374	$3,608	$5,905
Total	$5,796	$9,677	$14,051	$17,659	$23,564

Ownership Costs (5yr)

Average: $23,280 — This Car: $23,564
Cost/Mile: 33¢ — Cost/Mile: 34¢

Ownership Cost Rating: ○ Average

---

page 44

* Includes shaded options
** Other purchase requirements apply

● Poor ◐ Worse Than Average ○ Average ○ Better Than Average ○ Excellent ⊖ Insufficient Information

©1993 by IntelliChoice, Inc. (408) 554-8711 All Rights Reserved. Reproduction Prohibited.
Refer to *Section 3: Annotated Vehicle Charts* for an explanation of these charts.

Chevrolet Cavalier RS
2 Door Coupe

Compact

 2.2L 110 hp Gas Fuel Inject. | 4 Cylinder In-Line | Manual 5 Speed | 2 Wheel Front | Automatic Seatbelts

Purchase Price

Car Item	Dealer Cost	List
Base Price	**$8,901**	**$9,520**
Anti-Lock Brakes	Std	Std
Automatic 3 Speed	$426	$495
3.1L 140 hp Gas	$525	$610
Air Conditioning	$641	$745
Power Steering	Std	Std
Cruise Control	$194	$225
All Wheel Drive	N/A	N/A
AM/FM Stereo Cassette	$120	$140
Steering Wheel, Tilt	$125	$145
Power Windows	$228	$265
***Options Price**	**$1,712**	**$1,990**
***Total Price**	**$10,613**	**$11,510**
Target Price	$11,310	
Destination Charge	$475	
Avg. Tax & Fees	$709	
Total Target $	**$12,494**	
Average Dealer Option Cost:	**86%**	

Ownership Costs

Cost Area	5 Year Cost	Rate
Depreciation	$7,447	●
Financing ($251/month)	$2,057	
Insurance (Rating 8)	$7,223	●
State Fees	$480	
Fuel (Hwy 28 City 20)	$3,612	◐
Maintenance	$4,232	○
Repairs	$651	○

Warranty/Maintenance Info

Major Tune-Up	$157	○
Minor Tune-Up	$109	○
Brake Service	$216	○
Overall Warranty	3 yr/36k	○
Drivetrain Warranty	3 yr/36k	○
Rust Warranty	6 yr/100k	○
Maintenance Warranty	N/A	
Roadside Assistance	3 yr/36k	

Ownership Cost By Year
1993 / 1994 / 1995 / 1996 / 1997

Resale Value
1993	1994	1995	1996	1997
$8,978	$7,782	$6,877	$5,962	$5,047

Cumulative Costs
	1993	1994	1995	1996	1997
Annual	$6,543	$4,274	$4,732	$3,943	$6,210
Total	$6,543	$10,817	$15,549	$19,492	$25,702

Ownership Costs (5yr)
Average	This Car
$24,377	$25,702
Cost/Mile 35¢	Cost/Mile 37¢

Ownership Cost Rating
● Worse Than Average

The 1993 Chevrolet Cavalier is available in nine models - Z24 coupe, RS and Z24 convertibles, and VL and RS wagons, coupes and sedans. New for 1993, the RS coupe features Z24 Sport cloth trim with seat-back storage net, dual visor vanity mirrors, wind noise improvements and a manual outside right hand mirror. The RS upgrades the VL coupe with standard bodyside moldings, and the optional Performance Handling Package. Standard features include self-aligning steering wheel and radial tires.

Chevrolet Cavalier Z24
2 Door Coupe

Compact

 3.1L 140 hp Gas Fuel Inject. | 6 Cylinder "V" | Manual 5 Speed | 2 Wheel Front | Automatic Seatbelts

Purchase Price

Car Item	Dealer Cost	List
Base Price	**$11,313**	**$12,500**
Anti-Lock Brakes	Std	Std
Automatic 3 Speed	$426	** $495
Optional Engine	N/A	N/A
Air Conditioning	$641	$745
Power Steering	Std	Std
Cruise Control	$194	$225
All Wheel Drive	N/A	N/A
AM/FM Stereo Cassette	$120	$140
Steering Wheel, Tilt	$125	$145
Power Windows	$228	$265
***Options Price**	**$761**	**$885**
***Total Price**	**$12,074**	**$13,385**
Target Price	$12,931	
Destination Charge	$475	
Avg. Tax & Fees	$809	
Total Target $	**$14,215**	
Average Dealer Option Cost:	**86%**	

Ownership Costs

Cost Area	5 Year Cost	Rate
Depreciation	$8,399	●
Financing ($286/month)	$2,341	
Insurance (Rating 10)	$7,479	●
State Fees	$555	
Fuel (Hwy 28 City 19)	$3,704	◐
Maintenance	$5,068	●
Repairs	$700	○

Warranty/Maintenance Info

Major Tune-Up	$158	○
Minor Tune-Up	$109	○
Brake Service	$216	○
Overall Warranty	3 yr/36k	○
Drivetrain Warranty	3 yr/36k	○
Rust Warranty	6 yr/100k	○
Maintenance Warranty	N/A	
Roadside Assistance	3 yr/36k	

Resale Value
1993	1994	1995	1996	1997
$9,693	$8,560	$7,664	$6,761	$5,816

Cumulative Costs
	1993	1994	1995	1996	1997
Annual	$7,750	$4,383	$5,274	$4,056	$6,783
Total	$7,750	$12,133	$17,407	$21,463	$28,246

Ownership Costs (5yr)
Average	This Car
$25,987	$28,246
Cost/Mile 37¢	Cost/Mile 40¢

Ownership Cost Rating
● Worse Than Average

The 1993 Chevrolet Cavalier is available in nine models - Z24 coupe, RS and Z24 convertibles, and VL and RS wagons, coupes and sedans. New for 1993, the Z24 coupe features a concentric slave clutch for improved shifting, low-oil level sensor with warning light, and wind noise improvements. The Z24 upgrades the RS with a more powerful engine, cast aluminum wheels and a 4-way manual driver's seat adjuster.

---

* Includes shaded options
** Other purchase requirements apply

● Poor | ◐ Worse Than Average | ◯ Average | ○ Better Than Average | ○ Excellent | ⊖ Insufficient Information

©1993 by IntelliChoice, Inc. (408) 554-8711 All Rights Reserved. Reproduction Prohibited.
Refer to *Section 3: Annotated Vehicle Charts* for an explanation of these charts.

Chevrolet Cavalier VL
4 Door Sedan

 2.2L 110 hp Gas Fuel Inject.
 4 Cylinder In-Line
Manual 5 Speed
 2 Wheel Front
 Automatic Seatbelts

Compact — 2 Door Shown

Purchase Price
Car Item	Dealer Cost	List
Base Price	**$8,146**	**$8,620**
Anti-Lock Brakes	Std	Std
Automatic 3 Speed	$426	$495
Optional Engine	N/A	N/A
Air Conditioning	$641	$745
Power Steering	Std	Std
Cruise Control	Pkg	Pkg
All Wheel Drive	N/A	N/A
AM/FM Stereo Cassette	$406	$472
Steering Wheel, Tilt	$125	$145
Power Windows	$284	$330
*Options Price	$1,473	$1,712
*Total Price	$9,619	$10,332
Target Price	$10,234	
Destination Charge	$475	
Avg. Tax & Fees	$643	
Total Target $	**$11,352**	
Average Dealer Option Cost:	86%	

The 1993 Chevrolet Cavalier is available in nine models - Z24 coupe, RS and Z24 convertibles, and VL and RS wagons, coupes and sedans. New for 1993, the VL sedan features three new exterior colors, wind noise improvements, and dual visor vanity mirrors as optional equipment. Standard features include a self-aligning steering wheel, cloth reclining front bucket seats with integral head restraints, and steel wheels with bolt-on full wheel covers. Bodyside moldings are optional.

Ownership Costs
Cost Area	5 Year Cost	Rate
Depreciation	$6,640	O
Financing ($228/month)	$1,871	
Insurance (Rating 3)	$6,520	O
State Fees	$432	
Fuel (Hwy 32 City 23)	$3,152	O
Maintenance	$4,060	O
Repairs	$651	O

Warranty/Maintenance Info
Major Tune-Up	$128	◐
Minor Tune-Up	$84	O
Brake Service	$216	O
Overall Warranty	3 yr/36k	O
Drivetrain Warranty	3 yr/36k	O
Rust Warranty	6 yr/100k	O
Maintenance Warranty	N/A	
Roadside Assistance	3 yr/36k	

Ownership Cost By Year
1993 / 1994 / 1995 / 1996 / 1997

Resale Value
1993	1994	1995	1996	1997
$8,284	$7,218	$6,391	$5,567	$4,712

Cumulative Costs
	1993	1994	1995	1996	1997
Annual	$5,787	$3,820	$4,338	$3,556	$5,825
Total	$5,787	$9,607	$13,945	$17,501	$23,326

Ownership Costs (5yr)
Average $23,366 | This Car $23,326
Cost/Mile 33¢ | Cost/Mile 33¢

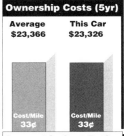
Ownership Cost Rating

Average

Chevrolet Cavalier RS
4 Door Sedan

 2.2L 110 hp Gas Fuel Inject.
 4 Cylinder In-Line
 Manual 5 Speed
 2 Wheel Front
 Automatic Seatbelts

Compact

Purchase Price
Car Item	Dealer Cost	List
Base Price	**$8,995**	**$9,620**
Anti-Lock Brakes	Std	Std
Automatic 3 Speed	$426	$495
3.1L 140 hp Gas	$525	$610
Air Conditioning	$641	$745
Power Steering	Std	Std
Cruise Control	$194	$225
All Wheel Drive	N/A	N/A
AM/FM Stereo Cassette	$120	$140
Steering Wheel, Tilt	$125	$145
Power Windows	$284	$330
*Options Price	$1,712	$1,990
*Total Price	$10,707	$11,610
Target Price	$11,412	
Destination Charge	$475	
Avg. Tax & Fees	$715	
Total Target $	**$12,602**	
Average Dealer Option Cost:	86%	

The 1993 Chevrolet Cavalier is available in nine models - Z24 coupe, RS and Z24 convertibles, and VL and RS wagons, coupes and sedans. New for 1993, the RS sedan features dual visor vanity mirrors, wind noise improvements, Z24 Sport cloth trim with seat-back storage net and a manual outside right hand mirror. The RS upgrades the VL sedan with standard bodyside moldings and the optional Performance Handling Package. Steel wheels with bolt-on full wheel covers are a standard feature.

Ownership Costs
Cost Area	5 Year Cost	Rate
Depreciation	$7,548	●
Financing ($253/month)	$2,077	
Insurance (Rating 5)	$6,786	O
State Fees	$485	
Fuel (Hwy 32 City 23)	$3,152	O
Maintenance	$4,232	O
Repairs	$651	O

Warranty/Maintenance Info
Major Tune-Up	$157	O
Minor Tune-Up	$109	O
Brake Service	$216	O
Overall Warranty	3 yr/36k	O
Drivetrain Warranty	3 yr/36k	O
Rust Warranty	6 yr/100k	O
Maintenance Warranty	N/A	
Roadside Assistance	3 yr/36k	

Resale Value
1993	1994	1995	1996	1997
$9,028	$7,822	$6,909	$5,988	$5,054

Cumulative Costs
	1993	1994	1995	1996	1997
Annual	$6,444	$4,118	$4,566	$3,766	$6,037
Total	$6,444	$10,562	$15,128	$18,894	$24,931

Ownership Costs (5yr)
Average $24,463 | This Car $24,931
Cost/Mile 35¢ | Cost/Mile 36¢

Ownership Cost Rating
O Average

---

page 46 | * Includes shaded options | ** Other purchase requirements apply

● Poor | ◐ Worse Than Average | O Average | O Better Than Average | O Excellent | ⊖ Insufficient Information

©1993 by IntelliChoice, Inc. (408) 554-8711 All Rights Reserved. Reproduction Prohibited.
Refer to *Section 3: Annotated Vehicle Charts* for an explanation of these charts.

Chevrolet Cavalier RS
2 Door Convertible

Compact

 2.2L 110 hp Gas Fuel Inject.
 4 Cylinder In-Line
 Manual 5 Speed
 2 Wheel Front
Automatic Seatbelts

Purchase Price

Car Item	Dealer Cost	List
Base Price	**$14,394**	**$15,395**
Anti-Lock Brakes	Std	Std
Automatic 3 Speed	$426	$495
3.1L 140 hp Gas	$525	$610
Air Conditioning	$641	$745
Power Steering	Std	Std
Cruise Control	$194	$225
All Wheel Drive	N/A	N/A
AM/FM Stereo Cassette	$120	$140
Steering Wheel, Tilt	$125	$145
Power Windows	Std	Std
***Options Price**	**$1,712**	**$1,990**
***Total Price**	**$16,106**	**$17,385**
Target Price	$17,359	
Destination Charge	$475	
Avg. Tax & Fees	$1,071	
Total Target $	**$18,905**	
Average Dealer Option Cost:	*86%*	

The 1993 Chevrolet Cavalier is available in nine models - Z24 coupe, RS and Z24 convertibles, and VL and RS wagons, coupes and sedans. New for 1993, the RS convertible features a glass backlight, Z24 Sport cloth trim with seat-back storage net, dual visor vanity mirrors, and optional rear window defogger. Standard features include bodyside moldings, a power operated top and cast aluminum wheels. Optional items include a sporty white vinyl interior trim, and the Performance Handling Package.

Ownership Costs

Cost Area	5 Year Cost	Rate
Depreciation	$10,880	○
Financing ($380/month)	$3,113	
Insurance (Rating 13)	$8,061	●
State Fees	$715	
Fuel (Hwy 28 City 20)	$3,612	○
Maintenance	$4,232	○
Repairs	$651	○

Warranty/Maintenance Info

Major Tune-Up	$157	○
Minor Tune-Up	$109	○
Brake Service	$216	○
Overall Warranty	3 yr/36k	○
Drivetrain Warranty	3 yr/36k	○
Rust Warranty	6 yr/100k	○
Maintenance Warranty	N/A	
Roadside Assistance	3 yr/36k	

Ownership Cost By Year

1993 $9,122 / 1994 $5,109 / 1995 $5,621 / 1996 $4,652 / 1997 $6,760

Resale Value

1993	1994	1995	1996	1997
$13,463	$11,976	$10,606	$9,274	$8,025

Cumulative Costs

	1993	1994	1995	1996	1997
Annual	$9,122	$5,109	$5,621	$4,652	$6,760
Total	$9,122	$14,231	$19,852	$24,504	$31,264

Ownership Costs (5yr)

Average	This Car
$29,422	$31,264
Cost/Mile 42¢	Cost/Mile 45¢

Ownership Cost Rating
Worse Than Average

Chevrolet Cavalier Z24
2 Door Convertible

Compact

 3.1L 140 hp Gas Fuel Inject.
 6 Cylinder "V"
 Manual 5 Speed
 2 Wheel Front
 Automatic Seatbelts

RS Model Shown

Purchase Price

Car Item	Dealer Cost	List
Base Price	**$16,566**	**$18,305**
Anti-Lock Brakes	Std	Std
Automatic 3 Speed	$426	** $495
Optional Engine	N/A	N/A
Air Conditioning	$641	$745
Power Steering	Std	Std
Cruise Control	$194	$225
All Wheel Drive	N/A	N/A
AM/FM Stereo Cassette	$120	$140
Steering Wheel, Tilt	$125	$145
Power Windows	Std	Std
***Options Price**	**$761**	**$885**
***Total Price**	**$17,327**	**$19,190**
Target Price	$18,759	
Destination Charge	$475	
Avg. Tax & Fees	$1,159	
Total Target $	**$20,393**	
Average Dealer Option Cost:	*86%*	

The 1993 Chevrolet Cavalier is available in nine models - Z24 coupe, RS and Z24 convertibles, and VL and RS wagons, coupes and sedans. New for 1993, the Z24 convertible features dual visor vanity mirrors, a glass backlight, 4-way manual adjustable driver's seat, and an optional rear window defogger. Standard features include bodyside moldings, tinted glass, radial touring tires, and 15 inch cast aluminum wheels. Optional features include a deck-lid spoiler and sporty white vinyl interior trim.

Ownership Costs

Cost Area	5 Year Cost	Rate
Depreciation	$11,495	○
Financing ($410/month)	$3,359	
Insurance (Rating 15)	$8,507	●
State Fees	$787	
Fuel (Hwy 28 City 19)	$3,704	●
Maintenance	$5,068	●
Repairs	$700	○

Warranty/Maintenance Info

Major Tune-Up	$158	○
Minor Tune-Up	$109	○
Brake Service	$216	○
Overall Warranty	3 yr/36k	○
Drivetrain Warranty	3 yr/36k	○
Rust Warranty	6 yr/100k	○
Maintenance Warranty	N/A	
Roadside Assistance	3 yr/36k	

Resale Value

1993	1994	1995	1996	1997
$14,808	$13,200	$11,711	$10,274	$8,898

Cumulative Costs

	1993	1994	1995	1996	1997
Annual	$9,485	$5,425	$6,320	$4,919	$7,471
Total	$9,485	$14,910	$21,230	$26,149	$33,620

Ownership Costs (5yr)

Average	This Car
$30,972	$33,620
Cost/Mile 44¢	Cost/Mile 48¢

Ownership Cost Rating
Worse Than Average

---

* Includes shaded options
** Other purchase requirements apply

● Poor ◐ Worse Than Average ○ Average ◯ Better Than Average ◯ Excellent ⊖ Insufficient Information

©1993 by IntelliChoice, Inc. (408) 554-8711   All Rights Reserved. Reproduction Prohibited.
Refer to *Section 3: Annotated Vehicle Charts* for an explanation of these charts.

Chevrolet Cavalier VL
4 Door Wagon

 2.2L 110 hp Gas Fuel Inject.
 4 Cylinder In-Line
 Automatic 3 Speed
 2 Wheel Front
Automatic Seatbelts

Compact Wagon — RS Model Shown

Purchase Price

Car Item	Dealer Cost	List
Base Price	$9,200	$9,735
Anti-Lock Brakes	Std	Std
Manual Transmission	N/A	N/A
Optional Engine	N/A	N/A
Air Conditioning	$641	$745
Power Steering	Std	Std
Cruise Control	Pkg	Pkg
All Wheel Drive	N/A	N/A
AM/FM Stereo Cassette	$406	$472
Steering Wheel, Tilt	$125	$145
Power Windows	N/A	N/A
*Options Price	$1,047	$1,217
*Total Price	$10,247	$10,952
Target Price	$10,926	
Destination Charge	$475	
Avg. Tax & Fees	$684	
Total Target $	**$12,085**	
Average Dealer Option Cost:	86%	

The 1993 Chevrolet Cavalier is available in nine models - Z24 coupe, RS and Z24 convertibles, and VL and RS wagons, coupes and sedans. New for 1993, the VL wagon features wind noise improvements, three new exterior colors, and dual visor vanity mirrors as an option. Standard features include a self-aligning steering wheel, 14 inch steel wheels with bolt-on full wheel covers, bucket seats and a stainless steel exhaust system. Optional features include a roof rack and bodyside moldings.

Ownership Costs

Cost Area	5 Year Cost	Rate
Depreciation	$7,295	●
Financing ($243/month)	$1,991	
Insurance (Rating 4)	$6,658	○
State Fees	$457	
Fuel (Hwy 32 City 23)	$3,152	○
Maintenance	$4,060	○
Repairs	$651	○

Warranty/Maintenance Info

Major Tune-Up	$128	○
Minor Tune-Up	$84	○
Brake Service	$216	○
Overall Warranty	3 yr/36k	○
Drivetrain Warranty	3 yr/36k	○
Rust Warranty	6 yr/100k	○
Maintenance Warranty	N/A	
Roadside Assistance	3 yr/36k	

Ownership Cost By Year

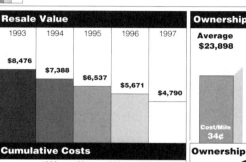

Legend: 1993, 1994, 1995, 1996, 1997

Resale Value

1993	1994	1995	1996	1997
$8,476	$7,388	$6,537	$5,671	$4,790

Cumulative Costs

	1993	1994	1995	1996	1997
Annual	$6,409	$3,911	$4,419	$3,640	$5,885
Total	$6,409	$10,320	$14,739	$18,379	$24,264

Ownership Costs (5yr)

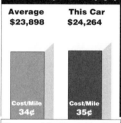

Average: $23,898 — Cost/Mile 34¢
This Car: $24,264 — Cost/Mile 35¢

Ownership Cost Rating
○ Average

Chevrolet Cavalier RS
4 Door Wagon

 2.2L 110 hp Gas Fuel Inject.
 4 Cylinder In-Line
 Automatic 3 Speed
 2 Wheel Front
Automatic Seatbelts

Compact Wagon

Purchase Price

Car Item	Dealer Cost	List
Base Price	$10,084	$10,785
Anti-Lock Brakes	Std	Std
Manual Transmission	N/A	N/A
3.1L 140 hp Gas	$525	$610
Air Conditioning	$641	$745
Power Steering	Std	Std
Cruise Control	$194	$225
All Wheel Drive	N/A	N/A
AM/FM Stereo Cassette	$120	$140
Steering Wheel, Tilt	$125	$145
Power Windows	$284	$330
*Options Price	$1,286	$1,495
*Total Price	$11,370	$12,280
Target Price	$12,146	
Destination Charge	$475	
Avg. Tax & Fees	$759	
Total Target $	**$13,380**	
Average Dealer Option Cost:	86%	

The 1993 Chevrolet Cavalier is available in nine models - Z24 coupe, RS and Z24 convertibles, and VL and RS wagons, coupes and sedans. New for 1993, the RS wagon features a standard split-folding rear seat, Z24 Sport cloth trim with seat-back storage pockets, dual visor vanity mirrors and a manual right hand outside rearview mirror. The RS upgrades the VL wagon with standard bodyside moldings, tinted glass, and the optional Performance Handling Package with includes a faster steering ratio.

Ownership Costs

Cost Area	5 Year Cost	Rate
Depreciation	$8,199	●
Financing ($269/month)	$2,204	
Insurance (Rating 5)	$6,786	○
State Fees	$511	
Fuel (Hwy 28 City 20)	$3,612	○
Maintenance	$4,232	○
Repairs	$651	○

Warranty/Maintenance Info

Major Tune-Up	$157	○
Minor Tune-Up	$109	○
Brake Service	$216	○
Overall Warranty	3 yr/36k	○
Drivetrain Warranty	3 yr/36k	○
Rust Warranty	6 yr/100k	○
Maintenance Warranty	N/A	
Roadside Assistance	3 yr/36k	

Resale Value

1993	1994	1995	1996	1997
$9,354	$8,099	$7,123	$6,156	$5,181

Cumulative Costs

	1993	1994	1995	1996	1997
Annual	$7,041	$4,302	$4,751	$3,920	$6,181
Total	$7,041	$11,343	$16,094	$20,014	$26,195

Ownership Costs (5yr)

Average: $25,038 — Cost/Mile 36¢
This Car: $26,195 — Cost/Mile 37¢

Ownership Cost Rating
◉ Worse Than Average

---

* Includes shaded options
** Other purchase requirements apply

● Poor ◉ Worse Than Average ○ Average ◯ Better Than Average ◯ Excellent ⊖ Insufficient Information

©1993 by IntelliChoice, Inc. (408) 554-8711 All Rights Reserved. Reproduction Prohibited.

Refer to *Section 3: Annotated Vehicle Charts* for an explanation of these charts.

Chevrolet Corsica LT
4 Door Sedan

2.2L 110 hp Gas Fuel Inject. | 4 Cylinder In-Line | Manual 5 Speed | 2 Wheel Front | Driver Airbag Psngr Belts

Compact

Purchase Price

Car Item	Dealer Cost	List
Base Price	**$10,312**	**$11,395**
Anti-Lock Brakes	Std	Std
Automatic 3 Speed	$477	$555
3.1L 140 hp Gas	$516	$600
Air Conditioning	$692	$805
Power Steering	Std	Std
Cruise Control	$194	** $225
All Wheel Drive	N/A	N/A
AM/FM Stereo Cassette	$120	$140
Steering Wheel, Tilt	$125	$145
Power Windows	$292	$340
***Options Price**	**$1,805**	**$2,100**
***Total Price**	**$12,117**	**$13,495**
Target Price	$13,140	
Destination Charge	$475	
Avg. Tax & Fees	$821	
Total Target $	**$14,436**	
Average Dealer Option Cost:	86%	

The 1993 Chevrolet Corsica is available as one model edition. New for 1993, the Corsica LT sedan features improved ride quality, noise reduction and enhanced exhaust noise isolation and tuning. Standard features include contoured front reclining bucket seats with head restraints and driver tilt mechanism, rear seat ducts, side window defoggers, tinted glass, and a stainless-steel exhaust system. Optional features include Comfort Convenience and Sport Handling Packages, and a full center console with armrest.

Ownership Costs

Cost Area	5 Year Cost	Rate
Depreciation	$8,494	●
Financing ($290/month)	$2,378	
Insurance (Rating 7)	$7,035	○
State Fees	$560	
Fuel (Hwy 28 City 20)	$3,612	○
Maintenance	$4,438	○
Repairs	$651	○

Warranty/Maintenance Info

Major Tune-Up	$147	○
Minor Tune-Up	$106	○
Brake Service	$216	○
Overall Warranty	3 yr/36k	○
Drivetrain Warranty	3 yr/36k	○
Rust Warranty	6 yr/100k	○
Maintenance Warranty	N/A	
Roadside Assistance	3 yr/36k	

Ownership Cost By Year

1993 / 1994 / 1995 / 1996 / 1997

Resale Value

1993	1994	1995	1996	1997
$9,935	$8,733	$7,691	$6,692	$5,942

Cumulative Costs

	1993	1994	1995	1996	1997
Annual	$7,647	$4,364	$4,883	$4,006	$6,268
Total	$7,647	$12,011	$16,894	$20,900	$27,168

Ownership Costs (5yr)

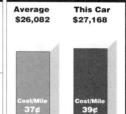

Average $26,082 / This Car $27,168
Cost/Mile 37¢ / Cost/Mile 39¢

Ownership Cost Rating
Worse Than Average

Chevrolet Corvette
2 Door Coupe

5.7L 300 hp Gas Fuel Inject. | 8 Cylinder "V" | Automatic 4 Speed | 2 Wheel Rear | Driver Airbag Psngr Belts

Sport

Purchase Price

Car Item	Dealer Cost	List
Base Price	**$29,579**	**$34,595**
Anti-Lock Brakes	Std	Std
Manual 6 Speed	N/C	N/C
Optional Engine	N/A	N/A
Air Conditioning	Std	Std
Power Steering	Std	Std
Cruise Control	Std	Std
All Wheel Drive	N/A	N/A
AM/FM Stereo Cassette	Std	Std
Steering Wheel, Scope	Std	Std
Power Windows	Std	Std
***Options Price**	**$0**	**$0**
***Total Price**	**$29,579**	**$34,595**
Target Price	$32,278	
Destination Charge	$550	
Avg. Tax & Fees	$1,992	
Luxury Tax	$283	
Total Target $	**$35,103**	

The 1993 Chevrolet Corvette is available in three models - (Base) Corvette and ZR-1 coupes, and (Base) Corvette convertible. New for 1993, the Corvette coupe features an optional 40th Anniversary Package, passive keyless entry system which engages a security system, and new machined wheels. Standard features include reclining cloth bucket seats with lateral support and back angle adjustment, power-operated retractable headlamps, and dual electrically adjustable and heated outside rearview mirrors.

Ownership Costs

Cost Area	5 Year Cost	Rate
Depreciation	$16,115	○
Financing ($706/month)	$5,782	
Insurance (Rating 20 Sport+)	$13,703	●
State Fees	$1,405	
Fuel (Hwy 25 City 17 -Prem.)	$4,578	○
Maintenance	$7,013	●
Repairs	$925	○

Warranty/Maintenance Info

Major Tune-Up	$213	○
Minor Tune-Up	$138	○
Brake Service	$308	○
Overall Warranty	3 yr/36k	○
Drivetrain Warranty	3 yr/36k	○
Rust Warranty	6 yr/100k	○
Maintenance Warranty	N/A	
Roadside Assistance	3 yr/36k	

Ownership Cost By Year

1993 / 1994 / 1995 / 1996 / 1997

Resale Value

1993	1994	1995	1996	1997
$27,078	$24,774	$22,844	$20,938	$18,988

Cumulative Costs

	1993	1994	1995	1996	1997
Annual	$14,213	$8,221	$9,191	$7,046	$10,851
Total	$14,213	$22,434	$31,625	$38,671	$49,522

Ownership Costs (5yr)

Average $51,442 / This Car $49,522
Cost/Mile 73¢ / Cost/Mile 71¢

Ownership Cost Rating
Average

---

* Includes shaded options
** Other purchase requirements apply

● Poor ◐ Worse Than Average ○ Average ○ Better Than Average ○ Excellent ⊖ Insufficient Information

©1993 by IntelliChoice, Inc. (408) 554-8711 All Rights Reserved. Reproduction Prohibited.
Refer to *Section 3: Annotated Vehicle Charts* for an explanation of these charts.

Chevrolet Corvette ZR-1
2 Door Coupe

Sport

 5.7L 405 hp Gas Fuel Inject.
 8 Cylinder "V"
 Manual 6 Speed
 2 Wheel Rear
Driver Airbag Psngr Belts

Purchase Price

Car Item	Dealer Cost	List
Base Price	**$56,192**	**$66,278**
Anti-Lock Brakes	Std	Std
Automatic Transmission	N/A	N/A
Optional Engine	N/A	N/A
Auto Climate Control	Std	Std
Power Steering	Std	Std
Cruise Control	Std	Std
All Wheel Drive	N/A	N/A
AM/FM Stereo CD	Std	Std
Steering Wheel, Scope	Std	Std
Power Windows	Std	Std
***Options Price**	$0	$0
***Total Price**	**$56,192**	**$66,278**
Target Price	$61,249	
Destination Charge	$550	
Avg. Tax & Fees	$3,758	
Luxury Tax	$3,180	
Total Target $	**$68,737**	

Ownership Costs

Cost Area	5 Year Cost	Rate
Depreciation		⊖
Financing ($1,382/month)	$11,322	
Insurance (Rating 25 Sport+)	$17,284	⦿
State Fees	$2,673	
Fuel (Hwy 25 City 17 -Prem.)	$4,578	○
Maintenance	$9,521	●
Repairs	$1,076	⦿

Warranty/Maintenance Info

Major Tune-Up	$218	○
Minor Tune-Up	$163	⦿
Brake Service	$308	○
Overall Warranty	3 yr/36k	○
Drivetrain Warranty	3 yr/36k	○
Rust Warranty	6 yr/100k	○
Maintenance Warranty	N/A	
Roadside Assistance	3 yr/36k	

Ownership Cost By Year

Insufficient Depreciation Information

- 1993
- 1994
- 1995
- 1996
- 1997

Resale Value

Insufficient Information

Ownership Costs (5yr)

Insufficient Information

Cumulative Costs

	1993	1994	1995	1996	1997
Annual	Insufficient Information				
Total	Insufficient Information				

Ownership Cost Rating

⊖ Insufficient Information

The 1993 Chevrolet Corvette is available in three models - (Base) Corvette and ZR-1 coupes, and (Base) Corvette convertible. New for 1993, the ZR-1 coupe features a more powerful engine, the optional 40th Anniversary Package, a passive keyless entry system which engages a security system, and larger rear tires. Standard features include a wider body to accommodate larger Goodyear tires, sport leather seating with six-way power adjustment and power lumbar support and "ZR-1" emblems.

Chevrolet Corvette
2 Door Convertible

Sport

 5.7L 300 hp Gas Fuel Inject.
 8 Cylinder "V"
 Automatic 4 Speed
 2 Wheel Rear
 Driver Airbag Psngr Belts

Purchase Price

Car Item	Dealer Cost	List
Base Price	**$35,222**	**$41,195**
Anti-Lock Brakes	Std	Std
Manual 6 Speed	N/C	N/C
Optional Engine	N/A	N/A
Air Conditioning	Std	Std
Power Steering	Std	Std
Cruise Control	Std	Std
All Wheel Drive	N/A	N/A
AM/FM Stereo Cassette	Std	Std
Steering Wheel, Scope	Std	Std
Power Windows	Std	Std
***Options Price**	$0	$0
***Total Price**	**$35,222**	**$41,195**
Target Price	$38,392	
Destination Charge	$550	
Avg. Tax & Fees	$2,364	
Luxury Tax	$894	
Total Target $	**$42,200**	

Ownership Costs

Cost Area	5 Year Cost	Rate
Depreciation	$19,055	○
Financing ($848/month)	$6,952	
Insurance (Rating 22 Sport+)	$14,896	⦿
State Fees	$1,669	
Fuel (Hwy 25 City 17 -Prem.)	$4,578	○
Maintenance	$7,013	⦿
Repairs	$925	○

Warranty/Maintenance Info

Major Tune-Up	$213	○
Minor Tune-Up	$138	○
Brake Service	$308	○
Overall Warranty	3 yr/36k	○
Drivetrain Warranty	3 yr/36k	○
Rust Warranty	6 yr/100k	○
Maintenance Warranty	N/A	
Roadside Assistance	3 yr/36k	

Resale Value

1993	1994	1995	1996	1997
$32,277	$29,685	$27,509	$25,336	$23,145

Ownership Costs (5yr)

Average	This Car
$57,774	$55,088
Cost/Mile 83¢	Cost/Mile 79¢

Cumulative Costs

	1993	1994	1995	1996	1997
Annual	$16,884	$9,164	$9,960	$7,692	$11,388
Total	$16,884	$26,048	$36,008	$43,700	$55,088

Ownership Cost Rating

○ Better Than Average

The 1993 Chevrolet Corvette is available in three models - (Base) Corvette and ZR-1 coupes, and (Base) Corvette convertible. New for 1993, the Corvette convertible features an optional 40th Anniversary Package, passive keyless entry system which engages a security system, and new machined wheels. Standard features include power-operated retractable headlamps, and dual electrically adjustable and heated outside rearview mirrors. A removable hardtop with rear window defogger is an optional feature.

---

* Includes shaded options
** Other purchase requirements apply

● Poor ⦿ Worse Than Average ○ Average ○ Better Than Average ○ Excellent ⊖ Insufficient Information

©1993 by IntelliChoice, Inc. (408) 554-8711 All Rights Reserved. Reproduction Prohibited.
Refer to *Section 3: Annotated Vehicle Charts* for an explanation of these charts.

Chevrolet Lumina
2 Door Coupe

 3.1L 140 hp Gas Fuel Inject.
 6 Cylinder "V"
 Automatic 3 Speed
 2 Wheel Front
 Automatic Seatbelts

Midsize — Z34 Model Shown

Purchase Price

Car Item	Dealer Cost	List
Base Price	**$12,854**	**$14,690**
Anti-Lock Brakes	$387	$450
4 Spd Elec.	$172	$200
Optional Engine	N/A	N/A
Air Conditioning	Std	Std
Power Steering	Std	Std
Cruise Control	$194	$225
All Wheel Drive	N/A	N/A
AM/FM Stereo Cassette	$120	$140
Steering Wheel, Tilt	$125	$145
Power Windows	$228	$265
*Options Price	$542	$630
*Total Price	$13,396	$15,320
Target Price		$14,302
Destination Charge		$505
Avg. Tax & Fees		$898
Total Target $		**$15,705**
Average Dealer Option Cost:		*86%*

The 1993 Chevrolet Lumina is available in five models - (Base) Lumina and Euro sedans, and (Base) Lumina, Euro and Z34 coupes. New for 1993, the (Base) Lumina coupe features a more powerful engine with improved horsepower, provisions for cellular phone installation, and a variety of new exterior colors. Standard features include reclining seats, tinted glass, energy-absorbing steering column, and tires with built-in tread wear indicators. The Special Service Package is an optional feature.

Ownership Costs

Cost Area	5 Year Cost	Rate
Depreciation	$7,977	○
Financing ($316/month)	$2,588	
Insurance (Rating 7)	$7,035	○
State Fees	$633	
Fuel (Hwy 27 City 19)	$3,775	○
Maintenance	$5,113	○
Repairs	$659	○

Warranty/Maintenance Info

Major Tune-Up	$179	○
Minor Tune-Up	$131	○
Brake Service	$277	●
Overall Warranty	3 yr/36k	○
Drivetrain Warranty	3 yr/36k	○
Rust Warranty	6 yr/100k	○
Maintenance Warranty	N/A	
Roadside Assistance	3 yr/36k	

Ownership Cost By Year
1993, 1994, 1995, 1996, 1997

Resale Value
1993	1994	1995	1996	1997
$11,566	$10,573	$9,828	$8,758	$7,728

Cumulative Costs
	1993	1994	1995	1996	1997
Annual	$7,420	$4,289	$4,843	$4,192	$7,036
Total	$7,420	$11,709	$16,552	$20,744	$27,780

Ownership Costs (5yr)
Average	This Car
$26,933	$27,780
Cost/Mile 38¢	Cost/Mile 40¢

Ownership Cost Rating
○ Average

Chevrolet Lumina Euro
2 Door Coupe

 3.1L 140 hp Gas Fuel Inject.
 6 Cylinder "V"
 Automatic 3 Speed
 2 Wheel Front
 Automatic Seatbelts

Midsize

Purchase Price

Car Item	Dealer Cost	List
Base Price	**$13,650**	**$15,600**
Anti-Lock Brakes	Std	Std
4 Spd Elec.	$172	$200
Optional Engine	N/A	N/A
Air Conditioning	Std	Std
Power Steering	Std	Std
Cruise Control	$194	$225
All Wheel Drive	N/A	N/A
AM/FM Stereo Cassette	$120	$140
Steering Wheel, Tilt	$125	$145
Power Windows	$228	$265
*Options Price	$542	$630
*Total Price	$14,192	$16,230
Target Price		$15,175
Destination Charge		$505
Avg. Tax & Fees		$951
Total Target $		**$16,631**
Average Dealer Option Cost:		*86%*

The 1993 Chevrolet Lumina is available in five models - (Base) Lumina and Euro sedans, and (Base) Lumina, Euro and Z34 coupes. New for 1993, the Euro coupe features provisions for cellular car phone installation, a more powerful engine with improved horsepower, and a variety of new exterior colors. The Euro upgrades the (Base) Lumina with a stronger engine, and Euro trim. Standard features include cloth upholstery, split-back front bench seat, tinted glass and styled aluminum wheels.

Ownership Costs

Cost Area	5 Year Cost	Rate
Depreciation	$7,876	○
Financing ($334/month)	$2,739	
Insurance (Rating 7)	$7,035	○
State Fees	$669	
Fuel (Hwy 27 City 19)	$3,775	○
Maintenance	$5,253	●
Repairs	$659	○

Warranty/Maintenance Info

Major Tune-Up	$179	○
Minor Tune-Up	$131	○
Brake Service	$277	●
Overall Warranty	3 yr/36k	○
Drivetrain Warranty	3 yr/36k	○
Rust Warranty	6 yr/100k	○
Maintenance Warranty	N/A	
Roadside Assistance	3 yr/36k	

Resale Value
1993	1994	1995	1996	1997
$12,502	$11,712	$11,074	$9,884	$8,755

Cumulative Costs
	1993	1994	1995	1996	1997
Annual	$7,482	$4,142	$4,835	$4,329	$7,218
Total	$7,482	$11,624	$16,459	$20,788	$28,006

Ownership Costs (5yr)
Average	This Car
$27,864	$28,006
Cost/Mile 40¢	Cost/Mile 40¢

Ownership Cost Rating
○ Average

Legend: ● Poor · ◐ Worse Than Average · ○ Average · ○ Better Than Average · ○ Excellent · ⊖ Insufficient Information

* Includes shaded options
** Other purchase requirements apply

©1993 by IntelliChoice, Inc. (408) 554-8711 All Rights Reserved. Reproduction Prohibited.

Refer to *Section 3: Annotated Vehicle Charts* for an explanation of these charts.

Chevrolet Lumina Z34
2 Door Coupe

Midsize

 3.4L 210 hp Gas Fuel Inject.
 6 Cylinder "V"
 Manual 5 Speed
 2 Wheel Front
Automatic Seatbelts

Purchase Price

Car Item	Dealer Cost	List
Base Price	**$16,100**	**$18,400**
Anti-Lock Brakes	Std	Std
Automatic 4 Speed	$172	$200
Optional Engine	N/A	N/A
Air Conditioning	Std	Std
Power Steering	Std	Std
Cruise Control	Std	Std
All Wheel Drive	N/A	N/A
AM/FM Stereo Cassette	Std	Std
Steering Wheel, Tilt	Std	Std
Power Windows	$228	$265
***Options Price**	**$400**	**$465**
***Total Price**	**$16,500**	**$18,865**
Target Price	$17,724	
Destination Charge	$505	
Avg. Tax & Fees	$1,105	
Total Target $	**$19,334**	
Average Dealer Option Cost:	*86%*	

The 1993 Chevrolet Lumina is available in five models - (Base) Lumina and Euro sedans, and (Base) Lumina, Euro and Z34 coupes. New for 1993, the Z34 coupe features a more powerful engine with improved horsepower, provisions for cellular phone installation and a variety of new exterior colors. The Z34 upgrades the Euro with tilt wheel, dual mirrors, and a full gauge package. Standard features include a rear spoiler, blackwall radial tires, and tinted glass. The Special Service Package is an option.

Ownership Costs

Cost Area	5 Year Cost	Rate
Depreciation	$10,443	O
Financing ($389/month)	$3,184	
Insurance (Rating 10+)	$8,975	O
State Fees	$775	
Fuel (Hwy 26 City 17)	$4,064	⊙
Maintenance	$5,382	⊙
Repairs	$659	O

Warranty/Maintenance Info

Major Tune-Up	$161	O
Minor Tune-Up	$111	O
Brake Service	$277	●
Overall Warranty	3 yr/36k	O
Drivetrain Warranty	3 yr/36k	O
Rust Warranty	6 yr/100k	O
Maintenance Warranty	N/A	
Roadside Assistance	3 yr/36k	

Ownership Cost By Year

1993, 1994, 1995, 1996, 1997

Resale Value

1993	1994	1995	1996	1997
$13,891	$12,516	$11,505	$10,148	$8,891

Cumulative Costs

	1993	1994	1995	1996	1997
Annual	$9,420	$5,295	$5,819	$4,988	$7,960
Total	$9,420	$14,715	$20,534	$25,522	$33,482

Ownership Costs (5yr)

Average	This Car
$30,558	$33,482
Cost/Mile 44¢	Cost/Mile 48¢

Ownership Cost Rating

● Poor

Chevrolet Lumina
4 Door Sedan

Midsize

 2.2L 110 hp Gas Fuel Inject.
 4 Cylinder "V"
 Automatic 3 Speed
 2 Wheel Front
 Automatic Seatbelts

Purchase Price

Car Item	Dealer Cost	List
Base Price	**$11,725**	**$13,400**
Anti-Lock Brakes	$387	$450
4 Spd Elec.	$172	$200
3.1L 140 hp Gas	$525	$610
Air Conditioning	$714	$830
Power Steering	Std	Std
Cruise Control	$194	$225
All Wheel Drive	N/A	N/A
AM/FM Stereo Cassette	$120	$140
Steering Wheel, Tilt	$125	$145
Power Windows	$284	$330
***Options Price**	**$1,837**	**$2,135**
***Total Price**	**$13,562**	**$15,535**
Target Price	$14,449	
Destination Charge	$505	
Avg. Tax & Fees	$908	
Total Target $	**$15,862**	
Average Dealer Option Cost:	*86%*	

The 1993 Chevrolet Lumina is available in five models - (Base) Lumina and Euro sedans, and (Base) Lumina, Euro and Z34 coupes. New for 1993, the (Base) Lumina sedan features provisions for cellular car phone installation, a variety of new exterior colors, and a more powerful engine with improved horsepower. Standard features include fully independent front and rear suspension, reclining seats, tinted glass, and auxiliary lighting. The Special Service Package is an optional feature.

Ownership Costs

Cost Area	5 Year Cost	Rate
Depreciation	$8,507	O
Financing ($319/month)	$2,613	
Insurance (Rating 4)	$6,658	○
State Fees	$640	
Fuel (Hwy 27 City 19)	$3,775	O
Maintenance	$5,133	O
Repairs	$659	O

Warranty/Maintenance Info

Major Tune-Up	$179	O
Minor Tune-Up	$131	O
Brake Service	$277	●
Overall Warranty	3 yr/36k	O
Drivetrain Warranty	3 yr/36k	O
Rust Warranty	6 yr/100k	O
Maintenance Warranty	N/A	
Roadside Assistance	3 yr/36k	

Resale Value

1993	1994	1995	1996	1997
$11,512	$10,218	$9,463	$8,381	$7,355

Cumulative Costs

	1993	1994	1995	1996	1997
Annual	$7,573	$4,527	$4,794	$4,129	$6,962
Total	$7,573	$12,100	$16,894	$21,023	$27,985

Ownership Costs (5yr)

Average	This Car
$27,153	$27,985
Cost/Mile 39¢	Cost/Mile 40¢

Ownership Cost Rating

O Average

---

* Includes shaded options
** Other purchase requirements apply

● Poor | ⊙ Worse Than Average | O Average | ○ Better Than Average | ◯ Excellent | ⊖ Insufficient Information

©1993 by IntelliChoice, Inc. (408) 554-8711 All Rights Reserved. Reproduction Prohibited.
Refer to *Section 3: Annotated Vehicle Charts* for an explanation of these charts.

Chevrolet Lumina Euro
4 Door Sedan

Midsize

 3.1L 140 hp Gas Fuel Inject.
 6 Cylinder "V"
 Automatic 3 Speed
 2 Wheel Front
 Automatic Seatbelts

Purchase Price

Car Item	Dealer Cost	List
Base Price	**$13,825**	**$15,800**
Anti-Lock Brakes	Std	Std
4 Spd Elec.	$172	$200
3.4L 200 hp Gas	Pkg	Pkg
Air Conditioning	Std	Std
Power Steering	Std	Std
Cruise Control	$194	$225
All Wheel Drive	N/A	N/A
AM/FM Stereo Cassette	$120	$140
Steering Wheel, Tilt	$125	$145
Power Windows	$284	$330
***Options Price**	**$598**	**$695**
***Total Price**	**$14,423**	**$16,495**
Target Price	$15,427	
Destination Charge	$505	
Avg. Tax & Fees	$967	
Total Target $	**$16,899**	
Average Dealer Option Cost:	*86%*	

Ownership Costs

Cost Area	5 Year Cost	Rate
Depreciation	$8,444	O
Financing ($340/month)	$2,783	
Insurance (Rating 5)	$6,786	O
State Fees	$680	
Fuel (Hwy 27 City 19)	$3,775	O
Maintenance	$5,253	●
Repairs	$659	O

Warranty/Maintenance Info

Major Tune-Up	$179	O
Minor Tune-Up	$131	O
Brake Service	$277	●
Overall Warranty	3 yr/36k	O
Drivetrain Warranty	3 yr/36k	O
Rust Warranty	6 yr/100k	O
Maintenance Warranty	N/A	
Roadside Assistance	3 yr/36k	

Ownership Cost By Year

1993, 1994, 1995, 1996, 1997

Resale Value

1993	1994	1995	1996	1997
$12,493	$11,386	$10,789	$9,599	$8,455

Cumulative Costs

	1993	1994	1995	1996	1997
Annual	$7,734	$4,427	$4,755	$4,282	$7,182
Total	$7,734	$12,161	$16,916	$21,198	$28,380

Ownership Costs (5yr)

Average	This Car
$28,135	$28,380
Cost/Mile 40¢	Cost/Mile 41¢

Ownership Cost Rating

O Average

The 1993 Chevrolet Lumina is available in five models - (Base) Lumina and Euro sedans, and (Base) Lumina, Euro and Z34 coupes. New for 1993, the Lumina Euro sedan features a variety of new exterior colors, a more powerful engine with improved horsepower, and provisions for cellular phone installation. The Euro upgrades the (Base) Lumina with Euro trim, and a stronger engine. Standard features include cast-aluminum wheels, rear spoiler, and tinted glass. The Special Service Package is an option.

Chrysler Concorde
4 Door Sedan

Large

 3.3L 153 hp Gas Fuel Inject.
 6 Cylinder "V"
 Automatic 4 Speed
 2 Wheel Front
 Driver/Psngr Airbags Std

Purchase Price

Car Item	Dealer Cost	List
Base Price	**$16,080**	**$18,341**
Anti-Lock Brakes	Std	Std
Manual Transmission	N/A	N/A
3.5L 214 hp Gas	Pkg	Pkg
Air Conditioning	Std	Std
Power Steering	Std	Std
Cruise Control	$190	$224
All Wheel Drive	N/A	N/A
AM/FM Stereo Cassette	$166	$195
Steering Wheel, Tilt	Std	Std
Power Windows	Pkg	Pkg
***Options Price**	**$356**	**$419**
***Total Price**	**$16,436**	**$18,760**
Target Price	$18,081	
Destination Charge	$490	
Avg. Tax & Fees	$1,122	
Total Target $	**$19,693**	
Average Dealer Option Cost:	*85%*	

Ownership Costs

Cost Area	5 Year Cost	Rate
Depreciation		⊖
Financing ($396/month)	$3,244	
Insurance (Rating 12)	$7,898	O
State Fees	$771	
Fuel (Hwy 26 City 18)	$3,949	O
Maintenance		⊖
Repairs	$740	O

Warranty/Maintenance Info

Major Tune-Up		⊖
Minor Tune-Up		⊖
Brake Service		⊖
Overall Warranty	1 yr/12k	●
Drivetrain Warranty	7 yr/70k	O
Rust Warranty	7 yr/100k	O
Maintenance Warranty	N/A	
Roadside Assistance	N/A	

Ownership Cost By Year

Insufficient Depreciation Information

Resale Value

Insufficient Information

Cumulative Costs

	1993	1994	1995	1996	1997
Annual	*Insufficient Information*				
Total	*Insufficient Information*				

Ownership Costs (5yr)

Insufficient Information

Ownership Cost Rating

⊖ Insufficient Information

The all-new Chrysler Concorde sedan makes its debut in 1993 with a one model line up. The Concorde is the Chrysler division's version of the Chrysler Corporation's new LH model whose production incorporated new factory habits such as the melding of diverse departments with fewer job classifications and lean production practices. The Concorde comes equipped with dual airbags, stainless steel exhaust system, floor console with rear ducts and a tilt steering column. A larger engine is optional.

---

* Includes shaded options
** Other purchase requirements apply

● Poor | ◉ Worse Than Average | O Average | ○ Better Than Average | ○ Excellent | ⊖ Insufficient Information

©1993 by IntelliChoice, Inc. (408) 554-8711 All Rights Reserved. Reproduction Prohibited.
Refer to *Section 3: Annotated Vehicle Charts* for an explanation of these charts.

page 53

Chrysler Imperial
4 Door Sedan

Luxury

 3.8L 150 hp Gas Fuel Inject.
 6 Cylinder "V"
 Automatic 4 Speed
 2 Wheel Front
 Driver Airbag Psngr Belts

Purchase Price

Car Item	Dealer Cost	List
Base Price	$25,320	$29,381
Anti-Lock Brakes	Std	Std
Manual Transmission	N/A	N/A
Optional Engine	N/A	N/A
Auto Climate Control	Std	Std
Power Steering	Std	Std
Cruise Control	Std	Std
All Wheel Drive	N/A	N/A
AM/FM Stereo Cassette	Std	Std
Steering Wheel, Tilt	Std	Std
Power Windows	Std	Std
*Options Price	$0	$0
*Total Price	$25,320	$29,381
Target Price	$28,492	
Destination Charge	$610	
Avg. Tax & Fees	$1,755	
Total Target $	$30,857	
Average Dealer Option Cost:	85%	

Ownership Costs

Cost Area	5 Year Cost	Rate
Depreciation	$20,899	●
Financing ($620/month)	$5,083	
Insurance (Rating 11)	$7,693	○
State Fees	$1,200	
Fuel (Hwy 26 City 19)	$3,847	○
Maintenance	$4,504	○
Repairs	$759	○

Warranty/Maintenance Info

Major Tune-Up	$146	○
Minor Tune-Up	$96	○
Brake Service	$272	●
Overall Warranty	1 yr/12k	●
Drivetrain Warranty	7 yr/70k	○
Rust Warranty	7 yr/100k	○
Maintenance Warranty	N/A	
Roadside Assistance	N/A	

The 1993 Chrysler Imperial is available in one model edition. New for 1993, the Imperial has a premium AM/FM stereo radio with cassette or CD player and graphic equalizer and a tamper resistant odometer. Features include an optional electrochromatic rear view mirror with headlamp, an optional front console with storage and map lights, optional automatic air suspension, electronic speed control, white sidewall tires and an electronic instrument cluster.

Resale Value

1993	1994	1995	1996	1997
$19,944	$16,927	$14,384	$12,128	$9,958

Cumulative Costs

	1993	1994	1995	1996	1997
Annual	$15,530	$7,514	$7,507	$5,647	$7,787
Total	$15,530	$23,044	$30,551	$36,198	$43,985

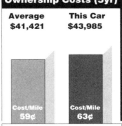

Ownership Costs (5yr)

Average: $41,421 — This Car: $43,985
Cost/Mile 59¢ — Cost/Mile 63¢

Ownership Cost Rating
Worse Than Average

Chrysler LeBaron
2 Door Coupe

Compact

 2.5L 100 hp Gas Fuel Inject.
 4 Cylinder In-Line
 Automatic 3 Speed
 2 Wheel Front
 Driver Airbag Psngr Belts

LX Model Shown

Purchase Price

Car Item	Dealer Cost	List
Base Price	$12,629	$13,999
Anti-Lock Brakes	$764	$899
Manual 5 Speed	($473)	** ($557)
3.0L 141 hp Gas	$590	** $694
Air Conditioning	Pkg	Pkg
Power Steering	Std	Std
Cruise Control	Pkg	Pkg
All Wheel Drive	N/A	N/A
AM/FM Stereo Cassette	$140	$165
Steering Wheel, Tilt	Pkg	Pkg
Power Windows	Std	Std
*Options Price	$140	$165
*Total Price	$12,769	$14,164
Target Price	$13,627	
Destination Charge	$510	
Avg. Tax & Fees	$854	
Total Target $	$14,991	
Average Dealer Option Cost:	85%	

Ownership Costs

Cost Area	5 Year Cost	Rate
Depreciation	$9,113	●
Financing ($301/month)	$2,470	
Insurance (Rating 7)	$7,035	○
State Fees	$587	
Fuel (Hwy 28 City 23)	$3,385	○
Maintenance	$4,074	○
Repairs	$731	○

Warranty/Maintenance Info

Major Tune-Up	$145	○
Minor Tune-Up	$81	○
Brake Service	$268	●
Overall Warranty	1 yr/12k	●
Drivetrain Warranty	7 yr/70k	○
Rust Warranty	7 yr/100k	○
Maintenance Warranty	N/A	
Roadside Assistance	N/A	

The 1993 LeBaron is available in five models-(Base) LeBaron, LX, and GTC coupes; and LE and Landau sedans. New for 1993, the LeBaron offer a V-6 engine and optional four-wheel anti-lock disc brakes. Features include rear window defroster, stainless steel exhaust system, and 14-inch "Preceptor" wheel covers. The LeBaron features three new colors for 1993: Wildberry, Light Driftwood, and Emerald Green. The option list includes wide bodyside moldings, power locks, and cast aluminum wheels.

Resale Value

1993	1994	1995	1996	1997
$11,119	$9,572	$8,287	$7,028	$5,878

Cumulative Costs

	1993	1994	1995	1996	1997
Annual	$7,038	$4,821	$5,476	$4,434	$5,626
Total	$7,038	$11,859	$17,335	$21,769	$27,395

Ownership Costs (5yr)

Average: $26,656 — This Car: $27,395
Cost/Mile 38¢ — Cost/Mile 39¢

Ownership Cost Rating
Average

---

page 54

* Includes shaded options
** Other purchase requirements apply

● Poor ◐ Worse Than Average ○ Average ○ Better Than Average ○ Excellent ⊖ Insufficient Information

©1993 by IntelliChoice, Inc. (408) 554-8711 All Rights Reserved. Reproduction Prohibited.
Refer to *Section 3: Annotated Vehicle Charts* for an explanation of these charts.

Chrysler LeBaron LX
2 Door Coupe

 3.0L 141 hp Gas Fuel Inject.
 6 Cylinder "V"
 Automatic 4 Speed
 2 Wheel Front
 Driver Airbag Psngr Belts

Compact

Purchase Price

Car Item	Dealer Cost	List
Base Price	$14,958	$16,676
Anti-Lock Brakes	$764	$899
Manual Transmission	N/A	N/A
Optional Engine	N/A	N/A
Air Conditioning	Std	Std
Power Steering	Std	Std
Cruise Control	Std	Std
All Wheel Drive	N/A	N/A
AM/FM Stereo Cassette	$140	$165
Steering Wheel, Tilt	Std	Std
Power Windows	Std	Std
***Options Price**	$140	$165
***Total Price**	$15,098	$16,841
Target Price	$16,183	
Destination Charge	$510	
Avg. Tax & Fees	$1,009	
Total Target $	$17,702	
Average Dealer Option Cost:	85%	

Ownership Costs

Cost Area	5 Year Cost	Rate
Depreciation	$10,849	●
Financing ($356/month)	$2,916	
Insurance (Rating 10)	$7,479	○
State Fees	$694	
Fuel (Hwy 27 City 20)	$3,682	●
Maintenance	$5,267	●
Repairs	$731	○

Warranty/Maintenance Info

Major Tune-Up	$173	○
Minor Tune-Up	$99	○
Brake Service	$268	●
Overall Warranty	1 yr/12k	●
Drivetrain Warranty	7 yr/70k	○
Rust Warranty	7 yr/100k	○
Maintenance Warranty	N/A	
Roadside Assistance	N/A	

Ownership Cost By Year

Resale Value
1993: $12,372 | 1994: $10,735 | 1995: $9,404 | 1996: $8,072 | 1997: $6,853

Ownership Costs (5yr)
Average: $28,955 (Cost/Mile 41¢) | This Car: $31,618 (Cost/Mile 45¢)

Cumulative Costs

	1993	1994	1995	1996	1997
Annual	$8,849	$5,242	$6,301	$4,739	$6,487
Total	$8,849	$14,091	$20,392	$25,131	$31,618

Ownership Cost Rating

Worse Than Average

The 1993 LeBaron is available in five models–(Base) LeBaron, LX, and GTC coupes; and LE and Landau sedans. New for 1993, the LX is powered by a 3.0-liter V6 engine with a 4-speed automatic transmission. It features a driver's-side air bag, power windows and a stainless steel exhaust system as standard equipment. The option list includes four-wheel anti-lock brake system and leather upholstery with a 12-way power driver's seat. Also featured is a redesigned front and grille with aero headlights.

Chrysler LeBaron GTC
2 Door Coupe

 3.0L 141 hp Gas Fuel Inject.
 6 Cylinder "V"
 Manual 5 Speed
 2 Wheel Front
 Driver Airbag Psngr Belts

Compact

Purchase Price

Car Item	Dealer Cost	List
Base Price	$15,101	$16,840
Anti-Lock Brakes	$764	$899
Automatic 4 Speed	$587	$690
Optional Engine	N/A	N/A
Air Conditioning	Std	Std
Power Steering	Std	Std
Cruise Control	Std	Std
All Wheel Drive	N/A	N/A
AM/FM Stereo Cassette	$450	$529
Steering Wheel, Tilt	Std	Std
Power Windows	Std	Std
***Options Price**	$1,037	$1,219
***Total Price**	$16,138	$18,059
Target Price	$17,303	
Destination Charge	$510	
Avg. Tax & Fees	$1,077	
Total Target $	$18,890	
Average Dealer Option Cost:	85%	

Ownership Costs

Cost Area	5 Year Cost	Rate
Depreciation	$11,466	●
Financing ($380/month)	$3,112	
Insurance (Rating 8)	$7,223	○
State Fees	$743	
Fuel (Hwy 28 City 21)	$3,530	○
Maintenance	$4,285	○
Repairs	$731	○

Warranty/Maintenance Info

Major Tune-Up	$173	○
Minor Tune-Up	$99	○
Brake Service	$272	●
Overall Warranty	1 yr/12k	●
Drivetrain Warranty	7 yr/70k	○
Rust Warranty	7 yr/100k	○
Maintenance Warranty	N/A	
Roadside Assistance	N/A	

Resale Value
1993: $12,877 | 1994: $11,261 | 1995: $9,966 | 1996: $8,660 | 1997: $7,424

Ownership Costs (5yr)
Average: $30,001 (Cost/Mile 43¢) | This Car: $31,090 (Cost/Mile 44¢)

Cumulative Costs

	1993	1994	1995	1996	1997
Annual	$9,550	$5,216	$5,760	$4,651	$5,913
Total	$9,550	$14,766	$20,526	$25,177	$31,090

Ownership Cost Rating
○ Average

The 1993 LeBaron is available in five models–(Base) LeBaron, LX, and GTC coupes; and LE and Landau sedans. New for 1993, the GTC is powered by a 3.0-liter V6 engine with a 5-speed manual transmission. It features a driver's-side air bag, air conditioning and power window and door locks. The option list includes a four-wheel anti-lock brake system and leather upholstery with a 12-way power driver's seat. Also featured are three new exterior colors and a redesigned front and grille with aero headlights.

---

* Includes shaded options
** Other purchase requirements apply

● Poor | Worse Than Average | Average | Better Than Average | Excellent | ⊖ Insufficient Information

©1993 by IntelliChoice, Inc. (408) 554-8711 All Rights Reserved. Reproduction Prohibited.
Refer to *Section 3: Annotated Vehicle Charts* for an explanation of these charts.

Chrysler LeBaron LE
4 Door Sedan

 2.5L 100 hp Gas Fuel Inject.
 4 Cylinder In-Line
 Automatic 3 Speed
 2 Wheel Front
Driver Airbag Psngr Belts

Base Model Shown — Compact

Purchase Price

Car Item	Dealer Cost	List
Base Price	$13,072	$14,497
Anti-Lock Brakes	$764	$899
4 Spd Auto	$113	** $133
3.0L 141 hp Gas	$590	** $694
Air Conditioning	Pkg	Pkg
Power Steering	Std	Std
Cruise Control	Std	Std
All Wheel Drive	N/A	N/A
AM/FM Stereo Cassette	$140	$165
Steering Wheel, Tilt	Std	Std
Power Windows	Pkg	Pkg
*Options Price	$140	$165
*Total Price	$13,212	$14,662
Target Price		$14,426
Destination Charge		$485
Avg. Tax & Fees		$897
Total Target $		$15,808
Average Dealer Option Cost:	85%	

The 1993 LeBaron is available in five models-(Base) LeBaron, LX, and GTC coupes; and LE and Landau sedans. New for 1993, the LE has a lower overall top gear ratio with the 3-speed automatic transaxle. Standard features include a driver's-side air bag, rear window defroster, reclining bucket seats and a stainless steel exhaust system. The option list includes four-wheel anti-lock disc brakes, air conditioning, and power windows and door locks. Exterior features include four new colors for 1993.

Ownership Costs

Cost Area	5 Year Cost	Rate
Depreciation	$10,294	●
Financing ($318/month)	$2,604	
Insurance (Rating 5)	$6,786	○
State Fees	$605	
Fuel (Hwy 28 City 23)	$3,385	◐
Maintenance	$4,057	○
Repairs	$731	◐

Warranty/Maintenance Info

Major Tune-Up	$138	○
Minor Tune-Up	$81	○
Brake Service	$268	●
Overall Warranty	1 yr/12k	●
Drivetrain Warranty	7 yr/70k	○
Rust Warranty	7 yr/100k	○
Maintenance Warranty	N/A	
Roadside Assistance	N/A	

Ownership Cost By Year

Resale Value

1993	1994	1995	1996	1997
$10,745	$9,243	$7,885	$6,649	$5,514

Cumulative Costs

	1993	1994	1995	1996	1997
Annual	$8,243	$4,774	$5,521	$4,372	$5,552
Total	$8,243	$13,017	$18,538	$22,910	$28,462

Ownership Costs (5yr)

Average: $27,084 — This Car: $28,462
Cost/Mile 39¢ — Cost/Mile 41¢

Ownership Cost Rating

Worse Than Average

Chrysler LeBaron Landau
4 Door Sedan

 3.0L 141 hp Gas Fuel Inject.
 6 Cylinder "V"
 Automatic 4 Speed
 2 Wheel Front
 Driver Airbag Psngr Belts

Compact

Purchase Price

Car Item	Dealer Cost	List
Base Price	$15,354	$17,119
Anti-Lock Brakes	$764	$899
Manual Transmission	N/A	N/A
Optional Engine	N/A	N/A
Air Conditioning	Std	Std
Power Steering	Std	Std
Cruise Control	Std	Std
All Wheel Drive	N/A	N/A
AM/FM Stereo Cassette	Std	Std
Steering Wheel, Tilt	Std	Std
Power Windows	Pkg	Pkg
*Options Price	$0	$0
*Total Price	$15,354	$17,119
Target Price		$16,861
Destination Charge		$485
Avg. Tax & Fees		$1,043
Total Target $		$18,389
Average Dealer Option Cost:	85%	

The 1993 LeBaron is available in five models-(Base) LeBaron, LX, and GTC coupes; and LE and Landau sedans. New for 1993, the Landau has a lower overall top gear ratio with the 3.0-liter engine. The Landau is equipped with a driver's-side air bag, air conditioning, rear window defroster, and a stainless steel exhaust system. The option list includes four-wheel anti-lock disc brakes and leather seats. Exterior features include four new colors: Wildberry, Emerald Green, Radiant Fire, and Light Champagne Metallic.

Ownership Costs

Cost Area	5 Year Cost	Rate
Depreciation	$12,085	●
Financing ($370/month)	$3,030	
Insurance (Rating 6)	$6,919	○
State Fees	$704	
Fuel (Hwy 27 City 20)	$3,682	◐
Maintenance	$4,258	○
Repairs	$750	◐

Warranty/Maintenance Info

Major Tune-Up	$166	○
Minor Tune-Up	$99	○
Brake Service	$268	●
Overall Warranty	1 yr/12k	●
Drivetrain Warranty	7 yr/70k	○
Rust Warranty	7 yr/100k	○
Maintenance Warranty	N/A	
Roadside Assistance	N/A	

Resale Value

1993	1994	1995	1996	1997
$12,093	$10,397	$8,871	$7,521	$6,304

Cumulative Costs

	1993	1994	1995	1996	1997
Annual	$9,761	$5,236	$5,928	$4,656	$5,847
Total	$9,761	$14,997	$20,925	$25,581	$31,428

Ownership Costs (5yr)

Average: $29,194 — This Car: $31,428
Cost/Mile 42¢ — Cost/Mile 45¢

Ownership Cost Rating

Worse Than Average

Legend: ● Poor ◐ Worse Than Average ○ Average ◯ Better Than Average ○ Excellent ⊖ Insufficient Information

* Includes shaded options
** Other purchase requirements apply

©1993 by IntelliChoice, Inc. (408) 554-8711 All Rights Reserved. Reproduction Prohibited.
Refer to *Section 3: Annotated Vehicle Charts* for an explanation of these charts.

Chrysler LeBaron
2 Door Convertible

Compact

 2.5L 100 hp Gas Fuel Inject.
 4 Cylinder In-Line
 Automatic 3 Speed
 2 Wheel Front
 Driver Airbag Psngr Belts

Purchase Price

Car Item	Dealer Cost	List
Base Price	**$15,607**	**$17,399**
Anti-Lock Brakes	$764	$899
Manual 5 Speed	($473) **	($557)
3.0L 141 hp Gas	$590	$694
Air Conditioning	Pkg	Pkg
Power Steering	Std	Std
Cruise Control	Pkg	Pkg
All Wheel Drive	N/A	N/A
AM/FM Stereo Cassette	$140	$165
Steering Wheel, Tilt	Pkg	Pkg
Power Windows	Std	Std
***Options Price**	**$140**	**$165**
***Total Price**	**$15,747**	**$17,564**
Target Price		$16,900
Destination Charge		$510
Avg. Tax & Fees		$1,052
Total Target $		**$18,462**
Average Dealer Option Cost:		**85%**

The 1993 Chrysler LeBaron Convertible is available in three models-(Base) LeBaron, GTC and LX. New for 1993, the LeBaron Convertible is powered by a V-6 engine. Exterior features includes a redesigned front and grille with aero headlights and a palette of new colors. Interior features include reclining bucket seats, power steering, and a center console. The option list includes leather seats, air conditioning, four-wheel anti-lock disc brakes, and 14" cast aluminum wheels.

Ownership Costs

Cost Area	5 Year Cost	Rate
Depreciation	$10,754	●
Financing ($371/month)	$3,042	
Insurance (Rating 11)	$7,693	○
State Fees	$724	
Fuel (Hwy 27 City 21)	$3,598	●
Maintenance	$4,074	◐
Repairs	$731	◐

Warranty/Maintenance Info

Major Tune-Up	$145	◐
Minor Tune-Up	$81	○
Brake Service	$268	●
Overall Warranty	1 yr/12k	●
Drivetrain Warranty	7 yr/70k	○
Rust Warranty	7 yr/100k	○
Maintenance Warranty	N/A	
Roadside Assistance	N/A	

Ownership Cost By Year

(1993, 1994, 1995, 1996, 1997)

Resale Value

1993	1994	1995	1996	1997
$13,912	$12,119	$10,642	$9,169	$7,708

Cumulative Costs

	1993	1994	1995	1996	1997
Annual	$8,148	$5,444	$5,983	$4,895	$6,146
Total	$8,148	$13,592	$19,575	$24,470	$30,616

Ownership Costs (5yr)

Average	This Car
$29,576	$30,616
Cost/Mile 42¢	Cost/Mile 44¢

Ownership Cost Rating

○ Average

Chrysler LeBaron GTC
2 Door Convertible

Compact

 3.0L 141 hp Gas Fuel Inject.
 6 Cylinder "V"
 Manual 5 Speed
 2 Wheel Front
 Driver Airbag Psngr Belts

Purchase Price

Car Item	Dealer Cost	List
Base Price	**$17,709**	**$19,815**
Anti-Lock Brakes	$764	$899
Automatic 4 Speed	$587	$690
Optional Engine	N/A	N/A
Air Conditioning	Std	Std
Power Steering	Std	Std
Cruise Control	Std	Std
All Wheel Drive	N/A	N/A
AM/FM Stereo Cassette	Std	Std
Steering Wheel, Tilt	Std	Std
Power Windows	Std	Std
***Options Price**	**$587**	**$690**
***Total Price**	**$18,296**	**$20,505**
Target Price		$19,713
Destination Charge		$510
Avg. Tax & Fees		$1,221
Total Target $		**$21,444**
Average Dealer Option Cost:		**85%**

The 1993 Chrysler LeBaron Convertible is available in three models-(Base) LeBaron, GTC and LX. New for 1993, the LeBaron GTC is powered by a V-6 engine and includes a power convertible top. Interior features includes air conditioning, reclining bucket seats, and power steering. Exterior features includes wide bodyside moldings and a redesigned front and grille with aero headlights. Other features inlcudes a driver's-side air bag and 15" cast aluminum wheels.

Ownership Costs

Cost Area	5 Year Cost	Rate
Depreciation	$12,424	●
Financing ($431/month)	$3,532	
Insurance (Rating 12)	$7,898	○
State Fees	$840	
Fuel (Hwy 28 City 21)	$3,530	○
Maintenance	$5,276	●
Repairs	$731	○

Warranty/Maintenance Info

Major Tune-Up	$173	○
Minor Tune-Up	$99	○
Brake Service	$272	●
Overall Warranty	1 yr/12k	●
Drivetrain Warranty	7 yr/70k	○
Rust Warranty	7 yr/100k	○
Maintenance Warranty	N/A	
Roadside Assistance	N/A	

Resale Value

1993	1994	1995	1996	1997
$14,700	$13,219	$11,837	$10,414	$9,020

Cumulative Costs

	1993	1994	1995	1996	1997
Annual	$10,604	$5,364	$6,560	$4,956	$6,747
Total	$10,604	$15,968	$22,528	$27,484	$34,231

Ownership Costs (5yr)

Average	This Car
$32,101	$34,231
Cost/Mile 46¢	Cost/Mile 49¢

Ownership Cost Rating

◐ Worse Than Average

---

* Includes shaded options
** Other purchase requirements apply

● Poor   ◐ Worse Than Average   ○ Average   ○ Better Than Average   ○ Excellent   ⊖ Insufficient Information

©1993 by IntelliChoice, Inc. (408) 554-8711   All Rights Reserved. Reproduction Prohibited.
Refer to *Section 3: Annotated Vehicle Charts* for an explanation of these charts.

Chrysler LeBaron LX
2 Door Convertible

Compact

 3.0L 141 hp Gas Fuel Inject.
 6 Cylinder "V"
 Automatic 4 Speed
 2 Wheel Front
 Driver Airbag Psngr Belts

Purchase Price

Car Item	Dealer Cost	List
Base Price	**$18,884**	**$21,165**
Anti-Lock Brakes	$764	$899
Manual Transmission	N/A	N/A
Optional Engine	N/A	N/A
Air Conditioning	Std	Std
Power Steering	Std	Std
Cruise Control	Std	Std
All Wheel Drive	N/A	N/A
AM/FM Stereo Cassette	Std	Std
Steering Wheel, Tilt	Std	Std
Power Windows	Std	Std
*Options Price	$0	$0
*Total Price	$18,884	$21,165
Target Price	$20,391	
Destination Charge	$510	
Avg. Tax & Fees	$1,262	
Total Target $	**$22,163**	
Average Dealer Option Cost:	85%	

Ownership Costs

Cost Area	5 Year Cost	Rate
Depreciation	$12,528	◯
Financing ($445/month)	$3,651	
Insurance (Rating 13)	$8,061	◯
State Fees	$867	
Fuel (Hwy 28 City 21)	$3,530	◯
Maintenance	$5,267	●
Repairs	$731	◯

Warranty/Maintenance Info

Major Tune-Up	$173	◯
Minor Tune-Up	$99	◯
Brake Service	$268	●
Overall Warranty	1 yr/12k	●
Drivetrain Warranty	7 yr/70k	◯
Rust Warranty	7 yr/100k	◯
Maintenance Warranty	N/A	
Roadside Assistance	N/A	

Ownership Cost By Year

1993, 1994, 1995, 1996, 1997

Resale Value

1993	1994	1995	1996	1997
$16,175	$14,434	$12,836	$11,175	$9,635

Cumulative Costs

	1993	1994	1995	1996	1997
Annual	$9,935	$5,698	$6,834	$5,241	$6,927
Total	$9,935	$15,633	$22,467	$27,708	$34,635

Ownership Costs (5yr)

Average: $32,668 — Cost/Mile 47¢
This Car: $34,635 — Cost/Mile 49¢

Ownership Cost Rating

● Worse Than Average

The 1993 Chrysler LeBaron Convertible is available in three models-(Base) LeBaron, GTC, and LX. New for 1993, the LeBaron LX is powered by a V-6 engine and includes a power convertible top. The interior features air conditioning, and reclining bucket seats. Exterior features include wide bodyside moldings and 15" cast aluminum wheels. Also featured is a driver's-side air bag and electronic speed control. The option list includes leather seats and the road touring suspension.

Chrysler New Yorker Salon
4 Door Sedan

Compact

 3.3L 147 hp Gas Fuel Inject.
 6 Cylinder "V"
 Automatic 4 Speed
 2 Wheel Front
 Driver Airbag Psngr Belts

Purchase Price

Car Item	Dealer Cost	List
Base Price	**$16,384**	**$18,705**
Anti-Lock Brakes	$764	$899
Manual Transmission	N/A	N/A
Optional Engine	N/A	N/A
Air Conditioning	Std	Std
Power Steering	Std	Std
Cruise Control	Pkg	Pkg
All Wheel Drive	N/A	N/A
AM/FM Stereo Cassette	$140	$165
Steering Wheel, Tilt	Pkg	Pkg
Power Windows	Std	Std
*Options Price	$140	$165
*Total Price	$16,524	$18,870
Target Price	$18,192	
Destination Charge	$565	
Avg. Tax & Fees	$1,132	
Total Target $	**$19,889**	
Average Dealer Option Cost:	85%	

Ownership Costs

Cost Area	5 Year Cost	Rate
Depreciation	$12,847	●
Financing ($400/month)	$3,277	
Insurance (Rating 5)	$6,786	◯
State Fees	$777	
Fuel (Hwy 26 City 20)	$3,756	●
Maintenance	$4,040	◯
Repairs	$771	●

Warranty/Maintenance Info

Major Tune-Up	$146	◯
Minor Tune-Up	$96	◯
Brake Service	$272	●
Overall Warranty	1 yr/12k	●
Drivetrain Warranty	7 yr/70k	◯
Rust Warranty	7 yr/100k	◯
Maintenance Warranty	N/A	
Roadside Assistance	N/A	

Resale Value

1993	1994	1995	1996	1997
$13,606	$11,705	$10,078	$8,501	$7,042

Cumulative Costs

	1993	1994	1995	1996	1997
Annual	$9,860	$5,523	$5,890	$4,925	$6,056
Total	$9,860	$15,383	$21,273	$26,198	$32,254

Ownership Costs (5yr)

Average: $30,697 — Cost/Mile 44¢
This Car: $32,254 — Cost/Mile 46¢

Ownership Cost Rating

● Worse Than Average

The 1993 New Yorker is available in two models - Salon and Fifth Avenue. New for 1993, the Salon features standard driver's-side air bag and available four-wheel anti-lock brakes. The Salon also offers the Owner's Choice Protection Plan. Other features for 1993 include a stainless steel exhaust system and a 50/50 front bench with dual folding arm rests and driver's side 6-way power seat. The option list includes 4-wheel disc anti-lock brakes, power door locks, leather upholstery and cast aluminum wheels.

---

page 58

* Includes shaded options
** Other purchase requirements apply

● Poor | ◐ Worse Than Average | ◯ Average | ◯ Better Than Average | ◯ Excellent | ⊖ Insufficient Information

©1993 by IntelliChoice, Inc. (408) 554-8711   All Rights Reserved. Reproduction Prohibited.
Refer to *Section 3: Annotated Vehicle Charts* for an explanation of these charts.

Chrysler New Yorker Fifth Avenue
4 Door Sedan

Midsize

3.3L 147 hp Gas Fuel Inject. | 6 Cylinder "V" | Automatic 4 Speed | 2 Wheel Front | Driver Airbag Psngr Belts

Purchase Price

Car Item	Dealer Cost	List
Base Price	**$19,261**	**$21,948**
Anti-Lock Brakes	$764	$899
Manual Transmission	N/A	N/A
3.8L 150 hp Gas	$223	$262
Air Conditioning	Std	Std
Power Steering	Std	Std
Cruise Control	Std	Std
All Wheel Drive	N/A	N/A
AM/FM Stereo Cassette	$140	$165
Steering Wheel, Tilt	Std	Std
Power Windows	Std	Std
***Options Price**	**$140**	**$165**
***Total Price**	**$19,401**	**$22,113**
Target Price		$21,511
Destination Charge		$610
Avg. Tax & Fees		$1,333
Total Target $		**$23,454**
Average Dealer Option Cost:		85%

The 1993 New Yorker is available in two models - Salon and Fifth Avenue. New for 1993, the Fifth Avenue features competitive levels of standard and optional equipment. Standard features for the Fifth Avenue includes air conditioning, electric rear window defroster, a stainless steel exhaust system and a driver's side 6-way power seat. It also comes fully equipped with a drivers'-side air bag and power windows and door locks. The option list includes anti-lock brakes, halogen headlights and leather upholstery.

Ownership Costs

Cost Area	5 Year Cost	Rate
Depreciation	$15,233	●
Financing ($471/month)	$3,864	
Insurance (Rating 6)	$6,919	○
State Fees	$909	
Fuel (Hwy 25 City 19)	$3,929	●
Maintenance	$4,086	○
Repairs	$731	○

Warranty/Maintenance Info

Major Tune-Up	$146	○
Minor Tune-Up	$96	○
Brake Service	$272	●
Overall Warranty	1 yr/12k	●
Drivetrain Warranty	7 yr/70k	○
Rust Warranty	7 yr/100k	○
Maintenance Warranty	N/A	
Roadside Assistance	N/A	

Ownership Cost By Year

1993, 1994, 1995, 1996, 1997

Resale Value

1993	1994	1995	1996	1997
$15,551	$13,442	$11,594	$9,821	$8,221

Cumulative Costs

	1993	1994	1995	1996	1997
Annual	$11,810	$5,996	$6,307	$5,310	$6,248
Total	$11,810	$17,806	$24,113	$29,423	$35,671

Ownership Costs (5yr)

Average	This Car
$33,879	$35,671
Cost/Mile 48¢	Cost/Mile 51¢

Ownership Cost Rating

● Worse Than Average

Dodge Colt
2 Door Coupe

Subcompact

1.5L 92 hp Gas Fuel Inject. | 4 Cylinder In-Line | Manual 5 Speed | 2 Wheel Front | Automatic Seatbelts

Purchase Price

Car Item	Dealer Cost	List
Base Price	**$7,488**	**$7,806**
Anti-Lock Brakes	N/A	N/A
Automatic Transmission	N/A	N/A
Optional Engine	N/A	N/A
Air Conditioning	$673	$783
Power Steering	N/A	N/A
Cruise Control	N/A	N/A
All Wheel Drive	N/A	N/A
AM/FM Stereo Cassette	N/A	N/A
Steering Wheel, Tilt	N/A	N/A
Power Windows	N/A	N/A
***Options Price**	**$0**	**$0**
***Total Price**	**$7,488**	**$7,806**
Target Price		$7,774
Destination Charge		$400
Avg. Tax & Fees		$491
Total Target $		**$8,665**
Average Dealer Option Cost:		86%

The 1993 Dodge Colt is available in four models - (Base) and GL coupe and sedan. The 1993 Base model has a coupe body style that replaces a previous hatchback model. This model is lighter than the one it replaces and features an independent rear suspension. Options on this model include tinted glass and an electric rear-window defroster. Black moldings, center console, and motorized passive shoulder belts and child seat anchors are standard.

Ownership Costs

Cost Area	5 Year Cost	Rate
Depreciation	$4,616	○
Financing ($174/month)	$1,428	
Insurance (Rating 8 [Est.])	$7,223	●
State Fees	$328	
Fuel (Hwy 40 City 32)	$2,398	○
Maintenance	$3,737	○
Repairs	$720	○

Warranty/Maintenance Info

Major Tune-Up	$215	●
Minor Tune-Up	$72	○
Brake Service	$169	○
Overall Warranty	1 yr/12k	●
Drivetrain Warranty	7 yr/70k	○
Rust Warranty	7 yr/100k	○
Maintenance Warranty	N/A	
Roadside Assistance	N/A	

Resale Value

1993	1994	1995	1996	1997
$6,944	$6,029	$5,378	$4,688	$4,049

Cumulative Costs

	1993	1994	1995	1996	1997
Annual	$4,238	$3,628	$4,079	$3,212	$5,293
Total	$4,238	$7,866	$11,945	$15,157	$20,450

Ownership Costs (5yr)

Average	This Car
$20,361	$20,450
Cost/Mile 29¢	Cost/Mile 29¢

Ownership Cost Rating

○ Average

---

* Includes shaded options
** Other purchase requirements apply

● Poor ◐ Worse Than Average ○ Average ○ Better Than Average ○ Excellent ⊖ Insufficient Information

©1993 by IntelliChoice, Inc. (408) 554-8711 All Rights Reserved. Reproduction Prohibited.
Refer to *Section 3: Annotated Vehicle Charts* for an explanation of these charts.

Dodge Colt GL
2 Door Coupe

 1.5L 92 hp Gas Fuel Inject. | 4 Cylinder In-Line | Manual 5 Speed | 2 Wheel Front | Automatic Seatbelts

Subcompact

Purchase Price

Car Item	Dealer Cost	List
Base Price	**$8,309**	**$8,705**
Anti-Lock Brakes	N/A	N/A
Automatic 3 Speed	$445	$518
Optional Engine	N/A	N/A
Air Conditioning	$673	$783
Power Steering	Pkg	Pkg
Cruise Control	N/A	N/A
All Wheel Drive	N/A	N/A
AM/FM Stereo Cassette	$156	$181
Steering Wheel, Tilt	N/A	N/A
Power Windows	N/A	N/A
*Options Price	$0	$0
*Total Price	**$8,309**	**$8,705**
Target Price		$8,665
Destination Charge		$400
Avg. Tax & Fees		$544
Total Target $		**$9,609**
Average Dealer Option Cost:		86%

The 1993 Dodge Colt is available in four models - (Base) and GL coupe and sedan. The Colt GL has a coupe body style that replaces a previous hatchback model. This model is lighter than the one it replaces and features an independent rear suspension. Color-keyed body-side moldings, door-handles, and bumpers, as well as full wheel covers are standard on this model. A rear spoiler and aluminum wheels are optional.

Ownership Costs

Cost Area	5 Year Cost	Rate
Depreciation	$5,283	○
Financing ($193/month)	$1,583	
Insurance (Rating 10 [Est.])	$7,479	●
State Fees	$365	
Fuel (Hwy 40 City 32)	$2,398	○
Maintenance	$3,842	○
Repairs	$720	○

Warranty/Maintenance Info

Major Tune-Up	$215	◉
Minor Tune-Up	$72	○
Brake Service	$169	○
Overall Warranty	1 yr/12k	●
Drivetrain Warranty	7 yr/70k	○
Rust Warranty	7 yr/100k	○
Maintenance Warranty	N/A	
Roadside Assistance	N/A	

Ownership Cost By Year

(1993, 1994, 1995, 1996, 1997)

Resale Value

1993	1994	1995	1996	1997
$7,232	$6,310	$5,678	$4,986	$4,326

Cumulative Costs

	1993	1994	1995	1996	1997
Annual	$5,014	$3,741	$4,200	$3,285	$5,430
Total	$5,014	$8,755	$12,955	$16,240	$21,670

Ownership Costs (5yr)

Average	This Car
$21,372	$21,670
Cost/Mile 31¢	Cost/Mile 31¢

Ownership Cost Rating
○ Average

Dodge Colt
4 Door Sedan

 1.5L 92 hp Gas Fuel Inject. | 4 Cylinder In-Line | Manual 5 Speed | 2 Wheel Front | Automatic Seatbelts

Subcompact

GL Model Shown

Purchase Price

Car Item	Dealer Cost	List
Base Price	**$9,001**	**$9,448**
Anti-Lock Brakes	N/A	N/A
Automatic 4 Speed	$603	$701
1.8L 113 hp Gas	$298	$346
Air Conditioning	$673	$783
Power Steering	Pkg	Pkg
Cruise Control	N/A	N/A
All Wheel Drive	N/A	N/A
AM/FM Stereo Cassette	$156	** $181
Steering Wheel, Tilt	N/A	N/A
Power Windows	N/A	N/A
*Options Price	$298	$346
*Total Price	**$9,299**	**$9,794**
Target Price		$9,772
Destination Charge		$400
Avg. Tax & Fees		$611
Total Target $		**$10,783**
Average Dealer Option Cost:		86%

The 1993 Dodge Colt is available in four models - (Base) and GL coupe and sedan. The four-door Colt is new to the line. Power brakes, motorized passive shoulder belt and three child seat anchors are standard items on this model. Options include fixed time intermittent windshield wipers, rear seat heater ducts, full wheel covers, and tinted glass.

Ownership Costs

Cost Area	5 Year Cost	Rate
Depreciation	$6,015	○
Financing ($217/month)	$1,776	
Insurance (Rating 10 [Est.])	$7,479	●
State Fees	$408	
Fuel (Hwy 34 City 27)	$2,831	○
Maintenance	$3,828	○
Repairs	$720	○

Warranty/Maintenance Info

Major Tune-Up	$210	◉
Minor Tune-Up	$72	○
Brake Service	$186	○
Overall Warranty	1 yr/12k	●
Drivetrain Warranty	7 yr/70k	○
Rust Warranty	7 yr/100k	○
Maintenance Warranty	N/A	
Roadside Assistance	N/A	

Resale Value

1993	1994	1995	1996	1997
$8,078	$7,067	$6,308	$5,484	$4,768

Cumulative Costs

	1993	1994	1995	1996	1997
Annual	$5,514	$3,984	$4,470	$3,528	$5,561
Total	$5,514	$9,498	$13,968	$17,496	$23,057

Ownership Costs (5yr)

Average	This Car
$22,598	$23,057
Cost/Mile 32¢	Cost/Mile 33¢

Ownership Cost Rating
○ Average

---

*Includes shaded options
**Other purchase requirements apply

● Poor | ◉ Worse Than Average | ○ Average | ○ Better Than Average | ○ Excellent | ⊖ Insufficient Information

©1993 by IntelliChoice, Inc. (408) 554-8711 All Rights Reserved. Reproduction Prohibited.
Refer to *Section 3: Annotated Vehicle Charts* for an explanation of these charts.

Dodge Colt GL
4 Door Sedan

 1.8L 113 hp Gas Fuel Inject.
 4 Cylinder In-Line
 Manual 5 Speed
 2 Wheel Front
Automatic Seatbelts

Subcompact

Purchase Price

Car Item	Dealer Cost	List
Base Price	**$9,846**	**$10,423**
Anti-Lock Brakes	$829	$964
Automatic 4 Speed	$551	$641
Optional Engine	N/A	N/A
Air Conditioning	$673	$783
Power Steering	Pkg	Pkg
Cruise Control	Pkg	Pkg
All Wheel Drive	N/A	N/A
AM/FM Stereo Cassette	$156	$181
Steering Wheel, Tilt	Pkg	Pkg
Power Windows	Pkg	Pkg
*Options Price	$0	$0
*Total Price	$9,846	$10,423
Target Price		$10,361
Destination Charge		$400
Avg. Tax & Fees		$646
Total Target $		**$11,407**
Average Dealer Option Cost:		86%

The 1993 Dodge Colt is available in four models - (Base) and GL coupe and sedan. The GL Sedan is new to the line. This is the only model that offers four-wheel disc anti-lock brakes as an option. Standard features include color-keyed body-side moldings, full wheel covers, and dual note horn. Optional items are a tilt steering column, power windows, power steering, and electronic speed control.

Ownership Costs

Cost Area	5 Year Cost	Rate
Depreciation	$6,558	○
Financing ($229/month)	$1,879	
Insurance (Rating 10 [Est.])	$7,479	●
State Fees	$433	
Fuel (Hwy 34 City 27)	$2,831	○
Maintenance	$3,955	○
Repairs	$720	○

Warranty/Maintenance Info

Major Tune-Up	$210	◉
Minor Tune-Up	$72	○
Brake Service	$186	○
Overall Warranty	1 yr/12k	●
Drivetrain Warranty	7 yr/70k	○
Rust Warranty	7 yr/100k	○
Maintenance Warranty	N/A	
Roadside Assistance	N/A	

Ownership Cost By Year

1993 / 1994 / 1995 / 1996 / 1997

Resale Value

1993	1994	1995	1996	1997
$8,080	$7,088	$6,368	$5,565	$4,849

Cumulative Costs

	1993	1994	1995	1996	1997
Annual	$6,185	$4,002	$4,518	$3,519	$5,631
Total	$6,185	$10,187	$14,705	$18,224	$23,855

Ownership Costs (5yr)

Average	This Car
$23,305	$23,855
Cost/Mile 33¢	Cost/Mile 34¢

Ownership Cost Rating
○ Average

Dodge Daytona
2 Door Hatchback

 2.5L 100 hp Gas Fuel Inject.
 4 Cylinder In-Line
 Manual 5 Speed
 2 Wheel Front
 Driver Airbag Psngr Belts

Subcompact

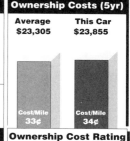
IROC Model Shown

Purchase Price

Car Item	Dealer Cost		List
Base Price	**$10,028**		**$10,874**
Anti-Lock Brakes	$764	**	$899
Automatic 3 Speed	$473		$557
3.0L 141 hp Gas	$590	**	$694
Air Conditioning	Pkg		Pkg
Power Steering	Std		Std
Cruise Control	$190	**	$224
All Wheel Drive	N/A		N/A
AM/FM Stereo Cassette	Pkg		Pkg
Steering Wheel, Tilt	Pkg		Pkg
Power Windows	$234	**	$275
*Options Price	$0		$0
*Total Price	$10,028		$10,874
Target Price			$10,587
Destination Charge			$499
Avg. Tax & Fees			$668
Total Target $			**$11,754**
Average Dealer Option Cost:			85%

The 1993 Dodge Daytona is available in four models-(Base) Daytona, ES, IROC and IROC R/T hatchbacks. New for 1993, the Daytona features available anti-lock brakes as well as a full stainless steel exhaust system. Standard features include driver's-side airbag and active 3-point belts for driver and front passenger, and aero-style halogen headlights. Key options include removable sunroof with storage bag and straps; CD player and Infinity speakers, power door lock, tilt steering and an anti-theft security system.

Ownership Costs

Cost Area	5 Year Cost	Rate
Depreciation	$7,636	●
Financing ($236/month)	$1,936	
Insurance (Rating 14)	$8,238	●
State Fees	$455	
Fuel (Hwy 31 City 23)	$3,204	◉
Maintenance	$3,550	○
Repairs	$731	○

Warranty/Maintenance Info

Major Tune-Up	$134	○
Minor Tune-Up	$73	○
Brake Service	$237	◉
Overall Warranty	1 yr/12k	●
Drivetrain Warranty	7 yr/70k	○
Rust Warranty	7 yr/100k	○
Maintenance Warranty	N/A	
Roadside Assistance	N/A	

Resale Value

1993	1994	1995	1996	1997
$7,880	$6,796	$5,903	$4,985	$4,118

Cumulative Costs

	1993	1994	1995	1996	1997
Annual	$6,970	$4,338	$4,803	$4,284	$5,355
Total	$6,970	$11,308	$16,111	$20,395	$25,750

Ownership Costs (5yr)

Average	This Car
$23,813	$25,750
Cost/Mile 34¢	Cost/Mile 37¢

Ownership Cost Rating
● Poor

---

* Includes shaded options
** Other purchase requirements apply

● Poor　◉ Worse Than Average　○ Average　○ Better Than Average　○ Excellent　⊖ Insufficient Information

©1993 by IntelliChoice, Inc. (408) 554-8711 All Rights Reserved. Reproduction Prohibited.
Refer to *Section 3: Annotated Vehicle Charts* for an explanation of these charts.

page 61

Dodge Daytona ES
2 Door Hatchback

 2.5L 100 hp Gas Fuel Inject.
 4 Cylinder In-Line
 Manual 5 Speed
 2 Wheel Front
 Driver Airbag Psngr Belts

Subcompact — IROC R/T Model Shown

Purchase Price

Car Item	Dealer Cost	List
Base Price	**$11,046**	**$12,018**
Anti-Lock Brakes	$764	** $899
Automatic 3 Speed	$473	$557
3.0L 141 hp Gas	$590	** $694
Air Conditioning	Pkg	Pkg
Power Steering	Std	Std
Cruise Control	Pkg	Pkg
All Wheel Drive	N/A	N/A
AM/FM Stereo Cassette	$286	$336
Steering Wheel, Tilt	Pkg	Pkg
Power Windows	$234	** $275
***Options Price**	$0	$0
***Total Price**	**$11,046**	**$12,018**
Target Price	$11,682	
Destination Charge	$499	
Avg. Tax & Fees	$734	
Total Target $	**$12,915**	
Average Dealer Option Cost:	*85%*	

Ownership Costs

Cost Area	5 Year Cost	Rate
Depreciation	$8,423	●
Financing ($260/month)	$2,128	
Insurance (Rating 15)	$8,507	●
State Fees	$501	
Fuel (Hwy 31 City 23)	$3,204	◐
Maintenance	$4,551	◐
Repairs	$731	○

Warranty/Maintenance Info

Major Tune-Up	$134	○
Minor Tune-Up	$73	○
Brake Service	$237	◐
Overall Warranty	1 yr/12k	●
Drivetrain Warranty	7 yr/70k	○
Rust Warranty	7 yr/100k	○
Maintenance Warranty	N/A	
Roadside Assistance	N/A	

Ownership Cost By Year
1993 / 1994 / 1995 / 1996 / 1997

Resale Value
1993	1994	1995	1996	1997
$8,431	$7,311	$6,365	$5,402	$4,492

Cumulative Costs
	1993	1994	1995	1996	1997
Annual	$7,721	$4,496	$5,438	$4,407	$5,983
Total	$7,721	$12,217	$17,655	$22,062	$28,045

Ownership Costs (5yr)
Average $25,100 — This Car $28,045
Cost/Mile 36¢ — Cost/Mile 40¢

Ownership Cost Rating
● Poor

The 1993 Daytona is available in four models-(Base) Daytona, ES, IROC and IROC R/T hatchbacks. New for 1993, the ES features front and rear integrated fascias, aero-style headlamps, integrated park and turn lamps, and wheel covers. Standard features include driver's-side airbag, deluxe intermittent windshield wipers, and aero-style halogen headlights with time delay. Key options include removable sunroof, CD player and Infinity speakers, power door locks, anti-theft security system, tilt steering and remote liftgate

---

Dodge Daytona IROC
2 Door Hatchback

 3.0L 141 hp Gas Fuel Inject.
 6 Cylinder "V"
 Manual 5 Speed
 2 Wheel Front
 Driver Airbag Psngr Belts

Sport

Purchase Price

Car Item	Dealer Cost	List
Base Price	**$12,195**	**$13,309**
Anti-Lock Brakes	$764	** $899
Automatic 4 Speed	$587	$690
Optional Engine	N/A	N/A
Air Conditioning	Pkg	Pkg
Power Steering	Std	Std
Cruise Control	Pkg	Pkg
All Wheel Drive	N/A	N/A
AM/FM Stereo Cassette	$286	$336
Steering Wheel, Tilt	Pkg	Pkg
Power Windows	Pkg	Pkg
***Options Price**	$286	$336
***Total Price**	**$12,481**	**$13,645**
Target Price	$13,226	
Destination Charge	$499	
Avg. Tax & Fees	$827	
Total Target $	**$14,552**	
Average Dealer Option Cost:	*85%*	

Ownership Costs

Cost Area	5 Year Cost	Rate
Depreciation	$9,307	●
Financing ($292/month)	$2,397	
Insurance (Rating 16)	$8,784	●
State Fees	$565	
Fuel (Hwy 28 City 19)	$3,704	○
Maintenance	$4,674	○
Repairs	$731	○

Warranty/Maintenance Info

Major Tune-Up	$161	○
Minor Tune-Up	$90	○
Brake Service	$262	○
Overall Warranty	1 yr/12k	●
Drivetrain Warranty	7 yr/70k	○
Rust Warranty	7 yr/100k	○
Maintenance Warranty	N/A	
Roadside Assistance	N/A	

Resale Value
1993	1994	1995	1996	1997
$9,008	$7,987	$7,114	$6,216	$5,245

Cumulative Costs
	1993	1994	1995	1996	1997
Annual	$9,056	$4,668	$5,531	$4,559	$6,348
Total	$9,056	$13,724	$19,255	$23,814	$30,162

Ownership Costs (5yr)
Average $31,343 — This Car $30,162
Cost/Mile 45¢ — Cost/Mile 43¢

Ownership Cost Rating
○ Average

The 1993 Daytona is available in four models-(Base) Daytona, ES, IROC and IROC R/T hatchbacks. New for 1993, the IROC features four wheel anti-lock brake system, 15-inch Triad wheel covers with white or Argent accent, aero-style headlamps and a full stainless steel exhaust system. Standard features include air conditioning, driver's-side airbag and active 3-point belts for driver and front passenger, outboard shoulder and lap belts for rear passengers and deluxe intermittent windshield wipers.

---

page 62

* Includes shaded options
** Other purchase requirements apply

● Poor ◐ Worse Than Average ○ Average ○ Better Than Average ○ Excellent ⊖ Insufficient Information

©1993 by IntelliChoice, Inc. (408) 554-8711   All Rights Reserved. Reproduction Prohibited.
Refer to *Section 3: Annotated Vehicle Charts* for an explanation of these charts.

Dodge Daytona IROC R/T
2 Door Hatchback

Sport

 2.2L 224 hp Turbo Gas Fuel Inject.
 4 Cylinder In-Line
 Manual 5 Speed
 2 Wheel Front
 Driver Airbag Psngr Belts

Purchase Price

Car Item	Dealer Cost	List
Base Price	**$17,425**	**$19,185**
Anti-Lock Brakes	Std	Std
Automatic Transmission	N/A	N/A
Optional Engine	N/A	N/A
Air Conditioning	Std	Std
Power Steering	Std	Std
Cruise Control	Pkg	Pkg
All Wheel Drive	N/A	N/A
AM/FM Stereo Cassette	$286	$336
Steering Wheel, Tilt	Pkg	Pkg
Power Windows	Pkg	Pkg
*Options Price	$286	$336
*Total Price	$17,711	$19,521
Target Price		$18,936
Destination Charge		$499
Avg. Tax & Fees		$1,172
Total Target $		**$20,607**
Average Dealer Option Cost:		85%

The 1993 Daytona is available in four models-(Base) Daytona, ES, IROC and IROC R/T hatchbacks. New for 1993, the IROC R/T features front and rear integrated fascias, integrated park and turn lamps, taillights and wheel covers. Standard features include driver's-side airbag, deluxe intermittent windshield wipers and aero-style halogen headlights with time delay. Other features include removable sunroof with storage bag and straps, CD player and Infinity speakers; power door locks and anti-theft security system.

Ownership Costs

Cost Area	5 Year Cost	Rate
Depreciation	$12,632	●
Financing ($414/month)	$3,395	
Insurance (Rating 21++)	$13,383	●
State Fees	$800	
Fuel (Hwy 27 City 19)	$3,775	○
Maintenance	$5,072	○
Repairs	$771	○

Warranty/Maintenance Info

Major Tune-Up	$133	○
Minor Tune-Up	$73	○
Brake Service	$262	○
Overall Warranty	1 yr/12k	●
Drivetrain Warranty	7 yr/70k	○
Rust Warranty	7 yr/100k	○
Maintenance Warranty	N/A	
Roadside Assistance	N/A	

Ownership Cost By Year

Bars for 1993–1997

Resale Value

1993	1994	1995	1996	1997
$12,868	$11,580	$10,395	$9,228	$7,975

Cumulative Costs

	1993	1994	1995	1996	1997
Annual	$12,624	$6,221	$7,230	$5,917	$7,836
Total	$12,624	$18,845	$26,075	$31,992	$39,828

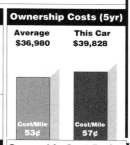

Ownership Costs (5yr)

Average	This Car
$36,980	$39,828
Cost/Mile 53¢	Cost/Mile 57¢

Ownership Cost Rating

● Poor

Dodge Dynasty
4 Door Sedan

Compact

 2.5L 100 hp Gas Fuel Inject.
 4 Cylinder In-Line
 Automatic 3 Speed
 2 Wheel Front
 Driver Airbag Psngr Belts

LE Model Shown

Purchase Price

Car Item	Dealer Cost	List
Base Price	**$12,956**	**$14,736**
Anti-Lock Brakes	$764	** $899
4 Spd Auto	$113	** $133
3.3L 147 hp Gas	$590	** $694
Air Conditioning	$706	$831
Power Steering	Std	Std
Cruise Control	Pkg	Pkg
All Wheel Drive	N/A	N/A
AM/FM Stereo Cassette	$140	$165
Steering Wheel, Tilt	Pkg	Pkg
Power Windows	Pkg	Pkg
*Options Price	$846	$996
*Total Price	$13,802	$15,732
Target Price		$15,066
Destination Charge		$565
Avg. Tax & Fees		$945
Total Target $		**$16,576**
Average Dealer Option Cost:		85%

The 1993 Dodge Dynasty is available in two models - (Base) and Dynasty LE. New for 1993, the Base model features a stainless steel exhaust system. Body-side moldings, cloth seats, power brakes, an overhead console, and tinted windows are standard features. Optional items include dual power heated mirrors, power windows, electronic speed control, and bodyside stripes.

Ownership Costs

Cost Area	5 Year Cost	Rate
Depreciation	$9,714	●
Financing ($333/month)	$2,729	
Insurance (Rating 3)	$6,520	○
State Fees	$652	
Fuel (Hwy 27 City 21)	$3,598	●
Maintenance	$3,837	○
Repairs	$731	○

Warranty/Maintenance Info

Major Tune-Up	$127	○
Minor Tune-Up	$73	○
Brake Service	$254	●
Overall Warranty	1 yr/12k	●
Drivetrain Warranty	7 yr/70k	○
Rust Warranty	7 yr/100k	○
Maintenance Warranty	N/A	
Roadside Assistance	N/A	

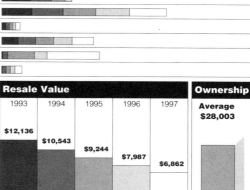

Resale Value

1993	1994	1995	1996	1997
$12,136	$10,543	$9,244	$7,987	$6,862

Cumulative Costs

	1993	1994	1995	1996	1997
Annual	$7,671	$4,886	$5,228	$4,364	$5,632
Total	$7,671	$12,557	$17,785	$22,149	$27,781

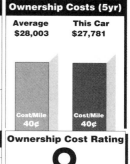

Ownership Costs (5yr)

Average	This Car
$28,003	$27,781
Cost/Mile 40¢	Cost/Mile 40¢

Ownership Cost Rating

○ Average

---

* Includes shaded options
** Other purchase requirements apply

● Poor ◐ Worse Than Average ○ Average ◑ Better Than Average ○ Excellent ⊖ Insufficient Information

©1993 by IntelliChoice, Inc. (408) 554-8711 All Rights Reserved. Reproduction Prohibited.
Refer to *Section 3: Annotated Vehicle Charts* for an explanation of these charts.

Dodge Dynasty LE
4 Door Sedan

Compact

 3.0L 141 hp Gas Fuel Inject.
 6 Cylinder "V"
 Automatic 4 Speed
 2 Wheel Front
 Driver Airbag Psngr Belts

Purchase Price

Car Item	Dealer Cost	List
Base Price	**$14,257**	**$16,267**
Anti-Lock Brakes	$764	$899
Manual Transmission	N/A	N/A
3.3L 147 hp Gas	N/C	N/C
Air Conditioning	Pkg	Pkg
Power Steering	Std	Std
Cruise Control	Pkg	Pkg
All Wheel Drive	N/A	N/A
AM/FM Stereo Cassette	$140	$165
Steering Wheel, Tilt	Pkg	Pkg
Power Windows	Pkg	Pkg
*Options Price	$140	$165
*Total Price	$14,397	$16,432
Target Price	$15,767	
Destination Charge	$565	
Avg. Tax & Fees	$987	
Total Target $	**$17,319**	
Average Dealer Option Cost:	85%	

The 1993 Dodge Dynasty is available in two models - (Base) and Dynasty LE. The Dynasty LE upgrades the Base model with illuminated entry system, electrochromatic automatic day/night mirrors, AM/FM stereo cassette with graphic equalizer and 10 speakers in 6 locations. Standard features include 14" wheel covers, body-side stripes, and 50/50 bench seats with dual folding armrests and cup holders.

Ownership Costs

Cost Area	5 Year Cost	Rate
Depreciation	$10,203	●
Financing ($348/month)	$2,853	
Insurance (Rating 5)	$6,786	○
State Fees	$680	
Fuel (Hwy 28 City 21)	$3,530	◐
Maintenance	$4,008	◐
Repairs	$731	◐

Warranty/Maintenance Info

Major Tune-Up	$162	○
Minor Tune-Up	$90	○
Brake Service	$254	●
Overall Warranty	1 yr/12k	●
Drivetrain Warranty	7 yr/70k	○
Rust Warranty	7 yr/100k	○
Maintenance Warranty	N/A	
Roadside Assistance	N/A	

Ownership Cost By Year

1993, 1994, 1995, 1996, 1997

Resale Value

1993	1994	1995	1996	1997
$12,684	$11,010	$9,604	$8,298	$7,116

Cumulative Costs

	1993	1994	1995	1996	1997
Annual	$7,964	$5,073	$5,432	$4,493	$5,829
Total	$7,964	$13,037	$18,469	$22,962	$28,791

Ownership Costs (5yr)

Average	This Car
$28,604	$28,791
Cost/Mile 41¢	Cost/Mile 41¢

Ownership Cost Rating
○ Average

Dodge Intrepid
4 Door Sedan

Large

 3.3L 153 hp Gas Fuel Inject.
 6 Cylinder "V"
 Automatic 4 Speed
 2 Wheel Front
 Driver/Psngr Airbags Std

Purchase Price

Car Item	Dealer Cost	List
Base Price	**$14,016**	**$15,930**
Anti-Lock Brakes	$530	$624
Manual Transmission	N/A	N/A
3.5L 214 hp Gas	$531	$625
Air Conditioning	Pkg	Pkg
Power Steering	Std	Std
Cruise Control	Pkg	Pkg
All Wheel Drive	N/A	N/A
AM/FM Stereo Cassette	$166	$195
Steering Wheel, Tilt	Std	Std
Power Windows	Pkg	Pkg
*Options Price	$697	$820
*Total Price	$14,713	$16,750
Target Price	$16,103	
Destination Charge	$490	
Avg. Tax & Fees	$1,002	
Total Target $	**$17,595**	
Average Dealer Option Cost:	85%	

The all-new Dodge Intrepid makes its debut in 1993 with a two model line up - (Base) and ES sedans. The Intrepid is the Dodge division's version of the Chrysler Corporation's new LH model whose production incorporated new factory habits such as the melding of diverse departments with fewer job clasifications and lean production practices. The standard Intrepid sedan features driver and passenger side airbags, thermoplastic front fenders, stainless steel exhaust system and a tilt steering column.

Ownership Costs

Cost Area	5 Year Cost	Rate
Depreciation		⊖
Financing ($354/month)	$2,899	
Insurance (Rating 10+)	$8,975	⊖
State Fees	$689	
Fuel (Hwy 26 City 18)	$3,949	◐
Maintenance		⊖
Repairs	$740	◐

Warranty/Maintenance Info

Major Tune-Up		⊖
Minor Tune-Up		⊖
Brake Service		⊖
Overall Warranty	1 yr/12k	●
Drivetrain Warranty	7 yr/70k	○
Rust Warranty	7 yr/100k	○
Maintenance Warranty	N/A	
Roadside Assistance	N/A	

Ownership Cost By Year

Insufficient Depreciation Information

Resale Value

Insufficient Information

Cumulative Costs

	1993	1994	1995	1996	1997
Annual	Insufficient Information				
Total	Insufficient Information				

Ownership Costs (5yr)

Insufficient Information

Ownership Cost Rating
⊖ Insufficient Information

---

page 64

* Includes shaded options
** Other purchase requirements apply

● Poor ◐ Worse Than Average ○ Average ◯ Better Than Average ○ Excellent ⊖ Insufficient Information

©1993 by IntelliChoice, Inc. (408) 554-8711 All Rights Reserved. Reproduction Prohibited.
Refer to *Section 3: Annotated Vehicle Charts* for an explanation of these charts.

Dodge Intrepid ES
4 Door Sedan

 3.3L 153 hp Gas Fuel Inject.
 6 Cylinder "V"
 Automatic 4 Speed
 2 Wheel Front
 Driver/Psngr Airbags Std

Large

Purchase Price

Car Item	Dealer Cost	List
Base Price	**$15,086**	**$17,189**
Anti-Lock Brakes	$658	$774
Manual Transmission	N/A	N/A
3.5L 214 hp Gas	$531	$625
Air Conditioning	$706	$831
Power Steering	Std	Std
Cruise Control	Pkg	Pkg
All Wheel Drive	N/A	N/A
AM/FM Stereo Cassette	$166	$195
Steering Wheel, Tilt	Std	Std
Power Windows	$290	** $341
*****Options Price**	**$1,403**	**$1,651**
*****Total Price**	**$16,489**	**$18,840**
Target Price		$18,095
Destination Charge		$490
Avg. Tax & Fees		$1,122
Total Target $		**$19,707**
Average Dealer Option Cost:		**85%**

The all-new Dodge Intrepid makes its debut in 1993 with a two model line up - (Base) and ES sedans. In addition to all the standard features of the (Base) sedan, the ES makes standard four wheel disc brakes, a cargo/convenience net, power remote decklid release, front fog lamps, touring tuned suspension and message center with warning lights. Options on the ES include an integrated child seat, AM/FM Cassette stereo, power windows and Performance Handling Group.

Ownership Costs

Cost Area	5 Year Cost	Rate
Depreciation		⊖
Financing ($396/month)	$3,247	
Insurance (Rating 11+)	$9,232	●
State Fees	$773	
Fuel (Hwy 26 City 18)	$3,949	○
Maintenance		⊖
Repairs	$740	○

Warranty/Maintenance Info

Major Tune-Up		⊖
Minor Tune-Up		⊖
Brake Service		⊖
Overall Warranty	1 yr/12k	●
Drivetrain Warranty	7 yr/70k	○
Rust Warranty	7 yr/100k	○
Maintenance Warranty	N/A	
Roadside Assistance	N/A	

Ownership Cost By Year

Insufficient Depreciation Information

■ 1993
■ 1994
■ 1995
□ 1996
□ 1997

Insufficient Maintenance Information

Resale Value

Insufficient Information

Cumulative Costs

	1993	1994	1995	1996	1997
Annual	Insufficient Information				
Total	Insufficient Information				

Ownership Costs (5yr)

Insufficient Information

Ownership Cost Rating

⊖ Insufficient Information

Dodge Shadow
2 Door Hatchback

 2.2L 93 hp Gas Fuel Inject.
 4 Cylinder In-Line
 Manual 5 Speed
 2 Wheel Front
 Driver Airbag Psngr Belts

Subcompact

ES Model Shown

Purchase Price

Car Item	Dealer Cost	List
Base Price	**$7,888**	**$8,397**
Anti-Lock Brakes	$764	$899
Automatic 3 Speed	$473	$557
2.5L 100 hp Gas	$243	$286
Air Conditioning	$765	$900
Power Steering	Std	Std
Cruise Control	$190	** $224
All Wheel Drive	N/A	N/A
AM/FM Stereo Cassette	$424	$499
Steering Wheel, Tilt	$126	** $148
Power Windows	N/A	N/A
*****Options Price**	**$0**	**$0**
*****Total Price**	**$7,888**	**$8,397**
Target Price		$8,343
Destination Charge		$485
Avg. Tax & Fees		$530
Total Target $		**$9,358**
Average Dealer Option Cost:		**85%**

The 1993 Shadow is available in six models - (Base) Shadow and ES two- and four-door hatchbacks; and Highline and ES convertibles. New for 1993, the (Base) Shadow two-door offers standard stainless steel exhaust, body color hood pentastar, champagne interior color, high performance torque converter, quieter final drive gearing on three speed automatic tansaxles, and a redesigned engine intake manifold. Two new exterior colors are available (Wildberry Pearlcoat and Light Driftwood Satin-glow).

Ownership Costs

Cost Area	5 Year Cost	Rate
Depreciation	$5,135	○
Financing ($188/month)	$1,542	
Insurance (Rating 8)	$7,223	●
State Fees	$355	
Fuel (Hwy 32 City 27)	$2,926	●
Maintenance	$3,648	○
Repairs	$731	○

Warranty/Maintenance Info

Major Tune-Up	$133	○
Minor Tune-Up	$73	○
Brake Service	$229	●
Overall Warranty	1 yr/12k	●
Drivetrain Warranty	7 yr/70k	○
Rust Warranty	7 yr/100k	○
Maintenance Warranty	N/A	
Roadside Assistance	N/A	

Ownership Cost By Year

■ 1993
■ 1994
■ 1995
□ 1996
□ 1997

Resale Value

1993	1994	1995	1996	1997
$7,227	$6,317	$5,611	$4,899	$4,223

Ownership Costs (5yr)

Average	This Car
$21,026	$21,560
Cost/Mile 30¢	Cost/Mile 31¢

Cumulative Costs

	1993	1994	1995	1996	1997
Annual	$4,799	$3,770	$4,243	$3,763	$4,985
Total	$4,799	$8,569	$12,812	$16,575	$21,560

Ownership Cost Rating

○ Average

---

* Includes shaded options
** Other purchase requirements apply

 Poor   Worse Than Average   Average   Better Than Average   Excellent   Insufficient Information

©1993 by IntelliChoice, Inc. (408) 554-8711   All Rights Reserved. Reproduction Prohibited.
Refer to *Section 3: Annotated Vehicle Charts* for an explanation of these charts.

Dodge Shadow
4 Door Hatchback

Subcompact

- 2.2L 93 hp Gas Fuel Inject.
- 4 Cylinder In-Line
- Manual 5 Speed
- 2 Wheel Front
- Driver Airbag Psngr Belts

ES Model Shown

Purchase Price

Car Item	Dealer Cost	List
Base Price	**$8,255**	**$8,797**
Anti-Lock Brakes	$764	$899
Automatic 3 Speed	$473	$557
2.5L 100 hp Gas	$243	$286
Air Conditioning	$765	$900
Power Steering	Std	Std
Cruise Control	$190	** $224
All Wheel Drive	N/A	N/A
AM/FM Stereo Cassette	$424	$499
Steering Wheel, Tilt	$126	** $148
Power Windows	N/A	N/A
*Options Price	$0	$0
*Total Price	**$8,255**	**$8,797**
Target Price		$8,737
Destination Charge		$485
Avg. Tax & Fees		$554
Total Target $		**$9,776**
Average Dealer Option Cost:		85%

The 1993 Shadow is available in six models - (Base) Shadow and ES two- and four-door hatchbacks; and Highline and ES convertibles. New for 1993, the Base Shadow four-door offers standard stainless steel exhaust, body color hood Pentastar, Champagne interior color, high performance torque converter, quieter final drive gearing on three speed automatic transaxles, and a redesigned engine intake manifold. Two new exterior colors are available (Wildberry Pearlcoat and Light Driftwood Satin-glow).

Ownership Costs

Cost Area	5 Year Cost	Rate
Depreciation	$5,408	○
Financing ($196/month)	$1,610	
Insurance (Rating 6)	$6,919	○
State Fees	$372	
Fuel (Hwy 32 City 27)	$2,926	◐
Maintenance	$3,648	○
Repairs	$731	○

Warranty/Maintenance Info

Major Tune-Up	$133	○
Minor Tune-Up	$73	○
Brake Service	$229	●
Overall Warranty	1 yr/12k	●
Drivetrain Warranty	7 yr/70k	○
Rust Warranty	7 yr/100k	○
Maintenance Warranty	N/A	
Roadside Assistance	N/A	

Ownership Cost By Year

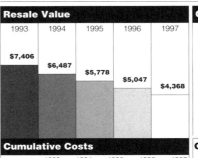

- 1993
- 1994
- 1995
- 1996
- 1997

Resale Value

1993	1994	1995	1996	1997
$7,406	$6,487	$5,778	$5,047	$4,368

Cumulative Costs

	1993	1994	1995	1996	1997
Annual	$5,014	$3,746	$4,202	$3,727	$4,925
Total	$5,014	$8,760	$12,962	$16,689	$21,614

Ownership Costs (5yr)

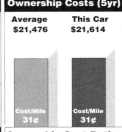

Average	This Car
$21,476	$21,614
Cost/Mile 31¢	Cost/Mile 31¢

Ownership Cost Rating

○ Average

Dodge Shadow ES
2 Door Hatchback

Subcompact

- 2.5L 100 hp Gas Fuel Inject.
- 4 Cylinder In-Line
- Manual 5 Speed
- 2 Wheel Front
- Driver Airbag Psngr Belts

Purchase Price

Car Item	Dealer Cost	List
Base Price	**$9,124**	**$9,804**
Anti-Lock Brakes	$764	$899
Automatic 3 Speed	$473	$557
3.0L 141 hp Gas	$590	** $694
Air Conditioning	$765	$900
Power Steering	Std	Std
Cruise Control	$190	** $224
All Wheel Drive	N/A	N/A
AM/FM Stereo Cassette	$183	$215
Steering Wheel, Tilt	$126	** $148
Power Windows	$225	** $265
*Options Price	$0	$0
*Total Price	**$9,124**	**$9,804**
Target Price		$9,673
Destination Charge		$485
Avg. Tax & Fees		$611
Total Target $		**$10,769**
Average Dealer Option Cost:		85%

The 1993 Shadow is available in six models - (Base) Shadow and ES two- and four-door hatchbacks; and Highline and ES convertibles. New for 1993, the ES Shadow two-door offers standard stainless steel exhaust, a redesigned timing belt cover, premium sound insulation, high performance torque converter, quieter final drive gearing on three speed automatic tansaxles, and a redesigned engine intake manifold. Two new exterior colors are available (Wildberry Pearlcoat and Light Driftwood Satin-glow).

Ownership Costs

Cost Area	5 Year Cost	Rate
Depreciation	$6,376	◐
Financing ($216/month)	$1,774	
Insurance (Rating 11)	$7,693	●
State Fees	$412	
Fuel (Hwy 32 City 25)	$3,030	◐
Maintenance	$3,751	○
Repairs	$731	○

Warranty/Maintenance Info

Major Tune-Up	$134	○
Minor Tune-Up	$73	○
Brake Service	$222	○
Overall Warranty	1 yr/12k	●
Drivetrain Warranty	7 yr/70k	○
Rust Warranty	7 yr/100k	○
Maintenance Warranty	N/A	
Roadside Assistance	N/A	

Ownership Cost By Year

- 1993
- 1994
- 1995
- 1996
- 1997

Resale Value

1993	1994	1995	1996	1997
$7,679	$6,710	$5,913	$5,124	$4,393

Cumulative Costs

	1993	1994	1995	1996	1997
Annual	$5,974	$4,025	$4,555	$3,986	$5,227
Total	$5,974	$9,999	$14,554	$18,540	$23,767

Ownership Costs (5yr)

Average	This Car
$22,609	$23,767
Cost/Mile 32¢	Cost/Mile 34¢

Ownership Cost Rating

◐ Worse Than Average

---

page 66

* Includes shaded options
** Other purchase requirements apply

● Poor ◐ Worse Than Average ○ Average ◯ Better Than Average ◯ Excellent ⊖ Insufficient Information

©1993 by IntelliChoice, Inc. (408) 554-8711 All Rights Reserved. Reproduction Prohibited.
Refer to *Section 3: Annotated Vehicle Charts* for an explanation of these charts.

Dodge Shadow ES
4 Door Hatchback

Subcompact

2.5L 100 hp Gas Fuel Inject. | 4 Cylinder In-Line | Manual 5 Speed | 2 Wheel Front | Driver Airbag Psngr Belts

Purchase Price

Car Item	Dealer Cost	List
Base Price	**$9,484**	**$10,204**
Anti-Lock Brakes	$764	$899
Automatic 3 Speed	$473	$557
3.0L 141 hp Gas	$590	** $694
Air Conditioning	$765	$900
Power Steering	Std	Std
Cruise Control	$190	** $224
All Wheel Drive	N/A	N/A
AM/FM Stereo Cassette	$183	$215
Steering Wheel, Tilt	$126	** $148
Power Windows	$281	** $331
*Options Price	$0	$0
*Total Price	$9,484	$10,204
Target Price		$10,061
Destination Charge		$485
Avg. Tax & Fees		$634
Total Target $		**$11,180**
Average Dealer Option Cost:		85%

Ownership Costs

Cost Area	5 Year Cost	Rate
Depreciation	$6,629	●
Financing ($225/month)	$1,842	
Insurance (Rating 7)	$7,035	○
State Fees	$428	
Fuel (Hwy 32 City 25)	$3,030	●
Maintenance	$3,768	○
Repairs	$731	○

Warranty/Maintenance Info

Major Tune-Up	$134	○
Minor Tune-Up	$73	○
Brake Service	$229	●
Overall Warranty	1 yr/12k	●
Drivetrain Warranty	7 yr/70k	○
Rust Warranty	7 yr/100k	○
Maintenance Warranty	N/A	
Roadside Assistance	N/A	

Ownership Cost By Year

1993, 1994, 1995, 1996, 1997

Resale Value

1993	1994	1995	1996	1997
$7,864	$6,883	$6,087	$5,292	$4,551

Cumulative Costs

	1993	1994	1995	1996	1997
Annual	$6,111	$3,936	$4,448	$3,863	$5,105
Total	$6,111	$10,047	$14,495	$18,358	$23,463

Ownership Costs (5yr)

Average $23,059 | This Car $23,463
Cost/Mile 33¢ | Cost/Mile 34¢

Ownership Cost Rating
○ Average

The 1993 Shadow is available in six models - (Base) Shadow and ES two- and four-door hatchbacks; and Highline and ES convertibles. New for 1993, the ES Shadow four-door offers standard stainless steel exhaust, new seat and door trim, premium sound insulation, high performance torque converter, quieter final drive gearing on three speed automatic transaxles, and a redesigned engine intake manifold. Two new exterior colors are available (Wildberry Pearlcoat and Light Driftwood Satin-glow).

Dodge Shadow Highline
2 Door Convertible

Subcompact

2.5L 100 hp Gas Fuel Inject. | 4 Cylinder In-Line | Manual 5 Speed | 2 Wheel Front | Driver Airbag Psngr Belts

Purchase Price

Car Item	Dealer Cost	List
Base Price	**$12,925**	**$14,028**
Anti-Lock Brakes	$764	$899
Automatic 3 Speed	$473	$557
3.0L 141 hp Gas	$769	** $905
Air Conditioning	$765	$900
Power Steering	Std	Std
Cruise Control	$190	** $224
All Wheel Drive	N/A	N/A
AM/FM Stereo Cassette	$183	$215
Steering Wheel, Tilt	$126	** $148
Power Windows	Std	Std
*Options Price	$0	$0
*Total Price	$12,925	$14,028
Target Price		$13,801
Destination Charge		$485
Avg. Tax & Fees		$859
Total Target $		**$15,145**
Average Dealer Option Cost:		85%

Ownership Costs

Cost Area	5 Year Cost	Rate
Depreciation	$8,957	●
Financing ($304/month)	$2,495	
Insurance (Rating 12 [Est.])	$7,898	●
State Fees	$581	
Fuel (Hwy 32 City 25)	$3,030	●
Maintenance	$3,656	○
Repairs	$731	○

Warranty/Maintenance Info

Major Tune-Up	$134	○
Minor Tune-Up	$73	○
Brake Service	$229	●
Overall Warranty	1 yr/12k	●
Drivetrain Warranty	7 yr/70k	○
Rust Warranty	7 yr/100k	○
Maintenance Warranty	N/A	
Roadside Assistance	N/A	

Resale Value

1993	1994	1995	1996	1997
$10,432	$9,163	$8,167	$7,168	$6,188

Cumulative Costs

	1993	1994	1995	1996	1997
Annual	$7,977	$4,629	$4,926	$4,320	$5,496
Total	$7,977	$12,606	$17,532	$21,852	$27,348

Ownership Costs (5yr)

Average $27,361 | This Car $27,348
Cost/Mile 39¢ | Cost/Mile 39¢

Ownership Cost Rating
○ Average

The 1993 Shadow is available in six models - (Base) shadow and ES two- and four-door hatchbacks; and Highline and ES convertibles. New for 1993, the Shadow Highline convertible is offered in two new exterior colors (Wildberry Pearlcoat and Light Driftwood Satin-glow). Also new is a standard stainless steel exhaust system, redesigned intake manifolds and a quieter final drive gearing on the three-speed auto. Other features include a water-proof vinyl top, 14-inch wheel covers and a bright grille.

---

* Includes shaded options
** Other purchase requirements apply

● Poor | ◐ Worse Than Average | ◑ Average | ○ Better Than Average | ○ Excellent | ⊖ Insufficient Information

©1993 by IntelliChoice, Inc. (408) 554-8711 All Rights Reserved. Reproduction Prohibited.
Refer to *Section 3: Annotated Vehicle Charts* for an explanation of these charts.

page 67

Dodge Shadow ES
2 Door Convertible

Subcompact

 2.5L 100 hp Gas Fuel Inject.
 4 Cylinder In-Line
 Manual 5 Speed
 2 Wheel Front
 Driver Airbag Psngr Belts

Highline Model Shown

Purchase Price

Car Item	Dealer Cost	List
Base Price	**$13,050**	**$14,167**
Anti-Lock Brakes	$764	$899
Automatic 3 Speed	$473	$557
3.0L 141 hp Gas	$769	** $905
Air Conditioning	$765	$900
Power Steering	Std	Std
Cruise Control	$190	** $224
All Wheel Drive	N/A	N/A
AM/FM Stereo Cassette	$183	$215
Steering Wheel, Tilt	$126	** $148
Power Windows	Std	Std
***Options Price**	**$0**	**$0**
***Total Price**	**$13,050**	**$14,167**
Target Price		$13,938
Destination Charge		$485
Avg. Tax & Fees		$868
Total Target $		**$15,291**
Average Dealer Option Cost:		*85%*

The 1993 Shadow is available in six models - (Base) Shadow and ES two- and four-door hatchbacks; and Highline and ES convertibles. New for 1993, the Shadow ES convertible is offered in two new exterior colors (Wildberry Pearlcoat and Light Driftwood Satin-glow). Also new is a standard stainless steel exhaust system, redesigned intake manifolds and quieter final drive gearing on the three-speed auto. Other features include a water-proof vinyl top, new seat and trim fabric and an optional V6 engine.

Ownership Costs

Cost Area	5 Year Cost	Rate
Depreciation	$9,247	●
Financing ($307/month)	$2,518	
Insurance (Rating 14)	$8,238	◐
State Fees	$587	
Fuel (Hwy 32 City 25)	$3,030	◐
Maintenance	$3,656	○
Repairs	$731	○

Warranty/Maintenance Info

Major Tune-Up	$134	○
Minor Tune-Up	$73	○
Brake Service	$229	●
Overall Warranty	1 yr/12k	●
Drivetrain Warranty	7 yr/70k	◐
Rust Warranty	7 yr/100k	◐
Maintenance Warranty	N/A	
Roadside Assistance	N/A	

Ownership Cost By Year

(bar chart: 1993–1997)

Resale Value

1993	1994	1995	1996	1997
$10,062	$8,855	$7,923	$6,973	$6,044

Cumulative Costs

	1993	1994	1995	1996	1997
Annual	$8,568	$4,640	$4,935	$4,345	$5,519
Total	$8,568	$13,208	$18,143	$22,488	$28,007

Ownership Costs (5yr)

Average	This Car
$27,518	$28,007
Cost/Mile 39¢	Cost/Mile 40¢

Ownership Cost Rating

◐ Average

Dodge Spirit Highline
4 Door Sedan

Compact

 2.5L 100 hp Gas Fuel Inject. · 4 Cylinder In-Line
 Manual 5 Speed · 2 Wheel Front
 Driver Airbag Psngr Belts

Purchase Price

Car Item	Dealer Cost	List
Base Price	**$10,863**	**$11,941**
Anti-Lock Brakes	$764	$899
Automatic 3 Speed	$473	$557
3.0L 141 hp Gas	$616	$725
Air Conditioning	Pkg	Pkg
Power Steering	Std	Std
Cruise Control	Pkg	Pkg
All Wheel Drive	N/A	N/A
AM/FM Stereo Cassette	$140	** $165
Steering Wheel, Tilt	Pkg	Pkg
Power Windows	Pkg	Pkg
***Options Price**	**$473**	**$557**
***Total Price**	**$11,336**	**$12,498**
Target Price		$12,238
Destination Charge		$485
Avg. Tax & Fees		$766
Total Target $		**$13,489**
Average Dealer Option Cost:		*85%*

The 1993 Dodge Spirit is available in two models - (Base) and ES. The Spirit is offered with a flexible fuel option for fleet applications as well as retail customers. The Flexible Fuel Vehicles operate on unleaded gasoline or any blend of gasoline and methanol containing up to 85% methanol (M85). Options for the Spirit include electric rear window defroster, dual power mirrors, and a full console with arm rests, bucket seats, and floor shift.

Ownership Costs

Cost Area	5 Year Cost	Rate
Depreciation	$7,951	◐
Financing ($271/month)	$2,223	
Insurance (Rating 2 [Est.])	$6,333	○
State Fees	$520	
Fuel (Hwy 28 City 23)	$3,385	○
Maintenance	$3,856	○
Repairs	$689	○

Warranty/Maintenance Info

Major Tune-Up	$127	○
Minor Tune-Up	$73	○
Brake Service	$259	●
Overall Warranty	1 yr/12k	●
Drivetrain Warranty	7 yr/70k	○
Rust Warranty	7 yr/100k	○
Maintenance Warranty	N/A	
Roadside Assistance	N/A	

Resale Value

1993	1994	1995	1996	1997
$8,778	$7,879	$7,417	$6,396	$5,538

Cumulative Costs

	1993	1994	1995	1996	1997
Annual	$7,623	$3,918	$4,181	$4,041	$5,194
Total	$7,623	$11,541	$15,722	$19,763	$24,957

Ownership Costs (5yr)

Average	This Car
$25,226	$24,957
Cost/Mile 36¢	Cost/Mile 36¢

Ownership Cost Rating

○ Average

---

page 68

* Includes shaded options
** Other purchase requirements apply

● Poor   ◐ Worse Than Average   ◐ Average   ○ Better Than Average   ○ Excellent   ⊖ Insufficient Information

©1993 by IntelliChoice, Inc. (408) 554-8711   All Rights Reserved. Reproduction Prohibited.
Refer to *Section 3: Annotated Vehicle Charts* for an explanation of these charts.

Dodge Spirit ES
4 Door Sedan

 2.5L 100 hp Gas Fuel Inject.
 4 Cylinder In-Line
 Automatic 3 Speed
 2 Wheel Front
Driver Airbag Psngr Belts

Compact

Purchase Price

Car Item	Dealer Cost	List
Base Price	**$13,304**	**$14,715**
Anti-Lock Brakes	$764	$899
4 Spd Elec.	$113	** $133
3.0L 141 hp Gas	$589	$694
Air Conditioning	Pkg	Pkg
Power Steering	Std	Std
Cruise Control	Std	Std
All Wheel Drive	N/A	N/A
AM/FM Stereo Cassette	$302	** $355
Steering Wheel, Tilt	Std	Std
Power Windows	Pkg	Pkg
*Options Price	$0	$0
*Total Price	$13,304	$14,715
Target Price		$14,444
Destination Charge		$485
Avg. Tax & Fees		$898
Total Target $		**$15,827**
Average Dealer Option Cost:		85%

The 1993 Dodge Spirit is available in two models - (Base) and ES. The Spirit ES comes with a flexible fuel option for fleet applications as well as retail customers. The Flexible Fuel Vehicles operate on unleaded gasoline or any blend of gasoline and methanol containing up to 85% methanol (M85). Standards on the Spirit ES include electric rear window defroster, a full console with arm rests, bucket seats, and floor shift, and remote trunk release.

Ownership Costs

Cost Area	5 Year Cost	Rate
Depreciation	$9,305	●
Financing ($318/month)	$2,607	
Insurance (Rating 5)	$6,786	○
State Fees	$608	
Fuel (Hwy 28 City 23)	$3,385	◐
Maintenance	$4,858	◐
Repairs	$689	◐

Warranty/Maintenance Info

Major Tune-Up	$127	○
Minor Tune-Up	$73	○
Brake Service	$259	●
Overall Warranty	1 yr/12k	●
Drivetrain Warranty	7 yr/70k	○
Rust Warranty	7 yr/100k	○
Maintenance Warranty	N/A	
Roadside Assistance	N/A	

Ownership Cost By Year

1993 / 1994 / 1995 / 1996 / 1997

Resale Value

1993	1994	1995	1996	1997
$10,250	$9,214	$8,672	$7,493	$6,522

Cumulative Costs

	1993	1994	1995	1996	1997
Annual	$8,755	$4,282	$4,926	$4,336	$5,939
Total	$8,755	$13,037	$17,963	$22,299	$28,238

Ownership Costs (5yr)

Average	This Car
$27,129	$28,238
Cost/Mile 39¢	Cost/Mile 40¢

Ownership Cost Rating
○ Average

Dodge Stealth
2 Door Coupe

 3.0L 164 hp Gas Fuel Inject.
 6 Cylinder "V"
 Manual 5 Speed
 2 Wheel Front
 Driver Airbag Psngr Belts

Sport

ES Model Shown

Purchase Price

Car Item	Dealer Cost	List
Base Price	**$16,900**	**$18,506**
Anti-Lock Brakes	Pkg	Pkg
Automatic 4 Speed	$742	$863
Optional Engine	N/A	N/A
Air Conditioning	Pkg	Pkg
Power Steering	Std	Std
Cruise Control	Pkg	Pkg
All Wheel Drive	N/A	N/A
AM/FM Stereo Cassette	Pkg	Pkg
Steering Wheel, Tilt	Std	Std
Power Windows	Pkg	Pkg
*Options Price	$0	$0
*Total Price	$16,900	$18,506
Target Price		$17,989
Destination Charge		$400
Avg. Tax & Fees		$1,108
Total Target $		**$19,497**
Average Dealer Option Cost:		86%

The 1993 Stealth is available as a 2+2 Sports Coupe with four trim levels - Base, ES, R/T and R/T Turbo All-Wheel Drive. New for 1993, the Stealth Base has five new exterior colors (White, Pearl White, Dark Gray Metallic, Pearl Blue and Medium Green). Other new features include standard ES sill molding, optional CD changer, remote keyless entry system, optional rear spoiler, leather-wrapped steering wheel and a standard 164 hp engine. The Stealth is built in Nagoya, Japan.

Ownership Costs

Cost Area	5 Year Cost	Rate
Depreciation	$9,507	○
Financing ($392/month)	$3,212	
Insurance (Rating 19)	$9,777	●
State Fees	$755	
Fuel (Hwy 24 City 18 -Prem.)	$4,547	●
Maintenance	$5,997	○
Repairs	$771	○

Warranty/Maintenance Info

Major Tune-Up	$383	●
Minor Tune-Up	$274	●
Brake Service	$261	○
Overall Warranty	1 yr/12k	●
Drivetrain Warranty	7 yr/70k	○
Rust Warranty	7 yr/100k	○
Maintenance Warranty	N/A	
Roadside Assistance	N/A	

Ownership Cost By Year

1993 / 1994 / 1995 / 1996 / 1997

Resale Value

1993	1994	1995	1996	1997
$17,027	$14,628	$12,997	$11,511	$9,990

Cumulative Costs

	1993	1994	1995	1996	1997
Annual	$6,712	$6,869	$7,002	$5,713	$8,270
Total	$6,712	$13,581	$20,583	$26,296	$34,566

Ownership Costs (5yr)

Average	This Car
$36,006	$34,566
Cost/Mile 51¢	Cost/Mile 49¢

Ownership Cost Rating
○ Average

---

* Includes shaded options
** Other purchase requirements apply

● Poor   ◐ Worse Than Average   ○ Average   ○ Better Than Average   ○ Excellent   ⊖ Insufficient Information

©1993 by IntelliChoice, Inc. (408) 554-8711 All Rights Reserved. Reproduction Prohibited.
Refer to *Section 3: Annotated Vehicle Charts* for an explanation of these charts.

Dodge Stealth ES
2 Door Coupe

Sport

 3.0L 222 hp Gas Fuel Inject.
 6 Cylinder "V"
 Manual 5 Speed
 2 Wheel Front
 Driver Airbag Psngr Belts

Purchase Price

Car Item	Dealer Cost	List
Base Price	**$18,498**	**$20,322**
Anti-Lock Brakes	Pkg	Pkg
Automatic 4 Speed	$742	$863
Optional Engine	N/A	N/A
Air Conditioning	Pkg	Pkg
Power Steering	Std	Std
Cruise Control	Pkg	Pkg
All Wheel Drive	N/A	N/A
AM/FM Stereo Cassette	Pkg	Pkg
Steering Wheel, Tilt	Std	Std
Power Windows	Pkg	Pkg
*Options Price	$0	$0
*Total Price	$18,498	$20,322
Target Price		$19,741
Destination Charge		$400
Avg. Tax & Fees		$1,214
Total Target $		**$21,355**
Average Dealer Option Cost:		86%

The 1993 Dodge Stealth is available as a 2+2 Sports Coupe with four trim levels - Base, ES, R/T and R/T Turbo All-Wheel Drive. New for 1993, the ES has five new exterior colors (White, Pearl White, Dark Gray, Pearl Blue and Medium Green). Other features include standard body-color headlight covers, mounted fog lamps, aluminum sport wheels, optional CD changer, leather wrapped steering wheel, remote keyless entry system and an available rear spoiler. The Stealth is built in Nagoya, Japan.

Ownership Costs

Cost Area	5 Year Cost	Rate
Depreciation	$10,748	○
Financing ($429/month)	$3,518	
Insurance (Rating 20++)	$12,805	●
State Fees	$828	
Fuel (Hwy 24 City 19 -Prem.)	$4,436	○
Maintenance	$6,613	○
Repairs	$771	○

Warranty/Maintenance Info

Major Tune-Up	$370	●
Minor Tune-Up	$261	●
Brake Service	$261	○
Overall Warranty	1 yr/12k	●
Drivetrain Warranty	7 yr/70k	○
Rust Warranty	7 yr/100k	○
Maintenance Warranty	N/A	
Roadside Assistance	N/A	

Ownership Cost By Year

1993, 1994, 1995, 1996, 1997

Resale Value

1993	1994	1995	1996	1997
$18,012	$15,588	$13,844	$12,246	$10,607

Cumulative Costs

	1993	1994	1995	1996	1997
Annual	$8,268	$7,554	$8,023	$6,452	$9,423
Total	$8,268	$15,822	$23,845	$30,297	$39,720

Ownership Costs (5yr)

Average	This Car
$37,748	$39,720
Cost/Mile 54¢	Cost/Mile 57¢

Ownership Cost Rating

● Worse Than Average

Dodge Stealth R/T
2 Door Coupe

Sport

 3.0L 222 hp Gas Fuel Inject.
 6 Cylinder "V"
 Manual 5 Speed
 2 Wheel Front
 Driver Airbag Psngr Belts

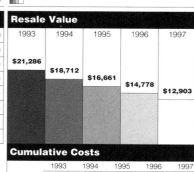

Turbo AWD Shown

Purchase Price

Car Item	Dealer Cost	List
Base Price	**$24,697**	**$27,366**
Anti-Lock Brakes	Std	Std
Automatic 4 Speed	$742	$863
Optional Engine	N/A	N/A
Auto Climate Control	Std	Std
Power Steering	Std	Std
Cruise Control	Std	Std
All Wheel Drive	N/A	N/A
AM/FM Stereo Cassette	Std	Std
Steering Wheel, Tilt	Std	Std
Power Windows	Std	Std
*Options Price	$0	$0
*Total Price	$24,697	$27,366
Target Price		$26,619
Destination Charge		$400
Avg. Tax & Fees		$1,629
Total Target $		**$28,648**
Average Dealer Option Cost:		86%

The 1993 Dodge Stealth is available as a 2+2 Sports Coupe with four trim levels - Base, ES, R/T and R/T Turbo All-Wheel Drive. New for 1993, the R/T Stealth has five new exterior colors (White, Pearl White, Dark Gray Metallic, Pearl Blue and Medium Green). Other features include an optional CD changer, remote keyless entry system, glass black roof, front air dam, black acrylic headlight covers, front and rear body-color bumpers and a rear spoiler. The Stealth is built in Nagoya, Japan.

Ownership Costs

Cost Area	5 Year Cost	Rate
Depreciation	$15,745	○
Financing ($576/month)	$4,720	
Insurance (Rating 22+)	$12,768	●
State Fees	$1,111	
Fuel (Hwy 24 City 19 -Prem.)	$4,436	○
Maintenance	$6,929	○
Repairs	$944	○

Warranty/Maintenance Info

Major Tune-Up	$370	●
Minor Tune-Up	$261	●
Brake Service	$261	○
Overall Warranty	1 yr/12k	●
Drivetrain Warranty	7 yr/70k	○
Rust Warranty	7 yr/100k	○
Maintenance Warranty	N/A	
Roadside Assistance	N/A	

Resale Value

1993	1994	1995	1996	1997
$21,286	$18,712	$16,661	$14,778	$12,903

Cumulative Costs

	1993	1994	1995	1996	1997
Annual	$12,857	$8,172	$8,809	$6,910	$9,906
Total	$12,857	$21,029	$29,838	$36,748	$46,654

Ownership Costs (5yr)

Average	This Car
$44,506	$46,654
Cost/Mile 64¢	Cost/Mile 67¢

Ownership Cost Rating

● Worse Than Average

---

page 70

* Includes shaded options
** Other purchase requirements apply

● Poor   ◐ Worse Than Average   ○ Average   ◑ Better Than Average   ○ Excellent   ⊖ Insufficient Information

©1993 by IntelliChoice, Inc. (408) 554-8711   All Rights Reserved. Reproduction Prohibited.
Refer to *Section 3: Annotated Vehicle Charts* for an explanation of these charts.

Dodge Stealth R/T Turbo AWD
2 Door Coupe

 3.0L 300 hp Turbo Gas Fuel Inject.
 6 Cylinder "V"
 Manual 5 Speed
 4 Wheel Full-Time
 Driver Airbag Psngr Belts

Sport

Purchase Price

Car Item	Dealer Cost	List
Base Price	**$29,749**	**$33,107**
Anti-Lock Brakes	Std	Std
Automatic Transmission	N/A	N/A
Optional Engine	N/A	N/A
Auto Climate Control	Std	Std
Power Steering	Std	Std
Cruise Control	Std	Std
4 Wheel Full-Time Drive	Std	Std
AM/FM Stereo Cassette	Std	Std
Steering Wheel, Tilt	Std	Std
Power Windows	Std	Std
*Options Price	$0	$0
*Total Price	$29,749	$33,107
Target Price		$32,321
Destination Charge		$400
Avg. Tax & Fees		$1,971
Luxury Tax		$272
Total Target $		**$34,964**

The 1993 Dodge Stealth is available as a 2+2 Sports Coupe with four trim levels - Base, ES, R/T and R/T Turbo All-Wheel Drive. New for 1993, the R/T Turbo has five new exterior colors (White, Pearl White, Dark Gray, Pearl Blue and Medium Green). Features include an available CD changer, optional chrome wheels, a forged crankshaft and four-bolt main bearings on the engine. The R/T Turbo features all-wheel drive with center differential and viscous coupling and 4-wheel hydraulic assist steering.

Ownership Costs

Cost Area	5 Year Cost	Rate
Depreciation	$18,591	O
Financing ($703/month)	$5,759	
Insurance (Rating 25++)	$15,433	●
State Fees	$1,341	
Fuel (Hwy 25 City 19 -Prem.)	$4,338	O
Maintenance	$8,228	●
Repairs	$944	O

Warranty/Maintenance Info

Major Tune-Up	$387	●
Minor Tune-Up	$261	●
Brake Service	$261	O
Overall Warranty	1 yr/12k	●
Drivetrain Warranty	7 yr/70k	O
Rust Warranty	7 yr/100k	O
Maintenance Warranty	N/A	
Roadside Assistance	N/A	

Resale Value

1993	1994	1995	1996	1997
$27,411	$24,062	$21,145	$18,717	$16,373

Cumulative Costs

	1993	1994	1995	1996	1997
Annual	$14,060	$9,867	$10,839	$8,131	$11,737
Total	$14,060	$23,927	$34,766	$42,897	$54,634

Ownership Costs (5yr)

Average	This Car
$50,014	$54,634
Cost/Mile 71¢	Cost/Mile 78¢

Ownership Cost Rating

● Poor

Dodge Viper RT/10
2 Door Convertible

 8.0L 400 hp Gas Fuel Inject.
 10 Cylinder "V"
 Manual 6 Speed
 2 Wheel Rear
 Manual Seatbelts Only

Sport

Purchase Price

Car Item	Dealer Cost	List
Base Price	**$43,625**	**$50,000**
Anti-Lock Brakes	N/A	N/A
Automatic Transmission	N/A	N/A
Optional Engine	N/A	N/A
Air Conditioning	N/A	N/A
Power Steering	Std	Std
Cruise Control	N/A	N/A
All Wheel Drive	N/A	N/A
AM/FM Stereo Cassette	Std	Std
Steering Wheel, Tilt	Std	Std
Power Windows	N/A	N/A
*Options Price	$0	$0
*Total Price	$43,625	$50,000
Target Price		$47,988
Destination Charge		$700
Avg. Tax & Fees		$2,941
Luxury Tax		$1,869
Total Target $		**$53,498**

The Dodge Viper RT/10 is only available as a two-seat open sports car. New for 1993, the Dodge Viper has three new colors (black in the fall, Emerald Green in the spring and yellow in the summer) and an extra interior color (Black-and-Tan w/Emerald Green exterior). Features include an optional air conditioning system, fog lamps w/covers, full analog gauge package, 3-point passive restraint system, remote control alarm system and power assisted steering.

Ownership Costs

Cost Area	5 Year Cost	Rate
Depreciation		⊖
Financing ($1,075/month)	$8,812	
Insurance (Rating 24 Sport+)	$17,546	◉
State Fees	$2,028	
Fuel (Hwy 22 City 13 [Est] Pr)	$5,589	●
Maintenance		⊖
Repairs		⊖

Warranty/Maintenance Info

Major Tune-Up		⊖
Minor Tune-Up		⊖
Brake Service		⊖
Overall Warranty	3 yr/36k	O
Drivetrain Warranty	3 yr/36k	O
Rust Warranty	7 yr/100k	O
Maintenance Warranty	N/A	
Roadside Assistance	N/A	

Resale Value

Insufficient Information

Cumulative Costs

	1993	1994	1995	1996	1997
Annual	Insufficient Information				
Total	Insufficient Information				

Ownership Costs (5yr)

Insufficient Information

Ownership Cost Rating

⊖ Insufficient Information

---

* Includes shaded options
** Other purchase requirements apply

● Poor ◉ Worse Than Average ◐ Average O Better Than Average ○ Excellent ⊖ Insufficient Information

©1993 by IntelliChoice, Inc. (408) 554-8711 All Rights Reserved. Reproduction Prohibited.
Refer to *Section 3: Annotated Vehicle Charts* for an explanation of these charts.

Eagle Summit DL
2 Door Coupe

ES 4 Door Shown
Subcompact

- 1.5L 92 hp Gas Fuel Inject.
- 4 Cylinder In-Line
- Manual 5 Speed
- 2 Wheel Front
- Automatic Seatbelts

Purchase Price

Car Item	Dealer Cost	List
Base Price	**$7,488**	**$7,806**
Anti-Lock Brakes	N/A	N/A
Automatic Transmission	N/A	N/A
Optional Engine	N/A	N/A
Air Conditioning	$673	$783
Power Steering	N/A	N/A
Cruise Control	N/A	N/A
All Wheel Drive	N/A	N/A
AM/FM Stereo Cassette	N/A	N/A
Steering Wheel, Tilt	N/A	N/A
Power Windows	N/A	N/A
*Options Price	$0	$0
*Total Price	$7,488	$7,806
Target Price		$7,774
Destination Charge		$400
Avg. Tax & Fees		$491
Total Target $		**$8,665**
Average Dealer Option Cost:		86%

The 1993 Eagle Summit is available in four models - the DL and ES two-door coupes and four-door sedans. This year's coupe body replaces the previous hatchback and offers a fully independent rear suspension. The DL coupe features anti-lock four-wheel disc brakes, 5-mph bumpers, optional tinted glass, tamper-resistant odometer, and a center-floor console with arm rest and storage. Its restraint system is equipped with three child seat anchors. Electronic rear-window defroster is also optional.

Ownership Costs

Cost Area	5 Year Cost	Rate
Depreciation	$5,084	○
Financing ($174/month)	$1,428	
Insurance (Rating 8 [Est.])	$7,223	●
State Fees	$328	
Fuel (Hwy 40 City 32)	$2,398	○
Maintenance	$4,426	○
Repairs	$789	◉

Warranty/Maintenance Info

Major Tune-Up	$260	●
Minor Tune-Up	$89	○
Brake Service	$197	○
Overall Warranty	1 yr/12k	●
Drivetrain Warranty	7 yr/70k	○
Rust Warranty	7 yr/100k	○
Maintenance Warranty	N/A	
Roadside Assistance	N/A	

Ownership Cost By Year
1993, 1994, 1995, 1996, 1997

Resale Value
- 1993: $6,254
- 1994: $5,553
- 1995: $5,231
- 1996: $4,081
- 1997: $3,581

Cumulative Costs

	1993	1994	1995	1996	1997
Annual	$4,937	$3,503	$4,020	$3,728	$5,488
Total	$4,937	$8,440	$12,460	$16,188	$21,676

Ownership Costs (5yr)

Average	This Car
$20,361	$21,676
Cost/Mile 29¢	Cost/Mile 31¢

Ownership Cost Rating
◉ Worse Than Average

Eagle Summit ES
2 Door Coupe

4 Door Model Shown
Subcompact

- 1.5L 92 hp Gas Fuel Inject.
- 4 Cylinder In-Line
- Manual 5 Speed
- 2 Wheel Front
- Automatic Seatbelts

Purchase Price

Car Item	Dealer Cost	List
Base Price	**$8,309**	**$8,705**
Anti-Lock Brakes	N/A	N/A
Automatic 3 Speed	$445	$518
Optional Engine	N/A	N/A
Air Conditioning	$673	$783
Power Steering	Pkg	Pkg
Cruise Control	Pkg	Pkg
All Wheel Drive	N/A	N/A
AM/FM Stereo Cassette	$156	$181
Steering Wheel, Tilt	N/A	N/A
Power Windows	N/A	N/A
*Options Price	$0	$0
*Total Price	$8,309	$8,705
Target Price		$8,665
Destination Charge		$400
Avg. Tax & Fees		$544
Total Target $		**$9,609**
Average Dealer Option Cost:		86%

The 1993 Eagle Summit is available in four models - the DL and ES two-door coupes and four-door sedans. The two-door ES features body-side moldings, split, folding seats in the rear and rear-seat easy entry on passenger side. Options include tinted glass, trunk lights, rear window defroster and power steering. A Sport Appearance Package, with rear spoiler and five-spoke aluminum wheels is also optional. The coupe features a fully independent rear suspension.

Ownership Costs

Cost Area	5 Year Cost	Rate
Depreciation	$5,365	○
Financing ($193/month)	$1,583	
Insurance (Rating 10 [Est.])	$7,479	●
State Fees	$365	
Fuel (Hwy 40 City 32)	$2,398	○
Maintenance	$4,523	○
Repairs	$789	◉

Warranty/Maintenance Info

Major Tune-Up	$260	●
Minor Tune-Up	$89	○
Brake Service	$197	○
Overall Warranty	1 yr/12k	●
Drivetrain Warranty	7 yr/70k	○
Rust Warranty	7 yr/100k	○
Maintenance Warranty	N/A	
Roadside Assistance	N/A	

Resale Value
- 1993: $7,032
- 1994: $6,219
- 1995: $6,069
- 1996: $4,759
- 1997: $4,244

Cumulative Costs

	1993	1994	1995	1996	1997
Annual	$5,223	$3,721	$3,985	$3,959	$5,614
Total	$5,223	$8,944	$12,929	$16,888	$22,502

Ownership Costs (5yr)

Average	This Car
$21,372	$22,502
Cost/Mile 31¢	Cost/Mile 32¢

Ownership Cost Rating
◉ Worse Than Average

---

* Includes shaded options
** Other purchase requirements apply

● Poor ◉ Worse Than Average ○ Average ○ Better Than Average ○ Excellent ⊖ Insufficient Information

©1993 by IntelliChoice, Inc. (408) 554-8711   All Rights Reserved. Reproduction Prohibited.
Refer to *Section 3: Annotated Vehicle Charts* for an explanation of these charts.

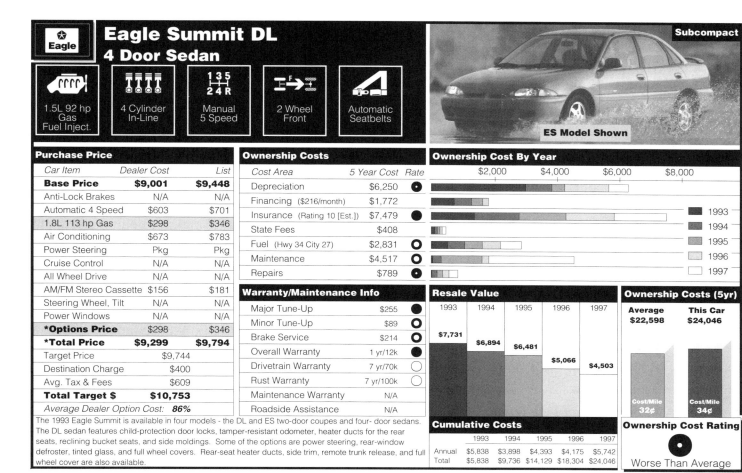

Eagle Summit DL
4 Door Sedan
Subcompact

- 1.5L 92 hp Gas Fuel Inject.
- 4 Cylinder In-Line
- Manual 5 Speed
- 2 Wheel Front
- Automatic Seatbelts

ES Model Shown

Purchase Price

Car Item	Dealer Cost	List
Base Price	**$9,001**	**$9,448**
Anti-Lock Brakes	N/A	N/A
Automatic 4 Speed	$603	$701
1.8L 113 hp Gas	$298	$346
Air Conditioning	$673	$783
Power Steering	Pkg	Pkg
Cruise Control	N/A	N/A
All Wheel Drive	N/A	N/A
AM/FM Stereo Cassette	$156	$181
Steering Wheel, Tilt	N/A	N/A
Power Windows	N/A	N/A
***Options Price**	**$298**	**$346**
***Total Price**	**$9,299**	**$9,794**
Target Price		$9,744
Destination Charge		$400
Avg. Tax & Fees		$609
Total Target $		**$10,753**
Average Dealer Option Cost:		**86%**

The 1993 Eagle Summit is available in four models - the DL and ES two-door coupes and four-door sedans. The DL sedan features child-protection door locks, tamper-resistant odometer, heater ducts for the rear seats, reclining bucket seats, and side moldings. Some of the options are power steering, rear-window defroster, tinted glass, and full wheel covers. Rear-seat heater ducts, side trim, remote trunk release, and full wheel cover are also available.

Ownership Costs

Cost Area	5 Year Cost	Rate
Depreciation	$6,250	●
Financing ($216/month)	$1,772	
Insurance (Rating 10 [Est.])	$7,479	●
State Fees	$408	
Fuel (Hwy 34 City 27)	$2,831	○
Maintenance	$4,517	◐
Repairs	$789	●

Warranty/Maintenance Info

Major Tune-Up	$255	●
Minor Tune-Up	$89	◐
Brake Service	$214	◐
Overall Warranty	1 yr/12k	●
Drivetrain Warranty	7 yr/70k	○
Rust Warranty	7 yr/100k	○
Maintenance Warranty	N/A	
Roadside Assistance	N/A	

Ownership Cost By Year
1993, 1994, 1995, 1996, 1997

Resale Value
- 1993: $7,731
- 1994: $6,894
- 1995: $6,481
- 1996: $5,066
- 1997: $4,503

Cumulative Costs

	1993	1994	1995	1996	1997
Annual	$5,838	$3,898	$4,393	$4,175	$5,742
Total	$5,838	$9,736	$14,129	$18,304	$24,046

Ownership Costs (5yr)
Average	This Car
$22,598	$24,046
Cost/Mile 32¢	Cost/Mile 34¢

Ownership Cost Rating: Worse Than Average

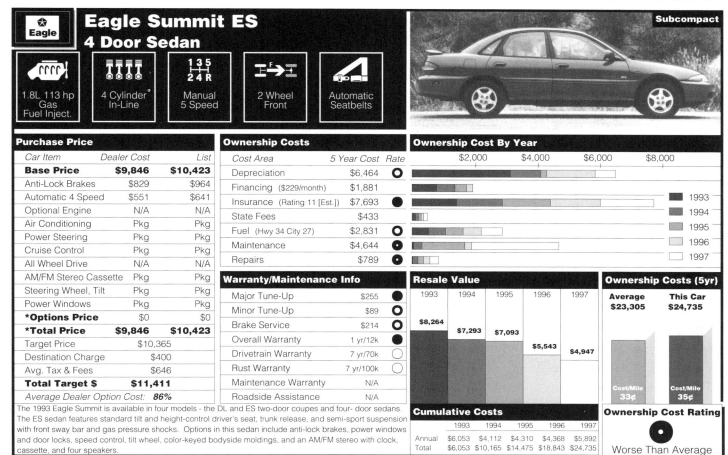

Eagle Summit ES
4 Door Sedan
Subcompact

- 1.8L 113 hp Gas Fuel Inject.
- 4 Cylinder In-Line
- Manual 5 Speed
- 2 Wheel Front
- Automatic Seatbelts

Purchase Price

Car Item	Dealer Cost	List
Base Price	**$9,846**	**$10,423**
Anti-Lock Brakes	$829	$964
Automatic 4 Speed	$551	$641
Optional Engine	N/A	N/A
Air Conditioning	Pkg	Pkg
Power Steering	Pkg	Pkg
Cruise Control	Pkg	Pkg
All Wheel Drive	N/A	N/A
AM/FM Stereo Cassette	Pkg	Pkg
Steering Wheel, Tilt	Pkg	Pkg
Power Windows	Pkg	Pkg
***Options Price**	**$0**	**$0**
***Total Price**	**$9,846**	**$10,423**
Target Price		$10,365
Destination Charge		$400
Avg. Tax & Fees		$646
Total Target $		**$11,411**
Average Dealer Option Cost:		**86%**

The 1993 Eagle Summit is available in four models - the DL and ES two-door coupes and four-door sedans. The ES sedan features standard tilt and height-control driver's seat, trunk release, and semi-sport suspension with front sway bar and gas pressure shocks. Options in this sedan include anti-lock brakes, power windows and door locks, speed control, tilt wheel, color-keyed bodyside moldings, and an AM/FM stereo with clock, cassette, and four speakers.

Ownership Costs

Cost Area	5 Year Cost	Rate
Depreciation	$6,464	○
Financing ($229/month)	$1,881	
Insurance (Rating 11 [Est.])	$7,693	●
State Fees	$433	
Fuel (Hwy 34 City 27)	$2,831	○
Maintenance	$4,644	◐
Repairs	$789	◐

Warranty/Maintenance Info

Major Tune-Up	$255	●
Minor Tune-Up	$89	◐
Brake Service	$214	◐
Overall Warranty	1 yr/12k	●
Drivetrain Warranty	7 yr/70k	○
Rust Warranty	7 yr/100k	○
Maintenance Warranty	N/A	
Roadside Assistance	N/A	

Resale Value
- 1993: $8,264
- 1994: $7,293
- 1995: $7,093
- 1996: $5,543
- 1997: $4,947

Cumulative Costs

	1993	1994	1995	1996	1997
Annual	$6,053	$4,112	$4,310	$4,368	$5,892
Total	$6,053	$10,165	$14,475	$18,843	$24,735

Ownership Costs (5yr)
Average	This Car
$23,305	$24,735
Cost/Mile 33¢	Cost/Mile 35¢

Ownership Cost Rating: Worse Than Average

---

* Includes shaded options
** Other purchase requirements apply

● Poor ◐ Worse Than Average ◑ Average ○ Better Than Average ○ Excellent ⊖ Insufficient Information

©1993 by IntelliChoice, Inc. (408) 554-8711 All Rights Reserved. Reproduction Prohibited.
Refer to *Section 3: Annotated Vehicle Charts* for an explanation of these charts.

Eagle Summit DL
3 Door Wagon

 1.8L 113 hp Gas Fuel Inject. | 4 Cylinder In-Line | Manual 5 Speed | 2 Wheel Front | Automatic Seatbelts

Compact Wagon — AWD Model Shown

Purchase Price

Car Item	Dealer Cost	List
Base Price	**$10,660**	**$11,455**
Anti-Lock Brakes	$829	** $964
Automatic 4 Speed	$622	$723
2.4L 136 hp Gas	$156	$181
Air Conditioning	$673	$783
Power Steering	Std	Std
Cruise Control	Pkg	Pkg
All Wheel Drive	N/A	N/A
AM/FM Stereo Cassette	$156	$181
Steering Wheel, Tilt	N/A	N/A
Power Windows	N/A	N/A
*Options Price	$1,607	$1,868
*Total Price	$12,267	$13,323
Target Price	$13,043	
Destination Charge	$400	
Avg. Tax & Fees	$809	
Total Target $	**$14,252**	
Average Dealer Option Cost:	**86%**	

Ownership Costs

Cost Area	5 Year Cost	Rate
Depreciation	$8,427	●
Financing ($286/month)	$2,348	
Insurance (Rating 6 [Est.])	$6,919	○
State Fees	$549	
Fuel (Hwy 26 City 20)	$3,756	●
Maintenance		⊖
Repairs	$789	●

Warranty/Maintenance Info

Major Tune-Up		⊖
Minor Tune-Up		⊖
Brake Service	$177	○
Overall Warranty	1 yr/12k	●
Drivetrain Warranty	7 yr/70k	○
Rust Warranty	7 yr/100k	○
Maintenance Warranty	N/A	
Roadside Assistance	N/A	

Ownership Cost By Year — 1993, 1994, 1995, 1996, 1997

Resale Value: 1993 $9,503 | 1994 $8,421 | 1995 $7,576 | 1996 $6,666 | 1997 $5,825

Ownership Costs (5yr): Insufficient Information

Cumulative Costs: Annual / Total — Insufficient Information

Ownership Cost Rating: ⊖ Insufficient Information

The 1993 Eagle Summit Wagon is available in three models - DL, LX, and AWD. The 1993 Eagle Summit Wagon DL offers for the first time, a cargo security cover on the outside. Other changes from the previous model include the addition of bodyside moldings, halogen headlamps, variable, intermittent windshield wipers, full wheel covers, and an optional luggage rack. Other options include anti-lock brakes, rear-window defroster, tinted glass, and rear wiper and washer.

Eagle Summit LX
3 Door Wagon

 2.4L 136 hp Gas Fuel Inject. | 4 Cylinder In-Line | Manual 5 Speed | 2 Wheel Front | Automatic Seatbelts

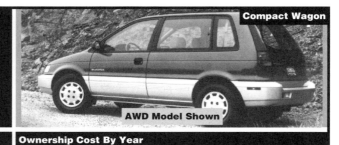
Compact Wagon — AWD Model Shown

Purchase Price

Car Item	Dealer Cost	List
Base Price	**$11,481**	**$12,368**
Anti-Lock Brakes	$829	$964
Automatic 4 Speed	$622	$723
Optional Engine	N/A	N/A
Air Conditioning	Pkg	Pkg
Power Steering	Std	Std
Cruise Control	Pkg	Pkg
All Wheel Drive	N/A	N/A
AM/FM Stereo Cassette	Pkg	Pkg
Steering Wheel, Tilt	Std	Std
Power Windows	Pkg	Pkg
*Options Price	$622	$723
*Total Price	$12,103	$13,091
Target Price	$12,888	
Destination Charge	$400	
Avg. Tax & Fees	$799	
Total Target $	**$14,087**	
Average Dealer Option Cost:	**86%**	

Ownership Costs

Cost Area	5 Year Cost	Rate
Depreciation	$8,405	●
Financing ($283/month)	$2,321	
Insurance (Rating 6 [Est.])	$6,919	○
State Fees	$539	
Fuel (Hwy 26 City 20)	$3,756	●
Maintenance		⊖
Repairs	$789	●

Warranty/Maintenance Info

Major Tune-Up		⊖
Minor Tune-Up		⊖
Brake Service	$177	○
Overall Warranty	1 yr/12k	●
Drivetrain Warranty	7 yr/70k	○
Rust Warranty	7 yr/100k	○
Maintenance Warranty	N/A	
Roadside Assistance	N/A	

Resale Value: 1993 $9,318 | 1994 $8,253 | 1995 $7,422 | 1996 $6,512 | 1997 $5,682

Ownership Costs (5yr): Insufficient Information

Cumulative Costs: Annual / Total — Insufficient Information

Ownership Cost Rating: ⊖ Insufficient Information

The 1993 Eagle Summit Wagon is available in three models - DL, LX, and AWD. The 1993 Eagle Summit Wagon LX features a new, full-cloth velour fabric. This model offers a standard child safety lock for the rear sliding door, power lock tailgate, dual electric mirrors, and rear, left side shelf for storage. The reclining, split-back bench in the rear can fold, tumble, and is removable. Options include remote keyless entry, luggage rack, electronic speed control, and rear window defroster.

---

* Includes shaded options
** Other purchase requirements apply

● Poor | ◐ Worse Than Average | ◑ Average | ○ Better Than Average | ○ Excellent | ⊖ Insufficient Information

©1993 by IntelliChoice, Inc. (408) 554-8711 All Rights Reserved. Reproduction Prohibited.
Refer to *Section 3: Annotated Vehicle Charts* for an explanation of these charts.

Eagle Summit AWD
3 Door Wagon

1.8L 113 hp Gas Fuel Inject. | 4 Cylinder In-Line | Manual 5 Speed | 4 Wheel Full-Time | Automatic Seatbelts

Compact Wagon

Purchase Price

Car Item	Dealer Cost	List
Base Price	**$12,535**	**$13,539**
Anti-Lock Brakes	$829	** $964
Automatic 4 Speed	$622	$723
2.4L 136 hp Gas	$156	$181
Air Conditioning	$673	$783
Power Steering	Std	Std
Cruise Control	Pkg	Pkg
4 Wheel Full-Time Drive	Std	Std
AM/FM Stereo Cassette	$156	$181
Steering Wheel, Tilt	Std	Std
Power Windows	Pkg	Pkg
*Options Price	$1,607	$1,868
*Total Price	$14,142	$15,407
Target Price	$15,090	
Destination Charge	$400	
Avg. Tax & Fees	$933	
Total Target $	**$16,423**	
Average Dealer Option Cost:	**86%**	

The 1993 Eagle Summit Wagon is available in three models - DL, LX, and AWD. The 1993 Eagle Summit Wagon AWD. The AWD model features a standard stainless steel exhaust system, leather-wrapped sport steering wheel, bodyside moldings, front and rear body color fascias, and full tinted glass with windshield sun shaded band. Leather front seats, sunroof, anti-lock brakes, rear lift gate wiper/washer, and power windows and door locks are optional.

Ownership Costs

Cost Area	5 Year Cost	Rate
Depreciation	$9,584	●
Financing ($330/month)	$2,706	
Insurance (Rating 7 [Est.])	$7,035	○
State Fees	$632	
Fuel (Hwy 23 City 19)	$4,111	●
Maintenance		⊖
Repairs	$789	●

Warranty/Maintenance Info

Major Tune-Up		⊖
Minor Tune-Up		⊖
Brake Service	$185	○
Overall Warranty	1 yr/12k	●
Drivetrain Warranty	7 yr/70k	○
Rust Warranty	7 yr/100k	○
Maintenance Warranty	N/A	
Roadside Assistance	N/A	

Ownership Cost By Year

1993 / 1994 / 1995 / 1996 / 1997

Insufficient Maintenance Information

Resale Value

1993	1994	1995	1996	1997
$10,890	$9,706	$8,779	$7,762	$6,839

Cumulative Costs

	1993	1994	1995	1996	1997
Annual	*Insufficient Information*				
Total	*Insufficient Information*				

Ownership Costs (5yr)

Insufficient Information

Ownership Cost Rating

⊖

Insufficient Information

Eagle Talon DL
2 Door Coupe

1.8L 92 hp Gas Fuel Inject. | 4 Cylinder In-Line | Manual 5 Speed | 2 Wheel Front | Automatic Seatbelts

Subcompact

Purchase Price

Car Item	Dealer Cost	List
Base Price	**$10,957**	**$11,752**
Anti-Lock Brakes	N/A	N/A
Automatic 4 Speed	$609	** $716
Optional Engine	N/A	N/A
Air Conditioning	Pkg	Pkg
Power Steering	Pkg	Pkg
Cruise Control	Pkg	Pkg
All Wheel Drive	N/A	N/A
AM/FM Stereo Cassette	$168	** $198
Steering Wheel, Tilt	Std	Std
Power Windows	N/A	N/A
*Options Price	$0	$0
*Total Price	$10,957	$11,752
Target Price	$11,551	
Destination Charge	$400	
Avg. Tax & Fees	$720	
Total Target $	**$12,671**	
Average Dealer Option Cost:	**85%**	

The 1993 Eagle Talon is available in four trim levels - (Base) DL, ES, TSi FWD and TSi AWD. New for 1993, the Talon DL features a rear spoiler, improved door sound speaker quality, more ergonomic shifter knob for manual transmissions, and medium quartz-painted aluminum wheels (available with dark exterior colors). One new color is offered: Deep Green. Eagle Talons are produced by Diamond-Star Motors in a joint venture between Chrysler and Mitsubishi and are built in Normal, Illinois.

Ownership Costs

Cost Area	5 Year Cost	Rate
Depreciation	$5,935	○
Financing ($255/month)	$2,087	
Insurance (Rating 13)	$8,061	●
State Fees	$486	
Fuel (Hwy 32 City 23)	$3,152	●
Maintenance	$4,417	○
Repairs	$759	●

Warranty/Maintenance Info

Major Tune-Up	$249	●
Minor Tune-Up	$89	○
Brake Service	$239	●
Overall Warranty	1 yr/12k	●
Drivetrain Warranty	7 yr/70k	○
Rust Warranty	7 yr/100k	○
Maintenance Warranty	N/A	
Roadside Assistance	N/A	

Ownership Cost By Year

Resale Value

1993	1994	1995	1996	1997
$10,450	$9,761	$8,575	$7,649	$6,736

Cumulative Costs

	1993	1994	1995	1996	1997
Annual	$5,353	$3,962	$5,442	$4,498	$5,642
Total	$5,353	$9,315	$14,757	$19,255	$24,897

Ownership Costs (5yr)

Average	This Car
$24,801	$24,897
Cost/Mile 35¢	Cost/Mile 36¢

Ownership Cost Rating

○

Average

---

* Includes shaded options
** Other purchase requirements apply

● Poor | ◐ Worse Than Average | ◉ Average | ○ Better Than Average | ○ Excellent | ⊖ Insufficient Information

©1993 by IntelliChoice, Inc. (408) 554-8711   All Rights Reserved. Reproduction Prohibited.
Refer to *Section 3: Annotated Vehicle Charts* for an explanation of these charts.

Eagle Talon ES
2 Door Coupe

Subcompact

- 2.0L 135 hp Gas Fuel Inject.
- 4 Cylinder In-Line
- Manual 5 Speed
- 2 Wheel Front
- Automatic Seatbelts

DL Model Shown

Purchase Price		
Car Item	Dealer Cost	List
Base Price	**$13,177**	**$14,197**
Anti-Lock Brakes	$802	$943
Automatic 4 Speed	$609	$716
Optional Engine	N/A	N/A
Air Conditioning	Pkg	Pkg
Power Steering	Std	Std
Cruise Control	Pkg	Pkg
All Wheel Drive	N/A	N/A
AM/FM Stereo Cassette	Std	Std
Steering Wheel, Tilt	Std	Std
Power Windows	Pkg	Pkg
*Options Price	$0	$0
*Total Price	$13,177	$14,197
Target Price		$13,942
Destination Charge		$400
Avg. Tax & Fees		$863
Total Target $		**$15,205**
Average Dealer Option Cost:		85%

Ownership Costs		
Cost Area	5 Year Cost	Rate
Depreciation	$7,068	○
Financing ($306/month)	$2,505	
Insurance (Rating 15)	$8,507	●
State Fees	$584	
Fuel (Hwy 29 City 22)	$3,390	⊙
Maintenance	$5,307	●
Repairs	$759	⊙

Warranty/Maintenance Info		
Major Tune-Up	$310	●
Minor Tune-Up	$89	○
Brake Service	$239	●
Overall Warranty	1 yr/12k	●
Drivetrain Warranty	7 yr/70k	◌
Rust Warranty	7 yr/100k	◌
Maintenance Warranty	N/A	
Roadside Assistance	N/A	

Ownership Cost By Year — $2,000 $4,000 $6,000 $8,000 $10,000 — 1993, 1994, 1995, 1996, 1997

Resale Value
1993	1994	1995	1996	1997
$12,521	$11,712	$10,287	$9,195	$8,137

Cumulative Costs
	1993	1994	1995	1996	1997
Annual	$6,141	$4,366	$6,222	$4,853	$6,538
Total	$6,141	$10,507	$16,729	$21,582	$28,120

Ownership Costs (5yr)
- Average: $27,552
- This Car: $28,120
- Cost/Mile: 39¢ / 40¢

Ownership Cost Rating: ○ Average

The 1993 Eagle Talon is available in four trim levels - (Base) DL, ES, TSi FWD and TSi AWD. New for 1993, the ES features a rear spoiler, improved door sound speaker quality, more ergonomic shifter knob for manual transmissions, sixteen-inch polycast wheels, and medium quartz-painted aluminum wheels (available with dark exterior colors). One new color is offered: Deep Green. Eagle Talons are produced in Normal, Illinois by Diamond-Star Motors in a joint venture between Chrysler and Mitsubishi.

Eagle Talon TSi Turbo
2 Door Coupe

Subcompact

- 2.0L 195 hp Turbo Gas Fuel Inject.
- 4 Cylinder In-Line
- Manual 5 Speed
- 2 Wheel Front
- Automatic Seatbelts

AWD Model Shown

Purchase Price		
Car Item	Dealer Cost	List
Base Price	**$14,548**	**$15,703**
Anti-Lock Brakes	$802	$943
Automatic 4 Speed	$728	** $857
Optional Engine	N/A	N/A
Air Conditioning	Pkg	Pkg
Power Steering	Std	Std
Cruise Control	Pkg	Pkg
All Wheel Drive	N/A	N/A
AM/FM Stereo Cassette	Std	Std
Steering Wheel, Tilt	Std	Std
Power Windows	Pkg	Pkg
*Options Price	$0	$0
*Total Price	$14,548	$15,703
Target Price		$15,427
Destination Charge		$400
Avg. Tax & Fees		$952
Total Target $		**$16,779**
Average Dealer Option Cost:		85%

Ownership Costs		
Cost Area	5 Year Cost	Rate
Depreciation	$8,568	○
Financing ($337/month)	$2,764	
Insurance (Rating 16+)	$10,541	●
State Fees	$645	
Fuel (Hwy 28 City 21)	$3,530	⊙
Maintenance	$5,480	●
Repairs	$1,046	●

Warranty/Maintenance Info		
Major Tune-Up	$310	●
Minor Tune-Up	$89	○
Brake Service	$239	●
Overall Warranty	1 yr/12k	●
Drivetrain Warranty	7 yr/70k	◌
Rust Warranty	7 yr/100k	◌
Maintenance Warranty	N/A	
Roadside Assistance	N/A	

Ownership Cost By Year — $2,000 $4,000 $6,000 $8,000 $10,000 $12,000 — 1993, 1994, 1995, 1996, 1997

Resale Value
1993	1994	1995	1996	1997
$12,701	$12,077	$10,439	$9,293	$8,211

Cumulative Costs
	1993	1994	1995	1996	1997
Annual	$8,107	$4,792	$7,033	$5,483	$7,158
Total	$8,107	$12,899	$19,932	$25,415	$32,573

Ownership Costs (5yr)
- Average: $29,246
- This Car: $32,573
- Cost/Mile: 42¢ / 47¢

Ownership Cost Rating: ● Poor

The 1993 Eagle Talon is available in four trim levels - (Base) DL, ES, TSi FWD and TSi AWD. New for 1993, the TSi (FWD) gets larger front brakes for improved braking and receives gas-filled front and rear shock absorbers. Other features include a rear spoiler, improved door sound speaker quality, more ergonomic shifter knob for manual transmissions, sixteen-inch polycast wheels, and medium quartz-painted aluminum wheels (available with dark exterior colors). One new color is offered: Deep Green.

---

page 76

* Includes shaded options
** Other purchase requirements apply

● Poor | ⊙ Worse Than Average | ○ Average | ◌ Better Than Average | ○ Excellent | ⊖ Insufficient Information

©1993 by IntelliChoice, Inc. (408) 554-8711 All Rights Reserved. Reproduction Prohibited.
Refer to *Section 3: Annotated Vehicle Charts* for an explanation of these charts.

Eagle Talon TSi Turbo AWD
2 Door Coupe

Subcompact

- 2.0L 195 hp Turbo Gas Fuel Inject.
- 4 Cylinder In-Line
- Manual 5 Speed
- 4 Wheel Full-Time
- Automatic Seatbelts

Purchase Price

Car Item	Dealer Cost	List
Base Price	**$16,435**	**$17,772**
Anti-Lock Brakes	$802	$943
Automatic 4 Speed	$728	** $857
Optional Engine	N/A	N/A
Air Conditioning	Pkg	Pkg
Power Steering	Std	Std
Cruise Control	Pkg	Pkg
4 Wheel Full-Time Drive	Std	Std
AM/FM Stereo Cassette	Std	Std
Steering Wheel, Tilt	Std	Std
Power Windows	Pkg	Pkg
*Options Price	$0	$0
*Total Price	$16,435	$17,772
Target Price		$17,481
Destination Charge		$400
Avg. Tax & Fees		$1,076
Total Target $		**$18,957**
Average Dealer Option Cost:		85%

Ownership Costs

Cost Area	5 Year Cost	Rate
Depreciation	$9,952	○
Financing ($381/month)	$3,123	
Insurance (Rating 18+)	$11,344	●
State Fees	$727	
Fuel (Hwy 25 City 20)	$3,837	●
Maintenance	$5,455	●
Repairs	$1,085	●

Warranty/Maintenance Info

Major Tune-Up	$310 ●
Minor Tune-Up	$89 ○
Brake Service	$239 ⊙
Overall Warranty	1 yr/12k ●
Drivetrain Warranty	7 yr/70k ○
Rust Warranty	7 yr/100k ○
Maintenance Warranty	N/A
Roadside Assistance	N/A

Ownership Cost By Year

(1993, 1994, 1995, 1996, 1997)

Resale Value

1993	1994	1995	1996	1997
$13,895	$13,238	$11,445	$10,200	$9,005

Cumulative Costs

	1993	1994	1995	1996	1997
Annual	$9,467	$5,177	$7,506	$5,838	$7,535
Total	$9,467	$14,644	$22,150	$27,988	$35,523

Ownership Costs (5yr)

Average	This Car
$31,574	$35,523
Cost/Mile 45¢	Cost/Mile 51¢

Ownership Cost Rating

● Poor

The 1993 Eagle Talon is available in four trim levels - (Base) DL, ES, TSi FWD and TSi AWD. New for 1993, the TSi (AWD) has larger front brakes for improved braking and receives gas-filled front and rear shock absorbers. Other features include a rear spoiler, improved door sound speaker quality, more ergonomic shifter knob for manual transmissions, sixteen-inch polycast wheels, and medium quartz-painted aluminum wheels (available with dark exterior colors). One new color is offered: Deep Green.

Eagle Vision ESi
4 Door Sedan

Large

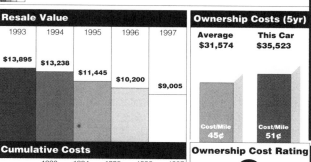

- 3.3L 153 hp Gas Fuel Inject.
- 6 Cylinder "V"
- Automatic 4 Speed
- 2 Wheel Front
- Driver/Psngr Airbags Std

Purchase Price

Car Item	Dealer Cost	List
Base Price	**$15,264**	**$17,387**
Anti-Lock Brakes	$509	$599
Manual Transmission	N/A	N/A
Optional Engine	N/A	N/A
Air Conditioning	Std	Std
Power Steering	Std	Std
Cruise Control	$190	$224
All Wheel Drive	N/A	N/A
AM/FM Stereo Cassette	$606	$713
Steering Wheel, Tilt	Std	Std
Power Windows	Pkg	Pkg
*Options Price	$796	$937
*Total Price	$16,060	$18,324
Target Price		$17,632
Destination Charge		$490
Avg. Tax & Fees		$1,094
Total Target $		**$19,216**
Average Dealer Option Cost:		85%

Ownership Costs

Cost Area	5 Year Cost	Rate
Depreciation		⊖
Financing ($386/month)	$3,166	
Insurance (Rating 10)	$7,479	○
State Fees	$753	
Fuel (Hwy 28 City 20)	$3,612	○
Maintenance		⊖
Repairs	$740	○

Warranty/Maintenance Info

Major Tune-Up	⊖
Minor Tune-Up	⊖
Brake Service	⊖
Overall Warranty	1 yr/12k ●
Drivetrain Warranty	7 yr/70k ○
Rust Warranty	7 yr/100k ○
Maintenance Warranty	N/A
Roadside Assistance	N/A

Ownership Cost By Year

Insufficient Depreciation Information

Resale Value

Insufficient Information

Cumulative Costs

	1993	1994	1995	1996	1997
Annual	Insufficient Information				
Total	Insufficient Information				

Ownership Costs (5yr)

Insufficient Information

Ownership Cost Rating

⊖ Insufficient Information

The all-new Eagle Vision makes its debut in 1993 with two editions - ESi and TSi sedan. The Vision is the Eagle division's version of the Chrysler Corporation's new LH model whose production incorporated new factory habits such as fewer job classifications, lean production practices, and the melding of diverse departments. The ESi sedan comes equipped with standard driver and passenger side airbags, thermoplastic front fenders, stainless steel exhaust system and a tilt steering column.

---

* Includes shaded options
** Other purchase requirements apply

● Poor ◐ Worse Than Average ○ Average ○ Better Than Average ○ Excellent ⊖ Insufficient Information

©1993 by IntelliChoice, Inc. (408) 554-8711 All Rights Reserved. Reproduction Prohibited.
Refer to *Section 3: Annotated Vehicle Charts* for an explanation of these charts.

Eagle Vision TSi
4 Door Sedan

 3.5L 214 hp Gas Fuel Inject.
 6 Cylinder "V"
 Automatic 4 Speed
 2 Wheel Front
 Driver/Psngr Airbags Std

Large

Purchase Price

Car Item	Dealer Cost	List
Base Price	**$18,458**	**$21,104**
Anti-Lock Brakes	Std	Std
Manual Transmission	N/A	N/A
Optional Engine	N/A	N/A
Auto Climate Control	Std	Std
Power Steering	Std	Std
Cruise Control	Std	Std
All Wheel Drive	N/A	N/A
AM/FM Stereo Cassette	Std	Std
Steering Wheel, Tilt	Std	Std
Power Windows	$290	$341
***Options Price**	$290	$341
***Total Price**	**$18,748**	**$21,445**
Target Price	$20,746	
Destination Charge	$490	
Avg. Tax & Fees	$1,281	
Total Target $	**$22,517**	
Average Dealer Option Cost:	85%	

The all-new Eagle Vision makes its debut in 1993 with two editions -ESi and TSi sedans. In addition to or replacing the standard features of the ESi, the TSi comes with automatic climate control, standard anti-lock braking system, cargo/convenience net, overhead console with compass, reading lamps, traveler computer and illuminated visor vanity mirrors; aluminum wheels and a larger engine. Options on the TSi include a built-in fold-away child seat, leather seating material and power windows.

Ownership Costs

Cost Area	5 Year Cost	Rate
Depreciation		⊖
Financing ($453/month)	$3,708	
Insurance (Rating 13+)	$9,673	●
State Fees	$877	
Fuel (Hwy 26 City 18)	$3,949	○
Maintenance		⊖
Repairs	$740	○

Warranty/Maintenance Info

Major Tune-Up		⊖
Minor Tune-Up		⊖
Brake Service		⊖
Overall Warranty	1 yr/12k	●
Drivetrain Warranty	7 yr/70k	○
Rust Warranty	7 yr/100k	○
Maintenance Warranty	N/A	
Roadside Assistance	N/A	

Ownership Cost By Year

Insufficient Depreciation Information

1993 / 1994 / 1995 / 1996 / 1997

Insufficient Maintenance Information

Resale Value
Insufficient Information

Cumulative Costs
	1993	1994	1995	1996	1997
Annual	Insufficient Information				
Total	Insufficient Information				

Ownership Costs (5yr)
Insufficient Information

Ownership Cost Rating
⊖ Insufficient Information

Ford Crown Victoria
4 Door Sedan

 4.6L 190 hp Gas Fuel Inject.
 8 Cylinder "V"
 Automatic 4 Speed
 2 Wheel Rear
Driver Airbag Psngr Opt

Large — LX Model Shown

Purchase Price

Car Item	Dealer Cost	List
Base Price	**$17,146**	**$19,972**
Anti-Lock Brakes	$591	$695
Manual Transmission	N/A	N/A
Optional Engine	N/A	N/A
Air Conditioning	Std	Std
Power Steering	Std	Std
Cruise Control	Pkg	Pkg
All Wheel Drive	N/A	N/A
AM/FM Stereo Cassette	$145	$171
Steering Wheel, Tilt	Std	Std
Power Windows	Std	Std
***Options Price**	$145	$171
***Total Price**	**$17,291**	**$20,143**
Target Price	$18,544	
Destination Charge	$545	
Avg. Tax & Fees	$1,161	
Total Target $	**$20,250**	
Average Dealer Option Cost:	85%	

The 1993 Ford Crown Victoria is available in two models - Crown Victoria (Base) and LX. New for 1993, the Crown Victoria (Base) features a new grille, color-coordinated body side moldings, driver's footrest, and decklid applique. Standard features include tinted glass, dual remote control power mirrors, deluxe wheel covers, gauge cluster, scuff plates, split bench seat, and a coolant recovery system. Optional features include a convenience group, and a cellular telephone with storage armrest.

Ownership Costs

Cost Area	5 Year Cost	Rate
Depreciation	$11,150	○
Financing ($407/month)	$3,335	
Insurance (Rating 8)	$7,223	○
State Fees	$828	
Fuel (Hwy 26 City 18)	$3,949	○
Maintenance	$4,003	○
Repairs	$761	●

Warranty/Maintenance Info

Major Tune-Up	$147	○
Minor Tune-Up	$89	○
Brake Service	$286	●
Overall Warranty	3 yr/36k	○
Drivetrain Warranty	3 yr/36k	○
Rust Warranty	6 yr/100k	○
Maintenance Warranty	N/A	
Roadside Assistance	N/A	

Ownership Cost By Year
1993 / 1994 / 1995 / 1996 / 1997

Resale Value

1993	1994	1995	1996	1997
$15,025	$13,308	$11,918	$10,550	$9,100

Cumulative Costs

	1993	1994	1995	1996	1997
Annual	$8,932	$5,320	$5,897	$4,938	$6,162
Total	$8,932	$14,252	$20,149	$25,087	$31,249

Ownership Costs (5yr)
Average	This Car
$32,446	$31,249
Cost/Mile 46¢	Cost/Mile 45¢

Ownership Cost Rating
○ Better Than Average

---

page 78 — * Includes shaded options ** Other purchase requirements apply

● Poor ◐ Worse Than Average ○ Average ○ Better Than Average ○ Excellent ⊖ Insufficient Information

©1993 by *IntelliChoice, Inc.* (408) 554-8711   All Rights Reserved. Reproduction Prohibited.
Refer to *Section 3: Annotated Vehicle Charts* for an explanation of these charts.

Ford Crown Victoria LX
4 Door Sedan

Large

 4.6L 190 hp Gas Fuel Inject.
 8 Cylinder "V"
 Automatic 4 Speed
 2 Wheel Rear
 Driver Airbag Psngr Opt

Purchase Price

Car Item	Dealer Cost	List
Base Price	**$18,495**	**$21,559**
Anti-Lock Brakes	$591	$695
Manual Transmission	N/A	N/A
Optional Engine	N/A	N/A
Air Conditioning	Std	Std
Power Steering	Std	Std
Cruise Control	Pkg	Pkg
All Wheel Drive	N/A	N/A
AM/FM Stereo Cassette	$282	** $331
Steering Wheel, Tilt	Std	Std
Power Windows	Std	Std
***Options Price**	**$0**	**$0**
***Total Price**	**$18,495**	**$21,559**
Target Price		$19,883
Destination Charge		$545
Avg. Tax & Fees		$1,242
Total Target $		**$21,670**
Average Dealer Option Cost:		87%

The 1993 Ford Crown Victoria is available in two models - Crown Victoria (Base) and LX. New for 1993, the LX features a decklid applique, driver's footrest, new cloth seat style and a new grille. Standard features include dual remote control power mirrors, luxury cloth power drivers seat, gauge cluster, deluxe wheel covers, and scuff plates. Optional features include cellular telephone with storage armrest, convenience group, cornering lamps, and a handling and performance package.

Ownership Costs

Cost Area	5 Year Cost	Rate
Depreciation	$12,169	O
Financing ($436/month)	$3,570	
Insurance (Rating 8)	$7,223	O
State Fees	$885	
Fuel (Hwy 26 City 18)	$3,949	O
Maintenance	$4,003	O
Repairs	$761	⊙

Warranty/Maintenance Info

Major Tune-Up	$147	O
Minor Tune-Up	$89	O
Brake Service	$286	⊙
Overall Warranty	3 yr/36k	O
Drivetrain Warranty	3 yr/36k	O
Rust Warranty	6 yr/100k	O
Maintenance Warranty	N/A	
Roadside Assistance	N/A	

Ownership Cost By Year

1993, 1994, 1995, 1996, 1997

Resale Value

1993	1994	1995	1996	1997
$16,210	$14,321	$12,661	$11,119	$9,501

Cumulative Costs

	1993	1994	1995	1996	1997
Annual	$9,279	$5,578	$6,225	$5,139	$6,339
Total	$9,279	$14,857	$21,082	$26,221	$32,560

Ownership Costs (5yr)

Average: $33,851 — This Car: $32,560
Cost/Mile 48¢ — Cost/Mile 47¢

Ownership Cost Rating

O Better Than Average

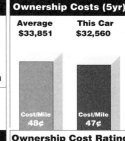

Ford Escort Pony
2 Door Hatchback

Subcompact

LX Model Shown

 1.9L 88 hp Gas Fuel Inject.
 4 Cylinder In-Line
 Manual 5 Speed
 2 Wheel Front
 Automatic Seatbelts

Purchase Price

Car Item	Dealer Cost	List
Base Price	**$7,688**	**$8,355**
Anti-Lock Brakes	N/A	N/A
Automatic 4 Speed	$622	$732
Optional Engine	N/A	N/A
Air Conditioning	Pkg	Pkg
Power Steering	Pkg	Pkg
Cruise Control	N/A	N/A
All Wheel Drive	N/A	N/A
AM/FM Stereo Cassette	$397	$467
Steering Wheel, Tilt	N/A	N/A
Power Windows	N/A	N/A
***Options Price**	**$0**	**$0**
***Total Price**	**$7,688**	**$8,355**
Target Price		$8,084
Destination Charge		$375
Avg. Tax & Fees		$510
Total Target $		**$8,969**
Average Dealer Option Cost:		85%

The 1993 Escort is available in seven models - Pony, LX and GT two-door hatchbacks; LX and LX-E sedans; and the LX four-door hatchback and wagon. New for 1993, the Pony offers a new grille, and several new exterior colors including Caymen Green, Mocha Frost and Silver Clearcoat Metallic. The Pony comes equipped with cargo cover, tinted glass, and intermittent wipers. Options include a Comfort Group, and AM/FM stereo cassette.

Ownership Costs

Cost Area	5 Year Cost	Rate
Depreciation	$5,180	⊙
Financing ($180/month)	$1,477	
Insurance (Rating 10 [Est.])	$7,479	●
State Fees	$349	
Fuel (Hwy 39 City 31)	$2,467	O
Maintenance	$3,417	O
Repairs	$680	O

Warranty/Maintenance Info

Major Tune-Up	$120	O
Minor Tune-Up	$68	O
Brake Service	$206	O
Overall Warranty	3 yr/36k	O
Drivetrain Warranty	3 yr/36k	O
Rust Warranty	6 yr/100k	O
Maintenance Warranty	N/A	
Roadside Assistance	N/A	

Ownership Cost By Year

1993, 1994, 1995, 1996, 1997

Resale Value

1993	1994	1995	1996	1997
$6,670	$5,817	$5,176	$4,483	$3,789

Cumulative Costs

	1993	1994	1995	1996	1997
Annual	$4,880	$3,494	$4,037	$3,403	$5,235
Total	$4,880	$8,374	$12,411	$15,814	$21,049

Ownership Costs (5yr)

Average: $20,979 — This Car: $21,049
Cost/Mile 30¢ — Cost/Mile 30¢

Ownership Cost Rating

O Average

---

* Includes shaded options
** Other purchase requirements apply

● Poor ⊙ Worse Than Average O Average O Better Than Average O Excellent ⊖ Insufficient Information

©1993 by IntelliChoice, Inc. (408) 554-8711 All Rights Reserved. Reproduction Prohibited.
Refer to *Section 3: Annotated Vehicle Charts* for an explanation of these charts.

Ford Escort LX — 2 Door Hatchback

Subcompact

- 1.9L 88 hp Gas Fuel Inject.
- 4 Cylinder In-Line
- Manual 5 Speed
- 2 Wheel Front
- Automatic Seatbelts

Purchase Price

Car Item	Dealer Cost	List
Base Price	**$8,419**	**$9,364**
Anti-Lock Brakes	N/A	N/A
Automatic 4 Speed	$622	** $732
Optional Engine	N/A	N/A
Air Conditioning	$645	** $759
Power Steering	$222	$261
Cruise Control	Pkg	Pkg
All Wheel Drive	N/A	N/A
AM/FM Stereo Cassette	$132	$155
Steering Wheel, Tilt	Pkg	Pkg
Power Windows	N/A	N/A
*Options Price	$222	$261
*Total Price	$8,641	$9,625
Target Price	$9,097	
Destination Charge	$375	
Avg. Tax & Fees	$574	
Total Target $	**$10,046**	
Average Dealer Option Cost:	85%	

The 1993 Escort is available in seven models - Pony, LX and GT two-door hatchbacks; LX and LX-E sedans; and the LX four-door hatchback and wagon. New for 1993, the two-door LX offers a new grille, and several new exterior colors including Caymen Green and Mocha Frost, as well as an optional Sport Appearance Package. LX models come equipped with body side moldings, luxury wheel covers, and a split folding rear seat. Ford will continue to market its "One-Price" LX Advantage for the 1993 model year.

Ownership Costs

Cost Area	5 Year Cost	Rate
Depreciation	$5,673	◉
Financing ($202/month)	$1,654	
Insurance (Rating 11)	$7,693	●
State Fees	$400	
Fuel (Hwy 37 City 30)	$2,578	○
Maintenance	$3,417	◯
Repairs	$680	○

Warranty/Maintenance Info

Major Tune-Up	$120	◯
Minor Tune-Up	$68	◯
Brake Service	$206	○
Overall Warranty	3 yr/36k	○
Drivetrain Warranty	3 yr/36k	○
Rust Warranty	6 yr/100k	○
Maintenance Warranty	N/A	
Roadside Assistance	N/A	

Ownership Cost By Year

Resale Value

1993	1994	1995	1996	1997
$7,597	$6,644	$5,929	$5,159	$4,373

Cumulative Costs

	1993	1994	1995	1996	1997
Annual	$5,177	$3,722	$4,220	$3,570	$5,406
Total	$5,177	$8,899	$13,119	$16,689	$22,095

Ownership Costs (5yr)

Average	This Car
$22,408	$22,095
Cost/Mile 32¢	Cost/Mile 32¢

Ownership Cost Rating: ◯ Better Than Average

Ford Escort LX — 4 Door Hatchback

Subcompact

- 1.9L 88 hp Gas Fuel Inject.
- 4 Cylinder In-Line
- Manual 5 Speed
- 2 Wheel Front
- Automatic Seatbelts

2 Door Model Shown

Purchase Price

Car Item	Dealer Cost	List
Base Price	**$8,804**	**$9,797**
Anti-Lock Brakes	N/A	N/A
Automatic 4 Speed	$622	** $732
Optional Engine	N/A	N/A
Air Conditioning	$645	** $759
Power Steering	$222	$261
Cruise Control	Pkg	Pkg
All Wheel Drive	N/A	N/A
AM/FM Stereo Cassette	$132	$155
Steering Wheel, Tilt	Pkg	Pkg
Power Windows	Pkg	Pkg
*Options Price	$222	$261
*Total Price	$9,026	$10,058
Target Price	$9,509	
Destination Charge	$375	
Avg. Tax & Fees	$598	
Total Target $	**$10,482**	
Average Dealer Option Cost:	85%	

The 1993 Escort is available in seven models - Pony, LX and GT two-door hatchbacks; LX and LX-E sedans; and the LX four-door hatchback and wagon. New for 1993, the four-door LX hatchback offers a new grille, wheel covers, taillamp graphics, exterior appliques, and several new exterior colors including Caymen Green, Mocha Frost, and Silver Clearcoat Metallic. Ford will continue to market its "One-Price" LX Advantage for the 1993 model year.

Ownership Costs

Cost Area	5 Year Cost	Rate
Depreciation	$5,983	◉
Financing ($211/month)	$1,726	
Insurance (Rating 8)	$7,223	◉
State Fees	$417	
Fuel (Hwy 37 City 30)	$2,578	○
Maintenance	$3,417	◯
Repairs	$680	○

Warranty/Maintenance Info

Major Tune-Up	$120	◯
Minor Tune-Up	$68	◯
Brake Service	$206	○
Overall Warranty	3 yr/36k	○
Drivetrain Warranty	3 yr/36k	○
Rust Warranty	6 yr/100k	○
Maintenance Warranty	N/A	
Roadside Assistance	N/A	

Resale Value

1993	1994	1995	1996	1997
$7,836	$6,876	$6,131	$5,311	$4,499

Cumulative Costs

	1993	1994	1995	1996	1997
Annual	$5,322	$3,666	$4,173	$3,530	$5,333
Total	$5,322	$8,988	$13,161	$16,691	$22,024

Ownership Costs (5yr)

Average	This Car
$22,895	$22,024
Cost/Mile 33¢	Cost/Mile 31¢

Ownership Cost Rating: ◯ Better Than Average

---

* Includes shaded options
** Other purchase requirements apply

● Poor ◉ Worse Than Average ○ Average ◯ Better Than Average ○ Excellent ⊖ Insufficient Information

©1993 by IntelliChoice, Inc. (408) 554-8711 All Rights Reserved. Reproduction Prohibited.
Refer to *Section 3: Annotated Vehicle Charts* for an explanation of these charts.

Ford Escort GT
2 Door Hatchback

Subcompact

 1.8L 127 hp Gas Fuel Inject. | 4 Cylinder In-Line | Manual 5 Speed | 2 Wheel Front | Automatic Seatbelts

Purchase Price

Car Item	Dealer Cost	List
Base Price	**$10,650**	**$11,871**
Anti-Lock Brakes	N/A	N/A
Automatic 4 Speed	$622	$732
Optional Engine	N/A	N/A
Air Conditioning	$645	$759
Power Steering	Std	Std
Cruise Control	Pkg	Pkg
All Wheel Drive	N/A	N/A
AM/FM Stereo Cassette	Std	Std
Steering Wheel, Tilt	Pkg	Pkg
Power Windows	N/A	N/A
*Options Price	$0	$0
*Total Price	$10,650	$11,871
Target Price	$11,256	
Destination Charge	$375	
Avg. Tax & Fees	$704	
Total Target $	**$12,335**	
Average Dealer Option Cost:	85%	

The 1993 Escort is available in seven models - Pony, LX and GT two-door hatchbacks; LX and LX-E sedans; and the LX four-door hatchback and wagon. New for 1993, styling enhancements on the GT include directional aluminum wheels, rocker cladding, new seat trim, and a rear spoiler. Among the standard features are: fog lights, sport bucket seats, and a sport handling suspension.

Ownership Costs

Cost Area	5 Year Cost	Rate
Depreciation	$7,161	●
Financing ($248/month)	$2,032	
Insurance (Rating 13)	$8,061	●
State Fees	$489	
Fuel (Hwy 31 City 26)	$3,029	●
Maintenance	$4,114	○
Repairs	$761	●

Warranty/Maintenance Info

Major Tune-Up	$138	○
Minor Tune-Up	$72	○
Brake Service	$228	●
Overall Warranty	3 yr/36k	○
Drivetrain Warranty	3 yr/36k	○
Rust Warranty	6 yr/100k	○
Maintenance Warranty	N/A	
Roadside Assistance	N/A	

Ownership Cost By Year

Resale Value

1993	1994	1995	1996	1997
$9,158	$8,048	$7,160	$6,178	$5,174

Cumulative Costs

	1993	1994	1995	1996	1997
Annual	$6,235	$4,181	$4,917	$4,034	$6,280
Total	$6,235	$10,416	$15,333	$19,367	$25,647

Ownership Costs (5yr)

Average	This Car
$24,935	$25,647
Cost/Mile 36¢	Cost/Mile 37¢

Ownership Cost Rating
● Worse Than Average

Ford Escort LX
4 Door Sedan

Subcompact

 1.9L 88 hp Gas Fuel Inject. | 4 Cylinder In-Line | Manual 5 Speed | 2 Wheel Front | Automatic Seatbelts

Purchase Price

Car Item	Dealer Cost	List
Base Price	**$9,021**	**$10,041**
Anti-Lock Brakes	N/A	N/A
Automatic 4 Speed	$622	** $732
Optional Engine	N/A	N/A
Air Conditioning	$645	** $759
Power Steering	$222	$261
Cruise Control	Pkg	Pkg
All Wheel Drive	N/A	N/A
AM/FM Stereo Cassette	$132	$155
Steering Wheel, Tilt	Pkg	Pkg
Power Windows	Pkg	Pkg
*Options Price	$222	$261
*Total Price	$9,243	$10,302
Target Price	$9,741	
Destination Charge	$375	
Avg. Tax & Fees	$613	
Total Target $	**$10,729**	
Average Dealer Option Cost:	85%	

The 1993 Escort is available in seven models - Pony, LX and GT two-door hatchbacks; LX and LX-E sedans; and the LX four-door hatchback and wagon. New for 1993, the four-door LX sedan offers a new grille, and several new exterior colors including Caymen Green, Mocha Frost and Silver Clearcoat Metallic. The LX features a tachometer, and dual visor vanity mirrors, and variable intermittent wipers. Ford will continue to market its "One-Price" LX Advantage for the 1993 model year.

Ownership Costs

Cost Area	5 Year Cost	Rate
Depreciation	$6,279	●
Financing ($216/month)	$1,766	
Insurance (Rating 8)	$7,223	●
State Fees	$427	
Fuel (Hwy 37 City 30)	$2,578	○
Maintenance	$3,667	○
Repairs	$680	○

Warranty/Maintenance Info

Major Tune-Up	$120	○
Minor Tune-Up	$68	○
Brake Service	$206	○
Overall Warranty	3 yr/36k	○
Drivetrain Warranty	3 yr/36k	○
Rust Warranty	6 yr/100k	○
Maintenance Warranty	N/A	
Roadside Assistance	N/A	

Resale Value

1993	1994	1995	1996	1997
$7,930	$6,929	$6,134	$5,297	$4,450

Cumulative Costs

	1993	1994	1995	1996	1997
Annual	$5,495	$3,721	$4,353	$3,552	$5,499
Total	$5,495	$9,216	$13,569	$17,121	$22,620

Ownership Costs (5yr)

Average	This Car
$23,169	$22,620
Cost/Mile 33¢	Cost/Mile 32¢

Ownership Cost Rating
○ Better Than Average

---

* Includes shaded options
** Other purchase requirements apply

● Poor | ◐ Worse Than Average | ◑ Average | ○ Better Than Average | ○ Excellent | ⊖ Insufficient Information

©1993 by IntelliChoice, Inc. (408) 554-8711 All Rights Reserved. Reproduction Prohibited.
Refer to *Section 3: Annotated Vehicle Charts* for an explanation of these charts.

page 81

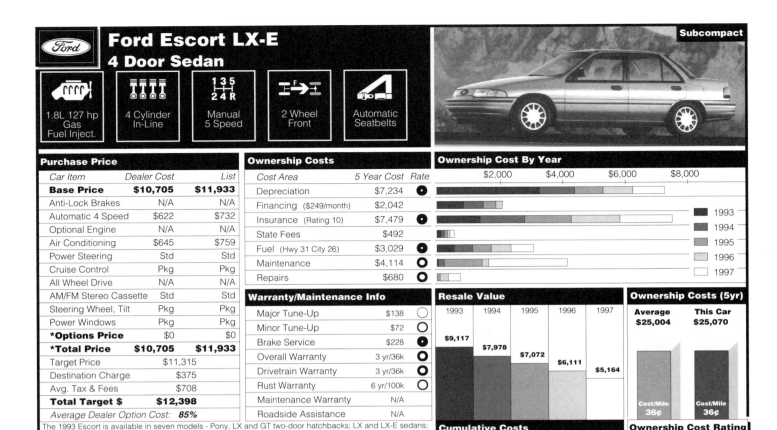

Ford Escort LX-E — 4 Door Sedan

Subcompact

- 1.8L 127 hp Gas Fuel Inject.
- 4 Cylinder In-Line
- Manual 5 Speed
- 2 Wheel Front
- Automatic Seatbelts

Purchase Price

Car Item	Dealer Cost	List
Base Price	**$10,705**	**$11,933**
Anti-Lock Brakes	N/A	N/A
Automatic 4 Speed	$622	$732
Optional Engine	N/A	N/A
Air Conditioning	$645	$759
Power Steering	Std	Std
Cruise Control	Pkg	Pkg
All Wheel Drive	N/A	N/A
AM/FM Stereo Cassette	Std	Std
Steering Wheel, Tilt	Pkg	Pkg
Power Windows	Pkg	Pkg
*Options Price	$0	$0
*Total Price	$10,705	$11,933
Target Price		$11,315
Destination Charge		$375
Avg. Tax & Fees		$708
Total Target $		**$12,398**
Average Dealer Option Cost:		85%

Ownership Costs

Cost Area	5 Year Cost	Rate
Depreciation	$7,234	●
Financing ($249/month)	$2,042	
Insurance (Rating 10)	$7,479	●
State Fees	$492	
Fuel (Hwy 31 City 26)	$3,029	●
Maintenance	$4,114	○
Repairs	$680	○

Warranty/Maintenance Info

Major Tune-Up	$138	○
Minor Tune-Up	$72	○
Brake Service	$228	●
Overall Warranty	3 yr/36k	○
Drivetrain Warranty	3 yr/36k	○
Rust Warranty	6 yr/100k	○
Maintenance Warranty	N/A	
Roadside Assistance	N/A	

Resale Value

1993	1994	1995	1996	1997
$9,117	$7,978	$7,072	$6,111	$5,164

Cumulative Costs

	1993	1994	1995	1996	1997
Annual	$6,238	$4,102	$4,809	$3,860	$6,061
Total	$6,238	$10,340	$15,149	$19,009	$25,070

Ownership Costs (5yr)

Average	This Car
$25,004	$25,070
Cost/Mile 36¢	Cost/Mile 36¢

Ownership Cost Rating: Average

The 1993 Escort is available in seven models - Pony, LX and GT two-door hatchbacks; LX and LX-E sedans; and the LX four-door hatchback and wagon. New for 1993, the LX-E sedan offers a new grille, and several new exterior colors including Caymen Green, Mocha Frost and Silver Clearcoat Metallic. The LX-E features a standard Light/Convenience Group, styled aluminum wheels, power steering, and dual power mirrors. Options include a premium sound system, and power sunroof.

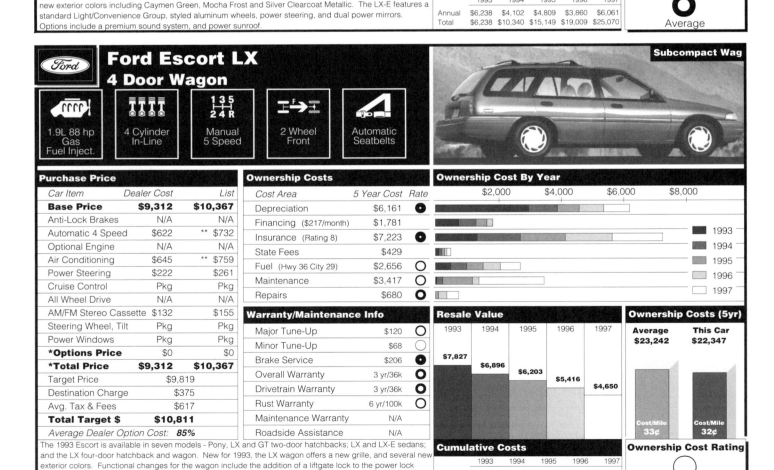

Ford Escort LX — 4 Door Wagon

Subcompact Wag

- 1.9L 88 hp Gas Fuel Inject.
- 4 Cylinder In-Line
- Manual 5 Speed
- 2 Wheel Front
- Automatic Seatbelts

Purchase Price

Car Item	Dealer Cost	List
Base Price	**$9,312**	**$10,367**
Anti-Lock Brakes	N/A	N/A
Automatic 4 Speed	$622	** $732
Optional Engine	N/A	N/A
Air Conditioning	$645	** $759
Power Steering	$222	$261
Cruise Control	Pkg	Pkg
All Wheel Drive	N/A	N/A
AM/FM Stereo Cassette	$132	$155
Steering Wheel, Tilt	Pkg	Pkg
Power Windows	Pkg	Pkg
*Options Price	$0	$0
*Total Price	$9,312	$10,367
Target Price		$9,819
Destination Charge		$375
Avg. Tax & Fees		$617
Total Target $		**$10,811**
Average Dealer Option Cost:		85%

Ownership Costs

Cost Area	5 Year Cost	Rate
Depreciation	$6,161	●
Financing ($217/month)	$1,781	
Insurance (Rating 8)	$7,223	●
State Fees	$429	
Fuel (Hwy 36 City 29)	$2,656	○
Maintenance	$3,417	○
Repairs	$680	○

Warranty/Maintenance Info

Major Tune-Up	$120	○
Minor Tune-Up	$68	○
Brake Service	$206	●
Overall Warranty	3 yr/36k	○
Drivetrain Warranty	3 yr/36k	○
Rust Warranty	6 yr/100k	○
Maintenance Warranty	N/A	
Roadside Assistance	N/A	

Resale Value

1993	1994	1995	1996	1997
$7,827	$6,896	$6,203	$5,416	$4,650

Cumulative Costs

	1993	1994	1995	1996	1997
Annual	$5,700	$3,672	$4,150	$3,520	$5,305
Total	$5,700	$9,372	$13,522	$17,042	$22,347

Ownership Costs (5yr)

Average	This Car
$23,242	$22,347
Cost/Mile 33¢	Cost/Mile 32¢

Ownership Cost Rating: Excellent

The 1993 Escort is available in seven models - Pony, LX and GT two-door hatchbacks; LX and LX-E sedans; and the LX four-door hatchback and wagon. New for 1993, the LX wagon offers a new grille, and several new exterior colors. Functional changes for the wagon include the addition of a liftgate lock to the power lock option. The LX features a tachometer, and dual visor vanity mirrors, and variable intermittent wipers. Ford will continue to market its "One-Price" LX Advantage for the 1993 model year.

---

* Includes shaded options
** Other purchase requirements apply

● Poor ◐ Worse Than Average ○ Average ◯ Better Than Average ○ Excellent ⊖ Insufficient Information

©1993 by IntelliChoice, Inc. (408) 554-8711   All Rights Reserved. Reproduction Prohibited.
Refer to *Section 3: Annotated Vehicle Charts* for an explanation of these charts.

Ford Festiva L
2 Door Hatchback

Subcompact

 1.3L 63 hp Gas Fuel Inject.
 4 Cylinder In-Line
 Manual 5 Speed
 2 Wheel Front
Automatic Seatbelts

Purchase Price

Car Item	Dealer Cost	List
Base Price	**$6,515**	**$6,941**
Anti-Lock Brakes	N/A	N/A
Automatic Transmission	N/A	N/A
Optional Engine	N/A	N/A
Air Conditioning	N/A	N/A
Power Steering	N/A	N/A
Cruise Control	N/A	N/A
All Wheel Drive	N/A	N/A
AM/FM Stereo Cassette	$397	$467
Steering Wheel, Tilt	N/A	N/A
Power Windows	N/A	N/A
*Options Price	$0	$0
*Total Price	**$6,515**	**$6,941**
Target Price	$6,765	
Destination Charge	$295	
Avg. Tax & Fees	$425	
Total Target $	**$7,485**	
Average Dealer Option Cost:	85%	

The 1993 Ford Festiva is available in two models - L, and GL hatchbacks. For 1993, standard equipment includes luggage compartment, door trim, courtesy lamps, inside hood release, front high-back reclining bucket seats, flip-fold rear seat, front and rear ashtrays, and narrow bodyside protection moldings. Optional features include rear window defroster, flip-up open air roof, sport option package, and automatic transaxle.

Ownership Costs

Cost Area	5 Year Cost	Rate
Depreciation	$4,325	O
Financing ($150/month)	$1,233	
Insurance (Rating 8)	$7,223	●
State Fees	$289	
Fuel (Hwy 42 City 35)	$2,243	○
Maintenance	$3,644	O
Repairs	$680	O

Warranty/Maintenance Info

Major Tune-Up	$233	◉
Minor Tune-Up	$75	O
Brake Service	$195	O
Overall Warranty	3 yr/36k	O
Drivetrain Warranty	3 yr/36k	O
Rust Warranty	6 yr/100k	O
Maintenance Warranty	N/A	
Roadside Assistance	N/A	

Ownership Cost By Year
1993, 1994, 1995, 1996, 1997

Resale Value
1993: $5,201 | 1994: $4,615 | 1995: $4,215 | 1996: $3,718 | 1997: $3,160

Cumulative Costs
	1993	1994	1995	1996	1997
Annual	$4,661	$3,044	$3,654	$3,768	$4,510
Total	$4,661	$7,705	$11,359	$15,127	$19,637

Ownership Costs (5yr)
Average $19,388 (Cost/Mile 28¢) | This Car $19,637 (Cost/Mile 28¢)

Ownership Cost Rating
O Average

Ford Festiva GL
2 Door Hatchback

Subcompact

 1.3L 63 hp Gas Fuel Inject.
 4 Cylinder In-Line
 Manual 5 Speed
 2 Wheel Front
 Automatic Seatbelts

Purchase Price

Car Item	Dealer Cost	List
Base Price	**$7,378**	**$7,869**
Anti-Lock Brakes	N/A	N/A
Automatic 3 Speed	$437	$515
Optional Engine	N/A	N/A
Air Conditioning	$734	$863
Power Steering	N/A	N/A
Cruise Control	N/A	N/A
All Wheel Drive	N/A	N/A
AM/FM Stereo Cassette	$132	$155
Steering Wheel, Tilt	N/A	N/A
Power Windows	N/A	N/A
*Options Price	$0	$0
*Total Price	**$7,378**	**$7,869**
Target Price	$7,670	
Destination Charge	$295	
Avg. Tax & Fees	$480	
Total Target $	**$8,445**	
Average Dealer Option Cost:	85%	

The 1993 Ford Festiva is available in two models - L, and GL hatchbacks. For 1993 standard equipment includes luggage compartment, door trim, courtesy lamps, inside hood release, front high-back reclining bucket seats, flip-fold rear seat, front and rear ashtrays, and narrow bodyside protection moldings. Optional features include rear window defroster, flip-up open air roof, sport option package, and automatic transaxle.

Ownership Costs

Cost Area	5 Year Cost	Rate
Depreciation	$4,884	◉
Financing ($170/month)	$1,391	
Insurance (Rating 10)	$7,479	●
State Fees	$326	
Fuel (Hwy 42 City 35)	$2,243	○
Maintenance	$3,798	O
Repairs	$680	O

Warranty/Maintenance Info

Major Tune-Up	$233	◉
Minor Tune-Up	$75	O
Brake Service	$195	O
Overall Warranty	3 yr/36k	O
Drivetrain Warranty	3 yr/36k	O
Rust Warranty	6 yr/100k	O
Maintenance Warranty	N/A	
Roadside Assistance	N/A	

Resale Value
1993: $5,660 | 1994: $5,068 | 1995: $4,641 | 1996: $4,133 | 1997: $3,561

Cumulative Costs
	1993	1994	1995	1996	1997
Annual	$5,284	$3,157	$3,846	$3,849	$4,665
Total	$5,284	$8,441	$12,287	$16,136	$20,801

Ownership Costs (5yr)
Average $20,432 (Cost/Mile 29¢) | This Car $20,801 (Cost/Mile 30¢)

Ownership Cost Rating
O Average

---

* Includes shaded options
** Other purchase requirements apply

● Poor | ◉ Worse Than Average | O Average | ○ Better Than Average | ○ Excellent | ⊖ Insufficient Information

©1993 by IntelliChoice, Inc. (408) 554-8711 All Rights Reserved. Reproduction Prohibited.
Refer to *Section 3: Annotated Vehicle Charts* for an explanation of these charts.

page 83

Ford Mustang LX
2 Door Coupe

 2.3L 105 hp Gas Fuel Inject.
 4 Cylinder In-Line
 Manual 5 Speed
 2 Wheel Rear
Driver Airbag Psngr Belts

Compact
5.0 Hatchback Shown

Purchase Price

Car Item	Dealer Cost	List
Base Price	**$9,670**	**$10,719**
Anti-Lock Brakes	N/A	N/A
Automatic 4 Speed	$506	$595
Optional Engine	N/A	N/A
Air Conditioning	$695	$817
Power Steering	Std	Std
Cruise Control	Pkg	Pkg
All Wheel Drive	N/A	N/A
AM/FM Stereo Cassette	$535	$629
Steering Wheel, Tilt	N/A	N/A
Power Windows	Pkg	Pkg
*Options Price	$1,736	$2,041
*Total Price	$11,406	$12,760
Target Price	$12,104	
Destination Charge	$440	
Avg. Tax & Fees	$759	
Total Target $	**$13,303**	
Average Dealer Option Cost:	85%	

Ownership Costs

Cost Area	5 Year Cost	Rate
Depreciation	$7,220	○
Financing ($267/month)	$2,191	
Insurance (Rating 16)	$8,784	●
State Fees	$528	
Fuel (Hwy 29 City 22)	$3,390	○
Maintenance	$3,642	○
Repairs	$761	◉

Warranty/Maintenance Info

Major Tune-Up	$111	○
Minor Tune-Up	$67	○
Brake Service	$229	○
Overall Warranty	3 yr/36k	○
Drivetrain Warranty	3 yr/36k	○
Rust Warranty	6 yr/100k	○
Maintenance Warranty	N/A	
Roadside Assistance	N/A	

Ownership Cost By Year

1993, 1994, 1995, 1996, 1997

Resale Value

1993	1994	1995	1996	1997
$9,965	$8,855	$7,976	$7,031	$6,083

Cumulative Costs

	1993	1994	1995	1996	1997
Annual	$6,673	$4,471	$5,082	$4,654	$5,636
Total	$6,673	$11,144	$16,226	$20,880	$26,516

Ownership Costs (5yr)

Average	This Car
$25,451	$26,516
Cost/Mile 36¢	Cost/Mile 38¢

Ownership Cost Rating
◉ Worse Than Average

The 1993 Ford Mustang is available in five models - LX hatchback and coupe, LX 5.0 Sport hatchback and coupe, and the GT hatchback. New for 1993, the LX coupe features an upgraded stereo with increased power, ergonomics, and improved display. Standard features include instrumentation, complete tinted glass, body color body-side moldings, dual remote mirrors, and low back reclining cloth bucket seats. Optional features include 4-way power driver seat, power equipment group and a convenience group.

Ford Mustang LX 5.0L
2 Door Coupe

 5.0L 205 hp Gas Fuel Inject.
 8 Cylinder "V"
 Manual 5 Speed
 2 Wheel Rear
 Driver Airbag Psngr Belts

Sport
Hatchback Model Shown

Purchase Price

Car Item	Dealer Cost	List
Base Price	**$12,524**	**$13,926**
Anti-Lock Brakes	N/A	N/A
Automatic 4 Speed	$506	$595
Optional Engine	N/A	N/A
Air Conditioning	$695	$817
Power Steering	Std	Std
Cruise Control	Pkg	Pkg
All Wheel Drive	N/A	N/A
AM/FM Stereo Cassette	$535	$629
Steering Wheel, Tilt	N/A	N/A
Power Windows	Pkg	Pkg
*Options Price	$1,230	$1,446
*Total Price	$13,754	$15,372
Target Price	$14,675	
Destination Charge	$440	
Avg. Tax & Fees	$914	
Total Target $	**$16,029**	
Average Dealer Option Cost:	85%	

Ownership Costs

Cost Area	5 Year Cost	Rate
Depreciation	$8,974	○
Financing ($322/month)	$2,641	
Insurance (Rating 18)	$9,453	●
State Fees	$632	
Fuel (Hwy 24 City 17)	$4,232	◉
Maintenance	$5,360	○
Repairs	$761	○

Warranty/Maintenance Info

Major Tune-Up	$154	○
Minor Tune-Up	$106	○
Brake Service	$285	○
Overall Warranty	3 yr/36k	○
Drivetrain Warranty	3 yr/36k	○
Rust Warranty	6 yr/100k	○
Maintenance Warranty	N/A	
Roadside Assistance	N/A	

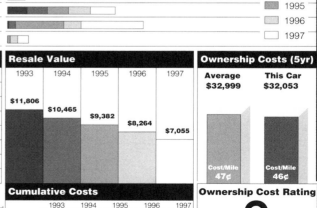

Resale Value

1993	1994	1995	1996	1997
$11,806	$10,465	$9,382	$8,264	$7,055

Cumulative Costs

	1993	1994	1995	1996	1997
Annual	$8,055	$5,208	$6,512	$5,249	$7,029
Total	$8,055	$13,263	$19,775	$25,024	$32,053

Ownership Costs (5yr)

Average	This Car
$32,999	$32,053
Cost/Mile 47¢	Cost/Mile 46¢

Ownership Cost Rating
○ Average

The 1993 Ford Mustang is available in five models - LX hatchback and coupe, LX 5.0 Sport hatchback and coupe, and the GT hatchback. New for 1993, the LX 5.0 Sport coupe features an upgraded stereo with improved display and ergonomics. Standard features include bodyside moldings, articulated sport seats and power lumbar support, tinted glass, a light group, instrumentation and stalk mounted controls. Optional features include power equipment group, rear window defroster and leather seating.

---

page 84

* Includes shaded options
** Other purchase requirements apply

● Poor   ◉ Worse Than Average   ○ Average   ○ Better Than Average   ○ Excellent   ⊖ Insufficient Information

©1993 by IntelliChoice, Inc. (408) 554-8711   All Rights Reserved. Reproduction Prohibited.
Refer to Section 3: Annotated Vehicle Charts for an explanation of these charts.

Ford Mustang LX
2 Door Hatchback

 2.3L 105 hp Gas Fuel Inject.
 4 Cylinder In-Line
 Manual 5 Speed
 2 Wheel Rear
 Driver Airbag Psngr Belts

Compact

5.0 Model Shown

Purchase Price

Car Item	Dealer Cost	List
Base Price	**$10,120**	**$11,224**
Anti-Lock Brakes	N/A	N/A
Automatic 4 Speed	$506	$595
Optional Engine	N/A	N/A
Air Conditioning	$695	$817
Power Steering	Std	Std
Cruise Control	Pkg	Pkg
All Wheel Drive	N/A	N/A
AM/FM Stereo Cassette	$535	$629
Steering Wheel, Tilt	N/A	N/A
Power Windows	Pkg	Pkg
*Options Price	$1,736	$2,041
*Total Price	$11,856	$13,265
Target Price	$12,593	
Destination Charge	$440	
Avg. Tax & Fees	$789	
Total Target $	**$13,822**	
Average Dealer Option Cost:	**85%**	

The 1993 Ford Mustang is available in five models - LX hatchback and coupe, LX 5.0 Sport hatchback and coupe, and the GT hatchback. New for 1993, the LX hatchback features an upgraded stereo with improved display, ergonomics, and increased power. Standard features include a rear spoiler, low back reclining cloth bucket seats, cargo area cover, split/fold down rear seat, and finned wheel covers. Optional features include a convenience group, power equipment group and a flip-up roof.

Ownership Costs

Cost Area	5 Year Cost	Rate
Depreciation	$7,511	○
Financing ($278/month)	$2,276	
Insurance (Rating 16)	$8,784	●
State Fees	$548	
Fuel (Hwy 29 City 22)	$3,390	○
Maintenance	$3,642	◐
Repairs	$761	◉

Warranty/Maintenance Info

Major Tune-Up	$111	○
Minor Tune-Up	$67	○
Brake Service	$229	○
Overall Warranty	3 yr/36k	○
Drivetrain Warranty	3 yr/36k	○
Rust Warranty	6 yr/100k	○
Maintenance Warranty	N/A	
Roadside Assistance	N/A	

Ownership Cost By Year
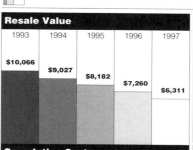

1993 / 1994 / 1995 / 1996 / 1997

Resale Value

1993	1994	1995	1996	1997
$10,066	$9,027	$8,182	$7,260	$6,311

Cumulative Costs

	1993	1994	1995	1996	1997
Annual	$7,131	$4,431	$5,069	$4,641	$5,640
Total	$7,131	$11,562	$16,631	$21,272	$26,912

Ownership Costs (5yr)

Average	This Car
$25,884	$26,912
Cost/Mile 37¢	Cost/Mile 38¢

Ownership Cost Rating

Average

Ford Mustang LX 5.0L
2 Door Hatchback

 5.0L 205 hp Gas Fuel Inject.
 8 Cylinder "V"
 Manual 5 Speed
 2 Wheel Rear
 Driver Airbag Psngr Belts

Sport

Purchase Price

Car Item	Dealer Cost	List
Base Price	**$13,222**	**$14,710**
Anti-Lock Brakes	N/A	N/A
Automatic 4 Speed	$506	$595
Optional Engine	N/A	N/A
Air Conditioning	$695	$817
Power Steering	Std	Std
Cruise Control	Pkg	Pkg
All Wheel Drive	N/A	N/A
AM/FM Stereo Cassette	$535	$629
Steering Wheel, Tilt	N/A	N/A
Power Windows	Pkg	Pkg
*Options Price	$1,230	$1,446
*Total Price	$14,452	$16,156
Target Price	$15,440	
Destination Charge	$440	
Avg. Tax & Fees	$960	
Total Target $	**$16,840**	
Average Dealer Option Cost:	**85%**	

The 1993 Ford Mustang is available in five models - LX hatchback and coupe, LX 5.0 Sport hatchback and coupe, and the GT hatchback. New for 1993, the LX 5.0 Sport hatchback features an upgraded stereo with improved display and ergonomics. Standard features include articulated sport seats with cloth trim and power lumbar support, stalk mounted controls, and a leather wrapped steering wheel. Optional features include leather seating surfaces, power equipment group, and a convenience group.

Ownership Costs

Cost Area	5 Year Cost	Rate
Depreciation	$9,403	○
Financing ($338/month)	$2,774	
Insurance (Rating 19++)	$12,221	●
State Fees	$664	
Fuel (Hwy 24 City 17)	$4,232	◉
Maintenance	$5,360	○
Repairs	$761	○

Warranty/Maintenance Info

Major Tune-Up	$154	○
Minor Tune-Up	$106	○
Brake Service	$285	○
Overall Warranty	3 yr/36k	○
Drivetrain Warranty	3 yr/36k	○
Rust Warranty	6 yr/100k	○
Maintenance Warranty	N/A	
Roadside Assistance	N/A	

Ownership Cost By Year

1993 / 1994 / 1995 / 1996 / 1997

Resale Value

1993	1994	1995	1996	1997
$12,045	$10,763	$9,727	$8,641	$7,437

Cumulative Costs

	1993	1994	1995	1996	1997
Annual	$9,201	$5,729	$7,050	$5,808	$7,627
Total	$9,201	$14,930	$21,980	$27,788	$35,415

Ownership Costs (5yr)

Average	This Car
$33,752	$35,415
Cost/Mile 48¢	Cost/Mile 51¢

Ownership Cost Rating

Worse Than Average

---

* Includes shaded options
** Other purchase requirements apply

 Poor
 Worse Than Average
 Average
 Better Than Average
 Excellent
 Insufficient Information

©1993 by IntelliChoice, Inc. (408) 554-8711 All Rights Reserved. Reproduction Prohibited.
Refer to *Section 3: Annotated Vehicle Charts* for an explanation of these charts.

Ford Mustang GT
2 Door Hatchback

Sport

- 5.0L 205 hp Gas Fuel Inject.
- 8 Cylinder "V"
- Manual 5 Speed
- 2 Wheel Rear
- Driver Airbag Psngr Belts

Purchase Price

Car Item	Dealer Cost	List
Base Price	**$14,144**	**$15,747**
Anti-Lock Brakes	N/A	N/A
Automatic 4 Speed	$506	$595
Optional Engine	N/A	N/A
Air Conditioning	Pkg	Pkg
Power Steering	Std	Std
Cruise Control	Pkg	Pkg
All Wheel Drive	N/A	N/A
AM/FM Stereo Cassette	$535	$629
Steering Wheel, Tilt	N/A	N/A
Power Windows	Pkg	Pkg
*Options Price	$535	$629
*Total Price	$14,679	$16,376
Target Price		$15,710
Destination Charge		$440
Avg. Tax & Fees		$976
Total Target $		**$17,126**
Average Dealer Option Cost:		85%

The 1993 Ford Mustang is available in five models - LX hatchback and coupe, LX 5.0 Sport hatchback and coupe, and the GT hatchback. New for 1993, the GT hatchback features an upgraded stereo with improved display and ergonomics. The GT upgrades the LX with an air dam, louvered wraparound tail lamp, and flared rocker moldings. Standard features include a unique rear spoiler, articulated sport seats and power lumbar support. Optional features include a power equipment group and a flip-up roof.

Ownership Costs

Cost Area	5 Year Cost	Rate
Depreciation	$9,284	○
Financing ($344/month)	$2,821	
Insurance (Rating 20+)	$12,293	●
State Fees	$673	
Fuel (Hwy 24 City 17)	$4,232	◉
Maintenance	$5,360	○
Repairs	$761	○

Warranty/Maintenance Info

Major Tune-Up	$154	○
Minor Tune-Up	$106	○
Brake Service	$285	○
Overall Warranty	3 yr/36k	○
Drivetrain Warranty	3 yr/36k	○
Rust Warranty	6 yr/100k	○
Maintenance Warranty	N/A	
Roadside Assistance	N/A	

Ownership Cost By Year

Legend: 1993, 1994, 1995, 1996, 1997

Resale Value
- 1993: $12,234
- 1994: $11,089
- 1995: $10,107
- 1996: $9,042
- 1997: $7,842

Cumulative Costs

	1993	1994	1995	1996	1997
Annual	$9,333	$5,623	$7,022	$5,805	$7,640
Total	$9,333	$14,956	$21,978	$27,783	$35,423

Ownership Costs (5yr)
- Average: $33,963
- This Car: $35,423
- Cost/Mile: 49¢ / 51¢

Ownership Cost Rating
Worse Than Average

Ford Mustang LX
2 Door Convertible

Compact

5.0 Model Shown

- 2.3L 105 hp Gas Fuel Inject.
- 4 Cylinder In-Line
- Manual 5 Speed
- 2 Wheel Rear
- Driver Airbag Psngr Belts

Purchase Price

Car Item	Dealer Cost	List
Base Price	**$15,747**	**$17,548**
Anti-Lock Brakes	N/A	N/A
Automatic 4 Speed	$506	$595
Optional Engine	N/A	N/A
Air Conditioning	$695	$817
Power Steering	Std	Std
Cruise Control	Pkg	Pkg
All Wheel Drive	N/A	N/A
AM/FM Stereo Cassette	$535	$629
Steering Wheel, Tilt	N/A	N/A
Power Windows	Std	Std
*Options Price	$1,736	$2,041
*Total Price	$17,483	$19,589
Target Price		$18,768
Destination Charge		$440
Avg. Tax & Fees		$1,160
Total Target $		**$20,368**
Average Dealer Option Cost:		85%

The 1993 Ford Mustang convertible is available in three models - LX, LX 5.0 Sport, and GT. New for 1993, the LX convertible features an upgraded stereo with improved display and ergonomics. Standard features include a convertible top cover, luggage rack, a light group, a power equipment group, instrumentation and low back reclining bucket seats. Optional features include a convenience group, rear window defroster, 4-way power driver seat, styled road wheels, and a leather seating trim package.

Ownership Costs

Cost Area	5 Year Cost	Rate
Depreciation	$10,821	○
Financing ($409/month)	$3,356	
Insurance (Rating 21)	$10,706	●
State Fees	$800	
Fuel (Hwy 29 City 22)	$3,390	○
Maintenance	$3,642	○
Repairs	$761	◉

Warranty/Maintenance Info

Major Tune-Up	$111	○
Minor Tune-Up	$67	○
Brake Service	$229	○
Overall Warranty	3 yr/36k	○
Drivetrain Warranty	3 yr/36k	○
Rust Warranty	6 yr/100k	○
Maintenance Warranty	N/A	
Roadside Assistance	N/A	

Resale Value
- 1993: $14,398
- 1994: $13,123
- 1995: $11,986
- 1996: $10,765
- 1997: $9,547

Cumulative Costs

	1993	1994	1995	1996	1997
Annual	$10,213	$5,430	$6,009	$5,462	$6,362
Total	$10,213	$15,643	$21,652	$27,114	$33,476

Ownership Costs (5yr)
- Average: $31,315
- This Car: $33,476
- Cost/Mile: 45¢ / 48¢

Ownership Cost Rating
Worse Than Average

---

* Includes shaded options
** Other purchase requirements apply

Legend: ● Poor | ◉ Worse Than Average | ○ Average | ○ Better Than Average | ○ Excellent | ⊖ Insufficient Information

©1993 by IntelliChoice, Inc. (408) 554-8711 All Rights Reserved. Reproduction Prohibited.
Refer to *Section 3: Annotated Vehicle Charts* for an explanation of these charts.

Ford Mustang LX 5.0L
2 Door Convertible

Sport

 5.0L 205 hp Gas Fuel Inject.
 8 Cylinder "V"
 Manual 5 Speed
 2 Wheel Rear
Driver Airbag Psngr Belts

Purchase Price

Car Item	Dealer Cost	List
Base Price	**$18,191**	**$20,293**
Anti-Lock Brakes	N/A	N/A
Automatic 4 Speed	$506	$595
Optional Engine	N/A	N/A
Air Conditioning	$695	$817
Power Steering	Std	Std
Cruise Control	Pkg	Pkg
All Wheel Drive	N/A	N/A
AM/FM Stereo Cassette	$535	$629
Steering Wheel, Tilt	N/A	N/A
Power Windows	Std	Std
***Options Price**	$1,230	$1,446
***Total Price**	**$19,421**	**$21,739**
Target Price	$20,944	
Destination Charge	$440	
Avg. Tax & Fees	$1,291	
Total Target $	**$22,675**	
Average Dealer Option Cost:	85%	

The 1993 Ford Mustang convertible is available in three models - LX, LX 5.0 Sport, and GT. New for 1993, the LX 5.0 Sport convertible features an upgraded stereo with improved display and ergonomics. The LX 5.0 Sport upgrades the LX model with articulated sport seats and headrests with cloth trim and power lumbar support, and a leather wrapped steering wheel. Optional features include 4-way power driver seat, a leather seating trim package, rear window defroster, and styled road wheels.

Ownership Costs

Cost Area	5 Year Cost	Rate
Depreciation	$11,699	○
Financing ($456/month)	$3,735	
Insurance (Rating 22+)	$13,400	●
State Fees	$887	
Fuel (Hwy 24 City 17)	$4,232	⊙
Maintenance	$5,360	○
Repairs	$761	○

Warranty/Maintenance Info

Major Tune-Up	$154	○
Minor Tune-Up	$106	○
Brake Service	$285	○
Overall Warranty	3 yr/36k	○
Drivetrain Warranty	3 yr/36k	○
Rust Warranty	6 yr/100k	○
Maintenance Warranty	N/A	
Roadside Assistance	N/A	

Ownership Cost By Year

Resale Value

1993	1994	1995	1996	1997
$16,064	$14,769	$13,614	$12,354	$10,976

Cumulative Costs

	1993	1994	1995	1996	1997
Annual	$11,692	$6,320	$7,640	$6,334	$8,088
Total	$11,692	$18,012	$25,652	$31,986	$40,074

Ownership Costs (5yr)

Average	This Car
$39,108	$40,074
Cost/Mile 56¢	Cost/Mile 57¢

Ownership Cost Rating

 Average

Ford Mustang GT
2 Door Convertible

Sport

 5.0L 205 hp Gas Fuel Inject.
 8 Cylinder "V"
 Manual 5 Speed
 2 Wheel Rear
Driver Airbag Psngr Belts

LX 5.0 Model Shown

Purchase Price

Car Item	Dealer Cost	List
Base Price	**$18,684**	**$20,848**
Anti-Lock Brakes	N/A	N/A
Automatic 4 Speed	$506	$595
Optional Engine	N/A	N/A
Air Conditioning	Pkg	Pkg
Power Steering	Std	Std
Cruise Control	Pkg	Pkg
All Wheel Drive	N/A	N/A
AM/FM Stereo Cassette	$535	$629
Steering Wheel, Tilt	N/A	N/A
Power Windows	Std	Std
***Options Price**	$535	$629
***Total Price**	**$19,219**	**$21,477**
Target Price	$20,745	
Destination Charge	$440	
Avg. Tax & Fees	$1,278	
Total Target $	**$22,463**	
Average Dealer Option Cost:	85%	

The 1993 Ford Mustang convertible is available in three models - LX, LX 5.0 Sport, and GT. New for 1993, the GT convertible features an upgraded stereo with improved display and ergonomics. The GT upgrades the LX models with a unique rear spoiler, an air dam, louvered wraparound tail lamp and flared rocker moldings. Standard features include luggage rack, instrumentation, power equipment group and a convertible top cover. Optional features include a convenience group and leather seating package.

Ownership Costs

Cost Area	5 Year Cost	Rate
Depreciation	$10,428	○
Financing ($451/month)	$3,701	
Insurance (Rating 23+)	$13,955	●
State Fees	$876	
Fuel (Hwy 24 City 17)	$4,232	⊙
Maintenance	$5,360	○
Repairs	$761	○

Warranty/Maintenance Info

Major Tune-Up	$154	○
Minor Tune-Up	$106	○
Brake Service	$285	○
Overall Warranty	3 yr/36k	○
Drivetrain Warranty	3 yr/36k	○
Rust Warranty	6 yr/100k	○
Maintenance Warranty	N/A	
Roadside Assistance	N/A	

Resale Value

1993	1994	1995	1996	1997
$16,748	$15,596	$14,545	$13,404	$12,035

Cumulative Costs

	1993	1994	1995	1996	1997
Annual	$10,880	$6,271	$7,637	$6,326	$8,198
Total	$10,880	$17,151	$24,788	$31,114	$39,312

Ownership Costs (5yr)

Average	This Car
$38,857	$39,312
Cost/Mile 56¢	Cost/Mile 56¢

Ownership Cost Rating

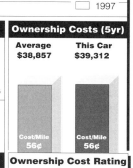 Average

---

* Includes shaded options
** Other purchase requirements apply

● Poor   ◐ Worse Than Average   ○ Average   ◯ Better Than Average   ○ Excellent   ⊖ Insufficient Information

©1993 by IntelliChoice, Inc. (408) 554-8711 All Rights Reserved. Reproduction Prohibited.
Refer to *Section 3: Annotated Vehicle Charts* for an explanation of these charts.

Ford Probe
2 Door Hatchback

Compact

 2.0L 115 hp Gas Fuel Inject.
 4 Cylinder In-Line
 Manual 5 Speed
 2 Wheel Front
Driver Airbag Psngr Belts

Purchase Price

Car Item	Dealer Cost	List
Base Price	**$11,557**	**$12,845**
Anti-Lock Brakes	$658	** $774
Automatic 4 Speed	$622	** $732
Optional Engine	N/A	N/A
Air Conditioning	$695	** $817
Power Steering	Std	Std
Cruise Control	$191	** $224
All Wheel Drive	N/A	N/A
AM/FM Stereo Cassette	$288	** $339
Steering Wheel, Tilt	Pkg	Pkg
Power Windows	Pkg	Pkg
*Options Price	$0	$0
*Total Price	**$11,557**	**$12,845**
Target Price	$12,271	
Destination Charge	$330	
Avg. Tax & Fees	$762	
Total Target $	**$13,363**	
Average Dealer Option Cost:	85%	

The 1993 Probe is available in two models- (Base) Probe and GT hatchbacks. New for 1993, the Probe has been redesigned in both the interior and exterior. Four wheel independent suspension modifications were made for improved handling and steering. Also featuring an all-new driver-oriented instrument panel with integrated door panels. The Probe is larger in almost every dimension, yet lighter than last year's model. The Preferred Equipment Package (PEP 253) is available as an option.

Ownership Costs

Cost Area	5 Year Cost	Rate
Depreciation	$6,607	○
Financing ($269/month)	$2,201	
Insurance (Rating 12)	$7,898	⦿
State Fees	$527	
Fuel (Hwy 33 City 26)	$2,927	○
Maintenance	$4,448	○
Repairs	$709	○

Warranty/Maintenance Info

Major Tune-Up	$165	○
Minor Tune-Up	$78	○
Brake Service	$178	○
Overall Warranty	3 yr/36k	○
Drivetrain Warranty	3 yr/36k	○
Rust Warranty	6 yr/100k	○
Maintenance Warranty	N/A	
Roadside Assistance	N/A	

Ownership Cost By Year

1993, 1994, 1995, 1996, 1997

Resale Value

1993	1994	1995	1996	1997
$10,315	$9,339	$8,686	$7,584	$6,756

Cumulative Costs

	1993	1994	1995	1996	1997
Annual	$6,138	$4,045	$4,665	$4,103	$6,366
Total	$6,138	$10,183	$14,848	$18,951	$25,317

Ownership Costs (5yr)

Average	This Car
$25,524	$25,317
Cost/Mile 36¢	Cost/Mile 36¢

Ownership Cost Rating
○ Average

Ford Probe GT
2 Door Hatchback

Compact

 2.5L 164 hp Gas Fuel Inject.
 6 Cylinder "V"
 Manual 5 Speed
 2 Wheel Front
 Driver Airbag Psngr Belts

Purchase Price

Car Item	Dealer Cost	List
Base Price	**$13,630**	**$15,174**
Anti-Lock Brakes	$506	** $595
Automatic 4 Speed	$622	** $732
Optional Engine	N/A	N/A
Air Conditioning	Pkg	Pkg
Power Steering	Std	Std
Cruise Control	$191	** $224
All Wheel Drive	N/A	N/A
AM/FM Stereo CD	$426	** $501
Steering Wheel, Tilt	Pkg	Pkg
Power Windows	Pkg	Pkg
*Options Price	$0	$0
*Total Price	**$13,630**	**$15,174**
Target Price	$14,526	
Destination Charge	$330	
Avg. Tax & Fees	$898	
Total Target $	**$15,754**	
Average Dealer Option Cost:	85%	

The 1993 Probe is available in two models- (Base) Probe and GT hatchbacks. New for 1993, the GT model has been redesigned in both the exterior and interior. The Probe GT is larger in almost every dimension, yet lighter than last year's model. The GT has a unique taillight treatment and front fascia that includes larger ducts for higher air flow. The GT has an improved center console, power lumbar and bolster controls on the front seats and offers a Preferred Equipment Package (PEP 263).

Ownership Costs

Cost Area	5 Year Cost	Rate
Depreciation	$8,554	○
Financing ($317/month)	$2,596	
Insurance (Rating 14)	$8,238	⦿
State Fees	$621	
Fuel (Hwy 26 City 21 -Prem.)	$4,056	●
Maintenance	$5,096	⦿
Repairs	$830	⦿

Warranty/Maintenance Info

Major Tune-Up	$186	○
Minor Tune-Up	$108	○
Brake Service	$177	○
Overall Warranty	3 yr/36k	○
Drivetrain Warranty	3 yr/36k	○
Rust Warranty	6 yr/100k	○
Maintenance Warranty	N/A	
Roadside Assistance	N/A	

Resale Value

1993	1994	1995	1996	1997
$11,088	$10,048	$9,434	$8,171	$7,200

Cumulative Costs

	1993	1994	1995	1996	1997
Annual	$8,218	$4,572	$5,446	$4,702	$7,053
Total	$8,218	$12,790	$18,236	$22,938	$29,991

Ownership Costs (5yr)

Average	This Car
$27,523	$29,991
Cost/Mile 39¢	Cost/Mile 43¢

Ownership Cost Rating
⦿ Worse Than Average

● Poor ⦿ Worse Than Average ○ Average ◐ Better Than Average ○ Excellent ⊖ Insufficient Information

* Includes shaded options
** Other purchase requirements apply

©1993 by IntelliChoice, Inc. (408) 554-8711 All Rights Reserved. Reproduction Prohibited.
Refer to *Section 3: Annotated Vehicle Charts* for an explanation of these charts.

Ford Taurus GL
4 Door Sedan

Midsize

- 3.0L 140 hp Gas Fuel Inject.
- 6 Cylinder "V"
- Automatic 4 Speed
- 2 Wheel Front
- Driver Airbag Psngr Opt

LX Model Shown

Purchase Price

Car Item	Dealer Cost	List
Base Price	**$13,308**	**$15,491**
Anti-Lock Brakes	$506	$595
Manual Transmission	N/A	N/A
3.8L 140 hp Gas	$472	** $555
Air Conditioning	Pkg	Pkg
Power Steering	Std	Std
Cruise Control	$191	$224
All Wheel Drive	N/A	N/A
AM/FM Stereo Cassette	$145	$171
Steering Wheel, Tilt	Std	Std
Power Windows	$303	$356
***Options Price**	**$639**	**$751**
***Total Price**	**$13,947**	**$16,242**
Target Price		$15,094
Destination Charge		$490
Avg. Tax & Fees		$946
Total Target $		**$16,530**
Average Dealer Option Cost:		85%

Ownership Costs

Cost Area	5 Year Cost	Rate
Depreciation	$8,938	◉
Financing ($332/month)	$2,723	
Insurance (Rating 3)	$6,520	○
State Fees	$669	
Fuel (Hwy 30 City 21)	$3,405	○
Maintenance	$4,253	○
Repairs	$709	○

Warranty/Maintenance Info

Major Tune-Up	$137	○
Minor Tune-Up	$87	○
Brake Service	$271	●
Overall Warranty	3 yr/36k	○
Drivetrain Warranty	3 yr/36k	○
Rust Warranty	6 yr/100k	○
Maintenance Warranty	N/A	
Roadside Assistance	N/A	

Ownership Cost By Year

Legend: 1993, 1994, 1995, 1996, 1997

Resale Value

1993	1994	1995	1996	1997
$13,155	$11,485	$10,172	$8,870	$7,592

Cumulative Costs

	1993	1994	1995	1996	1997
Annual	$6,555	$4,785	$5,422	$4,852	$5,603
Total	$6,555	$11,340	$16,762	$21,614	$27,217

Ownership Costs (5yr)

Average	This Car
$27,876	$27,217
Cost/Mile 40¢	Cost/Mile 39¢

Ownership Cost Rating

○ Better Than Average

The 1993 Ford Taurus sedan is available in three models - GL, LX and SHO sedans. New for 1993, the GL features body-colored bumpers, all new seat trim, and a variety of new colors. Standard features include a decklid spoiler, an exterior accent group, dual electric remote control mirrors, door trim panel and split bench seats with dual recliner. Optional features include 6-way power adjustable driver seat, remote keyless entry system, leather seating trim, a power moonroof, and cellular phone.

Ford Taurus LX
4 Door Sedan

Midsize

- 3.0L 140 hp Gas Fuel Inject.
- 6 Cylinder "V"
- Automatic 4 Speed
- 2 Wheel Front
- Driver Airbag Psngr Opt

Purchase Price

Car Item	Dealer Cost	List
Base Price	**$15,695**	**$18,300**
Anti-Lock Brakes	$506	$595
Manual Transmission	N/A	N/A
3.8L 140 hp Gas	$472	$555
Air Conditioning	Std	Std
Power Steering	Std	Std
Cruise Control	Pkg	Pkg
All Wheel Drive	N/A	N/A
AM/FM Stereo Cassette	$282	** $332
Steering Wheel, Tilt	Std	Std
Power Windows	Std	Std
***Options Price**	**$0**	**$0**
***Total Price**	**$15,695**	**$18,300**
Target Price		$17,077
Destination Charge		$490
Avg. Tax & Fees		$1,066
Total Target $		**$18,633**
Average Dealer Option Cost:		85%

Ownership Costs

Cost Area	5 Year Cost	Rate
Depreciation	$10,466	◉
Financing ($374/month)	$3,069	
Insurance (Rating 4)	$6,658	○
State Fees	$752	
Fuel (Hwy 30 City 21)	$3,405	○
Maintenance	$4,352	○
Repairs	$709	○

Warranty/Maintenance Info

Major Tune-Up	$137	○
Minor Tune-Up	$87	○
Brake Service	$271	●
Overall Warranty	3 yr/36k	○
Drivetrain Warranty	3 yr/36k	○
Rust Warranty	6 yr/100k	○
Maintenance Warranty	N/A	
Roadside Assistance	N/A	

Resale Value

1993	1994	1995	1996	1997
$13,802	$12,144	$10,806	$9,430	$8,167

Cumulative Costs

	1993	1994	1995	1996	1997
Annual	$8,201	$4,925	$5,607	$4,995	$5,683
Total	$8,201	$13,126	$18,733	$23,728	$29,411

Ownership Costs (5yr)

Average	This Car
$29,980	$29,411
Cost/Mile 43¢	Cost/Mile 42¢

Ownership Cost Rating

○ Better Than Average

The 1993 Ford Taurus sedan is available in three models - GL, LX and SHO sedans. New for 1993, the LX features a new floor console, and a variety of new colors. The LX sedan upgrades the GL with a console floor shift, convenience kit, bucket seats with dual recliners, and a 6-way power driver seat. Standard features include an exterior accent group, dual electric remote control mirrors and color-keyed rocker panel moldings. Optional features include a cellular phone and power moonroof.

---

* Includes shaded options
** Other purchase requirements apply

● Poor | ◉ Worse Than Average | ◐ Average | ○ Better Than Average | ⊘ Excellent | ⊖ Insufficient Information

©1993 by IntelliChoice, Inc. (408) 554-8711 All Rights Reserved. Reproduction Prohibited.
Refer to *Section 3: Annotated Vehicle Charts* for an explanation of these charts.

Ford Taurus SHO
4 Door Sedan

Midsize

 3.0L 220 hp Gas Fuel Inject.
 6 Cylinder "V"
Manual 5 Speed
 2 Wheel Front
 Driver Airbag Psngr Opt

Purchase Price

Car Item	Dealer Cost	List
Base Price	**$21,245**	**$24,829**
Anti-Lock Brakes	Std	Std
Automatic 4 Speed	$545	$641
3.2L 220 hp Gas	Grp	Grp
Auto Climate Control	Std	Std
Power Steering	Std	Std
Cruise Control	Std	Std
All Wheel Drive	N/A	N/A
AM/FM Stereo Cassette	Std	Std
Steering Wheel, Tilt	Std	Std
Power Windows	Std	Std
***Options Price**	**$545**	**$641**
***Total Price**	**$21,790**	**$25,470**
Target Price		$24,001
Destination Charge		$490
Avg. Tax & Fees		$1,485
Total Target $		**$25,976**
Average Dealer Option Cost:		*85%*

The 1993 Ford Taurus sedan is available in three models - GL, LX and SHO sedans. New for 1993, the SHO features a new steering wheel, and new floor console. The SHO sedan upgrades the LX with cornering lamps, floor mats, a leather wrapped steering wheel, leather seats with power lumbar support, rear window defroster and dual exhaust. Other standard features include a decklid spoiler, exterior accent group and convenience kit. Options include a cellular phone and power moonroof.

Ownership Costs

Cost Area	5 Year Cost	Rate
Depreciation	$15,331	●
Financing ($522/month)	$4,278	
Insurance (Rating 8+)	$8,668	◐
State Fees	$1,040	
Fuel (Hwy 30 City 21 -Prem.)	$3,760	◐
Maintenance	$4,904	◐
Repairs	$799	◐

Warranty/Maintenance Info

Major Tune-Up	$151	◐
Minor Tune-Up	$101	◐
Brake Service	$302	●
Overall Warranty	3 yr/36k	◐
Drivetrain Warranty	3 yr/36k	◐
Rust Warranty	6 yr/100k	◐
Maintenance Warranty	N/A	
Roadside Assistance	N/A	

Ownership Cost By Year

Bars for 1993, 1994, 1995, 1996, 1997

Resale Value

1993	1994	1995	1996	1997
$17,811	$15,683	$13,933	$12,213	$10,645

Cumulative Costs

	1993	1994	1995	1996	1997
Annual	$12,548	$6,307	$6,837	$6,012	$7,075
Total	$12,548	$18,855	$25,692	$31,704	$38,779

Ownership Costs (5yr)

	Average	This Car
	$37,312	$38,779
Cost/Mile	53¢	55¢

Ownership Cost Rating

● Worse Than Average

Ford Taurus GL
4 Door Wagon

Midsize Wagon

 3.0L 140 hp Gas Fuel Inject.
 6 Cylinder "V"
 Automatic 4 Speed
 2 Wheel Front
 Driver Airbag Psngr Opt

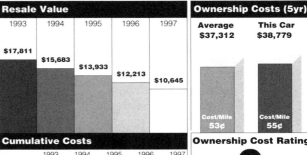

Purchase Price

Car Item	Dealer Cost	List
Base Price	**$14,298**	**$16,656**
Anti-Lock Brakes	$506	$595
Manual Transmission	N/A	N/A
3.8L 140 hp Gas	$472	** $555
Air Conditioning	Pkg	Pkg
Power Steering	Std	Std
Cruise Control	$191	$224
All Wheel Drive	N/A	N/A
AM/FM Stereo Cassette	$145	$171
Steering Wheel, Tilt	Std	Std
Power Windows	$303	$356
***Options Price**	**$639**	**$751**
***Total Price**	**$14,937**	**$17,407**
Target Price		$16,202
Destination Charge		$490
Avg. Tax & Fees		$1,014
Total Target $		**$17,706**
Average Dealer Option Cost:		*85%*

The 1993 Ford Taurus wagon is available in two models - GL and LX. New for 1993, the GL wagon features body-colored bumpers, all new seat trim, and a variety of new colors. Standard features include an exterior accent group, luggage rack, dual electric remote control mirrors, door trim panels, split bench seat with dual recliners and tinted glass. Optional features include a GL decor/equipment group, a light group, cellular phone, Luxury Convenience group and power moonroof.

Ownership Costs

Cost Area	5 Year Cost	Rate
Depreciation	$8,148	○
Financing ($356/month)	$2,917	
Insurance (Rating 4)	$6,658	◐
State Fees	$715	
Fuel (Hwy 30 City 21)	$3,405	◐
Maintenance	$4,173	○
Repairs	$709	○

Warranty/Maintenance Info

Major Tune-Up	$137	○
Minor Tune-Up	$87	○
Brake Service	$271	●
Overall Warranty	3 yr/36k	○
Drivetrain Warranty	3 yr/36k	○
Rust Warranty	6 yr/100k	○
Maintenance Warranty	N/A	
Roadside Assistance	N/A	

Resale Value

1993	1994	1995	1996	1997
$14,768	$13,253	$12,087	$10,805	$9,558

Cumulative Costs

	1993	1994	1995	1996	1997
Annual	$6,236	$4,727	$5,350	$4,803	$5,609
Total	$6,236	$10,963	$16,313	$21,116	$26,725

Ownership Costs (5yr)

	Average	This Car
	$29,067	$26,725
Cost/Mile	42¢	38¢

Ownership Cost Rating

○ Better Than Average

---

page 90

* Includes shaded options
** Other purchase requirements apply

● Poor   ◐ Worse Than Average   ◯ Average   ○ Better Than Average   ○ Excellent   ⊖ Insufficient Information

©1993 by IntelliChoice, Inc. (408) 554-8711   All Rights Reserved. Reproduction Prohibited.
Refer to *Section 3: Annotated Vehicle Charts* for an explanation of these charts.

Ford Taurus LX
4 Door Wagon

Midsize Wagon

- 3.8L 140 hp Gas Fuel Inject.
- 6 Cylinder "V"
- Automatic 4 Speed
- 2 Wheel Front
- Driver Airbag Psngr Opt

Purchase Price

Car Item	Dealer Cost	List
Base Price	**$17,131**	**$19,989**
Anti-Lock Brakes	$506	$595
Manual Transmission	N/A	N/A
Optional Engine	N/A	N/A
Air Conditioning	Std	Std
Power Steering	Std	Std
Cruise Control	Pkg	Pkg
All Wheel Drive	N/A	N/A
AM/FM Stereo Cassette	$282	$332
Steering Wheel, Tilt	Std	Std
Power Windows	Std	Std
*Options Price	$282	$332
*Total Price	$17,413	$20,321
Target Price		$19,007
Destination Charge		$490
Avg. Tax & Fees		$1,183
Total Target $		**$20,680**
Average Dealer Option Cost:		85%

The 1993 Ford Taurus wagon is available in two models - GL and LX. New for 1993, the LX wagon features a new floor console and a variety of new exterior colors. The LX wagon upgrades the GL with a console floor shift, convenience kit, cast aluminum wheels, leather seat trim, and a light group. Other standard features include an exterior accent group, luggage rack, and dual electric remote control mirrors. Optional features include a cellular phone, power moonroof and a Luxury Convenience group.

Ownership Costs

Cost Area	5 Year Cost	Rate
Depreciation	$9,903	○
Financing ($416/month)	$3,406	
Insurance (Rating 6)	$6,919	○
State Fees	$832	
Fuel (Hwy 27 City 19)	$3,775	○
Maintenance	$4,225	○
Repairs	$709	○

Warranty/Maintenance Info

Major Tune-Up	$138	○
Minor Tune-Up	$87	○
Brake Service	$271	●
Overall Warranty	3 yr/36k	○
Drivetrain Warranty	3 yr/36k	○
Rust Warranty	6 yr/100k	○
Maintenance Warranty	N/A	
Roadside Assistance	N/A	

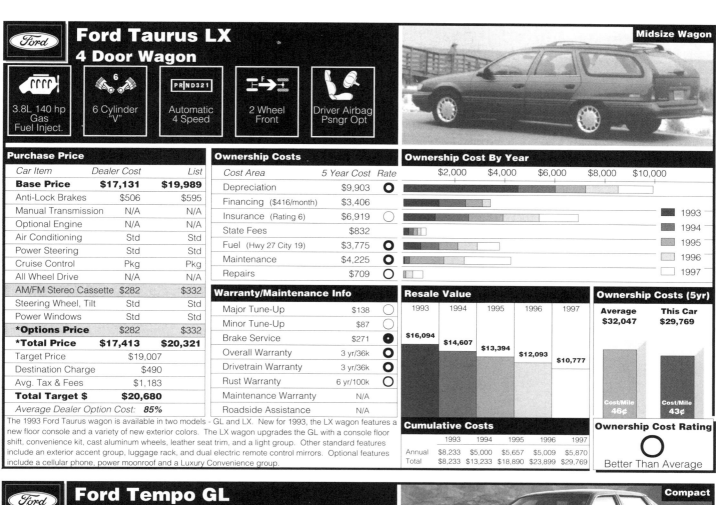

Ownership Cost By Year (1993–1997)

Resale Value: 1993 $16,094 | 1994 $14,607 | 1995 $13,394 | 1996 $12,093 | 1997 $10,777

Cumulative Costs

	1993	1994	1995	1996	1997
Annual	$8,233	$5,000	$5,657	$5,009	$5,870
Total	$8,233	$13,233	$18,890	$23,899	$29,769

Ownership Costs (5yr): Average $32,047 | This Car $29,769
Cost/Mile 46¢ / 43¢

Ownership Cost Rating: Better Than Average

Ford Tempo GL
2 Door Sedan

Compact

- 2.3L 96 hp Gas Fuel Inject.
- 4 Cylinder In-Line
- Manual 5 Speed
- 2 Wheel Front
- Belts Std, Driv Air Opt

4 Door Model Shown

Purchase Price

Car Item	Dealer Cost	List
Base Price	**$9,228**	**$10,267**
Anti-Lock Brakes	N/A	N/A
Automatic 3 Speed	$479	$563
3.0L 135 hp Gas	$583	$685
Air Conditioning	Pkg	Pkg
Power Steering	Std	Std
Cruise Control	$191	$224
All Wheel Drive	N/A	N/A
AM/FM Stereo Cassette	$132	$155
Steering Wheel, Tilt	Pkg	Pkg
Power Windows	N/A	N/A
*Options Price	$1,194	$1,403
*Total Price	$10,422	$11,670
Target Price		$11,208
Destination Charge		$465
Avg. Tax & Fees		$705
Total Target $		**$12,378**
Average Dealer Option Cost:		85%

The 1993 Ford Tempo is available in three models - GL two-door and four-door, and LX four-door sedan. For 1993, the GL two-door features color-keyed bumpers, interval windshield wipers, wheel covers, low back front bucket seats with cloth seat trim, and complete tinted glass. Optional features include rear window defroster, sport instrumentation cluster, polycast wheels, 6-way power adjustable driver seat, dual electric remote control mirrors and a light group.

Ownership Costs

Cost Area	5 Year Cost	Rate
Depreciation	$7,605	●
Financing ($249/month)	$2,038	
Insurance (Rating 8)	$7,223	○
State Fees	$485	
Fuel (Hwy 24 City 20)	$3,925	●
Maintenance	$3,937	○
Repairs	$709	○

Warranty/Maintenance Info

Major Tune-Up	$137	○
Minor Tune-Up	$87	○
Brake Service	$210	○
Overall Warranty	3 yr/36k	○
Drivetrain Warranty	3 yr/36k	○
Rust Warranty	6 yr/100k	○
Maintenance Warranty	N/A	
Roadside Assistance	N/A	

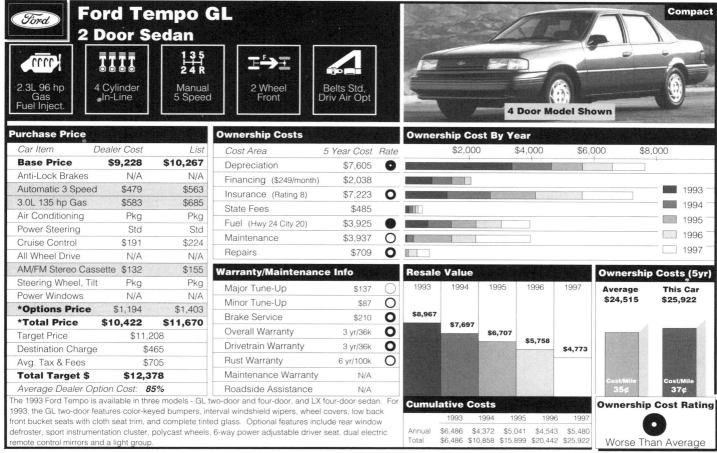

Ownership Cost By Year (1993–1997)

Resale Value: 1993 $8,967 | 1994 $7,697 | 1995 $6,707 | 1996 $5,758 | 1997 $4,773

Cumulative Costs

	1993	1994	1995	1996	1997
Annual	$6,486	$4,372	$5,041	$4,543	$5,480
Total	$6,486	$10,858	$15,899	$20,442	$25,922

Ownership Costs (5yr): Average $24,515 | This Car $25,922
Cost/Mile 35¢ / 37¢

Ownership Cost Rating: Worse Than Average

---

* Includes shaded options
** Other purchase requirements apply

Legend: ● Poor | ◐ Worse Than Average | ○ Average | ○ Better Than Average | ○ Excellent | ⊖ Insufficient Information

©1993 by IntelliChoice, Inc. (408) 554-8711 All Rights Reserved. Reproduction Prohibited.
Refer to *Section 3: Annotated Vehicle Charts* for an explanation of these charts.

Ford Tempo GL
4 Door Sedan

 2.3L 96 hp Gas Fuel Inject.
 4 Cylinder In-Line
 Manual 5 Speed
 2 Wheel Front
 Belts Std, Driv Air Opt

Compact

Purchase Price

Car Item	Dealer Cost	List
Base Price	**$9,228**	**$10,267**
Anti-Lock Brakes	N/A	N/A
Automatic 3 Speed	$479	$563
3.0L 135 hp Gas	$583	$685
Air Conditioning	Pkg	Pkg
Power Steering	Std	Std
Cruise Control	$191	$224
All Wheel Drive	N/A	N/A
AM/FM Stereo Cassette	$132	$155
Steering Wheel, Tilt	Pkg	Pkg
Power Windows	$281	** $330
***Options Price**	**$1,194**	**$1,403**
***Total Price**	**$10,422**	**$11,670**
Target Price		$11,208
Destination Charge		$465
Avg. Tax & Fees		$705
Total Target $		**$12,378**
Average Dealer Option Cost:		**85%**

The 1993 Ford Tempo is available in three models - GL two-door and four-door, and LX four-door sedan. For 1993, the GL four-door features low back front bucket seats with cloth seat trim, complete tinted glass, passenger assist grips, color-keyed bumpers, and interval windshield wipers. Optional features include rear window defroster, sport instrumentation cluster, polycast wheels, 6-way power adjustable driver seat, a light group and dual electric remote mirrors.

Ownership Costs

Cost Area	5 Year Cost	Rate
Depreciation	$7,263	●
Financing ($249/month)	$2,038	
Insurance (Rating 4)	$6,658	○
State Fees	$485	
Fuel (Hwy 24 City 20)	$3,925	●
Maintenance	$3,937	○
Repairs	$709	○

Warranty/Maintenance Info

Major Tune-Up	$137	◯
Minor Tune-Up	$87	○
Brake Service	$210	○
Overall Warranty	3 yr/36k	○
Drivetrain Warranty	3 yr/36k	○
Rust Warranty	6 yr/100k	○
Maintenance Warranty	N/A	
Roadside Assistance	N/A	

Ownership Cost By Year

1993, 1994, 1995, 1996, 1997

Resale Value

1993	1994	1995	1996	1997
$9,315	$8,062	$7,051	$6,084	$5,115

Cumulative Costs

	1993	1994	1995	1996	1997
Annual	$6,033	$4,246	$4,950	$4,444	$5,342
Total	$6,033	$10,279	$15,229	$19,673	$25,015

Ownership Costs (5yr)

Average	This Car
$24,515	$25,015
Cost/Mile 35¢	Cost/Mile 36¢

Ownership Cost Rating

○ Average

Ford Tempo LX
4 Door Sedan

 2.3L 96 hp Gas Fuel Inject.
4 Cylinder In-Line
 Manual 5 Speed
 2 Wheel Front
Belts Std, Driv Air Opt

Compact

Purchase Price

Car Item	Dealer Cost	List
Base Price	**$10,890**	**$12,135**
Anti-Lock Brakes	N/A	N/A
Automatic 3 Speed	Pkg	Pkg
3.0L 135 hp Gas	$583	$685
Air Conditioning	Pkg	Pkg
Power Steering	Std	Std
Cruise Control	$191	$224
All Wheel Drive	N/A	N/A
AM/FM Stereo Cassette	Pkg	Pkg
Steering Wheel, Tilt	Std	Std
Power Windows	$281	$330
***Options Price**	**$583**	**$685**
***Total Price**	**$11,473**	**$12,820**
Target Price		$12,386
Destination Charge		$465
Avg. Tax & Fees		$776
Total Target $		**$13,627**
Average Dealer Option Cost:		**85%**

The 1993 Ford Tempo is available in three models - GL two-door and four-door, and LX four-door sedan. For 1993, the LX features complete tinted glass, passenger assist grips, color-keyed bumpers, low back front bucket seats, and interval windshield wipers. The LX upgrades the GL with an illuminated entry system, power lock group, and sport instrumentation cluster. Optional features include a 6-way power adjustable driver seat, dual electric remote mirrors and a rear window defrost.

Ownership Costs

Cost Area	5 Year Cost	Rate
Depreciation	$8,619	●
Financing ($274/month)	$2,245	
Insurance (Rating 5)	$6,786	○
State Fees	$532	
Fuel (Hwy 28 City 21)	$3,530	●
Maintenance	$3,735	○
Repairs	$709	○

Warranty/Maintenance Info

Major Tune-Up	$137	◯
Minor Tune-Up	$87	○
Brake Service	$210	○
Overall Warranty	3 yr/36k	○
Drivetrain Warranty	3 yr/36k	○
Rust Warranty	6 yr/100k	○
Maintenance Warranty	N/A	
Roadside Assistance	N/A	

Resale Value

1993	1994	1995	1996	1997
$9,164	$7,928	$6,937	$5,987	$5,008

Cumulative Costs

	1993	1994	1995	1996	1997
Annual	$7,482	$4,255	$4,829	$4,394	$5,196
Total	$7,482	$11,737	$16,566	$20,960	$26,156

Ownership Costs (5yr)

Average	This Car
$25,502	$26,156
Cost/Mile 36¢	Cost/Mile 37¢

Ownership Cost Rating

○ Average

---

page 92

* Includes shaded options
** Other purchase requirements apply

● Poor | Worse Than Average | ○ Average | Better Than Average | ◯ Excellent | Insufficient Information

©1993 by IntelliChoice, Inc. (408) 554-8711 All Rights Reserved. Reproduction Prohibited.
Refer to *Section 3: Annotated Vehicle Charts* for an explanation of these charts.

Ford Thunderbird LX
2 Door Coupe

Large

- 3.8L 140 hp Gas Fuel Inject.
- 6 Cylinder "V"
- Automatic 4 Speed
- 2 Wheel Rear
- Automatic Seatbelts

Purchase Price

Car Item	Dealer Cost	List
Base Price	**$14,189**	**$15,797**
Anti-Lock Brakes	$591	$695
Manual Transmission	N/A	N/A
5.0L 185 hp Gas	$923	$1,086
Air Conditioning	Std	Std
Power Steering	Std	Std
Cruise Control	Std	Std
All Wheel Drive	N/A	N/A
AM/FM Stereo Cassette	Std	Std
Steering Wheel, Tilt	Std	Std
Power Windows	Std	Std
*Options Price	$923	$1,086
*Total Price	$15,112	$16,883
Target Price		$16,122
Destination Charge		$495
Avg. Tax & Fees		$1,005
Total Target $		**$17,622**
Average Dealer Option Cost:		85%

The 1993 Ford Thunderbird is available in two models- LX and Super Coupe. New for 1993, the LX comes equipped with dual electric mirrors and optional 15 inch cast aluminum wheels. The LX comes standard with anti-lock brakes, air conditioning, automatic suspension, rack-and-pinion steering, five-speed manual transmission, AM/FM cassette stereo, and full analog instrumentation. Options include speed control, moonroof, electric door locks, a 5.0L V-8 engine and a special option package.

Ownership Costs

Cost Area	5 Year Cost	Rate
Depreciation	$10,915	●
Financing ($354/month)	$2,903	
Insurance (Rating 8 [Est.])	$7,223	○
State Fees	$695	
Fuel (Hwy 24 City 17)	$4,232	●
Maintenance	$4,001	○
Repairs	$709	○

Warranty/Maintenance Info

Major Tune-Up	$156	○
Minor Tune-Up	$106	○
Brake Service	$296	●
Overall Warranty	3 yr/36k	○
Drivetrain Warranty	3 yr/36k	○
Rust Warranty	6 yr/100k	○
Maintenance Warranty	N/A	
Roadside Assistance	N/A	

Ownership Cost By Year
1993, 1994, 1995, 1996, 1997

Resale Value

1993	1994	1995	1996	1997
$12,008	$10,396	$9,138	$7,955	$6,707

Cumulative Costs

	1993	1994	1995	1996	1997
Annual	$9,160	$5,099	$5,642	$4,904	$5,873
Total	$9,160	$14,259	$19,901	$24,805	$30,678

Ownership Costs (5yr)
- Average: $29,213 — Cost/Mile 42¢
- This Car: $30,678 — Cost/Mile 44¢

Ownership Cost Rating
Worse Than Average

Ford Thunderbird Super Coupe
2 Door Coupe

Large

- 3.8L 210 hp Suprchrged Fuel Inject.
- 6 Cylinder "V"
- Manual 5 Speed
- 2 Wheel Rear
- Automatic Seatbelts

Purchase Price

Car Item	Dealer Cost	List
Base Price	**$18,856**	**$22,030**
Anti-Lock Brakes	Std	Std
Automatic 4 Speed	$506	$595
Optional Engine	N/A	N/A
Air Conditioning	Std	Std
Power Steering	Std	Std
Cruise Control	Pkg	Pkg
All Wheel Drive	N/A	N/A
AM/FM Stereo Cassette	Std	Std
Steering Wheel, Tilt	Pkg	Pkg
Power Windows	Std	Std
*Options Price	$506	$595
*Total Price	$19,362	$22,625
Target Price		$20,829
Destination Charge		$495
Avg. Tax & Fees		$1,297
Total Target $		**$22,621**
Average Dealer Option Cost:		85%

The 1993 Ford Thunderbird is available in two models-LX and Super Coupe. New for 1993 the Super Coupe comes equipped with 16 inch cast aluminum wheels and dual electric mirrors. The Super Coupe comes standard with a Supercharged 3.8L V-6 engine, full analog instrumentation, air conditioning, articulated sport seats, power windows, automatic suspension, 5-speed manual, rack-and-pinion steering and an AM/FM cassette stereo. Options include speed control, CD player, moonroof, and electric door locks.

Ownership Costs

Cost Area	5 Year Cost	Rate
Depreciation	$12,900	●
Financing ($455/month)	$3,727	
Insurance (Rating 13 [Est.])	$8,061	○
State Fees	$925	
Fuel (Hwy 23 City 17 -Prem.)	$4,777	●
Maintenance	$5,379	●
Repairs	$995	●

Warranty/Maintenance Info

Major Tune-Up	$141	○
Minor Tune-Up	$88	○
Brake Service	$298	●
Overall Warranty	3 yr/36k	○
Drivetrain Warranty	3 yr/36k	○
Rust Warranty	6 yr/100k	○
Maintenance Warranty	N/A	
Roadside Assistance		

Resale Value

1993	1994	1995	1996	1997
$16,202	$14,374	$12,800	$11,261	$9,721

Cumulative Costs

	1993	1994	1995	1996	1997
Annual	$10,618	$5,866	$7,080	$5,264	$7,936
Total	$10,618	$16,484	$23,564	$28,828	$36,764

Ownership Costs (5yr)
- Average: $34,908 — Cost/Mile 50¢
- This Car: $36,764 — Cost/Mile 53¢

Ownership Cost Rating
Worse Than Average

---

* Includes shaded options
** Other purchase requirements apply

- ● Poor
- ◐ Worse Than Average
- ◑ Average
- ○ Better Than Average
- ○ Excellent
- ⊖ Insufficient Information

©1993 by IntelliChoice, Inc. (408) 554-8711 All Rights Reserved. Reproduction Prohibited.
Refer to *Section 3: Annotated Vehicle Charts* for an explanation of these charts.

Geo Metro
2 Door Hatchback

 1.0L 55 hp Gas Fuel Inject.
 3 Cylinder In-Line
 Manual 5 Speed
 2 Wheel Front
 Automatic Seatbelts

Subcompact

Purchase Price

Car Item	Dealer Cost	List
Base Price	**$6,254**	**$6,710**
Anti-Lock Brakes	N/A	N/A
Automatic 3 Speed	$441	$495
Optional Engine	N/A	N/A
Air Conditioning	$641	$720
Power Steering	N/A	N/A
Cruise Control	N/A	N/A
All Wheel Drive	N/A	N/A
AM/FM Stereo Cassette	$441	$496
Steering Wheel, Tilt	N/A	N/A
Power Windows	N/A	N/A
*Options Price	$0	$0
*Total Price	$6,254	$6,710
Target Price	$6,577	
Destination Charge	$285	
Avg. Tax & Fees	$413	
Total Target $	**$7,275**	
Average Dealer Option Cost:	89%	

The 1993 Geo Metro is available in six models - (Base) Metro, XFi, and LSi two-door hatchbacks; (Base) Metro and LSi four-door hatchbacks; and the LSi convertible. New for 1993, the Base Metro two-door hatchback features new automatic front door locks and upgraded sound systems. Standard features include a self-aligning steering wheel, fold-down rear seat for cargo-carrying capability, and a laminated windshield glass with urethane bonding. Rear window defogger is optional.

Ownership Costs

Cost Area	5 Year Cost	Rate
Depreciation	$3,207	○
Financing ($146/month)	$1,199	
Insurance (Rating 11 [Est.])	$7,693	●
State Fees	$280	
Fuel (Hwy 50 City 46)	$1,805	○
Maintenance	$4,199	◐
Repairs	$570	○

Warranty/Maintenance Info

Major Tune-Up	$256 ●
Minor Tune-Up	$100 ◐
Brake Service	$214 ◐
Overall Warranty	3 yr/36k ◐
Drivetrain Warranty	3 yr/36k ◐
Rust Warranty	6 yr/100k ◐
Maintenance Warranty	N/A
Roadside Assistance	3 yr/36k

Ownership Cost By Year

1993, 1994, 1995, 1996, 1997

Resale Value

1993	1994	1995	1996	1997
$5,529	$5,044	$4,966	$4,565	$4,068

Cumulative Costs

	1993	1994	1995	1996	1997
Annual	$4,113	$2,980	$3,373	$3,197	$5,290
Total	$4,113	$7,093	$10,466	$13,663	$18,953

Ownership Costs (5yr)

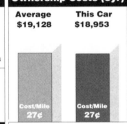

Average $19,128 | This Car $18,953
Cost/Mile 27¢ | Cost/Mile 27¢

Ownership Cost Rating

○ Average

Geo Metro XFi
2 Door Hatchback

 1.0L 49 hp Gas Fuel Inject.
3 Cylinder In-Line
 Manual 5 Speed
 2 Wheel Front
Automatic Seatbelts

Subcompact

Purchase Price

Car Item	Dealer Cost	List
Base Price	**$6,254**	**$6,710**
Anti-Lock Brakes	N/A	N/A
Automatic Transmission	N/A	N/A
Optional Engine	N/A	N/A
Air Conditioning	N/A	N/A
Power Steering	N/A	N/A
Cruise Control	N/A	N/A
All Wheel Drive	N/A	N/A
AM/FM Stereo Cassette	$174	** $195
Steering Wheel, Tilt	N/A	N/A
Power Windows	N/A	N/A
*Options Price	$0	$0
*Total Price	$6,254	$6,710
Target Price	$6,577	
Destination Charge	$285	
Avg. Tax & Fees	$413	
Total Target $	**$7,275**	
Average Dealer Option Cost:	89%	

The 1993 Geo Metro is available in six models - (Base) Metro, XFi, and LSi two-door hatchbacks; (Base) Metro and LSi four-door hatchbacks; and the LSi convertible. New for 1993, the Metro XFi two-door hatchback features new automatic front door locks and upgraded sound systems. Standard features include self-aligning steering wheel and radial tires. The Metro XFi has been the most fuel efficient vehicle sold in the United States for the past four consecutive years.

Ownership Costs

Cost Area	5 Year Cost	Rate
Depreciation	$3,305	○
Financing ($146/month)	$1,199	
Insurance (Rating 11 [Est.])	$7,693	●
State Fees	$280	
Fuel (Hwy 58 City 53)	$1,561	○
Maintenance	$4,199	◐
Repairs	$570	○

Warranty/Maintenance Info

Major Tune-Up	$256 ●
Minor Tune-Up	$100 ◐
Brake Service	$214 ◐
Overall Warranty	3 yr/36k ◐
Drivetrain Warranty	3 yr/36k ◐
Rust Warranty	6 yr/100k ◐
Maintenance Warranty	N/A
Roadside Assistance	3 yr/36k

Ownership Cost By Year

Resale Value

1993	1994	1995	1996	1997
$5,613	$5,233	$4,910	$4,488	$3,970

Cumulative Costs

	1993	1994	1995	1996	1997
Annual	$3,984	$2,828	$3,570	$3,167	$5,258
Total	$3,984	$6,812	$10,382	$13,549	$18,807

Ownership Costs (5yr)

Average $19,128 | This Car $18,807
Cost/Mile 27¢ | Cost/Mile 27¢

Ownership Cost Rating

○ Better Than Average

---

page 94

* Includes shaded options
** Other purchase requirements apply

● Poor | ◐ Worse Than Average | ○ Average | ○ Better Than Average | ○ Excellent | ⊖ Insufficient Information

©1993 by IntelliChoice, Inc. (408) 554-8711 All Rights Reserved. Reproduction Prohibited.
Refer to *Section 3: Annotated Vehicle Charts* for an explanation of these charts.

Geo Metro 4 Door Hatchback

Subcompact

- 1.0L 55 hp Gas Fuel Inject.
- 3 Cylinder In-Line
- Manual 5 Speed
- 2 Wheel Front
- Automatic Seatbelts

Purchase Price

Car Item	Dealer Cost	List
Base Price	**$6,709**	**$7,199**
Anti-Lock Brakes	N/A	N/A
Automatic 3 Speed	$441	$495
Optional Engine	N/A	N/A
Air Conditioning	$641	$720
Power Steering	N/A	N/A
Cruise Control	N/A	N/A
All Wheel Drive	N/A	N/A
AM/FM Stereo Cassette	$441	$496
Steering Wheel, Tilt	N/A	N/A
Power Windows	N/A	N/A
*Options Price	$0	$0
*Total Price	$6,709	$7,199
Target Price		$7,061
Destination Charge		$285
Avg. Tax & Fees		$442
Total Target $		**$7,788**
Average Dealer Option Cost:	**89%**	

The 1993 Geo Metro is available in six models - (Base) Metro, XFi, and LSi two-door hatchbacks; (Base) Metro and LSi four-door hatchbacks; and the LSi convertible. New for 1993, the Base Metro four-door hatchback features upgraded sound systems and new automatic front door locks. Standard features include rear door security locks, self-aligning steering wheel, laminated windshield glass with urethane bonding and blackwall radial tires. Cargo cover and rear defogger are optional.

Ownership Costs

Cost Area	5 Year Cost	Rate
Depreciation	$3,405	○
Financing ($157/month)	$1,282	
Insurance (Rating 8 [Est.])	$7,223	●
State Fees	$299	
Fuel (Hwy 50 City 46)	$1,805	○
Maintenance	$4,243	◐
Repairs	$570	○

Warranty/Maintenance Info

Major Tune-Up	$256	●
Minor Tune-Up	$100	◉
Brake Service	$214	○
Overall Warranty	3 yr/36k	○
Drivetrain Warranty	3 yr/36k	○
Rust Warranty	6 yr/100k	○
Maintenance Warranty	N/A	
Roadside Assistance	3 yr/36k	

Ownership Cost By Year
(1993–1997)

Resale Value

1993	1994	1995	1996	1997
$5,905	$5,385	$5,308	$4,894	$4,383

Cumulative Costs

	1993	1994	1995	1996	1997
Annual	$4,204	$2,955	$3,298	$3,165	$5,205
Total	$4,204	$7,159	$10,457	$13,622	$18,827

Ownership Costs (5yr)

Average	This Car
$19,678	$18,827
Cost/Mile 28¢	Cost/Mile 27¢

Ownership Cost Rating: Excellent

Geo Metro LSi 2 Door Hatchback

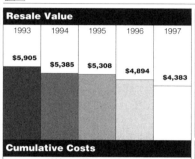
Base Model Shown

Subcompact

- 1.0L 55 hp Gas Fuel Inject.
- 3 Cylinder In-Line
- Manual 5 Speed
- 2 Wheel Front
- Automatic Seatbelts

Purchase Price

Car Item	Dealer Cost	List
Base Price	**$7,641**	**$8,199**
Anti-Lock Brakes	N/A	N/A
Automatic 3 Speed	$441	$495
Optional Engine	N/A	N/A
Air Conditioning	$641	$720
Power Steering	N/A	N/A
Cruise Control	N/A	N/A
All Wheel Drive	N/A	N/A
AM/FM Stereo Cassette	$441	$496
Steering Wheel, Tilt	N/A	N/A
Power Windows	N/A	N/A
*Options Price	$0	$0
*Total Price	$7,641	$8,199
Target Price		$8,056
Destination Charge		$285
Avg. Tax & Fees		$502
Total Target $		**$8,843**
Average Dealer Option Cost:	**89%**	

The 1993 Geo Metro is available in six models - (Base) Metro, XFi, and LSi two-door hatchbacks; (Base) Metro and LSi four-door hatchbacks; and the LSi convertible. New for 1993, the Metro LSi two-door hatchback features upgraded sound systems and new automatic front door locks. The LSi upgrades the Base edition with standard cloth reclining bucket seats with head restraints, split-folding rear seat, tachometer, and a rear cargo security cover. The Metros are built in Kosai, Japan.

Ownership Costs

Cost Area	5 Year Cost	Rate
Depreciation	$4,370	○
Financing ($178/month)	$1,457	
Insurance (Rating 12 [Est.])	$7,898	●
State Fees	$339	
Fuel (Hwy 50 City 46)	$1,805	○
Maintenance	$4,199	◐
Repairs	$570	○

Warranty/Maintenance Info

Major Tune-Up	$256	●
Minor Tune-Up	$100	◉
Brake Service	$214	○
Overall Warranty	3 yr/36k	○
Drivetrain Warranty	3 yr/36k	○
Rust Warranty	6 yr/100k	○
Maintenance Warranty	N/A	
Roadside Assistance	3 yr/36k	

Resale Value

1993	1994	1995	1996	1997
$6,240	$5,739	$5,516	$5,026	$4,473

Cumulative Costs

	1993	1994	1995	1996	1997
Annual	$5,131	$3,129	$3,622	$3,357	$5,399
Total	$5,131	$8,260	$11,882	$15,239	$20,638

Ownership Costs (5yr)

Average	This Car
$20,803	$20,638
Cost/Mile 30¢	Cost/Mile 29¢

Ownership Cost Rating: Average

---

* Includes shaded options
** Other purchase requirements apply

● Poor ◉ Worse Than Average ◐ Average ○ Better Than Average ○ Excellent ⊖ Insufficient Information

©1993 by IntelliChoice, Inc. (408) 554-8711 All Rights Reserved. Reproduction Prohibited.
Refer to *Section 3: Annotated Vehicle Charts* for an explanation of these charts.

Geo Metro LSi
4 Door Hatchback

- 1.0L 55 hp Gas Fuel Inject.
- 3 Cylinder In-Line
- Manual 5 Speed
- 2 Wheel Front
- Automatic Seatbelts

Subcompact — Base Model Shown

Purchase Price

Car Item	Dealer Cost	List
Base Price	**$8,014**	**$8,599**
Anti-Lock Brakes	N/A	N/A
Automatic 3 Speed	$441	$495
Optional Engine	N/A	N/A
Air Conditioning	$641	$720
Power Steering	N/A	N/A
Cruise Control	N/A	N/A
All Wheel Drive	N/A	N/A
AM/FM Stereo Cassette	$441	$496
Steering Wheel, Tilt	N/A	N/A
Power Windows	N/A	N/A
*Options Price	$0	$0
*Total Price	$8,014	$8,599
Target Price		$8,455
Destination Charge		$285
Avg. Tax & Fees		$526
Total Target $		**$9,266**
Average Dealer Option Cost:	**89%**	

The 1993 Geo Metro is available in six models - (Base) Metro, XFi, and LSi two-door hatchbacks; (Base) Metro and LSi four-door hatchbacks; and the LSi convertible. New for 1993, the Metro LSi four-door hatchback features new automatic front and rear door locks and upgraded sound systems. The LSi upgrades the Base edition with a split folding rear seat, cloth reclining bucket seats with head restraints, rear cargo security cover and tachometer. The Metros are built in Kosai, Japan.

Ownership Costs

Cost Area	5 Year Cost	Rate
Depreciation	$4,642	○
Financing ($186/month)	$1,527	
Insurance (Rating 10 [Est.])	$7,479	●
State Fees	$355	
Fuel (Hwy 50 City 46)	$1,805	○
Maintenance	$4,243	◐
Repairs	$570	○

Warranty/Maintenance Info

Major Tune-Up	$256	●
Minor Tune-Up	$100	◉
Brake Service	$214	○
Overall Warranty	3 yr/36k	○
Drivetrain Warranty	3 yr/36k	○
Rust Warranty	6 yr/100k	○
Maintenance Warranty	N/A	
Roadside Assistance	3 yr/36k	

Ownership Cost By Year

1993, 1994, 1995, 1996, 1997

Resale Value

1993	1994	1995	1996	1997
$6,492	$5,954	$5,719	$5,193	$4,624

Cumulative Costs

	1993	1994	1995	1996	1997
Annual	$5,258	$3,111	$3,568	$3,358	$5,326
Total	$5,258	$8,369	$11,937	$15,295	$20,621

Ownership Costs (5yr)

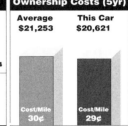

Average	This Car
$21,253	$20,621
Cost/Mile 30¢	Cost/Mile 29¢

Ownership Cost Rating
○ Better Than Average

Geo Metro LSi
2 Door Convertible

- 1.0L 55 hp Gas Fuel Inject.
- 3 Cylinder In-Line
- Manual 5 Speed
- 2 Wheel Front
- Driver Airbag Psngr Belts

Subcompact

Purchase Price

Car Item	Dealer Cost	List
Base Price	**$9,319**	**$9,999**
Anti-Lock Brakes	N/A	N/A
Automatic 3 Speed	$441	$495
Optional Engine	N/A	N/A
Air Conditioning	$641	$720
Power Steering	N/A	N/A
Cruise Control	N/A	N/A
All Wheel Drive	N/A	N/A
AM/FM Stereo Cassette	$464	$521
Steering Wheel, Tilt	N/A	N/A
Power Windows	N/A	N/A
*Options Price	$0	$0
*Total Price	$9,319	$9,999
Target Price		$9,855
Destination Charge		$255
Avg. Tax & Fees		$609
Total Target $		**$10,719**
Average Dealer Option Cost:	**89%**	

The 1993 Geo Metro is available in six models - (Base) Metro, XFi, and LSi two-door hatchbacks; (Base) Metro and LSi four-door hatchbacks; and the LSi convertible. New for 1993, the Metro LSi convertible features optional upgraded sound systems. Standard features include special custom cloth seat trim, intermittent wipers, 5-mph front and rear bumpers and blackwall radial tires. Safety features include energy-absorbing steering column, and tempered safety side glass windows.

Ownership Costs

Cost Area	5 Year Cost	Rate
Depreciation	$5,409	○
Financing ($215/month)	$1,765	
Insurance (Rating 13 [Est.])	$8,061	●
State Fees	$411	
Fuel (Hwy 46 City 41)	$1,988	○
Maintenance	$4,481	◐
Repairs	$570	○

Warranty/Maintenance Info

Major Tune-Up	$256	●
Minor Tune-Up	$100	◉
Brake Service	$214	○
Overall Warranty	3 yr/36k	○
Drivetrain Warranty	3 yr/36k	○
Rust Warranty	6 yr/100k	○
Maintenance Warranty	N/A	
Roadside Assistance	3 yr/36k	

Ownership Cost By Year

Resale Value

1993	1994	1995	1996	1997
$8,114	$7,231	$6,652	$5,999	$5,310

Cumulative Costs

	1993	1994	1995	1996	1997
Annual	$5,343	$3,690	$4,238	$3,627	$5,787
Total	$5,343	$9,033	$13,271	$16,898	$22,685

Ownership Costs (5yr)

Average	This Car
$22,828	$22,685
Cost/Mile 33¢	Cost/Mile 32¢

Ownership Cost Rating
○ Average

---

page 96

* Includes shaded options
** Other purchase requirements apply

● Poor   ◉ Worse Than Average   ◐ Average   ○ Better Than Average   ○ Excellent   ⊖ Insufficient Information

©1993 by *IntelliChoice, Inc.* (408) 554-8711   All Rights Reserved. Reproduction Prohibited.
Refer to *Section 3: Annotated Vehicle Charts* for an explanation of these charts.

Geo Prizm
4 Door Sedan

Subcompact

 1.6L 108 hp Gas Fuel Inject.
 4 Cylinder In-Line
 Manual 5 Speed
 2 Wheel Front
 Driver Airbag Psngr Belts

Purchase Price

Car Item	Dealer Cost	List
Base Price	**$9,515**	**$9,995**
Anti-Lock Brakes	$434	$505
Automatic 3 Speed	$426	$495
Optional Engine	N/A	N/A
Air Conditioning	$641	$745
Power Steering	$215	$250
Cruise Control	N/A	N/A
All Wheel Drive	N/A	N/A
AM/FM Stereo Cassette	$432	$502
Steering Wheel, Tilt	N/A	N/A
Power Windows	N/A	N/A
*Options Price	$215	$250
*Total Price	$9,730	$10,245
Target Price	$10,194	
Destination Charge	$365	
Avg. Tax & Fees	$634	
Total Target $	**$11,193**	
Average Dealer Option Cost:	**86%**	

The 1993 Geo Prizm is available in two models - (Base) Prizm and LSi. New for 1993, the (Base) Prizm features firmer bucket seats providing more room for the back seat passengers. The new design was built for wind noise reduction and more interior room with only one inch added to the exterior. Standard features include body-colored bumpers, tinted glass, and wrap-around rear side marker lamps. A power sunroof, rear defroster and dual electric outside mirrors are featured as options.

Ownership Costs

Cost Area	5 Year Cost	Rate
Depreciation	$5,818	○
Financing ($225/month)	$1,844	
Insurance (Rating 5 [Est.])	$6,786	○
State Fees	$424	
Fuel (Hwy 33 City 28)	$2,832	○
Maintenance	$4,731	●
Repairs	$570	○

Warranty/Maintenance Info

Major Tune-Up	$223	●
Minor Tune-Up	$91	○
Brake Service	$174	○
Overall Warranty	3 yr/36k	○
Drivetrain Warranty	3 yr/36k	○
Rust Warranty	6 yr/100k	○
Maintenance Warranty	N/A	
Roadside Assistance	3 yr/36k	

Ownership Cost By Year

1993 / 1994 / 1995 / 1996 / 1997

Resale Value

1993	1994	1995	1996	1997
$8,040	$7,098	$6,743	$6,059	$5,375

Cumulative Costs

	1993	1994	1995	1996	1997
Annual	$5,847	$3,683	$4,003	$3,561	$5,911
Total	$5,847	$9,530	$13,533	$17,094	$23,005

Ownership Costs (5yr)

Average	This Car
$23,105	$23,005
Cost/Mile 33¢	Cost/Mile 33¢

Ownership Cost Rating

Average

Geo Prizm LSi
4 Door Sedan

Subcompact

 1.6L 108 hp Gas Fuel Inject.
 4 Cylinder In-Line
 Manual 5 Speed
 2 Wheel Front
Driver Airbag Psngr Belts

Purchase Price

Car Item	Dealer Cost	List
Base Price	**$9,801**	**$10,630**
Anti-Lock Brakes	$434	$505
Automatic 4 Speed	$667	$775
1.8L 115 hp Gas	$303	$352
Air Conditioning	$641	$745
Power Steering	$215	$250
Cruise Control	Pkg	Pkg
All Wheel Drive	N/A	N/A
AM/FM Stereo Cassette	$432	$502
Steering Wheel, Tilt	Std	Std
Power Windows	Pkg	Pkg
*Options Price	$518	$602
*Total Price	$10,319	$11,232
Target Price	$10,890	
Destination Charge	$365	
Avg. Tax & Fees	$679	
Total Target $	**$11,934**	
Average Dealer Option Cost:	**86%**	

The 1993 Geo Prizm is available in two models - (Base) Prizm and LSi. New for 1993, the Prizm LSi features a new design which increases interior room with only one additional inch added to the exterior width. The LSi upgrades the base edition with specific seat and door trim, a split-folding rear seat with a "pass-through" feature to the trunk, a center console, and full wheel covers. Optional equipment includes a power sunroof, rear window defroster and dual electric outside mirrors.

Ownership Costs

Cost Area	5 Year Cost	Rate
Depreciation	$6,246	○
Financing ($240/month)	$1,966	
Insurance (Rating 10 [Est.])	$7,479	●
State Fees	$464	
Fuel (Hwy 34 City 28)	$2,784	○
Maintenance	$4,694	●
Repairs	$570	○

Warranty/Maintenance Info

Major Tune-Up	$210	●
Minor Tune-Up	$100	●
Brake Service	$174	○
Overall Warranty	3 yr/36k	○
Drivetrain Warranty	3 yr/36k	○
Rust Warranty	6 yr/100k	○
Maintenance Warranty	N/A	
Roadside Assistance	3 yr/36k	

Ownership Cost By Year

Resale Value

1993	1994	1995	1996	1997
$7,985	$7,300	$6,819	$6,267	$5,688

Cumulative Costs

	1993	1994	1995	1996	1997
Annual	$6,826	$3,613	$4,283	$3,596	$5,885
Total	$6,826	$10,439	$14,722	$18,318	$24,203

Ownership Costs (5yr)

Average	This Car
$24,216	$24,203
Cost/Mile 35¢	Cost/Mile 35¢

Ownership Cost Rating

Average

---

* Includes shaded options
** Other purchase requirements apply

● Poor ◐ Worse Than Average ○ Average ◯ Better Than Average ○ Excellent ⊖ Insufficient Information

©1993 by IntelliChoice, Inc. (408) 554-8711 All Rights Reserved. Reproduction Prohibited.
Refer to *Section 3: Annotated Vehicle Charts* for an explanation of these charts.

page **97**

Geo Storm 2+2
2 Door Coupe
Subcompact

- 1.6L 95 hp Gas Fuel Inject.
- 4 Cylinder In-Line
- Manual 5 Speed
- 2 Wheel Front
- Driver Airbag Psngr Belts

Purchase Price
Car Item	Dealer Cost	List
Base Price	**$10,515**	**$11,530**
Anti-Lock Brakes	N/A	N/A
Automatic 3 Speed	Pkg	Pkg
Optional Engine	N/A	N/A
Air Conditioning	$641	$745
Power Steering	Std	Std
Cruise Control	N/A	N/A
All Wheel Drive	N/A	N/A
AM/FM Stereo Cassette	$168	$195
Steering Wheel, Tilt	N/A	N/A
Power Windows	N/A	N/A
*Options Price	$0	$0
*Total Price	$10,515	$11,530
Target Price		$11,111
Destination Charge		$345
Avg. Tax & Fees		$692
Total Target $		**$12,148**
Average Dealer Option Cost:		**86%**

The 1993 Geo Storm is available in two models - (Base) 2+2 sport coupe and the GSi 2+2 sport coupe. New for 1993, the Storm sport coupe has retuned the engine for a lower torque peak to improve low-speed performance during air conditioning operation and to improve launch feel. Safety features include tempered safety glass for side and rear windows, and an energy-absorbing steering column. Tinted glass and full aero-style wheel covers are standard. New alloy wheels are offered as optional equipment.

Ownership Costs
Cost Area	5 Year Cost	Rate
Depreciation	$5,796	○
Financing ($244/month)	$2,000	
Insurance (Rating 12 [Est.])	$7,898	●
State Fees	$475	
Fuel (Hwy 36 City 30)	$2,615	○
Maintenance	$4,743	◐
Repairs	$580	○

Warranty/Maintenance Info
Major Tune-Up	$205	○
Minor Tune-Up	$100	◐
Brake Service	$175	○
Overall Warranty	3 yr/36k	○
Drivetrain Warranty	3 yr/36k	○
Rust Warranty	6 yr/100k	○
Maintenance Warranty	N/A	
Roadside Assistance	3 yr/36k	

Ownership Cost By Year

Resale Value
1993	1994	1995	1996	1997
$9,776	$8,507	$7,788	$7,082	$6,352

Cumulative Costs
	1993	1994	1995	1996	1997
Annual	$5,310	$4,254	$4,468	$3,808	$6,267
Total	$5,310	$9,564	$14,032	$17,840	$24,107

Ownership Costs (5yr)
- Average: $24,551
- This Car: $24,107
- Cost/Mile: 35¢ / 34¢

Ownership Cost Rating
○ Better Than Average

Geo Storm 2+2 GSi
2 Door Coupe
Subcompact

- 1.8L 140 hp Gas Fuel Inject.
- 4 Cylinder In-Line
- Manual 5 Speed
- 2 Wheel Front
- Driver Airbag Psngr Belts

Purchase Price
Car Item	Dealer Cost	List
Base Price	**$12,307**	**$13,495**
Anti-Lock Brakes	N/A	N/A
Automatic 4 Speed	Pkg	Pkg
Optional Engine	N/A	N/A
Air Conditioning	$641	$745
Power Steering	Std	Std
Cruise Control	N/A	N/A
All Wheel Drive	N/A	N/A
AM/FM Stereo Cassette	$168	$195
Steering Wheel, Tilt	N/A	N/A
Power Windows	N/A	N/A
*Options Price	$0	$0
*Total Price	$12,307	$13,495
Target Price		$13,044
Destination Charge		$345
Avg. Tax & Fees		$807
Total Target $		**$14,196**
Average Dealer Option Cost:		**86%**

The 1993 Geo Storm is available in two models - (Base) 2+2 sport coupe and the GSi 2+2 sport coupe. New for 1993, the Storm GSi sport coupe has retuned the engine for a lower torque peak to improve low-speed performance during air conditioning operation and to improve launch feel. The GSi upgrades the base model with a stronger horsepower, sport suspension, sport alloy wheels, and a wing-type rear spoiler. Safety features include tempered safety glass for side and rear windows and security locks.

Ownership Costs
Cost Area	5 Year Cost	Rate
Depreciation	$6,894	○
Financing ($285/month)	$2,339	
Insurance (Rating 13 [Est.])	$8,061	●
State Fees	$554	
Fuel (Hwy 31 City 23)	$3,204	◐
Maintenance	$5,640	●
Repairs	$570	○

Warranty/Maintenance Info
Major Tune-Up	$215	◐
Minor Tune-Up	$100	◐
Brake Service	$175	○
Overall Warranty	3 yr/36k	○
Drivetrain Warranty	3 yr/36k	○
Rust Warranty	6 yr/100k	○
Maintenance Warranty	N/A	
Roadside Assistance	3 yr/36k	

Resale Value
1993	1994	1995	1996	1997
$10,703	$9,503	$8,844	$8,080	$7,302

Cumulative Costs
	1993	1994	1995	1996	1997
Annual	$6,732	$4,458	$5,072	$4,071	$6,929
Total	$6,732	$11,190	$16,262	$20,333	$27,262

Ownership Costs (5yr)
- Average: $26,762
- This Car: $27,262
- Cost/Mile: 38¢ / 39¢

Ownership Cost Rating
○ Average

Legend: ● Poor · ◐ Worse Than Average · ○ Average · ○ Better Than Average · ○ Excellent · ⊖ Insufficient Information

* Includes shaded options
** Other purchase requirements apply

©1993 by *IntelliChoice, Inc.* (408) 554-8711   All Rights Reserved. Reproduction Prohibited.
Refer to *Section 3: Annotated Vehicle Charts* for an explanation of these charts.

Honda Accord DX
2 Door Coupe

 2.2L 125 hp Gas Fuel Inject.
 4 Cylinder In-Line
 Manual 5 Speed
 2 Wheel Front
 Driver Airbag Psngr Belts

Midsize — EX Model Shown

Purchase Price

Car Item	Dealer Cost	List
Base Price	$11,550	$13,750
Anti-Lock Brakes	N/A	N/A
Automatic 4 Speed	$630	$750
Optional Engine	N/A	N/A
Air Conditioning	Dlr	Dlr
Power Steering	Std	Std
Cruise Control	N/A	N/A
All Wheel Drive	N/A	N/A
AM/FM Stereo Cassette	Dlr	Dlr
Steering Wheel, Tilt	Std	Std
Power Windows	N/A	N/A
*Options Price	$630	$750
*Total Price	$12,180	$14,500
Target Price	$13,051	
Destination Charge	$310	
Avg. Tax & Fees	$816	
Total Target $	**$14,177**	
Average Dealer Option Cost:	**84%**	

Ownership Costs

Cost Area	5 Year Cost	Rate
Depreciation	$4,101	○
Financing ($285/month)	$2,336	
Insurance (Rating 7)	$7,035	◐
State Fees	$592	
Fuel (Hwy 28 City 22)	$3,454	◐
Maintenance	$4,820	◐
Repairs	$570	○

Warranty/Maintenance Info

Major Tune-Up	$254	●
Minor Tune-Up	$133	◐
Brake Service	$217	◐
Overall Warranty	3 yr/36k	◐
Drivetrain Warranty	3 yr/36k	◐
Rust Warranty	3 yr/unlim. mi	●
Maintenance Warranty	N/A	
Roadside Assistance	N/A	

Ownership Cost By Year

1993 / 1994 / 1995 / 1996 / 1997

Resale Value

1993	1994	1995	1996	1997
$13,491	$12,610	$11,796	$10,964	$10,076

Cumulative Costs

	1993	1994	1995	1996	1997
Annual	$3,789	$4,001	$4,940	$3,765	$6,413
Total	$3,789	$7,790	$12,730	$16,495	$22,908

Ownership Costs (5yr)

Average	This Car
$26,095	$22,908

Cost/Mile 37¢ — Cost/Mile 33¢

Ownership Cost Rating

Excellent

The 1993 Accord in available in ten models - LX and EX wagons; and DX, LX, EX, and SE sedans and coupes. The DX coupe features body side moldings, driver's side manual remote-operated mirror, maintenance interval indicator, tinted glass, adjustable steering column, electronic ignition, remote trunk and fuel filler door releases, tachometer and clear-lens multi-reflector headlights which provide the front view with a distinctive accent. The Honda Accord is built in Marysville, Ohio.

Honda Accord LX
2 Door Coupe

 2.2L 125 hp Gas Fuel Inject.
 4 Cylinder In-Line
 Manual 5 Speed
 2 Wheel Front
 Driver Airbag Psngr Belts

Midsize — SE Model Shown

Purchase Price

Car Item	Dealer Cost	List
Base Price	$13,734	$16,350
Anti-Lock Brakes	N/A	N/A
Automatic 4 Speed	$630	$750
Optional Engine	N/A	N/A
Air Conditioning	Std	Std
Power Steering	Std	Std
Cruise Control	Std	Std
All Wheel Drive	N/A	N/A
AM/FM Stereo Cassette	Std	Std
Steering Wheel, Tilt	Std	Std
Power Windows	Std	Std
*Options Price	$630	$750
*Total Price	$14,364	$17,100
Target Price	$15,461	
Destination Charge	$310	
Avg. Tax & Fees	$963	
Total Target $	**$16,734**	
Average Dealer Option Cost:	**84%**	

Ownership Costs

Cost Area	5 Year Cost	Rate
Depreciation	$5,980	○
Financing ($336/month)	$2,757	
Insurance (Rating 11)	$7,693	◐
State Fees	$696	
Fuel (Hwy 28 City 22)	$3,454	◐
Maintenance	$4,932	◐
Repairs	$570	○

Warranty/Maintenance Info

Major Tune-Up	$254	●
Minor Tune-Up	$133	◐
Brake Service	$217	◐
Overall Warranty	3 yr/36k	◐
Drivetrain Warranty	3 yr/36k	◐
Rust Warranty	3 yr/unlim. mi	●
Maintenance Warranty	N/A	
Roadside Assistance	N/A	

Resale Value

1993	1994	1995	1996	1997
$14,146	$13,295	$12,527	$11,677	$10,754

Cumulative Costs

	1993	1994	1995	1996	1997
Annual	$6,013	$4,253	$5,184	$3,968	$6,664
Total	$6,013	$10,266	$15,450	$19,418	$26,082

Ownership Costs (5yr)

Average	This Car
$28,753	$26,082

Cost/Mile 41¢ — Cost/Mile 37¢

Ownership Cost Rating

Excellent

The 1993 Accord in available in ten models - LX and EX wagons; and DX, LX, EX, and SE sedans and coupes. The LX coupe features body side moldings, driver's side manual remote-operated mirror, maintenance interval indicator, tinted glass, adjustable steering column, electronic ignition, remote trunk and fuel filler door releases, tachometer and clear-lens multi-reflector headlights which provide the front view with a distinctive accent. The Honda Accord is built in Marysville, Ohio.

---

* Includes shaded options
** Other purchase requirements apply

● Poor — ◐ Worse Than Average — ○ Average — ○ Better Than Average — ○ Excellent — ⊖ Insufficient Information

©1993 by IntelliChoice, Inc. (408) 554-8711   All Rights Reserved. Reproduction Prohibited.
Refer to *Section 3: Annotated Vehicle Charts* for an explanation of these charts.

Honda Accord EX
2 Door Coupe

Midsize

 2.2L 140 hp Gas Fuel Inject. 4 Cylinder In-Line Manual 5 Speed 2 Wheel Front — Driver Airbag Psngr Belts

Purchase Price

Car Item	Dealer Cost	List
Base Price	**$15,767**	**$18,770**
Anti-Lock Brakes	Std	Std
Automatic 4 Speed	$630	$750
Optional Engine	N/A	N/A
Air Conditioning	Std	Std
Power Steering	Std	Std
Cruise Control	Std	Std
All Wheel Drive	N/A	N/A
AM/FM Stereo Cassette	Std	Std
Steering Wheel, Tilt	Std	Std
Power Windows	Std	Std
*Options Price	$630	$750
*Total Price	$16,397	$19,520
Target Price		$17,723
Destination Charge		$310
Avg. Tax & Fees		$1,100
Total Target $		**$19,133**
Average Dealer Option Cost:	**84%**	

The 1993 Accord in available in ten models - LX and EX wagons; and DX, LX, EX, and SE sedans and coupes. The EX coupe features body side moldings, maintenance interval indicator, tinted glass, power moonroof, color-keyed dual power mirrors, remote trunk and fuel filler door releases, tachometer, and clear-lens multi-reflector headlights which provide the front view with a distinctive accent. The Honda Accord is built in Marysville, Ohio and features a 3-year/36,000-mile limited warranty.

Ownership Costs

Cost Area	5 Year Cost	Rate
Depreciation	$7,705	○
Financing ($385/month)	$3,152	
Insurance (Rating 13)	$8,061	◐
State Fees	$793	
Fuel (Hwy 28 City 22)	$3,454	○
Maintenance	$5,061	◐
Repairs	$570	○

Warranty/Maintenance Info

Major Tune-Up	$254	●
Minor Tune-Up	$133	◐
Brake Service	$280	◉
Overall Warranty	3 yr/36k	◐
Drivetrain Warranty	3 yr/36k	◐
Rust Warranty	3 yr/unlim. mi	◉
Maintenance Warranty	N/A	
Roadside Assistance	N/A	

Ownership Cost By Year

Resale Value

1993: $15,227 | 1994: $14,239 | 1995: $13,342 | 1996: $12,453 | 1997: $11,428

Cumulative Costs

	1993	1994	1995	1996	1997
Annual	$7,588	$4,605	$5,547	$4,129	$6,927
Total	$7,588	$12,193	$17,740	$21,869	$28,796

Ownership Costs (5yr)

Average: $31,228 | This Car: $28,796
Cost/Mile: 45¢ | Cost/Mile: 41¢

Ownership Cost Rating
○ Better Than Average

Honda Accord SE
2 Door Coupe

Midsize

 2.2L 140 hp Gas Fuel Inject. 4 Cylinder In-Line Automatic 4 Speed 2 Wheel Front — Driver Airbag Psngr Belts

Purchase Price

Car Item	Dealer Cost	List
Base Price	**$18,077**	**$21,520**
Anti-Lock Brakes	Std	Std
Manual Transmission	N/A	N/A
Optional Engine	N/A	N/A
Air Conditioning	Std	Std
Power Steering	Std	Std
Cruise Control	Std	Std
All Wheel Drive	N/A	N/A
AM/FM Stereo Cassette	Std	Std
Steering Wheel, Tilt	Std	Std
Power Windows	Std	Std
*Options Price	$0	$0
*Total Price	$18,077	$21,520
Target Price		$19,632
Destination Charge		$310
Avg. Tax & Fees		$1,215
Total Target $		**$21,157**
Average Dealer Option Cost:	**84%**	

The 1993 Accord in available in ten models - LX and EX wagons; and DX, LX, EX, and SE sedans and coupes. The SE coupe is new for 1993, and features leather-trimmed upholstery, exclusive 15-inch alloy wheels, body-colored side moldings, and a Honda/Bose music system specifically designed to match the acoustic environment of the Accord SE. Other features include a passenger-side airbag, power moonroof, windows and locks, and a choice of Atlantis Blue-Green Pearl or Cashmere Metallic paint.

Ownership Costs

Cost Area	5 Year Cost	Rate
Depreciation	$8,553	○
Financing ($425/month)	$3,485	
Insurance (Rating 14)	$8,238	◐
State Fees	$873	
Fuel (Hwy 28 City 22)	$3,454	○
Maintenance	$5,061	◐
Repairs	$570	○

Warranty/Maintenance Info

Major Tune-Up	$254	●
Minor Tune-Up	$133	◐
Brake Service	$280	◉
Overall Warranty	3 yr/36k	◐
Drivetrain Warranty	3 yr/36k	◐
Rust Warranty	3 yr/unlim. mi	◉
Maintenance Warranty	N/A	
Roadside Assistance	N/A	

Resale Value

1993: $16,739 | 1994: $15,667 | 1995: $14,678 | 1996: $13,713 | 1997: $12,604

Cumulative Costs

	1993	1994	1995	1996	1997
Annual	$8,292	$4,845	$5,756	$4,280	$7,061
Total	$8,292	$13,137	$18,893	$23,173	$30,234

Ownership Costs (5yr)

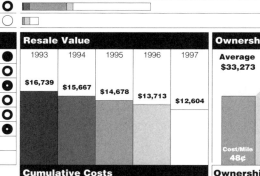

Average: $33,273 | This Car: $30,234
Cost/Mile: 48¢ | Cost/Mile: 43¢

Ownership Cost Rating
○ Excellent

---

page 100 | *Includes shaded options | **Other purchase requirements apply

● Poor | ◉ Worse Than Average | ◐ Average | ○ Better Than Average | ○ Excellent | ⊖ Insufficient Information

©1993 by IntelliChoice, Inc. (408) 554-8711 All Rights Reserved. Reproduction Prohibited.
Refer to *Section 3: Annotated Vehicle Charts* for an explanation of these charts.

Honda Accord DX
4 Door Sedan

Midsize

- 2.2L 125 hp Gas Fuel Inject.
- 4 Cylinder In-Line
- Manual 5 Speed
- 2 Wheel Front
- Driver Airbag Psngr Belts

EX Model Shown

Purchase Price

Car Item	Dealer Cost	List
Base Price	**$11,718**	**$13,950**
Anti-Lock Brakes	N/A	N/A
Automatic 4 Speed	$630	$750
Optional Engine	N/A	N/A
Air Conditioning	Dlr	Dlr
Power Steering	Std	Std
Cruise Control	N/A	N/A
All Wheel Drive	N/A	N/A
AM/FM Stereo Cassette	Dlr	Dlr
Steering Wheel, Tilt	Std	Std
Power Windows	N/A	N/A
***Options Price**	**$630**	**$750**
***Total Price**	**$12,348**	**$14,700**
Target Price	$13,236	
Destination Charge	$310	
Avg. Tax & Fees	$827	
Total Target $	**$14,373**	
Average Dealer Option Cost:	**84%**	

Ownership Costs

Cost Area	5 Year Cost	Rate
Depreciation	$4,638	○
Financing ($289/month)	$2,367	
Insurance (Rating 6)	$6,919	○
State Fees	$600	
Fuel (Hwy 28 City 22)	$3,454	○
Maintenance	$4,820	○
Repairs	$570	○

Warranty/Maintenance Info

Major Tune-Up	$254	●
Minor Tune-Up	$133	○
Brake Service	$217	○
Overall Warranty	3 yr/36k	○
Drivetrain Warranty	3 yr/36k	○
Rust Warranty	3 yr/unlim. mi	◉
Maintenance Warranty	N/A	
Roadside Assistance	N/A	

Ownership Cost By Year

Resale Value: 1993 $13,715 | 1994 $12,659 | 1995 $11,699 | 1996 $10,722 | 1997 $9,735

Cumulative Costs

	1993	1994	1995	1996	1997
Annual	$3,754	$4,166	$5,071	$3,889	$6,488
Total	$3,754	$7,920	$12,991	$16,880	$23,368

Ownership Costs (5yr)

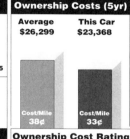

Average $26,299 | This Car $23,368
Cost/Mile 38¢ | Cost/Mile 33¢

Ownership Cost Rating: ○ Excellent

The 1993 Accord in available in ten models - LX and EX wagons; and DX, LX, EX, and SE sedans and coupes. The DX sedan features body side moldings, a driver's side manual remote-operated mirror, maintenance interval indicator, tinted glass, electronic ignition, adjustable steering column, remote trunk and fuel filler door releases, tachometer, and child-proof rear door locks. The Honda Accord is built in Marysville, Ohio. All 1993 Honda Accords feature a 3-year/36,000-mile limited warranty.

Honda Accord LX
4 Door Sedan

Midsize

- 2.2L 125 hp Gas Fuel Inject.
- 4 Cylinder In-Line
- Manual 5 Speed
- 2 Wheel Front
- Driver Airbag Psngr Belts

Purchase Price

Car Item	Dealer Cost	List
Base Price	**$13,902**	**$16,550**
Anti-Lock Brakes	N/A	N/A
Automatic 4 Speed	$630	$750
Optional Engine	N/A	N/A
Air Conditioning	Std	Std
Power Steering	Std	Std
Cruise Control	Std	Std
All Wheel Drive	N/A	N/A
AM/FM Stereo Cassette	Std	Std
Steering Wheel, Tilt	Std	Std
Power Windows	Std	Std
***Options Price**	**$630**	**$750**
***Total Price**	**$14,532**	**$17,300**
Target Price	$15,647	
Destination Charge	$310	
Avg. Tax & Fees	$974	
Total Target $	**$16,931**	
Average Dealer Option Cost:	**84%**	

Ownership Costs

Cost Area	5 Year Cost	Rate
Depreciation	$6,861	○
Financing ($340/month)	$2,790	
Insurance (Rating 8)	$7,223	○
State Fees	$704	
Fuel (Hwy 28 City 22)	$3,454	○
Maintenance	$4,932	○
Repairs	$570	○

Warranty/Maintenance Info

Major Tune-Up	$254	●
Minor Tune-Up	$133	○
Brake Service	$217	○
Overall Warranty	3 yr/36k	○
Drivetrain Warranty	3 yr/36k	○
Rust Warranty	3 yr/unlim. mi	◉
Maintenance Warranty	N/A	
Roadside Assistance	N/A	

Resale Value: 1993 $14,376 | 1994 $13,112 | 1995 $12,071 | 1996 $11,054 | 1997 $10,070

Cumulative Costs

	1993	1994	1995	1996	1997
Annual	$5,910	$4,587	$5,371	$4,041	$6,625
Total	$5,910	$10,497	$15,868	$19,909	$26,534

Ownership Costs (5yr)

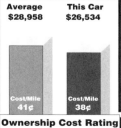

Average $28,958 | This Car $26,534
Cost/Mile 41¢ | Cost/Mile 38¢

Ownership Cost Rating: ○ Excellent

The 1993 Accord in available in ten models - LX and EX wagons; and DX, LX, EX and SE sedans and coupes. The LX sedan features body side moldings, driver's side manual remote-operated mirror, maintenance interval indicator, tinted glass, electronic ignition, adjustable steering column, remote trunk and fuel filler door releases, tachometer, and child-proof rear door locks. The Honda Accord is built in Marysville, Ohio. All 1993 Honda Accords feature a 3-year/36,000-mile limited warranty.

---

* Includes shaded options
** Other purchase requirements apply

● Poor | ◐ Worse Than Average | ○ Average | ○ Better Than Average | ○ Excellent | ⊖ Insufficient Information

©1993 by *IntelliChoice, Inc.* (408) 554-8711   All Rights Reserved. Reproduction Prohibited.
Refer to *Section 3: Annotated Vehicle Charts* for an explanation of these charts.

Honda Accord EX
4 Door Sedan

 2.2L 140 hp Gas Fuel Inject.
 4 Cylinder In-Line
 Manual 5 Speed
 2 Wheel Front
 Driver Airbag Psngr Belts

Midsize

Purchase Price

Car Item	Dealer Cost	List
Base Price	**$15,935**	**$18,970**
Anti-Lock Brakes	Std	Std
Automatic 4 Speed	$630	$750
Optional Engine	N/A	N/A
Air Conditioning	Std	Std
Power Steering	Std	Std
Cruise Control	Std	Std
All Wheel Drive	N/A	N/A
AM/FM Stereo Cassette	Std	Std
Steering Wheel, Tilt	Std	Std
Power Windows	Std	Std
***Options Price**	**$630**	**$750**
***Total Price**	**$16,565**	**$19,720**
Target Price		$17,911
Destination Charge		$310
Avg. Tax & Fees		$1,111
Total Target $		**$19,332**
Average Dealer Option Cost:		**84%**

The 1993 Accord in available in ten models - LX and EX wagons; and DX, LX, EX and SE sedans and coupes. The EX sedan includes body side moldings, maintenance interval indicator, tinted glass, power moonroof, color-keyed dual power mirrors, remote trunk and fuel filler door releases, clear-lens multi-reflector headlights which provide the front view with a distinctive accent, and child-proof rear door locks. All 1993 Honda Accords comes with a 3-year/36,000-mile limited warranty.

Ownership Costs

Cost Area	5 Year Cost	Rate
Depreciation	$8,087	○
Financing ($389/month)	$3,183	
Insurance (Rating 11)	$7,693	○
State Fees	$800	
Fuel (Hwy 28 City 22)	$3,454	○
Maintenance	$5,280	◉
Repairs	$570	○

Warranty/Maintenance Info

Major Tune-Up	$254	●
Minor Tune-Up	$133	○
Brake Service	$280	◉
Overall Warranty	3 yr/36k	○
Drivetrain Warranty	3 yr/36k	○
Rust Warranty	3 yr/unlim. mi	◉
Maintenance Warranty	N/A	
Roadside Assistance	N/A	

Ownership Cost By Year

1993, 1994, 1995, 1996, 1997

Resale Value

1993	1994	1995	1996	1997
$15,218	$14,162	$13,196	$12,258	$11,245

Cumulative Costs

	1993	1994	1995	1996	1997
Annual	$7,743	$4,614	$5,654	$4,105	$6,951
Total	$7,743	$12,357	$18,011	$22,116	$29,067

Ownership Costs (5yr)

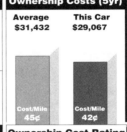

Average: $31,432 — This Car: $29,067
Cost/Mile: 45¢ — Cost/Mile: 42¢

Ownership Cost Rating

Better Than Average

Honda Accord SE
4 Door Sedan

 2.2L 140 hp Gas Fuel Inject.
 4 Cylinder In-Line
 Automatic 4 Speed
 2 Wheel Front
Driver/Psngr Airbags Std

Midsize

Purchase Price

Car Item	Dealer Cost	List
Base Price	**$18,245**	**$21,720**
Anti-Lock Brakes	Std	Std
Manual Transmission	N/A	N/A
Optional Engine	N/A	N/A
Air Conditioning	Std	Std
Power Steering	Std	Std
Cruise Control	Std	Std
All Wheel Drive	N/A	N/A
AM/FM Stereo Cassette	Std	Std
Steering Wheel, Tilt	Std	Std
Power Windows	Std	Std
***Options Price**	**$0**	**$0**
***Total Price**	**$18,245**	**$21,720**
Target Price		$19,821
Destination Charge		$310
Avg. Tax & Fees		$1,227
Total Target $		**$21,358**
Average Dealer Option Cost:		**84%**

The 1993 Accord in available in ten models - LX and EX wagons; and DX, LX, EX, and SE sedans and coupes. The SE sedan is new for 1993, and features leather-trimmed upholstery, exclusive 15-inch alloy wheels, body-colored side moldings, and a Honda/Bose music system specifically designed to match the acoustic environment of the Accord SE. Other features include a passenger-side airbag, power moonroof, windows and locks, and a choice of Geneva Green Pearl or Cashmere Metallic paint.

Ownership Costs

Cost Area	5 Year Cost	Rate
Depreciation	$9,077	○
Financing ($429/month)	$3,519	
Insurance (Rating 12)	$7,898	○
State Fees	$880	
Fuel (Hwy 28 City 22)	$3,454	○
Maintenance	$5,280	◉
Repairs	$570	○

Warranty/Maintenance Info

Major Tune-Up	$254	●
Minor Tune-Up	$133	○
Brake Service	$280	◉
Overall Warranty	3 yr/36k	○
Drivetrain Warranty	3 yr/36k	○
Rust Warranty	3 yr/unlim. mi	◉
Maintenance Warranty	N/A	
Roadside Assistance	N/A	

Resale Value

1993	1994	1995	1996	1997
$17,441	$15,965	$14,658	$13,464	$12,281

Cumulative Costs

	1993	1994	1995	1996	1997
Annual	$7,743	$5,197	$6,119	$4,442	$7,177
Total	$7,743	$12,940	$19,059	$23,501	$30,678

Ownership Costs (5yr)

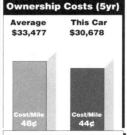

Average: $33,477 — This Car: $30,678
Cost/Mile: 48¢ — Cost/Mile: 44¢

Ownership Cost Rating

Excellent

---

* Includes shaded options
** Other purchase requirements apply

● Poor ◐ Worse Than Average ◉ Average ○ Better Than Average ○ Excellent ⊖ Insufficient Information

©1993 by IntelliChoice, Inc. (408) 554-8711 All Rights Reserved. Reproduction Prohibited.
Refer to *Section 3: Annotated Vehicle Charts* for an explanation of these charts.

Honda Accord LX
4 Door Wagon

 2.2L 125 hp Gas Fuel Inject.
 4 Cylinder In-Line
 Manual 5 Speed
 2 Wheel Front
 Driver Airbag Psngr Belts

Midsize Wagon

Purchase Price

Car Item	Dealer Cost	List
Base Price	$15,099	$17,975
Anti-Lock Brakes	N/A	N/A
Automatic 4 Speed	$630	$750
Optional Engine	N/A	N/A
Air Conditioning	Std	Std
Power Steering	Std	Std
Cruise Control	Std	Std
All Wheel Drive	N/A	N/A
AM/FM Stereo Cassette	Std	Std
Steering Wheel, Tilt	Std	Std
Power Windows	Std	Std
*Options Price	$630	$750
*Total Price	$15,729	$18,725
Target Price	$16,978	
Destination Charge	$310	
Avg. Tax & Fees	$1,054	
Total Target $	**$18,342**	
Average Dealer Option Cost:	84%	

The 1993 Accord in available in ten models - LX and EX wagons; and DX, LX, EX and SE sedans and coupes. The LX wagon includes body side moldings, integrated rear bumper step, full-size spare tire, air conditioning, power door locks and tailgate lock, rear window defroster with timer, cruise control, fold-down rear seatback (split 60/40), and a choice of five exterior colors. The Honda Accord is built in Marysville, Ohio. All 1993 Honda Accords feature a 3-year/36,000-mile limited warranty.

Ownership Costs

Cost Area	5 Year Cost	Rate
Depreciation	$6,411	◯
Financing ($369/month)	$3,021	
Insurance (Rating 10)	$7,479	◯
State Fees	$760	
Fuel (Hwy 27 City 21)	$3,598	◯
Maintenance	$4,921	●
Repairs	$570	◯

Warranty/Maintenance Info

Major Tune-Up	$254	●
Minor Tune-Up	$133	◉
Brake Service	$217	◯
Overall Warranty	3 yr/36k	◯
Drivetrain Warranty	3 yr/36k	◯
Rust Warranty	3 yr/unlim. mi	◉
Maintenance Warranty	N/A	
Roadside Assistance	N/A	

Ownership Cost By Year

- 1993
- 1994
- 1995
- 1996
- 1997

Resale Value

1993	1994	1995	1996	1997
$15,711	$14,776	$13,877	$12,904	$11,931

Cumulative Costs

	1993	1994	1995	1996	1997
Annual	$6,170	$4,420	$5,362	$4,106	$6,702
Total	$6,170	$10,590	$15,952	$20,058	$26,760

Ownership Costs (5yr)

Average	This Car
$30,415	$26,760
Cost/Mile 43¢	Cost/Mile 38¢

Ownership Cost Rating
Excellent

Honda Accord EX
4 Door Wagon

 2.2L 140 hp Gas Fuel Inject.
 4 Cylinder In-Line
 Manual 5 Speed
 2 Wheel Front
 Driver Airbag Psngr Belts

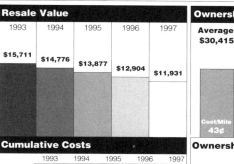

Midsize Wagon

Purchase Price

Car Item	Dealer Cost	List
Base Price	$17,157	$20,425
Anti-Lock Brakes	Std	Std
Automatic 4 Speed	$630	$750
Optional Engine	N/A	N/A
Air Conditioning	Std	Std
Power Steering	Std	Std
Cruise Control	Std	Std
All Wheel Drive	N/A	N/A
AM/FM Stereo Cassette	Std	Std
Steering Wheel, Tilt	Std	Std
Power Windows	Std	Std
*Options Price	$630	$750
*Total Price	$17,787	$21,175
Target Price	$19,280	
Destination Charge	$310	
Avg. Tax & Fees	$1,195	
Total Target $	**$20,785**	
Average Dealer Option Cost:	84%	

The 1993 Accord in available in ten models - LX and EX wagons; and DX, LX, EX and SE sedans and coupes. The EX wagon includes body side moldings, integrated rear bumper step, full-size spare tire, air conditioning, power door locks and tailgate lock, power moonroof, a remote entry system, cruise control, fold-down rear seatback (split 60/40), and a choice of five exterior colors. The Honda Accord is built in Marysville, Ohio. All 1993 Honda Accords feature a 3-year/36,000-mile limited warranty.

Ownership Costs

Cost Area	5 Year Cost	Rate
Depreciation	$8,642	◯
Financing ($418/month)	$3,424	
Insurance (Rating 12)	$7,898	◯
State Fees	$859	
Fuel (Hwy 27 City 21)	$3,598	◯
Maintenance	$5,061	●
Repairs	$570	◯

Warranty/Maintenance Info

Major Tune-Up	$254	●
Minor Tune-Up	$133	◉
Brake Service	$280	●
Overall Warranty	3 yr/36k	◯
Drivetrain Warranty	3 yr/36k	◯
Rust Warranty	3 yr/unlim. mi	◉
Maintenance Warranty	N/A	
Roadside Assistance	N/A	

Ownership Cost By Year

- 1993
- 1994
- 1995
- 1996
- 1997

Resale Value

1993	1994	1995	1996	1997
$16,371	$15,206	$14,212	$13,194	$12,143

Cumulative Costs

	1993	1994	1995	1996	1997
Annual	$8,222	$4,879	$5,707	$4,285	$6,959
Total	$8,222	$13,101	$18,808	$23,093	$30,052

Ownership Costs (5yr)

Average	This Car
$32,920	$30,052
Cost/Mile 47¢	Cost/Mile 43¢

Ownership Cost Rating
Excellent

---

* Includes shaded options
** Other purchase requirements apply

● Poor ◉ Worse Than Average ◐ Average ◯ Better Than Average ◯ Excellent ⊖ Insufficient Information

©1993 by IntelliChoice, Inc. (408) 554-8711   All Rights Reserved. Reproduction Prohibited.
Refer to *Section 3: Annotated Vehicle Charts* for an explanation of these charts.

Honda Civic DX
2 Door Coupe

Compact

 1.5L 102 hp Gas Fuel Inject.
 4 Cylinder In-Line
 Manual 5 Speed
 2 Wheel Front
Driver Airbag Psngr Belts

Purchase Price

Car Item	Dealer Cost	List
Base Price	**$8,798**	**$10,350**
Anti-Lock Brakes	N/A	N/A
Automatic 4 Speed	$823	$980
Optional Engine	N/A	N/A
Air Conditioning	Dlr	Dlr
Power Steering	Std	Std
Cruise Control	N/A	N/A
All Wheel Drive	N/A	N/A
AM/FM Stereo Cassette	Dlr	Dlr
Steering Wheel, Tilt	Std	Std
Power Windows	N/A	N/A
*Options Price	$823	$980
*Total Price	$9,621	$11,330
Target Price	$10,250	
Destination Charge	$310	
Avg. Tax & Fees	$644	
Total Target $	**$11,204**	
Average Dealer Option Cost:	84%	

The 1993 Honda Civic is available in nine models - DX, LX, and EX sedans; DX and EX coupes; and CX, DX, VX, and Si hatchbacks. New for 1993, the DX coupe is a new model featuring the same sleek exterior and interior as the rest of the Civic line-up. The DX has a double wishbone suspension front and rear that provides stable handling and a comfortable ride. Power-assisted brakes and 13-inch steel wheels with P175/70R13 radials are standard. A 3yr/36k-mile limited warranty has been provided by Honda.

Ownership Costs

Cost Area	5 Year Cost	Rate
Depreciation	$4,156	○
Financing ($225/month)	$1,846	
Insurance (Rating 5)	$6,786	O
State Fees	$465	
Fuel (Hwy 38 City 30)	$2,540	○
Maintenance	$4,539	O
Repairs	$570	○

Warranty/Maintenance Info

Major Tune-Up	$239	●
Minor Tune-Up	$129	●
Brake Service	$217	O
Overall Warranty	3 yr/36k	O
Drivetrain Warranty	3 yr/36k	O
Rust Warranty	3 yr/unlim. mi	●
Maintenance Warranty	N/A	
Roadside Assistance	N/A	

Ownership Cost By Year

Resale Value

1993	1994	1995	1996	1997
$10,389	$9,446	$8,694	$7,861	$7,048

Cumulative Costs

	1993	1994	1995	1996	1997
Annual	$3,467	$3,655	$4,398	$3,460	$5,922
Total	$3,467	$7,122	$11,520	$14,980	$20,902

Ownership Costs (5yr)

Average	This Car
$24,223	$20,902
Cost/Mile 35¢	Cost/Mile 30¢

Ownership Cost Rating
Excellent

Honda Civic EX
2 Door Coupe

Compact

 1.6L 125 hp Gas Fuel Inject.
4 Cylinder In-Line
 Manual 5 Speed
 2 Wheel Front
Driver Airbag Psngr Opt

Purchase Price

Car Item	Dealer Cost	List
Base Price	**$10,540**	**$12,400**
Anti-Lock Brakes	N/A	N/A
Automatic 4 Speed	$630	$750
Optional Engine	N/A	N/A
Air Conditioning	Dlr	Dlr
Power Steering	Std	Std
Cruise Control	Std	Std
All Wheel Drive	N/A	N/A
AM/FM Stereo Cassette	Dlr	Dlr
Steering Wheel, Tilt	Std	Std
Power Windows	Std	Std
*Options Price	$630	$750
*Total Price	$11,170	$13,150
Target Price	$11,944	
Destination Charge	$310	
Avg. Tax & Fees	$748	
Total Target $	**$13,002**	
Average Dealer Option Cost:	84%	

The 1993 Honda Civic is available in nine models - DX, LX, and EX sedans; DX and EX coupes; and CX, DX, VX, and Si hatchbacks. New for 1993, the EX coupe is a new model featuring the same sleek exterior and interior the rest of the Civic line-up has to offer. The EX coupe offers power-operated window and door locks, body-colored dual power side mirrors, power-operated moonroof with tilt feature, cruise control, and a cargo area light. A 3-year/36k-mile limited warranty has been provided by Honda.

Ownership Costs

Cost Area	5 Year Cost	Rate
Depreciation	$6,014	O
Financing ($261/month)	$2,142	
Insurance (Rating 6)	$6,919	O
State Fees	$539	
Fuel (Hwy 34 City 27)	$2,831	○
Maintenance	$4,636	O
Repairs	$570	○

Warranty/Maintenance Info

Major Tune-Up	$239	●
Minor Tune-Up	$129	●
Brake Service	$217	O
Overall Warranty	3 yr/36k	O
Drivetrain Warranty	3 yr/36k	O
Rust Warranty	3 yr/unlim. mi	●
Maintenance Warranty	N/A	
Roadside Assistance	N/A	

Resale Value

1993	1994	1995	1996	1997
$10,414	$9,418	$8,630	$7,790	$6,988

Cumulative Costs

	1993	1994	1995	1996	1997
Annual	$5,460	$3,898	$4,633	$3,591	$6,069
Total	$5,460	$9,358	$13,991	$17,582	$23,651

Ownership Costs (5yr)

Average	This Car
$25,785	$23,651
Cost/Mile 37¢	Cost/Mile 34¢

Ownership Cost Rating
Excellent

---

page 104

* Includes shaded options
** Other purchase requirements apply

● Poor ● Worse Than Average ○ Average ○ Better Than Average ○ Excellent ⊖ Insufficient Information

©1993 by IntelliChoice, Inc. (408) 554-8711 All Rights Reserved. Reproduction Prohibited.
Refer to *Section 3: Annotated Vehicle Charts* for an explanation of these charts.

Honda Civic del Sol S
2 Door Coupe

Subcompact

 1.5L 102 hp Gas Fuel Inject.   4 Cylinder In-Line   Manual 5 Speed   2 Wheel Front Driver Airbag Psngr Belts

Purchase Price

Car Item	Dealer Cost	List
Base Price	**$11,220**	**$13,200**
Anti-Lock Brakes	N/A	N/A
Automatic 4 Speed	$638	$750
Optional Engine	N/A	N/A
Air Conditioning	Dlr	Dlr
Power Steering	N/A	N/A
Cruise Control	Std	Std
All Wheel Drive	N/A	N/A
AM/FM Stereo Cassette	N/A	N/A
Steering Wheel, Tilt	Std	Std
Power Windows	Std	Std
***Options Price**	$0	$0
***Total Price**	**$11,220**	**$13,200**
Target Price		$12,014
Destination Charge		$310
Avg. Tax & Fees		$751
Total Target $		**$13,075**
Average Dealer Option Cost:		*85%*

The 1993 Honda Civic del Sol is available in two models - S and Si coupe. All new for 1993, the Civic del Sol S is an open-top and hard-top coupe combined in one package. A sensibly designed storage rack secures the roof panel in the trunk with a good amount of usable space still available. The S model includes a drivers air bag, five speed manual transmission, and power windows. Optional is a four speed automatic transmission.

Ownership Costs

Cost Area	5 Year Cost	Rate
Depreciation		⊖
Financing ($263/month)	$2,153	
Insurance (Rating 10 Sport)	$10,097	●
State Fees	$541	
Fuel (Hwy 38 City 34)	$2,403	○
Maintenance		⊖
Repairs	$570	◔

Warranty/Maintenance Info

Major Tune-Up		⊖
Minor Tune-Up		⊖
Brake Service		⊖
Overall Warranty	3 yr/36k	○
Drivetrain Warranty	3 yr/36k	○
Rust Warranty	3 yr/unlim. mi	◉
Maintenance Warranty	N/A	
Roadside Assistance	N/A	

Ownership Cost By Year

$2,000 $4,000 $6,000 $8,000 $10,000 $12,000

Insufficient Depreciation Information

Insufficient Maintenance Information

■ 1993
■ 1994
▨ 1995
▢ 1996
□ 1997

Resale Value

Insufficient Information

Ownership Costs (5yr)

Insufficient Information

Cumulative Costs

	1993	1994	1995	1996	1997
Annual	*Insufficient Information*				
Total	*Insufficient Information*				

Ownership Cost Rating

⊖

Insufficient Information

Honda Civic del Sol Si
2 Door Coupe

Subcompact

 1.6L 125 hp Gas Fuel Inject. 4 Cylinder In-Line   Manual 5 Speed   2 Wheel Front Driver Airbag Psngr Belts

Purchase Price

Car Item	Dealer Cost	List
Base Price	**$12,750**	**$15,000**
Anti-Lock Brakes	N/A	N/A
Automatic 4 Speed	$638	$750
Optional Engine	N/A	N/A
Air Conditioning	Dlr	Dlr
Power Steering	Std	Std
Cruise Control	Std	Std
All Wheel Drive	N/A	N/A
AM/FM Stereo Cassette	N/A	N/A
Steering Wheel, Tilt	Std	Std
Power Windows	Std	Std
***Options Price**	$0	$0
***Total Price**	**$12,750**	**$15,000**
Target Price		$13,696
Destination Charge		$310
Avg. Tax & Fees		$853
Total Target $		**$14,859**
Average Dealer Option Cost:		*85%*

The 1993 Honda Civic del Sol is available in two models - S and Si coupe. All new for 1993, the Civic del Sol Si is an open-top and hard-top coupe combined in one package. A sensibly designed storage rack secures the roof panel in the trunk with a good amount of usable space still available. The Si upgrades the S model with more horsepower, alloy wheels and power-operated side mirrors. Standard features include body-colored bumpers and side mirrors, and a center console storage space.

Ownership Costs

Cost Area	5 Year Cost	Rate
Depreciation		⊖
Financing ($299/month)	$2,448	
Insurance (Rating 11 Sport+)	$10,770	●
State Fees	$612	
Fuel (Hwy 33 City 29)	$2,788	○
Maintenance		⊖
Repairs	$570	◔

Warranty/Maintenance Info

Major Tune-Up		⊖
Minor Tune-Up		⊖
Brake Service		⊖
Overall Warranty	3 yr/36k	○
Drivetrain Warranty	3 yr/36k	○
Rust Warranty	3 yr/unlim. mi	◉
Maintenance Warranty	N/A	
Roadside Assistance	N/A	

Ownership Cost By Year

$2,000 $4,000 $6,000 $8,000 $10,000 $12,000

Insufficient Depreciation Information

Insufficient Maintenance Information

■ 1993
■ 1994
▨ 1995
▢ 1996
□ 1997

Resale Value

Insufficient Information

Ownership Costs (5yr)

Insufficient Information

Cumulative Costs

	1993	1994	1995	1996	1997
Annual	*Insufficient Information*				
Total	*Insufficient Information*				

Ownership Cost Rating

⊖

Insufficient Information

---

* Includes shaded options
** Other purchase requirements apply

● Poor   ◐ Worse Than Average   ○ Average   ◑ Better Than Average   ○ Excellent   ⊖ Insufficient Information

©1993 by IntelliChoice, Inc. (408) 554-8711 All Rights Reserved. Reproduction Prohibited.
Refer to *Section 3: Annotated Vehicle Charts* for an explanation of these charts.

Honda Civic CX
2 Door Hatchback

DX Model Shown — Compact

- 1.5L 70 hp Gas Fuel Inject.
- 4 Cylinder In-Line
- Manual 5 Speed
- 2 Wheel Front
- Driver Airbag Psngr Belts

Purchase Price

Car Item	Dealer Cost	List
Base Price	**$7,560**	**$8,400**
Anti-Lock Brakes	N/A	N/A
Automatic Transmission	N/A	N/A
Optional Engine	N/A	N/A
Air Conditioning	Dlr	Dlr
Power Steering	N/A	N/A
Cruise Control	N/A	N/A
All Wheel Drive	N/A	N/A
AM/FM Stereo CD	Dlr	Dlr
Steering Wheel, Tilt	N/A	N/A
Power Windows	N/A	N/A
*Options Price	$0	$0
*Total Price	$7,560	$8,400
Target Price	$8,034	
Destination Charge	$310	
Avg. Tax & Fees	$504	
Total Target $	**$8,848**	
Average Dealer Option Cost:	**84%**	

Ownership Costs

Cost Area	5 Year Cost	Rate
Depreciation	$2,769	○
Financing ($178/month)	$1,458	
Insurance (Rating 3)	$6,520	◐
State Fees	$348	
Fuel (Hwy 46 City 42)	$1,968	○
Maintenance	$4,452	◐
Repairs	$570	○

Warranty/Maintenance Info

Major Tune-Up	$239	●
Minor Tune-Up	$129	●
Brake Service	$217	◐
Overall Warranty	3 yr/36k	◐
Drivetrain Warranty	3 yr/36k	◐
Rust Warranty	3 yr/unlim. mi	◉
Maintenance Warranty	N/A	
Roadside Assistance	N/A	

The 1993 Honda Civic is available in nine models - DX, LX, and EX sedans; DX and EX coupes; and CX, DX, VX, and Si hatchbacks. New for 1993, the CX hatchback has cloth seat upholstery replacing the vinyl seat covering. Standard features on the CX hatchback include body-colored impact-absorbing bumpers, tinted glass, remote fuel filler door and rear hatch releases, and a rear window defroster.

Ownership Cost By Year

1993, 1994, 1995, 1996, 1997

Resale Value

1993	1994	1995	1996	1997
$8,414	$7,677	$7,155	$6,627	$6,079

Cumulative Costs

	1993	1994	1995	1996	1997
Annual	$2,740	$3,140	$3,846	$2,934	$5,425
Total	$2,740	$5,880	$9,726	$12,660	$18,085

Ownership Costs (5yr)

Average	This Car
$21,707	$18,085
Cost/Mile 31¢	Cost/Mile 26¢

Ownership Cost Rating
○ Excellent

Honda Civic DX
2 Door Hatchback

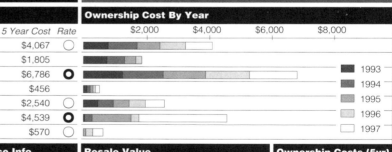
Compact

- 1.5L 102 hp Gas Fuel Inject.
- 4 Cylinder In-Line
- Manual 5 Speed
- 2 Wheel Front
- Driver Airbag Psngr Belts

Purchase Price

Car Item	Dealer Cost	List
Base Price	**$8,585**	**$10,100**
Anti-Lock Brakes	N/A	N/A
Automatic 4 Speed	$823	$980
Optional Engine	N/A	N/A
Air Conditioning	Dlr	Dlr
Power Steering	Grp	Grp
Cruise Control	N/A	N/A
All Wheel Drive	N/A	N/A
AM/FM Stereo Cassette	Dlr	Dlr
Steering Wheel, Tilt	Std	Std
Power Windows	N/A	N/A
*Options Price	$823	$980
*Total Price	$9,408	$11,080
Target Price	$10,019	
Destination Charge	$310	
Avg. Tax & Fees	$630	
Total Target $	**$10,959**	
Average Dealer Option Cost:	**84%**	

Ownership Costs

Cost Area	5 Year Cost	Rate
Depreciation	$4,067	○
Financing ($220/month)	$1,805	
Insurance (Rating 5)	$6,786	◐
State Fees	$456	
Fuel (Hwy 38 City 30)	$2,540	○
Maintenance	$4,539	◐
Repairs	$570	○

Warranty/Maintenance Info

Major Tune-Up	$239	●
Minor Tune-Up	$129	●
Brake Service	$217	◐
Overall Warranty	3 yr/36k	◐
Drivetrain Warranty	3 yr/36k	◐
Rust Warranty	3 yr/unlim. mi	◉
Maintenance Warranty	N/A	
Roadside Assistance	N/A	

The 1993 Honda Civic is available in nine models - DX, LX, and EX sedans; DX and EX coupes; and CX, DX, VX, and Si hatchbacks. The 1993 DX hatchback has added comfort and performance over the base hatchback. Standard features on the DX hatchback include body-colored impact-absorbing bumpers, power-assisted steering (when ordering the automatic transmission), and a rear window defroster.

Resale Value

1993	1994	1995	1996	1997
$10,169	$9,244	$8,509	$7,692	$6,892

Cumulative Costs

	1993	1994	1995	1996	1997
Annual	$3,423	$3,622	$4,371	$3,440	$5,907
Total	$3,423	$7,045	$11,416	$14,856	$20,763

Ownership Costs (5yr)

Average	This Car
$24,008	$20,763
Cost/Mile 34¢	Cost/Mile 30¢

Ownership Cost Rating
○ Excellent

---

* Includes shaded options
** Other purchase requirements apply

● Poor   ◉ Worse Than Average   ◐ Average   ○ Better Than Average   ○ Excellent   ⊖ Insufficient Information

©1993 by IntelliChoice, Inc. (408) 554-8711   All Rights Reserved. Reproduction Prohibited.
Refer to *Section 3: Annotated Vehicle Charts* for an explanation of these charts.

Honda Civic VX
2 Door Hatchback

Compact

 1.5L 92 hp Gas Fuel Inject.
 4 Cylinder In-Line
Manual 5 Speed
 2 Wheel Front
 Driver Airbag Psngr Belts

Purchase Price

Car Item	Dealer Cost	List
Base Price	**$9,180**	**$10,800**
Anti-Lock Brakes	N/A	N/A
Automatic Transmission	N/A	N/A
Optional Engine	N/A	N/A
Air Conditioning	Dlr	Dlr
Power Steering	N/A	N/A
Cruise Control	N/A	N/A
All Wheel Drive	N/A	N/A
AM/FM Stereo Cassette	Dlr	Dlr
Steering Wheel, Tilt	N/A	N/A
Power Windows	N/A	N/A
***Options Price**	**$0**	**$0**
***Total Price**	**$9,180**	**$10,800**
Target Price		$9,788
Destination Charge		$310
Avg. Tax & Fees		$616
Total Target $		**$10,714**
Average Dealer Option Cost:		84%

The 1993 Honda Civic is available in nine models - DX, LX, and EX sedans; DX and EX coupes; and CX, DX, VX, and Si hatchbacks. The 1993 VX hatchback has the increased fuel economy (VTEC-E) engine. Enhanced aerodynamic efficiency, including a special underbody rear panel to improve airflow, also contribute to the high fuel economy. Lightweight, 13-inch alloy wheels are mounted with P165/70 R13 tires designed to provide extra low rolling resistance. A special 5-speed transmission is standard.

Ownership Costs

Cost Area	5 Year Cost	Rate
Depreciation	$3,075	○
Financing ($215/month)	$1,765	
Insurance (Rating 5)	$6,786	O
State Fees	$445	
Fuel (Hwy 51 City 44)	$1,819	○
Maintenance	$4,452	O
Repairs	$570	○

Warranty/Maintenance Info

Major Tune-Up	$239	●
Minor Tune-Up	$129	●
Brake Service	$217	O
Overall Warranty	3 yr/36k	O
Drivetrain Warranty	3 yr/36k	O
Rust Warranty	3 yr/unlim. mi	⦿
Maintenance Warranty	N/A	
Roadside Assistance	N/A	

Ownership Cost By Year

1993 / 1994 / 1995 / 1996 / 1997

Resale Value

1993	1994	1995	1996	1997
$10,144	$9,351	$8,783	$8,217	$7,639

Cumulative Costs

	1993	1994	1995	1996	1997
Annual	$3,050	$3,336	$3,995	$3,035	$5,496
Total	$3,050	$6,386	$10,381	$13,416	$18,912

Ownership Costs (5yr)

Average	This Car
$23,767	$18,912
Cost/Mile 34¢	Cost/Mile 27¢

Ownership Cost Rating

○ Excellent

Honda Civic Si
2 Door Hatchback

Compact

 1.6L 125 hp Gas Fuel Inject.
 4 Cylinder In-Line
 Manual 5 Speed
2 Wheel Front
 Driver Airbag Psngr Belts

Purchase Price

Car Item	Dealer Cost	List
Base Price	**$10,370**	**$12,200**
Anti-Lock Brakes	N/A	N/A
Automatic Transmission	N/A	N/A
Optional Engine	N/A	N/A
Air Conditioning	Dlr	Dlr
Power Steering	Std	Std
Cruise Control	Std	Std
All Wheel Drive	N/A	N/A
AM/FM Stereo Cassette	Dlr	Dlr
Steering Wheel, Tilt	Std	Std
Power Windows	N/A	N/A
***Options Price**	**$0**	**$0**
***Total Price**	**$10,370**	**$12,200**
Target Price		$11,084
Destination Charge		$310
Avg. Tax & Fees		$695
Total Target $		**$12,089**
Average Dealer Option Cost:		84%

The 1993 Honda Civic is available in nine models - DX, LX, and EX sedans; DX and EX coupes; and CX, DX, VX, and Si hatchbacks. New for 1993, the Si hatchback has body-colored power side mirrors. Standard features on the Si hatchback include a sporty suspension, power-assisted steering, tachometer, all-season tires, cruise control, power moonroof with tilt feature, quartz digital clock, rear window wiper/washer, and a 5-speed manual transmission designed to match the engine's power characteristics.

Ownership Costs

Cost Area	5 Year Cost	Rate
Depreciation	$4,572	○
Financing ($243/month)	$1,991	
Insurance (Rating 6)	$6,919	O
State Fees	$501	
Fuel (Hwy 35 City 29)	$2,698	O
Maintenance	$4,855	O
Repairs	$570	○

Warranty/Maintenance Info

Major Tune-Up	$239	●
Minor Tune-Up	$129	●
Brake Service	$238	O
Overall Warranty	3 yr/36k	O
Drivetrain Warranty	3 yr/36k	O
Rust Warranty	3 yr/unlim. mi	⦿
Maintenance Warranty	N/A	
Roadside Assistance	N/A	

Resale Value

1993	1994	1995	1996	1997
$10,190	$9,546	$9,001	$8,251	$7,517

Cumulative Costs

	1993	1994	1995	1996	1997
Annual	$4,673	$3,465	$4,431	$3,456	$6,081
Total	$4,673	$8,138	$12,569	$16,025	$22,106

Ownership Costs (5yr)

Average	This Car
$24,970	$22,106
Cost/Mile 36¢	Cost/Mile 32¢

Ownership Cost Rating

○ Excellent

---

* Includes shaded options
** Other purchase requirements apply

● Poor ⦿ Worse Than Average O Average ○ Better Than Average ○ Excellent ⊖ Insufficient Information

©1993 by *IntelliChoice, Inc.* (408) 554-8711 All Rights Reserved. Reproduction Prohibited.
Refer to *Section 3: Annotated Vehicle Charts* for an explanation of these charts.

page 107

Honda Civic DX
4 Door Sedan

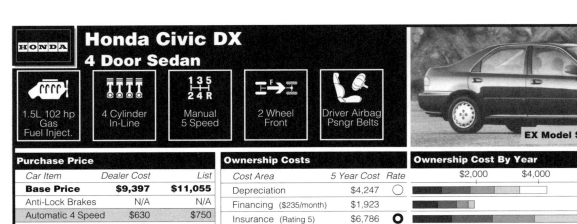

EX Model Shown

Compact

- 1.5L 102 hp Gas Fuel Inject.
- 4 Cylinder In-Line
- Manual 5 Speed
- 2 Wheel Front
- Driver Airbag Psngr Belts

Purchase Price

Car Item	Dealer Cost	List
Base Price	**$9,397**	**$11,055**
Anti-Lock Brakes	N/A	N/A
Automatic 4 Speed	$630	$750
Optional Engine	N/A	N/A
Air Conditioning	Dlr	Dlr
Power Steering	Std	Std
Cruise Control	N/A	N/A
All Wheel Drive	N/A	N/A
AM/FM Stereo Cassette	Dlr	Dlr
Steering Wheel, Tilt	Std	Std
Power Windows	N/A	N/A
***Options Price**	**$630**	**$750**
***Total Price**	**$10,027**	**$11,805**
Target Price		$10,696
Destination Charge		$310
Avg. Tax & Fees		$671
Total Target $		**$11,677**
Average Dealer Option Cost:		**84%**

The 1993 Honda Civic is available in nine models - DX, LX, and EX sedans; DX and EX coupes; and CX, DX, VX, and Si hatchbacks. New for 1993, the DX sedan retains the sporty new exterior and interior styling redesigns from last year. Features include power-assisted steering, impact-absorbing body-colored bumpers, tinted glass, and body side moldings as standard equipment. The DX also offers child-proof door locks, adjustable steering column, and remote fuel-filler door and trunk lid release.

Ownership Costs

Cost Area	5 Year Cost	Rate
Depreciation	$4,247	○
Financing ($235/month)	$1,923	
Insurance (Rating 5)	$6,786	O
State Fees	$485	
Fuel (Hwy 38 City 30)	$2,540	○
Maintenance	$4,539	O
Repairs	$570	○

Warranty/Maintenance Info

Major Tune-Up	$239	●
Minor Tune-Up	$129	●
Brake Service	$217	O
Overall Warranty	3 yr/36k	O
Drivetrain Warranty	3 yr/36k	O
Rust Warranty	3 yr/unlim. mi	◐
Maintenance Warranty	N/A	
Roadside Assistance	N/A	

Ownership Cost By Year

- 1993
- 1994
- 1995
- 1996
- 1997

Resale Value

1993	1994	1995	1996	1997
$10,746	$9,828	$9,058	$8,244	$7,430

Cumulative Costs

	1993	1994	1995	1996	1997
Annual	$3,620	$3,658	$4,435	$3,451	$5,926
Total	$3,620	$7,278	$11,713	$15,164	$21,090

Ownership Costs (5yr)

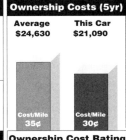

Average	This Car
$24,630	$21,090
Cost/Mile 35¢	Cost/Mile 30¢

Ownership Cost Rating

○ Excellent

Honda Civic LX
4 Door Sedan

Compact

- 1.5L 102 hp Gas Fuel Inject.
- 4 Cylinder In-Line
- Manual 5 Speed
- 2 Wheel Front
- Driver Airbag Psngr Belts

Purchase Price

Car Item	Dealer Cost	List
Base Price	**$10,102**	**$11,885**
Anti-Lock Brakes	N/A	N/A
Automatic 4 Speed	$630	$750
Optional Engine	N/A	N/A
Air Conditioning	Dlr	Dlr
Power Steering	Std	Std
Cruise Control	Std	Std
All Wheel Drive	N/A	N/A
AM/FM Stereo Cassette	Dlr	Dlr
Steering Wheel, Tilt	Std	Std
Power Windows	Std	Std
***Options Price**	**$630**	**$750**
***Total Price**	**$10,732**	**$12,635**
Target Price		$11,465
Destination Charge		$310
Avg. Tax & Fees		$718
Total Target $		**$12,493**
Average Dealer Option Cost:		**84%**

The 1993 Honda Civic is available in nine models - DX, LX, and EX sedans; DX and EX coupes; and CX, DX, VX, and Si hatchbacks. New for 1993, the LX sedan retains the sporty new exterior and interior styling redesigns from last year. Features include power-assisted steering, impact-absorbing body-colored bumpers, tinted glass, and body side moldings as standard equipment. The LX also offers full wheel covers, cruise control, power door locks and windows, tachometer and dual power mirrors.

Ownership Costs

Cost Area	5 Year Cost	Rate
Depreciation	$4,979	○
Financing ($251/month)	$2,057	
Insurance (Rating 6)	$6,919	O
State Fees	$519	
Fuel (Hwy 38 City 30)	$2,540	○
Maintenance	$4,539	O
Repairs	$570	○

Warranty/Maintenance Info

Major Tune-Up	$239	●
Minor Tune-Up	$129	●
Brake Service	$217	O
Overall Warranty	3 yr/36k	O
Drivetrain Warranty	3 yr/36k	O
Rust Warranty	3 yr/unlim. mi	◐
Maintenance Warranty	N/A	
Roadside Assistance	N/A	

Ownership Cost By Year

- 1993
- 1994
- 1995
- 1996
- 1997

Resale Value

1993	1994	1995	1996	1997
$10,707	$9,785	$9,053	$8,267	$7,514

Cumulative Costs

	1993	1994	1995	1996	1997
Annual	$4,563	$3,738	$4,458	$3,466	$5,898
Total	$4,563	$8,301	$12,759	$16,225	$22,123

Ownership Costs (5yr)

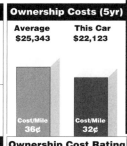

Average	This Car
$25,343	$22,123
Cost/Mile 36¢	Cost/Mile 32¢

Ownership Cost Rating

○ Excellent

---

* Includes shaded options
** Other purchase requirements apply

● Poor ◐ Worse Than Average O Average ○ Better Than Average ○ Excellent ⊖ Insufficient Information

©1993 by IntelliChoice, Inc. (408) 554-8711 All Rights Reserved. Reproduction Prohibited.
Refer to *Section 3: Annotated Vehicle Charts* for an explanation of these charts.

Honda Civic EX
4 Door Sedan

Compact

 1.6L 125 hp Gas Fuel Inject.
 4 Cylinder In-Line
 Manual 5 Speed
 2 Wheel Front
 Driver Airbag Psngr Belts

Purchase Price

Car Item	Dealer Cost	List
Base Price	**$12,835**	**$15,100**
Anti-Lock Brakes	Std	Std
Automatic 4 Speed	$630	$750
Optional Engine	N/A	N/A
Air Conditioning	Std	Std
Power Steering	Std	Std
Cruise Control	Std	Std
All Wheel Drive	N/A	N/A
AM/FM Stereo Cassette	Std	Std
Steering Wheel, Tilt	Std	Std
Power Windows	Std	Std
*Options Price	$630	$750
*Total Price	$13,465	$15,850
Target Price	$14,466	
Destination Charge	$310	
Avg. Tax & Fees	$901	
Total Target $	**$15,677**	
Average Dealer Option Cost:	*84%*	

Ownership Costs

Cost Area	5 Year Cost	Rate
Depreciation	$7,242	○
Financing ($315/month)	$2,583	
Insurance (Rating 10)	$7,479	○
State Fees	$646	
Fuel (Hwy 34 City 27)	$2,831	◐
Maintenance	$4,851	○
Repairs	$570	◐

Warranty/Maintenance Info

Major Tune-Up	$239	●
Minor Tune-Up	$129	●
Brake Service	$238	○
Overall Warranty	3 yr/36k	○
Drivetrain Warranty	3 yr/36k	○
Rust Warranty	3 yr/unlim. mi	◉
Maintenance Warranty	N/A	
Roadside Assistance	N/A	

Ownership Cost By Year
1993, 1994, 1995, 1996, 1997

Resale Value
1993	1994	1995	1996	1997
$12,471	$11,295	$10,347	$9,358	$8,435

Cumulative Costs
	1993	1994	1995	1996	1997
Annual	$6,393	$4,347	$5,116	$3,907	$6,439
Total	$6,393	$10,740	$15,856	$19,763	$26,202

Ownership Costs (5yr)
Average: $28,104 — This Car: $26,202
Cost/Mile 40¢ — Cost/Mile 37¢

Ownership Cost Rating
○ Better Than Average

The 1993 Honda Civic is available in nine models - DX, LX, and EX sedans; DX and EX coupes; and CX, DX, VX, and Si hatchbacks. New for 1993, the EX sedan retains the sporty new exterior and interior styling redesigns from last year. Features include power-assisted steering, impact-absorbing body-colored bumpers, standard air conditioning and stereo system, power 4-wheel disc brakes, P175/65 R14 radials on 14-inch (full-covered) wheels, and front and rear stabilizer bars for better handling.

Honda Prelude S
2 Door Coupe

Compact

 2.2L 135 hp Gas Fuel Inject.
 4 Cylinder In-Line
 Manual 5 Speed
 2 Wheel Front
Driver Airbag Psngr Belts

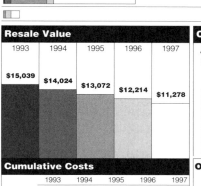

Purchase Price

Car Item	Dealer Cost	List
Base Price	**$14,280**	**$17,000**
Anti-Lock Brakes	N/A	N/A
Automatic 4 Speed	$630	$750
Optional Engine	N/A	N/A
Air Conditioning	Dlr	Dlr
Power Steering	Std	Std
Cruise Control	Std	Std
All Wheel Drive	N/A	N/A
AM/FM Stereo Cassette	Std	Std
Steering Wheel, Tilt	Std	Std
Power Windows	Std	Std
*Options Price	$630	$750
*Total Price	$14,910	$17,750
Target Price	$15,961	
Destination Charge	$310	
Avg. Tax & Fees	$995	
Total Target $	**$17,266**	
Average Dealer Option Cost:	*84%*	

Ownership Costs

Cost Area	5 Year Cost	Rate
Depreciation	$5,988	○
Financing ($347/month)	$2,845	
Insurance (Rating 16)	$8,784	●
State Fees	$722	
Fuel (Hwy 30 City 23)	$3,260	○
Maintenance	$5,209	◉
Repairs	$570	○

Warranty/Maintenance Info

Major Tune-Up	$277	●
Minor Tune-Up	$147	●
Brake Service	$272	●
Overall Warranty	3 yr/36k	○
Drivetrain Warranty	3 yr/36k	○
Rust Warranty	3 yr/unlim. mi	◉
Maintenance Warranty	N/A	
Roadside Assistance	N/A	

Resale Value
1993	1994	1995	1996	1997
$15,039	$14,024	$13,072	$12,214	$11,278

Cumulative Costs
	1993	1994	1995	1996	1997
Annual	$5,862	$4,638	$5,579	$4,189	$7,110
Total	$5,862	$10,500	$16,079	$20,268	$27,378

Ownership Costs (5yr)
Average: $29,735 — This Car: $27,378
Cost/Mile 42¢ — Cost/Mile 39¢

Ownership Cost Rating
○ Excellent

The 1993 Prelude is available in two models - S and Si. The Prelude S mechanics include a 16-valve 4-cylinder SOHC engine powered by 135 hp, variable-assist rack-and-pinion steering, power 4-wheel disc brakes, 5-speed manual transmission with overdrive and a fully independent suspension system. The S model also features a driver's side airbag, 4 speaker stereo system, cruise control, adjustable steering column, child safety-seat tether anchors and a rear window defroster with timer.

---

* Includes shaded options
** Other purchase requirements apply

● Poor ◐ Worse Than Average ○ Average ○ Better Than Average ○ Excellent ⊖ Insufficient Information

©1993 by IntelliChoice, Inc. (408) 554-8711 All Rights Reserved. Reproduction Prohibited.
Refer to *Section 3: Annotated Vehicle Charts* for an explanation of these charts.

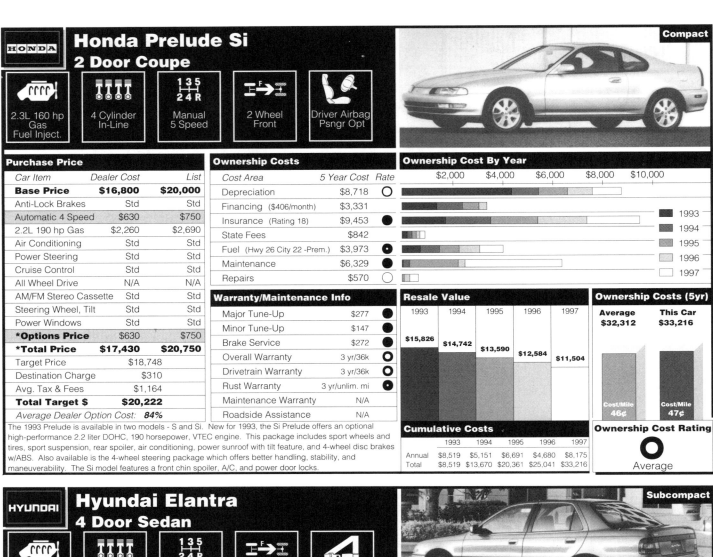

Honda Prelude Si
2 Door Coupe — Compact

- 2.3L 160 hp Gas Fuel Inject.
- 4 Cylinder In-Line
- Manual 5 Speed
- 2 Wheel Front
- Driver Airbag, Psngr Opt

Purchase Price

Car Item	Dealer Cost	List
Base Price	**$16,800**	**$20,000**
Anti-Lock Brakes	Std	Std
Automatic 4 Speed	$630	$750
2.2L 190 hp Gas	$2,260	$2,690
Air Conditioning	Std	Std
Power Steering	Std	Std
Cruise Control	Std	Std
All Wheel Drive	N/A	N/A
AM/FM Stereo Cassette	Std	Std
Steering Wheel, Tilt	Std	Std
Power Windows	Std	Std
*Options Price	$630	$750
*Total Price	$17,430	$20,750
Target Price	$18,748	
Destination Charge	$310	
Avg. Tax & Fees	$1,164	
Total Target $	**$20,222**	
Average Dealer Option Cost:	84%	

Ownership Costs

Cost Area	5 Year Cost	Rate
Depreciation	$8,718	○
Financing ($406/month)	$3,331	
Insurance (Rating 18)	$9,453	●
State Fees	$842	
Fuel (Hwy 26 City 22 -Prem.)	$3,973	⊙
Maintenance	$6,329	●
Repairs	$570	○

Warranty/Maintenance Info

Major Tune-Up	$277	●
Minor Tune-Up	$147	●
Brake Service	$272	●
Overall Warranty	3 yr/36k	○
Drivetrain Warranty	3 yr/36k	○
Rust Warranty	3 yr/unlim. mi	⊙
Maintenance Warranty	N/A	
Roadside Assistance	N/A	

Ownership Cost By Year (1993, 1994, 1995, 1996, 1997)

Resale Value

1993	1994	1995	1996	1997
$15,826	$14,742	$13,590	$12,584	$11,504

Cumulative Costs

	1993	1994	1995	1996	1997
Annual	$8,519	$5,151	$6,691	$4,680	$8,175
Total	$8,519	$13,670	$20,361	$25,041	$33,216

Ownership Costs (5yr)
- Average: $32,312
- This Car: $33,216
- Cost/Mile: 46¢ / 47¢

Ownership Cost Rating: ○ Average

The 1993 Prelude is available in two models - S and Si. New for 1993, the Si Prelude offers an optional high-performance 2.2 liter DOHC, 190 horsepower, VTEC engine. This package includes sport wheels and tires, sport suspension, rear spoiler, air conditioning, power sunroof with tilt feature, and 4-wheel disc brakes w/ABS. Also available is the 4-wheel steering package which offers better handling, stability, and maneuverability. The Si model features a front chin spoiler, A/C, and power door locks.

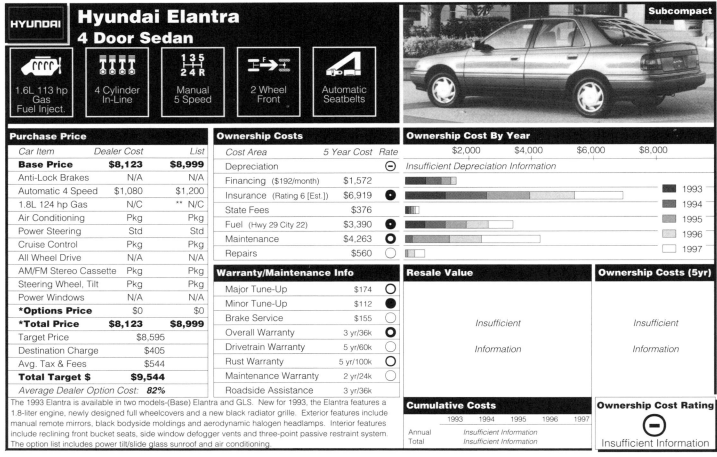

Hyundai Elantra
4 Door Sedan — Subcompact

- 1.6L 113 hp Gas Fuel Inject.
- 4 Cylinder In-Line
- Manual 5 Speed
- 2 Wheel Front
- Automatic Seatbelts

Purchase Price

Car Item	Dealer Cost	List
Base Price	**$8,123**	**$8,999**
Anti-Lock Brakes	N/A	N/A
Automatic 4 Speed	$1,080	$1,200
1.8L 124 hp Gas	N/C	** N/C
Air Conditioning	Pkg	Pkg
Power Steering	Std	Std
Cruise Control	Pkg	Pkg
All Wheel Drive	N/A	N/A
AM/FM Stereo Cassette	Pkg	Pkg
Steering Wheel, Tilt	Pkg	Pkg
Power Windows	N/A	N/A
*Options Price	$0	$0
*Total Price	$8,123	$8,999
Target Price	$8,595	
Destination Charge	$405	
Avg. Tax & Fees	$544	
Total Target $	**$9,544**	
Average Dealer Option Cost:	82%	

Ownership Costs

Cost Area	5 Year Cost	Rate
Depreciation		⊖
Financing ($192/month)	$1,572	
Insurance (Rating 6 [Est.])	$6,919	⊙
State Fees	$376	
Fuel (Hwy 29 City 22)	$3,390	⊙
Maintenance	$4,263	○
Repairs	$560	○

Warranty/Maintenance Info

Major Tune-Up	$174	○
Minor Tune-Up	$112	●
Brake Service	$155	○
Overall Warranty	3 yr/36k	○
Drivetrain Warranty	5 yr/60k	○
Rust Warranty	5 yr/100k	○
Maintenance Warranty	2 yr/24k	○
Roadside Assistance	3 yr/36k	

Ownership Cost By Year — Insufficient Depreciation Information

Resale Value: Insufficient Information

Cumulative Costs

	1993	1994	1995	1996	1997
Annual	Insufficient Information				
Total	Insufficient Information				

Ownership Costs (5yr): Insufficient Information

Ownership Cost Rating: ⊖ Insufficient Information

The 1993 Elantra is available in two models-(Base) Elantra and GLS. New for 1993, the Elantra features a 1.8-liter engine, newly designed full wheelcovers and a new black radiator grille. Exterior features include manual remote mirrors, black bodyside moldings and aerodynamic halogen headlamps. Interior features include reclining front bucket seats, side window defogger vents and three-point passive restraint system. The option list includes power tilt/slide glass sunroof and air conditioning.

---

* Includes shaded options
** Other purchase requirements apply

● Poor ◐ Worse Than Average ◑ Average ○ Better Than Average ○ Excellent ⊖ Insufficient Information

©1993 by IntelliChoice, Inc. (408) 554-8711 All Rights Reserved. Reproduction Prohibited.
Refer to *Section 3: Annotated Vehicle Charts* for an explanation of these charts.

Hyundai Elantra GLS
4 Door Sedan

 1.8L 124 hp Gas Fuel Inject.
 4 Cylinder In-Line
 Manual 5 Speed
 2 Wheel Front
Automatic Seatbelts

Subcompact — Base Model Shown

Purchase Price

Car Item	Dealer Cost	List
Base Price	**$9,087**	**$10,299**
Anti-Lock Brakes	N/A	N/A
Automatic 4 Speed	$628	$700
Optional Engine	N/A	N/A
Air Conditioning	Pkg	Pkg
Power Steering	Std	Std
Cruise Control	Pkg	Pkg
All Wheel Drive	N/A	N/A
AM/FM Stereo Cassette	Std	Std
Steering Wheel, Tilt	Std	Std
Power Windows	Std	Std
*Options Price	$0	$0
*Total Price	$9,087	$10,299
Target Price	$9,633	
Destination Charge	$405	
Avg. Tax & Fees	$609	
Total Target $	**$10,647**	
Average Dealer Option Cost:	82%	

The 1993 Elantra is available in two models-(Base) Elantra and GLS. New for 1993, the Elantra GLS features a 1.8-liter DOHC 16 valve dual balance shaft engine as standard equipment, newly designed deluxe wheelcovers, body color grille and soft grip three-spoke steering wheel. Functionally, the GLS includes power windows, door locks and remote control outside mirrors. Other features include driver's seat lumbar/tilt adjust, front seat headrests with tilt adjust and tilt steering wheel.

Ownership Costs

Cost Area	5 Year Cost	Rate
Depreciation		⊖
Financing ($214/month)	$1,754	
Insurance (Rating 8 [Est.])	$7,223	●
State Fees	$428	
Fuel (Hwy 28 City 21)	$3,530	●
Maintenance		⊖
Repairs	$560	○

Warranty/Maintenance Info

Major Tune-Up		⊖
Minor Tune-Up		⊖
Brake Service		⊖
Overall Warranty	3 yr/36k	●
Drivetrain Warranty	5 yr/60k	○
Rust Warranty	5 yr/100k	○
Maintenance Warranty	2 yr/24k	○
Roadside Assistance	3 yr/36k	

Ownership Cost By Year

Insufficient Depreciation Information
1993, 1994, 1995, 1996, 1997

Insufficient Maintenance Information

Resale Value
Insufficient Information

Cumulative Costs
	1993	1994	1995	1996	1997
Annual	Insufficient Information				
Total	Insufficient Information				

Ownership Costs (5yr)
Insufficient Information

Ownership Cost Rating
⊖ Insufficient Information

Hyundai Excel
2 Door Hatchback

 1.5L 81 hp Gas Fuel Inject.
4 Cylinder In-Line
 Manual 4 Speed
 2 Wheel Front
Automatic Seatbelts

Subcompact

Purchase Price

Car Item	Dealer Cost	List
Base Price	**$6,346**	**$6,799**
Anti-Lock Brakes	N/A	N/A
Automatic 4 Speed	$580	$625
Optional Engine	N/A	N/A
Air Conditioning	Pkg	Pkg
Power Steering	Pkg	Pkg
Cruise Control	N/A	N/A
All Wheel Drive	N/A	N/A
AM/FM Stereo Cassette	Pkg	Pkg
Steering Wheel, Tilt	N/A	N/A
Power Windows	N/A	N/A
*Options Price	$0	$0
*Total Price	$6,346	$6,799
Target Price	$6,692	
Destination Charge	$405	
Avg. Tax & Fees	$427	
Total Target $	**$7,524**	
Average Dealer Option Cost:	83%	

The 1993 Excel is available in four models-(Base) Excel and GS Hatchbacks, and (Base) Excel and GL Sedans. New for 1993, the Excel hatchback features a bold new grille and newly designed wheelcovers. Exterior features include dual manual remote mirrors, black bodyside moldings and aerodynamic halogen headlights. Interior features include reclining front bucket seats, 60/40 split fold-down rear seat and front passive restraint system. A pop-up sunroof is available on all Excel hatchbacks.

Ownership Costs

Cost Area	5 Year Cost	Rate
Depreciation	$4,739	●
Financing ($151/month)	$1,239	
Insurance (Rating 11 [Est.])	$7,693	●
State Fees	$288	
Fuel (Hwy 33 City 29)	$2,788	○
Maintenance	$3,703	○
Repairs	$630	○

Warranty/Maintenance Info

Major Tune-Up	$207	○
Minor Tune-Up	$123	●
Brake Service	$153	○
Overall Warranty	3 yr/36k	○
Drivetrain Warranty	5 yr/60k	○
Rust Warranty	5 yr/100k	○
Maintenance Warranty	2 yr/24k	○
Roadside Assistance	3 yr/36k	

Resale Value
1993	1994	1995	1996	1997
$5,740	$4,883	$4,197	$3,492	$2,785

Cumulative Costs
	1993	1994	1995	1996	1997
Annual	$4,307	$3,557	$4,194	$4,182	$4,840
Total	$4,307	$7,864	$12,058	$16,240	$21,080

Ownership Costs (5yr)
Average	This Car
$19,228	$21,080
Cost/Mile 27¢	Cost/Mile 30¢

Ownership Cost Rating
● Poor

---

* Includes shaded options
** Other purchase requirements apply

● Poor ● Worse Than Average ○ Average ○ Better Than Average ○ Excellent ⊖ Insufficient Information

©1993 by IntelliChoice, Inc. (408) 554-8711   All Rights Reserved. Reproduction Prohibited.
Refer to Section 3: Annotated Vehicle Charts for an explanation of these charts.

Hyundai Excel GS
2 Door Hatchback

 1.5L 81 hp Gas Fuel Inject.
 4 Cylinder In-Line
 Manual 5 Speed
 2 Wheel Front
 Automatic Seatbelts

Subcompact — Base Model Shown

Purchase Price

Car Item	Dealer Cost	List
Base Price	**$6,950**	**$7,699**
Anti-Lock Brakes	N/A	N/A
Automatic 4 Speed	$516	$575
Optional Engine	N/A	N/A
Air Conditioning	Pkg	Pkg
Power Steering	Pkg	Pkg
Cruise Control	N/A	N/A
All Wheel Drive	N/A	N/A
AM/FM Stereo Cassette	Std	Std
Steering Wheel, Tilt	N/A	N/A
Power Windows	N/A	N/A
*Options Price	$0	$0
*Total Price	$6,950	$7,699
Target Price		$7,337
Destination Charge		$405
Avg. Tax & Fees		$468
Total Target $		**$8,210**
Average Dealer Option Cost:		83%

Ownership Costs

Cost Area	5 Year Cost	Rate
Depreciation	$4,903	●
Financing ($165/month)	$1,352	
Insurance (Rating 11 [Est.])	$7,693	●
State Fees	$325	
Fuel (Hwy 36 City 29)	$2,656	○
Maintenance	$3,822	○
Repairs	$630	○

Warranty/Maintenance Info

Major Tune-Up	$207	○
Minor Tune-Up	$123	●
Brake Service	$153	○
Overall Warranty	3 yr/36k	○
Drivetrain Warranty	5 yr/60k	○
Rust Warranty	5 yr/100k	○
Maintenance Warranty	2 yr/24k	○
Roadside Assistance	3 yr/36k	

Ownership Cost By Year
(1993–1997 bar chart)

Resale Value
1993	1994	1995	1996	1997
$5,838	$5,096	$4,534	$3,940	$3,307

Cumulative Costs
	1993	1994	1995	1996	1997
Annual	$4,926	$3,461	$4,129	$4,059	$4,806
Total	$4,926	$8,387	$12,516	$16,575	$21,381

Ownership Costs (5yr)
- Average: $20,241 (Cost/Mile 29¢)
- This Car: $21,381 (Cost/Mile 31¢)

Ownership Cost Rating
● Worse Than Average

The 1993 Excel is available in four models-(Base) Excel and GS Hatchbacks, and (Base) Excel and GL Sedans. New for 1993, the GS hatchback features bodycolor trim, rear spoiler and rear window wiper/washer that integrates well with the functional rear hatch. Interior features include five-way adjustable sport contoured driver's seat with lumbar support and deluxe instrument cluster. The option list includes shift interlock feature, alloy wheels and power sliding sunroof with shade.

Hyundai Excel
4 Door Sedan

 1.5L 81 hp Gas Fuel Inject.
 4 Cylinder In-Line
 Manual 4 Speed
 2 Wheel Front / Automatic Seatbelts

Subcompact — GL Model Shown

Purchase Price

Car Item	Dealer Cost	List
Base Price	**$7,185**	**$7,699**
Anti-Lock Brakes	N/A	N/A
Automatic 4 Speed	$580	$625
Optional Engine	N/A	N/A
Air Conditioning	Pkg	Pkg
Power Steering	Pkg	Pkg
Cruise Control	N/A	N/A
All Wheel Drive	N/A	N/A
AM/FM Stereo Cassette	Pkg	Pkg
Steering Wheel, Tilt	N/A	N/A
Power Windows	N/A	N/A
*Options Price	$0	$0
*Total Price	$7,185	$7,699
Target Price		$7,589
Destination Charge		$405
Avg. Tax & Fees		$481
Total Target $		**$8,475**
Average Dealer Option Cost:		83%

Ownership Costs

Cost Area	5 Year Cost	Rate
Depreciation	$5,314	●
Financing ($170/month)	$1,396	
Insurance (Rating 10 [Est.])	$7,479	●
State Fees	$325	
Fuel (Hwy 33 City 29)	$2,788	○
Maintenance	$3,703	○
Repairs	$630	○

Warranty/Maintenance Info

Major Tune-Up	$207	○
Minor Tune-Up	$123	●
Brake Service	$153	○
Overall Warranty	3 yr/36k	○
Drivetrain Warranty	5 yr/60k	○
Rust Warranty	5 yr/100k	○
Maintenance Warranty	2 yr/24k	○
Roadside Assistance	3 yr/36k	

Resale Value
1993	1994	1995	1996	1997
$5,903	$5,072	$4,478	$3,851	$3,161

Cumulative Costs
	1993	1994	1995	1996	1997
Annual	$5,131	$3,547	$4,098	$4,077	$4,782
Total	$5,131	$8,678	$12,776	$16,853	$21,635

Ownership Costs (5yr)
- Average: $20,241 (Cost/Mile 29¢)
- This Car: $21,635 (Cost/Mile 31¢)

Ownership Cost Rating
● Worse Than Average

The 1993 Excel is available in four models-(Base) Excel and GS Hatchbacks, and (Base) Excel and GL Sedans. New for 1993, the Excel sedan features newly designed wheelcovers. Exterior features include dual manual remote mirrors, black bodyside moldings and aerodynamic halogen headlights. Interior features include reclining front bucket seats and front passive restraint system with lap belts. The option list includes alloy wheels, tinted windows and a power sliding sunroof with shade.

---

* Includes shaded options
** Other purchase requirements apply

● Poor   ◐ Worse Than Average   ○ Average   ◯ Better Than Average   ◯ Excellent   ⊖ Insufficient Information

©1993 by IntelliChoice, Inc. (408) 554-8711   All Rights Reserved. Reproduction Prohibited.
Refer to *Section 3: Annotated Vehicle Charts* for an explanation of these charts.

Hyundai Excel GL
4 Door Sedan

 1.5L 81 hp Gas Fuel Inject.  4 Cylinder In-Line Manual 5 Speed 2 Wheel Front Automatic Seatbelts

Subcompact

Purchase Price

Car Item	Dealer Cost	List
Base Price	**$7,762**	**$8,599**
Anti-Lock Brakes	N/A	N/A
Automatic 4 Speed	$516	$575
Optional Engine	N/A	N/A
Air Conditioning	Pkg	Pkg
Power Steering	Pkg	Pkg
Cruise Control	N/A	N/A
All Wheel Drive	N/A	N/A
AM/FM Stereo Cassette	Std	Std
Steering Wheel, Tilt	N/A	N/A
Power Windows	N/A	N/A
***Options Price**	**$0**	**$0**
***Total Price**	**$7,762**	**$8,599**
Target Price		$8,207
Destination Charge		$405
Avg. Tax & Fees		$521
Total Target $		**$9,133**
Average Dealer Option Cost:		84%

The 1993 Excel is available in four models-(Base) Excel and GS Hatchbacks, and (Base) Excel and GL Sedans. New for 1993, the GL sedan features bodycolor bumpers, mirrors, door handles and bodyside moldings. Interior features include deluxe instrument cluster, reclining front seatbacks, full cloth seat trim, cloth door insert panels and a convenient center console with cassette holder. Other features include a remote deck lid and fuel filler releases and a 60/40 split fold-down rear seatback.

Ownership Costs

Cost Area	5 Year Cost	Rate
Depreciation	$5,416	◐
Financing ($184/month)	$1,504	
Insurance (Rating 11 [Est.])	$7,693	●
State Fees	$360	
Fuel (Hwy 36 City 29)	$2,656	○
Maintenance	$3,703	○
Repairs	$630	○

Warranty/Maintenance Info

Major Tune-Up	$207	○
Minor Tune-Up	$123	●
Brake Service	$153	○
Overall Warranty	3 yr/36k	○
Drivetrain Warranty	5 yr/60k	○
Rust Warranty	5 yr/100k	○
Maintenance Warranty	2 yr/24k	○
Roadside Assistance	3 yr/36k	

Ownership Cost By Year

(bar chart by year 1993-1997)

Resale Value

1993	1994	1995	1996	1997
$6,856	$5,943	$5,210	$4,453	$3,717

Cumulative Costs

	1993	1994	1995	1996	1997
Annual	$4,904	$3,687	$4,280	$4,239	$4,852
Total	$4,904	$8,591	$12,871	$17,110	$21,962

Ownership Costs (5yr)

Average	This Car
$21,253	$21,962
Cost/Mile 30¢	Cost/Mile 31¢

Ownership Cost Rating

 Worse Than Average

Hyundai Scoupe
2 Door Coupe

 1.5L 92 hp Gas Fuel Inject. 4 Cylinder In-Line Manual 5 Speed 2 Wheel Front Automatic Seatbelts

Subcompact

LS Model Shown

Purchase Price

Car Item	Dealer Cost	List
Base Price	**$8,187**	**$9,069**
Anti-Lock Brakes	N/A	N/A
Automatic 4 Speed	$522	$575
Optional Engine	N/A	N/A
Air Conditioning	Pkg	Pkg
Power Steering	Pkg	Pkg
Cruise Control	N/A	N/A
All Wheel Drive	N/A	N/A
AM/FM Stereo Cassette	Pkg	Pkg
Steering Wheel, Tilt	N/A	N/A
Power Windows	N/A	N/A
***Options Price**	**$0**	**$0**
***Total Price**	**$8,187**	**$9,069**
Target Price		$8,664
Destination Charge		$405
Avg. Tax & Fees		$548
Total Target $		**$9,617**
Average Dealer Option Cost:		83%

The 1993 Scoupe is available in three models-(Base) Scoupe, LS and Turbo. New for 1993, the Scoupe features a redesigned hood and front fender sheetmetal, new front bumper fascia and compound-parabolic reflector headlights. Interior features include reclining front bucket seats, 60/40 split fold-down rear seat and three-spoke steering wheel. The option list includes pop-up sunroof, air conditioning and halogen driving lamps. A CD player is available in conjunction with the highline radio.

Ownership Costs

Cost Area	5 Year Cost	Rate
Depreciation	$5,543	◐
Financing ($193/month)	$1,584	
Insurance (Rating 11 [Est.])	$7,693	●
State Fees	$379	
Fuel (Hwy 35 City 27)	$2,788	○
Maintenance	$3,802	○
Repairs	$630	○

Warranty/Maintenance Info

Major Tune-Up	$207	○
Minor Tune-Up	$123	●
Brake Service	$163	○
Overall Warranty	3 yr/36k	○
Drivetrain Warranty	5 yr/60k	○
Rust Warranty	5 yr/100k	○
Maintenance Warranty	2 yr/24k	○
Roadside Assistance	3 yr/36k	

Resale Value

1993	1994	1995	1996	1997
$7,594	$6,430	$5,659	$4,867	$4,074

Cumulative Costs

	1993	1994	1995	1996	1997
Annual	$4,713	$3,992	$4,433	$4,267	$5,014
Total	$4,713	$8,705	$13,138	$17,405	$22,419

Ownership Costs (5yr)

Average	This Car
$21,782	$22,419
Cost/Mile 31¢	Cost/Mile 32¢

Ownership Cost Rating

Worse Than Average

---

* Includes shaded options
** Other purchase requirements apply

● Poor ◐ Worse Than Average ○ Average ◯ Better Than Average ◯ Excellent ⊖ Insufficient Information

©1993 by IntelliChoice, Inc. (408) 554-8711 All Rights Reserved. Reproduction Prohibited.
Refer to *Section 3: Annotated Vehicle Charts* for an explanation of these charts.

Hyundai Scoupe LS
2 Door Coupe

Subcompact

- 1.5L 92 hp Gas Fuel Inject.
- 4 Cylinder In-Line
- Manual 5 Speed
- 2 Wheel Front
- Automatic Seatbelts

Purchase Price

Car Item	Dealer Cost	List
Base Price	**$8,999**	**$10,199**
Anti-Lock Brakes	N/A	N/A
Automatic 4 Speed	$522	$575
Optional Engine	N/A	N/A
Air Conditioning	Pkg	Pkg
Power Steering	Std	Std
Cruise Control	N/A	N/A
All Wheel Drive	N/A	N/A
AM/FM Stereo Cassette	Std	Std
Steering Wheel, Tilt	Std	Std
Power Windows	Std	Std
*Options Price	$0	$0
*Total Price	$8,999	$10,199
Target Price		$9,538
Destination Charge		$405
Avg. Tax & Fees		$603
Total Target $		**$10,546**
Average Dealer Option Cost:		*83%*

The 1993 Scoupe is available in three models-(Base) Scoupe, LS and Turbo. New for 1993, the LS coupe exterior features bodycolor cladding as well as 14-inch steel wheels with deluxe full wheelcovers. New interior features include deluxe seat cloth trim with matching door and panel inserts. Other interior touches include a power package for window and outside mirror operation and 6-way adjustable driver's seat. The option list includes a removable pop-up sunroof and halogen driving lamps.

Ownership Costs

Cost Area	5 Year Cost	Rate
Depreciation	$6,298	●
Financing ($212/month)	$1,737	
Insurance (Rating 12 [Est.])	$7,898	●
State Fees	$424	
Fuel (Hwy 35 City 27)	$2,788	○
Maintenance	$3,885	○
Repairs	$630	○

Warranty/Maintenance Info

Major Tune-Up	$207	○
Minor Tune-Up	$123	●
Brake Service	$163	○
Overall Warranty	3 yr/36k	○
Drivetrain Warranty	5 yr/60k	○
Rust Warranty	5 yr/100k	○
Maintenance Warranty	2 yr/24k	○
Roadside Assistance	3 yr/36k	

Ownership Cost By Year

1993, 1994, 1995, 1996, 1997

Resale Value

1993	1994	1995	1996	1997
$8,163	$6,744	$5,912	$5,070	$4,248

Cumulative Costs

	1993	1994	1995	1996	1997
Annual	$5,186	$4,345	$4,614	$4,378	$5,137
Total	$5,186	$9,531	$14,145	$18,523	$23,660

Ownership Costs (5yr)

	Average	This Car
	$23,053	$23,660
Cost/Mile	33¢	34¢

Ownership Cost Rating

○ Average

Hyundai Scoupe Turbo
2 Door Coupe

Subcompact

- 1.5L 115 hp Turbo Gas Fuel Inject.
- 4 Cylinder In-Line
- Manual 5 Speed
- 2 Wheel Front
- Automatic Seatbelts

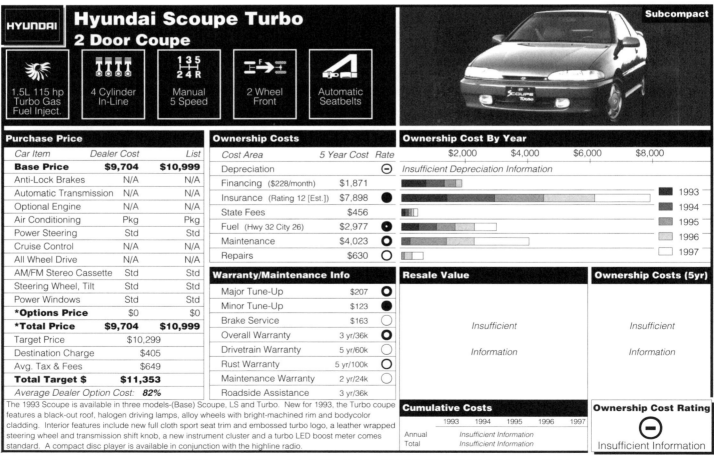

Purchase Price

Car Item	Dealer Cost	List
Base Price	**$9,704**	**$10,999**
Anti-Lock Brakes	N/A	N/A
Automatic Transmission	N/A	N/A
Optional Engine	N/A	N/A
Air Conditioning	Pkg	Pkg
Power Steering	Std	Std
Cruise Control	N/A	N/A
All Wheel Drive	N/A	N/A
AM/FM Stereo Cassette	Std	Std
Steering Wheel, Tilt	Std	Std
Power Windows	Std	Std
*Options Price	$0	$0
*Total Price	$9,704	$10,999
Target Price		$10,299
Destination Charge		$405
Avg. Tax & Fees		$649
Total Target $		**$11,353**
Average Dealer Option Cost:		*82%*

The 1993 Scoupe is available in three models-(Base) Scoupe, LS and Turbo. New for 1993, the Turbo coupe features a black-out roof, halogen driving lamps, alloy wheels with bright-machined rim and bodycolor cladding. Interior features include new full cloth sport seat trim and embossed turbo logo, a leather wrapped steering wheel and transmission shift knob, a new instrument cluster and a turbo LED boost meter comes standard. A compact disc player is available in conjunction with the highline radio.

Ownership Costs

Cost Area	5 Year Cost	Rate
Depreciation		⊖
Financing ($228/month)	$1,871	
Insurance (Rating 12 [Est.])	$7,898	●
State Fees	$456	
Fuel (Hwy 32 City 26)	$2,977	◉
Maintenance	$4,023	○
Repairs	$630	○

Warranty/Maintenance Info

Major Tune-Up	$207	○
Minor Tune-Up	$123	●
Brake Service	$163	○
Overall Warranty	3 yr/36k	○
Drivetrain Warranty	5 yr/60k	○
Rust Warranty	5 yr/100k	○
Maintenance Warranty	2 yr/24k	○
Roadside Assistance	3 yr/36k	

Ownership Cost By Year

Insufficient Depreciation Information

Resale Value

Insufficient Information

Cumulative Costs

	1993	1994	1995	1996	1997
Annual	*Insufficient Information*				
Total	*Insufficient Information*				

Ownership Costs (5yr)

Insufficient Information

Ownership Cost Rating

⊖ Insufficient Information

---

* Includes shaded options
** Other purchase requirements apply

● Poor | ◉ Worse Than Average | ○ Average | ○ Better Than Average | ○ Excellent | ⊖ Insufficient Information

©1993 by IntelliChoice, Inc. (408) 554-8711 All Rights Reserved. Reproduction Prohibited.
Refer to *Section 3: Annotated Vehicle Charts* for an explanation of these charts.

Hyundai Sonata
4 Door Sedan

Compact

2.0L 128 hp Gas Fuel Inject. | 4 Cylinder In-Line | Manual 5 Speed | 2 Wheel Front | Automatic Seatbelts

Purchase Price

Car Item	Dealer Cost	List
Base Price	**$11,061**	**$12,399**
Anti-Lock Brakes	N/A	N/A
Automatic 4 Speed	$734	$750
3.0L 142 hp Gas	$1,419	$1,540
Air Conditioning	Std	Std
Power Steering	Std	Std
Cruise Control	Pkg	Pkg
All Wheel Drive	N/A	N/A
AM/FM Stereo Cassette	Std	Std
Steering Wheel, Tilt	Std	Std
Power Windows	Pkg	Pkg
***Options Price**	**$2,153**	**$2,290**
***Total Price**	**$13,214**	**$14,689**
Target Price		$14,060
Destination Charge		$405
Avg. Tax & Fees		$874
Total Target $		**$15,339**
Average Dealer Option Cost:		89%

The 1993 Sonata is available in two models-(Base) Sonata and GLS. New for 1993, the Sonata features redesigned wheelcovers and a new grille. Exterior features include dual manual remote mirrors, bodyside molding and halogen lamps. Interior features include front bucket seats, driver seat with cushion height adjuster and two-point passive restraint system. The option list includes a power tilt/slide glass sunroof with inner sunshade, shift lock system, leather interior and aluminum alloy wheels.

Ownership Costs

Cost Area	5 Year Cost	Rate
Depreciation	$9,905	●
Financing ($308/month)	$2,527	
Insurance (Rating 11 [Est.])	$7,693	●
State Fees	$605	
Fuel (Hwy 24 City 18)	$4,117	●
Maintenance	$5,566	●
Repairs	$630	○

Warranty/Maintenance Info

Major Tune-Up	$307	●
Minor Tune-Up	$133	●
Brake Service	$196	○
Overall Warranty	3 yr/36k	◐
Drivetrain Warranty	5 yr/60k	○
Rust Warranty	5 yr/100k	○
Maintenance Warranty	2 yr/24k	◐
Roadside Assistance	3 yr/36k	

Ownership Cost By Year
1993, 1994, 1995, 1996, 1997

Resale Value
1993	1994	1995	1996	1997
$11,055	$9,422	$7,601	$6,475	$5,434

Cumulative Costs
	1993	1994	1995	1996	1997
Annual	$7,667	$5,082	$6,434	$4,673	$7,187
Total	$7,667	$12,749	$19,183	$23,856	$31,043

Ownership Costs (5yr)
Average	This Car
$27,107	$31,043
Cost/Mile 39¢	Cost/Mile 44¢

Ownership Cost Rating
● Poor

Hyundai Sonata GLS
4 Door Sedan

Compact

Base Model Shown

2.0L 128 hp Gas Fuel Inject. | 4 Cylinder In-Line | Manual 5 Speed | 2 Wheel Front | Automatic Seatbelts

Purchase Price

Car Item	Dealer Cost	List
Base Price	**$12,034**	**$13,799**
Anti-Lock Brakes	Pkg	Pkg
Automatic 4 Speed	$734	$750
3.0L 142 hp Gas	$1,419	$1,540
Air Conditioning	Std	Std
Power Steering	Std	Std
Cruise Control	Std	Std
All Wheel Drive	N/A	N/A
AM/FM Stereo Cassette	Std	Std
Steering Wheel, Tilt	Std	Std
Power Windows	Std	Std
***Options Price**	**$2,153**	**$2,290**
***Total Price**	**$14,187**	**$16,089**
Target Price		$15,123
Destination Charge		$405
Avg. Tax & Fees		$941
Total Target $		**$16,469**
Average Dealer Option Cost:		85%

The 1993 Sonata is available in two models-(Base) Sonata and GLS. New for 1993, the GLS sedan features a newly designed deluxe bodycolor and charcoal grille and deluxe wheelcovers. Other exterior equipment includes dual bodycolor power remote mirrors and bodycolor bumpers with bright accents. Functionally power windows, power mirrors, power antenna and power door locks are included. Interior features include driver's seat lumbar support, 60/40 split fold-down rear seat and full cloth seat trim.

Ownership Costs

Cost Area	5 Year Cost	Rate
Depreciation	$11,121	●
Financing ($331/month)	$2,713	
Insurance (Rating 14 [Est.])	$8,238	◉
State Fees	$659	
Fuel (Hwy 24 City 18)	$4,117	●
Maintenance	$5,566	●
Repairs	$630	○

Warranty/Maintenance Info

Major Tune-Up	$307	●
Minor Tune-Up	$133	●
Brake Service	$196	○
Overall Warranty	3 yr/36k	◐
Drivetrain Warranty	5 yr/60k	○
Rust Warranty	5 yr/100k	○
Maintenance Warranty	2 yr/24k	◐
Roadside Assistance	3 yr/36k	

Resale Value
1993	1994	1995	1996	1997
$11,109	$9,515	$7,805	$6,385	$5,348

Cumulative Costs
	1993	1994	1995	1996	1997
Annual	$8,936	$5,218	$6,480	$5,102	$7,308
Total	$8,936	$14,154	$20,634	$25,736	$33,044

Ownership Costs (5yr)
Average	This Car
$28,309	$33,044
Cost/Mile 40¢	Cost/Mile 47¢

Ownership Cost Rating
● Poor

---

* Includes shaded options
** Other purchase requirements apply

● Poor | ◉ Worse Than Average | ◐ Average | ○ Better Than Average | ○ Excellent | ⊖ Insufficient Information

©1993 by IntelliChoice, Inc. (408) 554-8711 All Rights Reserved. Reproduction Prohibited.
Refer to *Section 3: Annotated Vehicle Charts* for an explanation of these charts.

Infiniti G20
4 Door Sedan

Compact

 2.0L 140 hp Gas Fuel Inject. | 4 Cylinder In-Line | Manual 5 Speed | 2 Wheel Front | Automatic Seatbelts

Purchase Price

Car Item	Dealer Cost	List
Base Price	**$15,600**	**$19,500**
Anti-Lock Brakes	Std	Std
Automatic 4 Speed	$720	$900
Optional Engine	N/A	N/A
Air Conditioning	Std	Std
Power Steering	Std	Std
Cruise Control	Std	Std
All Wheel Drive	N/A	N/A
AM/FM Stereo Cassette	Std	Std
Steering Wheel, Tilt	Std	Std
Power Windows	Std	Std
*Options Price	$720	$900
*Total Price	$16,320	$20,400
Target Price		$17,694
Destination Charge		$450
Avg. Tax & Fees		$1,116
Total Target $		**$19,260**
Average Dealer Option Cost:		**80%**

The 1993 G20 is available as a four-door sedan. New for 1993, the G20 features four new exterior colors (Deep Blue Sapphire, Ruby Red, Crimson Pearl and Beige Sand) and revised availability of some exterior/interior color combinations. Sunroof availability has been expanded for models with cloth interiors, and the rear speakers for the premium AM/FM/cassette audio system have been upgraded for enhanced sound reproduction. Other features include air conditioning and optional leather seats.

Ownership Costs

Cost Area	5 Year Cost	Rate
Depreciation	$6,414	○
Financing ($387/month)	$3,173	
Insurance (Rating 16)	$8,449	●
State Fees	$835	
Fuel (Hwy 29 City 22)	$3,390	○
Maintenance	$3,324	○
Repairs	$600	○

Warranty/Maintenance Info

Major Tune-Up	$143	○
Minor Tune-Up	$73	○
Brake Service	$175	○
Overall Warranty	4 yr/60k	○
Drivetrain Warranty	6 yr/70k	○
Rust Warranty	7 yr/unlim. mi	○
Maintenance Warranty	N/A	
Roadside Assistance	4 yr/unlim. mi	

Ownership Cost By Year

1993, 1994, 1995, 1996, 1997

Resale Value

1993	1994	1995	1996	1997
$18,093	$16,395	$15,087	$13,918	$12,846

Cumulative Costs

	1993	1994	1995	1996	1997
Annual	$4,915	$5,280	$5,666	$4,109	$6,215
Total	$4,915	$10,195	$15,861	$19,970	$26,185

Ownership Costs (5yr)

Average	This Car
$32,011	$26,185
Cost/Mile 46¢	Cost/Mile 37¢

Ownership Cost Rating

Excellent

Infiniti J30
4 Door Sedan

Luxury

 3.0L 210 hp Gas Fuel Inject. | 6 Cylinder "V" | Automatic 4 Speed |  2 Wheel Rear | Driver/Psngr Airbags Std

Purchase Price

Car Item	Dealer Cost	List
Base Price	**$27,200**	**$34,000**
Anti-Lock Brakes	Std	Std
Manual Transmission	N/A	N/A
Optional Engine	N/A	N/A
Auto Climate Control	Std	Std
Power Steering	Std	Std
Cruise Control	Std	Std
All Wheel Drive	N/A	N/A
AM/FM Stereo CD	Std	Std
Steering Wheel, Tilt	Std	Std
Power Windows	Std	Std
*Options Price	$0	$0
*Total Price	$27,200	$34,000
Target Price		$30,221
Destination Charge		$450
Avg. Tax & Fees		$1,879
Luxury Tax		$67
Total Target $		**$32,617**

The 1993 J30 is available in a one model edition. New to the Infiniti lineup, the J30's exterior features twin projector beam headlights, a broad hood, and a wide grille. The rearend offers a sloping trunk, full-width taillights and an integrated bumper. Interior features include a leather-appointed seating, subtle wood accents, eight-way power driver's and front passenger's seats and a keyless remote. Standard features include a Bose Audio System and an in-dash compact disc player.

Ownership Costs

Cost Area	5 Year Cost	Rate
Depreciation		⊖
Financing ($656/month)	$5,372	
Insurance (Rating 20)	$9,788	○
State Fees	$1,379	
Fuel (Hwy 23 City 18 -Prem.)	$4,652	○
Maintenance	$3,493	○
Repairs	$600	○

Warranty/Maintenance Info

Major Tune-Up	$151	○
Minor Tune-Up	$72	○
Brake Service	$176	○
Overall Warranty	4 yr/60k	○
Drivetrain Warranty	6 yr/70k	○
Rust Warranty	7 yr/unlim. mi	○
Maintenance Warranty	N/A	
Roadside Assistance	4 yr/unlim. mi	

Ownership Cost By Year

Insufficient Depreciation Information

Resale Value

Insufficient Information

Cumulative Costs

	1993	1994	1995	1996	1997
Annual	*Insufficient Information*				
Total	*Insufficient Information*				

Ownership Costs (5yr)

Insufficient Information

Ownership Cost Rating

⊖ Insufficient Information

---

* Includes shaded options
** Other purchase requirements apply

 Poor | Worse Than Average | Average | Better Than Average | ○ Excellent | Insufficient Information

©1993 by IntelliChoice, Inc. (408) 554-8711 All Rights Reserved. Reproduction Prohibited.
Refer to *Section 3: Annotated Vehicle Charts* for an explanation of these charts.

Infiniti Q45
4 Door Sedan

Luxury

4.5L 278 hp Gas Fuel Inject. | 8 Cylinder "V" | Automatic 4 Speed | 2 Wheel Rear | Driver Airbag Psngr Belts

Purchase Price

Car Item	Dealer Cost	List
Base Price	**$35,520**	**$44,400**
Anti-Lock Brakes	Std	Std
Manual Transmission	N/A	N/A
Optional Engine	N/A	N/A
Auto Climate Control	Std	Std
Power Steering	Std	Std
Cruise Control	Std	Std
All Wheel Drive	N/A	N/A
AM/FM Stereo Cassette	Std	Std
Steering Wheel, Scope	Std	Std
Power Windows	Std	Std
***Options Price**	**$0**	**$0**
***Total Price**	**$35,520**	**$44,400**
Target Price	$39,605	
Destination Charge	$450	
Avg. Tax & Fees	$2,452	
Luxury/Gas Guzzler Tax	$2,006	
Total Target $	**$44,513**	

The 1993 Q45 is available as a four-door sedan. New for 1993, the Q45 has five new colors to choose from (Blue Lapis, Crimson Pearl, Grey Anthracite, Silver Crystal and Silver-Blue Ice). A new monochromatic interior treatment is available in a choice of grey, beige, white or black. Interior refinements include color-keyed air conditioning switches and steering wheel and instrument panel facing. Other features include automatic temperature control system and leather interior.

Ownership Costs

Cost Area	5 Year Cost	Rate
Depreciation	$16,874	O
Financing ($895/month)	$7,333	
Insurance (Rating 20+)	$11,746	O
State Fees	$1,795	
Fuel (Hwy 22 City 17 -Prem.)	$4,892	●
Maintenance	$3,783	O
Repairs	$660	O

Warranty/Maintenance Info

Major Tune-Up	$180	O
Minor Tune-Up	$110	O
Brake Service	$218	O
Overall Warranty	4 yr/60k	O
Drivetrain Warranty	6 yr/70k	O
Rust Warranty	7 yr/unlim. mi	O
Maintenance Warranty	N/A	
Roadside Assistance	4 yr/unlim. mi	

Ownership Cost By Year

1993, 1994, 1995, 1996, 1997

Resale Value

1993	1994	1995	1996	1997
$38,864	$36,535	$34,380	$30,942	$27,639

Cumulative Costs

	1993	1994	1995	1996	1997
Annual	$12,254	$8,354	$8,757	$7,818	$9,900
Total	$12,254	$20,608	$29,365	$37,183	$47,083

Ownership Costs (5yr)

Average	This Car
$52,664	$47,083
Cost/Mile 75¢	Cost/Mile 67¢

Ownership Cost Rating
O Excellent

Jaguar XJ6
4 Door Sedan

Luxury

4.0L 223 hp Gas Fuel Inject. | 6 Cylinder In-Line | Automatic 4 Speed | 2 Wheel Rear | Driver Airbag Psngr Belts

Purchase Price

Car Item	Dealer Cost	List
Base Price	**$40,596**	**$49,750**
Anti-Lock Brakes	Std	Std
Manual Transmission	N/A	N/A
Optional Engine	N/A	N/A
Auto Climate Control	Std	Std
Power Steering	Std	Std
Cruise Control	Std	Std
All Wheel Drive	N/A	N/A
AM/FM Stereo Cassette	Std	Std
Steering Wheel, Scope	Std	Std
Power Windows	Std	Std
***Options Price**	**$0**	**$0**
***Total Price**	**$40,596**	**$49,750**
Target Price	$45,873	
Destination Charge	$580	
Avg. Tax & Fees	$2,826	
Luxury Tax	$1,645	
Total Target $	**$50,924**	

The 1993 Jaguar XJ6 is available in two models - (Base) XJ6 four-door sedan and Vanden Plas four-door sedan. New for 1993, the Base sedan has many new features. Mechanical features include CFC-free air conditioning with recirculation, integrated fog lamps and new aerodynamic front spoiler, recalibrated transmission (designed for smoother shifting), improved cooling with twin electric fans and increased power assistance to steering at parking speed. An eight speaker stereo system is also standard.

Ownership Costs

Cost Area	5 Year Cost	Rate
Depreciation	$25,033	O
Financing ($1,023/month)	$8,388	
Insurance (Rating 21)	$10,706	O
State Fees	$2,013	
Fuel (Hwy 24 City 17)	$4,232	O
Maintenance	$6,809	●
Repairs	$1,505	●

Warranty/Maintenance Info

Major Tune-Up	$341	●
Minor Tune-Up	$155	◐
Brake Service	$262	O
Overall Warranty	4 yr/50k	O
Drivetrain Warranty	4 yr/50k	O
Rust Warranty	6 yr/unlim. mi	O
Maintenance Warranty	N/A	
Roadside Assistance	4 yr/50k	

Resale Value

1993	1994	1995	1996	1997
$41,333	$37,196	$33,183	$29,271	$25,891

Cumulative Costs

	1993	1994	1995	1996	1997
Annual	$16,403	$10,385	$11,362	$8,650	$11,886
Total	$16,403	$26,788	$38,150	$46,800	$58,686

Ownership Costs (5yr)

Average	This Car
$56,669	$58,686
Cost/Mile 81¢	Cost/Mile 84¢

Ownership Cost Rating
O Average

---

* Includes shaded options
** Other purchase requirements apply

● Poor | ◐ Worse Than Average | ◉ Average | O Better Than Average | ○ Excellent | ⊖ Insufficient Information

©1993 by IntelliChoice, Inc. (408) 554-8711 All Rights Reserved. Reproduction Prohibited.
Refer to *Section 3: Annotated Vehicle Charts* for an explanation of these charts.

Jaguar XJ6 Vanden Plas
4 Door Sedan

Luxury

 4.0L 223 hp Gas Fuel Inject.
 6 Cylinder In-Line
 Automatic 4 Speed
 2 Wheel Rear
Driver Airbag Psngr Belts

Purchase Price

Car Item	Dealer Cost	List
Base Price	**$46,308**	**$56,750**
Anti-Lock Brakes	Std	Std
Manual Transmission	N/A	N/A
Optional Engine	N/A	N/A
Auto Climate Control	Std	Std
Power Steering	Std	Std
Cruise Control	Std	Std
All Wheel Drive	N/A	N/A
AM/FM Stereo Cassette	Std	Std
Steering Wheel, Scope	Std	Std
Power Windows	Std	Std
*Options Price	$0	$0
***Total Price**	**$46,308**	**$56,750**
Target Price	$52,328	
Destination Charge	$580	
Avg. Tax & Fees	$3,218	
Luxury Tax	$2,291	
Total Target $	**$58,417**	

The 1993 XJ6 is available in two models - Base XJ6 four-door sedan and Vanden Plas four-door sedan. New for 1993, the Vanden Plas sedan includes integrated fog lamps and new aerodynamic front spoiler, recalibrated transmission (designed for smoother shifting), remote locking and alarm system, remote entry system, eight-speaker stereo system and new sheepskin footwell rugs. Safety features include driver's side airbag and first gear-inhibit feature to aid traction.

Ownership Costs

Cost Area	5 Year Cost	Rate
Depreciation	$31,408	O
Financing ($1,174/month)	$9,623	
Insurance (Rating 22)	$11,167	O
State Fees	$2,293	
Fuel (Hwy 24 City 17)	$4,232	O
Maintenance	$6,809	●
Repairs	$1,505	●

Warranty/Maintenance Info

Major Tune-Up	$341	●
Minor Tune-Up	$155	◉
Brake Service	$262	O
Overall Warranty	4 yr/50k	O
Drivetrain Warranty	4 yr/50k	O
Rust Warranty	6 yr/unlim. mi	O
Maintenance Warranty	N/A	
Roadside Assistance	4 yr/50k	

Ownership Cost By Year

Legend: 1993, 1994, 1995, 1996, 1997

Resale Value

1993	1994	1995	1996	1997
$43,690	$39,282	$34,931	$30,711	$27,009

Cumulative Costs

	1993	1994	1995	1996	1997
Annual	$22,208	$11,193	$12,093	$9,193	$12,350
Total	$22,208	$33,401	$45,494	$54,687	$67,037

Ownership Costs (5yr)

Average	This Car
$62,474	$67,037
Cost/Mile 89¢	Cost/Mile 96¢

Ownership Cost Rating

◉ Worse Than Average

Jaguar XJS
2 Door Coupe

Luxury

 4.0L 219 hp Gas Fuel Inject.
 6 Cylinder In-Line
 Automatic 4 Speed
 2 Wheel Rear
 Driver Airbag Psngr Belts

Purchase Price

Car Item	Dealer Cost	List
Base Price	**$41,611**	**$49,750**
Anti-Lock Brakes	Std	Std
Manual 5 Speed	N/C	N/C
Optional Engine	N/A	N/A
Auto Climate Control	Std	Std
Power Steering	Std	Std
Cruise Control	Std	Std
All Wheel Drive	N/A	N/A
AM/FM Stereo Cassette	Std	Std
Steering Wheel, Tilt	Std	Std
Power Windows	Std	Std
*Options Price	$0	$0
***Total Price**	**$41,611**	**$49,750**
Target Price	$47,020	
Destination Charge	$580	
Avg. Tax & Fees	$2,883	
Luxury Tax	$1,760	
Total Target $	**$52,243**	

The 1993 XJS is available in two bodystyles - coupe and convertible. New for 1993, the coupe features several new mechanical, comfort and convenience features. Mechanical features include a 4-speed electronic ZF automatic transmission which offers both sport and normal modes and a redesigned steering column positioned further from the driver. Other improved features include a theft deterrent vehicle alarm, radio frequency remote entry system and the seat slides are redesigned for increased legroom.

Ownership Costs

Cost Area	5 Year Cost	Rate
Depreciation	$29,856	O
Financing ($1,050/month)	$8,606	
Insurance (Rating 23 Sport+)	$15,492	◉
State Fees	$2,013	
Fuel (Hwy 23 City 17)	$4,328	O
Maintenance	$7,169	●
Repairs	$1,505	●

Warranty/Maintenance Info

Major Tune-Up	$445	●
Minor Tune-Up	$235	●
Brake Service	$234	O
Overall Warranty	4 yr/50k	O
Drivetrain Warranty	4 yr/50k	O
Rust Warranty	6 yr/unlim. mi	O
Maintenance Warranty	N/A	
Roadside Assistance	4 yr/50k	

Resale Value

1993	1994	1995	1996	1997
$36,719	$32,783	$29,072	$25,565	$22,387

Cumulative Costs

	1993	1994	1995	1996	1997
Annual	$23,328	$11,272	$12,153	$9,362	$12,854
Total	$23,328	$34,600	$46,753	$56,115	$68,969

Ownership Costs (5yr)

Average	This Car
$56,669	$68,969
Cost/Mile 81¢	Cost/Mile 99¢

Ownership Cost Rating

● Poor

---

* Includes shaded options
** Other purchase requirements apply

● Poor   ◉ Worse Than Average   O Average   ◯ Better Than Average   ◯ Excellent   ⊖ Insufficient Information

©1993 by IntelliChoice, Inc. (408) 554-8711   All Rights Reserved. Reproduction Prohibited.
Refer to *Section 3: Annotated Vehicle Charts* for an explanation of these charts.

Jaguar XJS
2 Door Convertible

Luxury

 4.0L 219 hp Gas Fuel Inject.
 6 Cylinder In-Line
 Automatic 4 Speed
 2 Wheel Rear
 Driver Airbag Psngr Belts

Purchase Price

Car Item	Dealer Cost	List
Base Price	**$47,466**	**$56,750**
Anti-Lock Brakes	Std	Std
Manual 5 Speed	N/C	N/C
Optional Engine	N/A	N/A
Auto Climate Control	Std	Std
Power Steering	Std	Std
Cruise Control	Std	Std
All Wheel Drive	N/A	N/A
AM/FM Stereo Cassette	Std	Std
Steering Wheel, Tilt	Std	Std
Power Windows	Std	Std
*Options Price	$0	$0
*Total Price	$47,466	$56,750
Target Price		$53,637
Destination Charge		$580
Avg. Tax & Fees		$3,284
Luxury Tax		$2,422
Total Target $		**$59,923**

The 1993 XJS is available in two bodystyles - coupe and convertible. New for 1993, the convertible features several new features including a 4-speed electronic ZF automatic transmission which offers both sport and normal modes, gearshift interlock with manual release, stainless steel front and rear crossbeams to increase body strength, a theft deterrent vehicle alarm, an improved seat belt design and a steering column that has been positioned 2-inches further from the driver, designed for more comfort.

Ownership Costs

Cost Area	5 Year Cost	Rate
Depreciation	$27,673	○
Financing ($1,204/month)	$9,871	
Insurance (Rating 24 Sport+)	$16,386	●
State Fees	$2,293	
Fuel (Hwy 23 City 17)	$4,328	○
Maintenance	$7,169	●
Repairs	$1,505	●

Warranty/Maintenance Info

Major Tune-Up	$445	●
Minor Tune-Up	$235	●
Brake Service	$234	○
Overall Warranty	4 yr/50k	○
Drivetrain Warranty	4 yr/50k	○
Rust Warranty	6 yr/unlim. mi	○
Maintenance Warranty	N/A	
Roadside Assistance	4 yr/50k	

Ownership Cost By Year

(1993, 1994, 1995, 1996, 1997)

Resale Value

1993	1994	1995	1996	1997
$47,134	$43,658	$39,946	$35,956	$32,250

Cumulative Costs

	1993	1994	1995	1996	1997
Annual	$21,355	$11,441	$12,639	$10,172	$13,617
Total	$21,355	$32,796	$45,435	$55,607	$69,224

Ownership Costs (5yr)

Average	This Car
$62,474	$69,224
Cost/Mile 89¢	Cost/Mile 99¢

Ownership Cost Rating

Worse Than Average

Lexus ES 300
4 Door Sedan

Luxury

 3.0L 185 hp Gas Fuel Inject.
 6 Cylinder "V"
 Manual 5 Speed
 2 Wheel Front
Driver Airbag Psngr Belts

Purchase Price

Car Item	Dealer Cost	List
Base Price	**$22,550**	**$27,500**
Anti-Lock Brakes	Std	Std
Automatic 4 Speed	$738	$900
Optional Engine	N/A	N/A
Auto Climate Control	Std	Std
Power Steering	Std	Std
Cruise Control	Std	Std
All Wheel Drive	N/A	N/A
AM/FM Stereo Cassette	Std	Std
Steering Wheel, Tilt	Std	Std
Power Windows	Std	Std
*Options Price	$738	$900
*Total Price	$23,288	$28,400
Target Price		$26,233
Destination Charge		$400
Avg. Tax & Fees		$1,620
Total Target $		**$28,253**
Average Dealer Option Cost:	79%	

The 1993 Lexus is available in four models - ES 300 and LS400 sedans, and SC 300 and SC 400 coupes. New for 1993, the ES 300 features an automatic locking retractor function which is incorporated into the front passenger and rear seat belts. This system assists in the use of installing child seats by eliminating the need for locking clips. The ES 300 features tilt wheel, an electroluminescent instrument panel and a remote entry system. The ES 300 is exempt from the federal gas-guzzler tax.

Ownership Costs

Cost Area	5 Year Cost	Rate
Depreciation	$9,429	○
Financing ($568/month)	$4,655	
Insurance (Rating 17)	$8,719	○
State Fees	$1,152	
Fuel (Hwy 23 City 18 -Prem.)	$4,652	○
Maintenance	$5,682	○
Repairs	$631	

Warranty/Maintenance Info

Major Tune-Up	$214	○
Minor Tune-Up	$121	○
Brake Service	$195	
Overall Warranty	4 yr/50k	○
Drivetrain Warranty	6 yr/70k	○
Rust Warranty	6 yr/unlim. mi	○
Maintenance Warranty	N/A	
Roadside Assistance	4 yr/unlim. mi	

Resale Value

1993	1994	1995	1996	1997
$26,452	$24,527	$22,717	$20,690	$18,824

Cumulative Costs

	1993	1994	1995	1996	1997
Annual	$6,547	$6,352	$7,204	$5,723	$9,094
Total	$6,547	$12,899	$20,103	$25,826	$34,920

Ownership Costs (5yr)

Average	This Car
$40,687	$34,920
Cost/Mile 58¢	Cost/Mile 50¢

Ownership Cost Rating

Excellent

---

* Includes shaded options
** Other purchase requirements apply

● Poor ◐ Worse Than Average ◯ Average ○ Better Than Average ○ Excellent ⊖ Insufficient Information

©1993 by IntelliChoice, Inc. (408) 554-8711 All Rights Reserved. Reproduction Prohibited.
Refer to *Section 3: Annotated Vehicle Charts* for an explanation of these charts.

Lexus LS 400
4 Door Sedan
Luxury

- 4.0L 250 hp Gas Fuel Inject.
- 8 Cylinder "V"
- Automatic 4 Speed
- 2 Wheel Rear
- Driver/Psngr Airbags Std

Purchase Price

Car Item	Dealer Cost	List
Base Price	**$37,280**	**$46,600**
Anti-Lock Brakes	Std	Std
Manual Transmission	N/A	N/A
Optional Engine	N/A	N/A
Auto Climate Control	Std	Std
Power Steering	Std	Std
Cruise Control	Std	Std
All Wheel Drive	N/A	N/A
AM/FM Stereo Cassette	Std	Std
Steering Wheel, Scope	Std	Std
Power Windows	Std	Std
***Options Price**	**$0**	**$0**
***Total Price**	**$37,280**	**$46,600**
Target Price		$42,686
Destination Charge		$400
Avg. Tax & Fees		$2,624
Luxury Tax		$1,309
Total Target $		**$47,019**

The 1993 Lexus is available in four models - ES 300 and LS400 sedans, and SC 300 and SC 400 coupes. New for 1993, the LS 400 adds a passenger side airbag and both front seat belts feature new pretentioners. Also new for 1993 are larger wheels, brakes, and tires. The available air-suspension system fine-tunes ride constantly. Other standard features include power seats and locks, a remote entry system and a tilt/telescopic steering wheel. The LS 400 is exempt from the federal gas guzzler tax.

Ownership Costs

Cost Area	5 Year Cost	Rate
Depreciation	$16,724	○
Financing ($945/month)	$7,746	
Insurance (Rating 19+)	$11,230	○
State Fees	$1,880	
Fuel (Hwy 23 City 18 -Prem.)	$4,652	○
Maintenance	$5,552	○
Repairs	$660	○

Warranty/Maintenance Info

Major Tune-Up	$270	●
Minor Tune-Up	$172	●
Brake Service	$221	○
Overall Warranty	4 yr/50k	○
Drivetrain Warranty	6 yr/70k	○
Rust Warranty	6 yr/unlim. mi	○
Maintenance Warranty	N/A	
Roadside Assistance	4 yr/unlim. mi	

Ownership Cost By Year
(1993–1997)

Resale Value

1993	1994	1995	1996	1997
$43,534	$41,860	$39,406	$34,561	$30,295

Cumulative Costs

	1993	1994	1995	1996	1997
Annual	$10,169	$7,734	$9,081	$9,882	$11,578
Total	$10,169	$17,903	$26,984	$36,866	$48,444

Ownership Costs (5yr)

Average	This Car
$54,311	$48,444
Cost/Mile 78¢	Cost/Mile 69¢

Ownership Cost Rating
Excellent

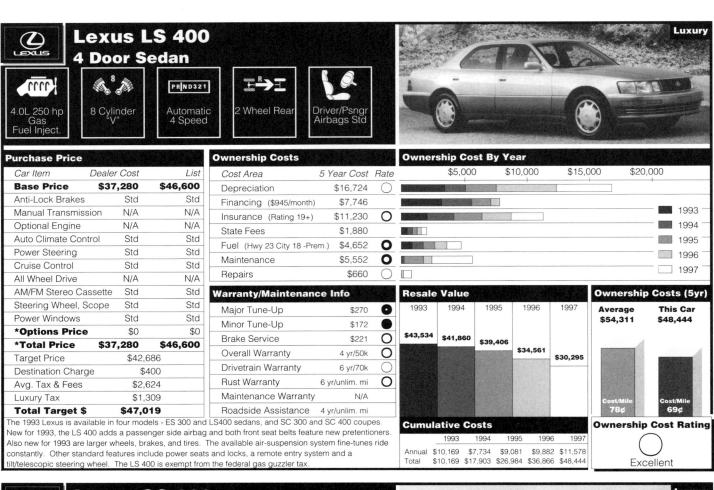

Lexus SC 300
2 Door Coupe
Luxury

- 3.0L 225 hp Gas Fuel Inject.
- 6 Cylinder In-Line
- Manual 5 Speed
- 2 Wheel Rear
- Driver/Psngr Airbags Std

Purchase Price

Car Item	Dealer Cost	List
Base Price	**$28,454**	**$34,700**
Anti-Lock Brakes	Std	Std
Automatic 4 Speed	$738	$900
Optional Engine	N/A	N/A
Auto Climate Control	Std	Std
Power Steering	Std	Std
Cruise Control	Std	Std
All Wheel Drive	N/A	N/A
AM/FM Stereo Cassette	Std	Std
Steering Wheel, Scope	Std	Std
Power Windows	Std	Std
***Options Price**	**$738**	**$900**
***Total Price**	**$29,192**	**$35,600**
Target Price		$33,387
Destination Charge		$400
Avg. Tax & Fees		$2,049
Luxury Tax		$379
Total Target $		**$36,215**

The 1993 Lexus is available in four models - ES 300 and LS400 sedans, and SC 300 and SC 400 coupes. New for 1993, the SC 300 features a standard passenger-side airbag and a locking retractor system for the rear seats. Also new is an automatic headlight control system that turns the headlights on and off as needed. The system uses a sensor behind the driver's side windshield to sense ambient light to determine when to activate the headlights. Also standard are aluminum wheels and fog lights.

Ownership Costs

Cost Area	5 Year Cost	Rate
Depreciation		⊖
Financing ($728/month)	$5,966	
Insurance (Rating 20+)	$11,746	●
State Fees	$1,440	
Fuel (Hwy 23 City 18 -Prem.)	$4,652	○
Maintenance	$5,713	○
Repairs	$631	○

Warranty/Maintenance Info

Major Tune-Up	$291	●
Minor Tune-Up	$188	●
Brake Service	$195	○
Overall Warranty	4 yr/50k	○
Drivetrain Warranty	6 yr/70k	○
Rust Warranty	6 yr/unlim. mi	○
Maintenance Warranty	N/A	
Roadside Assistance	4 yr/unlim. mi	

Ownership Cost By Year
Insufficient Depreciation Information

Resale Value
Insufficient Information

Cumulative Costs

	1993	1994	1995	1996	1997
Annual		Insufficient Information			
Total		Insufficient Information			

Ownership Costs (5yr)
Insufficient Information

Ownership Cost Rating
⊖ Insufficient Information

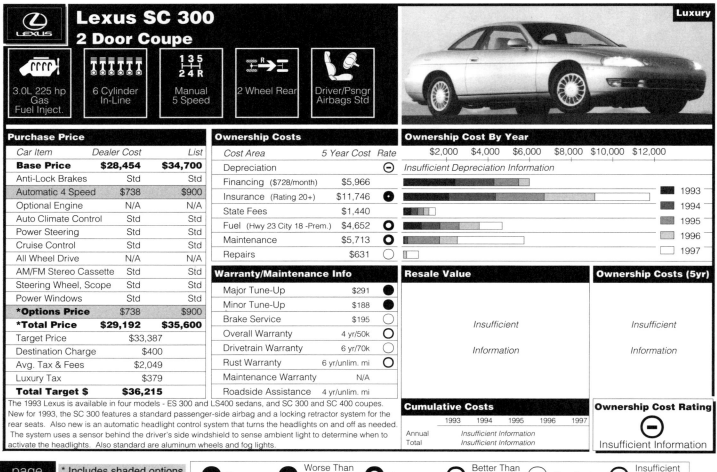

---

* Includes shaded options
** Other purchase requirements apply

● Poor ◐ Worse Than Average ○ Average ○ Better Than Average ○ Excellent ⊖ Insufficient Information

©1993 by IntelliChoice, Inc. (408) 554-8711 All Rights Reserved. Reproduction Prohibited.
Refer to *Section 3: Annotated Vehicle Charts* for an explanation of these charts.

Lexus SC 400
2 Door Coupe

Luxury

 4.0L 250 hp Gas Fuel Inject.
 8 Cylinder "V"
 Automatic 4 Speed
 2 Wheel Rear
 Driver/Psngr Airbags Std

Purchase Price

Car Item	Dealer Cost	List
Base Price	**$33,120**	**$41,400**
Anti-Lock Brakes	Std	Std
Manual Transmission	N/A	N/A
Optional Engine	N/A	N/A
Auto Climate Control	Std	Std
Power Steering	Std	Std
Cruise Control	Std	Std
All Wheel Drive	N/A	N/A
AM/FM Stereo Cassette	Std	Std
Steering Wheel, Tilt	Std	Std
Power Windows	Std	Std
***Options Price**	$0	$0
***Total Price**	**$33,120**	**$41,400**
Target Price	$37,922	
Destination Charge	$400	
Avg. Tax & Fees	$2,334	
Luxury Tax	$832	
Total Target $	**$41,488**	

Ownership Costs

Cost Area	5 Year Cost	Rate
Depreciation		⊖
Financing ($834/month)	$6,834	
Insurance (Rating 22+)	$12,768	●
State Fees	$1,672	
Fuel (Hwy 23 City 18 -Prem.)	$4,652	○
Maintenance	$6,360	●
Repairs	$631	○

Warranty/Maintenance Info

Major Tune-Up	$236	○
Minor Tune-Up	$127	○
Brake Service	$195	○
Overall Warranty	4 yr/50k	○
Drivetrain Warranty	6 yr/70k	○
Rust Warranty	6 yr/unlim. mi	○
Maintenance Warranty	N/A	
Roadside Assistance	4 yr/unlim. mi	

Ownership Cost By Year

Resale Value

Insufficient Information

Ownership Costs (5yr)

Insufficient Information

Cumulative Costs

	1993	1994	1995	1996	1997
Annual	*Insufficient Information*				
Total	*Insufficient Information*				

Ownership Cost Rating

⊖

Insufficient Information

The 1993 Lexus is available in four models - ES 300 and LS400 sedans, and SC 300 and SC 400 coupes. New for 1993, the SC 400 features a passenger-side airbag and seat belts with automatic locking retraction. An automatic headlight control system has been added which gauges ambient light intensity and operates the headlights accordingly. Also new for 1993 is graphite metallic paint. The SC 400 is faster off the line than the LS 400 due to a lower gearing of the differential and the transmission.

Lincoln Continental Executive
4 Door Sedan

Luxury

 3.8L 160 hp Gas Fuel Inject.
 6 Cylinder "V"
 Automatic 4 Speed
 2 Wheel Front
 Driver/Psngr Airbags Std

Purchase Price

Car Item	Dealer Cost	List
Base Price	**$28,256**	**$33,328**
Anti-Lock Brakes	Std	Std
Manual Transmission	N/A	N/A
Optional Engine	N/A	N/A
Auto Climate Control	Std	Std
Power Steering	Std	Std
Cruise Control	Std	Std
All Wheel Drive	N/A	N/A
AM/FM Stereo Cassette	Std	Std
Steering Wheel, Tilt	Std	Std
Power Windows	Std	Std
***Options Price**	$0	$0
***Total Price**	**$28,256**	**$33,328**
Target Price	$32,022	
Destination Charge	$590	
Avg. Tax & Fees	$1,970	
Luxury Tax	$261	
Total Target $	**$34,843**	

Ownership Costs

Cost Area	5 Year Cost	Rate
Depreciation	$23,089	●
Financing ($700/month)	$5,740	
Insurance (Rating 12)	$7,898	○
State Fees	$1,356	
Fuel (Hwy 26 City 17)	$4,064	○
Maintenance	$6,269	●
Repairs	$691	○

Warranty/Maintenance Info

Major Tune-Up	$162	○
Minor Tune-Up	$112	○
Brake Service	$277	●
Overall Warranty	4 yr/50k	○
Drivetrain Warranty	4 yr/50k	○
Rust Warranty	6 yr/100k	○
Maintenance Warranty	N/A	
Roadside Assistance	4 yr/unlim. mi	

Ownership Cost By Year

Resale Value

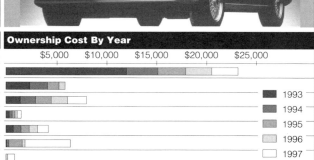

1993: $22,763; 1994: $19,604; 1995: $16,822; 1996: $14,210; 1997: $11,754

Ownership Costs (5yr)

Average	This Car
$44,376	$49,107
Cost/Mile 63¢	Cost/Mile 70¢

Cumulative Costs

	1993	1994	1995	1996	1997
Annual	$17,067	$7,799	$7,934	$6,209	$10,098
Total	$17,067	$24,866	$32,800	$39,009	$49,107

Ownership Cost Rating

●

Worse Than Average

The Lincoln Continental sedan is available in Executive Series and Signature Series editions. New for 1993, the Continental features additional standard items such as a remote keyless entry system, leather-wrapped steering wheel and Comfort Convenience Group. The Executive Series sedan features a standard 6-way power driver seat, styled aluminum wheels and leather seating material. Optional items include a power sun roof, cellular telephone, anti-theft alarm system and compact disc player.

---

* Includes shaded options
** Other purchase requirements apply

 Poor   Worse Than Average   Average   Better Than Average   Excellent   Insufficient Information

©1993 by IntelliChoice, Inc. (408) 554-8711   All Rights Reserved. Reproduction Prohibited.
Refer to *Section 3: Annotated Vehicle Charts* for an explanation of these charts.

Need More Facts? - All the Facts?

When you've decided on a model line, but need to know
— all feature and option pricing for each model in the line
— how the models in the line differ

Order:

Just the Facts™

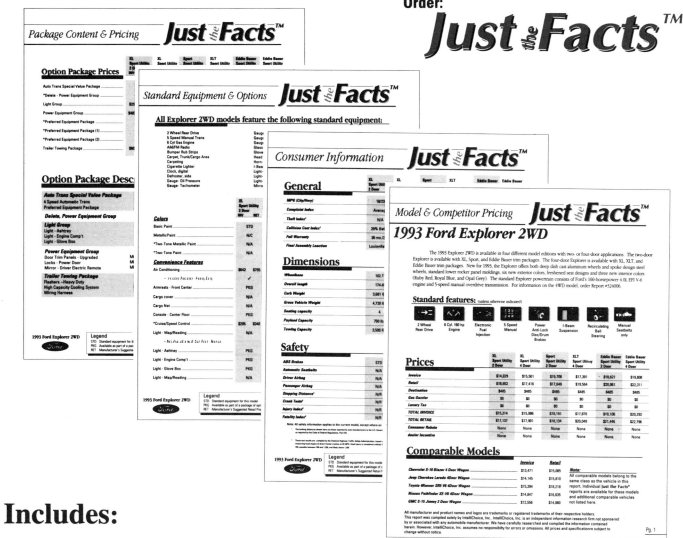

Includes:

■ Latest dealer and consumer rebates

■ Information provided for each car in the line, not just for <u>one</u> specific model!!

■ Complete pricing and feature information, selected safety and specification information

■ Competitor model prices also included

■ Latest information - updated daily

■ Easy to read and understand

■ Shipped 1st class within 24 hours, or faxed immediately

To order your copy of the *Just the Facts*™ report

call 1-800-227-2665 ext 1, or complete the request form in back of the book.

Just the Facts™ is a trademark of IntelliChoice, Inc.

✓ Up to 6 vehicles
All information in the report is provided for the model line (up to 6 vehicles), not just on a single vehicle.

✓ Complete Price Breakout
Dealer Cost (factory invoice), retail price, fees, gas and/or luxury car tax clearly shown for each car.

✓ Rebates, Incentives, Discount Financing
The latest factory-to-consumer and factory-to-dealer incentive programs.

✓ Competitor Model Pricing
Provided for up to 6 competitor vehicles. For instance, a Ford Taurus report also shows pricing for Buick Century Custom, Honda Accord LX, Mercury Sable GS, Chevrolet Euro Lumina, and 2 other models.

✓ All Available Equipment
Standard equipment is shown, plus availability, order code, dealer cost, and list price for all factory installed, and some dealer installed, optional equipment. Option package content is shown, as well as dealer cost and list price.

✓ Safety Information
Availability of anti-lock brakes, driver airbag, passenger airbag, and automatic seatbelts, plus U.S. Government and insurance industry results for crash tests, stopping distance, and injury ratings.

✓ Plus
U.S. Government complaint file, industry theft rating, collision cost rating, interior and exterior dimensions, warranty information, and where each vehicle is built.

✓ Designed to be Easy and Useful
Every element of the report has been designed to be simple to use, clear, understandable, and friendly.

Delivery Choices

Shipping	Price		Delivery
1st class mail	$14.95		Included
Instant Fax	$14.95	+	$6.00
Next-Day Air	$14.95	+	$12.50
2nd-Day Air	$14.95	+	$7.50

To order call: 1-800-227-2655

$14.⁹⁵

(VISA, Mastercard, Personal Check)

IntelliChoice Inc.

Before you've made your choice, compare with

The ArmChair Compare® Report

Compare any **two** vehicles *Side-by-Side* and see how they really **match up**.

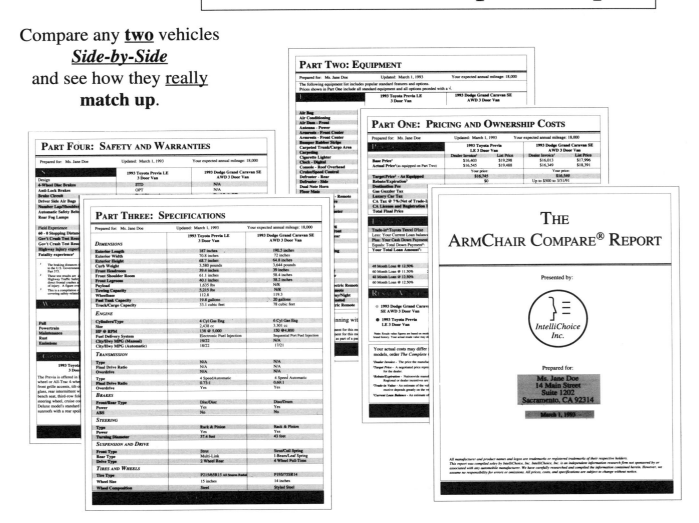

Your personalized automobile comparison report:

- ■ Compares two vehicles side-by-side
- ■ Provides full pricing, including dealer cost and Suggested List price, plus rebate information
- ■ Shows estimated monthly payments and projected five-year resale value
- ■ Provides customized ownership costs based on your driving profile
- ■ Includes selected specifications, safety, and warranty information

To order your copy of THE ARMCHAIR COMPARE® REPORT *call 1-800-227-2665 ext. 1, or complete the request form in back of the book.*

The ArmChair Compare® Report is a registered trademark of IntelliChoice, Inc.

The ArmChair Compare® Report

Includes everything from the Just-the-Facts report shown on the previous pages, PLUS:

✓ Direct Comparison

This report shows any two competitive models of your choice in a side-by-side comparison. Use it to see just how the cars "stack up" in features, prices, safety, ownership costs, and specifications.

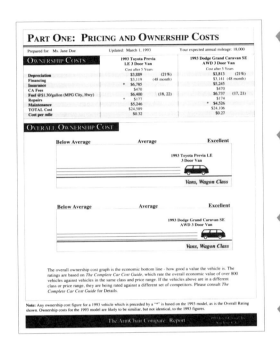

✓ Personalized Ownership Costs

The information in this report is customized to your state and how many miles per year you drive.

✓ Trade-In Value

The current value of your existing car.

✓ Financing Costs

Monthly payments based on latest rates, and your personal down payment, and your trade-in value.

✓ Expected Resale Value

The value of your new car in 5 years.

✓ Additional Dimensions and Specifications

See how the two vehicles compare on horsepower, brakes, turning diameter, and much more.

IntelliChoice Inc.

$19.⁹⁵

(VISA, Mastercard, Personal Check)

Delivery Choices

Shipping	Price	Delivery
1st class mail	$19.95	Included
Instant Fax	$19.95 +	$6.00
Next-Day Air	$19.95 +	$12.50
2nd-Day Air	$19.95 +	$7.50

To order call: 1-800-227-2655

Immediate Faxing Available

Lincoln Continental Signature
4 Door Sedan

Luxury

- 3.8L 160 hp Gas Fuel Inject.
- 6 Cylinder "V"
- Automatic 4 Speed
- 2 Wheel Front
- Driver/Psngr Airbags Std

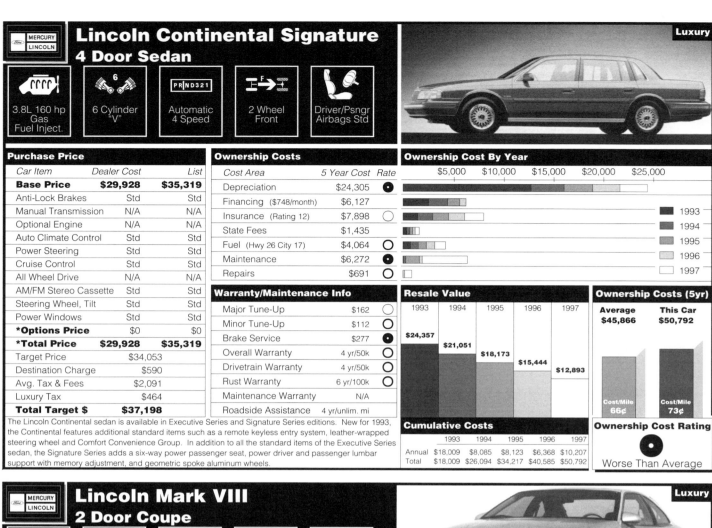

Purchase Price

Car Item	Dealer Cost	List
Base Price	**$29,928**	**$35,319**
Anti-Lock Brakes	Std	Std
Manual Transmission	N/A	N/A
Optional Engine	N/A	N/A
Auto Climate Control	Std	Std
Power Steering	Std	Std
Cruise Control	Std	Std
All Wheel Drive	N/A	N/A
AM/FM Stereo Cassette	Std	Std
Steering Wheel, Tilt	Std	Std
Power Windows	Std	Std
*Options Price	$0	$0
*Total Price	$29,928	$35,319
Target Price		$34,053
Destination Charge		$590
Avg. Tax & Fees		$2,091
Luxury Tax		$464
Total Target $		**$37,198**

The Lincoln Continental sedan is available in Executive Series and Signature Series editions. New for 1993, the Continental features additional standard items such as a remote keyless entry system, leather-wrapped steering wheel and Comfort Convenience Group. In addition to all the standard items of the Executive Series sedan, the Signature Series adds a six-way power passenger seat, power driver and passenger lumbar support with memory adjustment, and geometric spoke aluminum wheels.

Ownership Costs

Cost Area	5 Year Cost	Rate
Depreciation	$24,305	●
Financing ($748/month)	$6,127	
Insurance (Rating 12)	$7,898	○
State Fees	$1,435	
Fuel (Hwy 26 City 17)	$4,064	◐
Maintenance	$6,272	●
Repairs	$691	◐

Warranty/Maintenance Info

Major Tune-Up	$162	◐
Minor Tune-Up	$112	◐
Brake Service	$277	●
Overall Warranty	4 yr/50k	◐
Drivetrain Warranty	4 yr/50k	◐
Rust Warranty	6 yr/100k	◐
Maintenance Warranty	N/A	
Roadside Assistance	4 yr/unlim. mi	

Ownership Cost By Year
1993, 1994, 1995, 1996, 1997

Resale Value

1993	1994	1995	1996	1997
$24,357	$21,051	$18,173	$15,444	$12,893

Cumulative Costs

	1993	1994	1995	1996	1997
Annual	$18,009	$8,085	$8,123	$6,368	$10,207
Total	$18,009	$26,094	$34,217	$40,585	$50,792

Ownership Costs (5yr)
- Average: $45,866 (Cost/Mile 66¢)
- This Car: $50,792 (Cost/Mile 73¢)

Ownership Cost Rating
● Worse Than Average

Lincoln Mark VIII
2 Door Coupe

Luxury

- 4.6L 280 hp Gas Fuel Inject.
- 8 Cylinder "V"
- Automatic 4 Speed
- 2 Wheel Rear
- Driver/Psngr Airbags Std

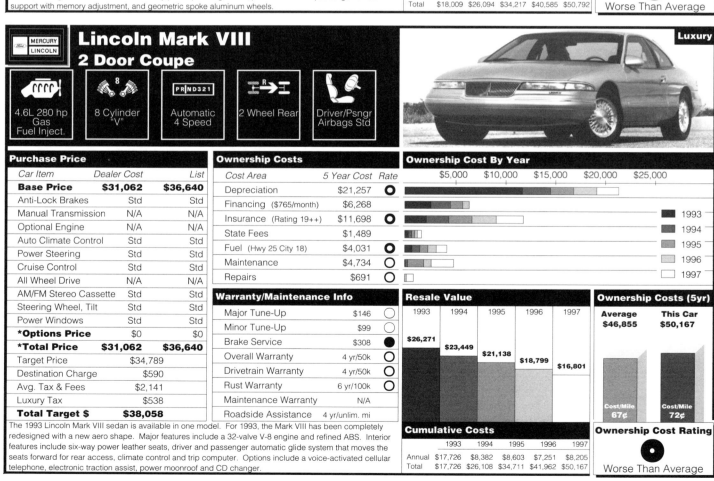

Purchase Price

Car Item	Dealer Cost	List
Base Price	**$31,062**	**$36,640**
Anti-Lock Brakes	Std	Std
Manual Transmission	N/A	N/A
Optional Engine	N/A	N/A
Auto Climate Control	Std	Std
Power Steering	Std	Std
Cruise Control	Std	Std
All Wheel Drive	N/A	N/A
AM/FM Stereo Cassette	Std	Std
Steering Wheel, Tilt	Std	Std
Power Windows	Std	Std
*Options Price	$0	$0
*Total Price	$31,062	$36,640
Target Price		$34,789
Destination Charge		$590
Avg. Tax & Fees		$2,141
Luxury Tax		$538
Total Target $		**$38,058**

The 1993 Lincoln Mark VIII sedan is available in one model. For 1993, the Mark VIII has been completely redesigned with a new aero shape. Major features include a 32-valve V-8 engine and refined ABS. Interior features include six-way power leather seats, driver and passenger automatic glide system that moves the seats forward for rear access, climate control and trip computer. Options include a voice-activated cellular telephone, electronic traction assist, power moonroof and CD changer.

Ownership Costs

Cost Area	5 Year Cost	Rate
Depreciation	$21,257	○
Financing ($765/month)	$6,268	
Insurance (Rating 19++)	$11,698	●
State Fees	$1,489	
Fuel (Hwy 25 City 18)	$4,031	○
Maintenance	$4,734	○
Repairs	$691	○

Warranty/Maintenance Info

Major Tune-Up	$146	○
Minor Tune-Up	$99	○
Brake Service	$308	●
Overall Warranty	4 yr/50k	○
Drivetrain Warranty	4 yr/50k	○
Rust Warranty	6 yr/100k	○
Maintenance Warranty	N/A	
Roadside Assistance	4 yr/unlim. mi	

Resale Value

1993	1994	1995	1996	1997
$26,271	$23,449	$21,138	$18,799	$16,801

Cumulative Costs

	1993	1994	1995	1996	1997
Annual	$17,726	$8,382	$8,603	$7,251	$8,205
Total	$17,726	$26,108	$34,711	$41,962	$50,167

Ownership Costs (5yr)
- Average: $46,855 (Cost/Mile 67¢)
- This Car: $50,167 (Cost/Mile 72¢)

Ownership Cost Rating
● Worse Than Average

---

* Includes shaded options
** Other purchase requirements apply

- ● Poor
- ◐ Worse Than Average
- ◑ Average
- ○ Better Than Average
- ○ Excellent
- ⊖ Insufficient Information

©1993 by IntelliChoice, Inc. (408) 554-8711 All Rights Reserved. Reproduction Prohibited.
Refer to *Section 3: Annotated Vehicle Charts* for an explanation of these charts.

Lincoln Town Car Executive
4 Door Sedan

Luxury

 4.6L 190 hp Gas Fuel Inject.
 8 Cylinder "V"
 Automatic 4 Speed
 2 Wheel Rear
Driver/Psngr Airbags Std

Purchase Price

Car Item	Dealer Cost	List
Base Price	**$28,990**	**$34,190**
Anti-Lock Brakes	Std	Std
Manual Transmission	N/A	N/A
Optional Engine	N/A	N/A
Auto Climate Control	Std	Std
Power Steering	Std	Std
Cruise Control	Std	Std
All Wheel Drive	N/A	N/A
AM/FM Stereo Cassette	Std	Std
Steering Wheel, Tilt	Std	Std
Power Windows	Std	Std
*Options Price	$0	$0
*Total Price	$28,990	$34,190
Target Price	$32,911	
Destination Charge	$590	
Avg. Tax & Fees	$2,023	
Luxury Tax	$350	
Total Target $	**$35,874**	

The Lincoln Town Car sedan is available in Executive, Signature, and Cartier Designer Series editions. New standard features for 1993 on all Town Cars include the Comfort Convenience Group, electronic instrument cluster, leather-wrapped steering wheel, express down driver side power window, and digital air conditioning controls. The Executive Series features color-keyed body side moldings, digital instrumentation and cloth seating material. The Town Car is a sister to the Mercury Grand Marquis.

Ownership Costs

Cost Area	5 Year Cost	Rate
Depreciation	$22,226	●
Financing ($721/month)	$5,909	
Insurance (Rating 13)	$8,061	○
State Fees	$1,391	
Fuel (Hwy 26 City 18)	$3,949	○
Maintenance	$4,291	○
Repairs	$691	○

Warranty/Maintenance Info

Major Tune-Up	$147	○
Minor Tune-Up	$99	○
Brake Service	$306	●
Overall Warranty	4 yr/50k	○
Drivetrain Warranty	4 yr/50k	○
Rust Warranty	6 yr/100k	○
Maintenance Warranty	N/A	
Roadside Assistance	4 yr/unlim. mi	

Ownership Cost By Year

1993, 1994, 1995, 1996, 1997

Resale Value

1993	1994	1995	1996	1997
$23,884	$20,995	$18,350	$15,951	$13,648

Cumulative Costs

	1993	1994	1995	1996	1997
Annual	$17,070	$7,605	$7,921	$6,367	$7,555
Total	$17,070	$24,675	$32,596	$38,963	$46,518

Ownership Costs (5yr)

Average	This Car
$45,021	$46,518
Cost/Mile 64¢	Cost/Mile 66¢

Ownership Cost Rating

○ Average

Lincoln Town Car Signature
4 Door Sedan

Luxury

 4.6L 190 hp Gas Fuel Inject.
 8 Cylinder "V"
 Automatic 4 Speed
 2 Wheel Rear
 Driver/Psngr Airbags Std

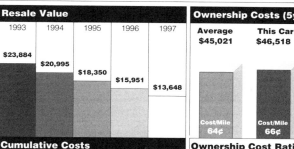

Executive Model Shown

Purchase Price

Car Item	Dealer Cost	List
Base Price	**$30,085**	**$35,494**
Anti-Lock Brakes	Std	Std
Manual Transmission	N/A	N/A
Optional Engine	N/A	N/A
Auto Climate Control	Std	Std
Power Steering	Std	Std
Cruise Control	Std	Std
All Wheel Drive	N/A	N/A
AM/FM Stereo Cassette	Std	Std
Steering Wheel, Tilt	Std	Std
Power Windows	Std	Std
*Options Price	$0	$0
*Total Price	$30,085	$35,494
Target Price	$34,146	
Destination Charge	$590	
Avg. Tax & Fees	$2,098	
Luxury Tax	$474	
Total Target $	**$37,308**	

The Lincoln Town Car sedan is available in Executive, Signature, and Cartier Designer Series editions. New standard features for 1993 on all Town Cars include the Comfort Convenience Group, electronic instrument cluster, leather-wrapped steering wheel, express down driver side power window, and digital air conditioning controls. The Signature sedan upgrades the Executive by adding front seatback map pockets, body side accent striping, color-coordinated body side moldings and dual shade paint.

Ownership Costs

Cost Area	5 Year Cost	Rate
Depreciation	$21,843	○
Financing ($750/month)	$6,145	
Insurance (Rating 13)	$8,061	○
State Fees	$1,445	
Fuel (Hwy 26 City 18)	$3,949	○
Maintenance	$4,291	○
Repairs	$691	○

Warranty/Maintenance Info

Major Tune-Up	$147	○
Minor Tune-Up	$99	○
Brake Service	$306	●
Overall Warranty	4 yr/50k	○
Drivetrain Warranty	4 yr/50k	○
Rust Warranty	6 yr/100k	○
Maintenance Warranty	N/A	
Roadside Assistance	4 yr/unlim. mi	

Resale Value

1993	1994	1995	1996	1997
$25,603	$22,863	$20,296	$17,845	$15,465

Cumulative Costs

	1993	1994	1995	1996	1997
Annual	$16,897	$7,542	$7,900	$6,446	$7,640
Total	$16,897	$24,439	$32,339	$38,785	$46,425

Ownership Costs (5yr)

Average	This Car
$45,997	$46,425
Cost/Mile 66¢	Cost/Mile 66¢

Ownership Cost Rating

○ Average

---

* Includes shaded options
** Other purchase requirements apply

● Poor   ◐ Worse Than Average   ○ Average   ◯ Better Than Average   ◯ Excellent   ⊖ Insufficient Information

©1993 by IntelliChoice, Inc. (408) 554-8711   All Rights Reserved. Reproduction Prohibited.
Refer to *Section 3: Annotated Vehicle Charts* for an explanation of these charts.

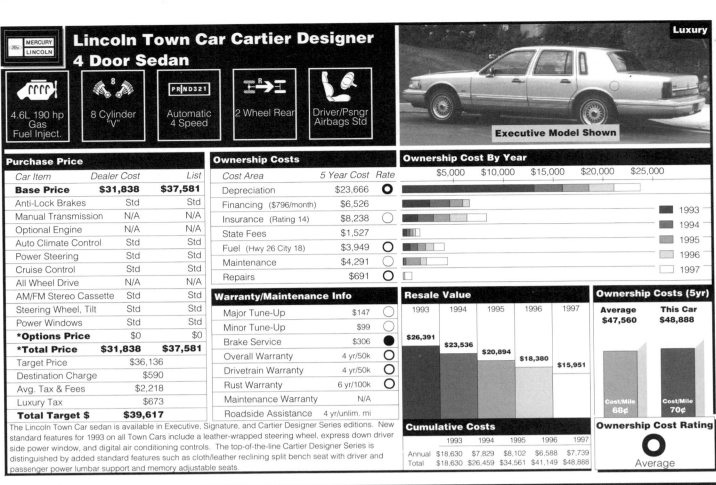
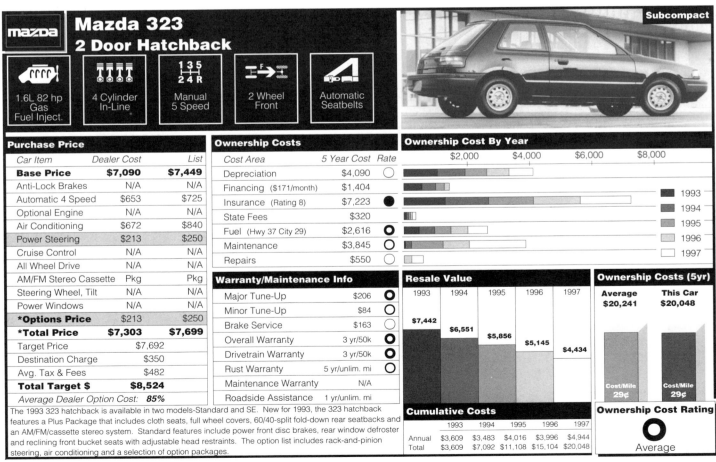

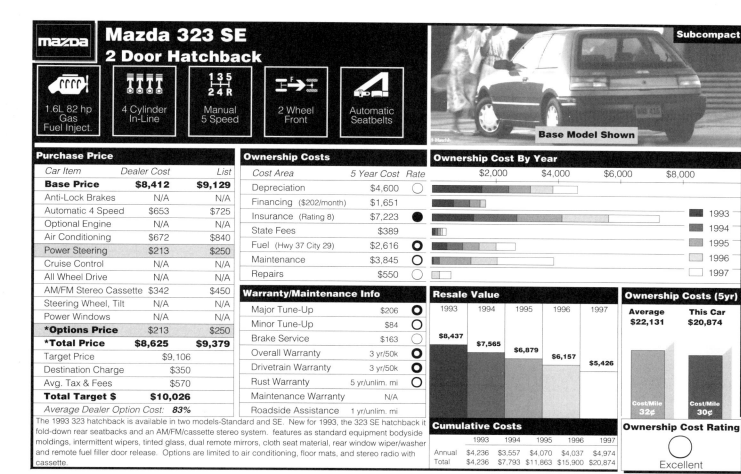

Mazda 323 SE
2 Door Hatchback
Subcompact — Base Model Shown

- 1.6L 82 hp Gas Fuel Inject.
- 4 Cylinder In-Line
- Manual 5 Speed
- 2 Wheel Front
- Automatic Seatbelts

Purchase Price

Car Item	Dealer Cost	List
Base Price	**$8,412**	**$9,129**
Anti-Lock Brakes	N/A	N/A
Automatic 4 Speed	$653	$725
Optional Engine	N/A	N/A
Air Conditioning	$672	$840
Power Steering	$213	$250
Cruise Control	N/A	N/A
All Wheel Drive	N/A	N/A
AM/FM Stereo Cassette	$342	$450
Steering Wheel, Tilt	N/A	N/A
Power Windows	N/A	N/A
*Options Price	$213	$250
*Total Price	$8,625	$9,379
Target Price		$9,106
Destination Charge		$350
Avg. Tax & Fees		$570
Total Target $		**$10,026**
Average Dealer Option Cost:		83%

Ownership Costs

Cost Area	5 Year Cost	Rate
Depreciation	$4,600	○
Financing ($202/month)	$1,651	
Insurance (Rating 8)	$7,223	●
State Fees	$389	
Fuel (Hwy 37 City 29)	$2,616	○
Maintenance	$3,845	○
Repairs	$550	○

Warranty/Maintenance Info

Major Tune-Up	$206	○
Minor Tune-Up	$84	○
Brake Service	$163	
Overall Warranty	3 yr/50k	○
Drivetrain Warranty	3 yr/50k	○
Rust Warranty	5 yr/unlim. mi	○
Maintenance Warranty	N/A	
Roadside Assistance	1 yr/unlim. mi	

Resale Value
- 1993: $8,437
- 1994: $7,565
- 1995: $6,879
- 1996: $6,157
- 1997: $5,426

Cumulative Costs

	1993	1994	1995	1996	1997
Annual	$4,236	$3,557	$4,070	$4,037	$4,974
Total	$4,236	$7,793	$11,863	$15,900	$20,874

Ownership Costs (5yr)
- Average: $22,131 (Cost/Mile 32¢)
- This Car: $20,874 (Cost/Mile 30¢)

Ownership Cost Rating
○ Excellent

The 1993 323 hatchback is available in two models—Standard and SE. New for 1993, the 323 SE hatchback features fold-down rear seatbacks and an AM/FM cassette stereo system. features as standard equipment bodyside moldings, intermittent wipers, tinted glass, dual remote mirrors, cloth seat material, rear window wiper/washer and remote fuel filler door release. Options are limited to air conditioning, floor mats, and stereo radio with cassette.

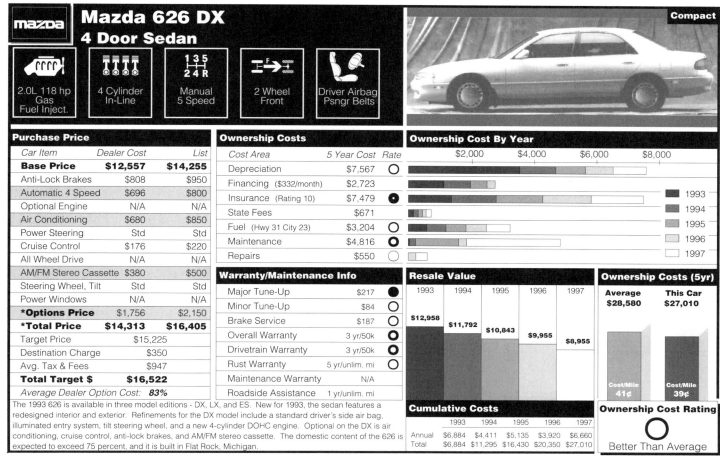

Mazda 626 DX
4 Door Sedan
Compact

- 2.0L 118 hp Gas Fuel Inject.
- 4 Cylinder In-Line
- Manual 5 Speed
- 2 Wheel Front
- Driver Airbag Psngr Belts

Purchase Price

Car Item	Dealer Cost	List
Base Price	**$12,557**	**$14,255**
Anti-Lock Brakes	$808	$950
Automatic 4 Speed	$696	$800
Optional Engine	N/A	N/A
Air Conditioning	$680	$850
Power Steering	Std	Std
Cruise Control	$176	$220
All Wheel Drive	N/A	N/A
AM/FM Stereo Cassette	$380	$500
Steering Wheel, Tilt	Std	Std
Power Windows	N/A	N/A
*Options Price	$1,756	$2,150
*Total Price	$14,313	$16,405
Target Price		$15,225
Destination Charge		$350
Avg. Tax & Fees		$947
Total Target $		**$16,522**
Average Dealer Option Cost:		83%

Ownership Costs

Cost Area	5 Year Cost	Rate
Depreciation	$7,567	○
Financing ($332/month)	$2,723	
Insurance (Rating 10)	$7,479	●
State Fees	$671	
Fuel (Hwy 31 City 23)	$3,204	○
Maintenance	$4,816	○
Repairs	$550	○

Warranty/Maintenance Info

Major Tune-Up	$217	●
Minor Tune-Up	$84	○
Brake Service	$187	○
Overall Warranty	3 yr/50k	○
Drivetrain Warranty	3 yr/50k	○
Rust Warranty	5 yr/unlim. mi	○
Maintenance Warranty	N/A	
Roadside Assistance	1 yr/unlim. mi	

Resale Value
- 1993: $12,958
- 1994: $11,792
- 1995: $10,843
- 1996: $9,955
- 1997: $8,955

Cumulative Costs

	1993	1994	1995	1996	1997
Annual	$6,884	$4,411	$5,135	$3,920	$6,660
Total	$6,884	$11,295	$16,430	$20,350	$27,010

Ownership Costs (5yr)
- Average: $28,580 (Cost/Mile 41¢)
- This Car: $27,010 (Cost/Mile 39¢)

Ownership Cost Rating
○ Better Than Average

The 1993 626 is available in three model editions – DX, LX, and ES. New for 1993, the sedan features a redesigned interior and exterior. Refinements for the DX model include a standard driver's side air bag, illuminated entry system, tilt steering wheel, and a new 4-cylinder DOHC engine. Optional on the DX is air conditioning, cruise control, anti-lock brakes, and AM/FM stereo cassette. The domestic content of the 626 is expected to exceed 75 percent, and it is built in Flat Rock, Michigan.

---

* Includes shaded options
** Other purchase requirements apply

● Poor ◉ Worse Than Average ○ Average ○ Better Than Average ○ Excellent ⊖ Insufficient Information

©1993 by IntelliChoice, Inc. (408) 554-8711 All Rights Reserved. Reproduction Prohibited.
Refer to *Section 3: Annotated Vehicle Charts* for an explanation of these charts.

Mazda 626 LX
4 Door Sedan

Compact

- 2.0L 118 hp Gas Fuel Inject.
- 4 Cylinder In-Line
- Manual 5 Speed
- 2 Wheel Front
- Driver Airbag Psngr Belts

Purchase Price

Car Item	Dealer Cost	List
Base Price	**$14,482**	**$16,440**
Anti-Lock Brakes	$808	$950
Automatic 4 Speed	$696	$800
Optional Engine	N/A	N/A
Air Conditioning	Std	Std
Power Steering	Std	Std
Cruise Control	Std	Std
All Wheel Drive	N/A	N/A
AM/FM Stereo Cassette	Std	Std
Steering Wheel, Tilt	Std	Std
Power Windows	Std	Std
***Options Price**	$696	$800
***Total Price**	**$15,178**	**$17,240**
Target Price		$16,201
Destination Charge		$350
Avg. Tax & Fees		$1,004
Total Target $		**$17,555**
Average Dealer Option Cost:		82%

The 1993 626 is available in three model editions - DX, LX, and ES. New for 1993, the sedan features a redesigned interior and exterior. Refinements for the LX model include standard power door locks, AM/FM stereo cassette, dual power mirrors, and air conditioning. Optional on the LX is a power moonroof, theft deterrent system and compact disc player. The domestic content of the 626 is expected to exceed 75 percent, and it is built in Flat Rock, Michigan.

Ownership Costs

Cost Area	5 Year Cost	Rate
Depreciation	$8,617	○
Financing ($353/month)	$2,892	
Insurance (Rating 10)	$7,479	◉
State Fees	$704	
Fuel (Hwy 31 City 23)	$3,204	○
Maintenance	$4,816	○
Repairs	$550	○

Warranty/Maintenance Info

Major Tune-Up	$217	●
Minor Tune-Up	$84	○
Brake Service	$187	○
Overall Warranty	3 yr/50k	○
Drivetrain Warranty	3 yr/50k	○
Rust Warranty	5 yr/unlim. mi	○
Maintenance Warranty	N/A	
Roadside Assistance	1 yr/unlim. mi	

Resale Value

1993	1994	1995	1996	1997
$12,928	$11,779	$10,834	$9,943	$8,938

Cumulative Costs

	1993	1994	1995	1996	1997
Annual	$8,026	$4,454	$5,171	$3,941	$6,670
Total	$8,026	$12,480	$17,651	$21,592	$28,262

Ownership Costs (5yr)

Average	This Car
$29,297	$28,262
Cost/Mile 42¢	Cost/Mile 40¢

Ownership Cost Rating
○ Better Than Average

Mazda 626 ES
4 Door Sedan

Compact

- 2.5L 164 hp Gas Fuel Inject.
- 6 Cylinder "V"
- Manual 5 Speed
- 2 Wheel Front
- Driver Airbag Psngr Belts

Purchase Price

Car Item	Dealer Cost	List
Base Price	**$16,305**	**$18,725**
Anti-Lock Brakes	$680	$800
Automatic 4 Speed	$696	$800
Optional Engine	N/A	N/A
Air Conditioning	Std	Std
Power Steering	Std	Std
Cruise Control	Std	Std
All Wheel Drive	N/A	N/A
AM/FM Stereo Cassette	Std	Std
Steering Wheel, Tilt	Std	Std
Power Windows	Std	Std
***Options Price**	$696	$800
***Total Price**	**$17,001**	**$19,525**
Target Price		$18,206
Destination Charge		$350
Avg. Tax & Fees		$1,127
Total Target $		**$19,683**
Average Dealer Option Cost:		83%

The 1993 626 is available in three model editions - DX, LX, and ES. New for 1993, the sedan features a redesigned interior and exterior. Refinements for the ES model include standard theft deterrent system, dual exhaust, and locking alloy wheels. Optional on the ES only, is a Leather Package which includes leather seat trim, power driver's seat, and leather-wrapped steering wheel. The domestic content of the 626 is expected to exceed 75 percent, and it is built in Flat Rock, Michigan.

Ownership Costs

Cost Area	5 Year Cost	Rate
Depreciation	$9,754	○
Financing ($396/month)	$3,243	
Insurance (Rating 13)	$7,797	○
State Fees	$795	
Fuel (Hwy 25 City 19)	$3,929	◉
Maintenance	$5,568	○
Repairs	$580	○

Warranty/Maintenance Info

Major Tune-Up	$212	◉
Minor Tune-Up	$129	●
Brake Service	$188	○
Overall Warranty	3 yr/50k	○
Drivetrain Warranty	3 yr/50k	○
Rust Warranty	5 yr/unlim. mi	○
Maintenance Warranty	N/A	
Roadside Assistance	1 yr/unlim. mi	

Resale Value

1993	1994	1995	1996	1997
$14,547	$13,221	$12,113	$11,079	$9,929

Cumulative Costs

	1993	1994	1995	1996	1997
Annual	$8,901	$5,017	$5,841	$4,418	$7,489
Total	$8,901	$13,918	$19,759	$24,177	$31,666

Ownership Costs (5yr)

Average	This Car
$31,260	$31,666
Cost/Mile 45¢	Cost/Mile 45¢

Ownership Cost Rating
○ Average

---

* Includes shaded options
** Other purchase requirements apply

● Poor ◉ Worse Than Average ● Average ○ Better Than Average ○ Excellent ⊖ Insufficient Information

©1993 by IntelliChoice, Inc. (408) 554-8711   All Rights Reserved. Reproduction Prohibited.
Refer to *Section 3: Annotated Vehicle Charts* for an explanation of these charts.

Mazda 929
4 Door Sedan

- 3.0L 195 hp Gas Fuel Inject.
- 6 Cylinder "V"
- Automatic 4 Speed
- 2 Wheel Rear
- Driver/Psngr Airbags Std

Large

Purchase Price

Car Item	Dealer Cost	List
Base Price	**$24,835**	**$29,200**
Anti-Lock Brakes	Std	Std
Manual Transmission	N/A	N/A
Optional Engine	N/A	N/A
Auto Climate Control	Std	Std
Power Steering	Std	Std
Cruise Control	Std	Std
All Wheel Drive	N/A	N/A
AM/FM Stereo Cassette	Std	Std
Steering Wheel, Tilt	Std	Std
Power Windows	Std	Std
*Options Price	$0	$0
*Total Price	$24,835	$29,200
Target Price	$27,001	
Destination Charge	$350	
Avg. Tax & Fees	$1,664	
Total Target $	**$29,015**	
Average Dealer Option Cost:	82%	

The 1993 929 is available as a one-model luxury sedan. New for 1993, the 929 sedan is equipped with an automatic transmission, dual airbags, four-wheel disc brakes, glass moonroof and automatic climate control. Two option packages are available: The Leather Package includes supple leather seating surfaces, an eight-way adjustable driver's seat and a four-way power passenger seat. The Premium Package adds wood trim on the center console, cellular phone rewiring and a Panasonic multi-disc CD player.

Ownership Costs

Cost Area	5 Year Cost	Rate
Depreciation	$14,494	○
Financing ($583/month)	$4,780	
Insurance (Rating 16)	$8,784	○
State Fees	$1,182	
Fuel (Hwy 24 City 19)	$4,015	●
Maintenance	$6,298	●
Repairs	$580	○

Warranty/Maintenance Info

Major Tune-Up	$341	●
Minor Tune-Up	$109	○
Brake Service	$244	○
Overall Warranty	3 yr/50k	○
Drivetrain Warranty	3 yr/50k	○
Rust Warranty	5 yr/unlim. mi	○
Maintenance Warranty	1 yr/15k	○
Roadside Assistance	1 yr/unlim. mi	

Ownership Cost By Year

1993, 1994, 1995, 1996, 1997

Resale Value

1993	1994	1995	1996	1997
$19,887	$18,223	$16,928	$15,728	$14,521

Cumulative Costs

	1993	1994	1995	1996	1997
Annual	$13,778	$6,106	$6,626	$4,931	$8,692
Total	$13,778	$19,884	$26,510	$31,441	$40,133

Ownership Costs (5yr)

Average	This Car
$41,430	$40,133
Cost/Mile 59¢	Cost/Mile 57¢

Ownership Cost Rating

○ Better Than Average

Mazda MX-3
2 Door Coupe

- 1.6L 88 hp Gas Fuel Inject.
- 4 Cylinder In-Line
- Manual 5 Speed
- 2 Wheel Front
- Automatic Seatbelts

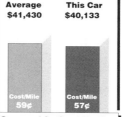

Subcompact

GS Model Shown

Purchase Price

Car Item	Dealer Cost	List
Base Price	**$10,701**	**$11,875**
Anti-Lock Brakes	N/A	N/A
Automatic 4 Speed	$653	$725
Optional Engine	N/A	N/A
Air Conditioning	$664	$830
Power Steering	Std	Std
Cruise Control	N/A	N/A
All Wheel Drive	N/A	N/A
AM/FM Stereo Cassette	Std	Std
Steering Wheel, Tilt	N/A	N/A
Power Windows	Pkg	Pkg
*Options Price	$0	$0
*Total Price	$10,701	$11,875
Target Price	$11,345	
Destination Charge	$350	
Avg. Tax & Fees	$707	
Total Target $	**$12,402**	
Average Dealer Option Cost:	85%	

The 1993 MX-3 is available in two models—Standard and GS. New for 1993, the MX-3 features an AM/FM/cassette stereo sound system with antitheft coding as standard equipment. Interior features include reclining bucket seats with a fold-down rear seatback and three-point lap and shoulder belts. The option list includes 14-inch aluminum alloy wheels with locks, power window and door locks and air conditioning. A new exterior color for 1993 is Laguna Blue Metallic.

Ownership Costs

Cost Area	5 Year Cost	Rate
Depreciation		⊖
Financing ($249/month)	$2,043	
Insurance (Rating 10)	$7,479	●
State Fees	$489	
Fuel (Hwy 35 City 28)	$2,740	○
Maintenance	$4,857	●
Repairs	$560	○

Warranty/Maintenance Info

Major Tune-Up	$201	○
Minor Tune-Up	$79	○
Brake Service	$187	○
Overall Warranty	3 yr/50k	○
Drivetrain Warranty	3 yr/50k	○
Rust Warranty	5 yr/unlim. mi	○
Maintenance Warranty	N/A	
Roadside Assistance	1 yr/unlim. mi	

Ownership Cost By Year

Insufficient Depreciation Information

Resale Value

Insufficient Information

Cumulative Costs

	1993	1994	1995	1996	1997
Annual	Insufficient Information				
Total	Insufficient Information				

Ownership Costs (5yr)

Insufficient Information

Ownership Cost Rating

⊖ Insufficient Information

---

* Includes shaded options
** Other purchase requirements apply

● Poor   ◉ Worse Than Average   ◐ Average   ○ Better Than Average   ○ Excellent   ⊖ Insufficient Information

©1993 by IntelliChoice, Inc. (408) 554-8711 All Rights Reserved. Reproduction Prohibited.
Refer to *Section 3: Annotated Vehicle Charts* for an explanation of these charts.

Mazda MX-3 GS
2 Door Coupe

 1.8L 130 hp Gas Fuel Inject.
 6 Cylinder "V"
 Manual 5 Speed
 2 Wheel Front
 Automatic Seatbelts

Subcompact

Purchase Price

Car Item	Dealer Cost	List
Base Price	**$13,049**	**$14,645**
Anti-Lock Brakes	$680	$800
Automatic 4 Speed	$653	$725
Optional Engine	N/A	N/A
Air Conditioning	$664	$830
Power Steering	Std	Std
Cruise Control	Pkg	Pkg
All Wheel Drive	N/A	N/A
AM/FM Stereo Cassette	Std	Std
Steering Wheel, Tilt	Std	Std
Power Windows	Pkg	Pkg
*Options Price	$0	$0
*Total Price	$13,049	$14,645
Target Price	$13,892	
Destination Charge	$350	
Avg. Tax & Fees	$862	
Total Target $	**$15,104**	
Average Dealer Option Cost:	**85%**	

The 1993 MX-3 is available in two models-Standard and GS. New for 1993, the GS is powered by a 1.8-liter DOHC 24-valve V6 engine. Standard features for the GS includes 15-inch aluminum alloy wheels, AM/FM/cassette stereo sound system with antitheft coding and rear spoiler. Interior features include reclining front bucket seats with a fold-down rear seatback and three-point lap and shoulder belts. Laguna Blue Metallic has been added to the color palette.

Ownership Costs

Cost Area	5 Year Cost	Rate
Depreciation		⊖
Financing ($304/month)	$2,488	
Insurance (Rating 12)	$7,898	●
State Fees	$600	
Fuel (Hwy 28 City 23)	$3,385	◉
Maintenance	$5,991	●
Repairs	$580	○

Warranty/Maintenance Info

Major Tune-Up	$181	○
Minor Tune-Up	$99	◉
Brake Service	$193	○
Overall Warranty	3 yr/50k	○
Drivetrain Warranty	3 yr/50k	○
Rust Warranty	5 yr/unlim. mi	○
Maintenance Warranty	N/A	
Roadside Assistance	1 yr/unlim. mi	

Ownership Cost By Year

Insufficient Depreciation Information

1993 / 1994 / 1995 / 1996 / 1997

Resale Value

Insufficient Information

Cumulative Costs

	1993	1994	1995	1996	1997
Annual	Insufficient Information				
Total	Insufficient Information				

Ownership Costs (5yr)

Insufficient Information

Ownership Cost Rating

⊖ Insufficient Information

Mazda MX-5 Miata
2 Door Convertible

 1.6L 116 hp Gas Fuel Inject.
4 Cylinder In-Line
Manual 5 Speed
 2 Wheel Rear
 Driver Airbag Psngr Belts

Sport

Purchase Price

Car Item	Dealer Cost	List
Base Price	**$13,632**	**$15,300**
Anti-Lock Brakes	$765	** $900
Automatic 4 Speed	$652	$750
Optional Engine	N/A	N/A
Air Conditioning	$664	$830
Power Steering	Pkg	Pkg
Cruise Control	Pkg	Pkg
All Wheel Drive	N/A	N/A
AM/FM Stereo Cassette	Std	Std
Steering Wheel, Tilt	N/A	N/A
Power Windows	Pkg	Pkg
*Options Price	$664	$830
*Total Price	$14,296	$16,130
Target Price	$15,236	
Destination Charge	$350	
Avg. Tax & Fees	$944	
Total Target $	**$16,530**	
Average Dealer Option Cost:	**83%**	

The 1993 MX-5 Miata is available in a one model edition. New for 1993, the MX-5 offers a tan interior with leather seating surfaces and tan vinyl top on Crystal White and Classic Red Miata. All Miatas now come standard with a driver's-side airbag and three new option packages. The option list includes power windows, aluminum wheels and power-assisted steering. Also new for 1993 is Mazda's Sensory Sound System and one new exterior color: Brilliant Black.

Ownership Costs

Cost Area	5 Year Cost	Rate
Depreciation	$6,807	○
Financing ($332/month)	$2,723	
Insurance (Rating 11 Sport+)	$10,770	◉
State Fees	$659	
Fuel (Hwy 30 City 24)	$3,198	○
Maintenance	$4,619	○
Repairs	$550	○

Warranty/Maintenance Info

Major Tune-Up	$244	○
Minor Tune-Up	$79	○
Brake Service	$179	○
Overall Warranty	3 yr/50k	◉
Drivetrain Warranty	3 yr/50k	◉
Rust Warranty	5 yr/unlim. mi	○
Maintenance Warranty	N/A	
Roadside Assistance	1 yr/unlim. mi	

Ownership Cost By Year

1993 / 1994 / 1995 / 1996 / 1997

Resale Value

1993	1994	1995	1996	1997
$15,523	$14,215	$12,626	$11,201	$9,723

Cumulative Costs

	1993	1994	1995	1996	1997
Annual	$4,929	$5,175	$6,194	$5,799	$7,229
Total	$4,929	$10,104	$16,298	$22,097	$29,326

Ownership Costs (5yr)

Average	This Car
$33,727	$29,326
Cost/Mile 48¢	Cost/Mile 42¢

Ownership Cost Rating

○ Excellent

---

page 128

* Includes shaded options
** Other purchase requirements apply

● Poor   ◉ Worse Than Average   ○ Average   ○ Better Than Average   ○ Excellent   ⊖ Insufficient Information

©1993 by IntelliChoice, Inc. (408) 554-8711 All Rights Reserved. Reproduction Prohibited.
Refer to *Section 3: Annotated Vehicle Charts* for an explanation of these charts.

Mazda MX-6 2 Door Coupe

Compact

Engine: 2.0L 118 hp Gas Fuel Inject. | 4 Cylinder In-Line | Manual 5 Speed | 2 Wheel Front | Driver Airbag Psngr Belts

LS Model Shown

Purchase Price

Car Item	Dealer Cost	List
Base Price	$14,358	$16,300
Anti-Lock Brakes	$808	$950
Automatic 4 Speed	$696	$800
Optional Engine	N/A	N/A
Air Conditioning	$680	$850
Power Steering	Std	Std
Cruise Control	Std	Std
All Wheel Drive	N/A	N/A
AM/FM Stereo Cassette	Std	Std
Steering Wheel, Tilt	Std	Std
Power Windows	Std	Std
*Options Price	$1,376	$1,650
*Total Price	$15,734	$17,950
Target Price	$16,790	
Destination Charge	$350	
Avg. Tax & Fees	$1,040	
Total Target $	**$18,180**	
Average Dealer Option Cost:	81%	

The 1993 MX-6 is available in two models - base and LS. New for 1993 the coupes feature a redesigned interior and exterior. The base model comes equipped with standard driver's side airbag, cruise control, power door locks, dual power mirrors, and power windows. Options include power sunroof, air conditioning, theft deterrent system, and anti-lock brakes. The domestic content of the MX-6 is expected to exceed 75 percent, and is built in Flat Rock, Michigan.

Ownership Costs

Cost Area	5 Year Cost	Rate
Depreciation	$9,092	○
Financing ($365/month)	$2,995	
Insurance (Rating 16)	$8,784	●
State Fees	$732	
Fuel (Hwy 31 City 23)	$3,204	○
Maintenance	$4,954	◉
Repairs	$550	○

Warranty/Maintenance Info

Major Tune-Up	$217	●
Minor Tune-Up	$84	○
Brake Service	$187	○
Overall Warranty	3 yr/50k	○
Drivetrain Warranty	3 yr/50k	○
Rust Warranty	5 yr/unlim. mi	○
Maintenance Warranty	N/A	
Roadside Assistance	1 yr/unlim. mi	

Resale Value: 1993 $12,748 | 1994 $11,776 | 1995 $10,884 | 1996 $9,998 | 1997 $9,088

Cumulative Costs

	1993	1994	1995	1996	1997
Annual	$9,121	$4,567	$5,520	$4,220	$6,883
Total	$9,121	$13,688	$19,208	$23,428	$30,311

Ownership Costs (5yr): Average $29,907 | This Car $30,311 | Cost/Mile 43¢ / 43¢

Ownership Cost Rating: Average

Mazda MX-6 LS 2 Door Coupe

Compact

Engine: 2.5L 164 hp Gas Fuel Inject. | 6 Cylinder "V" | Manual 5 Speed | 2 Wheel Front | Driver Airbag Psngr Belts

Purchase Price

Car Item	Dealer Cost	List
Base Price	$16,175	$18,575
Anti-Lock Brakes	$680	$800
Automatic 4 Speed	$696	$800
Optional Engine	N/A	N/A
Air Conditioning	$680	$850
Power Steering	Std	Std
Cruise Control	Std	Std
All Wheel Drive	N/A	N/A
AM/FM Stereo Cassette	Std	Std
Steering Wheel, Tilt	Std	Std
Power Windows	Std	Std
*Options Price	$1,376	$1,650
*Total Price	$17,551	$20,225
Target Price	$18,791	
Destination Charge	$350	
Avg. Tax & Fees	$1,163	
Total Target $	**$20,304**	
Average Dealer Option Cost:	82%	

The 1993 MX-6 is available in two models - base and LS. New for 1993 the coupes feature a redesigned interior and exterior. The LS model comes equipped with standard theft deterrent system, fog lights, dual exhaust, and leather-wrapped steering wheel. Options include power sunroof, air conditioning, rear spoiler, and anti-locks brakes. The domestic content of the MX-6 is expected to exceed 75 percent, and is built in Flat Rock, Michigan.

Ownership Costs

Cost Area	5 Year Cost	Rate
Depreciation	$10,387	○
Financing ($408/month)	$3,344	
Insurance (Rating 18)	$9,058	●
State Fees	$823	
Fuel (Hwy 25 City 19)	$3,929	◉
Maintenance	$5,420	◉
Repairs	$580	○

Warranty/Maintenance Info

Major Tune-Up	$212	◉
Minor Tune-Up	$109	○
Brake Service	$188	○
Overall Warranty	3 yr/50k	○
Drivetrain Warranty	3 yr/50k	○
Rust Warranty	5 yr/unlim. mi	○
Maintenance Warranty	N/A	
Roadside Assistance	1 yr/unlim. mi	

Resale Value: 1993 $13,449 | 1994 $12,470 | 1995 $11,591 | 1996 $10,751 | 1997 $9,917

Cumulative Costs

	1993	1994	1995	1996	1997
Annual	$10,901	$4,929	$5,890	$4,475	$7,346
Total	$10,901	$15,830	$21,720	$26,195	$33,541

Ownership Costs (5yr): Average $31,861 | This Car $33,541 | Cost/Mile 46¢ / 48¢

Ownership Cost Rating: Worse Than Average

---

* Includes shaded options
** Other purchase requirements apply

Legend: ● Poor | ◉ Worse Than Average | ○ Average | ◯ Better Than Average | ○ Excellent | ⊖ Insufficient Information

©1993 by IntelliChoice, Inc. (408) 554-8711 All Rights Reserved. Reproduction Prohibited.
Refer to *Section 3: Annotated Vehicle Charts* for an explanation of these charts.

Mazda Protege DX
4 Door Sedan

LX Model Shown — Subcompact

- 1.8L 103 hp Gas Fuel Inject.
- 4 Cylinder In-Line
- Manual 5 Speed
- 2 Wheel Front
- Automatic Seatbelts

Purchase Price

Car Item	Dealer Cost	List
Base Price	**$9,891**	**$10,854**
Anti-Lock Brakes	N/A	N/A
Automatic 4 Speed	$653	$725
Optional Engine	N/A	N/A
Air Conditioning	$672	$840
Power Steering	Std	Std
Cruise Control	N/A	N/A
All Wheel Drive	N/A	N/A
AM/FM Stereo Cassette	$342	$450
Steering Wheel, Tilt	Pkg	Pkg
Power Windows	N/A	N/A
*Options Price	$0	$0
*Total Price	$9,891	$10,854
Target Price	$10,470	
Destination Charge	$350	
Avg. Tax & Fees	$653	
Total Target $	**$11,473**	
Average Dealer Option Cost:	**82%**	

Ownership Costs

Cost Area	5 Year Cost	Rate
Depreciation	$4,921	○
Financing ($231/month)	$1,890	
Insurance (Rating 12)	$7,898	●
State Fees	$448	
Fuel (Hwy 36 City 28)	$2,699	○
Maintenance	$4,153	○
Repairs	$550	○

Warranty/Maintenance Info

Major Tune-Up	$206	○
Minor Tune-Up	$84	○
Brake Service	$192	○
Overall Warranty	3 yr/50k	○
Drivetrain Warranty	3 yr/50k	○
Rust Warranty	5 yr/unlim. mi	○
Maintenance Warranty	N/A	
Roadside Assistance	1 yr/unlim. mi	

Ownership Cost By Year
1993, 1994, 1995, 1996, 1997

Resale Value
1993	1994	1995	1996	1997
$9,132	$8,702	$8,158	$7,364	$6,552

Ownership Costs (5yr)
- Average: $23,790
- This Car: $22,559
- Cost/Mile: 34¢ / 32¢

Cumulative Costs
	1993	1994	1995	1996	1997
Annual	$5,241	$3,349	$4,289	$4,282	$5,398
Total	$5,241	$8,590	$12,879	$17,161	$22,559

Ownership Cost Rating: Excellent

The 1993 Mazda Protege is available in two models - DX and LX four-door sedans. The 1993 Mazda Protege DX comes with a larger center console with covered storage and a cup holder. Power steering has been added as a standard equipment. This model is available with a new option package that includes an AM/FM cassette stereo with four speakers, a tachometer, enhanced instrument warning lights, tilt steering, and a trunk light.

Mazda Protege LX
4 Door Sedan

Subcompact

- 1.8L 125 hp Gas Fuel Inject.
- 4 Cylinder In-Line
- Manual 5 Speed
- 2 Wheel Front
- Automatic Seatbelts

Purchase Price

Car Item	Dealer Cost	List
Base Price	**$11,128**	**$12,349**
Anti-Lock Brakes	N/A	N/A
Automatic 4 Speed	$653	$725
Optional Engine	N/A	N/A
Air Conditioning	$672	$840
Power Steering	Std	Std
Cruise Control	Std	Std
All Wheel Drive	N/A	N/A
AM/FM Stereo Cassette	Std	Std
Steering Wheel, Tilt	Std	Std
Power Windows	Std	Std
*Options Price	$0	$0
*Total Price	$11,128	$12,349
Target Price	$11,806	
Destination Charge	$350	
Avg. Tax & Fees	$735	
Total Target $	**$12,891**	
Average Dealer Option Cost:	**85%**	

Ownership Costs

Cost Area	5 Year Cost	Rate
Depreciation	$6,070	○
Financing ($259/month)	$2,124	
Insurance (Rating 12)	$7,898	●
State Fees	$508	
Fuel (Hwy 30 City 25)	$3,140	◉
Maintenance	$4,386	○
Repairs	$550	○

Warranty/Maintenance Info

Major Tune-Up	$207	○
Minor Tune-Up	$84	○
Brake Service	$148	
Overall Warranty	3 yr/50k	○
Drivetrain Warranty	3 yr/50k	○
Rust Warranty	5 yr/unlim. mi	○
Maintenance Warranty	N/A	
Roadside Assistance	1 yr/unlim. mi	

Resale Value
1993	1994	1995	1996	1997
$9,821	$9,283	$8,603	$7,716	$6,821

Ownership Costs (5yr)
- Average: $25,472
- This Car: $24,676
- Cost/Mile: 36¢ / 35¢

Cumulative Costs
	1993	1994	1995	1996	1997
Annual	$6,165	$3,628	$4,649	$4,494	$5,740
Total	$6,165	$9,793	$14,442	$18,936	$24,676

Ownership Cost Rating: Better Than Average

The 1993 Mazda Protege is available in two models - DX and LX four-door sedans. The 1993 Mazda Protege LX model features a new seat and door trim. The trunk can be accessed from the passenger compartment via the fold-down seatback, which is split 60/40. Standard equipment includes dual color-keyed power mirrors, velour upholstery, and dual visor vanity mirrors.

---

page 130

* Includes shaded options
** Other purchase requirements apply

● Poor ◉ Worse Than Average ○ Average ○ Better Than Average ○ Excellent ⊖ Insufficient Information

©1993 by IntelliChoice, Inc. (408) 554-8711   All Rights Reserved. Reproduction Prohibited.
Refer to *Section 3: Annotated Vehicle Charts* for an explanation of these charts.

Mazda RX-7
2 Door Coupe

Sport

 1.3L 255 hp Turbo Rotary Fuel Inject.
 2 Chamber In-Line
 Manual 5 Speed
 2 Wheel Rear
 Driver Airbag Psngr Belts

Purchase Price

Car Item	Dealer Cost	List
Base Price	**$27,641**	**$32,500**
Anti-Lock Brakes	Std	Std
Automatic 4 Speed	$740	$850
Optional Engine	N/A	N/A
Air Conditioning	Std	Std
Power Steering	Std	Std
Cruise Control	Std	Std
All Wheel Drive	N/A	N/A
AM/FM Stereo Cassette	Std	Std
Steering Wheel, Tilt	N/A	N/A
Power Windows	Std	Std
***Options Price**	**$0**	**$0**
***Total Price**	**$27,641**	**$32,500**
Target Price		$30,201
Destination Charge		$350
Avg. Tax & Fees		$1,857
Luxury Tax		$55
Total Target $		**$32,463**

The 1993 RX-7 is available in a one model edition with two option packages: the performance-oriented 'R1' package and the Touring package. The 'R1' package includes dual oil coolers, Z-rated tires and an aero package that reduces front and rear lift at higher speeds. The Touring package includes leather seating surfaces, cruise control, power sunroof and a special Bose Acoustic Wave speaker system. The RX-7 was developed jointly by Mazda's Yokohama, Japan and Irvine, Calif. design studios.

Ownership Costs

Cost Area	5 Year Cost	Rate
Depreciation	$16,995	○
Financing ($652/month)	$5,346	
Insurance (Rating 21 Sport+)	$14,300	●
State Fees	$1,315	
Fuel (Hwy 25 City 17 -Prem.)	$4,578	○
Maintenance	$5,715	○
Repairs	$650	○

Warranty/Maintenance Info

Major Tune-Up	$188	○
Minor Tune-Up	$136	○
Brake Service	$272	○
Overall Warranty	3 yr/50k	○
Drivetrain Warranty	3 yr/50k	○
Rust Warranty	5 yr/unlim. mi	○
Maintenance Warranty		
Roadside Assistance	1 yr/unlim. mi	

Ownership Cost By Year

1993 / 1994 / 1995 / 1996 / 1997

Resale Value

1993	1994	1995	1996	1997
$23,934	$21,479	$19,318	$17,305	$15,468

Cumulative Costs

	1993	1994	1995	1996	1997
Annual	$14,570	$8,325	$8,882	$7,140	$9,980
Total	$14,570	$22,895	$31,777	$38,917	$48,897

Ownership Costs (5yr)

Average	This Car
$49,432	$48,897
Cost/Mile 71¢	Cost/Mile 70¢

Ownership Cost Rating

○ Average

Mercedes Benz 190E 2.3
4 Door Sedan

Luxury

 2.3L 130 hp Gas Fuel Inject.
 4 Cylinder In-Line
 Manual 5 Speed
 2 Wheel Rear
Driver Airbag Psngr Belts

Purchase Price

Car Item	Dealer Cost	List
Base Price	**$24,030**	**$28,950**
Anti-Lock Brakes	Std	Std
Automatic 4 Speed	$750	$900
Optional Engine	N/A	N/A
Auto Climate Control	Std	Std
Power Steering	Std	Std
Cruise Control	Std	Std
All Wheel Drive	N/A	N/A
AM/FM Stereo Cassette	Std	Std
Steering Wheel, Tilt	N/A	N/A
Power Windows	Std	Std
***Options Price**	**$750**	**$900**
***Total Price**	**$24,780**	**$29,850**
Target Price		$27,350
Destination Charge		$400
Avg. Tax & Fees		$1,691
Total Target $		**$29,441**
Average Dealer Option Cost:	82%	

The 1993 Mercedes 190 Class is available in two models - 190E 2.3 and 190E 2.6 sedan. For 1993, the 2.3 comes equipped with electronic cruise control, power-assisted disc brakes with ABS, aluminum-alloy wheels, single-sweep windshield wiper; halogen headlights and fog lamps; and central locking. Other features include eight-way manually adjustable front seats, automatic climate control and a first-aid kit. A 24-hour roadside assistance program is provided by Mercedes on all models.

Ownership Costs

Cost Area	5 Year Cost	Rate
Depreciation	$10,713	○
Financing ($592/month)	$4,850	
Insurance (Rating 17)	$9,084	○
State Fees	$1,211	
Fuel (Hwy 26 City 20 -Prem.)	$4,149	○
Maintenance	$5,014	○
Repairs	$831	○

Warranty/Maintenance Info

Major Tune-Up	$263	◉
Minor Tune-Up	$168	●
Brake Service	$198	○
Overall Warranty	4 yr/50k	○
Drivetrain Warranty	4 yr/50k	○
Rust Warranty	4 yr/50k	◉
Maintenance Warranty	N/A	
Roadside Assistance	4 yr/50k	

Resale Value

1993	1994	1995	1996	1997
$23,712	$22,456	$21,253	$20,007	$18,728

Cumulative Costs

	1993	1994	1995	1996	1997
Annual	$10,545	$5,948	$6,860	$5,081	$7,418
Total	$10,545	$16,493	$23,353	$28,434	$35,852

Ownership Costs (5yr)

Average	This Car
$41,772	$35,852
Cost/Mile 60¢	Cost/Mile 51¢

Ownership Cost Rating

○ Excellent

---

* Includes shaded options
** Other purchase requirements apply

● Poor ◉ Worse Than Average ○ Average ○ Better Than Average ○ Excellent ⊖ Insufficient Information

©1993 by IntelliChoice, Inc. (408) 554-8711 All Rights Reserved. Reproduction Prohibited.
Refer to *Section 3: Annotated Vehicle Charts* for an explanation of these charts.

Mercedes Benz 190E 2.6
4 Door Sedan

Luxury

2.6L 158 hp Gas Fuel Inject. | 6 Cylinder In-Line | Manual 5 Speed | 2 Wheel Rear | Driver Airbag Psngr Belts

2.3 Model Shown

Purchase Price

Car Item	Dealer Cost	List
Base Price	**$28,220**	**$34,000**
Anti-Lock Brakes	Std	Std
Automatic 4 Speed	$750	$900
Optional Engine	N/A	N/A
Auto Climate Control	Std	Std
Power Steering	Std	Std
Cruise Control	Std	Std
All Wheel Drive	N/A	N/A
AM/FM Stereo Cassette	Std	Std
Steering Wheel, Tilt	N/A	N/A
Power Windows	Std	Std
***Options Price**	**$750**	**$900**
***Total Price**	**$28,970**	**$34,900**
Target Price	$32,256	
Destination Charge	$400	
Avg. Tax & Fees	$1,986	
Luxury Tax	$266	
Total Target $	**$34,908**	

Ownership Costs

Cost Area	5 Year Cost	Rate
Depreciation	$12,849	○
Financing ($702/month)	$5,750	
Insurance (Rating 19)	$9,777	O
State Fees	$1,412	
Fuel (Hwy 25 City 20 -Prem.)	$4,237	O
Maintenance	$5,065	O
Repairs	$831	O

Warranty/Maintenance Info

Major Tune-Up	$249	O
Minor Tune-Up	$150	●
Brake Service	$198	○
Overall Warranty	4 yr/50k	O
Drivetrain Warranty	4 yr/50k	O
Rust Warranty	4 yr/50k	●
Maintenance Warranty	N/A	
Roadside Assistance	4 yr/50k	

Ownership Cost By Year

1993 / 1994 / 1995 / 1996 / 1997

Resale Value

1993	1994	1995	1996	1997
$28,741	$27,056	$25,473	$23,823	$22,059

Cumulative Costs

	1993	1994	1995	1996	1997
Annual	$11,555	$6,841	$7,605	$5,734	$8,186
Total	$11,555	$18,396	$26,001	$31,735	$39,921

Ownership Costs (5yr)

Average	This Car
$45,553	$39,921
Cost/Mile 65¢	Cost/Mile 57¢

Ownership Cost Rating

○ Excellent

The 1993 Mercedes 190 Class is available in two models - 190E 2.3 and 190E 2.6 sedan. For 1993, the 2.6 comes equipped with electronic cruise control, power-assisted disc brakes with ABS, alloy wheels, single-sweep windshield wiper and central locking. Other features include 10-way electrically adjustable front seats, automatic climate control and a first-aid kit. The Sportline Package is offered on the 2.6 which includes larger alloy wheels, sport steering wheel, sport seats and sport suspension.

Mercedes Benz 300 CE
2 Door Coupe

Luxury

3.2L 217 hp Gas Fuel Inject. | 6 Cylinder In-Line | Automatic 4 Speed | 2 Wheel Rear | Driver/Psngr Airbags Std

Purchase Price

Car Item	Dealer Cost	List
Base Price	**$50,630**	**$61,000**
Anti-Lock Brakes	Std	Std
Manual Transmission	N/A	N/A
Optional Engine	N/A	N/A
Auto Climate Control	Std	Std
Power Steering	Std	Std
Cruise Control	Std	Std
All Wheel Drive	N/A	N/A
AM/FM Stereo Cassette	Std	Std
Steering, Tilt w/Memory	Std	Std
Power Windows	Std	Std
***Options Price**	**$0**	**$0**
***Total Price**	**$50,630**	**$61,000**
Target Price	$56,452	
Destination Charge	$400	
Avg. Tax & Fees	$3,457	
Luxury Tax	$2,685	
Total Target $	**$62,994**	

Ownership Costs

Cost Area	5 Year Cost	Rate
Depreciation	$24,330	O
Financing ($1,266/month)	$10,376	
Insurance (Rating 24)	$11,704	O
State Fees	$2,456	
Fuel (Hwy 25 City 19 -Prem.)	$4,338	O
Maintenance	$6,217	●
Repairs	$1,215	O

Warranty/Maintenance Info

Major Tune-Up	$345	●
Minor Tune-Up	$188	●
Brake Service	$218	O
Overall Warranty	4 yr/50k	O
Drivetrain Warranty	4 yr/50k	O
Rust Warranty	4 yr/50k	●
Maintenance Warranty	N/A	
Roadside Assistance	4 yr/50k	

Resale Value

1993	1994	1995	1996	1997
$49,612	$46,842	$44,121	$41,422	$38,664

Cumulative Costs

	1993	1994	1995	1996	1997
Annual	$21,336	$10,042	$10,683	$7,889	$10,686
Total	$21,336	$31,378	$42,061	$49,950	$60,636

Ownership Costs (5yr)

Average	This Car
$66,249	$60,636
Cost/Mile 95¢	Cost/Mile 87¢

Ownership Cost Rating

○ Better Than Average

The 1993 300 Class is available in ten models - 300D 2.5 Turbo, 300E 2.6, 300E 2.8, 300E 4MATIC, 400E and 500E sedans; 300TE and 300TE 4MATIC wagons; and the 300CE coupe and convertible. For 1993, the 300 CE features front fog lamps, 10-way electrically adjustable front bucket seats with two-position driver-side memory, high-performance sound system, automatic climate control, anti-theft vehicle alarm system and leather upholstery. The Sportline feature is optional.

---

page 132

* Includes shaded options
** Other purchase requirements apply

● Poor   ◐ Worse Than Average   ○ Average   ○ Better Than Average   ○ Excellent   ⊖ Insufficient Information

©1993 by IntelliChoice, Inc. (408) 554-8711 All Rights Reserved. Reproduction Prohibited.

Refer to *Section 3: Annotated Vehicle Charts* for an explanation of these charts.

Mercedes Benz 300 D 2.5 Turbo
4 Door Sedan

Luxury

 2.5L 121 hp Turbo Dsl Fuel Inject.
 5 Cylinder In-Line
 Automatic 4 Speed
 2 Wheel Rear
 Driver/Psngr Airbags Std

Purchase Price

Car Item	Dealer Cost	List
Base Price	**$36,350**	**$43,800**
Anti-Lock Brakes	Std	Std
Manual Transmission	N/A	N/A
Optional Engine	N/A	N/A
Auto Climate Control	Std	Std
Power Steering	Std	Std
Cruise Control	Std	Std
All Wheel Drive	N/A	N/A
AM/FM Stereo Cassette	Std	Std
Steering Wheel, Scope	Grp	Grp
Power Windows	Std	Std
***Options Price**	**$0**	**$0**
***Total Price**	**$36,350**	**$43,800**
Target Price	$40,530	
Destination Charge	$400	
Avg. Tax & Fees	$2,489	
Luxury Tax	$1,093	
Total Target $	**$44,512**	

The 300 Diesel is available in two models–the 300D 2.5 Turbo (300 Class) and 300SD Turbo (S Class) sedans. New for 1993, the 300D 2.5 Turbo features Supplemental Restraint System with dual air bags and 3-point front seat belts with height-adjustable shoulder-belt anchors. Interior features include 10-way electrically adjustable front bucket seats with remote retractable rear head restraints and includes Zebrano wood interior accents. The 300D 2.5 Turbo is not available in California.

Ownership Costs

Cost Area	5 Year Cost	Rate
Depreciation	$15,144	○
Financing ($895/month)	$7,333	
Insurance (Rating 18)	$9,058	○
State Fees	$1,768	
Fuel (Hwy 33 City 26)	$3,309	○
Maintenance	$4,440	◐
Repairs	$1,215	●

Warranty/Maintenance Info

Major Tune-Up	$143	○
Minor Tune-Up	$60	○
Brake Service	$205	○
Overall Warranty	4 yr/50k	○
Drivetrain Warranty	4 yr/50k	○
Rust Warranty	4 yr/50k	●
Maintenance Warranty	N/A	
Roadside Assistance	4 yr/50k	

Ownership Cost By Year

Resale Value
1993	1994	1995	1996	1997
$36,112	$34,617	$32,893	$31,066	$29,368

Cumulative Costs
	1993	1994	1995	1996	1997
Annual	$14,266	$6,846	$7,748	$5,771	$7,636
Total	$14,266	$21,112	$28,860	$34,631	$42,267

Ownership Costs (5yr)
Average: $52,215 — Cost/Mile 75¢
This Car: $42,267 — Cost/Mile 60¢

Ownership Cost Rating
Excellent

Mercedes Benz 300 E 2.8
4 Door Sedan

Luxury

 2.8L 194 hp Gas Fuel Inject.
 6 Cylinder In-Line
 Automatic 4 Speed
 2 Wheel Rear
 Driver/Psngr Airbags Std

Base Model Shown

Purchase Price

Car Item	Dealer Cost	List
Base Price	**$36,350**	**$43,800**
Anti-Lock Brakes	Std	Std
Manual Transmission	N/A	N/A
Optional Engine	N/A	N/A
Auto Climate Control	Std	Std
Power Steering	Std	Std
Cruise Control	Std	Std
All Wheel Drive	N/A	N/A
AM/FM Stereo Cassette	Std	Std
Steering Wheel, Scope	$291	$350
Power Windows	Std	Std
***Options Price**	**$291**	**$350**
***Total Price**	**$36,641**	**$44,150**
Target Price	$40,855	
Destination Charge	$400	
Avg. Tax & Fees	$2,509	
Luxury Tax	$1,126	
Total Target $	**$44,890**	

The 1993 300 Class is available in ten models - 300D 2.5 Turbo, 300E 2.6, 300E 2.8, 300E 4MATIC, 400E and 500E sedans; 300TE and 300TE 4MATIC wagons; and the 300CE coupe and convertible. For 1993, the 300E 2.8 features 10-way electrically adjustable front bucket seats with driver-side memory, high-performance sound system, automatic climate control, anti-theft vehicle alarm system and leather upholstery. Options include a trunk-mounted CD changer, cellular telephone and Automatic Slip Control.

Ownership Costs

Cost Area	5 Year Cost	Rate
Depreciation	$15,730	○
Financing ($902/month)	$7,394	
Insurance (Rating 18)	$9,058	○
State Fees	$1,782	
Fuel (Hwy 25 City 19 -Prem.)	$4,338	◐
Maintenance	$5,186	◐
Repairs	$1,215	◐

Warranty/Maintenance Info

Major Tune-Up	$247	◐
Minor Tune-Up	$148	●
Brake Service	$237	◐
Overall Warranty	4 yr/50k	○
Drivetrain Warranty	4 yr/50k	○
Rust Warranty	4 yr/50k	●
Maintenance Warranty	N/A	
Roadside Assistance	4 yr/50k	

Resale Value
1993	1994	1995	1996	1997
$36,450	$34,628	$32,935	$31,004	$29,160

Cumulative Costs
	1993	1994	1995	1996	1997
Annual	$14,492	$7,454	$8,046	$6,183	$8,528
Total	$14,492	$21,946	$29,992	$36,175	$44,703

Ownership Costs (5yr)
Average: $52,477 — Cost/Mile 75¢
This Car: $44,703 — Cost/Mile 64¢

Ownership Cost Rating
Excellent

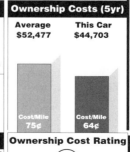

---

* Includes shaded options
** Other purchase requirements apply

● Poor Worse Than Average Average Better Than Average Excellent Insufficient Information

©1993 by IntelliChoice, Inc. (408) 554-8711 All Rights Reserved. Reproduction Prohibited.
Refer to *Section 3: Annotated Vehicle Charts* for an explanation of these charts.

Mercedes Benz 300 E
4 Door Sedan

Luxury

- 3.2L 217 hp Gas Fuel Inject.
- 6 Cylinder In-Line
- Automatic 4 Speed
- 2 Wheel Rear
- Driver/Psngr Airbags Std

Purchase Price

Car Item	Dealer Cost	List
Base Price	**$41,420**	**$49,900**
Anti-Lock Brakes	Std	Std
Manual Transmission	N/A	N/A
Optional Engine	N/A	N/A
Auto Climate Control	Std	Std
Power Steering	Std	Std
Cruise Control	Std	Std
All Wheel Drive	N/A	N/A
AM/FM Stereo Cassette	Std	Std
Steering Wheel, Scope	Std	Std
Power Windows	Std	Std
*Options Price	$0	$0
*Total Price	$41,420	$49,900
Target Price		$46,183
Destination Charge		$400
Avg. Tax & Fees		$2,832
Luxury Tax		$1,658
Total Target $		**$51,073**

Ownership Costs

Cost Area	5 Year Cost	Rate
Depreciation	$18,547	○
Financing ($1,026/month)	$8,413	
Insurance (Rating 19)	$9,358	○
State Fees	$2,012	
Fuel (Hwy 25 City 19 -Prem.)	$4,338	◐
Maintenance	$5,192	◐
Repairs	$1,215	◐

Warranty/Maintenance Info

Major Tune-Up	$247	○
Minor Tune-Up	$148	●
Brake Service	$237	○
Overall Warranty	4 yr/50k	○
Drivetrain Warranty	4 yr/50k	○
Rust Warranty	4 yr/50k	●
Maintenance Warranty	N/A	
Roadside Assistance	4 yr/50k	

Ownership Cost By Year

1993, 1994, 1995, 1996, 1997

Resale Value

1993	1994	1995	1996	1997
$40,547	$38,662	$36,798	$34,728	$32,526

Cumulative Costs

	1993	1994	1995	1996	1997
Annual	$17,115	$7,946	$8,521	$6,498	$8,995
Total	$17,115	$25,061	$33,582	$40,080	$49,075

Ownership Costs (5yr)

Average: $56,782
This Car: $49,075
Cost/Mile: 81¢ / 70¢

Ownership Cost Rating

Excellent

The 1993 300 Class is available in ten models - 300D 2.5 Turbo, 300E 2.6, 300E 2.8, 300E 4MATIC, 400E and 500E sedans; 300TE and 300TE 4MATIC wagons; and the 300CE coupe and convertible. For 1993, the 300E features front fog lights, 10-way electrically adjustable front bucket seats with two-position driver-side memory, high-performance sound system, automatic climate control, anti-theft vehicle alarm system and leather upholstery. The Sportline feature is optional.

Mercedes Benz 300 SD
4 Door Sedan

Luxury
SE Model Shown

- 3.4L 148 hp Turbo Dsl Fuel Inject.
- 6 Cylinder In-Line
- Automatic 4 Speed
- 2 Wheel Rear
- Driver/Psngr Airbags Std

Purchase Price

Car Item	Dealer Cost	List
Base Price	**$58,020**	**$69,900**
Anti-Lock Brakes	Std	Std
Manual Transmission	N/A	N/A
Optional Engine	N/A	N/A
Auto Climate Control	Std	Std
Power Steering	Std	Std
Cruise Control	Std	Std
All Wheel Drive	N/A	N/A
AM/FM Stereo Cassette	Std	Std
Steering, Tilt w/Memory	Std	Std
Power Windows	Std	Std
*Options Price	$0	$0
*Total Price	$58,020	$69,900
Target Price		$64,692
Destination Charge		$400
Avg. Tax & Fees		$3,958
Luxury Tax		$3,509
Total Target $		**$72,559**

Ownership Costs

Cost Area	5 Year Cost	Rate
Depreciation	$27,968	○
Financing ($1,458/month)	$11,953	
Insurance (Rating 22)	$10,640	○
State Fees	$2,812	
Fuel (Hwy 23 City 20)	$4,543	○
Maintenance	$5,080	○
Repairs	$1,275	●

Warranty/Maintenance Info

Major Tune-Up	$191	○
Minor Tune-Up	$143	○
Brake Service	$293	●
Overall Warranty	4 yr/50k	○
Drivetrain Warranty	4 yr/50k	○
Rust Warranty	4 yr/50k	●
Maintenance Warranty	N/A	
Roadside Assistance	4 yr/50k	

Resale Value

1993	1994	1995	1996	1997
$57,283	$53,698	$50,360	$47,424	$44,591

Cumulative Costs

	1993	1994	1995	1996	1997
Annual	$23,853	$11,249	$11,257	$8,098	$9,814
Total	$23,853	$35,102	$46,359	$54,457	$64,271

Ownership Costs (5yr)

Average: $74,154
This Car: $64,271
Cost/Mile: $1.06 / 92¢

Ownership Cost Rating

Excellent

The 300 Diesel is available in two models - the 300D 2.5 Turbo (300 Class) and 300SD Turbo (S Class) sedans. New for 1993, the 300 SD Turbo features dual air bags and 3-point front bucket seat belts with automatic shoulder-belt anchor adjustment. Interior features include leather-upholstered seats, zebrano wood accents, pre-wiring for CD changer and telephone and 12-way electrically adjustable front bucket seats with 3-position memory and adjustable lumbar seats. The 300SD is not avaialble in California.

---

page 134

* Includes shaded options
** Other purchase requirements apply

● Poor ◐ Worse Than Average ○ Average ○ Better Than Average ○ Excellent ⊖ Insufficient Information

©1993 by IntelliChoice, Inc. (408) 554-8711 All Rights Reserved. Reproduction Prohibited.
Refer to *Section 3: Annotated Vehicle Charts* for an explanation of these charts.

Mercedes Benz 300 SE
4 Door Sedan

Luxury

 3.2L 228 hp Gas Fuel Inject.
 6 Cylinder In-Line
 Automatic 5 Speed
 2 Wheel Rear
Driver/Psngr Airbags Std

Purchase Price

Car Item	Dealer Cost	List
Base Price	**$58,020**	**$69,900**
Anti-Lock Brakes	Std	Std
Manual Transmission	N/A	N/A
Optional Engine	N/A	N/A
Auto Climate Control	Std	Std
Power Steering	Std	Std
Cruise Control	Std	Std
All Wheel Drive	N/A	N/A
AM/FM Stereo Cassette	Std	Std
Steering, Tilt w/Memory	Std	Std
Power Windows	Std	Std
*Options Price	$0	$0
*Total Price	$58,020	$69,900
Target Price		$64,692
Destination Charge		$400
Avg. Tax & Fees		$3,958
Luxury/Gas Guzzler Tax		$5,209
Total Target $		**$74,259**

The 1993 S Class is available in five models - 300SD, 300SE, 400SEL, 500SEL and 600SEL sedans. For 1993, the 300SE features Bose-Beta sound system, halogen headlamps and front fog lights, remote central locking system, 12-way power adjustable front bucket seats, anti-theft alarm system and leather upholstery. The S-Class cars are completely CFC-free, including no ozone-depleting chlorofluorocarbons in their air conditioning systems. All models are covered by a 24-hour roadside assistance program.

Ownership Costs

Cost Area	5 Year Cost	Rate
Depreciation	$28,587	○
Financing ($1,492/month)	$12,232	
Insurance (Rating 22)	$10,640	○
State Fees	$2,812	
Fuel (Hwy 19 City 15 -Prem.)	$5,610	●
Maintenance	$5,897	◐
Repairs	$1,215	○

Warranty/Maintenance Info

Major Tune-Up	$323 ●
Minor Tune-Up	$166 ◐
Brake Service	$293 ◐
Overall Warranty	4 yr/50k ○
Drivetrain Warranty	4 yr/50k ○
Rust Warranty	4 yr/50k ●
Maintenance Warranty	N/A
Roadside Assistance	4 yr/50k

Ownership Cost By Year

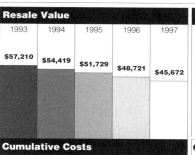

1993 / 1994 / 1995 / 1996 / 1997

Resale Value

1993	1994	1995	1996	1997
$57,210	$54,419	$51,729	$48,721	$45,672

Cumulative Costs

	1993	1994	1995	1996	1997
Annual	$25,903	$10,740	$11,042	$8,400	$10,908
Total	$25,903	$36,643	$47,685	$56,085	$66,993

Ownership Costs (5yr)

Average $74,154 / This Car $66,993

Cost/Mile $1.06 / Cost/Mile 96¢

Ownership Cost Rating

○ Better Than Average

Mercedes Benz 300 CE
2 Door Convertible

Luxury

 3.2L 217 hp Gas Fuel Inject.
 6 Cylinder In-Line
 Automatic 4 Speed
 2 Wheel Rear
 Driver/Psngr Airbags Std

Purchase Price

Car Item	Dealer Cost	List
Base Price	**$63,500**	**$76,500**
Anti-Lock Brakes	Std	Std
Manual Transmission	N/A	N/A
Optional Engine	N/A	N/A
Auto Climate Control	Std	Std
Power Steering	Std	Std
Cruise Control	Std	Std
All Wheel Drive	N/A	N/A
AM/FM Stereo Cassette	Std	Std
Steering, Tilt w/Memory	Std	Std
Power Windows	Std	Std
*Options Price	$0	$0
*Total Price	$63,500	$76,500
Target Price		$70,802
Destination Charge		$400
Avg. Tax & Fees		$4,329
Luxury Tax		$4,120
Total Target $		**$79,651**

The 1993 300 Class is available in ten models - 300D 2.5 Turbo, 300E 2.6, 300E 2.8, 300E 4MATIC, 400E and 500E sedans; 300TE and 300TE 4MATIC wagons; and the 300CE coupe and convertible. For 1993, the 300CE Cabriolet introduces a safety system featuring a rollbar integrated into the two rear-seat head restraints. These pop up within a fraction of a second should the car start to roll over, or if there is a severe front, side or rear impact.

Ownership Costs

Cost Area	5 Year Cost	Rate
Depreciation		⊖
Financing ($1,601/month)	$13,120	
Insurance (Rating 25)	$12,346	○
State Fees	$3,075	
Fuel (Hwy 23 City 18 -Prem.)	$4,652	○
Maintenance	$6,217	◐
Repairs	$1,215	○

Warranty/Maintenance Info

Major Tune-Up	$345 ●
Minor Tune-Up	$188 ●
Brake Service	$218 ○
Overall Warranty	4 yr/50k ○
Drivetrain Warranty	4 yr/50k ○
Rust Warranty	4 yr/50k ●
Maintenance Warranty	N/A
Roadside Assistance	4 yr/50k

Ownership Cost By Year

Insufficient Depreciation Information

1993 / 1994 / 1995 / 1996 / 1997

Resale Value

Insufficient Information

Cumulative Costs

	1993	1994	1995	1996	1997
Annual	Insufficient Information				
Total	Insufficient Information				

Ownership Costs (5yr)

Insufficient Information

Ownership Cost Rating

⊖ Insufficient Information

---

* Includes shaded options
** Other purchase requirements apply

● Poor ◐ Worse Than Average ◑ Average ○ Better Than Average ◯ Excellent ⊖ Insufficient Information

©1993 by IntelliChoice, Inc. (408) 554-8711 All Rights Reserved. Reproduction Prohibited.
Refer to *Section 3: Annotated Vehicle Charts* for an explanation of these charts.

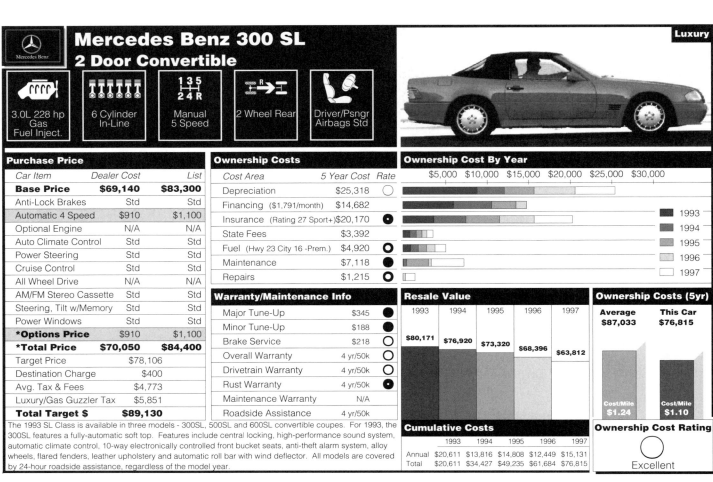

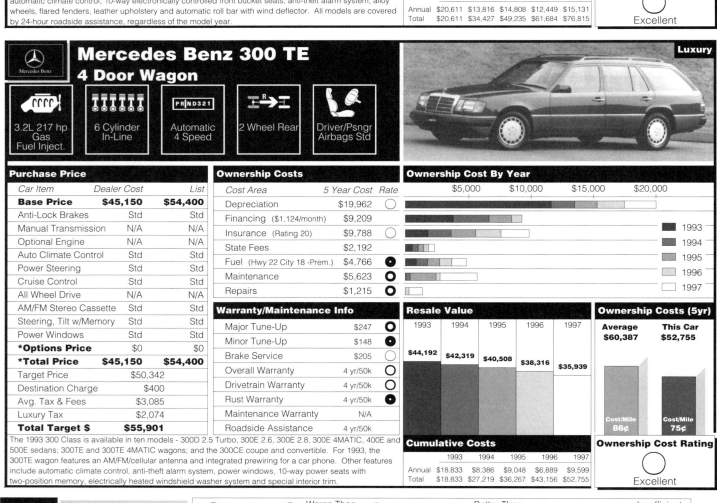

Mercedes Benz 300 E 4Matic
4 Door Sedan

 3.0L 177 hp Gas Fuel Inject. | 6 Cylinder In-Line | Automatic 4 Speed | 4 Wheel Full-Time | Driver/Psngr Airbags Std

2WD Model Shown — Luxury

Purchase Price

Car Item	Dealer Cost	List
Base Price	**$47,890**	**$57,700**
Anti-Lock Brakes	Std	Std
Manual Transmission	N/A	N/A
Optional Engine	N/A	N/A
Auto Climate Control	Std	Std
Power Steering	Std	Std
Cruise Control	Std	Std
4 Wheel Full-Time Drive	Std	Std
AM/FM Stereo Cassette	Std	Std
Steering Wheel, Scope	Std	Std
Power Windows	Std	Std
*Options Price	$0	$0
*Total Price	$47,890	$57,700
Target Price		$53,397
Destination Charge		$400
Avg. Tax & Fees		$3,271
Luxury/Gas Guzzler Tax		$3,380
Total Target $		**$60,448**

The 1993 300 Class is available in ten models - 300D 2.5 Turbo, 300E 2.6, 300E 2.8, 300E 4MATIC, 400E and 500E sedans; 300TE and 300TE 4MATIC wagons; and the 300CE coupe and convertible. For 1993, the 300E 4MATIC features an all-wheel drive system which distributes power between the front and rear wheels as needed. Other features include central locking, traction control, high-performance sound system, automatic climate control, 10-way power front bucket seats and leather upholstery.

Ownership Costs

Cost Area	5 Year Cost	Rate
Depreciation	$21,734	○
Financing ($1,215/month)	$9,957	
Insurance (Rating 20)	$9,788	○
State Fees	$2,325	
Fuel (Hwy 21 City 17 -Prem.)	$5,017	⊙
Maintenance	$5,695	○
Repairs	$1,215	○

Warranty/Maintenance Info

Major Tune-Up	$247	○
Minor Tune-Up	$148	⊙
Brake Service	$237	○
Overall Warranty	4 yr/50k	○
Drivetrain Warranty	4 yr/50k	○
Rust Warranty	4 yr/50k	●
Maintenance Warranty	N/A	
Roadside Assistance	4 yr/50k	

Ownership Cost By Year

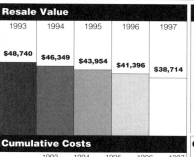

Resale Value

1993	1994	1995	1996	1997
$48,740	$46,349	$43,954	$41,396	$38,714

Cumulative Costs

	1993	1994	1995	1996	1997
Annual	$19,223	$9,216	$9,892	$7,384	$10,016
Total	$19,223	$28,439	$38,331	$45,715	$55,731

Ownership Costs (5yr)

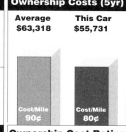

Average $63,318 | This Car $55,731
Cost/Mile 90¢ | Cost/Mile 80¢

Ownership Cost Rating

Excellent

Mercedes Benz 300 TE 4Matic
4 Door Wagon

 3.0L 177 hp Gas Fuel Inject. | 6 Cylinder In-Line | Automatic 4 Speed | 4 Wheel Full-Time | Driver/Psngr Airbags Std

2WD Model Shown — Luxury

Purchase Price

Car Item	Dealer Cost	List
Base Price	**$51,210**	**$61,700**
Anti-Lock Brakes	Std	Std
Manual Transmission	N/A	N/A
Optional Engine	N/A	N/A
Auto Climate Control	Std	Std
Power Steering	Std	Std
Cruise Control	Std	Std
4 Wheel Full-Time Drive	Std	Std
AM/FM Stereo Cassette	Std	Std
Steering, Tilt w/Memory	Std	Std
Power Windows	Std	Std
*Options Price	$0	$0
*Total Price	$51,210	$61,700
Target Price		$57,099
Destination Charge		$400
Avg. Tax & Fees		$3,496
Luxury/Gas Guzzler Tax		$3,750
Total Target $		**$64,745**

The 1993 300 Class is available in ten models - 300D 2.5 Turbo, 300E 2.6, 300E 2.8, 300E 4MATIC, 400E and 500E sedans; 300TE and 300TE 4MATIC wagons; and the 300CE coupe and convertible. For 1993, the 300TE 4MATIC features an AM/FM/cellular antenna and integrated prewiring for a car phone. Other features include automatic climate control, anti-theft alarm system, power windows, 10-way power seats with two-position memory, electrically heated windshield washer system and an active-bass sound system.

Ownership Costs

Cost Area	5 Year Cost	Rate
Depreciation	$22,706	○
Financing ($1,301/month)	$10,666	
Insurance (Rating 21)	$10,214	○
State Fees	$2,485	
Fuel (Hwy 21 City 17 -Prem.)	$5,017	⊙
Maintenance	$5,646	○
Repairs	$1,215	○

Warranty/Maintenance Info

Major Tune-Up	$247	○
Minor Tune-Up	$148	⊙
Brake Service	$205	○
Overall Warranty	4 yr/50k	○
Drivetrain Warranty	4 yr/50k	○
Rust Warranty	4 yr/50k	●
Maintenance Warranty	N/A	
Roadside Assistance	4 yr/50k	

Resale Value

1993	1994	1995	1996	1997
$52,092	$49,799	$47,554	$44,889	$42,039

Cumulative Costs

	1993	1994	1995	1996	1997
Annual	$20,581	$9,458	$9,965	$7,659	$10,286
Total	$20,581	$30,039	$40,004	$47,663	$57,949

Ownership Costs (5yr)

Average $66,870 | This Car $57,949
Cost/Mile 96¢ | Cost/Mile 83¢

Ownership Cost Rating

Excellent

---

* Includes shaded options
** Other purchase requirements apply

● Poor | ⊙ Worse Than Average | ◐ Average | ○ Better Than Average | ○ Excellent | ⊖ Insufficient Information

©1993 by IntelliChoice, Inc. (408) 554-8711 All Rights Reserved. Reproduction Prohibited.
Refer to *Section 3: Annotated Vehicle Charts* for an explanation of these charts.

Mercedes Benz 400 E
4 Door Sedan

 4.2L 275 hp Gas Fuel Inject.
 8 Cylinder "V"
 Automatic 4 Speed
 2 Wheel Rear
Driver/Psngr Airbags Std

Luxury

Purchase Price

Car Item	Dealer Cost	List
Base Price	**$46,810**	**$56,400**
Anti-Lock Brakes	Std	Std
Manual Transmission	N/A	N/A
Optional Engine	N/A	N/A
Auto Climate Control	Std	Std
Power Steering	Std	Std
Cruise Control	Std	Std
All Wheel Drive	N/A	N/A
AM/FM Stereo Cassette	Std	Std
Steering Wheel, Scope	Std	Std
Power Windows	Std	Std
*Options Price	$0	$0
*Total Price	$46,810	$56,400
Target Price		$52,193
Destination Charge		$400
Avg. Tax & Fees		$3,198
Luxury Tax		$2,259
Total Target $		**$58,050**

The 1993 300 Class is available in ten models - 300D 2.5 Turbo, 300E 2.6, 300E 2.8, 300E 4MATIC, 400E and 500E sedans; 300TE and 300TE 4MATIC wagons; and the 300CE coupe and convertible. For 1993, the 400E features central locking, high-performance sound system, automatic climate control, 10-way electronically controlled front bucket seats, anti-theft vehicle alarm system, and Supplemental Restraint System with dual airbags and three-point safety belts with height adjustable shoulder belt anchors.

Ownership Costs

Cost Area	5 Year Cost	Rate
Depreciation	$22,490	O
Financing ($1,167/month)	$9,563	
Insurance (Rating 20++)	$12,235	O
State Fees	$2,272	
Fuel (Hwy 24 City 18 -Prem.)	$4,547	O
Maintenance	$5,573	O
Repairs	$1,215	O

Warranty/Maintenance Info

Major Tune-Up	$303	●
Minor Tune-Up	$152	◉
Brake Service	$218	O
Overall Warranty	4 yr/50k	O
Drivetrain Warranty	4 yr/50k	O
Rust Warranty	4 yr/50k	◉
Maintenance Warranty	N/A	
Roadside Assistance	4 yr/50k	

Ownership Cost By Year

Bars across years 1993-1997 against scale $5,000–$25,000.

Resale Value

1993	1994	1995	1996	1997
$45,313	$43,030	$40,689	$38,187	$35,560

Cumulative Costs

	1993	1994	1995	1996	1997
Annual	$20,450	$9,378	$10,032	$7,715	$10,321
Total	$20,450	$29,828	$39,860	$47,575	$57,896

Ownership Costs (5yr)

Average	This Car
$62,163	$57,896
Cost/Mile 89¢	Cost/Mile 83¢

Ownership Cost Rating
O Better Than Average

Mercedes Benz 400 SEL
4 Door Sedan

 4.2L 275 hp Gas Fuel Inject.
 8 Cylinder "V"
 Automatic 4 Speed
 2 Wheel Rear
 Driver/Psngr Airbags Std

Luxury

Purchase Price

Car Item	Dealer Cost	List
Base Price	**$65,320**	**$78,700**
Anti-Lock Brakes	Std	Std
Manual Transmission	N/A	N/A
Optional Engine	N/A	N/A
Auto Climate Control	Std	Std
Power Steering	Std	Std
Cruise Control	Std	Std
All Wheel Drive	N/A	N/A
AM/FM Stereo Cassette	Std	Std
Steering, Tilt w/Memory	Std	Std
Power Windows	Std	Std
*Options Price	$0	$0
*Total Price	$65,320	$78,700
Target Price		$72,832
Destination Charge		$400
Avg. Tax & Fees		$4,453
Luxury/Gas Guzzler Tax		$6,423
Total Target $		**$84,108**

The 1993 S Class is available in five models - 300SD, 300SE, 400SEL, 500SEL and 600SEL sedans. New for 1993, the 400SEL features a longer wheelbase and additional rear legroom. Other features include remote central locking system, 12-way electrically adjustable front bucket seats with 3-position memory, Bose sound system, anti-theft alarm system and leather upholstery. The S-Class cars are completely CFC-free, including no ozone-depleting chlorofluorocarbons in their air conditioning systems.

Ownership Costs

Cost Area	5 Year Cost	Rate
Depreciation	$38,982	O
Financing ($1,690/month)	$13,854	
Insurance (Rating 23)	$11,629	O
State Fees	$3,165	
Fuel (Hwy 18 City 14 -Prem.)	$5,962	●
Maintenance	$6,521	●
Repairs	$1,215	O

Warranty/Maintenance Info

Major Tune-Up	$303	●
Minor Tune-Up	$152	◉
Brake Service	$342	●
Overall Warranty	4 yr/50k	O
Drivetrain Warranty	4 yr/50k	O
Rust Warranty	4 yr/50k	◉
Maintenance Warranty	N/A	
Roadside Assistance	4 yr/50k	

Ownership Cost By Year

Bars across years 1993-1997 against scale $5K–$40K.

Resale Value

1993	1994	1995	1996	1997
$58,868	$55,257	$51,821	$48,362	$45,126

Cumulative Costs

	1993	1994	1995	1996	1997
Annual	$35,113	$12,404	$12,776	$9,307	$11,728
Total	$35,113	$47,517	$60,293	$69,600	$81,328

Ownership Costs (5yr)

Average	This Car
$81,970	$81,328
Cost/Mile $1.17	Cost/Mile $1.16

Ownership Cost Rating
O Average

---

* Includes shaded options
** Other purchase requirements apply

● Poor ◉ Worse Than Average O Average ◯ Better Than Average ○ Excellent ⊖ Insufficient Information

©1993 by IntelliChoice, Inc. (408) 554-8711 All Rights Reserved. Reproduction Prohibited.
Refer to *Section 3: Annotated Vehicle Charts* for an explanation of these charts.

Mercedes Benz 500 E
4 Door Sedan

 5.0L 315 hp Gas Fuel Inject.
 8 Cylinder "V"
 Automatic 4 Speed
 2 Wheel Rear
 Driver/Psngr Airbags Std

Luxury

Purchase Price

Car Item	Dealer Cost	List
Base Price	**$66,400**	**$80,000**
Anti-Lock Brakes	Std	Std
Manual Transmission	N/A	N/A
Optional Engine	N/A	N/A
Auto Climate Control	Std	Std
Power Steering	Std	Std
Cruise Control	Std	Std
All Wheel Drive	N/A	N/A
AM/FM Stereo Cassette	Std	Std
Steering Wheel, Scope	Std	Std
Power Windows	Std	Std
*Options Price	$0	$0
*Total Price	$66,400	$80,000
Target Price	$74,036	
Destination Charge	$400	
Avg. Tax & Fees	$4,526	
Luxury/Gas Guzzler Tax	$6,144	
Total Target $	**$85,106**	

The 1993 300 Class is available in ten models - 300D 2.5 Turbo, 300E 2.6, 300E 2.8, 300E 4MATIC, 400E and 500E sedans; 300TE and 300TE 4MATIC wagons; and the 300CE coupe and convertible. For 1993, the 500E features central locking, high-performance sound system, automatic climate control, 10-way power front bucket seats, anti-theft vehicle alarm system, alloy wheels, flared fenders, leather upholstery and Automatic Slip Control. All models are covered by 24-hour roadside assistance.

Ownership Costs

Cost Area	5 Year Cost	Rate
Depreciation	$39,136	○
Financing ($1,711/month)	$14,020	
Insurance (Rating 22++)	$13,300	○
State Fees	$3,216	
Fuel (Hwy 19 City 16 -Prem.)	$5,449	●
Maintenance	$6,962	●
Repairs	$1,215	○

Warranty/Maintenance Info

Major Tune-Up	$332	●
Minor Tune-Up	$169	●
Brake Service	$218	○
Overall Warranty	4 yr/50k	○
Drivetrain Warranty	4 yr/50k	○
Rust Warranty	4 yr/50k	●
Maintenance Warranty	N/A	
Roadside Assistance	4 yr/50k	

Ownership Cost By Year

$5K $10K $15K $20K $25K $30K $35K $40K

- 1993
- 1994
- 1995
- 1996
- 1997

Resale Value

1993	1994	1995	1996	1997
$62,326	$58,493	$54,834	$50,081	$45,970

Cumulative Costs

	1993	1994	1995	1996	1997
Annual	$32,948	$12,930	$13,464	$10,884	$13,073
Total	$32,948	$45,878	$59,342	$70,226	$83,299

Ownership Costs (5yr)

Average	This Car
$83,125	$83,299
Cost/Mile $1.19	Cost/Mile $1.19

Ownership Cost Rating

○ Average

Mercedes Benz 500 SEL
4 Door Sedan

 5.0L 315 hp Gas Fuel Inject.
 8 Cylinder "V"
 Automatic 4 Speed
 2 Wheel Rear
 Driver/Psngr Airbags Std

Luxury

Purchase Price

Car Item	Dealer Cost	List
Base Price	**$78,350**	**$94,400**
Anti-Lock Brakes	Std	Std
Manual Transmission	N/A	N/A
Optional Engine	N/A	N/A
Auto Climate Control	Std	Std
Power Steering	Std	Std
Cruise Control	Std	Std
All Wheel Drive	N/A	N/A
AM/FM Stereo Cassette	Std	Std
Steering, Tilt w/Memory	Std	Std
Power Windows	Std	Std
*Options Price	$0	$0
*Total Price	$78,350	$94,400
Target Price	$87,360	
Destination Charge	$400	
Avg. Tax & Fees	$5,336	
Luxury/Gas Guzzler Tax	$8,776	
Total Target $	**$101,872**	

The 1993 S Class is available in five models - 300SD, 300SE, 400SEL, 500SEL and 600SEL sedans. For 1993, the 500SEL features 11-speaker Bose sound system, soft-close power door latches, remote central locking system, 12-way power adjustable front bucket seats, anti-theft alarm system and leather upholstery. The S-Class cars are completely CFC-free, including no ozone-depleting chlorofluorocarbons in their air conditioning systems. All models are covered by a 24-hour roadside assistance program.

Ownership Costs

Cost Area	5 Year Cost	Rate
Depreciation	$47,466	○
Financing ($2,047/month)	$16,781	
Insurance (Rating 27+)	$17,288	●
State Fees	$3,792	
Fuel (Hwy 17 City 13 -Prem.)	$6,362	●
Maintenance	$6,527	●
Repairs	$1,215	○

Warranty/Maintenance Info

Major Tune-Up	$332	●
Minor Tune-Up	$169	●
Brake Service	$218	○
Overall Warranty	4 yr/50k	○
Drivetrain Warranty	4 yr/50k	○
Rust Warranty	4 yr/50k	●
Maintenance Warranty	N/A	
Roadside Assistance	4 yr/50k	

Ownership Cost By Year

$5K $10K $15K $20K $25K $30K $35K $40K $45K $50K

- 1993
- 1994
- 1995
- 1996
- 1997

Resale Value

1993	1994	1995	1996	1997
$71,086	$66,883	$62,268	$58,429	$54,406

Cumulative Costs

	1993	1994	1995	1996	1997
Annual	$43,154	$15,233	$15,851	$11,288	$13,904
Total	$43,154	$58,387	$74,238	$85,526	$99,430

Ownership Costs (5yr)

Average	This Car
$95,915	$99,430
Cost/Mile $1.37	Cost/Mile $1.42

Ownership Cost Rating

○ Average

---

* Includes shaded options
** Other purchase requirements apply

● Poor   ◐ Worse Than Average   ○ Average   ◯ Better Than Average   ✦ Excellent   ⊖ Insufficient Information

©1993 by IntelliChoice, Inc. (408) 554-8711   All Rights Reserved. Reproduction Prohibited.
Refer to *Section 3: Annotated Vehicle Charts* for an explanation of these charts.

Mercedes Benz 500 SL
2 Door Convertible — Luxury

 5.0L 315 hp Gas Fuel Inject.   8 Cylinder "V"   Automatic 4 Speed   2 Wheel Rear   Driver/Psngr Airbags Std

Purchase Price

Car Item	Dealer Cost	List
Base Price	N/R	$98,500
Anti-Lock Brakes	Std	Std
Manual Transmission	N/A	N/A
Optional Engine	N/A	N/A
Auto Climate Control	Std	Std
Power Steering	Std	Std
Cruise Control	Std	Std
All Wheel Drive	N/A	N/A
AM/FM Stereo Cassette	Std	Std
Steering, Tilt w/Memory	Std	Std
Power Windows	Std	Std
*Options Price	$0	$0
*Total Price	N/R	$98,500
Target Price		$98,500
Destination Charge		$400
Avg. Tax & Fees		$5,934
Luxury/Gas Guzzler Tax		$8,990
Total Target $		**$113,824**

Ownership Costs

Cost Area	5 Year Cost	Rate
Depreciation		⊖
Financing ($2,288/month)	$18,751	
Insurance (27 Sport+ [Est.])	$20,170	●
State Fees	$3,955	
Fuel (Hwy 17 City 13 -Prem.)	$6,362	●
Maintenance	$6,962	●
Repairs	$1,215	○

Warranty/Maintenance Info

Major Tune-Up	$332	●
Minor Tune-Up	$169	●
Brake Service	$218	○
Overall Warranty	4 yr/50k	○
Drivetrain Warranty	4 yr/50k	○
Rust Warranty	4 yr/50k	●
Maintenance Warranty	N/A	
Roadside Assistance	4 yr/50k	

Ownership Cost By Year — Insufficient Depreciation Information
■ 1993 ■ 1994 ■ 1995 ■ 1996 □ 1997

Resale Value: Insufficient Information

Ownership Costs (5yr): Insufficient Information

Cumulative Costs (1993 1994 1995 1996 1997): Annual — Insufficient Information; Total — Insufficient Information

Ownership Cost Rating: ⊖ Insufficient Information

The 1993 SL Class is available in three models - 300SL, 500SL and 600SL convertible coupes. For 1993, the 500SL features a fully-automatic soft top. Features include remote central locking, high-performance sound system, automatic climate control, 10-way electronically controlled front bucket seats, Automatic Slip Control, alloy wheels, flared fenders, leather upholstery and automatic roll bar with wind deflector. All models are covered by 24-hour roadside assistance, regardless of the model year.

Mercedes Benz 600 SEL
4 Door Sedan — Luxury

 6.0L 389 hp Gas Fuel Inject.   12 Cylinder "V"   Automatic 4 Speed   2 Wheel Rear   Driver/Psngr Airbags Std

Purchase Price

Car Item	Dealer Cost	List
Base Price	N/R	$129,000
Anti-Lock Brakes	Std	Std
Manual Transmission	N/A	N/A
Optional Engine	N/A	N/A
Auto Climate Control	Std	Std
Power Steering	Std	Std
Cruise Control	Std	Std
All Wheel Drive	N/A	N/A
AM/FM Stereo Cassette	Std	Std
Steering, Tilt w/Memory	Std	Std
Power Windows	Std	Std
*Options Price	$0	$0
*Total Price	N/R	$129,000
Target Price		$129,000
Destination Charge		N/R
Avg. Tax & Fees		$7,740
Luxury/Gas Guzzler Tax		$13,600
Total Target $		**$150,340**

Ownership Costs

Cost Area	5 Year Cost	Rate
Depreciation		⊖
Financing ($3,022/month)	$24,766	
Insurance (27 Sport+ [Est.])	$19,071	●
State Fees	$5,160	
Fuel (Hwy 16 City 12 -Prem.)	$6,821	●
Maintenance	$7,005	●
Repairs	$1,505	●

Warranty/Maintenance Info

Major Tune-Up	$356	●
Minor Tune-Up	$187	●
Brake Service	$342	●
Overall Warranty	4 yr/50k	○
Drivetrain Warranty	4 yr/50k	○
Rust Warranty	4 yr/50k	●
Maintenance Warranty	N/A	
Roadside Assistance	4 yr/50k	

Ownership Cost By Year — Insufficient Depreciation Information
■ 1993 ■ 1994 ■ 1995 ■ 1996 □ 1997

Resale Value: Insufficient Information

Ownership Costs (5yr): Insufficient Information

Cumulative Costs (1993 1994 1995 1996 1997): Annual — Insufficient Information; Total — Insufficient Information

Ownership Cost Rating: ⊖ Insufficient Information

The 1993 S Class is available in five models - 300SD, 300SE, 400SEL, 500SEL and 600SEL sedans. For 1993, the 600SEL features 11-speaker Bose sound system, soft-close power door latches, remote central locking system, 12-way power adjustable front bucket seats, anti-theft alarm system and leather upholstery. The S-Class cars are completely CFC-free, including no ozone-depleting chlorofluorocarbons in their air conditioning systems. All models are covered by a 24-hour roadside assistance program.

---

*Includes shaded options   ** Other purchase requirements apply

● Poor   ◐ Worse Than Average   ◓ Average   ◔ Better Than Average   ○ Excellent   ⊖ Insufficient Information

©1993 by IntelliChoice, Inc. (408) 554-8711   All Rights Reserved. Reproduction Prohibited.
Refer to *Section 3: Annotated Vehicle Charts* for an explanation of these charts.

Mercury Capri
2 Door Convertible

Subcompact

- 1.6L 100 hp Gas Fuel Inject.
- 4 Cylinder In-Line
- Manual 5 Speed
- 2 Wheel Front
- Driver Airbag Psngr Belts

Purchase Price

Car Item	Dealer Cost	List
Base Price	**$12,429**	**$14,452**
Anti-Lock Brakes	N/A	N/A
Automatic 4 Speed	$622	$732
Optional Engine	N/A	N/A
Air Conditioning	$659	$817
Power Steering	Std	Std
Cruise Control	$191	$224
All Wheel Drive	N/A	N/A
AM/FM Stereo Cassette	$145	$171
Steering Wheel, Tilt	N/A	N/A
Power Windows	Std	Std
*Options Price	$0	$0
*Total Price	$12,429	$14,452
Target Price		$13,135
Destination Charge		$375
Avg. Tax & Fees		$824
Total Target $		**$14,334**
Average Dealer Option Cost:		83%

The Capri convertible is available in (Base) Capri and XR2 editions. New for 1993, the Capri features new corporate radios. The (Base) Capri features styled steel wheels with center caps, driver side airbag, power driver and passenger side mirrors, electronic AM/FM stereo, body color bumpers and cloth sport bucket seats. Options on the base model include air conditioning, cruise control, removable hardtop roof and leather seating material.

Ownership Costs

Cost Area	5 Year Cost	Rate
Depreciation	$6,910	○
Financing ($288/month)	$2,362	
Insurance (Rating 14)	$8,238	●
State Fees	$593	
Fuel (Hwy 31 City 25)	$3,083	●
Maintenance	$4,914	●
Repairs	$709	○

Warranty/Maintenance Info

Major Tune-Up	$245	●
Minor Tune-Up	$154	●
Brake Service	$192	○
Overall Warranty	3 yr/36k	○
Drivetrain Warranty	3 yr/36k	○
Rust Warranty	6 yr/100k	○
Maintenance Warranty	N/A	
Roadside Assistance	N/A	

Ownership Cost By Year

1993, 1994, 1995, 1996, 1997

Resale Value

1993	1994	1995	1996	1997
$12,790	$10,798	$9,662	$8,525	$7,424

Cumulative Costs

	1993	1994	1995	1996	1997
Annual	$4,813	$5,332	$5,428	$4,998	$6,238
Total	$4,813	$10,145	$15,573	$20,571	$26,809

Ownership Costs (5yr)

Average	This Car
$27,838	$26,809
Cost/Mile 40¢	Cost/Mile 38¢

Ownership Cost Rating

○ Better Than Average

Mercury Capri XR2
2 Door Convertible

Subcompact

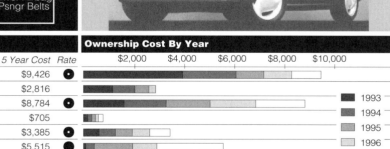

- 1.6L 132 hp Turbo Gas Fuel Inject.
- 4 Cylinder In-Line
- Manual 5 Speed
- 2 Wheel Front
- Driver Airbag Psngr Belts

Purchase Price

Car Item	Dealer Cost	List
Base Price	**$14,835**	**$17,250**
Anti-Lock Brakes	N/A	N/A
Automatic Transmission	N/A	N/A
Optional Engine	N/A	N/A
Air Conditioning	Std	Std
Power Steering	Std	Std
Cruise Control	Std	Std
All Wheel Drive	N/A	N/A
AM/FM Stereo Cassette	Std	Std
Steering Wheel, Tilt	N/A	N/A
Power Windows	Std	Std
*Options Price	$0	$0
*Total Price	$14,835	$17,250
Target Price		$15,738
Destination Charge		$375
Avg. Tax & Fees		$982
Total Target $		**$17,095**
Average Dealer Option Cost:		85%

The Capri convertible is available in (Base) Capri and XR2 editions. New for 1993, the Capri features new corporate radios. The upgraded XR2 model features 3-spoke aluminum wheels, electronic AM/FM stereo with cassette tape player, front fog lamps, decklid spoiler and unique cloth interior trim and seating material as standard equipment. Optional on the XR2 is a removable hardtop roof, tonneau cover, premium stereo system and leather seating material.

Ownership Costs

Cost Area	5 Year Cost	Rate
Depreciation	$9,426	○
Financing ($344/month)	$2,816	
Insurance (Rating 16)	$8,784	●
State Fees	$705	
Fuel (Hwy 28 City 23)	$3,385	●
Maintenance	$5,515	●
Repairs	$954	●

Warranty/Maintenance Info

Major Tune-Up	$250	●
Minor Tune-Up	$154	●
Brake Service	$192	○
Overall Warranty	3 yr/36k	○
Drivetrain Warranty	3 yr/36k	○
Rust Warranty	6 yr/100k	○
Maintenance Warranty	N/A	
Roadside Assistance	N/A	

Ownership Cost By Year

1993, 1994, 1995, 1996, 1997

Resale Value

1993	1994	1995	1996	1997
$13,129	$10,991	$9,886	$8,764	$7,669

Cumulative Costs

	1993	1994	1995	1996	1997
Annual	$7,654	$5,855	$5,897	$5,332	$6,847
Total	$7,654	$13,509	$19,406	$24,738	$31,585

Ownership Costs (5yr)

Average	This Car
$30,986	$31,585
Cost/Mile 44¢	Cost/Mile 45¢

Ownership Cost Rating

○ Average

---

* Includes shaded options
** Other purchase requirements apply

● Poor ◐ Worse Than Average ◑ Average ○ Better Than Average ○ Excellent ⊖ Insufficient Information

©1993 by IntelliChoice, Inc. (408) 554-8711 All Rights Reserved. Reproduction Prohibited.
Refer to *Section 3: Annotated Vehicle Charts* for an explanation of these charts.

page 141

Mercury Cougar XR7
2 Door Coupe

 3.8L 140 hp Gas Fuel Inject.
 6 Cylinder "V"
 Automatic 4 Speed
 2 Wheel Rear
 Automatic Seatbelts
 Large

Purchase Price

Car Item	Dealer Cost	List
Base Price	**$13,351**	**$14,855**
Anti-Lock Brakes	$591	$695
Manual Transmission	N/A	N/A
5.0L 200 hp Gas	$1,012	$1,190
Air Conditioning	Std	Std
Power Steering	Std	Std
Cruise Control	$191	** $224
All Wheel Drive	N/A	N/A
AM/FM Stereo Cassette	$132	$155
Steering Wheel, Tilt	$123	** $145
Power Windows	Std	Std
*Options Price	$1,144	$1,345
*Total Price	$14,495	$16,200
Target Price	$15,440	
Destination Charge	$495	
Avg. Tax & Fees	$964	
Total Target $	**$16,899**	
Average Dealer Option Cost:	**85%**	

Ownership Costs

Cost Area	5 Year Cost	Rate
Depreciation	$10,251	●
Financing ($340/month)	$2,783	
Insurance (Rating 8 [Est.])	$7,223	○
State Fees	$668	
Fuel (Hwy 24 City 17)	$4,232	●
Maintenance	$4,375	○
Repairs	$709	○

Warranty/Maintenance Info

Major Tune-Up	$201	○
Minor Tune-Up	$170	●
Brake Service	$319	●
Overall Warranty	3 yr/36k	○
Drivetrain Warranty	3 yr/36k	○
Rust Warranty	6 yr/100k	○
Maintenance Warranty	N/A	
Roadside Assistance	N/A	

Ownership Cost By Year
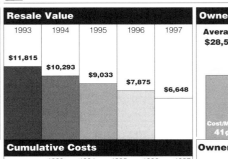

Resale Value
1993	1994	1995	1996	1997
$11,815	$10,293	$9,033	$7,875	$6,648

Cumulative Costs
	1993	1994	1995	1996	1997
Annual	$8,577	$5,043	$5,714	$4,972	$5,935
Total	$8,577	$13,620	$19,334	$24,306	$30,241

Ownership Costs (5yr)
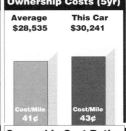
Average $28,535 — This Car $30,241
Cost/Mile 41¢ — Cost/Mile 43¢

Ownership Cost Rating

Poor

The 1993 Mercury Cougar XR7 is available in one model. New for 1993, the Cougar XR7 features a sport instrument cluster, leather-wrapped shift knob, color-keyed steering wheel and column, new wheel covers and color-keyed roof moldings and outside mirrors. Options include a 5.0L V8 engine, Ford JBL audio system, CD player, power locks, electric rear window defroster, tilt steering wheel, 6-way dual power seats, speed control, electronic climate control and a remote-control keyless entry system.

Mercury Grand Marquis GS
4 Door Sedan

 4.6L 190 hp Gas Fuel Inject.
 8 Cylinder "V"
 Automatic 4 Speed
 2 Wheel Rear
Driver/Psngr Airbags Std

LS Model Shown

Purchase Price

Car Item	Dealer Cost	List
Base Price	**$18,857**	**$21,973**
Anti-Lock Brakes	$591	$695
Manual Transmission	N/A	N/A
Optional Engine	N/A	N/A
Air Conditioning	Std	Std
Power Steering	Std	Std
Cruise Control	Pkg	Pkg
All Wheel Drive	N/A	N/A
AM/FM Stereo Cassette	Std	Std
Steering Wheel, Tilt	Std	Std
Power Windows	Std	Std
*Options Price	$0	$0
*Total Price	$18,857	$21,973
Target Price	$20,211	
Destination Charge	$545	
Avg. Tax & Fees	$1,263	
Total Target $	**$22,019**	
Average Dealer Option Cost:	**85%**	

Ownership Costs

Cost Area	5 Year Cost	Rate
Depreciation	$11,794	○
Financing ($443/month)	$3,627	
Insurance (Rating 10)	$7,479	◔
State Fees	$901	
Fuel (Hwy 26 City 18)	$3,949	○
Maintenance	$4,335	○
Repairs	$761	●

Warranty/Maintenance Info

Major Tune-Up	$157	○
Minor Tune-Up	$101	○
Brake Service	$316	●
Overall Warranty	3 yr/36k	○
Drivetrain Warranty	3 yr/36k	○
Rust Warranty	6 yr/100k	○
Maintenance Warranty	N/A	
Roadside Assistance	N/A	

Resale Value
1993	1994	1995	1996	1997
$16,508	$14,745	$13,235	$11,778	$10,225

Cumulative Costs
	1993	1994	1995	1996	1997
Annual	$9,409	$5,549	$6,243	$5,171	$6,474
Total	$9,409	$14,958	$21,201	$26,372	$32,846

Ownership Costs (5yr)
Average $34,262 — This Car $32,846
Cost/Mile 49¢ — Cost/Mile 47¢

Ownership Cost Rating
○ Better Than Average

The Grand Marquis four-door sedan is available in GS and LS editions. New features on the GS sedan include dual air bags, electronically controlled overdrive transmission, a stainless steel exhaust system and power driver seat. Other standard features include wide bodyside moldings, a backlighted analog instrumentation panel and the Luxury Sound Insulation Package. Optional on the GS are a keyless entry system, leather steering wheel and cast aluminum wheels.

---

* Includes shaded options
** Other purchase requirements apply

● Poor ◐ Worse Than Average ◑ Average ○ Better Than Average ○ Excellent ⊖ Insufficient Information

©1993 by IntelliChoice, Inc. (408) 554-8711   All Rights Reserved. Reproduction Prohibited.
Refer to *Section 3: Annotated Vehicle Charts* for an explanation of these charts.

Mercury Grand Marquis LS
4 Door Sedan

Large

 4.6L 190 hp Gas Fuel Inject.
 8 Cylinder "V"
 Automatic 4 Speed
 2 Wheel Rear
Driver/Psngr Airbags Std

Purchase Price

Car Item	Dealer Cost	List
Base Price	$19,305	$22,500
Anti-Lock Brakes	$591	$695
Manual Transmission	N/A	N/A
Optional Engine	N/A	N/A
Air Conditioning	Std	Std
Power Steering	Std	Std
Cruise Control	Pkg	Pkg
All Wheel Drive	N/A	N/A
AM/FM Stereo Cassette	Std	Std
Steering Wheel, Tilt	Std	Std
Power Windows	Std	Std
***Options Price**	$0	$0
***Total Price**	$19,305	$22,500
Target Price	$20,706	
Destination Charge	$545	
Avg. Tax & Fees	$1,293	
Total Target $	$22,544	
Average Dealer Option Cost:	**85%**	

The Grand Marquis four-door sedan is avaialable in GS and LS editions. New features on the LS sedan include dual air bags, electronically controlled overdrive transmission, a stainless steel exhaust system, brake shift interlock, and power driver seat. The LS upgrades the GS by offering a vinyl roof, 6-way power passenger seat, leather seating material and Performance/Handling Package as optional equipment. The Grand Marquis is a sister to the Lincoln Town Car.

Ownership Costs

Cost Area	5 Year Cost	Rate
Depreciation	$11,911	O
Financing ($453/month)	$3,714	
Insurance (Rating 10)	$7,479	O
State Fees	$920	
Fuel (Hwy 26 City 18)	$3,949	O
Maintenance	$4,335	O
Repairs	$761	●

Warranty/Maintenance Info

Major Tune-Up	$157	O
Minor Tune-Up	$101	O
Brake Service	$316	●
Overall Warranty	3 yr/36k	O
Drivetrain Warranty	3 yr/36k	O
Rust Warranty	6 yr/100k	O
Maintenance Warranty	N/A	
Roadside Assistance	N/A	

Ownership Cost By Year

1993 / 1994 / 1995 / 1996 / 1997

Resale Value

1993	1994	1995	1996	1997
$17,095	$15,295	$13,727	$12,215	$10,633

Cumulative Costs

	1993	1994	1995	1996	1997
Annual	$9,388	$5,617	$6,322	$5,236	$6,506
Total	$9,388	$15,005	$21,327	$26,563	$33,069

Ownership Costs (5yr)

Average: $34,784 (Cost/Mile 50¢)
This Car: $33,069 (Cost/Mile 47¢)

Ownership Cost Rating

O **Better Than Average**

Mercury Sable GS
4 Door Sedan

Midsize

 3.0L 140 hp Gas Fuel Inject.
 6 Cylinder "V"
 Automatic 4 Speed
 2 Wheel Front
 Driver/Psngr Airbags Std

Purchase Price

Car Item	Dealer Cost	List
Base Price	$14,891	$17,349
Anti-Lock Brakes	$506	$595
Manual Transmission	N/A	N/A
3.8L 140 hp Gas	$472	$555
Air Conditioning	Std	Std
Power Steering	Std	Std
Cruise Control	Pkg	Pkg
All Wheel Drive	N/A	N/A
AM/FM Stereo Cassette	$145	$171
Steering Wheel, Tilt	Std	Std
Power Windows	Pkg	** Pkg
***Options Price**	$145	$171
***Total Price**	$15,036	$17,520
Target Price	$16,277	
Destination Charge	$490	
Avg. Tax & Fees	$1,018	
Total Target $	$17,785	
Average Dealer Option Cost:	**85%**	

Sable is offered as a four-door sedan and four-door wagon in GS and LS editions. The 1993 GS sedan features new body color bumpers and bodyside moldings, new integrated console, dual illuminated visor mirrors and new seat fabrics. Other standard features include a driver side air bag, rear heat ducts, backlighted instrument cluster with tachometer, temperature and fuel gauges, and luggage rack. A larger engine and extended range fuel tank are optional. The Sable is a sister to the Ford Taurus.

Ownership Costs

Cost Area	5 Year Cost	Rate
Depreciation	$9,462	O
Financing ($357/month)	$2,930	
Insurance (Rating 4)	$6,658	○
State Fees	$720	
Fuel (Hwy 30 City 21)	$3,405	O
Maintenance	$4,503	O
Repairs	$709	O

Warranty/Maintenance Info

Major Tune-Up	$149	O
Minor Tune-Up	$99	O
Brake Service	$253	O
Overall Warranty	3 yr/36k	O
Drivetrain Warranty	3 yr/36k	O
Rust Warranty	6 yr/100k	O
Maintenance Warranty	N/A	
Roadside Assistance	N/A	

Resale Value

1993	1994	1995	1996	1997
$13,617	$12,022	$10,812	$9,541	$8,323

Cumulative Costs

	1993	1994	1995	1996	1997
Annual	$7,476	$4,832	$5,452	$4,946	$5,681
Total	$7,476	$12,308	$17,760	$22,706	$28,387

Ownership Costs (5yr)

Average: $29,183 (Cost/Mile 42¢)
This Car: $28,387 (Cost/Mile 41¢)

Ownership Cost Rating

O **Better Than Average**

---

* Includes shaded options
** Other purchase requirements apply

● Poor ◐ Worse Than Average ○ Average ○ Better Than Average ○ Excellent ⊖ Insufficient Information

©1993 by IntelliChoice, Inc. (408) 554-8711 All Rights Reserved. Reproduction Prohibited.
Refer to *Section 3: Annotated Vehicle Charts* for an explanation of these charts.

Mercury Sable LS — 4 Door Sedan — Midsize

- 3.8L 140 hp Gas Fuel Inject.
- 6 Cylinder "V"
- Automatic 4 Speed
- 2 Wheel Front
- Driver/Psngr Airbags Std

Purchase Price

Car Item	Dealer Cost	List
Base Price	**$15,811**	**$18,430**
Anti-Lock Brakes	$506	$595
Manual Transmission	N/A	N/A
3.0L 140 hp Gas	($472)	($555)
Air Conditioning	Std	Std
Power Steering	Std	Std
Cruise Control	Pkg	Pkg
All Wheel Drive	N/A	N/A
AM/FM Stereo Cassette	$139	** $163
Steering Wheel, Tilt	Std	Std
Power Windows	Std	Std
***Options Price**	**$0**	**$0**
***Total Price**	**$15,811**	**$18,430**
Target Price		$17,149
Destination Charge		$490
Avg. Tax & Fees		$1,071
Total Target $		**$18,710**
Average Dealer Option Cost:		85%

Sable is offered as a four-door sedan and four-door wagon in GS and LS editions. The 1993 LS sedan features new body color bumpers and bodyside moldings, new integrated console, dual illuminated visor mirrors and new seat fabrics. The LS sedan upgrades the GS by offering rear center armrests, a cargo/convenience net, remote fuel and trunk releases, Light Group and a larger engine as standard equipment. Options on the LS include a power sun roof and keyless remote entry system.

Ownership Costs

Cost Area	5 Year Cost	Rate
Depreciation	$10,050	O
Financing ($376/month)	$3,082	
Insurance (Rating 5)	$6,786	◯
State Fees	$756	
Fuel (Hwy 27 City 19)	$3,775	O
Maintenance	$4,572	O
Repairs	$709	O

Warranty/Maintenance Info

Major Tune-Up	$159	O
Minor Tune-Up	$109	O
Brake Service	$253	O
Overall Warranty	3 yr/36k	O
Drivetrain Warranty	3 yr/36k	O
Rust Warranty	6 yr/100k	O
Maintenance Warranty	N/A	
Roadside Assistance	N/A	

Ownership Cost By Year

1993, 1994, 1995, 1996, 1997

Resale Value

1993	1994	1995	1996	1997
$14,369	$12,682	$11,353	$9,968	$8,660

Cumulative Costs

	1993	1994	1995	1996	1997
Annual	$7,814	$5,085	$5,719	$5,192	$5,920
Total	$7,814	$12,899	$18,618	$23,810	$29,730

Ownership Costs (5yr)

- Average: $30,113
- This Car: $29,730
- Cost/Mile: 43¢ / 42¢

Ownership Cost Rating: Average

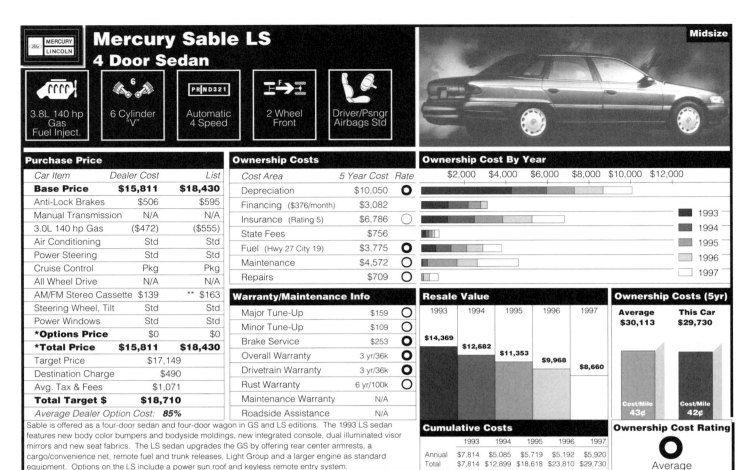

Mercury Sable GS — 4 Door Wagon — Midsize Wagon

- 3.0L 140 hp Gas Fuel Inject.
- 6 Cylinder "V"
- Automatic 4 Speed
- 2 Wheel Front
- Driver/Psngr Airbags Std

LS Model Shown

Purchase Price

Car Item	Dealer Cost	List
Base Price	**$15,835**	**$18,459**
Anti-Lock Brakes	$506	$595
Manual Transmission	N/A	N/A
3.8L 140 hp Gas	$472	$555
Air Conditioning	Std	Std
Power Steering	Std	Std
Cruise Control	Pkg	Pkg
All Wheel Drive	N/A	N/A
AM/FM Stereo Cassette	$145	$171
Steering Wheel, Tilt	Std	Std
Power Windows	Pkg	** Pkg
***Options Price**	**$145**	**$171**
***Total Price**	**$15,980**	**$18,630**
Target Price		$17,334
Destination Charge		$490
Avg. Tax & Fees		$1,082
Total Target $		**$18,906**
Average Dealer Option Cost:		85%

Sable is offered as a four-door sedan and four-door wagon in GS and LS editions. The 1993 GS wagon features new body color bumpers and bodyside moldings, new integrated console, dual illuminated visor mirrors and new seat fabrics. Other standard features include a driver side air bag, rear heat ducts, backlighted instrument cluster with tachometer, temperature and fuel gauges, and luggage rack. A cargo cover and picnic tray are optional. The Sable is a sister to the Ford Taurus.

Ownership Costs

Cost Area	5 Year Cost	Rate
Depreciation	$8,565	O
Financing ($380/month)	$3,113	
Insurance (Rating 5)	$6,786	◯
State Fees	$765	
Fuel (Hwy 30 City 21)	$3,405	◯
Maintenance	$4,369	O
Repairs	$709	O

Warranty/Maintenance Info

Major Tune-Up	$149	◯
Minor Tune-Up	$99	O
Brake Service	$271	●
Overall Warranty	3 yr/36k	O
Drivetrain Warranty	3 yr/36k	O
Rust Warranty	6 yr/100k	O
Maintenance Warranty	N/A	
Roadside Assistance	N/A	

Resale Value

1993	1994	1995	1996	1997
$15,455	$13,981	$12,849	$11,585	$10,341

Cumulative Costs

	1993	1994	1995	1996	1997
Annual	$6,871	$4,803	$5,463	$4,813	$5,762
Total	$6,871	$11,674	$17,137	$21,950	$27,712

Ownership Costs (5yr)

- Average: $30,318
- This Car: $27,712
- Cost/Mile: 43¢ / 40¢

Ownership Cost Rating: Excellent

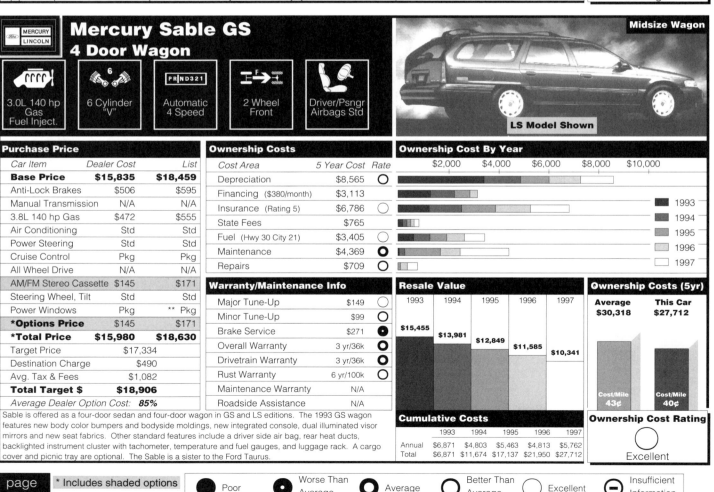

---

page 144

* Includes shaded options
** Other purchase requirements apply

Legend: ● Poor · ◐ Worse Than Average · ◯ Average · ○ Better Than Average · ○ Excellent · ⊖ Insufficient Information

©1993 by IntelliChoice, Inc. (408) 554-8711 All Rights Reserved. Reproduction Prohibited.
Refer to *Section 3: Annotated Vehicle Charts* for an explanation of these charts.

Mercury Sable LS
4 Door Wagon

Midsize Wagon

- 3.8L 140 hp Gas Fuel Inject.
- 6 Cylinder "V"
- Automatic 4 Speed
- 2 Wheel Front
- Driver/Psngr Airbags Std

Purchase Price

Car Item	Dealer Cost	List
Base Price	**$16,684**	**$19,457**
Anti-Lock Brakes	$506	$595
Manual Transmission	N/A	N/A
3.0L 140 hp Gas	($472)	($555)
Air Conditioning	Std	Std
Power Steering	Std	Std
Cruise Control	Pkg	Pkg
All Wheel Drive	N/A	N/A
AM/FM Stereo Cassette	$139	** $163
Steering Wheel, Tilt	Std	Std
Power Windows	Std	Std
*Options Price	$0	$0
*Total Price	$16,684	$19,457
Target Price	$18,130	
Destination Charge	$490	
Avg. Tax & Fees	$1,130	
Total Target $	**$19,750**	
Average Dealer Option Cost:	85%	

Sable is offered as a four-door sedan and four-door wagon in GS and LS editions. The 1993 LS wagon features new body color bumpers and bodyside moldings, new integrated console, dual illuminated visor mirrors and new seat fabrics. In addition to or in place of the base GS standard equipment, the LS wagon offers a larger standard engine, and a keyless remote entry system, power sun roof and electronic instrument cluster as optional. The Sable is a sister to the Ford Taurus.

Ownership Costs

Cost Area	5 Year Cost	Rate
Depreciation	$8,692	○
Financing ($397/month)	$3,253	
Insurance (Rating 6)	$6,919	○
State Fees	$799	
Fuel (Hwy 27 City 19)	$3,775	○
Maintenance	$4,436	○
Repairs	$709	○

Warranty/Maintenance Info

Major Tune-Up	$159	○
Minor Tune-Up	$109	○
Brake Service	$271	●
Overall Warranty	3 yr/36k	○
Drivetrain Warranty	3 yr/36k	○
Rust Warranty	6 yr/100k	○
Maintenance Warranty	N/A	
Roadside Assistance	N/A	

Ownership Cost By Year

1993, 1994, 1995, 1996, 1997

Resale Value

1993	1994	1995	1996	1997
$16,472	$14,941	$13,740	$12,407	$11,058

Cumulative Costs

	1993	1994	1995	1996	1997
Annual	$6,856	$5,019	$5,679	$5,014	$6,015
Total	$6,856	$11,875	$17,554	$22,568	$28,583

Ownership Costs (5yr)

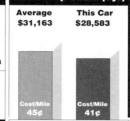

Average	This Car
$31,163	$28,583
Cost/Mile 45¢	Cost/Mile 41¢

Ownership Cost Rating

Excellent

Mercury Topaz GS
2 Door Sedan

Compact

- 2.3L 96 hp Gas Fuel Inject.
- 4 Cylinder In-Line
- Manual 5 Speed
- 2 Wheel Front
- Belts Std, Driv Air Opt

Purchase Price

Car Item	Dealer Cost	List
Base Price	**$9,703**	**$10,801**
Anti-Lock Brakes	N/A	N/A
Automatic 3 Speed	$479	$563
3.0L 130 hp Gas	$583	$685
Air Conditioning	Pkg	Pkg
Power Steering	Std	Std
Cruise Control	$191	$224
All Wheel Drive	N/A	N/A
AM/FM Stereo Cassette	$132	$155
Steering Wheel, Tilt	$123	$145
Power Windows	N/A	N/A
*Options Price	$1,194	$1,403
*Total Price	$10,897	$12,204
Target Price	$11,732	
Destination Charge	$465	
Avg. Tax & Fees	$737	
Total Target $	**$12,934**	
Average Dealer Option Cost:	85%	

The Topaz is offered as a two-door and four-door GS sedan. Refinements for 1993 include an improved interior with a console-mounted removable cupholder and leather-wrapped shift knob for the manual transaxle. Standard features on the two-door GS sedan include turbine style wheel covers, sport instrumentation with tachometer, temperature and fuel gauges, and trip odometer, and tinted window glass. The Comfort/Convenience and Power Lock Groups are optional. The Topaz is a sister to the Ford Tempo.

Ownership Costs

Cost Area	5 Year Cost	Rate
Depreciation	$8,017	●
Financing ($260/month)	$2,130	
Insurance (Rating 8)	$7,223	○
State Fees	$507	
Fuel (Hwy 24 City 20)	$3,925	●
Maintenance	$4,197	○
Repairs	$709	○

Warranty/Maintenance Info

Major Tune-Up	$154	○
Minor Tune-Up	$99	○
Brake Service	$221	○
Overall Warranty	3 yr/36k	○
Drivetrain Warranty	3 yr/36k	○
Rust Warranty	6 yr/100k	○
Maintenance Warranty	N/A	
Roadside Assistance	N/A	

Resale Value

1993	1994	1995	1996	1997
$9,102	$7,842	$6,847	$5,898	$4,917

Cumulative Costs

	1993	1994	1995	1996	1997
Annual	$6,955	$4,423	$5,131	$4,611	$5,588
Total	$6,955	$11,378	$16,509	$21,120	$26,708

Ownership Costs (5yr)

Average	This Car
$24,973	$26,708
Cost/Mile 36¢	Cost/Mile 38¢

Ownership Cost Rating

Worse Than Average

---

* Includes shaded options
** Other purchase requirements apply

● Poor ◐ Worse Than Average ◑ Average ○ Better Than Average ○ Excellent ⊖ Insufficient Information

©1993 by IntelliChoice, Inc. (408) 554-8711 All Rights Reserved. Reproduction Prohibited.
Refer to *Section 3: Annotated Vehicle Charts* for an explanation of these charts.

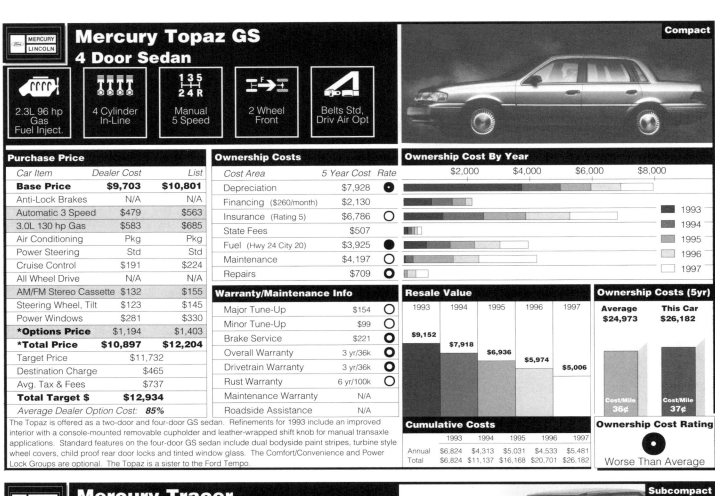
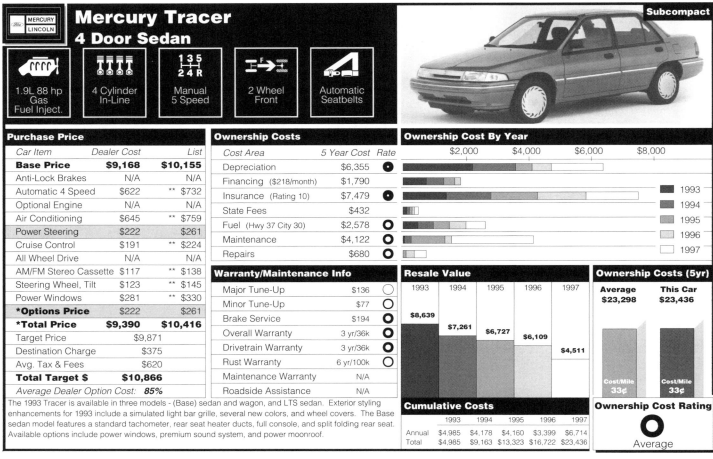

Mercury Tracer LTS
4 Door Sedan

Subcompact

 1.8L 127 hp Gas Fuel Inject. 4 Cylinder In-Line Manual 5 Speed 2 Wheel Front Automatic Seatbelts

Purchase Price

Car Item	Dealer Cost	List
Base Price	**$10,831**	**$12,023**
Anti-Lock Brakes	N/A	N/A
Automatic 4 Speed	$622	$732
Optional Engine	N/A	N/A
Air Conditioning	$645	$759
Power Steering	Std	Std
Cruise Control	Std	Std
All Wheel Drive	N/A	N/A
AM/FM Stereo Cassette	Std	Std
Steering Wheel, Tilt	Std	Std
Power Windows	$281	$330
***Options Price**	$0	$0
***Total Price**	**$10,831**	**$12,023**
Target Price		$11,416
Destination Charge		$375
Avg. Tax & Fees		$714
Total Target $		**$12,505**
Average Dealer Option Cost:		85%

The 1993 Tracer is available in three models - (Base) sedan and wagon, and LTS sedan. Exterior styling enhancements for 1993 include a rear spoiler with stop lamp, several new colors, and new seat and door trim. The LTS sedan features a standard air dam, 4-way adjustable driver's seat, tilt steering wheel, and cruise control. Available options include power windows, premium sound system, and power moonroof.

Ownership Costs

Cost Area	5 Year Cost	Rate
Depreciation	$7,525	●
Financing ($251/month)	$2,060	
Insurance (Rating 11)	$7,693	◉
State Fees	$496	
Fuel (Hwy 31 City 26)	$3,029	◉
Maintenance	$4,391	○
Repairs	$680	○

Warranty/Maintenance Info

Major Tune-Up	$147	○
Minor Tune-Up	$81	○
Brake Service	$205	○
Overall Warranty	3 yr/36k	○
Drivetrain Warranty	3 yr/36k	○
Rust Warranty	6 yr/100k	○
Maintenance Warranty	N/A	
Roadside Assistance	N/A	

Ownership Cost By Year

1993, 1994, 1995, 1996, 1997

Resale Value

1993	1994	1995	1996	1997
$9,431	$8,036	$7,421	$6,739	$4,980

Cumulative Costs

	1993	1994	1995	1996	1997
Annual	$6,082	$4,429	$4,493	$3,639	$7,231
Total	$6,082	$10,511	$15,004	$18,643	$25,874

Ownership Costs (5yr)

Average	This Car
$25,106	$25,874
Cost/Mile 36¢	Cost/Mile 37¢

Ownership Cost Rating

● Worse Than Average

Mercury Tracer
4 Door Wagon

Subcompact Wag

 1.9L 88 hp Gas Fuel Inject. 4 Cylinder In-Line Manual 5 Speed 2 Wheel Front Automatic Seatbelts

Purchase Price

Car Item	Dealer Cost	List
Base Price	**$9,903**	**$10,982**
Anti-Lock Brakes	N/A	N/A
Automatic 4 Speed	$622	$732
Optional Engine	N/A	N/A
Air Conditioning	$645	** $759
Power Steering	Std	Std
Cruise Control	$191	** $224
All Wheel Drive	N/A	N/A
AM/FM Stereo Cassette	$117	** $138
Steering Wheel, Tilt	$123	** $145
Power Windows	$281	** $330
***Options Price**	$0	$0
***Total Price**	**$9,903**	**$10,982**
Target Price		$10,422
Destination Charge		$375
Avg. Tax & Fees		$654
Total Target $		**$11,451**
Average Dealer Option Cost:		85%

The 1993 Tracer is available in three models - (Base) sedan and wagon, and LTS sedan. Exterior styling enhancements for 1993 include a simulated light bar grille, several new colors, and wheel covers. The Base wagon model features a standard cargo cover, power steering, rear window wiper/washer, and dual power mirrors. Available options include power windows, premium sound system, luggage rack, and power moonroof.

Ownership Costs

Cost Area	5 Year Cost	Rate
Depreciation	$6,658	◉
Financing ($230/month)	$1,887	
Insurance (Rating 11)	$7,693	◉
State Fees	$454	
Fuel (Hwy 36 City 29)	$2,656	○
Maintenance	$4,374	○
Repairs	$680	○

Warranty/Maintenance Info

Major Tune-Up	$136	○
Minor Tune-Up	$77	○
Brake Service	$194	○
Overall Warranty	3 yr/36k	○
Drivetrain Warranty	3 yr/36k	○
Rust Warranty	6 yr/100k	○
Maintenance Warranty	N/A	
Roadside Assistance	N/A	

Resale Value

1993	1994	1995	1996	1997
$8,970	$7,582	$7,054	$6,443	$4,793

Cumulative Costs

	1993	1994	1995	1996	1997
Annual	$5,338	$4,280	$4,356	$3,464	$6,964
Total	$5,338	$9,618	$13,974	$17,438	$24,402

Ownership Costs (5yr)

Average	This Car
$23,934	$24,402
Cost/Mile 34¢	Cost/Mile 35¢

Ownership Cost Rating

○ Average

---

* Includes shaded options
** Other purchase requirements apply

● Poor ◐ Worse Than Average ◉ Average ○ Better Than Average ○ Excellent ⊖ Insufficient Information

©1993 by IntelliChoice, Inc. (408) 554-8711 All Rights Reserved. Reproduction Prohibited.
Refer to *Section 3: Annotated Vehicle Charts* for an explanation of these charts.

Mitsubishi 3000GT
2 Door Coupe

Sport

 3.0L 222 hp Gas Fuel Inject.
 6 Cylinder "V"
 Manual 5 Speed
 2 Wheel Front
Driver Airbag Psngr Belts

VR-4 Model Shown

Purchase Price

Car Item	Dealer Cost	List
Base Price	**$20,111**	**$23,659**
Anti-Lock Brakes	N/A	N/A
Automatic 4 Speed	$712	$840
Optional Engine	N/A	N/A
Air Conditioning	Std	Std
Power Steering	Std	Std
Cruise Control	Std	Std
All Wheel Drive	N/A	N/A
AM/FM Stereo Cassette	Std	Std
Steering Wheel, Tilt	Std	Std
Power Windows	Std	Std
*Options Price	$0	$0
*Total Price	$20,111	$23,659
Target Price		$21,600
Destination Charge		$443
Avg. Tax & Fees		$1,343
Total Target $		**$23,386**
Average Dealer Option Cost:		78%

The 1993 Mitsubishi 3000GT is available in a one model edition with three trim levels - (Base), SL and VR-4. New for 1993, the (Base) 3000GT features new standard equipment, including air conditioning, power windows/locks, a six-speaker AM/FM cassette with graphic equalizer, cruise control and rear spoiler. The 3000GT receives a new audio system anti-theft feature and an optional cargo area-mounted compact disc auto changer that holds up to six discs. Cloth upholstery is standard.

Ownership Costs

Cost Area	5 Year Cost	Rate
Depreciation	$10,757	◯
Financing ($470/month)	$3,853	
Insurance (Rating 20+)	$12,293	●
State Fees	$965	
Fuel (Hwy 24 City 19)	$4,015	◯
Maintenance	$6,619	◯
Repairs	$680	◯

Warranty/Maintenance Info

Major Tune-Up	$250	◯
Minor Tune-Up	$197	●
Brake Service	$233	◯
Overall Warranty	3 yr/36k	◯
Drivetrain Warranty	5 yr/60k	◯
Rust Warranty	7 yr/100k	◯
Maintenance Warranty	N/A	
Roadside Assistance	3 yr/36k	

Ownership Cost By Year

$2,000 $4,000 $6,000 $8,000 $10,000 $12,000 $14,000

■ 1993
■ 1994
■ 1995
□ 1996
□ 1997

Resale Value

1993	1994	1995	1996	1997
$20,468	$18,065	$16,184	$14,439	$12,629

Cumulative Costs

	1993	1994	1995	1996	1997
Annual	$7,825	$7,277	$8,101	$6,505	$9,473
Total	$7,825	$15,102	$23,203	$29,708	$39,181

Ownership Costs (5yr)

Average	This Car
$40,950	$39,181
Cost/Mile 59¢	Cost/Mile 56¢

Ownership Cost Rating

◯ Better Than Average

Mitsubishi 3000GT SL
2 Door Coupe

Sport

 3.0L 222 hp Gas Fuel Inject.
 6 Cylinder "V"
 Manual 5 Speed
 2 Wheel Front
 Driver Airbag Psngr Belts

Purchase Price

Car Item	Dealer Cost	List
Base Price	**$23,539**	**$28,709**
Anti-Lock Brakes	Std	Std
Automatic 4 Speed	$712	$860
Optional Engine	N/A	N/A
Auto Climate Control	Std	Std
Power Steering	Std	Std
Cruise Control	Std	Std
All Wheel Drive	N/A	N/A
AM/FM Stereo Cassette	Std	Std
Steering Wheel, Tilt	Std	Std
Power Windows	Std	Std
*Options Price	$0	$0
*Total Price	$23,539	$28,709
Target Price		$25,429
Destination Charge		$443
Avg. Tax & Fees		$1,586
Total Target $		**$27,458**
Average Dealer Option Cost:		77%

The 1993 Mitsubishi 3000GT is available in a one model edition with three trim levels - (Base), SL and VR-4. New for 1993, the 3000GT SL features new standard equipment, including air conditioning, power windows/locks, AM/FM cassette with graphic equalizer, cruise control and rear spoiler. The SL now includes remote keyless entry and an upgraded stereo system with a visual audio display that allows custom selections to be made from a variety of surround sound effects such as concert hall.

Ownership Costs

Cost Area	5 Year Cost	Rate
Depreciation	$13,341	◯
Financing ($552/month)	$4,524	
Insurance (Rating 23)	$11,066	●
State Fees	$1,166	
Fuel (Hwy 24 City 19)	$4,015	◯
Maintenance	$6,750	◯
Repairs	$680	◯

Warranty/Maintenance Info

Major Tune-Up	$250	◯
Minor Tune-Up	$197	●
Brake Service	$233	◯
Overall Warranty	3 yr/36k	◯
Drivetrain Warranty	5 yr/60k	◯
Rust Warranty	7 yr/100k	◯
Maintenance Warranty	N/A	
Roadside Assistance	3 yr/36k	

Ownership Cost By Year

$2,000 $4,000 $6,000 $8,000 $10,000 $12,000 $14,000

■ 1993
■ 1994
■ 1995
□ 1996
□ 1997

Resale Value

1993	1994	1995	1996	1997
$22,363	$19,932	$17,896	$16,012	$14,117

Cumulative Costs

	1993	1994	1995	1996	1997
Annual	$10,106	$7,325	$8,184	$6,474	$9,453
Total	$10,106	$17,431	$25,615	$32,089	$41,542

Ownership Costs (5yr)

Average	This Car
$45,795	$41,542
Cost/Mile 65¢	Cost/Mile 59¢

Ownership Cost Rating

◯ Excellent

---

page 148 | *Includes shaded options ** Other purchase requirements apply

● Poor   ◔ Worse Than Average   ◑ Average   ◯ Better Than Average   ◯ Excellent   ⊖ Insufficient Information

©1993 by IntelliChoice, Inc. (408) 554-8711   All Rights Reserved. Reproduction Prohibited.

Refer to Section 3: Annotated Vehicle Charts for an explanation of these charts.

Mitsubishi 3000GT VR-4
2 Door Coupe

Sport

- 3.0L 300 hp Turbo Gas Fuel Inject.
- 6 Cylinder "V"
- Manual 5 Speed
- 4 Wheel Full-Time
- Driver Airbag Psngr Belts

Purchase Price

Car Item	Dealer Cost	List
Base Price	**$29,797**	**$37,250**
Anti-Lock Brakes	Std	Std
Automatic Transmission	N/A	N/A
Optional Engine	N/A	N/A
Auto Climate Control	Std	Std
Power Steering	Std	Std
Cruise Control	Std	Std
4 Wheel Full-Time Drive	Std	Std
AM/FM Stereo Cassette	Std	Std
Steering Wheel, Tilt	Std	Std
Power Windows	Std	Std
*Options Price	$0	$0
***Total Price**	**$29,797**	**$37,250**
Target Price	$32,528	
Destination Charge	$443	
Avg. Tax & Fees	$2,026	
Luxury Tax	$297	
Total Target $	**$35,294**	

Ownership Costs

Cost Area	5 Year Cost	Rate
Depreciation		⊖
Financing ($709/month)	$5,814	
Insurance (Rating 26++)	$16,231	●
State Fees	$1,508	
Fuel (Hwy 24 City 18)	$4,117	◓
Maintenance	$7,714	◓
Repairs	$680	○

Warranty/Maintenance Info

Major Tune-Up	$261	○
Minor Tune-Up	$214	●
Brake Service	$233	○
Overall Warranty	3 yr/36k	○
Drivetrain Warranty	5 yr/60k	○
Rust Warranty	7 yr/100k	○
Maintenance Warranty	N/A	
Roadside Assistance	3 yr/36k	

Ownership Cost By Year

Insufficient Depreciation Information

- 1993
- 1994
- 1995
- 1996
- 1997

Resale Value

Insufficient Information

Ownership Costs (5yr)

Insufficient Information

Cumulative Costs

	1993	1994	1995	1996	1997
Annual		Insufficient Information			
Total		Insufficient Information			

Ownership Cost Rating

⊖ Insufficient Information

The 1993 Mitsubishi 3000GT is available in a one model edition with three trim levels - (Base), SL and VR-4. New for 1993, the 3000GT VR-4 features air conditioning, power windows/locks, optional chrome wheels, optional leather seating, cruise control and rear spoiler. The VR-4 includes remote keyless entry and an upgraded stereo system with a visual audio display that allows custom selections to be made from a variety of surround sound effects such as cabin, studio, concert hall or stadium field.

Mitsubishi Diamante ES
4 Door Sedan

Midsize

- 3.0L 175 hp Gas Fuel Inject.
- 6 Cylinder "V"
- Automatic 4 Speed
- 2 Wheel Front
- Driver Airbag Psngr Belts

Purchase Price

Car Item	Dealer Cost	List
Base Price	**$18,814**	**$22,399**
Anti-Lock Brakes	$880	$1,100
Manual Transmission	N/A	N/A
Optional Engine	N/A	N/A
Auto Climate Control	Std	Std
Power Steering	Std	Std
Cruise Control	Std	Std
All Wheel Drive	N/A	N/A
AM/FM Stereo Cassette	Std	Std
Steering Wheel, Tilt	Std	Std
Power Windows	Std	Std
*Options Price	$0	$0
***Total Price**	**$18,814**	**$22,399**
Target Price	$20,163	
Destination Charge	$443	
Avg. Tax & Fees	$1,258	
Total Target $	**$21,864**	
Average Dealer Option Cost:	79%	

Ownership Costs

Cost Area	5 Year Cost	Rate
Depreciation	$10,150	○
Financing ($439/month)	$3,602	
Insurance (Rating 15)	$8,199	○
State Fees	$914	
Fuel (Hwy 25 City 18 -Prem.)	$4,451	●
Maintenance	$6,166	●
Repairs	$740	○

Warranty/Maintenance Info

Major Tune-Up	$211	●
Minor Tune-Up	$153	●
Brake Service	$233	○
Overall Warranty	3 yr/36k	○
Drivetrain Warranty	5 yr/60k	○
Rust Warranty	5 yr/unlim. mi	○
Maintenance Warranty	N/A	
Roadside Assistance	3 yr/36k	

Ownership Cost By Year

- 1993
- 1994
- 1995
- 1996
- 1997

Resale Value

1993	1994	1995	1996	1997
$19,057	$17,007	$15,354	$13,140	$11,714

Ownership Costs (5yr)

Average	This Car
$34,171	$34,222
Cost/Mile 49¢	Cost/Mile 49¢

Cumulative Costs

	1993	1994	1995	1996	1997
Annual	$6,922	$6,086	$6,971	$6,182	$8,061
Total	$6,922	$13,008	$19,979	$26,161	$34,222

Ownership Cost Rating

○ Average

The 1993 Mitsubishi Diamante is offered as a four-door sedan in ES and LS editions. New for 1993, the Diamante ES has five new colors to choose from, a dashboard-mounted remote electric trunk release switch, illuminated driver power window, door lock and mirror switches, carpeted floor mats and trunk cargo mat. The audio system has been improved with a new anti-theft feature. The ES offers automatic air conditioning, leather wrapped steering wheel and optional leather seating surfaces.

---

* Includes shaded options
** Other purchase requirements apply

● Poor   ◐ Worse Than Average   ○ Average   ◯ Better Than Average   ◯ Excellent   ⊖ Insufficient Information

©1993 by IntelliChoice, Inc. (408) 554-8711 All Rights Reserved. Reproduction Prohibited.
Refer to *Section 3: Annotated Vehicle Charts* for an explanation of these charts.

Mitsubishi Diamante LS
4 Door Sedan

Midsize

 3.0L 202 hp Gas Fuel Inject.
 6 Cylinder "V"
 Automatic 4 Speed
 2 Wheel Front
 Driver Airbag Psngr Belts

Purchase Price

Car Item	Dealer Cost	List
Base Price	**$23,868**	**$29,850**
Anti-Lock Brakes	Std	Std
Manual Transmission	N/A	N/A
Optional Engine	N/A	N/A
Auto Climate Control	Std	Std
Power Steering	Std	Std
Cruise Control	Std	Std
All Wheel Drive	N/A	N/A
AM/FM Stereo CD	Std	Std
Steering Wheel, Tilt	Std	Std
Power Windows	Std	Std
*Options Price	$0	$0
*Total Price	$23,868	$29,850
Target Price		$25,798
Destination Charge		$443
Avg. Tax & Fees		$1,615
Total Target $		**$27,856**
Average Dealer Option Cost:		77%

Ownership Costs

Cost Area	5 Year Cost	Rate
Depreciation		⊖
Financing ($560/month)	$4,588	
Insurance (Rating 18)	$9,058	○
State Fees	$1,212	
Fuel (Hwy 24 City 18 -Prem.)	$4,547	●
Maintenance	$6,296	●
Repairs	$740	○

Warranty/Maintenance Info

Major Tune-Up	$226	●
Minor Tune-Up	$169	●
Brake Service	$233	○
Overall Warranty	3 yr/36k	○
Drivetrain Warranty	5 yr/60k	○
Rust Warranty	5 yr/unlim. mi	○
Maintenance Warranty	N/A	
Roadside Assistance	3 yr/36k	

Ownership Cost By Year

Insufficient Depreciation Information

■ 1993
■ 1994
▨ 1995
▨ 1996
□ 1997

Resale Value

Insufficient Information

Ownership Costs (5yr)

Insufficient Information

Cumulative Costs

	1993	1994	1995	1996	1997
Annual	Insufficient Information				
Total	Insufficient Information				

Ownership Cost Rating

⊖ Insufficient Information

The 1993 Diamante is offered as a four-door sedan in ES and LS editions. New for 1993, the LS has five new colors to choose from (Zurich White Pearl, Reno Silver, Panama Green Pearl, Imperial Amethyst and Princeton Blue), a dash-mounted remote electric trunk release switch, illuminated driver power window, door lock and mirror switches, carpeted floor mats and trunk cargo mat. The LS includes the above plus alloy wheels and optional power sunroof and a new trunk mounted automatic compact disc changer.

Mitsubishi Eclipse
2 Door Coupe

Subcompact

 1.8L 92 hp Gas Fuel Inject.
 4 Cylinder In-Line
 Manual 5 Speed
 2 Wheel Front
Automatic Seatbelts

GS Model Shown

Purchase Price

Car Item	Dealer Cost	List
Base Price	**$10,252**	**$11,719**
Anti-Lock Brakes	N/A	N/A
Automatic 4 Speed	$593	$670
Optional Engine	N/A	N/A
Air Conditioning	$685	$835
Power Steering	$225	$274
Cruise Control	N/A	N/A
All Wheel Drive	N/A	N/A
AM/FM Stereo Cassette	$146	$178
Steering Wheel, Tilt	Std	Std
Power Windows	N/A	N/A
*Options Price	$225	$274
*Total Price	$10,477	$11,993
Target Price		$10,902
Destination Charge		$393
Avg. Tax & Fees		$689
Total Target $		**$11,984**
Average Dealer Option Cost:		84%

Ownership Costs

Cost Area	5 Year Cost	Rate
Depreciation	$4,965	○
Financing ($241/month)	$1,973	
Insurance (Rating 13)	$8,061	●
State Fees	$495	
Fuel (Hwy 32 City 23)	$3,152	◐
Maintenance	$4,604	◐
Repairs	$600	○

Warranty/Maintenance Info

Major Tune-Up	$220	●
Minor Tune-Up	$127	●
Brake Service	$189	○
Overall Warranty	3 yr/36k	○
Drivetrain Warranty	5 yr/60k	○
Rust Warranty	7 yr/100k	○
Maintenance Warranty	N/A	
Roadside Assistance	3 yr/36k	

Ownership Cost By Year

■ 1993
■ 1994
▨ 1995
▨ 1996
□ 1997

Resale Value

1993	1994	1995	1996	1997
$11,230	$10,629	$9,259	$8,107	$7,019

Ownership Costs (5yr)

Average	This Car
$25,072	$23,850
Cost/Mile 36¢	Cost/Mile 34¢

Cumulative Costs

	1993	1994	1995	1996	1997
Annual	$3,816	$3,723	$5,479	$4,922	$5,910
Total	$3,816	$7,539	$13,018	$17,940	$23,850

Ownership Cost Rating

○ Excellent

The 1993 Eclipse is available in five models - (Base) Eclipse, GS, GS 16V, GS 16V Turbo and GSX 16V Turbo AWD. New for 1993, the Base Eclipse has optional redesigned wheel covers. The exterior features aero headlamps, a color-keyed front airdam amd fog lamps. The interior features a console, digital clock, cloth reclining high back bucket seats and a remote control hatch release. The Eclipse features a 3yr/36K mile new car limited warranty and a 5yr/60K mile powertrain limited warranty.

---

page 150

*Includes shaded options
**Other purchase requirements apply

● Poor ◐ Worse Than Average ○ Average ○ Better Than Average ○ Excellent ⊖ Insufficient Information

©1993 by IntelliChoice, Inc. (408) 554-8711   All Rights Reserved. Reproduction Prohibited.
Refer to *Section 3: Annotated Vehicle Charts* for an explanation of these charts.

Mitsubishi Eclipse GS
2 Door Coupe

Subcompact

1.8L 92 hp Gas Fuel Inject. | 4 Cylinder In-Line | Manual 5 Speed | 2 Wheel Front | Automatic Seatbelts

Purchase Price

Car Item	Dealer Cost	List
Base Price	**$11,680**	**$13,429**
Anti-Lock Brakes	N/A	N/A
Automatic 4 Speed	$593	$680
Optional Engine	N/A	N/A
Air Conditioning	$685	$835
Power Steering	Std	Std
Cruise Control	$181	$221
All Wheel Drive	N/A	N/A
AM/FM Stereo Cassette	Std	Std
Steering Wheel, Tilt	Std	Std
Power Windows	Pkg	Pkg
*Options Price	$0	$0
*Total Price	$11,680	$13,429
Target Price		$12,176
Destination Charge		$393
Avg. Tax & Fees		$766
Total Target $		**$13,335**
Average Dealer Option Cost:		84%

The 1993 Eclipse is available in five models - (Base) Eclipse, GS, GS 16V, GS 16V Turbo and GSX 16V Turbo AWD. New for 1993, the Eclipse GS includes a new one-piece rear spoiler, new side graphics and sculptured lower body side sill garnish. The GS also adds an optional newly designed wheel cover. On the inside, the Eclipse cockpit offers a new seat stitching design and an upgraded audio system with improved door speakers. The Eclipse features a three year/36,000mile new car limited warranty.

Ownership Costs

Cost Area	5 Year Cost	Rate
Depreciation	$5,545	○
Financing ($268/month)	$2,198	
Insurance (Rating 14)	$8,238	●
State Fees	$553	
Fuel (Hwy 32 City 23)	$3,152	◉
Maintenance	$4,604	◉
Repairs	$600	○

Warranty/Maintenance Info

Major Tune-Up	$220	●
Minor Tune-Up	$127	●
Brake Service	$189	○
Overall Warranty	3 yr/36k	○
Drivetrain Warranty	5 yr/60k	
Rust Warranty	7 yr/100k	○
Maintenance Warranty	N/A	
Roadside Assistance	3 yr/36k	

Ownership Cost By Year
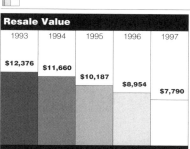
1993, 1994, 1995, 1996, 1997

Resale Value
1993	1994	1995	1996	1997
$12,376	$11,660	$10,187	$8,954	$7,790

Cumulative Costs
	1993	1994	1995	1996	1997
Annual	$4,161	$3,955	$5,674	$5,067	$6,033
Total	$4,161	$8,116	$13,790	$18,857	$24,890

Ownership Costs (5yr)

Average $26,687 — This Car $24,890
Cost/Mile 38¢ — Cost/Mile 36¢

Ownership Cost Rating: Excellent

Mitsubishi Eclipse GS 16V
2 Door Coupe

Subcompact

2.0L 135 hp Gas Fuel Inject. | 4 Cylinder In-Line | Manual 5 Speed | 2 Wheel Front | Automatic Seatbelts

Purchase Price

Car Item	Dealer Cost	List
Base Price	**$12,494**	**$14,359**
Anti-Lock Brakes	N/A	N/A
Automatic 4 Speed	$593	$680
Optional Engine	N/A	N/A
Air Conditioning	$685	$835
Power Steering	Std	Std
Cruise Control	$181	$221
All Wheel Drive	N/A	N/A
AM/FM Stereo Cassette	Std	Std
Steering Wheel, Tilt	Std	Std
Power Windows	Pkg	Pkg
*Options Price	$0	$0
*Total Price	$12,494	$14,359
Target Price		$13,037
Destination Charge		$393
Avg. Tax & Fees		$820
Total Target $		**$14,250**
Average Dealer Option Cost:		84%

The 1993 Eclipse is available in five models - (Base) Eclipse, GS, GS 16V, GS 16V Turbo and GSX 16V Turbo AWD. New for 1993, the Eclipse GS 16V includes a new one-piece rear spoiler, new side graphics and sculptured lower body side sill garnish. The GS 16V also receives a new alloy wheel design. On the inside, the Eclipse cockpit offers a new seat stitching design and an upgraded audio system with improved door speakers. The Eclipse features a 3year/36,000 mile new car limited warranty.

Ownership Costs

Cost Area	5 Year Cost	Rate
Depreciation	$5,832	○
Financing ($286/month)	$2,348	
Insurance (Rating 15)	$8,507	●
State Fees	$591	
Fuel (Hwy 29 City 22)	$3,390	◉
Maintenance	$5,302	●
Repairs	$600	○

Warranty/Maintenance Info

Major Tune-Up	$276	●
Minor Tune-Up	$139	●
Brake Service	$189	○
Overall Warranty	3 yr/36k	○
Drivetrain Warranty	5 yr/60k	
Rust Warranty	7 yr/100k	○
Maintenance Warranty	N/A	
Roadside Assistance	3 yr/36k	

Ownership Cost By Year
1993, 1994, 1995, 1996, 1997

Resale Value
1993	1994	1995	1996	1997
$13,080	$12,352	$10,971	$9,661	$8,418

Cumulative Costs
	1993	1994	1995	1996	1997
Annual	$4,539	$4,131	$6,033	$5,279	$6,588
Total	$4,539	$8,670	$14,703	$19,982	$26,570

Ownership Costs (5yr)
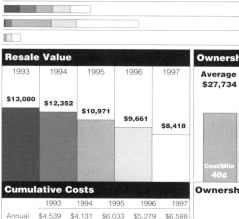
Average $27,734 — This Car $26,570
Cost/Mile 40¢ — Cost/Mile 38¢

Ownership Cost Rating: Excellent

---

* Includes shaded options
** Other purchase requirements apply

● Poor ◉ Worse Than Average ○ Average ◯ Better Than Average ○ Excellent ⊖ Insufficient Information

©1993 by IntelliChoice, Inc. (408) 554-8711 All Rights Reserved. Reproduction Prohibited.
Refer to *Section 3: Annotated Vehicle Charts* for an explanation of these charts.

Mitsubishi Eclipse GS 16V Turbo
2 Door Coupe

Subcompact

- 2.0L 195 hp Turbo Gas Fuel Inject.
- 4 Cylinder In-Line
- Manual 5 Speed
- 2 Wheel Front
- Automatic Seatbelts

Purchase Price

Car Item	Dealer Cost	List
Base Price	**$15,700**	**$18,049**
Anti-Lock Brakes	$781	$952
Automatic 4 Speed	$710	$810
Optional Engine	N/A	N/A
Air Conditioning	Std	Std
Power Steering	Std	Std
Cruise Control	Std	Std
All Wheel Drive	N/A	N/A
AM/FM Stereo Cassette	Std	Std
Steering Wheel, Tilt	Std	Std
Power Windows	Std	Std
*Options Price	$0	$0
*Total Price	$15,700	$18,049
Target Price		$16,449
Destination Charge		$393
Avg. Tax & Fees		$1,026
Total Target $		**$17,868**
Average Dealer Option Cost:		84%

The 1993 Eclipse is available in five models - (Base) Eclipse, GS, GS 16V, GS 16V Turbo and GSX 16V Turbo AWD. New for 1993, the Eclipse GS 16V Turbo includes a new one-piece rear spoiler, new side graphics and sculptured lower body side sill garnish. On the inside, the Eclipse cockpit offers a new seat stitching design and an upgraded audio system with improved speakers. GS 16V Turbo includes a graphic equalizer, and air conditioning as standard equipment. Leather front seating is optional.

Ownership Costs

Cost Area	5 Year Cost	Rate
Depreciation	$8,431	○
Financing ($359/month)	$2,944	
Insurance (Rating 17+)	$10,901	●
State Fees	$738	
Fuel (Hwy 28 City 21)	$3,530	◐
Maintenance	$5,642	
Repairs	$740	◐

Warranty/Maintenance Info

Major Tune-Up	$276	●
Minor Tune-Up	$139	●
Brake Service	$189	○
Overall Warranty	3 yr/36k	◐
Drivetrain Warranty	5 yr/60k	○
Rust Warranty	7 yr/100k	○
Maintenance Warranty	N/A	
Roadside Assistance	3 yr/36k	

Ownership Cost By Year

Bars for 1993, 1994, 1995, 1996, 1997

Resale Value

1993	1994	1995	1996	1997
$14,800	$14,080	$12,351	$10,850	$9,437

Cumulative Costs

	1993	1994	1995	1996	1997
Annual	$7,225	$4,867	$7,196	$6,140	$7,497
Total	$7,225	$12,092	$19,288	$25,428	$32,925

Ownership Costs (5yr)

Average	This Car
$31,885	$32,925
Cost/Mile 46¢	Cost/Mile 47¢

Ownership Cost Rating

● Worse Than Average

Mitsubishi Eclipse GSX 16V Turbo AWD
2 Door Coupe

Subcompact

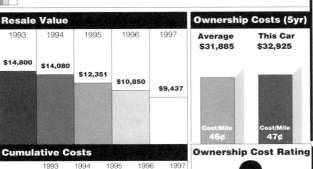

- 2.0L 195 hp Turbo Gas Fuel Inject.
- 4 Cylinder In-Line
- Manual 5 Speed
- 4 Wheel Full-Time
- Automatic Seatbelts

Purchase Price

Car Item	Dealer Cost	List
Base Price	**$18,068**	**$20,769**
Anti-Lock Brakes	Std	Std
Automatic 4 Speed	$710	$810
Optional Engine	N/A	N/A
Air Conditioning	Std	Std
Power Steering	Std	Std
Cruise Control	Std	Std
4 Wheel Full-Time Drive	Std	Std
AM/FM Stereo Cassette	Std	Std
Steering Wheel, Tilt	Std	Std
Power Windows	Std	Std
*Options Price	$0	$0
*Total Price	$18,068	$20,769
Target Price		$18,986
Destination Charge		$393
Avg. Tax & Fees		$1,181
Total Target $		**$20,560**
Average Dealer Option Cost:		86%

The 1993 Eclipse is available in five models - (Base) Eclipse, GS, GS 16V, GS 16V Turbo and GSX 16V Turbo AWD. New for 1993, the GSX 16V Turbo AWD includes a new one-piece rear spoiler, new side graphics, sculptured lower body side sill garnish and a new alloy wheel design. On the inside, the Eclipse cockpit offers a new seat stitching design and an upgraded audio system with improved speakers. GSX includes a graphic equalizer, and air conditioning as standard. Leather front seating is optional.

Ownership Costs

Cost Area	5 Year Cost	Rate
Depreciation	$10,058	○
Financing ($413/month)	$3,388	
Insurance (Rating 19+)	$11,732	●
State Fees	$846	
Fuel (Hwy 25 City 20)	$3,837	●
Maintenance	$5,642	●
Repairs	$780	◐

Warranty/Maintenance Info

Major Tune-Up	$276	●
Minor Tune-Up	$139	●
Brake Service	$189	○
Overall Warranty	3 yr/36k	◐
Drivetrain Warranty	5 yr/60k	○
Rust Warranty	7 yr/100k	○
Maintenance Warranty	N/A	
Roadside Assistance	3 yr/36k	

Resale Value

1993	1994	1995	1996	1997
$16,537	$15,660	$13,697	$12,053	$10,502

Cumulative Costs

	1993	1994	1995	1996	1997
Annual	$8,603	$5,405	$7,770	$6,587	$7,917
Total	$8,603	$14,008	$21,778	$28,365	$36,282

Ownership Costs (5yr)

Average	This Car
$34,946	$36,282
Cost/Mile 50¢	Cost/Mile 52¢

Ownership Cost Rating

● Worse Than Average

---

page 152 — * Includes shaded options   ** Other purchase requirements apply

● Poor   ◐ Worse Than Average   ◑ Average   ◯ Better Than Average   ○ Excellent   ⊖ Insufficient Information

©1993 by IntelliChoice, Inc. (408) 554-8711   All Rights Reserved. Reproduction Prohibited.
Refer to *Section 3: Annotated Vehicle Charts* for an explanation of these charts.

Mitsubishi Expo LRV
3 Door Wagon

 1.8L 113 hp Gas Fuel Inject.
 4 Cylinder In-Line
Manual 5 Speed
 2 Wheel Front
 Automatic Seatbelts

Compact Wagon

Purchase Price

Car Item	Dealer Cost	List
Base Price	**$10,291**	**$11,429**
Anti-Lock Brakes	N/A	N/A
Automatic 4 Speed	$600	$670
Optional Engine	N/A	N/A
Air Conditioning	$640	$780
Power Steering	Std	Std
Cruise Control	$160	$200
All Wheel Drive	N/A	N/A
AM/FM Stereo Cassette	$312	$446
Steering Wheel, Tilt	Std	Std
Power Windows	Pkg	Pkg
***Options Price**	$1,552	$1,896
***Total Price**	**$11,843**	**$13,325**
Target Price		$12,509
Destination Charge		$418
Avg. Tax & Fees		$783
Total Target $		**$13,710**
Average Dealer Option Cost:		*79%*

The 1993 Expo LRV is available in three models-LRV, LRV Sport and LRV Sport AWD. New for 1993, the Expo LRV features full wheel covers and a new inner tailgate handle. Interior features an optional convenience package that includes center armrest and upgraded door trim with cloth insert. The exterior of Expo LRV has a functional aerodynamic design for good fuel economy and low wind noise. The LRV also features a sliding curbside rear passenger door that provides easy access to the back seat.

Ownership Costs

Cost Area	5 Year Cost	Rate
Depreciation	$7,010	O
Financing ($276/month)	$2,259	
Insurance (Rating 6)	$6,919	O
State Fees	$549	
Fuel (Hwy 28 City 23)	$3,385	O
Maintenance	$5,041	O
Repairs	$680	O

Warranty/Maintenance Info

Major Tune-Up	$181	O
Minor Tune-Up	$125	●
Brake Service	$243	●
Overall Warranty	3 yr/36k	O
Drivetrain Warranty	5 yr/60k	○
Rust Warranty	7 yr/100k	○
Maintenance Warranty	N/A	
Roadside Assistance	3 yr/36k	

Ownership Cost By Year

1993 / 1994 / 1995 / 1996 / 1997

Resale Value

1993	1994	1995	1996	1997
$11,413	$10,069	$9,057	$7,652	$6,700

Cumulative Costs

	1993	1994	1995	1996	1997
Annual	$5,323	$4,391	$5,152	$4,599	$6,378
Total	$5,323	$9,714	$14,866	$19,465	$25,843

Ownership Costs (5yr)

Average	This Car
$25,936	$25,843
Cost/Mile 37¢	Cost/Mile 37¢

Ownership Cost Rating

O Average

Mitsubishi Expo LRV Sport
3 Door Wagon

 2.4L 136 hp Gas Fuel Inject.
 4 Cylinder In-Line
 Manual 5 Speed
2 Wheel Front
 Automatic Seatbelts

Compact Wagon

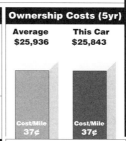

Purchase Price

Car Item	Dealer Cost	List
Base Price	**$12,413**	**$14,269**
Anti-Lock Brakes	$800	$976
Automatic 4 Speed	$600	$690
Optional Engine	N/A	N/A
Air Conditioning	$640	$780
Power Steering	Std	Std
Cruise Control	Std	Std
All Wheel Drive	N/A	N/A
AM/FM Stereo Cassette	Std	Std
Steering Wheel, Tilt	Std	Std
Power Windows	Std	Std
***Options Price**	$1,240	$1,470
***Total Price**	**$13,653**	**$15,739**
Target Price		$14,473
Destination Charge		$418
Avg. Tax & Fees		$907
Total Target $		**$15,798**
Average Dealer Option Cost:		*80%*

The 1993 Expo LRV is available in three models-LRV, LRV Sport and LRV Sport AWD. New for 1993, the Expo LRV Sport features a power package that includes power windows and door locks, power remote door mirrors and the power tailgate lock/unlock mechanism as standard equipment. The exterior of Expo LRV has a functional aerodynamic design for good fuel economy and low wind noise. The LRV also features a sliding curbside rear passenger door that provides easy access to the back seat.

Ownership Costs

Cost Area	5 Year Cost	Rate
Depreciation	$7,946	O
Financing ($318/month)	$2,602	
Insurance (Rating 8)	$7,223	O
State Fees	$646	
Fuel (Hwy 26 City 20)	$3,756	●
Maintenance	$5,173	●
Repairs	$680	O

Warranty/Maintenance Info

Major Tune-Up	$181	O
Minor Tune-Up	$125	●
Brake Service	$243	●
Overall Warranty	3 yr/36k	O
Drivetrain Warranty	5 yr/60k	○
Rust Warranty	7 yr/100k	○
Maintenance Warranty	N/A	
Roadside Assistance	3 yr/36k	

Resale Value

1993	1994	1995	1996	1997
$13,341	$11,810	$10,603	$8,953	$7,852

Cumulative Costs

	1993	1994	1995	1996	1997
Annual	$5,777	$4,835	$5,623	$5,026	$6,765
Total	$5,777	$10,612	$16,235	$21,261	$28,026

Ownership Costs (5yr)

Average	This Car
$28,009	$28,026
Cost/Mile 40¢	Cost/Mile 40¢

Ownership Cost Rating

O Average

---

* Includes shaded options
** Other purchase requirements apply

● Poor ◐ Worse Than Average O Average ○ Better Than Average ○ Excellent ⊖ Insufficient Information

©1993 by IntelliChoice, Inc. (408) 554-8711 All Rights Reserved. Reproduction Prohibited.
Refer to *Section 3: Annotated Vehicle Charts* for an explanation of these charts.

Mitsubishi Expo LRV AWD
3 Door Wagon

Compact Wagon

- 2.4L 136 hp Gas Fuel Inject.
- 4 Cylinder In-Line
- Manual 5 Speed
- 4 Wheel Full-Time
- Automatic Seatbelts

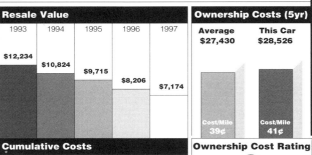
2WD Sport Shown

Purchase Price

Car Item	Dealer Cost	List
Base Price	**$11,857**	**$13,169**
Anti-Lock Brakes	N/A	N/A
Automatic 4 Speed	$600	$670
Optional Engine	N/A	N/A
Air Conditioning	$640	$780
Power Steering	Std	Std
Cruise Control	$160	$200
4 Wheel Full-Time Drive	Std	Std
AM/FM Stereo Cassette	$312	$446
Steering Wheel, Tilt	Std	Std
Power Windows	Pkg	Pkg
*Options Price	$1,552	$1,896
*Total Price	$13,409	$15,065
Target Price	$14,201	
Destination Charge	$418	
Avg. Tax & Fees	$886	
Total Target $	**$15,505**	
Average Dealer Option Cost:	79%	

The 1993 Expo LRV is available in three models-LRV, LRV Sport and LRV Sport AWD. New for 1993, the Expo LRV AWD features a convenience package and a power package that includes power windows and door locks, power remote door mirrors and the power tailgate lock/unlock mechanism as standard equipment. The LRV AWD provides extra flexibility and cargo room with three rows of seating, four hinged doors and a rear liftgate for easy access and loading convenience.

Ownership Costs

Cost Area	5 Year Cost	Rate
Depreciation	$8,331	○
Financing ($312/month)	$2,554	
Insurance (Rating 7)	$7,035	○
State Fees	$619	
Fuel (Hwy 23 City 19)	$4,111	●
Maintenance	$5,196	◉
Repairs	$680	○

Warranty/Maintenance Info

Major Tune-Up	$181	○
Minor Tune-Up	$125	◉
Brake Service	$243	◉
Overall Warranty	3 yr/36k	○
Drivetrain Warranty	5 yr/60k	◯
Rust Warranty	7 yr/100k	◯
Maintenance Warranty	N/A	
Roadside Assistance	3 yr/36k	

Ownership Cost By Year

1993, 1994, 1995, 1996, 1997

Resale Value

1993	1994	1995	1996	1997
$12,234	$10,824	$9,715	$8,206	$7,174

Cumulative Costs

	1993	1994	1995	1996	1997
Annual	$6,594	$4,725	$5,555	$4,912	$6,740
Total	$6,594	$11,319	$16,874	$21,786	$28,526

Ownership Costs (5yr)

Average	This Car
$27,430	$28,526
Cost/Mile 39¢	Cost/Mile 41¢

Ownership Cost Rating
○ Average

Mitsubishi Expo
4 Door Wagon

Midsize Wagon

- 2.4L 136 hp Gas Fuel Inject.
- 4 Cylinder In-Line
- Manual 5 Speed
- 2 Wheel Front
- Automatic Seatbelts

SP Model Shown

Purchase Price

Car Item	Dealer Cost	List
Base Price	**$11,939**	**$13,569**
Anti-Lock Brakes	N/A	N/A
Automatic 4 Speed	$600	$680
Optional Engine	N/A	N/A
Air Conditioning	$640	$780
Power Steering	Std	Std
Cruise Control	$160	$200
All Wheel Drive	N/A	N/A
AM/FM Stereo Cassette	$312	$446
Steering Wheel, Tilt	Std	Std
Power Windows	Pkg	Pkg
*Options Price	$1,712	$2,106
*Total Price	$13,651	$15,675
Target Price	$14,459	
Destination Charge	$418	
Avg. Tax & Fees	$905	
Total Target $	**$15,782**	
Average Dealer Option Cost:	79%	

The 1993 Expo is available in four models- (Base) Expo and SP; Expo AWD and SP AWD. New for 1993, the Expo features remote keyless entry and Blue interior color availability. The interior features three rows of seats allowing a variety of different seating and cargo arrangements. The second and third rows feature 50/50 split bench seat backs for additional cargo storage. The option list includes a Convenience Package featuring cloth door trim, power tailgate, and dual power remote mirrors.

Ownership Costs

Cost Area	5 Year Cost	Rate
Depreciation		⊖
Financing ($317/month)	$2,600	
Insurance (Rating 7)	$7,035	○
State Fees	$645	
Fuel (Hwy 26 City 20)	$3,756	○
Maintenance	$5,176	◉
Repairs	$680	○

Warranty/Maintenance Info

Major Tune-Up	$181	○
Minor Tune-Up	$125	○
Brake Service	$243	○
Overall Warranty	3 yr/36k	○
Drivetrain Warranty	5 yr/60k	◯
Rust Warranty	7 yr/100k	◯
Maintenance Warranty	N/A	
Roadside Assistance	3 yr/36k	

Ownership Cost By Year

Insufficient Depreciation Information

Resale Value

Insufficient Information

Cumulative Costs

	1993	1994	1995	1996	1997
Annual	Insufficient Information				
Total	Insufficient Information				

Ownership Costs (5yr)

Insufficient Information

Ownership Cost Rating
⊖ Insufficient Information

---

* Includes shaded options
** Other purchase requirements apply

● Poor | ◉ Worse Than Average | ○ Average | ◯ Better Than Average | ◯ Excellent | ⊖ Insufficient Information

©1993 by IntelliChoice, Inc. (408) 554-8711 All Rights Reserved. Reproduction Prohibited.
Refer to *Section 3: Annotated Vehicle Charts* for an explanation of these charts.

Mitsubishi Expo SP
4 Door Wagon

Midsize Wagon

 2.4L 136 hp Gas Fuel Inject. 4 Cylinder In-Line Manual 5 Speed 2 Wheel Front — Automatic Seatbelts

Purchase Price

Car Item	Dealer Cost	List
Base Price	**$13,474**	**$15,669**
Anti-Lock Brakes	$800	$976
Automatic 4 Speed	$600	$680
Optional Engine	N/A	N/A
Air Conditioning	$640	$780
Power Steering	Std	Std
Cruise Control	Std	Std
All Wheel Drive	N/A	N/A
AM/FM Stereo Cassette	Std	Std
Steering Wheel, Tilt	Std	Std
Power Windows	Std	Std
***Options Price**	**$1,240**	**$1,460**
***Total Price**	**$14,714**	**$17,129**
Target Price		$15,626
Destination Charge		$418
Avg. Tax & Fees		$977
Total Target $		**$17,021**
Average Dealer Option Cost:		**81%**

The 1993 Expo is available in four models- (Base) Expo and SP; Expo AWD and SP AWD. New for 1993, the Expo SP features a remote keyless entry system, rear cargo cover and a Power Package that includes power windows and door locks. A Convenience Package featuring upgraded cloth door trim, power tailgate lock and front center armrest is standard on all Expo SP models. A new two-tone color scheme for 1993 is Norfolk Green with Silver Metallic and Oceanside Blue with Gray Metallic.

Ownership Costs

Cost Area	5 Year Cost	Rate
Depreciation		⊖
Financing ($342/month)	$2,804	
Insurance (Rating 10)	$7,479	○
State Fees	$701	
Fuel (Hwy 26 City 20)	$3,756	○
Maintenance	$5,173	◐
Repairs	$680	○

Warranty/Maintenance Info

Major Tune-Up	$181	○
Minor Tune-Up	$125	○
Brake Service	$243	○
Overall Warranty	3 yr/36k	○
Drivetrain Warranty	5 yr/60k	○
Rust Warranty	7 yr/100k	○
Maintenance Warranty	N/A	
Roadside Assistance	3 yr/36k	

Ownership Cost By Year

Insufficient Depreciation Information

- 1993
- 1994
- 1995
- 1996
- 1997

Resale Value

Insufficient Information

Cumulative Costs

	1993	1994	1995	1996	1997
Annual	*Insufficient Information*				
Total	*Insufficient Information*				

Ownership Costs (5yr)

Insufficient Information

Ownership Cost Rating

⊖ Insufficient Information

Mitsubishi Expo AWD
4 Door Wagon

Midsize Wagon

 2.4L 136 hp Gas Fuel Inject. — 4 Cylinder In-Line — Manual 5 Speed 4 Wheel Full-Time Automatic Seatbelts

SP Model Shown

Purchase Price

Car Item	Dealer Cost	List
Base Price	**$13,105**	**$14,889**
Anti-Lock Brakes	N/A	N/A
Automatic 4 Speed	$600	$680
Optional Engine	N/A	N/A
Air Conditioning	$640	$780
Power Steering	Std	Std
Cruise Control	$160	$200
4 Wheel Full-Time Drive	Std	Std
AM/FM Stereo Cassette	$312	$446
Steering Wheel, Tilt	Std	Std
Power Windows	Pkg	Pkg
***Options Price**	**$1,712**	**$2,106**
***Total Price**	**$14,817**	**$16,995**
Target Price		$15,726
Destination Charge		$418
Avg. Tax & Fees		$981
Total Target $		**$17,125**
Average Dealer Option Cost:		**79%**

The 1993 Expo is available in four models- (Base) Expo and SP; Expo AWD and SP AWD. New for 1993, the Expo AWD features remote keyless entry and Blue interior color availability. New options for the Expo AWD includes remote keyless entry and a power package that includes power windows and door locks. Standard features include protective bodyside moldings and dual manual remote door mirrors. Other features include three rows of seats allowing a variety of different seating and cargo arrangements.

Ownership Costs

Cost Area	5 Year Cost	Rate
Depreciation		⊖
Financing ($344/month)	$2,821	
Insurance (Rating 8)	$7,223	○
State Fees	$696	
Fuel (Hwy 23 City 19)	$4,111	●
Maintenance	$5,196	◐
Repairs	$680	○

Warranty/Maintenance Info

Major Tune-Up	$181	○
Minor Tune-Up	$125	○
Brake Service	$243	○
Overall Warranty	3 yr/36k	○
Drivetrain Warranty	5 yr/60k	○
Rust Warranty	7 yr/100k	○
Maintenance Warranty	N/A	
Roadside Assistance	3 yr/36k	

Ownership Cost By Year

Insufficient Depreciation Information

- 1993
- 1994
- 1995
- 1996
- 1997

Resale Value

Insufficient Information

Cumulative Costs

	1993	1994	1995	1996	1997
Annual	*Insufficient Information*				
Total	*Insufficient Information*				

Ownership Costs (5yr)

Insufficient Information

Ownership Cost Rating

⊖ Insufficient Information

---

* Includes shaded options
** Other purchase requirements apply

● Poor ◐ Worse Than Average ○ Average ○ Better Than Average ○ Excellent ⊖ Insufficient Information

©1993 by IntelliChoice, Inc. (408) 554-8711 All Rights Reserved. Reproduction Prohibited.
Refer to *Section 3: Annotated Vehicle Charts* for an explanation of these charts.

Mitsubishi Expo SP AWD
4 Door Wagon

- 2.4L 136 hp Gas Fuel Inject.
- 4 Cylinder In-Line
- Manual 5 Speed
- 4 Wheel Full-Time
- Automatic Seatbelts

Midsize Wagon — 2WD Model Shown

Purchase Price

Car Item	Dealer Cost	List
Base Price	$14,640	$17,019
Anti-Lock Brakes	$800	$976
Automatic 4 Speed	$600	$680
Optional Engine	N/A	N/A
Air Conditioning	$640	$780
Power Steering	Std	Std
Cruise Control	Std	Std
4 Wheel Full-Time Drive	Std	Std
AM/FM Stereo Cassette	Std	Std
Steering Wheel, Tilt	Std	Std
Power Windows	Std	Std
***Options Price**	$1,240	$1,460
***Total Price**	$15,880	$18,479
Target Price	$16,898	
Destination Charge	$418	
Avg. Tax & Fees	$1,055	
Total Target $	$18,371	
Average Dealer Option Cost:	81%	

Ownership Costs

Cost Area	5 Year Cost	Rate
Depreciation		⊖
Financing ($369/month)	$3,027	
Insurance (Rating 11)	$7,693	○
State Fees	$755	
Fuel (Hwy 23 City 19)	$4,111	●
Maintenance	$5,196	◐
Repairs	$680	○

Warranty/Maintenance Info

Major Tune-Up	$181	○
Minor Tune-Up	$125	○
Brake Service	$243	○
Overall Warranty	3 yr/36k	○
Drivetrain Warranty	5 yr/60k	◯
Rust Warranty	7 yr/100k	◯
Maintenance Warranty	N/A	
Roadside Assistance	3 yr/36k	

Ownership Cost By Year — Insufficient Depreciation Information (1993, 1994, 1995, 1996, 1997)

Resale Value — Insufficient Information

Cumulative Costs — Annual/Total: Insufficient Information (1993–1997)

Ownership Costs (5yr) — Insufficient Information

Ownership Cost Rating — ⊖ Insufficient Information

The 1993 Expo is available in four models- (Base) Expo and SP; Expo AWD and SP AWD. New for 1993, the Expo SP AWD features a remote keyless entry system and a power package that includes power windows and door locks. The interior features three rows of seats allowing a variety of different seating and cargo arrangements. The second seat features a sliding mechanism that allows access to the rear seat. The second and third rows feature special 50/50 split seat backs for additional cargo storage.

Mitsubishi Galant S
4 Door Sedan

- 2.0L 121 hp Gas Fuel Inject.
- 4 Cylinder In-Line
- Manual 5 Speed
- 2 Wheel Front
- Automatic Seatbelts

Compact — LS Model Shown

Purchase Price

Car Item	Dealer Cost	List
Base Price	$11,356	$12,599
Anti-Lock Brakes	N/A	N/A
Automatic 4 Speed	$658	$1,530
Optional Engine	N/A	N/A
Air Conditioning	$658	$802
Power Steering	Std	Std
Cruise Control	$160	$200
All Wheel Drive	N/A	N/A
AM/FM Stereo Cassette	Pkg	Pkg
Steering Wheel, Tilt	Std	Std
Power Windows	Pkg	Pkg
***Options Price**	$1,316	$2,332
***Total Price**	$12,672	$14,931
Target Price	$13,409	
Destination Charge	$393	
Avg. Tax & Fees	$843	
Total Target $	$14,645	
Average Dealer Option Cost:	65%	

Ownership Costs

Cost Area	5 Year Cost	Rate
Depreciation	$7,547	○
Financing ($294/month)	$2,412	
Insurance (Rating 8)	$7,223	○
State Fees	$613	
Fuel (Hwy 28 City 21)	$3,530	○
Maintenance	$5,005	◐
Repairs	$600	◯

Warranty/Maintenance Info

Major Tune-Up	$172	○
Minor Tune-Up	$105	○
Brake Service	$262	◐
Overall Warranty	3 yr/36k	○
Drivetrain Warranty	5 yr/60k	◯
Rust Warranty	7 yr/100k	◯
Maintenance Warranty	N/A	
Roadside Assistance	3 yr/36k	

Resale Value: 1993 $12,355 | 1994 $10,928 | 1995 $9,677 | 1996 $8,306 | 1997 $7,098

Cumulative Costs:

	1993	1994	1995	1996	1997
Annual	$5,480	$4,630	$5,485	$4,627	$6,708
Total	$5,480	$10,110	$15,595	$20,222	$26,930

Ownership Costs (5yr): Average $27,315 | This Car $26,930 | Cost/Mile 39¢ vs 38¢

Ownership Cost Rating: ○ Better Than Average

The 1993 Galant is available in three models - S, LS and ES sedans. New for 1993, all models feature a redesigned body with a more aerodynamic and streamlined exterior. The Base model is now the "S" which comes standard with improved cloth material for better comfort. Standard equipment includes tilt steering, center console, intermittent wipers, auto-off headlamps and on-board diagnostic system. Optional features include front mudguards, full wheel covers and AM/FM cassetter with six speakers.

---

* Includes shaded options
** Other purchase requirements apply

● Poor | ◐ Worse Than Average | ○ Average | ◯ Better Than Average | ◯ Excellent | ⊖ Insufficient Information

©1993 by IntelliChoice, Inc. (408) 554-8711   All Rights Reserved. Reproduction Prohibited.
Refer to *Section 3: Annotated Vehicle Charts* for an explanation of these charts.

Mitsubishi Galant ES
4 Door Sedan

 2.0L 121 hp Gas Fuel Inject. 4 Cylinder In-Line Automatic 4 Speed 2 Wheel Front Automatic Seatbelts

LS Model Shown — Compact

Purchase Price

Car Item	Dealer Cost	List
Base Price	**$13,182**	**$15,509**
Anti-Lock Brakes	N/A	N/A
Manual Transmission	N/A	N/A
Optional Engine	N/A	N/A
Air Conditioning	$658	$802
Power Steering	Std	Std
Cruise Control	Std	Std
All Wheel Drive	N/A	N/A
AM/FM Stereo Cassette	Pkg	Pkg
Steering Wheel, Tilt	Std	Std
Power Windows	Std	Std
*Options Price	$658	$802
*Total Price	**$13,840**	**$16,311**
Target Price	$14,691	
Destination Charge	$393	
Avg. Tax & Fees	$921	
Total Target $	**$16,005**	
Average Dealer Option Cost:	83%	

The 1993 Galant is available in three models - S, LS and ES sedans. New for 1993, all models feature a redesigned body with a more aerodynamic and streamlined exterior. The ES (formerly LS) upgrades the S with standard cruise control, power antenna, rear seat heater ducts, power mirrors and center armrest with trunk access. Optional equipment includes full wheel covers and AM/FM ETR cassette stereo with EQ.

Ownership Costs

Cost Area	5 Year Cost	Rate
Depreciation	$8,214	◯
Financing ($322/month)	$2,637	
Insurance (Rating 10)	$7,479	◯
State Fees	$668	
Fuel (Hwy 28 City 21)	$3,530	◯
Maintenance	$5,005	⊙
Repairs	$600	◯

Warranty/Maintenance Info

Major Tune-Up	$172	◯
Minor Tune-Up	$105	◯
Brake Service	$262	⊙
Overall Warranty	3 yr/36k	◯
Drivetrain Warranty	5 yr/60k	◯
Rust Warranty	7 yr/100k	◯
Maintenance Warranty	N/A	
Roadside Assistance	3 yr/36k	

Ownership Cost By Year

1993, 1994, 1995, 1996, 1997

Resale Value

1993	1994	1995	1996	1997
$13,353	$11,840	$10,492	$9,056	$7,791

Cumulative Costs

	1993	1994	1995	1996	1997
Annual	$5,996	$4,847	$5,690	$4,772	$6,828
Total	$5,996	$10,843	$16,533	$21,305	$28,133

Ownership Costs (5yr)

Average	This Car
$28,500	$28,133
Cost/Mile 41¢	Cost/Mile 40¢

Ownership Cost Rating
◯ Average

Mitsubishi Galant LS
4 Door Sedan

 2.0L 121 hp Gas Fuel Inject. 4 Cylinder In-Line Automatic 4 Speed 2 Wheel Front Automatic Seatbelts

Compact

Purchase Price

Car Item	Dealer Cost	List
Base Price	**$14,060**	**$16,539**
Anti-Lock Brakes	N/A	N/A
Manual Transmission	N/A	N/A
Optional Engine	N/A	N/A
Air Conditioning	$658	$802
Power Steering	Std	Std
Cruise Control	Std	Std
All Wheel Drive	N/A	N/A
AM/FM Stereo Cassette	Pkg	Pkg
Steering Wheel, Tilt	Std	Std
Power Windows	Std	Std
*Options Price	$658	$802
*Total Price	**$14,718**	**$17,341**
Target Price	$15,646	
Destination Charge	$393	
Avg. Tax & Fees	$979	
Total Target $	**$17,018**	
Average Dealer Option Cost:	83%	

The 1993 Galant is available in three models - S, LS and ES sedans. New for 1993, all models feature a redesigned body with a more aerodynamic and streamlined exterior. The LS (formerly the GS) represents the top-of-the-line and makes standard color-keyed rear spoiler, ETACS IV instrumentation control system, P195 H-rated tires and dual exhaust. Optional equipment includes power glass sunroof and wood door trim accents.

Ownership Costs

Cost Area	5 Year Cost	Rate
Depreciation	$8,857	◯
Financing ($342/month)	$2,804	
Insurance (Rating 10)	$7,479	◯
State Fees	$709	
Fuel (Hwy 28 City 21)	$3,530	◯
Maintenance	$5,404	⊙
Repairs	$600	◯

Warranty/Maintenance Info

Major Tune-Up	$172	◯
Minor Tune-Up	$105	◯
Brake Service	$262	⊙
Overall Warranty	3 yr/36k	◯
Drivetrain Warranty	5 yr/60k	◯
Rust Warranty	7 yr/100k	◯
Maintenance Warranty	N/A	
Roadside Assistance	3 yr/36k	

Resale Value

1993	1994	1995	1996	1997
$14,132	$12,495	$11,029	$9,492	$8,161

Cumulative Costs

	1993	1994	1995	1996	1997
Annual	$6,310	$5,033	$6,040	$4,892	$7,108
Total	$6,310	$11,343	$17,383	$22,275	$29,383

Ownership Costs (5yr)

Average	This Car
$29,384	$29,383
Cost/Mile 42¢	Cost/Mile 42¢

Ownership Cost Rating
◯ Average

---

* Includes shaded options
** Other purchase requirements apply

● Poor ⊙ Worse Than Average ◯ Average ◎ Better Than Average ○ Excellent ⊖ Insufficient Information

©1993 by IntelliChoice, Inc. (408) 554-8711 All Rights Reserved. Reproduction Prohibited.
Refer to *Section 3: Annotated Vehicle Charts* for an explanation of these charts.

Mitsubishi Mirage S
2 Door Coupe

LS Model Shown — Subcompact

 1.5L 92 hp Gas Fuel Inject.   4 Cylinder In-Line   Manual 5 Speed   2 Wheel Front Automatic Seatbelts

Purchase Price

Car Item	Dealer Cost	List
Base Price	**$7,042**	**$7,649**
Anti-Lock Brakes	N/A	N/A
Automatic Transmission	N/A	N/A
Optional Engine	N/A	N/A
Air Conditioning	$640	$780
Power Steering	N/A	N/A
Cruise Control	N/A	N/A
All Wheel Drive	N/A	N/A
AM/FM Stereo Cassette	N/A	N/A
Steering Wheel, Tilt	N/A	N/A
Power Windows	N/A	N/A
*Options Price	$0	$0
*Total Price	$7,042	$7,649
Target Price		$7,357
Destination Charge		$393
Avg. Tax & Fees		$468
Total Target $		**$8,218**
Average Dealer Option Cost:		*81%*

The 1993 Mitsubishi Mirage is available in six models - S, ES and LS Coupes, and S, ES and LS Sedans. New for 1993, the S Coupe has been designed with a much more spacious interior. Standard features include highback front bucket seats, steering column-mounted controls, on-board diagnostic system, black finished trim, and a soft-dash touch wraparound dash. Optional features offer an electric rear window defroster (w/ auto shut off), and a radio accommodation package.

Ownership Costs

Cost Area	5 Year Cost	Rate
Depreciation	$4,207	○
Financing ($165/month)	$1,353	
Insurance (Rating 8)	$7,223	●
State Fees	$320	
Fuel (Hwy 40 City 32)	$2,398	○
Maintenance	$4,344	○
Repairs	$600	○

Warranty/Maintenance Info

Major Tune-Up	$156	○
Minor Tune-Up	$88	○
Brake Service	$274	●
Overall Warranty	3 yr/36k	○
Drivetrain Warranty	5 yr/60k	○
Rust Warranty	7 yr/100k	○
Maintenance Warranty	N/A	
Roadside Assistance	3 yr/36k	

Ownership Cost By Year

Bars spanning $2,000–$8,000 for years 1993, 1994, 1995, 1996, 1997.

Resale Value

1993	1994	1995	1996	1997
$6,824	$5,967	$5,345	$4,746	$4,011

Cumulative Costs

	1993	1994	1995	1996	1997
Annual	$3,861	$3,428	$4,104	$3,428	$5,624
Total	$3,861	$7,289	$11,393	$14,821	$20,445

Ownership Costs (5yr)

Average: $20,184 (Cost/Mile 29¢)
This Car: $20,445 (Cost/Mile 29¢)

Ownership Cost Rating
○ Average

Mitsubishi Mirage ES
2 Door Coupe

LS Model Shown — Subcompact

 1.5L 92 hp Gas Fuel Inject.   4 Cylinder In-Line   Manual 5 Speed   2 Wheel Front   Automatic Seatbelts

Purchase Price

Car Item	Dealer Cost	List
Base Price	**$8,042**	**$8,939**
Anti-Lock Brakes	N/A	N/A
Automatic 3 Speed	$430	$470
Optional Engine	N/A	N/A
Air Conditioning	$640	$780
Power Steering	$215	$262
Cruise Control	N/A	N/A
All Wheel Drive	N/A	N/A
AM/FM Stereo Cassette	$312	$446
Steering Wheel, Tilt	N/A	N/A
Power Windows	N/A	N/A
*Options Price	$215	$262
*Total Price	$8,257	$9,201
Target Price		$8,640
Destination Charge		$393
Avg. Tax & Fees		$548
Total Target $		**$9,581**
Average Dealer Option Cost:		*79%*

The 1993 Mitsubishi Mirage is available in six models - S, ES and LS Coupes, and S, ES and LS Sedans. New for 1993, ES Coupe has been designed for more space and style. Standard features include lowback bucket seats, keyless locking, a soft-dash touch wraparound dash, and color-keyed front and rear bumpers, bodyside moldings and grille. Optional features offer a split fold down rear seat, electric rear window defroster (w/auto shut off), and 2-speed wipers with intermittent setting.

Ownership Costs

Cost Area	5 Year Cost	Rate
Depreciation	$5,366	●
Financing ($193/month)	$1,578	
Insurance (Rating 10 [Est.])	$7,479	●
State Fees	$384	
Fuel (Hwy 40 City 32)	$2,398	○
Maintenance	$4,441	○
Repairs	$600	○

Warranty/Maintenance Info

Major Tune-Up	$156	○
Minor Tune-Up	$88	○
Brake Service	$274	●
Overall Warranty	3 yr/36k	○
Drivetrain Warranty	5 yr/60k	○
Rust Warranty	7 yr/100k	○
Maintenance Warranty	N/A	
Roadside Assistance	3 yr/36k	

Resale Value

1993	1994	1995	1996	1997
$6,883	$6,175	$5,605	$4,872	$4,215

Cumulative Costs

	1993	1994	1995	1996	1997
Annual	$5,321	$3,412	$4,209	$3,643	$5,661
Total	$5,321	$8,733	$12,942	$16,585	$22,246

Ownership Costs (5yr)

Average: $21,931 (Cost/Mile 31¢)
This Car: $22,246 (Cost/Mile 32¢)

Ownership Cost Rating
○ Average

---

*Includes shaded options
**Other purchase requirements apply

Legend: ● Poor ◐ Worse Than Average ◑ Average ○ Better Than Average ○ Excellent ⊖ Insufficient Information

©1993 by IntelliChoice, Inc. (408) 554-8711 All Rights Reserved. Reproduction Prohibited.
Refer to *Section 3: Annotated Vehicle Charts* for an explanation of these charts.

Mitsubishi Mirage LS
2 Door Coupe

 1.5L 92 hp Gas Fuel Inject. 4 Cylinder In-Line Manual 5 Speed 2 Wheel Front Automatic Seatbelts

Subcompact

Purchase Price

Car Item	Dealer Cost	List
Base Price	**$9,266**	**$10,299**
Anti-Lock Brakes	N/A	N/A
Automatic 3 Speed	$430	$470
Optional Engine	N/A	N/A
Air Conditioning	$640	$780
Power Steering	Std	Std
Cruise Control	N/A	N/A
All Wheel Drive	N/A	N/A
AM/FM Stereo Cassette	Std	Std
Steering Wheel, Tilt	Std	Std
Power Windows	N/A	N/A
*Options Price	$0	$0
*Total Price	$9,266	$10,299
Target Price		$9,714
Destination Charge		$393
Avg. Tax & Fees		$612
Total Target $		**$10,719**
Average Dealer Option Cost:		80%

The 1993 Mitsubishi Mirage is available in six models - S, ES and LS Coupes, and S, LS and ES Sedans. New for 1993, the LS Coupe has been designed with a much more spacious interior. Standard features include lowback front bucket seats, a soft-dash touch wraparound dash, electric rear-window defroster (w/auto shut off), and color-keyed integrated front and rear bumpers and body side moldings. An on-board diagnostic system and steering column-mounted controls are also standard.

Ownership Costs

Cost Area	5 Year Cost	Rate
Depreciation	$5,482	○
Financing ($215/month)	$1,765	
Insurance (Rating 11 [Est.])	$7,693	●
State Fees	$428	
Fuel (Hwy 40 City 32)	$2,398	○
Maintenance	$4,561	◉
Repairs	$600	○

Warranty/Maintenance Info

Major Tune-Up	$156	○
Minor Tune-Up	$88	○
Brake Service	$274	●
Overall Warranty	3 yr/36k	○
Drivetrain Warranty	5 yr/60k	○
Rust Warranty	7 yr/100k	○
Maintenance Warranty	N/A	
Roadside Assistance	3 yr/36k	

Ownership Cost By Year

1993, 1994, 1995, 1996, 1997

Resale Value

1993	1994	1995	1996	1997
$8,561	$7,505	$6,762	$5,978	$5,237

Cumulative Costs

	1993	1994	1995	1996	1997
Annual	$4,909	$3,870	$4,528	$3,760	$5,860
Total	$4,909	$8,779	$13,307	$17,067	$22,927

Ownership Costs (5yr)

Average	This Car
$23,166	$22,927
Cost/Mile 33¢	Cost/Mile 33¢

Ownership Cost Rating

○ Average

Mitsubishi Mirage S
4 Door Sedan

 1.5L 92 hp Gas Fuel Inject. 4 Cylinder In-Line Manual 5 Speed 2 Wheel Front Automatic Seatbelts

Subcompact

LS Model Shown

Purchase Price

Car Item	Dealer Cost	List
Base Price	**$8,496**	**$9,439**
Anti-Lock Brakes	N/A	N/A
Automatic 3 Speed	$430	$480
Optional Engine	N/A	N/A
Air Conditioning	$640	$780
Power Steering	$215	$262
Cruise Control	N/A	N/A
All Wheel Drive	N/A	N/A
AM/FM Stereo Cassette	$312	$446
Steering Wheel, Tilt	N/A	N/A
Power Windows	N/A	N/A
*Options Price	$215	$262
*Total Price	$8,711	$9,701
Target Price		$9,121
Destination Charge		$393
Avg. Tax & Fees		$577
Total Target $		**$10,091**
Average Dealer Option Cost:		78%

The 1993 Mitsubishi Mirage is available in six models - S, ES and LS Coupes, and S, ES and LS Sedans. New for 1993, the S Sedan has been completely redesigned in the body and interior. Standard features include highback front bucket seats with full face cloth upholstery, steering column-mounted controls, keyless locking, a soft-dash touch wraparound dash, and an on-board diagnostic system. Electric rear-window defroster (w/auto shut off), and tinted glass are all optional items.

Ownership Costs

Cost Area	5 Year Cost	Rate
Depreciation	$5,422	○
Financing ($203/month)	$1,663	
Insurance (Rating 8)	$7,223	●
State Fees	$405	
Fuel (Hwy 40 City 32)	$2,398	○
Maintenance	$4,402	○
Repairs	$600	○

Warranty/Maintenance Info

Major Tune-Up	$156	○
Minor Tune-Up	$88	○
Brake Service	$274	●
Overall Warranty	3 yr/36k	○
Drivetrain Warranty	5 yr/60k	○
Rust Warranty	7 yr/100k	○
Maintenance Warranty	N/A	
Roadside Assistance	3 yr/36k	

Resale Value

1993	1994	1995	1996	1997
$7,778	$6,817	$6,105	$5,430	$4,669

Cumulative Costs

	1993	1994	1995	1996	1997
Annual	$4,929	$3,648	$4,320	$3,542	$5,674
Total	$4,929	$8,577	$12,897	$16,439	$22,113

Ownership Costs (5yr)

Average	This Car
$22,493	$22,113
Cost/Mile 32¢	Cost/Mile 32¢

Ownership Cost Rating

○ Better Than Average

---

* Includes shaded options
** Other purchase requirements apply

● Poor ◉ Worse Than Average ○ Average ○ Better Than Average ○ Excellent ⊖ Insufficient Information

©1993 by IntelliChoice, Inc. (408) 554-8711 All Rights Reserved. Reproduction Prohibited.
Refer to *Section 3: Annotated Vehicle Charts* for an explanation of these charts.

Mitsubishi Mirage ES
4 Door Sedan°

Subcompact — LS Model Shown

- 1.8L 113 hp Gas Fuel Inject.
- 4 Cylinder In-Line
- Manual 5 Speed
- 2 Wheel Front
- Automatic Seatbelts

Purchase Price

Car Item	Dealer Cost	List
Base Price	**$9,430**	**$10,479**
Anti-Lock Brakes	N/A	N/A
Automatic 4 Speed	$582	$640
Optional Engine	N/A	N/A
Air Conditioning	$640	$780
Power Steering	Std	Std
Cruise Control	$160	$200
All Wheel Drive	N/A	N/A
AM/FM Stereo Cassette	$312	$446
Steering Wheel, Tilt	Pkg	Pkg
Power Windows	Pkg	Pkg
*Options Price	$0	$0
*Total Price	$9,430	$10,479
Target Price		$9,888
Destination Charge		$393
Avg. Tax & Fees		$623
Total Target $		**$10,904**
Average Dealer Option Cost:		**79%**

The 1993 Mitsubishi Mirage is available in six models - S, ES and LS Coupes, and S, ES and LS Sedans. New for 1993, the ES Sedan has been completely redesigned in the body and interior. With environmental automotive engineering, recyclable components and asbestos-free materials now contribute to the contemporary design of the Mirage. Standard features include steering column-mounted controls, black finished trim, lowback front bucket seats and an on-board diagnostic system.

Ownership Costs

Cost Area	5 Year Cost	Rate
Depreciation	$6,314	●
Financing ($219/month)	$1,795	
Insurance (Rating 11 [Est.])	$7,693	●
State Fees	$435	
Fuel (Hwy 34 City 27)	$2,831	○
Maintenance	$4,662	◉
Repairs	$600	○

Warranty/Maintenance Info

Major Tune-Up	$155	○
Minor Tune-Up	$88	○
Brake Service	$274	●
Overall Warranty	3 yr/36k	○
Drivetrain Warranty	5 yr/60k	○
Rust Warranty	7 yr/100k	○
Maintenance Warranty	N/A	
Roadside Assistance	3 yr/36k	

Ownership Cost By Year

1993 through 1997 bar chart.

Resale Value

1993	1994	1995	1996	1997
$7,735	$6,902	$6,212	$5,351	$4,590

Cumulative Costs

	1993	1994	1995	1996	1997
Annual	$6,015	$3,742	$4,562	$3,956	$6,055
Total	$6,015	$9,757	$14,319	$18,275	$24,330

Ownership Costs (5yr)

	Average	This Car
	$23,368	$24,330
Cost/Mile	33¢	35¢

Ownership Cost Rating
● Worse Than Average

Mitsubishi Mirage LS
4 Door Sedan

Subcompact

- 1.8L 113 hp Gas Fuel Inject.
- 4 Cylinder In-Line
- Manual 5 Speed
- 2 Wheel Front
- Automatic Seatbelts

Purchase Price

Car Item	Dealer Cost	List
Base Price	**$10,632**	**$12,079**
Anti-Lock Brakes	$800	$976
Automatic 4 Speed	$582	$660
Optional Engine	N/A	N/A
Air Conditioning	$640	$780
Power Steering	Std	Std
Cruise Control	Std	Std
All Wheel Drive	N/A	N/A
AM/FM Stereo Cassette	Std	Std
Steering Wheel, Tilt	Std	Std
Power Windows	Std	Std
*Options Price	$0	$0
*Total Price	$10,632	$12,079
Target Price		$11,169
Destination Charge		$393
Avg. Tax & Fees		$703
Total Target $		**$12,265**
Average Dealer Option Cost:		**81%**

The 1993 Mitsubishi Mirage is available in six models - S, ES and LS Coupes, and S, ES and LS Sedans. New for 1993, the LS Sedan has received a completely redesigned body and interior as well as enhanced engineering refinements. Standard features include color-keyed front and rear bumpers, grille and bodyside moldings, black finished trim, lowback front luxury seats, split fold down rear seat, steering column-mounted controls and a soft-dash touch wraparound dash.

Ownership Costs

Cost Area	5 Year Cost	Rate
Depreciation	$6,368	○
Financing ($247/month)	$2,021	
Insurance (Rating 11 [Est.])	$7,693	●
State Fees	$499	
Fuel (Hwy 34 City 27)	$2,831	○
Maintenance	$4,662	◉
Repairs	$600	○

Warranty/Maintenance Info

Major Tune-Up	$155	○
Minor Tune-Up	$88	○
Brake Service	$274	●
Overall Warranty	3 yr/36k	○
Drivetrain Warranty	5 yr/60k	○
Rust Warranty	7 yr/100k	○
Maintenance Warranty	N/A	
Roadside Assistance	3 yr/36k	

Resale Value

1993	1994	1995	1996	1997
$9,910	$8,644	$7,727	$6,779	$5,897

Cumulative Costs

	1993	1994	1995	1996	1997
Annual	$5,311	$4,259	$4,847	$4,071	$6,186
Total	$5,311	$9,570	$14,417	$18,488	$24,674

Ownership Costs (5yr)

	Average	This Car
	$25,169	$24,674
Cost/Mile	36¢	35¢

Ownership Cost Rating
○ Better Than Average

---

* Includes shaded options
** Other purchase requirements apply

● Poor | ◉ Worse Than Average | ○ Average | ○ Better Than Average | ○ Excellent | ⊖ Insufficient Information

©1993 by IntelliChoice, Inc. (408) 554-8711 All Rights Reserved. Reproduction Prohibited.
Refer to *Section 3: Annotated Vehicle Charts* for an explanation of these charts.

Nissan 240SX
2 Door Coupe

Subcompact

 2.4L 155 hp Gas Fuel Inject.
 4 Cylinder In-Line
 Manual 5 Speed
 2 Wheel Rear
Automatic Seatbelts

Purchase Price

Car Item	Dealer Cost	List
Base Price	**$13,094**	**$14,755**
Anti-Lock Brakes	N/A	N/A
Automatic 4 Speed	$736	$830
Optional Engine	N/A	N/A
Air Conditioning	$720	$850
Power Steering	Std	Std
Cruise Control	N/A	N/A
All Wheel Drive	N/A	N/A
AM/FM Stereo Cassette	Std	Std
Steering Wheel, Tilt	Std	Std
Power Windows	N/A	N/A
*Options Price	$0	$0
*Total Price	$13,094	$14,755
Target Price		$13,897
Destination Charge		$350
Avg. Tax & Fees		$863
Total Target $		**$15,110**
Average Dealer Option Cost:		87%

The 1993 Nissan 240SX comes in five models - Coupe, Fastback and convertible in either base or SE trim. New for 1993, the Base SX comes with new exterior colors and mid-season non-CFC air conditioning. The SX comes standard with rear-wheel drive, full-length center console, four-wheel independent suspension, four-wheel disc brakes, rack-and-pinion steering, AM/FM cassette stereo, and contoured seats. Options include air conditioning and electronic four-speed transmission.

Ownership Costs

Cost Area	5 Year Cost	Rate
Depreciation	$6,591	○
Financing ($304/month)	$2,490	
Insurance (Rating 16)	$8,784	●
State Fees	$605	
Fuel (Hwy 28 City 22)	$3,454	◉
Maintenance	$4,512	○
Repairs	$550	○

Warranty/Maintenance Info

Major Tune-Up	$135	○
Minor Tune-Up	$75	○
Brake Service	$173	○
Overall Warranty	3 yr/36k	◉
Drivetrain Warranty	5 yr/60k	
Rust Warranty	5 yr/unlim. mi	○
Maintenance Warranty	N/A	
Roadside Assistance	3 yr/36k	

Ownership Cost By Year

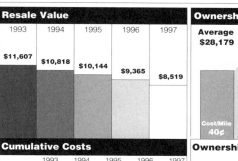

1993, 1994, 1995, 1996, 1997

Resale Value

1993	1994	1995	1996	1997
$11,607	$10,818	$10,144	$9,365	$8,519

Cumulative Costs

	1993	1994	1995	1996	1997
Annual	$6,998	$4,240	$4,967	$3,994	$6,787
Total	$6,998	$11,238	$16,205	$20,199	$26,986

Ownership Costs (5yr)

Average	This Car
$28,179	$26,986
Cost/Mile 40¢	Cost/Mile 39¢

Ownership Cost Rating

○ Excellent

Nissan 240SX SE
2 Door Coupe

Subcompact

 2.4L 155 hp Gas Fuel Inject.
 4 Cylinder In-Line
 Manual 5 Speed
 2 Wheel Rear
 Automatic Seatbelts

Purchase Price

Car Item	Dealer Cost	List
Base Price	**$15,105**	**$17,220**
Anti-Lock Brakes	N/A	N/A
Automatic 4 Speed	$736	$830
Optional Engine	N/A	N/A
Air Conditioning	$720	$850
Power Steering	Std	Std
Cruise Control	Std	Std
All Wheel Drive	N/A	N/A
AM/FM Stereo Cassette	Std	Std
Steering Wheel, Tilt	Std	Std
Power Windows	Std	Std
*Options Price	$0	$0
*Total Price	$15,105	$17,220
Target Price		$16,086
Destination Charge		$350
Avg. Tax & Fees		$998
Total Target $		**$17,434**
Average Dealer Option Cost:		86%

The 1993 Nissan 240SX comes in five models - Coupe, Fastback and convertible in either base or SE trim. New for 1993, the SE Coupe comes with new exterior colors and mid-season non-CFC air conditioning. The SE comes standard with rear-wheel drive, full-length center console, four-wheel independent suspension, four-wheel disc brakes, power rack-and-pinion steering, AM/FM cassette stereo, and contoured seats. Options include leather seating surfaces, power sunroof and air conditioning.

Ownership Costs

Cost Area	5 Year Cost	Rate
Depreciation	$8,718	○
Financing ($350/month)	$2,872	
Insurance (Rating 16)	$8,784	●
State Fees	$703	
Fuel (Hwy 28 City 22)	$3,454	◉
Maintenance	$4,512	○
Repairs	$550	○

Warranty/Maintenance Info

Major Tune-Up	$135	○
Minor Tune-Up	$75	○
Brake Service	$173	○
Overall Warranty	3 yr/36k	◉
Drivetrain Warranty	5 yr/60k	○
Rust Warranty	5 yr/unlim. mi	○
Maintenance Warranty	N/A	
Roadside Assistance	3 yr/36k	

Resale Value

1993	1994	1995	1996	1997
$12,421	$11,484	$10,611	$9,686	$8,716

Cumulative Costs

	1993	1994	1995	1996	1997
Annual	$8,692	$4,530	$5,261	$4,185	$6,925
Total	$8,692	$13,222	$18,483	$22,668	$29,593

Ownership Costs (5yr)

Average	This Car
$30,953	$29,593
Cost/Mile 44¢	Cost/Mile 42¢

Ownership Cost Rating

○ Excellent

---

* Includes shaded options
** Other purchase requirements apply

● Poor ◉ Worse Than Average ○ Average ○ Better Than Average ○ Excellent ⊖ Insufficient Information

©1993 by IntelliChoice, Inc. (408) 554-8711 All Rights Reserved. Reproduction Prohibited.
Refer to *Section 3: Annotated Vehicle Charts* for an explanation of these charts.

Nissan 240SX
2 Door Fastback

 2.4L 155 hp Gas Fuel Inject.
 4 Cylinder In-Line
 Manual 5 Speed
 2 Wheel Rear
 Automatic Seatbelts

Subcompact — SE Model Shown

Purchase Price

Car Item	Dealer Cost	List
Base Price	**$13,575**	**$15,475**
Anti-Lock Brakes	N/A	N/A
Automatic 4 Speed	$723	$825
Optional Engine	N/A	N/A
Air Conditioning	$720	$850
Power Steering	Std	Std
Cruise Control	N/A	N/A
All Wheel Drive	N/A	N/A
AM/FM Stereo Cassette	Std	Std
Steering Wheel, Tilt	Std	Std
Power Windows	N/A	N/A
*Options Price	$0	$0
***Total Price**	**$13,575**	**$15,475**
Target Price	$14,419	
Destination Charge	$350	
Avg. Tax & Fees	$896	
Total Target $	**$15,665**	
Average Dealer Option Cost:	86%	

Ownership Costs

Cost Area	5 Year Cost	Rate
Depreciation	$6,935	○
Financing ($315/month)	$2,581	
Insurance (Rating 17)	$9,084	●
State Fees	$633	
Fuel (Hwy 28 City 22)	$3,454	◉
Maintenance	$4,512	○
Repairs	$550	○

Warranty/Maintenance Info

Major Tune-Up	$135	○
Minor Tune-Up	$75	○
Brake Service	$173	○
Overall Warranty	3 yr/36k	◉
Drivetrain Warranty	5 yr/60k	○
Rust Warranty	5 yr/unlim. mi	○
Maintenance Warranty	N/A	
Roadside Assistance	3 yr/36k	

Ownership Cost By Year

(Bars for 1993–1997, scale $2,000–$10,000)

Resale Value

1993	1994	1995	1996	1997
$11,947	$11,125	$10,419	$9,607	$8,730

Cumulative Costs

	1993	1994	1995	1996	1997
Annual	$7,313	$4,365	$5,083	$4,101	$6,887
Total	$7,313	$11,678	$16,761	$20,862	$27,749

Ownership Costs (5yr)

Average	This Car
$28,989	$27,749
Cost/Mile 41¢	Cost/Mile 40¢

Ownership Cost Rating
○ Excellent

The 1993 Nissan 240SX comes in five models - Coupe, Fastback and convertible in either base or SE trim. New for 1993, the Base Fastback comes with new exterior colors and mid-season non-CFC air conditioning. Standard features include rear-wheel drive, full aerodynamic wheel covers, tilt steering column, four-wheel independent suspension, AM/FM cassette stereo, four-wheel disc brakes, fold-down rear seatback, and power rack-and-pinion steering. Air conditioning is optional.

Nissan 240SX SE
2 Door Fastback

 2.4L 155 hp Gas Fuel Inject.
 4 Cylinder In-Line
 Manual 5 Speed
 2 Wheel Rear
Automatic Seatbelts

Subcompact

Purchase Price

Car Item	Dealer Cost	List
Base Price	**$15,505**	**$17,675**
Anti-Lock Brakes	$843	** $995
Automatic 4 Speed	$727	$830
Optional Engine	N/A	N/A
Air Conditioning	$720	$850
Power Steering	Std	Std
Cruise Control	Std	Std
All Wheel Drive	N/A	N/A
AM/FM Stereo Cassette	Std	Std
Steering Wheel, Tilt	Std	Std
Power Windows	Std	Std
*Options Price	$0	$0
***Total Price**	**$15,505**	**$17,675**
Target Price	$16,524	
Destination Charge	$350	
Avg. Tax & Fees	$1,024	
Total Target $	**$17,898**	
Average Dealer Option Cost:	85%	

Ownership Costs

Cost Area	5 Year Cost	Rate
Depreciation	$8,986	○
Financing ($360/month)	$2,948	
Insurance (Rating 17)	$9,084	●
State Fees	$720	
Fuel (Hwy 28 City 22)	$3,454	◉
Maintenance	$4,512	○
Repairs	$550	○

Warranty/Maintenance Info

Major Tune-Up	$135	○
Minor Tune-Up	$75	○
Brake Service	$173	○
Overall Warranty	3 yr/36k	◉
Drivetrain Warranty	5 yr/60k	○
Rust Warranty	5 yr/unlim. mi	○
Maintenance Warranty	N/A	
Roadside Assistance	3 yr/36k	

Resale Value

1993	1994	1995	1996	1997
$12,703	$11,744	$10,848	$9,901	$8,912

Cumulative Costs

	1993	1994	1995	1996	1997
Annual	$8,965	$4,636	$5,363	$4,278	$7,012
Total	$8,965	$13,601	$18,964	$23,242	$30,254

Ownership Costs (5yr)

Average	This Car
$31,465	$30,254
Cost/Mile 45¢	Cost/Mile 43¢

Ownership Cost Rating
○ Excellent

The 1993 Nissan 240SX comes in five models - Coupe, Fastback and convertible in either base or SE trim. New for 1993, the SE Coupe comes with new exterior colors and mid-season non-CFC air conditioning. The SE comes standard with rear-wheel drive, full-length center console, four-wheel independent suspension, four-wheel disc brakes, power rack-and-pinion steering, AM/FM cassette stereo, and contoured seats. Options include leather seating surfaces, power sunroof and air conditioning.

---

page 162

* Includes shaded options
** Other purchase requirements apply

● Poor   ◉ Worse Than Average   ○ Average   ○ Better Than Average   ○ Excellent   ⊖ Insufficient Information

©1993 by IntelliChoice, Inc. (408) 554-8711   All Rights Reserved. Reproduction Prohibited.
Refer to *Section 3: Annotated Vehicle Charts* for an explanation of these charts.

Nissan 240SX
2 Door Convertible

Subcompact

 2.4L 155 hp Gas Fuel Inject.
 4 Cylinder In-Line
Automatic 4 Speed
 2 Wheel Rear
 Automatic Seatbelts

Purchase Price

Car Item	Dealer Cost	List
Base Price	$19,601	$22,345
Anti-Lock Brakes	N/A	N/A
Manual Transmission	N/A	N/A
Optional Engine	N/A	N/A
Air Conditioning	$720	$850
Power Steering	Std	Std
Cruise Control	Std	Std
All Wheel Drive	N/A	N/A
AM/FM Stereo Cassette	Std	Std
Steering Wheel, Tilt	Std	Std
Power Windows	Std	Std
*Options Price	$0	$0
*Total Price	$19,601	$22,345
Target Price		$21,034
Destination Charge		$350
Avg. Tax & Fees		$1,296
Total Target $		$22,680
Average Dealer Option Cost:	**85%**	

The 1993 Nissan 240SX comes in five models - Coupe, Fastback and convertible in either base or SE trim. New for 1993, the convertible offers a power operated fabric top, cruise control, four-speed automatic transmission, power front and rear windows and a 4-speaker audio system. A non-CFC air conditioner will be available mid-season. The convertible also comes equipped with a 2.4L four-cylinder engine, four-wheel independent suspension, four-wheel disc brakes, and rack-and-pinion steering.

Ownership Costs

Cost Area	5 Year Cost	Rate
Depreciation	$11,446	◯
Financing ($456/month)	$3,736	
Insurance (Rating 22)	$11,167	●
State Fees	$908	
Fuel (Hwy 26 City 21)	$3,673	◉
Maintenance	$4,738	◉
Repairs	$550	◯

Warranty/Maintenance Info

Major Tune-Up	$135	◯
Minor Tune-Up	$75	◯
Brake Service	$173	◯
Overall Warranty	3 yr/36k	◉
Drivetrain Warranty	5 yr/60k	◯
Rust Warranty	5 yr/unlim. mi	◯
Maintenance Warranty	N/A	
Roadside Assistance	3 yr/36k	

Ownership Cost By Year

1993, 1994, 1995, 1996, 1997

Resale Value

1993	1994	1995	1996	1997
$16,929	$15,315	$13,913	$12,520	$11,234

Cumulative Costs

	1993	1994	1995	1996	1997
Annual	$10,321	$6,022	$6,631	$5,292	$7,952
Total	$10,321	$16,343	$22,974	$28,266	$36,218

Ownership Costs (5yr)

Average	This Car
$36,719	$36,218
Cost/Mile 52¢	Cost/Mile 52¢

Ownership Cost Rating

◯ Better Than Average

Nissan 300ZX
2 Door Coupe

Sport

 3.0L 222 hp Gas Fuel Inject.
 6 Cylinder "V"
 Manual 5 Speed
 2 Wheel Rear
 Driver Airbag Psngr Belts

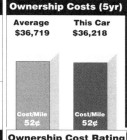

Purchase Price

Car Item	Dealer Cost	List
Base Price	$25,786	$30,095
Anti-Lock Brakes	Std	Std
Automatic 4 Speed	$814	** $950
Optional Engine	N/A	N/A
Auto Climate Control	Std	Std
Power Steering	Std	Std
Cruise Control	Std	Std
All Wheel Drive	N/A	N/A
AM/FM Stereo Cassette	Std	Std
Steering Wheel, Tilt	Std	Std
Power Windows	Std	Std
*Options Price	$0	$0
*Total Price	$25,786	$30,095
Target Price		$27,961
Destination Charge		$350
Avg. Tax & Fees		$1,720
Total Target $		$30,031
Average Dealer Option Cost:	**85%**	

The 1993 Nissan 300ZX comes in four models - ZX (Base), 2+2, turbo and convertible. New for 1993, the Base 300 ZX comes equipped with a Bose audio system and anti-lock brakes. Standard features include a 3.0L V-6 engine, five-speed manual transmission, two-way shock absorbers, driver's side airbag, rack-and-pinion steering, cruise control, power door locks and windows, air conditioning and an automatic theft deterrent system.

Ownership Costs

Cost Area	5 Year Cost	Rate
Depreciation	$16,827	◉
Financing ($604/month)	$4,947	
Insurance (Rating 21 Sport+)	$14,300	●
State Fees	$1,218	
Fuel (Hwy 24 City 18)	$4,117	◉
Maintenance	$5,034	◯
Repairs	$700	◯

Warranty/Maintenance Info

Major Tune-Up	$163	◯
Minor Tune-Up	$79	◯
Brake Service	$145	◯
Overall Warranty	3 yr/36k	◉
Drivetrain Warranty	5 yr/60k	◯
Rust Warranty	5 yr/unlim. mi	◯
Maintenance Warranty	N/A	
Roadside Assistance	3 yr/36k	

Resale Value

1993	1994	1995	1996	1997
$21,180	$18,832	$16,860	$15,018	$13,204

Cumulative Costs

	1993	1994	1995	1996	1997
Annual	$14,667	$7,904	$8,526	$6,698	$9,346
Total	$14,667	$22,571	$31,097	$37,795	$47,141

Ownership Costs (5yr)

Average	This Car
$47,125	$47,141
Cost/Mile 67¢	Cost/Mile 67¢

Ownership Cost Rating

◯ Average

---

* Includes shaded options
** Other purchase requirements apply

● Poor ◐ Worse Than Average ◑ Average ◯ Better Than Average ◎ Excellent ⊖ Insufficient Information

©1993 by IntelliChoice, Inc. (408) 554-8711   All Rights Reserved. Reproduction Prohibited.
Refer to *Section 3: Annotated Vehicle Charts* for an explanation of these charts.

Nissan 300ZX 2+2
2 Door Coupe

 3.0L 222 hp Gas Fuel Inject.
 6 Cylinder "V"
 Manual 5 Speed
 2 Wheel Rear
 Driver Airbag Psngr Belts

Base Model Shown — Sport

Purchase Price

Car Item	Dealer Cost	List
Base Price	**$28,724**	**$33,525**
Anti-Lock Brakes	Std	Std
Automatic 4 Speed	$814	$950
Optional Engine	N/A	N/A
Auto Climate Control	Std	Std
Power Steering	Std	Std
Cruise Control	Std	Std
All Wheel Drive	N/A	N/A
AM/FM Stereo Cassette	Std	Std
Steering Wheel, Tilt	Std	Std
Power Windows	Std	Std
*Options Price	$0	$0
*Total Price	$28,724	$33,525
Target Price		$31,300
Destination Charge		$350
Avg. Tax & Fees		$1,922
Luxury Tax		$165
Total Target $		**$33,737**

Ownership Costs

Cost Area	5 Year Cost	Rate
Depreciation	$19,415	●
Financing ($678/month)	$5,558	
Insurance (Rating 22 Sport+)	$14,896	●
State Fees	$1,355	
Fuel (Hwy 24 City 18)	$4,117	●
Maintenance	$5,034	○
Repairs	$700	○

Warranty/Maintenance Info

Major Tune-Up	$163	○
Minor Tune-Up	$79	○
Brake Service	$145	○
Overall Warranty	3 yr/36k	◉
Drivetrain Warranty	5 yr/60k	○
Rust Warranty	5 yr/unlim. mi	○
Maintenance Warranty	N/A	
Roadside Assistance	3 yr/36k	

Ownership Cost By Year

Resale Value
1993: $22,639 | 1994: $20,279 | 1995: $18,184 | 1996: $16,244 | 1997: $14,322

Ownership Costs (5yr)
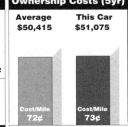
Average $50,415 | This Car $51,075
Cost/Mile 72¢ | Cost/Mile 73¢

Cumulative Costs

	1993	1994	1995	1996	1997
Annual	$17,313	$8,253	$8,916	$6,990	$9,603
Total	$17,313	$25,566	$34,482	$41,472	$51,075

Ownership Cost Rating
◯ Average

The 1993 Nissan 300ZX comes in four models - ZX (Base), 2+2, turbo, and convertible. New for 1993, the 2+2 includes a Bose audio system and anti-lock brakes. Standards include a driver's side airbag, a 3.0 liter V-6 engine, five-speed manual or four-speed automatic transmission, rack-and-pinion steering, side window defoggers, analog instrumentation, air conditioning and an automatic theft deterrent system. Leather seating is an available option.

Nissan 300ZX Turbo
2 Door Coupe

 3.0L 300 hp Turbo Gas Fuel Inject.
 6 Cylinder "V"
 Manual 5 Speed
 2 Wheel Rear
 Driver Airbag Psngr Belts

Sport

Purchase Price

Car Item	Dealer Cost	List
Base Price	**$31,779**	**$37,090**
Anti-Lock Brakes	Std	Std
Automatic 4 Speed	$814	$950
Optional Engine	N/A	N/A
Auto Climate Control	Std	Std
Power Steering	Std	Std
Cruise Control	Std	Std
All Wheel Drive	N/A	N/A
AM/FM Stereo Cassette	Std	Std
Steering Wheel, Tilt	Std	Std
Power Windows	Std	Std
*Options Price	$0	$0
*Total Price	$31,779	$37,090
Target Price		$34,639
Destination Charge		$350
Avg. Tax & Fees		$2,123
Luxury Tax		$499
Total Target $		**$37,611**

Ownership Costs

Cost Area	5 Year Cost	Rate
Depreciation	$22,057	●
Financing ($756/month)	$6,196	
Insurance (Rating 23 Sport+)	$15,492	●
State Fees	$1,498	
Fuel (Hwy 24 City 18)	$4,117	●
Maintenance	$5,720	○
Repairs	$810	○

Warranty/Maintenance Info

Major Tune-Up	$163	○
Minor Tune-Up	$79	○
Brake Service	$145	○
Overall Warranty	3 yr/36k	◉
Drivetrain Warranty	5 yr/60k	○
Rust Warranty	5 yr/unlim. mi	○
Maintenance Warranty	N/A	
Roadside Assistance	3 yr/36k	

Resale Value
1993: $23,813 | 1994: $21,479 | 1995: $19,419 | 1996: $17,504 | 1997: $15,554

Ownership Costs (5yr)

Average $53,836 | This Car $55,890
Cost/Mile 77¢ | Cost/Mile 80¢

Cumulative Costs

	1993	1994	1995	1996	1997
Annual	$20,471	$8,622	$9,341	$7,228	$10,228
Total	$20,471	$29,093	$38,434	$45,662	$55,890

Ownership Cost Rating
◯ Average

The 1993 Nissan 300ZX comes in four models - (Base), 2+2, turbo, and convertible. New for 1993, the Turbo features a Bose audio system and anti-lock brakes. Standard features include high speed Z-rated tires with alloy wheels, an engine oil cooler, integrated rear spoiler, vented front air dam, two-way shock absorbers, driver's side airbag, rack-and-pinion steering, air conditioning, cruise control, and an automatic theft deterrent system. Leather seating is an option.

---

* Includes shaded options
** Other purchase requirements apply

● Poor | ◐ Worse Than Average | ◯ Average | ◔ Better Than Average | ○ Excellent | ⊖ Insufficient Information

©1993 by IntelliChoice, Inc. (408) 554-8711 All Rights Reserved. Reproduction Prohibited.
Refer to *Section 3: Annotated Vehicle Charts* for an explanation of these charts.

Nissan 300ZX
2 Door Convertible

 3.0L 222 hp Gas Fuel Inject.
 6 Cylinder "V"
 Manual 5 Speed
 2 Wheel Rear
 Driver Airbag Psngr Belts

Sport

Purchase Price

Car Item	Dealer Cost	List
Base Price	**$31,333**	**$36,570**
Anti-Lock Brakes	Std	Std
Automatic 4 Speed	$857	$1,000
Optional Engine	N/A	N/A
Air Conditioning	Std	Std
Power Steering	Std	Std
Cruise Control	Std	Std
All Wheel Drive	N/A	N/A
AM/FM Stereo Cassette	Std	Std
Steering Wheel, Tilt	Std	Std
Power Windows	Std	Std
*Options Price	$0	$0
*Total Price	$31,333	$36,570
Target Price		$34,153
Destination Charge		$350
Avg. Tax & Fees		$2,094
Luxury Tax		$450
Total Target $		**$37,047**

The 1993 Nissan 300ZX Convertible is available in a one model edition. New for 1993, the 300ZX features a manually-operated convertible top that folds under a hard storage compartment cover. Nissan performs the convertible conversion at its 300ZX plant in Japan. Interior refinements include a leather-trimmed interior, driver-side air bag and full power-operated features. Functional features include low-profile high-performance radials mounted to 16-inch, five spoke alloy wheels.

Ownership Costs

Cost Area	5 Year Cost	Rate
Depreciation		⊖
Financing ($745/month)	$6,102	
Insurance (Rating 23 Sport+)	$15,492	●
State Fees	$1,476	
Fuel (Hwy 24 City 18 -Prem.)	$4,547	●
Maintenance	$5,034	○
Repairs	$700	○

Warranty/Maintenance Info

Major Tune-Up	$163	○
Minor Tune-Up	$79	○
Brake Service	$145	○
Overall Warranty	3 yr/36k	●
Drivetrain Warranty	5 yr/60k	○
Rust Warranty	5 yr/unlim. mi	○
Maintenance Warranty	N/A	
Roadside Assistance	3 yr/36k	

Ownership Cost By Year

Insufficient Depreciation Information

1993, 1994, 1995, 1996, 1997

Resale Value

Insufficient Information

Cumulative Costs

	1993	1994	1995	1996	1997
Annual	Insufficient Information				
Total	Insufficient Information				

Ownership Costs (5yr)

Insufficient Information

Ownership Cost Rating

⊖ Insufficient Information

Nissan Altima XE
4 Door Sedan

 2.4L 150 hp Gas Fuel Inject.
4 Cylinder In-Line
Manual 5 Speed
 2 Wheel Front
 Driver Airbag Psngr Belts

Compact

SE Model Shown

Purchase Price

Car Item	Dealer Cost	List
Base Price	**$11,469**	**$12,999**
Anti-Lock Brakes	$843	** $995
Automatic 4 Speed	$728	$825
Optional Engine	N/A	N/A
Air Conditioning	Pkg	Pkg
Power Steering	Std	Std
Cruise Control	$195	$230
All Wheel Drive	N/A	N/A
AM/FM Stereo Cassette	Pkg	Pkg
Steering Wheel, Tilt	Std	Std
Power Windows	N/A	N/A
*Options Price	$728	$825
*Total Price	$12,197	$13,824
Target Price		$12,988
Destination Charge		$300
Avg. Tax & Fees		$805
Total Target $		**$14,093**
Average Dealer Option Cost:		85%

The Nissan Altima is offered in four model editions - XE, GXE, SE and GLE four-door sedans. New for 1993, the Altima replaces the Stanza in the Nissan line-up. The XE makes standard a driver's side air bag, automatic seat belts, tilt steering wheel, remote trunk and fuel door releases, and child safety door locks. Functional features include a trip odometer and tachometer, and rear window defroster. Among the options available are cruise control, anti-lock braking system, and diversity antenna system.

Ownership Costs

Cost Area	5 Year Cost	Rate
Depreciation	$7,187	○
Financing ($283/month)	$2,322	
Insurance (Rating 11)	$7,693	●
State Fees	$565	
Fuel (Hwy 29 City 21)	$3,465	○
Maintenance	$4,005	○
Repairs	$560	○

Warranty/Maintenance Info

Major Tune-Up	$133	○
Minor Tune-Up	$75	○
Brake Service	$187	○
Overall Warranty	3 yr/36k	●
Drivetrain Warranty	5 yr/60k	○
Rust Warranty	5 yr/unlim. mi	○
Maintenance Warranty	N/A	
Roadside Assistance	3 yr/36k	

Ownership Cost By Year

$2,000 $4,000 $6,000 $8,000

1993, 1994, 1995, 1996, 1997

Resale Value

1993	1994	1995	1996	1997
$11,245	$10,078	$8,995	$7,935	$6,906

Cumulative Costs

	1993	1994	1995	1996	1997
Annual	$6,062	$4,348	$5,202	$4,658	$5,527
Total	$6,062	$10,410	$15,612	$20,270	$25,797

Ownership Costs (5yr)

Average	This Car
$26,364	$25,797
Cost/Mile 38¢	Cost/Mile 37¢

Ownership Cost Rating

○ Better Than Average

---

* Includes shaded options
** Other purchase requirements apply

 Poor
 Worse Than Average
 Average
 Better Than Average
○ Excellent
 Insufficient Information

©1993 by IntelliChoice, Inc. (408) 554-8711 All Rights Reserved. Reproduction Prohibited.
Refer to *Section 3: Annotated Vehicle Charts* for an explanation of these charts.

Nissan Altima GXE
4 Door Sedan

 2.4L 150 hp Gas Fuel Inject. | 4 Cylinder In-Line | Manual 5 Speed | 2 Wheel Front | Driver Airbag Psngr Belts

Compact

Purchase Price

Car Item	Dealer Cost	List
Base Price	**$12,231**	**$14,024**
Anti-Lock Brakes	$1,012	** $1,195
Automatic 4 Speed	$719	$825
Optional Engine	N/A	N/A
Air Conditioning	Pkg	Pkg
Power Steering	Std	Std
Cruise Control	$195	$230
All Wheel Drive	N/A	N/A
AM/FM Stereo Cassette	Pkg	Pkg
Steering Wheel, Tilt	Std	Std
Power Windows	Std	Std
*Options Price	$719	$825
*Total Price	$12,950	$14,849
Target Price	$13,810	
Destination Charge	$300	
Avg. Tax & Fees	$857	
Total Target $	**$14,967**	
Average Dealer Option Cost:	85%	

The Nissan Altima is offered in four model editions - XE, GXE, SE and GLE four-door sedans. New for 1993, the Altima replaces the Stanza in the Nissan line-up. Standard equipment on the GXE model includes power windows with driver's express down, power door locks, and a pass-through trunk with rear seat armrest. Options include front limited slip differential, anti-lock braking system, diversity antenna system, and a Value Option Package which includes air conditioning, cruise control, cassette stereo, and power antenna.

Ownership Costs

Cost Area	5 Year Cost	Rate
Depreciation	$8,649	●
Financing ($301/month)	$2,466	
Insurance (Rating 11)	$7,693	●
State Fees	$605	
Fuel (Hwy 29 City 21)	$3,465	○
Maintenance	$4,005	○
Repairs	$560	○

Warranty/Maintenance Info

Major Tune-Up	$133	○
Minor Tune-Up	$75	○
Brake Service	$187	○
Overall Warranty	3 yr/36k	●
Drivetrain Warranty	5 yr/60k	○
Rust Warranty	5 yr/unlim. mi	○
Maintenance Warranty	N/A	
Roadside Assistance	3 yr/36k	

Ownership Cost By Year

(1993, 1994, 1995, 1996, 1997)

Resale Value

1993	1994	1995	1996	1997
$10,762	$9,547	$8,462	$7,362	$6,318

Cumulative Costs

	1993	1994	1995	1996	1997
Annual	$7,489	$4,450	$5,241	$4,715	$5,548
Total	$7,489	$11,939	$17,180	$21,895	$27,443

Ownership Costs (5yr)

Average: $27,244 | This Car: $27,443
Cost/Mile 39¢ | Cost/Mile 39¢

Ownership Cost Rating

○ Average

Nissan Altima SE
4 Door Sedan

 2.4L 150 hp Gas Fuel Inject. | 4 Cylinder In-Line | Manual 5 Speed | 2 Wheel Front | Driver Airbag Psngr Belts

Compact

Purchase Price

Car Item	Dealer Cost	List
Base Price	**$14,326**	**$16,524**
Anti-Lock Brakes	$1,012	$1,195
Automatic 4 Speed	$698	$825
Optional Engine	N/A	N/A
Air Conditioning	Std	Std
Power Steering	Std	Std
Cruise Control	Std	Std
All Wheel Drive	N/A	N/A
AM/FM Stereo Cassette	Std	Std
Steering Wheel, Tilt	Std	Std
Power Windows	Std	Std
*Options Price	$698	$825
*Total Price	$15,024	$17,349
Target Price	$16,085	
Destination Charge	$300	
Avg. Tax & Fees	$995	
Total Target $	**$17,380**	
Average Dealer Option Cost:	85%	

The Nissan Altima is offered in four model editions - XE, GXE, SE and GLE four-door sedans. New for 1993, the Altima replaces the Stanza in the Nissan line-up. The sportier SE model features a standard AM/FM cassette stereo, cruise control, diversity antenna system, sport bucket seats, fog lights, leather wrapped steering wheel, and rear spoiler. Optional on the SE only is a power glass sunroof with sunshade. Anti-lock brakes, limited slip differential, and a CD player are also optional.

Ownership Costs

Cost Area	5 Year Cost	Rate
Depreciation	$9,806	●
Financing ($349/month)	$2,864	
Insurance (Rating 11)	$7,693	●
State Fees	$706	
Fuel (Hwy 29 City 21)	$3,465	○
Maintenance	$3,963	○
Repairs	$560	○

Warranty/Maintenance Info

Major Tune-Up	$133	○
Minor Tune-Up	$75	○
Brake Service	$169	○
Overall Warranty	3 yr/36k	●
Drivetrain Warranty	5 yr/60k	○
Rust Warranty	5 yr/unlim. mi	○
Maintenance Warranty	N/A	
Roadside Assistance	3 yr/36k	

Ownership Cost By Year

Resale Value

1993	1994	1995	1996	1997
$12,635	$11,253	$10,005	$8,759	$7,574

Cumulative Costs

	1993	1994	1995	1996	1997
Annual	$8,220	$4,764	$5,483	$4,908	$5,682
Total	$8,220	$12,984	$18,467	$23,375	$29,057

Ownership Costs (5yr)

Average: $29,391 | This Car: $29,057
Cost/Mile 42¢ | Cost/Mile 42¢

Ownership Cost Rating

○ Average

---

page 166

* Includes shaded options
** Other purchase requirements apply

● Poor | ◐ Worse Than Average | ◑ Average | ○ Better Than Average | ○ Excellent | ⊖ Insufficient Information

©1993 by IntelliChoice, Inc. (408) 554-8711 All Rights Reserved. Reproduction Prohibited.
Refer to *Section 3: Annotated Vehicle Charts* for an explanation of these charts.

Nissan Altima GLE
4 Door Sedan

Compact

 2.4L 150 hp Gas Fuel Inject.
 4 Cylinder In-Line
 Automatic 4 Speed
 2 Wheel Front
 Driver Airbag Psngr Belts

Purchase Price

Car Item	Dealer Cost	List
Base Price	**$15,909**	**$18,349**
Anti-Lock Brakes	$1,012	$1,195
Manual Transmission	N/A	N/A
Optional Engine	N/A	N/A
Auto Climate Control	Std	Std
Power Steering	Std	Std
Cruise Control	Std	Std
All Wheel Drive	N/A	N/A
AM/FM Stereo CD	Std	Std
Steering Wheel, Tilt	Std	Std
Power Windows	Std	Std
*Options Price	$0	$0
*Total Price	$15,909	$18,349
Target Price		$17,083
Destination Charge		$300
Avg. Tax & Fees		$1,055
Total Target $		**$18,438**
Average Dealer Option Cost:	**85%**	

The Nissan Altima is offered in four model editions - XE, GXE, SE and GLE four-door sedans. New for 1993, the Altima replaces the Stanza in the Nissan line-up. The top-of-the-line GLE model makes standard a theft deterrent system, automatic temperature control, CD player, power sunroof, adjustable lumbar support, and head-up display which projects driver information onto the windshield so the driver need not take his eyes off the road. Optional on the GLE is leather seating surfaces.

Ownership Costs

Cost Area	5 Year Cost	Rate
Depreciation		⊖
Financing ($371/month)	$3,037	
Insurance (Rating 11)	$7,468	●
State Fees	$746	
Fuel (Hwy 29 City 21)	$3,465	◐
Maintenance	$4,005	◯
Repairs	$560	◯

Warranty/Maintenance Info

Major Tune-Up	$133	◯
Minor Tune-Up	$75	◯
Brake Service	$187	◯
Overall Warranty	3 yr/36k	◐
Drivetrain Warranty	5 yr/60k	◯
Rust Warranty	5 yr/unlim. mi	◯
Maintenance Warranty	N/A	
Roadside Assistance	3 yr/36k	

Ownership Cost By Year

Insufficient Depreciation Information

1993, 1994, 1995, 1996, 1997

Resale Value

Insufficient Information

Cumulative Costs

	1993	1994	1995	1996	1997
Annual		Insufficient Information			
Total		Insufficient Information			

Ownership Costs (5yr)

Insufficient Information

Ownership Cost Rating

⊖ Insufficient Information

Nissan Maxima GXE
4 Door Sedan

Compact

 3.0L 160 hp Gas Fuel Inject.
 6 Cylinder "V"
 Automatic 4 Speed
 2 Wheel Front
 Driver Airbag Psngr Belts

Purchase Price

Car Item	Dealer Cost	List
Base Price	**$18,172**	**$20,960**
Anti-Lock Brakes	$843	$995
Manual Transmission	N/A	N/A
Optional Engine	N/A	N/A
Air Conditioning	Std	Std
Power Steering	Std	Std
Cruise Control	Std	Std
All Wheel Drive	N/A	N/A
AM/FM Stereo Cassette	Std	Std
Steering Wheel, Tilt	Std	Std
Power Windows	Std	Std
*Options Price	$0	$0
*Total Price	$18,172	$20,960
Target Price		$19,454
Destination Charge		$350
Avg. Tax & Fees		$1,203
Total Target $		**$21,007**
Average Dealer Option Cost:	**85%**	

The 1993 Maxima is available in two models-SE and GXE. New for 1993, the GXE features CFC-free air conditioning, limited slip differential and V-rated all season tires. The GXE also features driver's-side airbag, fore and aft headrest adjustments, an electric trunk release and electric defog exterior mirrors. The optional Luxury Package for the GXE includes a power glass sliding sunroof with rear-tilt feature. The GXE is also available with cloth or leather interior.

Ownership Costs

Cost Area	5 Year Cost	Rate
Depreciation	$9,449	◯
Financing ($422/month)	$3,460	
Insurance (Rating 13)	$7,797	◐
State Fees	$852	
Fuel (Hwy 26 City 19)	$3,847	◐
Maintenance	$5,206	◐
Repairs	$640	◯

Warranty/Maintenance Info

Major Tune-Up	$162	◯
Minor Tune-Up	$79	◯
Brake Service	$189	◯
Overall Warranty	3 yr/36k	◐
Drivetrain Warranty	5 yr/60k	◯
Rust Warranty	5 yr/unlim. mi	◯
Maintenance Warranty	N/A	
Roadside Assistance	3 yr/36k	

Resale Value

1993	1994	1995	1996	1997
$16,630	$15,279	$14,070	$12,841	$11,558

Cumulative Costs

	1993	1994	1995	1996	1997
Annual	$8,230	$5,057	$5,832	$4,479	$7,653
Total	$8,230	$13,287	$19,119	$23,598	$31,251

Ownership Costs (5yr)

Average	This Car
$32,492	$31,251
Cost/Mile 46¢	Cost/Mile 45¢

Ownership Cost Rating

◯ Better Than Average

---

* Includes shaded options
** Other purchase requirements apply

● Poor ◐ Worse Than Average ◯ Average ◯ Better Than Average ◯ Excellent ⊖ Insufficient Information

©1993 by IntelliChoice, Inc. (408) 554-8711 All Rights Reserved. Reproduction Prohibited.
Refer to *Section 3: Annotated Vehicle Charts* for an explanation of these charts.

page **167**

Nissan Maxima SE
4 Door Sedan

- 3.0L 190 hp Gas Fuel Inject.
- 6 Cylinder "V"
- Manual 5 Speed
- 2 Wheel Front
- Driver Airbag Psngr Belts

Compact

Purchase Price

Car Item	Dealer Cost	List
Base Price	**$19,095**	**$22,025**
Anti-Lock Brakes	$843	$995
Automatic 4 Speed	$811	$935
Optional Engine	N/A	N/A
Air Conditioning	Std	Std
Power Steering	Std	Std
Cruise Control	Std	Std
All Wheel Drive	N/A	N/A
AM/FM Stereo Cassette	N/A	N/A
Steering Wheel, Tilt	Std	Std
Power Windows	Std	Std
*Options Price	$811	$935
*Total Price	$19,906	$22,960
Target Price		$21,343
Destination Charge		$350
Avg. Tax & Fees		$1,318
Total Target $		**$23,011**
Average Dealer Option Cost:		85%

The 1993 Maxima is available in two models-SE and GXE. New for 1993, the SE features CFC-free air conditioning, limited slip differential and V-rated all-season tires. The SE also features as standard equipment unique analog gauges that reverse contrast from day to night for improved visability, a new child safety-seat seatbelt system and a Bose four-speaker electronically tuned audio system. Additionally, the SE receives new seat cloth, two new available colors and a medium beige interior.

Ownership Costs

Cost Area	5 Year Cost	Rate
Depreciation	$10,930	○
Financing ($462/month)	$3,791	
Insurance (Rating 14)	$7,956	○
State Fees	$932	
Fuel (Hwy 25 City 19)	$3,929	◉
Maintenance	$5,280	◉
Repairs	$640	○

Warranty/Maintenance Info

Major Tune-Up	$116	○
Minor Tune-Up	$67	○
Brake Service	$189	○
Overall Warranty	3 yr/36k	○
Drivetrain Warranty	5 yr/60k	○
Rust Warranty	5 yr/unlim. mi	○
Maintenance Warranty	N/A	
Roadside Assistance	3 yr/36k	

Ownership Cost By Year

1993, 1994, 1995, 1996, 1997

Resale Value

1993	1994	1995	1996	1997
$17,337	$15,961	$14,638	$13,256	$12,081

Cumulative Costs

	1993	1994	1995	1996	1997
Annual	$9,731	$5,241	$6,256	$4,712	$7,518
Total	$9,731	$14,972	$21,228	$25,940	$33,458

Ownership Costs (5yr)

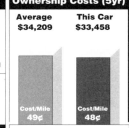

Average $34,209 / This Car $33,458
Cost/Mile 49¢ / Cost/Mile 48¢

Ownership Cost Rating: Better Than Average

Nissan NX 1600
2 Door Coupe

- 1.6L 110 hp Gas Fuel Inject.
- 4 Cylinder In-Line
- Manual 5 Speed
- 2 Wheel Front
- Driver Airbag Psngr Belts

Subcompact

2000 Model Shown

Purchase Price

Car Item	Dealer Cost	List
Base Price	**$10,444**	**$11,635**
Anti-Lock Brakes	N/A	N/A
Automatic 4 Speed	$740	$825
Optional Engine	N/A	N/A
Air Conditioning	$720	$850
Power Steering	Std	Std
Cruise Control	N/A	N/A
All Wheel Drive	N/A	N/A
AM/FM Stereo Cassette	$381	$450
Steering Wheel, Tilt	Std	Std
Power Windows	N/A	N/A
*Options Price	$0	$0
*Total Price	$10,444	$11,635
Target Price		$11,034
Destination Charge		$350
Avg. Tax & Fees		$689
Total Target $		**$12,073**
Average Dealer Option Cost:		86%

The 1993 NX is available in two models-1600 and 2000. New for 1993, the NX 1600 features an upgraded interior which includes sports-styled bucket seats and raised thigh bolsters. Removable T-bar panels are available as an option. Standard features include a driver's-side airbag, power mirrors, full instrumentation and front & rear stabilizer bars. Other features include molded door trim, full-length center console and cargo area with fold down rear seat back.

Ownership Costs

Cost Area	5 Year Cost	Rate
Depreciation	$6,199	○
Financing ($243/month)	$1,988	
Insurance (Rating 11)	$7,693	●
State Fees	$480	
Fuel (Hwy 38 City 28)	$2,623	○
Maintenance	$4,041	○
Repairs	$560	○

Warranty/Maintenance Info

Major Tune-Up	$135	○
Minor Tune-Up	$74	○
Brake Service	$205	○
Overall Warranty	3 yr/36k	○
Drivetrain Warranty	5 yr/60k	○
Rust Warranty	5 yr/unlim. mi	○
Maintenance Warranty	N/A	
Roadside Assistance	3 yr/36k	

Ownership Cost By Year

1993, 1994, 1995, 1996, 1997

Resale Value

1993	1994	1995	1996	1997
$9,074	$8,130	$7,428	$6,705	$5,874

Cumulative Costs

	1993	1994	1995	1996	1997
Annual	$5,897	$3,840	$4,348	$3,482	$6,017
Total	$5,897	$9,737	$14,085	$17,567	$23,584

Ownership Costs (5yr)

Average $24,669 / This Car $23,584
Cost/Mile 35¢ / Cost/Mile 34¢

Ownership Cost Rating: Excellent

---

* Includes shaded options
** Other purchase requirements apply

 Poor · Worse Than Average · Average · Better Than Average · Excellent · Insufficient Information

©1993 by IntelliChoice, Inc. (408) 554-8711 All Rights Reserved. Reproduction Prohibited.
Refer to *Section 3: Annotated Vehicle Charts* for an explanation of these charts.

Nissan NX 2000
2 Door Coupe

Subcompact

 2.0L 140 hp Gas Fuel Inject.
 4 Cylinder In-Line
 Manual 5 Speed
 2 Wheel Front
 Driver Airbag Psngr Belts

Purchase Price

Car Item	Dealer Cost	List
Base Price	**$13,062**	**$14,720**
Anti-Lock Brakes	$593	$700
Automatic 4 Speed	$732	$825
Optional Engine	N/A	N/A
Air Conditioning	$720	$850
Power Steering	Std	Std
Cruise Control	Pkg	Pkg
All Wheel Drive	N/A	N/A
AM/FM Stereo Cassette	Std	Std
Steering Wheel, Tilt	Std	Std
Power Windows	Pkg	Pkg
*Options Price	$0	$0
*Total Price	$13,062	$14,720
Target Price	$13,862	
Destination Charge	$350	
Avg. Tax & Fees	$862	
Total Target $	**$15,074**	
Average Dealer Option Cost:	86%	

The 1993 NX is available in two models—1600 and 2000. New for 1993, the NX 2000 features an upgraded interior with a standard T-bar roof. Interior features include sport-styled bucket seats, molded door trim, full-length center console and cargo area with fold down rear seat back. The NX 2000 also has an available Power Package, which includes power windows and door locks and cruise control.

Ownership Costs

Cost Area	5 Year Cost	Rate
Depreciation	$7,871	○
Financing ($303/month)	$2,482	
Insurance (Rating 13)	$8,061	●
State Fees	$603	
Fuel (Hwy 30 City 23)	$3,260	◔
Maintenance	$4,262	○
Repairs	$560	○

Warranty/Maintenance Info

Major Tune-Up	$117	○
Minor Tune-Up	$74	○
Brake Service	$205	○
Overall Warranty	3 yr/36k	○
Drivetrain Warranty	5 yr/60k	
Rust Warranty	5 yr/unlim. mi	○
Maintenance Warranty	N/A	
Roadside Assistance	3 yr/36k	

Resale Value

1993: $11,188 | 1994: $10,010 | 1995: $9,096 | 1996: $8,198 | 1997: $7,203

Cumulative Costs

	1993	1994	1995	1996	1997
Annual	$7,209	$4,452	$5,096	$3,926	$6,416
Total	$7,209	$11,661	$16,757	$20,683	$27,099

Ownership Costs (5yr)

Average: $28,140 | This Car: $27,099
Cost/Mile: 40¢ | Cost/Mile: 39¢

Ownership Cost Rating: ○ Better Than Average

Nissan Sentra E
2 Door Sedan

Subcompact

 1.6L 110 hp Gas Fuel Inject.
 4 Cylinder In-Line
 Manual 5 Speed
 2 Wheel Front
 Belts Std, Driv Air Opt

Purchase Price

Car Item	Dealer Cost	List
Base Price	**$8,177**	**$8,715**
Anti-Lock Brakes	N/A	N/A
Automatic 4 Speed	Pkg	Pkg
Optional Engine	N/A	N/A
Air Conditioning	$843	$995
Power Steering	Pkg	Pkg
Cruise Control	N/A	N/A
All Wheel Drive	N/A	N/A
AM/FM Stereo Cassette	$508	$600
Steering Wheel, Tilt	Pkg	Pkg
Power Windows	N/A	N/A
*Options Price	$0	$0
*Total Price	$8,177	$8,715
Target Price	$8,653	
Destination Charge	$350	
Avg. Tax & Fees	$541	
Total Target $	**$9,544**	
Average Dealer Option Cost:	89%	

The 1993 Nissan Sentra is available in seven models - E, XE, SE and SE-R two-door and E, XE and GXE four-door sedans. New for 1993, the two-door E offers an optional driver's side air bag. Standard features include a five-speed manual, reclining front bucket seats, black body side molding, four-wheel independent suspension, and front disc/rear drum brakes. Options include air conditioning, AM/FM cassette stereo, and a power steering package.

Ownership Costs

Cost Area	5 Year Cost	Rate
Depreciation	$3,942	○
Financing ($192/month)	$1,572	
Insurance (Rating 11)	$7,693	●
State Fees	$363	
Fuel (Hwy 38 City 29)	$2,579	○
Maintenance	$3,935	○
Repairs	$550	○

Warranty/Maintenance Info

Major Tune-Up	$135	○
Minor Tune-Up	$74	○
Brake Service	$205	○
Overall Warranty	3 yr/36k	○
Drivetrain Warranty	5 yr/60k	○
Rust Warranty	5 yr/unlim. mi	○
Maintenance Warranty	N/A	
Roadside Assistance	3 yr/36k	

Resale Value

1993: $8,760 | 1994: $7,841 | 1995: $7,122 | 1996: $6,351 | 1997: $5,602

Cumulative Costs

	1993	1994	1995	1996	1997
Annual	$3,472	$3,649	$4,193	$3,466	$5,854
Total	$3,472	$7,121	$11,314	$14,780	$20,634

Ownership Costs (5yr)

Average: $21,384 | This Car: $20,634
Cost/Mile: 31¢ | Cost/Mile: 29¢

Ownership Cost Rating: ○ Better Than Average

---

* Includes shaded options
** Other purchase requirements apply

● Poor | ◔ Worse Than Average | ◐ Average | ○ Better Than Average | ○ Excellent | ⊖ Insufficient Information

©1993 by IntelliChoice, Inc. (408) 554-8711 All Rights Reserved. Reproduction Prohibited.
Refer to *Section 3: Annotated Vehicle Charts* for an explanation of these charts.

Nissan Sentra E
4 Door Sedan

Subcompact

 1.6L 110 hp Gas Fuel Inject.
 4 Cylinder In-Line
 Manual 5 Speed
 2 Wheel Front
 Belts Std, Driv Air Opt

2 Door Model Shown

Purchase Price

Car Item	Dealer Cost	List
Base Price	**$9,020**	**$10,165**
Anti-Lock Brakes	N/A	N/A
Automatic 4 Speed	Pkg	Pkg
Optional Engine	N/A	N/A
Air Conditioning	$843	$995
Power Steering	Pkg	Pkg
Cruise Control	N/A	N/A
All Wheel Drive	N/A	N/A
AM/FM Stereo Cassette	$508	$600
Steering Wheel, Tilt	Pkg	Pkg
Power Windows	N/A	N/A
***Options Price**	$0	$0
***Total Price**	**$9,020**	**$10,165**
Target Price		$9,560
Destination Charge		$350
Avg. Tax & Fees		$601
Total Target $		**$10,511**
Average Dealer Option Cost:		87%

The 1993 Nissan Sentra is available in seven models - E, XE, SE and SE-R two-door and E, XE and GXE four-door sedans. New for 1993, the four-door E offers an optional driver's side air bag. Standard features include a five-speed manual transmission, four-wheel independent suspension, front motorized shoulder belts with manual lap belts and front disc/rear drum brakes. Options include air conditioning, AM/FM cassette stereo, and a power steering package.

Ownership Costs

Cost Area	5 Year Cost	Rate
Depreciation	$4,513	○
Financing ($211/month)	$1,731	
Insurance (Rating 10)	$7,479	●
State Fees	$421	
Fuel (Hwy 38 City 29)	$2,579	○
Maintenance	$3,935	○
Repairs	$550	○

Warranty/Maintenance Info

Major Tune-Up	$135	○
Minor Tune-Up	$74	○
Brake Service	$205	○
Overall Warranty	3 yr/36k	○
Drivetrain Warranty	5 yr/60k	○
Rust Warranty	5 yr/unlim. mi	○
Maintenance Warranty	N/A	
Roadside Assistance	3 yr/36k	

Ownership Cost By Year
(bars for 1993, 1994, 1995, 1996, 1997)

Resale Value
1993	1994	1995	1996	1997
$9,631	$8,591	$7,735	$6,846	$5,998

Cumulative Costs
	1993	1994	1995	1996	1997
Annual	$3,609	$3,792	$4,330	$3,562	$5,915
Total	$3,609	$7,401	$11,731	$15,293	$21,208

Ownership Costs (5yr)
Average	This Car
$23,015	$21,208
Cost/Mile 33¢	Cost/Mile 30¢

Ownership Cost Rating
○ Excellent

Nissan Sentra XE
2 Door Sedan

Subcompact

 1.6L 110 hp Gas Fuel Inject.
 4 Cylinder In-Line
 Manual 5 Speed
 2 Wheel Front
 Belts Std, Driv Air Opt

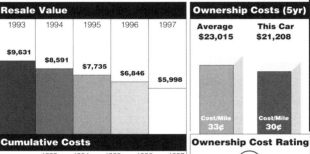
E Model Shown

Purchase Price

Car Item	Dealer Cost	List
Base Price	**$9,318**	**$10,500**
Anti-Lock Brakes	N/A	N/A
Automatic 4 Speed	$643	$725
Optional Engine	N/A	N/A
Air Conditioning	$843	$995
Power Steering	Std	Std
Cruise Control	$195	$230
All Wheel Drive	N/A	N/A
AM/FM Stereo Cassette	$508	$600
Steering Wheel, Tilt	Std	Std
Power Windows	N/A	N/A
***Options Price**	$0	$0
***Total Price**	**$9,318**	**$10,500**
Target Price		$9,882
Destination Charge		$350
Avg. Tax & Fees		$621
Total Target $		**$10,853**
Average Dealer Option Cost:		86%

The 1993 Nissan Sentra is available in seven models - E, XE, SE and SE-R two-door and E, XE and GXE four-door sedans. New for 1993, the XE two-door offers a driver's side air bag and cruise control. Standard features include rack-and-pinion power steering, a five-speed manual transmission, a four-wheel independent suspension, and front disc/rear drum brakes. Options include an AM/FM stereo cassette and a four-speed automatic transmission.

Ownership Costs

Cost Area	5 Year Cost	Rate
Depreciation	$4,982	○
Financing ($218/month)	$1,788	
Insurance (Rating 12)	$7,898	●
State Fees	$435	
Fuel (Hwy 38 City 29)	$2,579	○
Maintenance	$4,054	○
Repairs	$550	○

Warranty/Maintenance Info

Major Tune-Up	$135	○
Minor Tune-Up	$74	○
Brake Service	$205	○
Overall Warranty	3 yr/36k	○
Drivetrain Warranty	5 yr/60k	○
Rust Warranty	5 yr/unlim. mi	○
Maintenance Warranty	N/A	
Roadside Assistance	3 yr/36k	

Ownership Cost By Year
(bars for 1993, 1994, 1995, 1996, 1997)

Resale Value
1993	1994	1995	1996	1997
$9,033	$8,104	$7,360	$6,620	$5,871

Cumulative Costs
	1993	1994	1995	1996	1997
Annual	$4,654	$3,783	$4,372	$3,506	$5,971
Total	$4,654	$8,437	$12,809	$16,315	$22,286

Ownership Costs (5yr)
Average	This Car
$23,392	$22,286
Cost/Mile 33¢	Cost/Mile 32¢

Ownership Cost Rating
○ Excellent

---

* Includes shaded options
** Other purchase requirements apply

● Poor ◉ Worse Than Average ○ Average ○ Better Than Average ○ Excellent ⊖ Insufficient Information

©1993 by IntelliChoice, Inc. (408) 554-8711   All Rights Reserved. Reproduction Prohibited.
Refer to *Section 3: Annotated Vehicle Charts* for an explanation of these charts.

Nissan Sentra SE
2 Door Sedan

Subcompact

- 1.6L 110 hp Gas Fuel Inject.
- 4 Cylinder In-Line
- Manual 5 Speed
- 2 Wheel Front
- Belts Std, Driv Air Opt

SE-R Model Shown

Purchase Price

Car Item	Dealer Cost	List
Base Price	$9,617	$10,900
Anti-Lock Brakes	N/A	N/A
Automatic 4 Speed	$639	$725
Optional Engine	N/A	N/A
Air Conditioning	$843	$995
Power Steering	Std	Std
Cruise Control	$195	$230
All Wheel Drive	N/A	N/A
AM/FM Stereo Cassette	$508	$600
Steering Wheel, Tilt	Std	Std
Power Windows	N/A	N/A
*Options Price	$0	$0
*Total Price	$9,617	$10,900
Target Price		$10,205
Destination Charge		$350
Avg. Tax & Fees		$641
Total Target $		**$11,196**
Average Dealer Option Cost:		85%

The 1993 Nissan Sentra is available in seven models - E, XE, SE and SE-R two-door and E, XE and GXE four-door sedans. New for 1993, the Base SE includes an optional driver's side airbag and new exterior modifications. Standard features include four-wheel independent suspension, front and rear stabilizer bars and refined front-end fascia. Transmission choices include a five-speed manual or an optional four-speed automatic. Options include cruise control and air conditioning.

Ownership Costs

Cost Area	5 Year Cost	Rate
Depreciation	$5,258	○
Financing ($225/month)	$1,844	
Insurance (Rating 12)	$7,898	●
State Fees	$451	
Fuel (Hwy 38 City 29)	$2,579	○
Maintenance	$4,054	○
Repairs	$550	○

Warranty/Maintenance Info

Major Tune-Up	$135	○
Minor Tune-Up	$74	○
Brake Service	$205	○
Overall Warranty	3 yr/36k	○
Drivetrain Warranty	5 yr/60k	○
Rust Warranty	5 yr/unlim. mi	○
Maintenance Warranty	N/A	
Roadside Assistance	3 yr/36k	

Ownership Cost By Year

1993, 1994, 1995, 1996, 1997

Resale Value

1993	1994	1995	1996	1997
$8,853	$7,995	$7,359	$6,652	$5,938

Cumulative Costs

	1993	1994	1995	1996	1997
Annual	$5,204	$3,732	$4,279	$3,480	$5,939
Total	$5,204	$8,936	$13,215	$16,695	$22,634

Ownership Costs (5yr)

Average	This Car
$23,842	$22,634
Cost/Mile 34¢	Cost/Mile 32¢

Ownership Cost Rating

○ Excellent

Nissan Sentra XE
4 Door Sedan

Subcompact

- 1.6L 110 hp Gas Fuel Inject.
- 4 Cylinder In-Line
- Manual 5 Speed
- 2 Wheel Front
- Belts Std, Driv Air Opt

GXE Model Shown

Purchase Price

Car Item	Dealer Cost	List
Base Price	$9,930	$11,190
Anti-Lock Brakes	N/A	N/A
Automatic 4 Speed	$643	$725
Optional Engine	N/A	N/A
Air Conditioning	$843	$995
Power Steering	Std	Std
Cruise Control	Pkg	Pkg
All Wheel Drive	N/A	N/A
AM/FM Stereo Cassette	$508	$600
Steering Wheel, Tilt	Std	Std
Power Windows	N/A	N/A
*Options Price	$0	$0
*Total Price	$9,930	$11,190
Target Price		$10,543
Destination Charge		$350
Avg. Tax & Fees		$660
Total Target $		**$11,553**
Average Dealer Option Cost:		86%

The 1993 Nissan Sentra is available in seven models - E, XE, SE and SE-R two-door and E, XE and GXE four-door sedans. New for 1993, the four-door XE offers an optional driver's side air bag and cruise control. Standard features include power rack-and-pinion steering, a five-speed manual transmission, a four-wheel independent suspension, and front disc/rear drum brakes. Options include an AM/FM stereo cassette and a four-speed automatic transmission.

Ownership Costs

Cost Area	5 Year Cost	Rate
Depreciation	$5,375	○
Financing ($232/month)	$1,903	
Insurance (Rating 10)	$7,479	●
State Fees	$461	
Fuel (Hwy 38 City 29)	$2,579	○
Maintenance	$4,054	○
Repairs	$550	○

Warranty/Maintenance Info

Major Tune-Up	$135	○
Minor Tune-Up	$74	○
Brake Service	$205	○
Overall Warranty	3 yr/36k	○
Drivetrain Warranty	5 yr/60k	○
Rust Warranty	5 yr/unlim. mi	○
Maintenance Warranty	N/A	
Roadside Assistance	3 yr/36k	

Resale Value

1993	1994	1995	1996	1997
$9,216	$8,315	$7,616	$6,893	$6,178

Cumulative Costs

	1993	1994	1995	1996	1997
Annual	$5,148	$3,716	$4,272	$3,415	$5,850
Total	$5,148	$8,864	$13,136	$16,551	$22,401

Ownership Costs (5yr)

Average	This Car
$24,168	$22,401
Cost/Mile 35¢	Cost/Mile 32¢

Ownership Cost Rating

○ Excellent

---

* Includes shaded options
** Other purchase requirements apply

● Poor ◐ Worse Than Average ○ Average ◯ Better Than Average ◯ Excellent ⊖ Insufficient Information

©1993 by IntelliChoice, Inc. (408) 554-8711 All Rights Reserved. Reproduction Prohibited.
Refer to *Section 3: Annotated Vehicle Charts* for an explanation of these charts.

Nissan Sentra SE-R
2 Door Sedan

 2.0L 140 hp Gas Fuel Inject.
 4 Cylinder In-Line
 Manual 5 Speed
 2 Wheel Front
Belts Std, Driv Air Opt

Subcompact

Purchase Price

Car Item	Dealer Cost	List
Base Price	**$10,989**	**$12,455**
Anti-Lock Brakes	$593	$700
Automatic 4 Speed	$729	$825
Optional Engine	N/A	N/A
Air Conditioning	$843	$995
Power Steering	Std	Std
Cruise Control	$195	$230
All Wheel Drive	N/A	N/A
AM/FM Stereo Cassette	$508	$600
Steering Wheel, Tilt	Std	Std
Power Windows	N/A	N/A
*Options Price	$0	$0
*Total Price	$10,989	$12,455
Target Price	$11,691	
Destination Charge	$350	
Avg. Tax & Fees	$730	
Total Target $	**$12,771**	
Average Dealer Option Cost:	**85%**	

The 1993 Sentra is available in seven models - E, XE, SE and SE-R two-door and E, XE and GXE four-door sedans. New for 1993, the SE-R includes an optional driver's side airbag and new exterior modifications. Features include four-wheel independent suspension with front and rear stabilizer bars, a 5-speed manual or an optional four-speed automatic, sport-tuned suspension, rack-and-pinion steering, four-wheel disc brakes, and fog lamps. Options include anti-lock brakes and cruise control.

Ownership Costs

Cost Area	5 Year Cost	Rate
Depreciation	$6,185	O
Financing ($257/month)	$2,104	
Insurance (Rating 14)	$8,238	●
State Fees	$512	
Fuel (Hwy 32 City 24)	$3,088	◉
Maintenance	$4,055	O
Repairs	$550	O

Warranty/Maintenance Info

Major Tune-Up	$117	O
Minor Tune-Up	$74	O
Brake Service	$162	O
Overall Warranty	3 yr/36k	O
Drivetrain Warranty	5 yr/60k	O
Rust Warranty	5 yr/unlim. mi	O
Maintenance Warranty	N/A	
Roadside Assistance	3 yr/36k	

Ownership Cost By Year

1993 / 1994 / 1995 / 1996 / 1997

Resale Value

1993	1994	1995	1996	1997
$10,030	$9,027	$8,264	$7,435	$6,586

Cumulative Costs

	1993	1994	1995	1996	1997
Annual	$5,884	$4,139	$4,751	$3,812	$6,146
Total	$5,884	$10,023	$14,774	$18,586	$24,732

Ownership Costs (5yr)

Average	This Car
$25,592	$24,732
Cost/Mile 37¢	Cost/Mile 35¢

Ownership Cost Rating
Better Than Average

Nissan Sentra GXE
4 Door Sedan

 1.6L 110 hp Gas Fuel Inject.
 4 Cylinder In-Line
 Manual 5 Speed
2 Wheel Front
Driver Airbag Psngr Belts

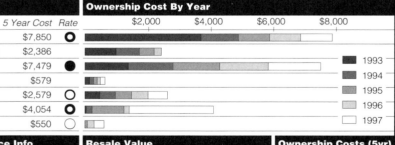
Subcompact

Purchase Price

Car Item	Dealer Cost	List
Base Price	**$12,481**	**$14,145**
Anti-Lock Brakes	$593	$700
Automatic 4 Speed	$639	$725
Optional Engine	N/A	N/A
Air Conditioning	Std	Std
Power Steering	Std	Std
Cruise Control	Std	Std
All Wheel Drive	N/A	N/A
AM/FM Stereo Cassette	Std	Std
Steering Wheel, Tilt	Std	Std
Power Windows	Std	Std
*Options Price	$0	$0
*Total Price	$12,481	$14,145
Target Price	$13,316	
Destination Charge	$350	
Avg. Tax & Fees	$828	
Total Target $	**$14,494**	
Average Dealer Option Cost:	**86%**	

The 1993 Nissan Sentra is available in seven models - E, XE, SE and SE-R two-door and E, XE and GXE four-door sedans. New for 1993, the GXE four-door offers a driver's side air bag, power windows, door locks, air conditioning, rack-and-pinion steering, and AM/FM cassette stereo. Other features include a five-speed manual transmission, four-wheel independent suspension, full-analog instrumentation, split rear seatbacks and front and rear stabilizer bars. Options include an anti-lock brake system.

Ownership Costs

Cost Area	5 Year Cost	Rate
Depreciation	$7,850	O
Financing ($291/month)	$2,386	
Insurance (Rating 10)	$7,479	●
State Fees	$579	
Fuel (Hwy 38 City 29)	$2,579	O
Maintenance	$4,054	O
Repairs	$550	O

Warranty/Maintenance Info

Major Tune-Up	$135	O
Minor Tune-Up	$74	O
Brake Service	$205	O
Overall Warranty	3 yr/36k	O
Drivetrain Warranty	5 yr/60k	O
Rust Warranty	5 yr/unlim. mi	O
Maintenance Warranty	N/A	
Roadside Assistance	3 yr/36k	

Resale Value

1993	1994	1995	1996	1997
$10,782	$9,587	$8,592	$7,594	$6,644

Cumulative Costs

	1993	1994	1995	1996	1997
Annual	$6,754	$4,187	$4,688	$3,745	$6,103
Total	$6,754	$10,941	$15,629	$19,374	$25,477

Ownership Costs (5yr)

Average	This Car
$27,493	$25,477
Cost/Mile 39¢	Cost/Mile 36¢

Ownership Cost Rating
Excellent

---

page 172

* Includes shaded options
** Other purchase requirements apply

● Poor | ◉ Worse Than Average | ○ Average | O Better Than Average | O Excellent | ⊖ Insufficient Information

©1993 by IntelliChoice, Inc. (408) 554-8711 All Rights Reserved. Reproduction Prohibited.
Refer to *Section 3: Annotated Vehicle Charts* for an explanation of these charts.

Oldsmobile Achieva S
2 Door Coupe
Compact

- 2.3L 120 hp Gas Fuel Inject.
- 4 Cylinder In-Line
- Manual 5 Speed
- 2 Wheel Front
- Automatic Seatbelts

Purchase Price

Car Item	Dealer Cost	List
Base Price	**$11,809**	**$13,049**
Anti-Lock Brakes	Std	Std
Automatic 3 Speed	$477	$555
3.3L 160 hp Gas	$396	$460
Air Conditioning	$714	$830
Power Steering	Std	Std
Cruise Control	$194	$225
All Wheel Drive	N/A	N/A
AM/FM Stereo Cassette	$194	$225
Steering Wheel, Tilt	$125	$145
Power Windows	$237	$275
***Options Price**	**$1,781**	**$2,070**
***Total Price**	**$13,590**	**$15,119**
Target Price		$14,392
Destination Charge		$475
Avg. Tax & Fees		$899
Total Target $		**$15,766**
Average Dealer Option Cost:		86%

The 1993 Oldsmobile Achieva is available in four models - S and SL coupes, and S and SL sedans. New for 1993, the Achieva S coupe has a major noise reduction program, standard Battery Run Down Protection to prevent inadvertent power drainage, and an advanced entertainment system. Standard features include rear seat heat ducts, self-aligning urethane steering wheel, and two-way adjustable contour front bucket seats. The Sport Coupe Package and the SCX Package are optional features.

Ownership Costs

Cost Area	5 Year Cost	Rate
Depreciation	$8,929	○
Financing ($317/month)	$2,598	
Insurance (Rating 11)	$7,693	○
State Fees	$624	
Fuel (Hwy 29 City 20)	$3,548	○
Maintenance		⊖
Repairs	$630	○

Warranty/Maintenance Info

Major Tune-Up		⊖
Minor Tune-Up		⊖
Brake Service		⊖
Overall Warranty	3 yr/36k	○
Drivetrain Warranty	3 yr/36k	○
Rust Warranty	6 yr/100k	○
Maintenance Warranty	N/A	
Roadside Assistance	3 yr/36k	

Ownership Cost By Year
1993, 1994, 1995, 1996, 1997

Resale Value
- 1993: $11,011
- 1994: $9,826
- 1995: $8,833
- 1996: $7,782
- 1997: $6,837

Cumulative Costs
Annual: Insufficient Information
Total: Insufficient Information

Ownership Costs (5yr)
Insufficient Information

Ownership Cost Rating
⊖ Insufficient Information

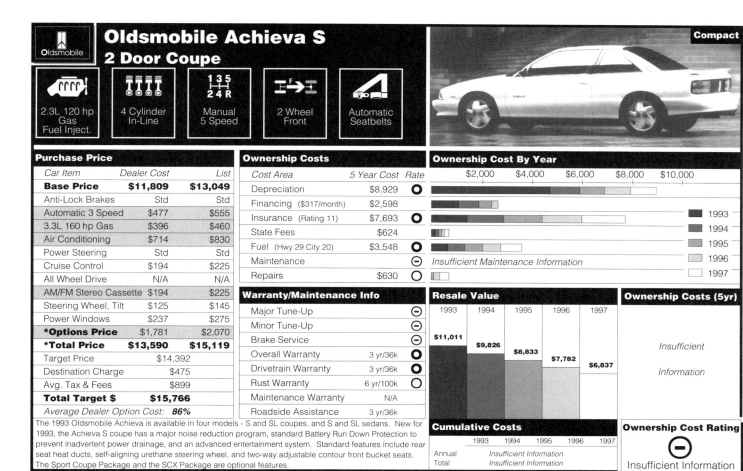

Oldsmobile Achieva SL
2 Door Coupe
Compact

- 2.3L 160 hp Gas Fuel Inject.
- 4 Cylinder In-Line
- Automatic 3 Speed
- 2 Wheel Front
- Automatic Seatbelts

SCX Model Shown

Purchase Price

Car Item	Dealer Cost	List
Base Price	**$13,438**	**$14,849**
Anti-Lock Brakes	Std	Std
Manual Transmission	N/A	N/A
3.3L 160 hp Gas	$43	$50
Air Conditioning	$714	$830
Power Steering	Std	Std
Cruise Control	$194	$225
All Wheel Drive	N/A	N/A
AM/FM Stereo Cassette	Std	Std
Steering Wheel, Tilt	$125	$145
Power Windows	$237	$275
***Options Price**	**$714**	**$830**
***Total Price**	**$14,152**	**$15,679**
Target Price		$15,029
Destination Charge		$475
Avg. Tax & Fees		$937
Total Target $		**$16,441**
Average Dealer Option Cost:		86%

The 1993 Oldsmobile Achieva is available in four models - S and SL coupes, and S and SL sedans. New for 1993, the Achieva SL coupe improves the quietness of the engine with a major noise reduction program. The SL upgrades the S with more horsepower, standard styled wheels and a fold-down rear seat. Standard features include 4-way adjustable driver side front bucket seat, side window defoggers and deluxe bodyside moldings. The Sport Coupe Package and SCX Package are optional.

Ownership Costs

Cost Area	5 Year Cost	Rate
Depreciation	$9,721	●
Financing ($330/month)	$2,709	
Insurance (Rating 11)	$7,693	○
State Fees	$646	
Fuel (Hwy 29 City 22)	$3,390	○
Maintenance		⊖
Repairs	$630	○

Warranty/Maintenance Info

Major Tune-Up		⊖
Minor Tune-Up		⊖
Brake Service		⊖
Overall Warranty	3 yr/36k	○
Drivetrain Warranty	3 yr/36k	○
Rust Warranty	6 yr/100k	○
Maintenance Warranty	N/A	
Roadside Assistance	3 yr/36k	

Resale Value
- 1993: $10,625
- 1994: $9,546
- 1995: $8,598
- 1996: $7,642
- 1997: $6,720

Cumulative Costs
Annual: Insufficient Information
Total: Insufficient Information

Ownership Costs (5yr)
Insufficient Information

Ownership Cost Rating
⊖ Insufficient Information

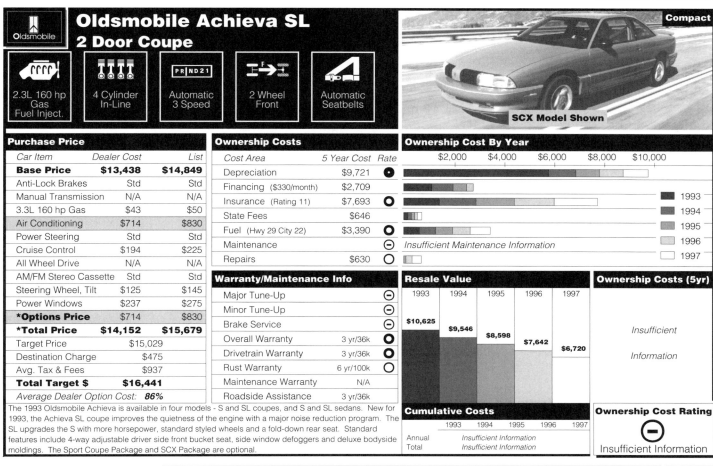

---

* Includes shaded options
** Other purchase requirements apply

● Poor ◐ Worse Than Average ◑ Average ○ Better Than Average ◯ Excellent ⊖ Insufficient Information

©1993 by IntelliChoice, Inc. (408) 554-8711   All Rights Reserved.   Reproduction Prohibited.
Refer to *Section 3: Annotated Vehicle Charts* for an explanation of these charts.

Oldsmobile Achieva S
4 Door Sedan

- 2.3L 120 hp Gas Fuel Inject.
- 4 Cylinder In-Line
- Manual 5 Speed
- 2 Wheel Front
- Automatic Seatbelts

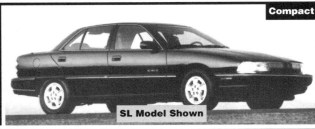
SL Model Shown — Compact

Purchase Price

Car Item	Dealer Cost	List
Base Price	**$11,900**	**$13,149**
Anti-Lock Brakes	Std	Std
Automatic 3 Speed	$477	$555
2.3L 160 hp Gas	$353	$410
Air Conditioning	$714	$830
Power Steering	Std	Std
Cruise Control	$194	$225
All Wheel Drive	N/A	N/A
AM/FM Stereo Cassette	$194	$225
Steering Wheel, Tilt	$125	$145
Power Windows	$292	$340
*Options Price	$1,738	$2,020
*Total Price	$13,638	$15,169
Target Price	$14,445	
Destination Charge	$475	
Avg. Tax & Fees	$902	
Total Target $	**$15,822**	
Average Dealer Option Cost:	**86%**	

Ownership Costs

Cost Area	5 Year Cost	Rate
Depreciation	$8,821	○
Financing ($318/month)	$2,607	
Insurance (Rating 7)	$7,035	○
State Fees	$626	
Fuel (Hwy 29 City 22)	$3,390	○
Maintenance		⊖
Repairs	$630	○

Warranty/Maintenance Info

Major Tune-Up		⊖
Minor Tune-Up		⊖
Brake Service		⊖
Overall Warranty	3 yr/36k	○
Drivetrain Warranty	3 yr/36k	○
Rust Warranty	6 yr/100k	○
Maintenance Warranty	N/A	
Roadside Assistance	3 yr/36k	

Ownership Cost By Year

Resale Value
1993: $11,264 — 1994: $10,076 — 1995: $9,033 — 1996: $8,010 — 1997: $7,001

Ownership Costs (5yr)
Insufficient Information

Cumulative Costs
Annual: Insufficient Information
Total: Insufficient Information

Ownership Cost Rating
⊖ Insufficient Information

The 1993 Oldsmobile Achieva is available in four models - S and SL coupes, and S and SL sedans. New for 1993, the Achieva S sedan has a major noise reduction program, and a standard Battery Run Down Protection to prevent inadvertent power drainage. Standard features include rear child-security door locks, self-aligning sport steering wheel, and deluxe bodyside moldings. An electric rear window defogger, rear split folding seat, and a luggage carrier are all optional features.

Oldsmobile Achieva SL
4 Door Sedan

- 2.3L 160 hp Gas Fuel Inject.
- 4 Cylinder In-Line
- Automatic 3 Speed
- 2 Wheel Front
- Automatic Seatbelts

Compact

Purchase Price

Car Item	Dealer Cost	List
Base Price	**$13,529**	**$14,949**
Anti-Lock Brakes	Std	Std
Manual Transmission	N/A	N/A
3.3L 160 hp Gas	$43	$50
Air Conditioning	$714	$830
Power Steering	Std	Std
Cruise Control	$194	$225
All Wheel Drive	N/A	N/A
AM/FM Stereo Cassette	Std	Std
Steering Wheel, Tilt	$125	$145
Power Windows	$292	$340
*Options Price	$714	$830
*Total Price	$14,243	$15,779
Target Price	$15,128	
Destination Charge	$475	
Avg. Tax & Fees	$943	
Total Target $	**$16,546**	
Average Dealer Option Cost:	**86%**	

Ownership Costs

Cost Area	5 Year Cost	Rate
Depreciation	$9,166	○
Financing ($333/month)	$2,725	
Insurance (Rating 7)	$7,035	○
State Fees	$651	
Fuel (Hwy 29 City 22)	$3,390	○
Maintenance		⊖
Repairs	$630	○

Warranty/Maintenance Info

Major Tune-Up		⊖
Minor Tune-Up		⊖
Brake Service		⊖
Overall Warranty	3 yr/36k	○
Drivetrain Warranty	3 yr/36k	○
Rust Warranty	6 yr/100k	○
Maintenance Warranty	N/A	
Roadside Assistance	3 yr/36k	

Resale Value
1993: $11,194 — 1994: $10,157 — 1995: $9,236 — 1996: $8,324 — 1997: $7,380

Ownership Costs (5yr)
Insufficient Information

Cumulative Costs
Annual: Insufficient Information
Total: Insufficient Information

Ownership Cost Rating
⊖ Insufficient Information

The 1993 Oldsmobile Achieva is available in four models - S and SL coupes, and the S and SL sedans. New for 1993, the Achieva SL sedan improves the quietness of the engine with a major noise reduction program. The SL upgrades the S with standard styled wheels, a fold-down rear seat and more horsepower. Standard features include deluxe bodyside moldings, self-aligning steering wheel and front and rear body color fascias. Electric rear window defogger and a luggage carrier are optional.

---

* Includes shaded options
** Other purchase requirements apply

● Poor   ◐ Worse Than Average   ◑ Average   ◐ Better Than Average   ○ Excellent   ⊖ Insufficient Information

©1993 by IntelliChoice, Inc. (408) 554-8711 All Rights Reserved. Reproduction Prohibited.
Refer to *Section 3: Annotated Vehicle Charts* for an explanation of these charts.

Oldsmobile Cutlass Ciera S
4 Door Sedan

SL Model Shown

Midsize

- 2.2L 110 hp Gas Fuel Inject.
- 4 Cylinder In-Line
- Automatic 3 Speed
- 2 Wheel Front
- Belts Std, Driv Air Opt

Purchase Price

Car Item	Dealer Cost	List
Base Price	**$12,708**	**$14,199**
Anti-Lock Brakes	N/A	N/A
4 Spd Auto	$172	$200
3.3L 160 hp Gas	$568	$660
Air Conditioning	Std	Std
Power Steering	Std	Std
Cruise Control	$194	$225
All Wheel Drive	N/A	N/A
AM/FM Stereo Cassette	$142	$165
Steering Wheel, Tilt	$125	** $145
Power Windows	$292	$340
*Options Price	$1,196	$1,390
*Total Price	$13,904	$15,589
Target Price	$14,794	
Destination Charge	$500	
Avg. Tax & Fees	$926	
Total Target $	**$16,220**	
Average Dealer Option Cost:	86%	

The 1993 Ciera is available in four models- S and SL sedans and Cruiser wagons. New for 1993, the Ciera S sedan has a 2.2-liter multiport-injected engine and air conditioning as standard equipment. It features automatic door locks with an available remote package, air conditioning and new exterior body-color body side moldings. The Ciera S model now offers the option of a driver's-side air bag. Other optional features include leather seating surfaces and engine block heater.

Ownership Costs

Cost Area	5 Year Cost	Rate
Depreciation	$8,826	●
Financing ($326/month)	$2,672	
Insurance (Rating 4)	$6,658	◐
State Fees	$645	
Fuel (Hwy 26 City 19)	$3,847	◉
Maintenance	$4,170	◐
Repairs	$709	◐

Warranty/Maintenance Info

Major Tune-Up	$180	◐
Minor Tune-Up	$127	◐
Brake Service	$207	◐
Overall Warranty	3 yr/36k	◐
Drivetrain Warranty	3 yr/36k	◐
Rust Warranty	6 yr/100k	◐
Maintenance Warranty	N/A	
Roadside Assistance	3 yr/36k	

Ownership Cost By Year
1993 / 1994 / 1995 / 1996 / 1997

Resale Value
- 1993: $12,206
- 1994: $10,828
- 1995: $9,669
- 1996: $8,519
- 1997: $7,394

Cumulative Costs

	1993	1994	1995	1996	1997
Annual	$7,277	$4,647	$5,125	$4,646	$5,832
Total	$7,277	$11,924	$17,049	$21,695	$27,527

Ownership Costs (5yr)
- Average: $27,208
- This Car: $27,527
- Cost/Mile: 39¢ / 39¢

Ownership Cost Rating
Average

Oldsmobile Cutlass Ciera SL
4 Door Sedan

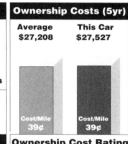

Midsize

- 3.3L 160 hp Gas Fuel Inject.
- 6 Cylinder "V"
- Automatic 3 Speed
- 2 Wheel Front
- Driver Airbag Psngr Belts

Purchase Price

Car Item	Dealer Cost	List
Base Price	**$15,662**	**$17,899**
Anti-Lock Brakes	N/A	N/A
4 Spd Auto	$172	$200
Optional Engine	N/A	N/A
Air Conditioning	Std	Std
Power Steering	Std	Std
Cruise Control	$194	$225
All Wheel Drive	N/A	N/A
AM/FM Stereo Cassette	Std	Std
Steering Wheel, Tilt	Std	Std
Power Windows	$292	$340
*Options Price	$486	$565
*Total Price	$16,148	$18,464
Target Price	$17,273	
Destination Charge	$500	
Avg. Tax & Fees	$1,079	
Total Target $	**$18,852**	
Average Dealer Option Cost:	86%	

The 1993 Ciera is available in four models- S and SL sedans and Cruiser Wagons. New for 1993, the Ciera SL sedan offers as a standard feature a driver's-side air bag. It also has a new 2.2-liter multiport-injected engine (instead of throttle-body injection) that produces 110 horsepower at 5200 rpm. The SL also features an added convenience group, split-bench front seat with power reclining feature and a power trunk-lid release. Added options include leather seats and an upgraded sound system.

Ownership Costs

Cost Area	5 Year Cost	Rate
Depreciation	$11,119	◉
Financing ($379/month)	$3,107	
Insurance (Rating 6)	$6,919	○
State Fees	$760	
Fuel (Hwy 29 City 19)	$3,640	◐
Maintenance	$4,187	◐
Repairs	$709	◐

Warranty/Maintenance Info

Major Tune-Up	$180	◐
Minor Tune-Up	$127	◐
Brake Service	$207	◐
Overall Warranty	3 yr/36k	◐
Drivetrain Warranty	3 yr/36k	◐
Rust Warranty	6 yr/100k	◐
Maintenance Warranty	N/A	
Roadside Assistance	3 yr/36k	

Resale Value
- 1993: $13,472
- 1994: $11,782
- 1995: $10,375
- 1996: $9,024
- 1997: $7,733

Cumulative Costs

	1993	1994	1995	1996	1997
Annual	$8,863	$5,132	$5,501	$4,910	$6,035
Total	$8,863	$13,995	$19,496	$24,406	$30,441

Ownership Costs (5yr)
- Average: $30,148
- This Car: $30,441
- Cost/Mile: 43¢ / 43¢

Ownership Cost Rating
Average

---

* Includes shaded options
** Other purchase requirements apply

Legend: ● Poor | ◐ Worse Than Average | ◑ Average | ○ Better Than Average | ○ Excellent | ⊖ Insufficient Information

©1993 by IntelliChoice, Inc. (408) 554-8711 All Rights Reserved. Reproduction Prohibited.
Refer to *Section 3: Annotated Vehicle Charts* for an explanation of these charts.

Oldsmobile Cutlass Ciera Cruiser S
4 Door Wagon

Midsize Wagon

 2.2L 110 hp Gas Fuel Inject. | 4 Cylinder In-Line | Automatic 3 Speed | 2 Wheel Front | Belts Std, Driv Air Opt

Purchase Price

Car Item	Dealer Cost	List
Base Price	**$13,335**	**$14,899**
Anti-Lock Brakes	N/A	N/A
4 Spd Auto	$172	$200
3.3L 160 hp Gas	$568	$660
Air Conditioning	Std	Std
Power Steering	Std	Std
Cruise Control	$194	$225
All Wheel Drive	N/A	N/A
AM/FM Stereo Cassette	$142	$165
Steering Wheel, Tilt	$125	** $145
Power Windows	$292	$340
***Options Price**	**$1,196**	**$1,390**
***Total Price**	**$14,531**	**$16,289**
Target Price		$15,478
Destination Charge		$500
Avg. Tax & Fees		$967
Total Target $		**$16,945**
Average Dealer Option Cost:		**86%**

The 1993 Ciera is available in four models- S and SL sedans and Cruiser Wagons. New for 1993, the S wagon now offers the option of a driver's-side air bag. It also features a 55/45 split-bench seat with power recliners and air conditioning as standard equipment. Other features include dual mirrors and bodyside molding. Options include aluminum wheels, leather seats, and woodgrain paneling. The S wagon also offers option packages which include power accessories and extra equipment.

Ownership Costs

Cost Area	5 Year Cost	Rate
Depreciation	$9,639	●
Financing ($341/month)	$2,791	
Insurance (Rating 5)	$6,786	○
State Fees	$672	
Fuel (Hwy 29 City 19)	$3,640	O
Maintenance	$4,171	O
Repairs	$709	O

Warranty/Maintenance Info

Major Tune-Up	$181	O
Minor Tune-Up	$127	◉
Brake Service	$207	O
Overall Warranty	3 yr/36k	O
Drivetrain Warranty	3 yr/36k	O
Rust Warranty	6 yr/100k	O
Maintenance Warranty	N/A	
Roadside Assistance	3 yr/36k	

Ownership Cost By Year
1993 / 1994 / 1995 / 1996 / 1997

Resale Value
1993	1994	1995	1996	1997
$11,535	$10,221	$9,296	$8,321	$7,306

Cumulative Costs
	1993	1994	1995	1996	1997
Annual	$8,716	$4,611	$4,903	$4,468	$5,710
Total	$8,716	$13,327	$18,230	$22,698	$28,408

Ownership Costs (5yr)
Average $27,924 | This Car $28,408
Cost/Mile 40¢ | Cost/Mile 41¢

Ownership Cost Rating
Average

Oldsmobile Cutlass Ciera Cruiser SL
4 Door Wagon

Midsize Wagon

 3.3L 160 hp Gas Fuel Inject. | 6 Cylinder "V" | Automatic 4 Speed | 2 Wheel Front | Driver Airbag Psngr Belts

Purchase Price

Car Item	Dealer Cost	List
Base Price	**$16,099**	**$18,399**
Anti-Lock Brakes	N/A	N/A
Manual Transmission	N/A	N/A
Optional Engine	N/A	N/A
Air Conditioning	Std	Std
Power Steering	Std	Std
Cruise Control	$194	$225
All Wheel Drive	N/A	N/A
AM/FM Stereo Cassette	Std	Std
Steering Wheel, Tilt	Std	Std
Power Windows	$292	$340
***Options Price**	**$486**	**$565**
***Total Price**	**$16,585**	**$18,964**
Target Price		$17,754
Destination Charge		$500
Avg. Tax & Fees		$1,108
Total Target $		**$19,362**
Average Dealer Option Cost:		**86%**

The 1993 Ciera is available in four models- S and SL sedans and Cruiser Wagons. New for 1993, the SL wagon now offers as a standard feature a driver's-side air bag. It also has a new multiport-injected engine (instead of throttle-body injection) that produces 110 horsepower at 5200 rpm. In addition to the features of the S model, the SL offers a convenience group and split-bench front seat with power recline feature. Added options include leather seats and an upgraded sound system.

Ownership Costs

Cost Area	5 Year Cost	Rate
Depreciation	$10,991	●
Financing ($389/month)	$3,190	
Insurance (Rating 7)	$7,035	○
State Fees	$779	
Fuel (Hwy 27 City 19)	$3,775	O
Maintenance	$4,187	O
Repairs	$709	O

Warranty/Maintenance Info

Major Tune-Up	$180	O
Minor Tune-Up	$127	◉
Brake Service	$207	O
Overall Warranty	3 yr/36k	O
Drivetrain Warranty	3 yr/36k	O
Rust Warranty	6 yr/100k	O
Maintenance Warranty	N/A	
Roadside Assistance	3 yr/36k	

Resale Value
1993	1994	1995	1996	1997
$13,160	$11,681	$10,587	$9,497	$8,371

Cumulative Costs
	1993	1994	1995	1996	1997
Annual	$9,772	$4,999	$5,258	$4,710	$5,927
Total	$9,772	$14,771	$20,029	$24,739	$30,666

Ownership Costs (5yr)
Average $30,659 | This Car $30,666
Cost/Mile 44¢ | Cost/Mile 44¢

Ownership Cost Rating
Average

---

* Includes shaded options
** Other purchase requirements apply

● Poor | ◉ Worse Than Average | O Average | ○ Better Than Average | ○ Excellent | ⊖ Insufficient Information

©1993 by IntelliChoice, Inc. (408) 554-8711   All Rights Reserved. Reproduction Prohibited.
Refer to *Section 3: Annotated Vehicle Charts* for an explanation of these charts.

Oldsmobile Cutlass Supreme S
2 Door Coupe

Midsize

 3.1L 140 hp Gas Fuel Inject.   6 Cylinder "V"   Automatic 3 Speed 2 Wheel Front Automatic Seatbelts

Purchase Price

Car Item	Dealer Cost	List
Base Price	**$13,733**	**$15,695**
Anti-Lock Brakes	$387	$450
4 Spd Auto $1	$151	$175
3.4L 200 hp Gas	$1,170	** $1,360
Air Conditioning	Std	Std
Power Steering	Std	Std
Cruise Control	$194	** $225
All Wheel Drive	N/A	N/A
AM/FM Stereo Cassette	$142	$165
Steering Wheel, Tilt	$125	** $145
Power Windows	$237	$275
*Options Price	$379	$440
*Total Price	$14,112	$16,135
Target Price	$15,043	
Destination Charge	$505	
Avg. Tax & Fees	$943	
Total Target $	**$16,491**	
Average Dealer Option Cost:	86%	

Ownership Costs

Cost Area	5 Year Cost	Rate
Depreciation	$9,401	●
Financing ($331/month)	$2,717	
Insurance (Rating 7)	$7,035	○
State Fees	$665	
Fuel (Hwy 27 City 19)	$3,775	○
Maintenance	$5,061	○
Repairs	$659	○

Warranty/Maintenance Info

Major Tune-Up	$179	○
Minor Tune-Up	$131	○
Brake Service	$287	●
Overall Warranty	3 yr/36k	○
Drivetrain Warranty	3 yr/36k	○
Rust Warranty	6 yr/100k	○
Maintenance Warranty	N/A	
Roadside Assistance	3 yr/36k	

Ownership Cost By Year

1993, 1994, 1995, 1996, 1997

Resale Value

1993	1994	1995	1996	1997
$12,183	$10,647	$9,448	$8,213	$7,090

Cumulative Costs

	1993	1994	1995	1996	1997
Annual	$7,651	$4,880	$5,363	$4,371	$7,048
Total	$7,651	$12,531	$17,894	$22,265	$29,313

Ownership Costs (5yr)

Average: $27,767   This Car: $29,313
Cost/Mile 40¢   Cost/Mile 42¢

Ownership Cost Rating

● Worse Than Average

The 1993 Oldsmobile Cutlass Supreme is available in seven models - S, SL, and International coupes and sedans, and the convertible. New for 1993, the Cutlass Supreme S coupe adds graphics to the climate-control system and improvements to the touring suspension system. Standard features include an instrument panel with electronic speedometer, 4-way manual adjustable front bucket seats, and dual engine cooling fans. Optional items include a Sport Luxury Package, bodyside moldings and leather trim.

Oldsmobile Cutlass Supreme International
2 Door Coupe

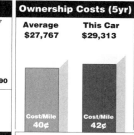

Midsize

 3.4L 200 hp Gas Fuel Inject.   6 Cylinder "V"   Automatic 4 Speed   2 Wheel Front   Automatic Seatbelts

Purchase Price

Car Item	Dealer Cost	List
Base Price	**$19,949**	**$22,799**
Anti-Lock Brakes	Std	Std
Manual Transmission	N/A	N/A
3.1L 140 hp Gas	($856)	($995)
Auto Climate Control	Std	Std
Power Steering	Std	Std
Cruise Control	Std	Std
All Wheel Drive	N/A	N/A
AM/FM Stereo Cassette	Std	Std
Steering Wheel, Tilt	Std	Std
Power Windows	$237	$275
*Options Price	$237	$275
*Total Price	$20,186	$23,074
Target Price	$21,758	
Destination Charge	$505	
Avg. Tax & Fees	$1,349	
Total Target $	**$23,612**	
Average Dealer Option Cost:	86%	

Ownership Costs

Cost Area	5 Year Cost	Rate
Depreciation	$15,096	●
Financing ($475/month)	$3,890	
Insurance (Rating 12)	$7,898	○
State Fees	$943	
Fuel (Hwy 26 City 17)	$4,064	●
Maintenance	$5,132	○
Repairs	$659	○

Warranty/Maintenance Info

Major Tune-Up	$161	○
Minor Tune-Up	$111	○
Brake Service	$287	●
Overall Warranty	3 yr/36k	○
Drivetrain Warranty	3 yr/36k	○
Rust Warranty	6 yr/100k	○
Maintenance Warranty	N/A	
Roadside Assistance	3 yr/36k	

Ownership Cost By Year

$5,000, $10,000, $15,000, $20,000

Resale Value

1993	1994	1995	1996	1997
$14,878	$12,979	$11,438	$9,886	$8,516

Cumulative Costs

	1993	1994	1995	1996	1997
Annual	$12,847	$5,872	$6,278	$5,040	$7,645
Total	$12,847	$18,719	$24,997	$30,037	$37,682

Ownership Costs (5yr)

Average: $34,862   This Car: $37,682
Cost/Mile 50¢   Cost/Mile 54¢

Ownership Cost Rating

● Poor

The 1993 Oldsmobile Cutlass Supreme is available in seven models - S, SL, and International coupes and sedans, and the convertible. New for 1993, the Cutlass Supreme International coupe adds graphics to the climate-control system and improvements to the touring suspension system. The International coupe upgrades the SL with power adjustable bucket seats, a twin dual cam engine, touring car suspension and new aluminum wheels. Optional items include bodyside moldings, astroroof and a luggage carrier.

---

* Includes shaded options
** Other purchase requirements apply

● Poor   ◐ Worse Than Average   ○ Average   ◯ Better Than Average   ○ Excellent   ⊖ Insufficient Information

©1993 by IntelliChoice, Inc. (408) 554-8711   All Rights Reserved. Reproduction Prohibited.
Refer to *Section 3: Annotated Vehicle Charts* for an explanation of these charts.

page 177

Oldsmobile Cutlass Supreme S
4 Door Sedan

Midsize

3.1L 140 hp Gas Fuel Inject. | 6 Cylinder "V" | Automatic 3 Speed | 2 Wheel Front | Automatic Seatbelts

2 Door Coupe Shown

Purchase Price

Car Item	Dealer Cost	List
Base Price	**$13,821**	**$15,795**
Anti-Lock Brakes	$387	$450
4 Spd Auto $1	$151	$175
3.4L 200 hp Gas	$1,415	** $1,645
Air Conditioning	Std	Std
Power Steering	Std	Std
Cruise Control	$194	** $225
All Wheel Drive	N/A	N/A
AM/FM Stereo Cassette	$142	$165
Steering Wheel, Tilt	$125	** $145
Power Windows	$292	$340
*Options Price	$434	$505
*Total Price	$14,255	$16,300
Target Price		$15,197
Destination Charge		$505
Avg. Tax & Fees		$953
Total Target $		**$16,655**
Average Dealer Option Cost:		*86%*

Ownership Costs

Cost Area	5 Year Cost	Rate
Depreciation	$8,961	O
Financing ($335/month)	$2,744	
Insurance (Rating 5)	$6,786	◯
State Fees	$672	
Fuel (Hwy 27 City 19)	$3,775	O
Maintenance	$5,061	O
Repairs	$659	O

Warranty/Maintenance Info

Major Tune-Up	$179	O
Minor Tune-Up	$131	O
Brake Service	$287	●
Overall Warranty	3 yr/36k	O
Drivetrain Warranty	3 yr/36k	O
Rust Warranty	6 yr/100k	O
Maintenance Warranty	N/A	
Roadside Assistance	3 yr/36k	

Ownership Cost By Year

1993, 1994, 1995, 1996, 1997

Resale Value

1993	1994	1995	1996	1997
$12,667	$11,176	$10,008	$8,815	$7,694

Cumulative Costs

	1993	1994	1995	1996	1997
Annual	$7,298	$4,797	$5,288	$4,281	$6,994
Total	$7,298	$12,095	$17,383	$21,664	$28,658

Ownership Costs (5yr)

Average: $27,935 — This Car: $28,658
Cost/Mile 40¢ — Cost/Mile 41¢

Ownership Cost Rating

O — Average

The 1993 Cutlass Supreme is available in five models- S and International coupes and sedans and (Base) convertible. New for 1993, the Supreme S sedan comes equipped with a 140-horsepower, 3.1-liter V-6 engine, four-wheel disc brakes, power rack-and-pinion steering and 15-inch wheels and tires. In addition, the options list offers ABS brakes, leather upholstery and rear fold-down seat back. The S sedan also offers the electronically-controlled four-speed automatic transmission (optional).

Oldsmobile Cutlass Supreme International
4 Door Sedan

Midsize

3.4L 200 hp Gas Fuel Inject. | 6 Cylinder "V" | Automatic 4 Speed | 2 Wheel Front | Automatic Seatbelts

Purchase Price

Car Item	Dealer Cost	List
Base Price	**$20,037**	**$22,899**
Anti-Lock Brakes	Std	Std
Manual Transmission	N/A	N/A
3.1L 140 hp Gas	($856)	($995)
Auto Climate Control	Std	Std
Power Steering	Std	Std
Cruise Control	Std	Std
All Wheel Drive	N/A	N/A
AM/FM Stereo Cassette	Std	Std
Steering Wheel, Tilt	Std	Std
Power Windows	$292	$340
*Options Price	$292	$340
*Total Price	$20,329	$23,239
Target Price		$21,915
Destination Charge		$505
Avg. Tax & Fees		$1,358
Total Target $		**$23,778**
Average Dealer Option Cost:		*86%*

Ownership Costs

Cost Area	5 Year Cost	Rate
Depreciation	$14,940	●
Financing ($478/month)	$3,917	
Insurance (Rating 10)	$7,479	◯
State Fees	$949	
Fuel (Hwy 26 City 19)	$3,847	◉
Maintenance	$5,132	O
Repairs	$659	O

Warranty/Maintenance Info

Major Tune-Up	$161	O
Minor Tune-Up	$111	O
Brake Service	$287	●
Overall Warranty	3 yr/36k	O
Drivetrain Warranty	3 yr/36k	O
Rust Warranty	6 yr/100k	O
Maintenance Warranty	N/A	
Roadside Assistance	3 yr/36k	

Resale Value

1993	1994	1995	1996	1997
$15,242	$13,330	$11,778	$10,217	$8,838

Cumulative Costs

	1993	1994	1995	1996	1997
Annual	$12,544	$5,773	$6,169	$4,920	$7,517
Total	$12,544	$18,317	$24,486	$29,406	$36,923

Ownership Costs (5yr)

Average: $35,030 — This Car: $36,923
Cost/Mile 50¢ — Cost/Mile 53¢

Ownership Cost Rating

● — Worse Than Average

The 1993 Cutlass Supreme is available in five models- S and International coupes and sedans, and (Base) convertible. New for 1993, the Supreme International sedan comes equipped with the Twin Dual Cam engine, a four-speed automatic transmission, and aluminum wheels. Convenience features include a tilt steering wheel with controls for AC and radio functions, and power-adjustable bucket seats. The option list presents leather trim, a power sun roof, and two equipment packages.

---

page **178**

* Includes shaded options
** Other purchase requirements apply

● Poor   ◉ Worse Than Average   ◯ Average   ◯ Better Than Average   ◯ Excellent   ⊖ Insufficient Information

©1993 by IntelliChoice, Inc. (408) 554-8711 All Rights Reserved. Reproduction Prohibited.
Refer to *Section 3: Annotated Vehicle Charts* for an explanation of these charts.

Oldsmobile Cutlass Supreme Convertible
2 Door Convertible — Midsize

 3.1L 140 hp Gas Fuel Inject.
 6 Cylinder "V"
 Automatic 4 Speed
 2 Wheel Front
 Automatic Seatbelts

Purchase Price

Car Item	Dealer Cost	List
Base Price	**$19,862**	**$22,699**
Anti-Lock Brakes	$387	$450
Manual Transmission	N/A	N/A
3.4L 200 hp Gas	$856	** $995
Air Conditioning	Std	Std
Power Steering	Std	Std
Cruise Control	$194	** $225
All Wheel Drive	N/A	N/A
AM/FM Stereo Cassette	Std	Std
Steering Wheel, Tilt	$125	** $145
Power Windows	Std	Std
***Options Price**	**$0**	**$0**
***Total Price**	**$19,862**	**$22,699**
Target Price		$21,405
Destination Charge		$505
Avg. Tax & Fees		$1,328
Total Target $		**$23,238**
Average Dealer Option Cost:		*86%*

The 1993 Oldsmobile Cutlass Supreme is available in seven models - S, SL, and International coupes and sedans, and the convertible. New for 1993, the Cutlass Supreme Convertible has an improved touring suspension and adds graphics to the climate-control system. Standard features include body-color front and rear bumper fascias, Convenience Group, front-end panel rocket emblem and 4-way manual adjustable front bucket seats. Optional items include leather trim and bodyside moldings.

Ownership Costs

Cost Area	5 Year Cost	Rate
Depreciation	$12,634	○
Financing ($467/month)	$3,827	
Insurance (Rating 12)	$7,898	○
State Fees	$928	
Fuel (Hwy 27 City 19)	$3,775	○
Maintenance	$5,192	○
Repairs	$659	○

Warranty/Maintenance Info

Major Tune-Up	$179	○
Minor Tune-Up	$131	○
Brake Service	$277	●
Overall Warranty	3 yr/36k	○
Drivetrain Warranty	3 yr/36k	○
Rust Warranty	6 yr/100k	○
Maintenance Warranty	N/A	
Roadside Assistance	3 yr/36k	

Ownership Cost By Year

Years: 1993, 1994, 1995, 1996, 1997

Resale Value

1993	1994	1995	1996	1997
$18,172	$15,883	$14,018	$12,235	$10,604

Cumulative Costs

	1993	1994	1995	1996	1997
Annual	$9,096	$6,205	$6,537	$5,225	$7,850
Total	$9,096	$15,301	$21,838	$27,063	$34,913

Ownership Costs (5yr)

Average	This Car
$34,478	$34,913
Cost/Mile 49¢	Cost/Mile 50¢

Ownership Cost Rating
○ Average

Oldsmobile Eighty-Eight Royale
4 Door Sedan — Large

 3.8L 170 hp Gas Fuel Inject.
 6 Cylinder "V"
 Automatic 4 Speed
 2 Wheel Front
 Driver Airbag Psngr Belts

Purchase Price

Car Item	Dealer Cost	List
Base Price	**$17,105**	**$19,549**
Anti-Lock Brakes	Std	Std
Manual Transmission	N/A	N/A
Optional Engine	N/A	N/A
Air Conditioning	Std	Std
Power Steering	Std	Std
Cruise Control	Pkg	Pkg
All Wheel Drive	N/A	N/A
AM/FM Stereo Cassette	$206	$240
Steering Wheel, Tilt	Std	Std
Power Windows	Std	Std
***Options Price**	**$206**	**$240**
***Total Price**	**$17,311**	**$19,789**
Target Price		$18,564
Destination Charge		$555
Avg. Tax & Fees		$1,159
Total Target $		**$20,278**
Average Dealer Option Cost:		*86%*

The 1993 Eighty-Eight is available in two models - Royale and Royale LS. New for 1993, the Royale is powered by a 3800 V-6 engine. It is fully equipped with a driver's-side air bag, ABS brakes, and air conditioning. Other features include tilt steering wheel, a 55/45 split-bench front seat, Pass Key anti-theft system, and all-season radial tires. In addition, various options are available such as a touring suspension system, aluminum wheels, and a 3000-pound trailer-towing package.

Ownership Costs

Cost Area	5 Year Cost	Rate
Depreciation	$10,745	○
Financing ($408/month)	$3,340	
Insurance (Rating 8)	$7,223	○
State Fees	$813	
Fuel (Hwy 28 City 19)	$3,704	○
Maintenance	$5,003	○
Repairs	$709	○

Warranty/Maintenance Info

Major Tune-Up	$179	○
Minor Tune-Up	$127	○
Brake Service	$243	○
Overall Warranty	3 yr/36k	○
Drivetrain Warranty	3 yr/36k	○
Rust Warranty	6 yr/100k	○
Maintenance Warranty	N/A	
Roadside Assistance	3 yr/36k	

Resale Value

1993	1994	1995	1996	1997
$14,861	$13,332	$12,108	$10,883	$9,533

Cumulative Costs

	1993	1994	1995	1996	1997
Annual	$9,078	$5,120	$5,545	$4,457	$7,337
Total	$9,078	$14,198	$19,743	$24,200	$31,537

Ownership Costs (5yr)

Average	This Car
$32,095	$31,537
Cost/Mile 46¢	Cost/Mile 45¢

Ownership Cost Rating
○ Average

---

* Includes shaded options
** Other purchase requirements apply

● Poor | ◐ Worse Than Average | ○ Average | ◯ Better Than Average | ○ Excellent | ⊖ Insufficient Information

©1993 by IntelliChoice, Inc. (408) 554-8711   All Rights Reserved. Reproduction Prohibited.
Refer to *Section 3: Annotated Vehicle Charts* for an explanation of these charts.

Oldsmobile Eighty-Eight Royale LS
4 Door Sedan

Large

 3.8L 170 hp Gas Fuel Inject. 6 Cylinder "V" Automatic 4 Speed 2 Wheel Front Driver Airbag Psngr Belts

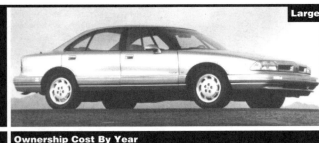

Purchase Price

Car Item	Dealer Cost	List
Base Price	**$19,205**	**$21,949**
Anti-Lock Brakes	Std	Std
Manual Transmission	N/A	N/A
Optional Engine	N/A	N/A
Air Conditioning	Std	Std
Power Steering	Std	Std
Cruise Control	Std	Std
All Wheel Drive	N/A	N/A
AM/FM Stereo Cassette	Std	Std
Steering Wheel, Tilt	Std	Std
Power Windows	Std	Std
*Options Price	$0	$0
*Total Price	$19,205	$21,949
Target Price	$20,673	
Destination Charge	$555	
Avg. Tax & Fees	$1,286	
Total Target $	**$22,514**	
Average Dealer Option Cost:	86%	

The 1993 Eighty-Eight is available in two models- Royale and Royale LS. New for 1993, the Royale LS is powered by a 3800 V-6 engine. It features air conditioning, a 55/45 split-bench front seat, Pass Key anti-theft system, and all-season radial tires. Also standard is a driver's-side air bag. The option list presents leather upholstery, a CD player, 16-inch aluminum wheels touring suspension with variable assist steering, and electronic instrumentation with a driver information center.

Ownership Costs

Cost Area	5 Year Cost	Rate
Depreciation	$12,813	●
Financing ($452/month)	$3,708	
Insurance (Rating 8)	$7,223	○
State Fees	$901	
Fuel (Hwy 28 City 19)	$3,704	○
Maintenance	$5,076	●
Repairs	$709	○

Warranty/Maintenance Info

Major Tune-Up	$179	●
Minor Tune-Up	$127	●
Brake Service	$243	●
Overall Warranty	3 yr/36k	●
Drivetrain Warranty	3 yr/36k	●
Rust Warranty	6 yr/100k	○
Maintenance Warranty	N/A	
Roadside Assistance	3 yr/36k	

Ownership Cost By Year

1993, 1994, 1995, 1996, 1997

Resale Value

1993	1994	1995	1996	1997
$15,373	$13,723	$12,417	$11,073	$9,701

Cumulative Costs

	1993	1994	1995	1996	1997
Annual	$10,978	$5,376	$5,752	$4,618	$7,410
Total	$10,978	$16,354	$22,106	$26,724	$34,134

Ownership Costs (5yr)

Average	This Car
$34,238	$34,134
Cost/Mile 49¢	Cost/Mile 49¢

Ownership Cost Rating
○ Average

Oldsmobile Ninety-Eight Regency
4 Door Sedan

Luxury

 3.8L 170 hp Gas Fuel Inject. 6 Cylinder "V" Automatic 4 Speed 2 Wheel Front Driver Airbag Psngr Belts

Elite Model Shown

Purchase Price

Car Item	Dealer Cost	List
Base Price	**$21,594**	**$24,595**
Anti-Lock Brakes	Std	Std
Manual Transmission	N/A	N/A
Optional Engine	N/A	N/A
Auto Climate Control	Std	Std
Power Steering	Std	Std
Cruise Control	Std	Std
All Wheel Drive	N/A	N/A
AM/FM Stereo Cassette	$206	$240
Steering Wheel, Tilt	Std	Std
Power Windows	Std	Std
*Options Price	$206	$240
*Total Price	$21,800	$24,835
Target Price	$23,659	
Destination Charge	$600	
Avg. Tax & Fees	$1,467	
Total Target $	**$25,726**	
Average Dealer Option Cost:	86%	

The 1993 Ninety-Eight is available in three models- Regency, Regency Elite, and Touring sedans. New for 1993, the Regency is powered by a V-6 engine and is equipped with a driver's-side air bag. Standard features include front-wheel drive, four-speed electronically controlled automatic transmission, and ABS brakes. Other features include automatic air conditioning, and cruise control. The option list presents leather upholstery, engine block, and a Towing Package that tows up to 3,000 lbs.

Ownership Costs

Cost Area	5 Year Cost	Rate
Depreciation	$14,655	●
Financing ($517/month)	$4,238	
Insurance (Rating 7)	$7,035	○
State Fees	$1,017	
Fuel (Hwy 27 City 19)	$3,775	○
Maintenance	$5,109	●
Repairs	$730	○

Warranty/Maintenance Info

Major Tune-Up	$177	○
Minor Tune-Up	$125	○
Brake Service	$243	●
Overall Warranty	3 yr/36k	●
Drivetrain Warranty	3 yr/36k	●
Rust Warranty	6 yr/100k	○
Maintenance Warranty	N/A	
Roadside Assistance	3 yr/36k	

Ownership Cost By Year

Resale Value

1993	1994	1995	1996	1997
$17,782	$15,804	$14,255	$12,729	$11,071

Cumulative Costs

	1993	1994	1995	1996	1997
Annual	$12,008	$5,877	$6,131	$4,840	$7,703
Total	$12,008	$17,885	$24,016	$28,856	$36,559

Ownership Costs (5yr)

Average	This Car
$38,018	$36,559
Cost/Mile 54¢	Cost/Mile 52¢

Ownership Cost Rating
○ Better Than Average

---

* Includes shaded options
** Other purchase requirements apply

● Poor ◐ Worse Than Average ○ Average ○ Better Than Average ○ Excellent ⊖ Insufficient Information

©1993 by IntelliChoice, Inc. (408) 554-8711 All Rights Reserved. Reproduction Prohibited.
Refer to *Section 3: Annotated Vehicle Charts* for an explanation of these charts.

Oldsmobile Ninety-Eight Regency Elite
4 Door Sedan

Luxury

Purchase Price

Car Item	Dealer Cost	List
Base Price	**$23,624**	**$26,999**
Anti-Lock Brakes	Std	Std
Manual Transmission	N/A	N/A
Optional Engine	N/A	N/A
Auto Climate Control	Std	Std
Power Steering	Std	Std
Cruise Control	Std	Std
All Wheel Drive	N/A	N/A
AM/FM Stereo Cassette	Std	Std
Steering Wheel, Tilt	Std	Std
Power Windows	Std	Std
*Options Price	$0	$0
***Total Price**	**$23,624**	**$26,999**
Target Price		$25,735
Destination Charge		$600
Avg. Tax & Fees		$1,593
Total Target $		**$27,928**
Average Dealer Option Cost:		86%

The 1993 Ninety-Eight is available in three models- Regency, Regency Elite, and Touring sedans. New for 1993, the Regency Elite is powered by a V-6 engine, it is also equipped with a driver's-side air bag and an overhead storage console. Standard features include ABS brakes, automatic air conditioning, and cruise control. Other features include aluminum wheels, a six-way power adjuster for the front passenger's seat, and steering wheel touch controls for HVAC and entertainment systems.

Ownership Costs

Cost Area	5 Year Cost	Rate
Depreciation	$15,835	○
Financing ($561/month)	$4,600	
Insurance (Rating 8)	$7,223	○
State Fees	$1,104	
Fuel (Hwy 27 City 19)	$3,775	○
Maintenance	$5,109	○
Repairs	$730	○

Warranty/Maintenance Info

Major Tune-Up	$177	○
Minor Tune-Up	$125	○
Brake Service	$243	○
Overall Warranty	3 yr/36k	○
Drivetrain Warranty	3 yr/36k	○
Rust Warranty	6 yr/100k	○
Maintenance Warranty	N/A	
Roadside Assistance	3 yr/36k	

Ownership Cost By Year
1993, 1994, 1995, 1996, 1997

Resale Value
1993	1994	1995	1996	1997
$19,292	$17,191	$15,511	$13,866	$12,093

Cumulative Costs
	1993	1994	1995	1996	1997
Annual	$12,908	$6,168	$6,389	$5,039	$7,872
Total	$12,908	$19,076	$25,465	$30,504	$38,376

Ownership Costs (5yr)
Average $39,638 / This Car $38,376
Cost/Mile 57¢ / Cost/Mile 55¢

Ownership Cost Rating
Better Than Average

Oldsmobile Ninety-Eight Touring Sedan
4 Door Sedan

Luxury

Purchase Price

Car Item	Dealer Cost	List
Base Price	**$25,987**	**$29,699**
Anti-Lock Brakes	Std	Std
Manual Transmission	N/A	N/A
Suprchrg 3.8L 205 hp	$879	$1,022
Auto Climate Control	Std	Std
Power Steering	Std	Std
Cruise Control	Std	Std
All Wheel Drive	N/A	N/A
AM/FM Stereo Cassette	Std	Std
Steering Wheel, Tilt	Std	Std
Power Windows	Std	Std
*Options Price	$0	$0
***Total Price**	**$25,987**	**$29,699**
Target Price		$28,433
Destination Charge		$600
Avg. Tax & Fees		$1,755
Total Target $		**$30,788**
Average Dealer Option Cost:		86%

The 1993 Ninety-Eight is available in three models- Regency, Regency Elite, and Touring sedans. New for 1993, the Touring Sedan is powered by a 3800 V-6 engine or the option of a 205 horsepower supercharged engine. Standard features includes adjustable bucket seats, leather upholstery, and 7.0x16 cast aluminum wheels. Other features include ABS brakes, FE3 touring suspension and full analog instrumentation. The option list presents heated bucket seats and a Towing Package.

Ownership Costs

Cost Area	5 Year Cost	Rate
Depreciation	$17,452	○
Financing ($619/month)	$5,071	
Insurance (Rating 10)	$7,479	○
State Fees	$1,212	
Fuel (Hwy 27 City 19)	$3,775	○
Maintenance	$5,263	○
Repairs	$730	○

Warranty/Maintenance Info

Major Tune-Up	$177	○
Minor Tune-Up	$125	○
Brake Service	$243	○
Overall Warranty	3 yr/36k	○
Drivetrain Warranty	3 yr/36k	○
Rust Warranty	6 yr/100k	○
Maintenance Warranty	N/A	
Roadside Assistance	3 yr/36k	

Resale Value
1993	1994	1995	1996	1997
$20,327	$18,311	$16,681	$15,070	$13,336

Cumulative Costs
	1993	1994	1995	1996	1997
Annual	$15,003	$6,304	$6,580	$5,111	$7,984
Total	$15,003	$21,307	$27,887	$32,998	$40,982

Ownership Costs (5yr)
Average $41,659 / This Car $40,982
Cost/Mile 60¢ / Cost/Mile 59¢

Ownership Cost Rating
Average

---

* Includes shaded options
** Other purchase requirements apply

● Poor | ◐ Worse Than Average | ○ Average | ○ Better Than Average | ○ Excellent | ⊖ Insufficient Information

©1993 by IntelliChoice, Inc. (408) 554-8711 All Rights Reserved. Reproduction Prohibited.
Refer to *Section 3: Annotated Vehicle Charts* for an explanation of these charts.

Plymouth Acclaim
4 Door Sedan

Compact

 2.5L 100 hp Gas Fuel Inject.   4 Cylinder In-Line   Manual 5 Speed   2 Wheel Front   Driver Airbag Psngr Belts

Purchase Price

Car Item	Dealer Cost	List
Base Price	**$10,863**	**$11,941**
Anti-Lock Brakes	$764	$899
Automatic 3 Speed	$473	$557
3.0L 141 hp Gas	$616	** $725
Air Conditioning	Pkg	Pkg
Power Steering	Std	Std
Cruise Control	Pkg	Pkg
All Wheel Drive	N/A	N/A
AM/FM Stereo Cassette	$140	** $165
Steering Wheel, Tilt	Pkg	Pkg
Power Windows	Pkg	Pkg
*Options Price	$473	$557
*Total Price	$11,336	$12,498
Target Price		$12,165
Destination Charge		$485
Avg. Tax & Fees		$763
Total Target $		**$13,413**
Average Dealer Option Cost:		*85%*

The 1993 Plymouth Acclaim is available one model edition. For 1993, the Acclaim is available to fleet and retail buyers with a flexible fuel package. The Flexible Fuel Vehicles operate on unleaded gasoline or any blend of gasoline and methanol containing up to 85% methanol (M85). Standard features include tinted glass all around, power-assisted rack-and-pinion steering, windshield wipers with variable intermittent feature, and childproof rear door locks.

Ownership Costs

Cost Area	5 Year Cost	Rate
Depreciation	$7,252	◐
Financing ($270/month)	$2,209	
Insurance (Rating 3)	$6,520	○
State Fees	$520	
Fuel (Hwy 27 City 23)	$3,454	◐
Maintenance	$4,111	○
Repairs	$689	◐

Warranty/Maintenance Info

Major Tune-Up	$138	○
Minor Tune-Up	$80	○
Brake Service	$267	●
Overall Warranty	1 yr/12k	●
Drivetrain Warranty	7 yr/70k	○
Rust Warranty	7 yr/100k	○
Maintenance Warranty	N/A	
Roadside Assistance	N/A	

Ownership Cost By Year

1993 / 1994 / 1995 / 1996 / 1997

Resale Value

1993	1994	1995	1996	1997
$9,544	$8,451	$7,794	$6,824	$6,161

Cumulative Costs

	1993	1994	1995	1996	1997
Annual	$6,827	$4,180	$4,465	$4,098	$5,185
Total	$6,827	$11,007	$15,472	$19,570	$24,755

Ownership Costs (5yr)

Average	This Car
$25,226	$24,755
Cost/Mile 36¢	Cost/Mile 35¢

Ownership Cost Rating
○ Better Than Average

Plymouth Colt
2 Door Coupe

Subcompact

 1.5L 92 hp Gas Fuel Inject. 4 Cylinder In-Line   Manual 5 Speed   2 Wheel Front   Automatic Seatbelts

Purchase Price

Car Item	Dealer Cost	List
Base Price	**$7,488**	**$7,806**
Anti-Lock Brakes	N/A	N/A
Automatic Transmission	N/A	N/A
Optional Engine	N/A	N/A
Air Conditioning	$673	$783
Power Steering	N/A	N/A
Cruise Control	N/A	N/A
All Wheel Drive	N/A	N/A
AM/FM Stereo Cassette	N/A	N/A
Steering Wheel, Tilt	N/A	N/A
Power Windows	N/A	N/A
*Options Price	$0	$0
*Total Price	$7,488	$7,806
Target Price		$7,774
Destination Charge		$400
Avg. Tax & Fees		$491
Total Target $		**$8,665**
Average Dealer Option Cost:		*86%*

The 1993 Plymouth Colt is available in four models - (Base) and GL coupes and sedans. The (Base) model replaces an earlier hatchback. Standard features on this model include front arm rests, power brakes, center console, stainless steel exhaust system, and three child seat anchors. The model features optional electric rear window defroster, tinted glass, and air-conditioning.

Ownership Costs

Cost Area	5 Year Cost	Rate
Depreciation	$4,510	○
Financing ($174/month)	$1,428	
Insurance (Rating 8 [Est.])	$7,223	●
State Fees	$328	
Fuel (Hwy 35 City 29)	$2,698	○
Maintenance	$4,092	○
Repairs	$630	○

Warranty/Maintenance Info

Major Tune-Up	$241	●
Minor Tune-Up	$82	○
Brake Service	$185	○
Overall Warranty	1 yr/12k	●
Drivetrain Warranty	7 yr/70k	○
Rust Warranty	7 yr/100k	○
Maintenance Warranty	N/A	
Roadside Assistance	N/A	

Resale Value

1993	1994	1995	1996	1997
$6,804	$5,956	$5,362	$4,753	$4,155

Cumulative Costs

	1993	1994	1995	1996	1997
Annual	$4,434	$3,626	$4,167	$3,192	$5,490
Total	$4,434	$8,060	$12,227	$15,419	$20,909

Ownership Costs (5yr)

Average	This Car
$20,361	$20,909
Cost/Mile 29¢	Cost/Mile 30¢

Ownership Cost Rating
○ Average

---

* Includes shaded options ** Other purchase requirements apply

● Poor   ◉ Worse Than Average   ○ Average   ○ Better Than Average   ○ Excellent   ⊖ Insufficient Information

©1993 by IntelliChoice, Inc. (408) 554-8711   All Rights Reserved. Reproduction Prohibited.
Refer to *Section 3: Annotated Vehicle Charts* for an explanation of these charts.

Plymouth Colt GL
2 Door Coupe

Subcompact

- 1.5L 92 hp Gas Fuel Inject.
- 4 Cylinder In-Line
- Manual 5 Speed
- 2 Wheel Front
- Automatic Seatbelts

Purchase Price

Car Item	Dealer Cost	List
Base Price	**$8,309**	**$8,705**
Anti-Lock Brakes	N/A	N/A
Automatic 3 Speed	$445	$518
Optional Engine	N/A	N/A
Air Conditioning	$673	$783
Power Steering	Pkg	Pkg
Cruise Control	N/A	N/A
All Wheel Drive	N/A	N/A
AM/FM Stereo Cassette	$156	$181
Steering Wheel, Tilt	N/A	N/A
Power Windows	N/A	N/A
***Options Price**	$0	$0
***Total Price**	**$8,309**	**$8,705**
Target Price		$8,665
Destination Charge		$400
Avg. Tax & Fees		$544
Total Target $		**$9,609**
Average Dealer Option Cost:		**86%**

The 1993 Plymouth Colt is available in four models - (Base) and GL coupes and sedans. The coupe model replaces an earlier hatchback model. Colt GL offers standard color-keyed body-side moldings, full wheel covers, passenger and driver side visors, and stainless steel exhaust system. Options include tinted glass, rear-window electric defroster, color-keyed rear spoiler, power steering, 13" aluminum wheel, and fixed time, intermittent windshield wipers.

Ownership Costs

Cost Area	5 Year Cost	Rate
Depreciation	$5,219	○
Financing ($193/month)	$1,583	
Insurance (Rating 10 [Est.])	$7,479	●
State Fees	$365	
Fuel (Hwy 35 City 29)	$2,698	○
Maintenance	$4,190	○
Repairs	$630	○

Warranty/Maintenance Info

Major Tune-Up	$241	●
Minor Tune-Up	$82	○
Brake Service	$185	○
Overall Warranty	1 yr/12k	●
Drivetrain Warranty	7 yr/70k	○
Rust Warranty	7 yr/100k	○
Maintenance Warranty	N/A	
Roadside Assistance	N/A	

Ownership Cost By Year
1993, 1994, 1995, 1996, 1997

Resale Value
1993	1994	1995	1996	1997
$7,259	$6,355	$5,723	$5,041	$4,390

Cumulative Costs
	1993	1994	1995	1996	1997
Annual	$5,043	$3,788	$4,342	$3,336	$5,655
Total	$5,043	$8,831	$13,173	$16,509	$22,164

Ownership Costs (5yr)
Average	This Car
$21,372	$22,164
Cost/Mile 31¢	Cost/Mile 32¢

Ownership Cost Rating
Worse Than Average

Plymouth Colt
4 Door Sedan

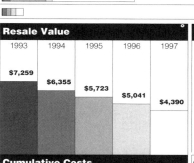
2 Door Coupe Shown

Subcompact

- 1.5L 92 hp Gas Fuel Inject.
- 4 Cylinder In-Line
- Manual 5 Speed
- 2 Wheel Front
- Automatic Seatbelts

Purchase Price

Car Item	Dealer Cost	List
Base Price	**$9,001**	**$9,448**
Anti-Lock Brakes	N/A	N/A
Automatic 4 Speed	$603	$701
1.8L 113 hp Gas	$298	$346
Air Conditioning	$673	$783
Power Steering	Pkg	Pkg
Cruise Control	N/A	N/A
All Wheel Drive	N/A	N/A
AM/FM Stereo Cassette	$156	** $181
Steering Wheel, Tilt	N/A	N/A
Power Windows	N/A	N/A
***Options Price**	$298	$346
***Total Price**	**$9,299**	**$9,794**
Target Price		$9,772
Destination Charge		$400
Avg. Tax & Fees		$611
Total Target $		**$10,783**
Average Dealer Option Cost:		**86%**

The 1993 Plymouth Colt is available in four models - (Base) and GL coupes and sedans. The Colt sedan is a new addition to the line for 1993. Standard features on this model include body-side moldings, reclining cloth-face bucket seats, half wheel covers, and stainless steel exhaust system. Options include tinted glass, trunk light, passenger side visor, electric rear window defroster, and power steering.

Ownership Costs

Cost Area	5 Year Cost	Rate
Depreciation	$5,668	○
Financing ($217/month)	$1,776	
Insurance (Rating 12 [Est.])	$7,898	●
State Fees	$408	
Fuel (Hwy 34 City 27)	$2,831	○
Maintenance	$4,183	○
Repairs	$630	○

Warranty/Maintenance Info

Major Tune-Up	$236	●
Minor Tune-Up	$82	○
Brake Service	$202	○
Overall Warranty	1 yr/12k	●
Drivetrain Warranty	7 yr/70k	○
Rust Warranty	7 yr/100k	○
Maintenance Warranty	N/A	
Roadside Assistance	N/A	

Resale Value
1993	1994	1995	1996	1997
$8,394	$7,342	$6,583	$5,831	$5,115

Cumulative Costs
	1993	1994	1995	1996	1997
Annual	$5,276	$4,113	$4,638	$3,542	$5,825
Total	$5,276	$9,389	$14,027	$17,569	$23,394

Ownership Costs (5yr)
Average	This Car
$22,598	$23,394
Cost/Mile 32¢	Cost/Mile 33¢

Ownership Cost Rating
Worse Than Average

---

* Includes shaded options
** Other purchase requirements apply

● Poor ◐ Worse Than Average ○ Average ◑ Better Than Average ○ Excellent ⊖ Insufficient Information

©1993 by IntelliChoice, Inc. (408) 554-8711 All Rights Reserved. Reproduction Prohibited.
Refer to *Section 3: Annotated Vehicle Charts* for an explanation of these charts.

page 183

Plymouth Colt GL
4 Door Sedan

Subcompact

 1.8L 113 hp Gas Fuel Inject.
 4 Cylinder In-Line
 Manual 5 Speed
 2 Wheel Front
Automatic Seatbelts

2 Door Coupe Shown

Purchase Price

Car Item	Dealer Cost	List
Base Price	**$9,846**	**$10,423**
Anti-Lock Brakes	$829	$964
Automatic 4 Speed	$551	$641
Optional Engine	N/A	N/A
Air Conditioning	Pkg	Pkg
Power Steering	Pkg	Pkg
Cruise Control	Pkg	Pkg
All Wheel Drive	N/A	N/A
AM/FM Stereo Cassette	Pkg	Pkg
Steering Wheel, Tilt	Pkg	Pkg
Power Windows	Pkg	Pkg
*Options Price	$0	$0
*Total Price	$9,846	$10,423
Target Price	$10,361	
Destination Charge	$400	
Avg. Tax & Fees	$646	
Total Target $	**$11,407**	
Average Dealer Option Cost:	*86%*	

The 1993 Plymouth Colt is available in four models - (Base) and GL coupes and sedans. The Colt GL sedan is a new addition to the line for 1993. Standard items on this model include reclining, full-cloth bucket seats, child protection door locks, power brakes, and color-keyed 5 mph bumpers. Options include tinted glass, electronic speed control (new for 1993), 13" aluminum wheels, and tilt steering column.

Ownership Costs

Cost Area	5 Year Cost	Rate
Depreciation	$6,255	○
Financing ($229/month)	$1,879	
Insurance (Rating 12 [Est.])	$7,898	●
State Fees	$433	
Fuel (Hwy 34 City 27)	$2,831	○
Maintenance	$4,434	○
Repairs	$630	○

Warranty/Maintenance Info

Major Tune-Up	$236	●
Minor Tune-Up	$82	○
Brake Service	$232	◉
Overall Warranty	1 yr/12k	●
Drivetrain Warranty	7 yr/70k	○
Rust Warranty	7 yr/100k	○
Maintenance Warranty	N/A	
Roadside Assistance	N/A	

Ownership Cost By Year

1993, 1994, 1995, 1996, 1997

Resale Value

1993	1994	1995	1996	1997
$8,578	$7,499	$6,725	$5,911	$5,152

Cumulative Costs

	1993	1994	1995	1996	1997
Annual	$5,765	$4,177	$4,799	$3,616	$6,003
Total	$5,765	$9,942	$14,741	$18,357	$24,360

Ownership Costs (5yr)

Average	This Car
$23,305	$24,360
Cost/Mile 33¢	Cost/Mile 35¢

Ownership Cost Rating

◉ Worse Than Average

Plymouth Colt Vista
3 Door Wagon

Compact Wagon

 1.8L 113 hp Gas Fuel Inject.
 4 Cylinder In-Line | Manual 5 Speed | 2 Wheel Front | Automatic Seatbelts

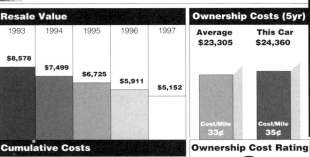

Purchase Price

Car Item	Dealer Cost	List
Base Price	**$10,660**	**$11,455**
Anti-Lock Brakes	$829	** $964
Automatic 4 Speed	$622	$723
2.4L 136 hp Gas	$156	$181
Air Conditioning	$673	$783
Power Steering	Std	Std
Cruise Control	Pkg	Pkg
All Wheel Drive	N/A	N/A
AM/FM Stereo Cassette	$156	** $181
Steering Wheel, Tilt	N/A	N/A
Power Windows	N/A	N/A
*Options Price	$1,451	$1,687
*Total Price	$12,111	$13,142
Target Price	$12,877	
Destination Charge	$400	
Avg. Tax & Fees	$799	
Total Target $	**$14,076**	
Average Dealer Option Cost:	*86%*	

The 1993 Plymouth Colt Vista wagon is available in three models - (Base), SE, and AWD. New for 1993, the Base model offers optional cargo security cover. Other options include anti-lock brakes, rear window defroster, tinted glass, luggage rack for the roof, dual electric mirrors, electronic speed control, tailgate power lock, and rear stabilizer bar. Standard features include child protection on rear sliding door lock, remote hood release, and vented side windows.

Ownership Costs

Cost Area	5 Year Cost	Rate
Depreciation	$7,667	○
Financing ($283/month)	$2,319	
Insurance (Rating 7)	$7,035	○
State Fees	$541	
Fuel (Hwy 26 City 20)	$3,756	◉
Maintenance		⊖
Repairs	$630	○

Insufficient Maintenance Information

Warranty/Maintenance Info

Major Tune-Up		⊖
Minor Tune-Up		⊖
Brake Service	$165	○
Overall Warranty	1 yr/12k	●
Drivetrain Warranty	7 yr/70k	○
Rust Warranty	7 yr/100k	○
Maintenance Warranty	N/A	
Roadside Assistance	N/A	

Ownership Cost By Year

1993, 1994, 1995, 1996, 1997

Resale Value

1993	1994	1995	1996	1997
$9,799	$8,825	$8,059	$7,201	$6,409

Cumulative Costs

	1993	1994	1995	1996	1997
Annual	*Insufficient Information*				
Total	*Insufficient Information*				

Ownership Costs (5yr)

Insufficient Information

Ownership Cost Rating

⊖ Insufficient Information

---

page 184

*Includes shaded options
** Other purchase requirements apply

● Poor | ◉ Worse Than Average | ○ Average | ○ Better Than Average | ○ Excellent | ⊖ Insufficient Information

©1993 by IntelliChoice, Inc. (408) 554-8711   All Rights Reserved. Reproduction Prohibited.
Refer to *Section 3: Annotated Vehicle Charts* for an explanation of these charts.

Plymouth Colt Vista SE
3 Door Wagon

 2.4L 136 hp Gas Fuel Inject.
 4 Cylinder In-Line
 Manual 5 Speed
 2 Wheel Front
Automatic Seatbelts

Compact Wagon — Base Model Shown

Purchase Price

Car Item	Dealer Cost	List
Base Price	**$11,481**	**$12,368**
Anti-Lock Brakes	$829	** $964
Automatic 4 Speed	$622	$723
Optional Engine	N/A	N/A
Air Conditioning	Pkg	Pkg
Power Steering	Std	Std
Cruise Control	Pkg	Pkg
All Wheel Drive	N/A	N/A
AM/FM Stereo Cassette	Pkg	Pkg
Steering Wheel, Tilt	Std	Std
Power Windows	Pkg	Pkg
***Options Price**	**$622**	**$723**
***Total Price**	**$12,103**	**$13,091**
Target Price		$12,888
Destination Charge		$400
Avg. Tax & Fees		$799
Total Target $		**$14,087**
Average Dealer Option Cost:		**86%**

The 1993 Plymouth Colt Vista wagon is available in three models - (Base), SE, and AWD. The Colt Vista SE features standard power brakes, child protection on rear sliding door lock, a stainless steel exhaust system, power lock on tailgate, fixed intermittent wiper/washer in rear, tilt steering column, and dual electric mirrors. Options include keyless remote entry, electronic speed control, power windows and door locks, roof luggage rack, and two tone pain with accent colored bumpers.

Ownership Costs

Cost Area	5 Year Cost	Rate
Depreciation	$7,730	○
Financing ($283/month)	$2,321	
Insurance (Rating 7)	$7,035	○
State Fees	$539	
Fuel (Hwy 26 City 20)	$3,756	◉
Maintenance		⊖
Repairs	$630	○

Warranty/Maintenance Info

Major Tune-Up		⊖
Minor Tune-Up		⊖
Brake Service	$165	○
Overall Warranty	1 yr/12k	●
Drivetrain Warranty	7 yr/70k	○
Rust Warranty	7 yr/100k	○
Maintenance Warranty	N/A	
Roadside Assistance	N/A	

Ownership Cost By Year

1993 / 1994 / 1995 / 1996 / 1997

Insufficient Maintenance Information

Resale Value

1993	1994	1995	1996	1997
$9,749	$8,765	$8,002	$7,147	$6,357

Cumulative Costs

Annual — Insufficient Information
Total — Insufficient Information

Ownership Costs (5yr)

Insufficient Information

Ownership Cost Rating

⊖ Insufficient Information

Plymouth Colt Vista AWD
3 Door Wagon

 1.8L 113 hp Gas Fuel Inject.
 4 Cylinder In-Line
 Manual 5 Speed
 4 Wheel Full-Time
Automatic Seatbelts

Compact Wagon — 2WD Base Shown

Purchase Price

Car Item	Dealer Cost	List
Base Price	**$12,535**	**$13,539**
Anti-Lock Brakes	$829	** $964
Automatic 4 Speed	$622	$723
2.4L 136 hp Gas	$156	$181
Air Conditioning	$673	$783
Power Steering	Std	Std
Cruise Control	Pkg	Pkg
4 Wheel Full-Time Drive	Std	Std
AM/FM Stereo Cassette	$156	** $181
Steering Wheel, Tilt	Std	Std
Power Windows	Pkg	Pkg
***Options Price**	**$1,451**	**$1,687**
***Total Price**	**$13,986**	**$15,226**
Target Price		$14,923
Destination Charge		$400
Avg. Tax & Fees		$922
Total Target $		**$16,245**
Average Dealer Option Cost:		**86%**

The 1993 Plymouth Colt Vista wagon is available in three models - (Base), SE, and AWD. New this year, the AWD model offers full cloth velour interior fabric with custom package. Exterior light blue and two-tone colors have been added to the choices. Other options include tachometer and low fuel warning light on the instrument cluster, keyless entry, power windows and locks, and electronic speed control. Standard features include mudguards and four speakers with radio, cassette, and clock.

Ownership Costs

Cost Area	5 Year Cost	Rate
Depreciation	$9,049	○
Financing ($327/month)	$2,676	
Insurance (Rating 10)	$7,479	○
State Fees	$625	
Fuel (Hwy 23 City 19)	$4,111	●
Maintenance		⊖
Repairs	$740	◉

Warranty/Maintenance Info

Major Tune-Up		⊖
Minor Tune-Up		⊖
Brake Service	$165	○
Overall Warranty	1 yr/12k	●
Drivetrain Warranty	7 yr/70k	○
Rust Warranty	7 yr/100k	○
Maintenance Warranty	N/A	
Roadside Assistance	N/A	

Ownership Cost By Year

1993 / 1994 / 1995 / 1996 / 1997

Insufficient Maintenance Information

Resale Value

1993	1994	1995	1996	1997
$11,049	$9,939	$9,054	$8,095	$7,196

Cumulative Costs

Annual — Insufficient Information
Total — Insufficient Information

Ownership Costs (5yr)

Insufficient Information

Ownership Cost Rating

⊖ Insufficient Information

---

* Includes shaded options
** Other purchase requirements apply

● Poor   ◉ Worse Than Average   ○ Average   ○ Better Than Average   ○ Excellent   ⊖ Insufficient Information

©1993 by IntelliChoice, Inc. (408) 554-8711 All Rights Reserved. Reproduction Prohibited.
Refer to *Section 3: Annotated Vehicle Charts* for an explanation of these charts.

Plymouth Laser
2 Door Hatchback

Subcompact

- 1.8L 92 hp Gas Fuel Inject.
- 4 Cylinder In-Line
- Manual 5 Speed
- 2 Wheel Front
- Automatic Seatbelts

Purchase Price

Car Item	Dealer Cost	List
Base Price	**$10,687**	**$11,406**
Anti-Lock Brakes	N/A	N/A
Automatic 4 Speed	$609	$716
Optional Engine	N/A	N/A
Air Conditioning	Pkg	Pkg
Power Steering	Pkg	Pkg
Cruise Control	Pkg	Pkg
All Wheel Drive	N/A	N/A
AM/FM Stereo Cassette	$168	$198
Steering Wheel, Tilt	Std	Std
Power Windows	N/A	N/A
*Options Price	$0	$0
*Total Price	$10,687	$11,406
Target Price		$11,262
Destination Charge		$400
Avg. Tax & Fees		$701
Total Target $		**$12,363**
Average Dealer Option Cost:		85%

The 1993 Plymouth Laser is available in four models - (Base), RS, RS Turbo, and RS Turbo AWD. The Laser is available in deep green exterior and gold colored wheels, stripes, and badging. Stainless steel exhaust system, body-side moldings, and dual visor vanity mirrors with covers are standard for this model. Remote release fuel fill and liftgate and passive shoulder belts for the front are also standard. Options include electronic speed control, tinted windows, and sun-roof.

Ownership Costs

Cost Area	5 Year Cost	Rate
Depreciation	$5,187	○
Financing ($248/month)	$2,037	
Insurance (Rating 13)	$8,061	●
State Fees	$472	
Fuel (Hwy 26 City 23)	$3,530	●
Maintenance	$3,941	○
Repairs	$649	○

Warranty/Maintenance Info

Major Tune-Up	$230	●
Minor Tune-Up	$82	○
Brake Service	$228	●
Overall Warranty	1 yr/12k	●
Drivetrain Warranty	7 yr/70k	○
Rust Warranty	7 yr/100k	○
Maintenance Warranty	N/A	
Roadside Assistance	N/A	

Ownership Cost By Year

(Bar chart showing costs for 1993-1997)

Resale Value

1993	1994	1995	1996	1997
$10,582	$9,945	$8,817	$7,997	$7,176

Cumulative Costs

	1993	1994	1995	1996	1997
Annual	$4,952	$3,927	$5,286	$4,313	$5,399
Total	$4,952	$8,879	$14,165	$18,478	$23,877

Ownership Costs (5yr)

Average	This Car
$24,411	$23,877
Cost/Mile 35¢	Cost/Mile 34¢

Ownership Cost Rating
○ Better Than Average

Plymouth Laser RS
2 Door Hatchback

Subcompact

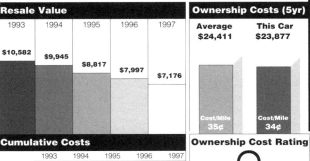

- 2.0L 135 hp Gas Fuel Inject.
- 4 Cylinder In-Line
- Manual 5 Speed
- 2 Wheel Front
- Automatic Seatbelts

Purchase Price

Car Item	Dealer Cost	List
Base Price	**$12,739**	**$13,749**
Anti-Lock Brakes	$802	$943
Automatic 4 Speed	$609	$716
Optional Engine	N/A	N/A
Air Conditioning	Pkg	Pkg
Power Steering	Std	Std
Cruise Control	Pkg	Pkg
All Wheel Drive	N/A	N/A
AM/FM Stereo Cassette	Std	Std
Steering Wheel, Tilt	Std	Std
Power Windows	Pkg	Pkg
*Options Price	$0	$0
*Total Price	$12,739	$13,749
Target Price		$13,469
Destination Charge		$400
Avg. Tax & Fees		$834
Total Target $		**$14,703**
Average Dealer Option Cost:		85%

The 1993 Plymouth Laser is available in four models - (Base), RS, RS Turbo, and RS Turbo AWD. The Laser RS is available in the new deep green exterior with gold colored wheels, stripes, and badging. Standard features for this model include a removable Tonneau cover, a stereo cassette player with six speakers and a clock, and a sport tuned stainless steel exhaust system with dual chrome tips. Options include electronic speed control, rear liftgate wiper/washer, and sun-roof.

Ownership Costs

Cost Area	5 Year Cost	Rate
Depreciation	$6,299	○
Financing ($295/month)	$2,422	
Insurance (Rating 14)	$8,238	●
State Fees	$565	
Fuel (Hwy 29 City 22)	$3,390	◉
Maintenance	$4,871	◉
Repairs	$649	○

Warranty/Maintenance Info

Major Tune-Up	$285	●
Minor Tune-Up	$82	○
Brake Service	$228	●
Overall Warranty	1 yr/12k	●
Drivetrain Warranty	7 yr/70k	○
Rust Warranty	7 yr/100k	○
Maintenance Warranty	N/A	
Roadside Assistance	N/A	

Resale Value

1993	1994	1995	1996	1997
$12,057	$11,524	$10,214	$9,322	$8,404

Cumulative Costs

	1993	1994	1995	1996	1997
Annual	$6,008	$3,971	$5,881	$4,437	$6,137
Total	$6,008	$9,979	$15,860	$20,297	$26,434

Ownership Costs (5yr)

Average	This Car
$27,048	$26,434
Cost/Mile 39¢	Cost/Mile 38¢

Ownership Cost Rating
○ Better Than Average

---

*Includes shaded options
**Other purchase requirements apply

● Poor ◉ Worse Than Average ○ Average ○ Better Than Average ○ Excellent ⊖ Insufficient Information

©1993 by IntelliChoice, Inc. (408) 554-8711 All Rights Reserved. Reproduction Prohibited.
Refer to *Section 3: Annotated Vehicle Charts* for an explanation of these charts.

page 186

Plymouth Laser RS Turbo
2 Door Hatchback

Subcompact

 2.0L 195 hp Gas Fuel Inject. | 4 Cylinder In-Line | Manual 5 Speed | 2 Wheel Front | Automatic Seatbelts

Base Model Shown

Purchase Price

Car Item	Dealer Cost	List
Base Price	**$14,105**	**$15,267**
Anti-Lock Brakes	$802	$943
Automatic 4 Speed	$728	$857
Optional Engine	N/A	N/A
Air Conditioning	Pkg	Pkg
Power Steering	Std	Std
Cruise Control	Pkg	Pkg
All Wheel Drive	N/A	N/A
AM/FM Stereo Cassette	Std	Std
Steering Wheel, Tilt	Std	Std
Power Windows	Pkg	Pkg
***Options Price**	**$0**	**$0**
***Total Price**	**$14,105**	**$15,267**
Target Price	$14,946	
Destination Charge	$400	
Avg. Tax & Fees	$924	
Total Target $	**$16,270**	
Average Dealer Option Cost:	85%	

The 1993 Plymouth Laser is available in four models - (Base), RS, RS Turbo, and RS Turbo AWD. The Laser RS Turbo is available in the new deep green exterior with gold colored wheels, stripes, and badging. Gas-filled shock absorbers at all four wheels are also new. Standard features in this model include a turbo boost gauge on the instrument panel, 16" turbine wheels, and sill applique body-side moldings with front stone guard.

Ownership Costs

Cost Area	5 Year Cost	Rate
Depreciation	$7,770	○
Financing ($327/month)	$2,680	
Insurance (Rating 16+)	$10,541	●
State Fees	$627	
Fuel (Hwy 28 City 21)	$3,530	◉
Maintenance	$5,079	●
Repairs	$935	●

Warranty/Maintenance Info

Major Tune-Up	$285	●
Minor Tune-Up	$82	○
Brake Service	$228	◉
Overall Warranty	1 yr/12k	●
Drivetrain Warranty	7 yr/70k	○
Rust Warranty	7 yr/100k	○
Maintenance Warranty	N/A	
Roadside Assistance	N/A	

Ownership Cost By Year

$2,000 – $12,000

■ 1993 ■ 1994 ■ 1995 □ 1996 □ 1997

Resale Value

1993	1994	1995	1996	1997
$12,503	$11,673	$10,300	$9,401	$8,500

Cumulative Costs

	1993	1994	1995	1996	1997
Annual	$7,712	$4,891	$6,661	$5,054	$6,843
Total	$7,712	$12,603	$19,264	$24,318	$31,161

Ownership Costs (5yr)

Average	This Car
$28,755	$31,161
Cost/Mile 41¢	Cost/Mile 45¢

Ownership Cost Rating

● Poor

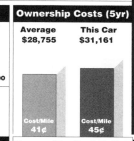

Plymouth Laser RS Turbo AWD
2 Door Hatchback

Subcompact

 2.0L 195 hp Gas Fuel Inject. | 4 Cylinder In-Line | Manual 5 Speed | 2 Wheel Front | Automatic Seatbelts

2WD Base Shown

Purchase Price

Car Item	Dealer Cost	List
Base Price	**$16,039**	**$17,371**
Anti-Lock Brakes	$802	$943
Automatic 4 Speed	$728	$857
Optional Engine	N/A	N/A
Air Conditioning	Pkg	Pkg
Power Steering	Std	Std
Cruise Control	Pkg	Pkg
All Wheel Drive	N/A	N/A
AM/FM Stereo Cassette	Std	Std
Steering Wheel, Tilt	Std	Std
Power Windows	Pkg	Pkg
***Options Price**	**$0**	**$0**
***Total Price**	**$16,039**	**$17,371**
Target Price	$17,049	
Destination Charge	$400	
Avg. Tax & Fees	$1,050	
Total Target $	**$18,499**	
Average Dealer Option Cost:	85%	

The 1993 Plymouth Laser is available in four models - (Base), RS, RS Turbo, and RS Turbo AWD. The RS Turbo AWD features the new gas-filled shock absorbers at all four wheels. A deep green exterior color with gold colored wheels, stripes, and badging are also new. Standards on this model include 16", seven-spoke, argent or gold wheels, cloth seats, leather-wrapped steering wheel, and a turbo boost gauge on the instrument panel. Sunroof and electronic speed control are optional.

Ownership Costs

Cost Area	5 Year Cost	Rate
Depreciation	$8,863	○
Financing ($372/month)	$3,046	
Insurance (Rating 17+)	$10,901	●
State Fees	$711	
Fuel (Hwy 25 City 20)	$3,837	●
Maintenance	$5,037	◉
Repairs	$975	●

Warranty/Maintenance Info

Major Tune-Up	$285	●
Minor Tune-Up	$82	○
Brake Service	$228	◉
Overall Warranty	1 yr/12k	●
Drivetrain Warranty	7 yr/70k	○
Rust Warranty	7 yr/100k	○
Maintenance Warranty	N/A	
Roadside Assistance	N/A	

Resale Value

1993	1994	1995	1996	1997
$13,894	$13,339	$11,727	$10,685	$9,636

Cumulative Costs

	1993	1994	1995	1996	1997
Annual	$8,847	$4,885	$7,132	$5,344	$7,161
Total	$8,847	$13,732	$20,864	$26,208	$33,369

Ownership Costs (5yr)

Average	This Car
$31,123	$33,369
Cost/Mile 44¢	Cost/Mile 48¢

Ownership Cost Rating

● Poor

---

* Includes shaded options
** Other purchase requirements apply

● Poor | ◉ Worse Than Average | ○ Average | ○ Better Than Average | ○ Excellent | ⊖ Insufficient Information

©1993 by IntelliChoice, Inc. (408) 554-8711 All Rights Reserved. Reproduction Prohibited.
Refer to *Section 3: Annotated Vehicle Charts* for an explanation of these charts.

Plymouth Sundance
2 Door Hatchback

Subcompact

 2.2L 93 hp Gas Fuel Inject.
 4 Cylinder In-Line
 Manual 5 Speed
 2 Wheel Front
 Driver Airbag Psngr Belts

Duster Model Shown

Purchase Price

Car Item	Dealer Cost	List
Base Price	**$7,888**	**$8,397**
Anti-Lock Brakes	$764	$899
Automatic 3 Speed	$473	$557
2.5L 100 hp Gas	$243	$286
Air Conditioning	$765	$900
Power Steering	Std	Std
Cruise Control	$190	** $224
All Wheel Drive	N/A	N/A
AM/FM Stereo Cassette	$424	$499
Steering Wheel, Tilt	$126	** $148
Power Windows	N/A	N/A
*Options Price	$0	$0
*Total Price	$7,888	$8,397
Target Price		$8,343
Destination Charge		$485
Avg. Tax & Fees		$530
Total Target $		**$9,358**
Average Dealer Option Cost:		85%

The 1993 Plymouth Sundance is available in four models - (Base) and Duster two-door hatchbacks and four-door sedans. For 1993, "America" has been dropped from the Base name. The two-door hatchback now has a body-color hood "Pentastar". Stainless steel exhaust system and sound insulation are standard. Options on this model include a tilt steering column, body-side stripes, power locks, and intermittent windshield wipers.

Ownership Costs

Cost Area	5 Year Cost	Rate
Depreciation	$5,090	◯
Financing ($188/month)	$1,542	
Insurance (Rating 10)	$7,479	●
State Fees	$355	
Fuel (Hwy 32 City 27)	$2,926	◉
Maintenance	$3,895	◯
Repairs	$731	◯

Warranty/Maintenance Info

Major Tune-Up	$149	○
Minor Tune-Up	$81	○
Brake Service	$242	●
Overall Warranty	1 yr/12k	●
Drivetrain Warranty	7 yr/70k	○
Rust Warranty	7 yr/100k	○
Maintenance Warranty	N/A	
Roadside Assistance	N/A	

Ownership Cost By Year

1993, 1994, 1995, 1996, 1997

Resale Value

1993	1994	1995	1996	1997
$7,236	$6,334	$5,647	$4,934	$4,268

Cumulative Costs

	1993	1994	1995	1996	1997
Annual	$4,841	$3,836	$4,488	$3,876	$4,977
Total	$4,841	$8,677	$13,165	$17,041	$22,018

Ownership Costs (5yr)

Average	This Car
$21,026	$22,018
Cost/Mile 30¢	Cost/Mile 31¢

Ownership Cost Rating

● Worse Than Average

Plymouth Sundance
4 Door Hatchback

Subcompact

 2.2L 93 hp Gas Fuel Inject.
 4 Cylinder In-Line
 Manual 5 Speed
 2 Wheel Front
 Driver Airbag Psngr Belts

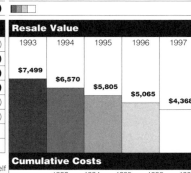
Duster Model Shown

Purchase Price

Car Item	Dealer Cost	List
Base Price	**$8,255**	**$8,797**
Anti-Lock Brakes	$764	$899
Automatic 3 Speed	$473	$557
2.5L 100 hp Gas	$243	$286
Air Conditioning	$765	$900
Power Steering	Std	Std
Cruise Control	$190	** $224
All Wheel Drive	N/A	N/A
AM/FM Stereo Cassette	$424	$499
Steering Wheel, Tilt	$126	** $148
Power Windows	N/A	N/A
*Options Price	$0	$0
*Total Price	$8,255	$8,797
Target Price		$8,737
Destination Charge		$485
Avg. Tax & Fees		$554
Total Target $		**$9,776**
Average Dealer Option Cost:		85%

The 1993 Plymouth Sundance is available in four models - (Base) and Duster two-door hatchbacks and four-door sedans. The Sundance sedan features standard 5 mph bumpers, sound insulation, removable shelf panel, driver-side airbag, and steel wheels. This model also features front and rear fascias in accent colors. Options include tinted glass, anti-lock brakes, power locks, electronic speed control, and a tilt steering column. Stainless steel exhaust system and sound insulation are standard.

Ownership Costs

Cost Area	5 Year Cost	Rate
Depreciation	$5,408	◯
Financing ($196/month)	$1,610	
Insurance (Rating 6)	$6,919	◯
State Fees	$372	
Fuel (Hwy 32 City 27)	$2,926	◉
Maintenance	$3,895	◯
Repairs	$731	◯

Warranty/Maintenance Info

Major Tune-Up	$149	○
Minor Tune-Up	$81	○
Brake Service	$242	●
Overall Warranty	1 yr/12k	●
Drivetrain Warranty	7 yr/70k	○
Rust Warranty	7 yr/100k	○
Maintenance Warranty	N/A	
Roadside Assistance	N/A	

Resale Value

1993	1994	1995	1996	1997
$7,499	$6,570	$5,805	$5,065	$4,368

Cumulative Costs

	1993	1994	1995	1996	1997
Annual	$4,925	$3,781	$4,470	$3,795	$4,890
Total	$4,925	$8,706	$13,176	$16,971	$21,861

Ownership Costs (5yr)

Average	This Car
$21,476	$21,861
Cost/Mile 31¢	Cost/Mile 31¢

Ownership Cost Rating

◯ Average

---

page 188

* Includes shaded options
** Other purchase requirements apply

● Poor ◉ Worse Than Average ◯ Average ○ Better Than Average ○ Excellent ⊖ Insufficient Information

©1993 by IntelliChoice, Inc. (408) 554-8711   All Rights Reserved. Reproduction Prohibited.
Refer to *Section 3: Annotated Vehicle Charts* for an explanation of these charts.

Plymouth Sundance Duster
2 Door Hatchback

 3.0L 141 hp Gas Fuel Inject.
 6 Cylinder "V"
 Manual 5 Speed
 2 Wheel Front
 Driver Airbag Psngr Belts

Subcompact

Purchase Price

Car Item	Dealer Cost	List
Base Price	**$9,748**	**$10,498**
Anti-Lock Brakes	$764	$899
Automatic 4 Speed	$587	$690
2.5L 100 hp Gas	($590)	($694)
Air Conditioning	$765	$900
Power Steering	Std	Std
Cruise Control	$190	** $224
All Wheel Drive	N/A	N/A
AM/FM Stereo Cassette	$183	$215
Steering Wheel, Tilt	$126	** $148
Power Windows	$225	** $265
*Options Price	$0	$0
*Total Price	$9,748	$10,498
Target Price	$10,346	
Destination Charge	$485	
Avg. Tax & Fees	$652	
Total Target $	**$11,483**	
Average Dealer Option Cost:	85%	

Ownership Costs

Cost Area	5 Year Cost	Rate
Depreciation	$6,614	●
Financing ($231/month)	$1,891	
Insurance (Rating 11)	$7,693	●
State Fees	$440	
Fuel (Hwy 28 City 19)	$3,704	●
Maintenance	$4,487	○
Repairs	$740	●

Warranty/Maintenance Info

Major Tune-Up	$177	○
Minor Tune-Up	$102	●
Brake Service	$242	●
Overall Warranty	1 yr/12k	●
Drivetrain Warranty	7 yr/70k	○
Rust Warranty	7 yr/100k	○
Maintenance Warranty	N/A	
Roadside Assistance	N/A	

Ownership Cost By Year
(Bar chart showing years 1993–1997)

Resale Value

1993	1994	1995	1996	1997
$8,757	$7,433	$6,549	$5,680	$4,869

Cumulative Costs

	1993	1994	1995	1996	1997
Annual	$5,799	$4,606	$5,199	$4,311	$5,654
Total	$5,799	$10,405	$15,604	$19,915	$25,569

Ownership Costs (5yr)

Average	This Car
$23,390	$25,569
Cost/Mile 33¢	Cost/Mile 37¢

Ownership Cost Rating
● Poor

The 1993 Plymouth Sundance is available in four models - (Base) and Duster two-door hatchbacks and four-door sedans. The Plymouth Duster offers as standard a stainless steel exhaust system, sound insulation, and seat and door trim fabric. Anti-lock brakes are optional. Other options include power door locks, fog lamps, a tachometer, and power, dual outside mirrors. A decor package that provides a monochromatic exterior appearance is new for 1993.

Plymouth Sundance Duster
4 Door Hatchback

 3.0L 141 hp Gas Fuel Inject.
 6 Cylinder "V"
 Manual 5 Speed
 2 Wheel Front
 Driver Airbag Psngr Belts

Subcompact

Purchase Price

Car Item	Dealer Cost	List
Base Price	**$10,108**	**$10,898**
Anti-Lock Brakes	$764	$899
Automatic 4 Speed	$587	$690
2.5L 100 hp Gas	($590)	($694)
Air Conditioning	$765	$900
Power Steering	Std	Std
Cruise Control	$190	** $224
All Wheel Drive	N/A	N/A
AM/FM Stereo Cassette	$183	$215
Steering Wheel, Tilt	$126	** $148
Power Windows	$281	** $331
*Options Price	$0	$0
*Total Price	$10,108	$10,898
Target Price	$10,736	
Destination Charge	$485	
Avg. Tax & Fees	$675	
Total Target $	**$11,896**	
Average Dealer Option Cost:	85%	

Ownership Costs

Cost Area	5 Year Cost	Rate
Depreciation	$6,659	○
Financing ($239/month)	$1,960	
Insurance (Rating 8)	$7,223	○
State Fees	$455	
Fuel (Hwy 28 City 19)	$3,704	●
Maintenance	$4,487	○
Repairs	$740	●

Warranty/Maintenance Info

Major Tune-Up	$177	○
Minor Tune-Up	$102	●
Brake Service	$242	●
Overall Warranty	1 yr/12k	●
Drivetrain Warranty	7 yr/70k	○
Rust Warranty	7 yr/100k	○
Maintenance Warranty	N/A	
Roadside Assistance	N/A	

Resale Value

1993	1994	1995	1996	1997
$9,264	$8,111	$7,206	$6,230	$5,237

Cumulative Costs

	1993	1994	1995	1996	1997
Annual	$5,651	$4,370	$5,143	$4,328	$5,736
Total	$5,651	$10,021	$15,164	$19,492	$25,228

Ownership Costs (5yr)

Average	This Car
$23,840	$25,228
Cost/Mile 34¢	Cost/Mile 36¢

Ownership Cost Rating
◐ Worse Than Average

The 1993 Plymouth Sundance is available in four models - (Base) and Duster two-door hatchbacks and four-door sedans. The Duster four-door offers standard stainless steel exhaust system, sound insulation, and seat and door trim fabric. Other standard features include a sport suspension, "Triad" wheel covers, power rack and pinion steering, wood grain instrument panel bezel, and color-keyed bumpers. An optional decor package that provides a monochromatic exterior appearance is new for 1993.

---

* Includes shaded options
** Other purchase requirements apply

● Poor   ◐ Worse Than Average   ◐ Average   ○ Better Than Average   ○ Excellent   ⊖ Insufficient Information

©1993 by IntelliChoice, Inc. (408) 554-8711 All Rights Reserved. Reproduction Prohibited.
Refer to *Section 3: Annotated Vehicle Charts* for an explanation of these charts.

Pontiac Bonneville SE
4 Door Sedan

3.8L 170 hp Gas Fuel Inject. | 6 Cylinder "V" | Automatic 4 Speed | 2 Wheel Front | Driver Airbag Psngr Belts

Large

Purchase Price

Car Item	Dealer Cost	List
Base Price	**$17,014**	**$19,444**
Anti-Lock Brakes	Std	Std
Manual Transmission	N/A	N/A
Optional Engine	N/A	N/A
Air Conditioning	Std	Std
Power Steering	Std	Std
Cruise Control	Pkg	Pkg
All Wheel Drive	N/A	N/A
AM/FM Stereo Cassette	$120	$140
Steering Wheel, Tilt	Std	Std
Power Windows	Std	Std
***Options Price**	$120	$140
***Total Price**	**$17,134**	**$19,584**
Target Price		$18,632
Destination Charge		$555
Avg. Tax & Fees		$1,160
Total Target $		**$20,347**
Average Dealer Option Cost:		86%

The 1993 Pontiac Bonneville is available in three models - SE, SSE, and SSEi Sedans. New for 1993, the SE offers a Sport Luxury Package, Exterior Appearance Package, Performance and Handling Package, traction control and an energy-absorbing steering column. Standard features include composite halogen headlamps, PASS-Key II theft-deterrent system, fog lamps, rear door child security locks and Solar Ray tinted glass that compliments the Bonneville's contemporary styling.

Ownership Costs

Cost Area	5 Year Cost	Rate
Depreciation	$11,215	○
Financing ($409/month)	$3,352	
Insurance (Rating 7)	$7,035	○
State Fees	$805	
Fuel (Hwy 28 City 19)	$3,704	○
Maintenance	$5,320	◉
Repairs	$709	○

Warranty/Maintenance Info

Major Tune-Up	$183	○
Minor Tune-Up	$131	◉
Brake Service	$263	○
Overall Warranty	3 yr/36k	○
Drivetrain Warranty	3 yr/36k	○
Rust Warranty	6 yr/100k	○
Maintenance Warranty	N/A	
Roadside Assistance	3 yr/36k	

Ownership Cost By Year

1993, 1994, 1995, 1996, 1997

Resale Value

1993	1994	1995	1996	1997
$14,090	$12,675	$11,543	$10,390	$9,132

Cumulative Costs

	1993	1994	1995	1996	1997
Annual	$9,887	$4,978	$5,516	$4,352	$7,407
Total	$9,887	$14,865	$20,381	$24,733	$32,140

Ownership Costs (5yr)

Average	This Car
$31,892	$32,140
Cost/Mile 46¢	Cost/Mile 46¢

Ownership Cost Rating

Average

Pontiac Bonneville SSE
4 Door Sedan

3.8L 170 hp Gas Fuel Inject. | 6 Cylinder "V" | Automatic 4 Speed | 2 Wheel Front | Driver Airbag Psngr Opt

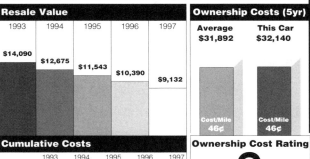

Large — SE Model Shown

Purchase Price

Car Item	Dealer Cost	List
Base Price	**$21,739**	**$24,844**
Anti-Lock Brakes	Std	Std
Manual Transmission	N/A	N/A
Suprchrg 3.8L 205 hp	$900	$1,047
Air Conditioning	Std	Std
Power Steering	Std	Std
Cruise Control	Std	Std
All Wheel Drive	N/A	N/A
AM/FM Stereo Cassette	Std	Std
Steering Wheel, Tilt	Std	Std
Power Windows	Std	Std
***Options Price**	$0	$0
***Total Price**	**$21,739**	**$24,844**
Target Price		$23,878
Destination Charge		$555
Avg. Tax & Fees		$1,476
Total Target $		**$25,909**
Average Dealer Option Cost:		86%

The 1993 Pontiac Bonneville is available in three models - SE, SSE, and SSEi Sedans. New for 1993, the SSE upgrades the SE edition by adding front bucket seats, specific body fascias and grille, decklid spoiler, deluxe acoustical insulation, center console storage and a driver information center. Other standard features include, PASS-Key II theft-deterrent system, accessory emergency road kit, and Solar Ray tinted glass that compliments the Bonneville's contemporary styling.

Ownership Costs

Cost Area	5 Year Cost	Rate
Depreciation	$15,606	◉
Financing ($521/month)	$4,268	
Insurance (Rating 11)	$7,693	○
State Fees	$1,016	
Fuel (Hwy 28 City 19)	$3,704	○
Maintenance	$5,382	◉
Repairs	$730	○

Warranty/Maintenance Info

Major Tune-Up	$183	○
Minor Tune-Up	$131	◉
Brake Service	$263	○
Overall Warranty	3 yr/36k	○
Drivetrain Warranty	3 yr/36k	○
Rust Warranty	6 yr/100k	○
Maintenance Warranty	N/A	
Roadside Assistance	3 yr/36k	

Ownership Cost By Year

1993, 1994, 1995, 1996, 1997

Resale Value

1993	1994	1995	1996	1997
$17,643	$15,475	$13,571	$11,890	$10,303

Cumulative Costs

	1993	1994	1995	1996	1997
Annual	$12,452	$6,191	$6,675	$5,129	$7,952
Total	$12,452	$18,643	$25,318	$30,447	$38,399

Ownership Costs (5yr)

Average	This Car
$37,109	$38,399
Cost/Mile 53¢	Cost/Mile 55¢

Ownership Cost Rating

Worse Than Average

---

page 190

* Includes shaded options
** Other purchase requirements apply

● Poor | ◉ Worse Than Average | ● Average | ○ Better Than Average | ○ Excellent | ⊖ Insufficient Information

©1993 by IntelliChoice, Inc. (408) 554-8711 All Rights Reserved. Reproduction Prohibited.
Refer to *Section 3: Annotated Vehicle Charts* for an explanation of these charts.

Pontiac Bonneville SSEi
4 Door Sedan

Large

 3.8L 205 hp Suprchrged Fuel Inject.
 6 Cylinder "V"
 Automatic 4 Speed
 2 Wheel Front
 Driver/Psngr Airbags Std

Purchase Price

Car Item	Dealer Cost	List
Base Price	$25,764	$29,444
Anti-Lock Brakes	Std	Std
Manual Transmission	N/A	N/A
Optional Engine	N/A	N/A
Auto Climate Control	Std	Std
Power Steering	Std	Std
Cruise Control	Std	Std
All Wheel Drive	N/A	N/A
AM/FM Stereo Cassette	Std	Std
Steering Wheel, Tilt	Std	Std
Power Windows	Std	Std
*Options Price	$0	$0
*Total Price	$25,764	$29,444
Target Price		$28,540
Destination Charge		$555
Avg. Tax & Fees		$1,755
Total Target $		$30,850
Average Dealer Option Cost:		86%

The 1993 Pontiac Bonneville is available in three models - SE, SSE, and SSEi Sedans. New for 1993, the SSEi upgrades the SSE with a passenger side air bag, traction control, leather seating areas, remote keyless entry, Head-Up Display, and automatic electrochromic mirror. The SSEi also offers a power glass sunroof as an optional feature. Standard items include a full-feature anti-theft system, power antenna, a driver information center and a six-way power driver's seat.

Ownership Costs

Cost Area	5 Year Cost	Rate
Depreciation	$18,654	○
Financing ($620/month)	$5,082	
Insurance (Rating 13)	$7,797	○
State Fees	$1,200	
Fuel (Hwy 25 City 17)	$4,145	●
Maintenance	$6,474	●
Repairs	$730	○

Warranty/Maintenance Info

Major Tune-Up	$183	○
Minor Tune-Up	$131	●
Brake Service	$263	○
Overall Warranty	3 yr/36k	○
Drivetrain Warranty	3 yr/36k	○
Rust Warranty	6 yr/100k	○
Maintenance Warranty	N/A	
Roadside Assistance	3 yr/36k	

Ownership Cost By Year

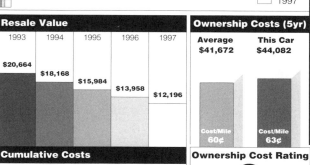

1993 / 1994 / 1995 / 1996 / 1997

Resale Value

1993	1994	1995	1996	1997
$20,664	$18,168	$15,984	$13,958	$12,196

Cumulative Costs

	1993	1994	1995	1996	1997
Annual	$14,858	$6,919	$7,790	$5,678	$8,837
Total	$14,858	$21,777	$29,567	$35,245	$44,082

Ownership Costs (5yr)

Average	This Car
$41,672	$44,082
Cost/Mile 60¢	Cost/Mile 63¢

Ownership Cost Rating

Worse Than Average

Pontiac Firebird
2 Door Coupe

Compact

 3.4L 160 hp Gas Fuel Inject.
 6 Cylinder "V"
 Manual 5 Speed
 2 Wheel Rear
 Driver/Psngr Airbags Std

Trans Am Model Shown

Purchase Price

Car Item	Dealer Cost	List
Base Price	N/R	N/R
Anti-Lock Brakes	Std	Std
Automatic 4 Speed	N/R	N/R
Optional Engine	N/A	N/A
Air Conditioning	N/R	N/R
Power Steering	Std	Std
Cruise Control	N/R	N/R
All Wheel Drive	N/A	N/A
AM/FM Stereo Cassette	Std	Std
Steering Wheel, Tilt	Std	Std
Power Windows	N/R	N/R
*Options Price	$0	$0
*Total Price	N/R	N/R
Target Price		N/R
Destination Charge		N/R
Avg. Tax & Fees		N/R
Total Target $		N/R
Average Dealer Option Cost:		N/A

The 1993 Pontiac Firebird is available in three models-(Base) Firebird, Formula and Trans Am. New for 1993, the Firebird features a completely redesigned body with four new exterior colors. Standard equipment includes aero rear deck spoiler, 2-way manual adjustment with a 6-way power seat as an available option, folding rear seat and an all new analog instrumentation. Safety equipment includes dual airbags, energy-absorbing steering column and interlocking door latches. Pricing was not available at time of printing.

Ownership Costs

Cost Area	5 Year Cost	Rate
Depreciation		⊖
Financing ($0/month)		
Insurance (Rating N/R)		⊖
State Fees		
Fuel (Hwy 28 City 19)	$3,704	○
Maintenance		⊖
Repairs	$709	○

Warranty/Maintenance Info

Major Tune-Up		⊖
Minor Tune-Up		⊖
Brake Service		⊖
Overall Warranty	3 yr/36k	○
Drivetrain Warranty	3 yr/36k	○
Rust Warranty	6 yr/100k	○
Maintenance Warranty	N/A	
Roadside Assistance	3 yr/36k	

Ownership Cost By Year

Insufficient Depreciation Information
Insufficient Financing Information
Insufficient Insurance Information
Insufficient State Fee Information

Insufficient Maintenance Information

1993 / 1994 / 1995 / 1996 / 1997

Resale Value

Insufficient Information

Cumulative Costs

	1993	1994	1995	1996	1997
Annual	Insufficient Information				
Total	Insufficient Information				

Ownership Costs (5yr)

Insufficient Information

Ownership Cost Rating

⊖

Insufficient Information

---

* Includes shaded options
** Other purchase requirements apply

● Poor ◉ Worse Than Average ○ Average ○ Better Than Average ○ Excellent ⊖ Insufficient Information

©1993 by IntelliChoice, Inc. (408) 554-8711 All Rights Reserved. Reproduction Prohibited.
Refer to *Section 3: Annotated Vehicle Charts* for an explanation of these charts.

page 191

Pontiac Firebird Formula
2 Door Coupe

Sport

5.7L 275 hp Gas Fuel Inject. | 8 Cylinder "V" | Manual 6 Speed | 2 Wheel Rear | Driver/Psngr Airbags Std

Purchase Price

Car Item	Dealer Cost	List
Base Price	**N/R**	**N/R**
Anti-Lock Brakes	Std	Std
Automatic 4 Speed	N/R	N/R
Optional Engine	N/A	N/A
Air Conditioning	N/R	N/R
Power Steering	Std	Std
Cruise Control	N/R	N/R
All Wheel Drive	N/A	N/A
AM/FM Stereo Cassette	N/A	N/A
Steering Wheel, Tilt	Std	Std
Power Windows	N/R	N/R
***Options Price**	$0	$0
***Total Price**	**N/R**	**N/R**
Target Price	N/R	
Destination Charge	N/R	
Avg. Tax & Fees	N/R	
Total Target $	**N/R**	
Average Dealer Option Cost:	N/A	

Ownership Costs

Cost Area	5 Year Cost	Rate
Depreciation		⊖
Financing ($0/month)		
Insurance (Rating N/R)		⊖
State Fees		
Fuel (Hwy 25 City 17)	$4,145	○
Maintenance		⊖
Repairs	$750	○

Warranty/Maintenance Info

Major Tune-Up		⊖
Minor Tune-Up		⊖
Brake Service		⊖
Overall Warranty	3 yr/36k	○
Drivetrain Warranty	3 yr/36k	○
Rust Warranty	6 yr/100k	○
Maintenance Warranty	N/A	
Roadside Assistance	3 yr/36k	

Ownership Cost By Year

$1,000 $2,000 $3,000 $4,000 $5,000

Insufficient Depreciation Information
Insufficient Financing Information
Insufficient Insurance Information
Insufficient State Fee Information
Insufficient Maintenance Information

■ 1993
■ 1994
■ 1995
▨ 1996
□ 1997

Resale Value

Insufficient Information

Ownership Costs (5yr)

Insufficient Information

Cumulative Costs

1993 1994 1995 1996 1997

Annual — Insufficient Information
Total — Insufficient Information

Ownership Cost Rating

⊖ Insufficient Information

The 1993 Firebird is available in three models-(Base) Firebird, Formula and Trans Am. New for 1993, the Firebird Formula features a completely redesigned body with four new exterior colors. Standard equipment includes aero rear deck spoiler, 2-way manual adjustment with a 6-way power seat as an available option and sport cast aluminum wheels. Other optional equipment includes power mirrors, bodycolor bodyside moldings and a removable hatch roof. Pricing was not available at time of printing.

Pontiac Firebird Trans Am
2 Door Coupe

Sport

5.7L 275 hp Gas Fuel Inject. | 8 Cylinder "V" | Manual 6 Speed | 2 Wheel Rear | Driver/Psngr Airbags Std

Purchase Price

Car Item	Dealer Cost	List
Base Price	**N/R**	**N/R**
Anti-Lock Brakes	Std	Std
Automatic 4 Speed	N/R	N/R
Optional Engine	N/A	N/A
Air Conditioning	Std	Std
Power Steering	Std	Std
Cruise Control	Std	Std
All Wheel Drive	N/A	N/A
AM/FM Stereo Cassette	Std	Std
Steering Wheel, Tilt	Std	Std
Power Windows	Std	Std
***Options Price**	$0	$0
***Total Price**	**N/R**	**N/R**
Target Price	N/R	
Destination Charge	N/R	
Avg. Tax & Fees	N/R	
Total Target $	**N/R**	
Average Dealer Option Cost:	N/A	

Ownership Costs

Cost Area	5 Year Cost	Rate
Depreciation		⊖
Financing ($0/month)		
Insurance (Rating N/R)		⊖
State Fees		
Fuel (Hwy 25 City 17)	$4,145	○
Maintenance		⊖
Repairs	$750	○

Warranty/Maintenance Info

Major Tune-Up		⊖
Minor Tune-Up		⊖
Brake Service		⊖
Overall Warranty	3 yr/36k	○
Drivetrain Warranty	3 yr/36k	○
Rust Warranty	6 yr/100k	○
Maintenance Warranty	N/A	
Roadside Assistance	3 yr/36k	

Ownership Cost By Year

$1,000 $2,000 $3,000 $4,000 $5,000

Insufficient Depreciation Information
Insufficient Financing Information
Insufficient Insurance Information
Insufficient State Fee Information
Insufficient Maintenance Information

■ 1993
■ 1994
■ 1995
▨ 1996
□ 1997

Resale Value

Insufficient Information

Ownership Costs (5yr)

Insufficient Information

Cumulative Costs

1993 1994 1995 1996 1997

Annual — Insufficient Information
Total — Insufficient Information

Ownership Cost Rating

⊖ Insufficient Information

The 1993 Firebird is available in three models-(Base) Firebird, Formula and Trans Am. New for 1993, the Firebird Trans AM features a completely redesigned body with four new exterior colors. Standard equipment includes a TA-specific rear decklid spoiler, fog lamps, and cloth seats with an adjustable six-way head restraint. Convenience options are limited to compact disc player, removable hatch roof, and sunshade. Pricing was not available at time of printing.

---

page **192**

* Includes shaded options
** Other purchase requirements apply

● Poor   ◐ Worse Than Average   ◑ Average   ◯ Better Than Average   ○ Excellent   ⊖ Insufficient Information

©1993 by IntelliChoice, Inc. (408) 554-8711   All Rights Reserved. Reproduction Prohibited.

Refer to *Section 3: Annotated Vehicle Charts* for an explanation of these charts.

Pontiac Grand Am SE
2 Door Coupe

 2.3L 115 hp Gas Fuel Inject. | 4 Cylinder In-Line | Manual 5 Speed | 2 Wheel Front | Automatic Seatbelts

1992 Model Shown — Compact

Purchase Price

Car Item	Dealer Cost	List
Base Price	**$11,334**	**$12,524**
Anti-Lock Brakes	Std	Std
Automatic 3 Speed	$477	$555
3.3L 160 hp Gas	$396	** $460
Air Conditioning	$714	$830
Power Steering	Std	Std
Cruise Control	$194	$225
All Wheel Drive	N/A	N/A
AM/FM Stereo Cassette	$120	$140
Steering Wheel, Tilt	$125	$145
Power Windows	$237	** $275
***Options Price**	**$1,311**	**$1,525**
***Total Price**	**$12,645**	**$14,049**
Target Price	$13,462	
Destination Charge	$475	
Avg. Tax & Fees	$842	
Total Target $	**$14,779**	
Average Dealer Option Cost:	**86%**	

Ownership Costs

Cost Area	5 Year Cost	Rate
Depreciation	$7,860	O
Financing ($297/month)	$2,435	
Insurance (Rating 10)	$7,479	◉
State Fees	$581	
Fuel (Hwy 32 City 22)	$3,221	O
Maintenance	$4,036	O
Repairs	$651	O

Warranty/Maintenance Info

Major Tune-Up	$137	○
Minor Tune-Up	$85	O
Brake Service	$214	O
Overall Warranty	3 yr/36k	O
Drivetrain Warranty	3 yr/36k	O
Rust Warranty	6 yr/100k	O
Maintenance Warranty	N/A	
Roadside Assistance	3 yr/36k	

Ownership Cost By Year
(1993–1997 bar chart, scale $2,000–$8,000)

Resale Value
1993	1994	1995	1996	1997
$10,208	$9,315	$8,551	$7,754	$6,919

Cumulative Costs
	1993	1994	1995	1996	1997
Annual	$7,753	$4,058	$4,628	$3,820	$6,004
Total	$7,753	$11,811	$16,439	$20,259	$26,263

Ownership Costs (5yr)
Average	This Car
$26,557	$26,263
Cost/Mile 38¢	Cost/Mile 38¢

Ownership Cost Rating

Average

The 1993 Pontiac Grand Am is available in four models - SE Coupe, SE Sedan, GT Coupe, and GT Sedan. New for 1993, the SE Coupe comes standard with a tuned absorber on the engine mount, and a redesigned exhaust system, both reducing vibration and noise. Other standard features include anti-lock brakes, an acoustical insulation package, wraparound headlamps, and an extensive anti-corrosion protection. The Grand Am has been Pontiac's sales volume leader since 1985.

Pontiac Grand Am GT
2 Door Coupe

 2.3L 175 hp Gas Fuel Inject. | 4 Cylinder In-Line | Manual 5 Speed | 2 Wheel Front | Automatic Seatbelts

Compact

Purchase Price

Car Item	Dealer Cost	List
Base Price	**$12,601**	**$13,924**
Anti-Lock Brakes	Std	Std
Automatic 3 Speed	$477	** $555
2.3L 160 hp Gas	($120)	** ($140)
Air Conditioning	$714	$830
Power Steering	Std	Std
Cruise Control	$194	$225
All Wheel Drive	N/A	N/A
AM/FM Stereo Cassette	$120	$140
Steering Wheel, Tilt	$125	$145
Power Windows	$237	** $275
***Options Price**	**$834**	**$970**
***Total Price**	**$13,435**	**$14,894**
Target Price	$14,337	
Destination Charge	$475	
Avg. Tax & Fees	$895	
Total Target $	**$15,707**	
Average Dealer Option Cost:	**86%**	

Ownership Costs

Cost Area	5 Year Cost	Rate
Depreciation	$8,480	O
Financing ($316/month)	$2,588	
Insurance (Rating 11+)	$9,232	◉
State Fees	$615	
Fuel (Hwy 30 City 21)	$3,405	O
Maintenance	$4,352	O
Repairs	$651	O

Warranty/Maintenance Info

Major Tune-Up	$137	○
Minor Tune-Up	$85	O
Brake Service	$214	O
Overall Warranty	3 yr/36k	O
Drivetrain Warranty	3 yr/36k	O
Rust Warranty	6 yr/100k	O
Maintenance Warranty	N/A	
Roadside Assistance	3 yr/36k	

Ownership Cost By Year
(1993–1997 bar chart, scale $2,000–$10,000)

Resale Value
1993	1994	1995	1996	1997
$11,592	$10,361	$9,321	$8,269	$7,227

Cumulative Costs
	1993	1994	1995	1996	1997
Annual	$7,727	$4,823	$5,479	$4,495	$6,798
Total	$7,727	$12,550	$18,029	$22,524	$29,322

Ownership Costs (5yr)
Average	This Car
$27,283	$29,322
Cost/Mile 39¢	Cost/Mile 42¢

Ownership Cost Rating

Worse Than Average

The 1993 Pontiac Grand Am is available in four models - SE Coupe, SE Sedan, GT Coupe, and GT Sedan. The GT Coupe upgrades the SE edition by adding front, side and rear aero-body skirts, dual chrome-tipped exhaust outlets and neutral density wraparound taillamps. Other standard features include an acoustical insulation package, analog gauges, automatic transmission shift interlock, and an extensive anti-corrosion protection. The Grand Am has been Pontiac's sales volume leader since 1985.

---

* Includes shaded options
** Other purchase requirements apply

● Poor | ◉ Worse Than Average | ◎ Average | ○ Better Than Average | ○ Excellent | ⊖ Insufficient Information

©1993 by IntelliChoice, Inc. (408) 554-8711 All Rights Reserved. Reproduction Prohibited.
Refer to *Section 3: Annotated Vehicle Charts* for an explanation of these charts.

Pontiac Grand Am SE
4 Door Sedan

Compact

 2.3L 115 hp Gas Fuel Inject.
 4 Cylinder In-Line
 Manual 5 Speed
 2 Wheel Front
Automatic Seatbelts

Purchase Price

Car Item	Dealer Cost	List
Base Price	**$11,425**	**$12,624**
Anti-Lock Brakes	Std	Std
Automatic 3 Speed	$477	$555
3.3L 160 hp Gas	$396	** $460
Air Conditioning	$714	$830
Power Steering	Std	Std
Cruise Control	$194	$225
All Wheel Drive	N/A	N/A
AM/FM Stereo Cassette	$120	$140
Steering Wheel, Tilt	$125	$145
Power Windows	$292	** $340
*Options Price	$1,311	$1,525
*Total Price	$12,736	$14,149
Target Price		$13,561
Destination Charge		$475
Avg. Tax & Fees		$848
Total Target $		**$14,884**
Average Dealer Option Cost:		87%

Ownership Costs

Cost Area	5 Year Cost	Rate
Depreciation	$8,077	O
Financing ($299/month)	$2,451	
Insurance (Rating 6)	$6,919	O
State Fees	$585	
Fuel (Hwy 32 City 22)	$3,221	O
Maintenance	$4,036	O
Repairs	$651	O

Warranty/Maintenance Info

Major Tune-Up	$137	O
Minor Tune-Up	$85	O
Brake Service	$214	O
Overall Warranty	3 yr/36k	O
Drivetrain Warranty	3 yr/36k	O
Rust Warranty	6 yr/100k	O
Maintenance Warranty	N/A	
Roadside Assistance	3 yr/36k	

Ownership Cost By Year

1993 / 1994 / 1995 / 1996 / 1997

Resale Value

1993	1994	1995	1996	1997
$10,248	$9,291	$8,507	$7,662	$6,807

Cumulative Costs

	1993	1994	1995	1996	1997
Annual	$7,722	$4,021	$4,540	$3,753	$5,904
Total	$7,722	$11,743	$16,283	$20,036	$25,940

Ownership Costs (5yr)

Average	This Car
$26,643	$25,940
Cost/Mile 38¢	Cost/Mile 37¢

Ownership Cost Rating
Better Than Average

The 1993 Pontiac Grand Am is available in four models - SE Coupe, SE Sedan, GT Coupe, and GT Sedan. New for 1993, the SE Sedan comes standard with a redesigned exhaust system, and a tuned absorber on the engine mount, both reducing vibration and noise. Other standard features include wraparound headlamps, anti-lock brakes, an acoustical insulation package, and an extensive anti-corrosion protection. The Grand Am has been Pontiac's sales volume leader since 1985.

Pontiac Grand Am GT
4 Door Sedan

Compact

 2.3L 175 hp Gas Fuel Inject.
 4 Cylinder In-Line
 Manual 5 Speed
 2 Wheel Front
Automatic Seatbelts

Purchase Price

Car Item	Dealer Cost	List
Base Price	**$12,692**	**$14,024**
Anti-Lock Brakes	Std	Std
Automatic 3 Speed	$477	** $555
2.3L 160 hp Gas	($120)	** ($140)
Air Conditioning	$714	$830
Power Steering	Std	Std
Cruise Control	$194	$225
All Wheel Drive	N/A	N/A
AM/FM Stereo Cassette	$120	$140
Steering Wheel, Tilt	$125	$145
Power Windows	$292	** $340
*Options Price	$834	$970
*Total Price	$13,526	$14,994
Target Price		$14,437
Destination Charge		$475
Avg. Tax & Fees		$901
Total Target $		**$15,813**
Average Dealer Option Cost:		86%

Ownership Costs

Cost Area	5 Year Cost	Rate
Depreciation	$8,676	O
Financing ($318/month)	$2,604	
Insurance (Rating 7)	$7,035	O
State Fees	$619	
Fuel (Hwy 30 City 21)	$3,405	O
Maintenance	$4,339	O
Repairs	$651	O

Warranty/Maintenance Info

Major Tune-Up	$137	O
Minor Tune-Up	$85	O
Brake Service	$214	O
Overall Warranty	3 yr/36k	O
Drivetrain Warranty	3 yr/36k	O
Rust Warranty	6 yr/100k	O
Maintenance Warranty	N/A	
Roadside Assistance	3 yr/36k	

Resale Value

1993	1994	1995	1996	1997
$11,589	$10,335	$9,273	$8,199	$7,137

Cumulative Costs

	1993	1994	1995	1996	1997
Annual	$7,439	$4,431	$5,061	$4,061	$6,337
Total	$7,439	$11,870	$16,931	$20,992	$27,329

Ownership Costs (5yr)

Average	This Car
$27,369	$27,329
Cost/Mile 39¢	Cost/Mile 39¢

Ownership Cost Rating
Average

The 1993 Pontiac Grand Am is available in four models - SE Coupe, SE Sedan, GT Coupe, and GT Sedan. The GT Sedan upgrades the SE edition by adding front, side and rear aero-body skirts, neutral density wraparound taillamps, composite wraparound headlamps and dual chrome-tipped exhaust outlets. Other standard features include an automatic transmission shift interlock, analog gauges, and an acoustical insulation package. The Grand Am has been Pontiac's sales volume leader since 1985.

---

* Includes shaded options
** Other purchase requirements apply

● Poor ◐ Worse Than Average ○ Average ○ Better Than Average ○ Excellent ⊖ Insufficient Information

©1993 by IntelliChoice, Inc. (408) 554-8711 All Rights Reserved. Reproduction Prohibited.
Refer to *Section 3: Annotated Vehicle Charts* for an explanation of these charts.

Pontiac Grand Prix SE
2 Door Coupe

 3.1L 140 hp Gas Fuel Inject.
 6 Cylinder "V"
 Automatic 3 Speed
 2 Wheel Front
 Automatic Seatbelts

Midsize

Purchase Price

Car Item	Dealer Cost	List
Base Price	**$13,466**	**$15,390**
Anti-Lock Brakes	$387	$450
Manual 5 Speed	N/C	** N/C
3.4L 210 hp Gas	$1,041	** $1,210
Air Conditioning	Std	Std
Power Steering	Std	Std
Cruise Control	$194	$225
All Wheel Drive	N/A	N/A
AM/FM Stereo Cassette	$120	$140
Steering Wheel, Tilt	$125	$145
Power Windows	$237	$275
***Options Price**	**$551**	**$640**
***Total Price**	**$14,017**	**$16,030**
Target Price		$15,127
Destination Charge		$505
Avg. Tax & Fees		$947
Total Target $		**$16,579**
Average Dealer Option Cost:		**87%**

The 1993 Pontiac Grand Prix is available in five models - SE and GT Coupe, SE, LE, and STE Sedan. New for 1993, the SE Coupe has made optional a Cellular Telephone wiring package, an Aero Performance Package and a Sport Appearance Package. Standard features include acoustical insulation, split bench seats, analog gauges, and an extensive anti-corrosion protection. Additional optional items are the Generation II Head-Up Display, and a Cold Climate Package that includes a rear defogger.

Ownership Costs

Cost Area	5 Year Cost	Rate
Depreciation	$9,486	●
Financing ($333/month)	$2,731	
Insurance (Rating 7)	$7,035	○
State Fees	$661	
Fuel (Hwy 27 City 19)	$3,775	◐
Maintenance	$4,954	◐
Repairs	$709	◐

Warranty/Maintenance Info

Major Tune-Up	$183	◐
Minor Tune-Up	$134	◐
Brake Service	$281	●
Overall Warranty	3 yr/36k	◐
Drivetrain Warranty	3 yr/36k	◐
Rust Warranty	6 yr/100k	○
Maintenance Warranty	N/A	
Roadside Assistance	3 yr/36k	

Ownership Cost By Year

1993, 1994, 1995, 1996, 1997

Resale Value

1993	1994	1995	1996	1997
$12,025	$10,548	$9,405	$8,227	$7,093

Cumulative Costs

	1993	1994	1995	1996	1997
Annual	$7,903	$4,831	$5,320	$4,342	$6,955
Total	$7,903	$12,734	$18,054	$22,396	$29,351

Ownership Costs (5yr)

Average	This Car
$27,659	$29,351
Cost/Mile 40¢	Cost/Mile 42¢

Ownership Cost Rating

Worse Than Average

Pontiac Grand Prix GT
2 Door Coupe

 3.1L 140 hp Gas Fuel Inject.
 6 Cylinder "V"
 Automatic 4 Speed
 2 Wheel Front
 Automatic Seatbelts

Midsize

SE Model Shown

Purchase Price

Car Item	Dealer Cost	List
Base Price	**$17,798**	**$20,340**
Anti-Lock Brakes	Std	Std
Manual 5 Speed	($172)	** ($200)
3.4L 210 hp Gas	$920	$1,070
Air Conditioning	Std	Std
Power Steering	Std	Std
Cruise Control	Std	Std
All Wheel Drive	N/A	N/A
AM/FM Stereo Cassette	Std	Std
Steering Wheel, Tilt	Std	Std
Power Windows	Std	Std
***Options Price**	**$0**	**$0**
***Total Price**	**$17,798**	**$20,340**
Target Price		$19,387
Destination Charge		$505
Avg. Tax & Fees		$1,203
Total Target $		**$21,095**
Average Dealer Option Cost:		**87%**

The 1993 Pontiac Grand Prix is available in five models - SE and GT Coupe, SE, LE, and STE Sedan. New for 1993, the GT Coupe upgrades the SE Coupe by adding lower aero ground effects, front and rear bucket seats, front overhead console with storage and reading lamps and a 6-way power driver's seat. Other optional features include a power glass sunroof, Generation II Head-Up Display, remote keyless entry, and a Cold Climate Package that includes a rear window defogger.

Ownership Costs

Cost Area	5 Year Cost	Rate
Depreciation	$12,989	●
Financing ($424/month)	$3,474	
Insurance (Rating 11)	$7,693	○
State Fees	$834	
Fuel (Hwy 29 City 19)	$3,640	◐
Maintenance	$5,119	◐
Repairs	$709	◐

Warranty/Maintenance Info

Major Tune-Up	$183	◐
Minor Tune-Up	$134	◐
Brake Service	$281	●
Overall Warranty	3 yr/36k	◐
Drivetrain Warranty	3 yr/36k	◐
Rust Warranty	6 yr/100k	○
Maintenance Warranty	N/A	
Roadside Assistance	3 yr/36k	

Resale Value

1993	1994	1995	1996	1997
$14,410	$12,504	$10,998	$9,507	$8,106

Cumulative Costs

	1993	1994	1995	1996	1997
Annual	$10,482	$5,631	$6,044	$4,848	$7,453
Total	$10,482	$16,113	$22,157	$27,005	$34,458

Ownership Costs (5yr)

Average	This Car
$32,066	$34,458
Cost/Mile 46¢	Cost/Mile 49¢

Ownership Cost Rating

● Poor

---

* Includes shaded options
** Other purchase requirements apply

● Poor   ◐ Worse Than Average   ○ Average   ○ Better Than Average   ○ Excellent   ⊖ Insufficient Information

©1993 by IntelliChoice, Inc. (408) 554-8711   All Rights Reserved. Reproduction Prohibited.
Refer to *Section 3: Annotated Vehicle Charts* for an explanation of these charts.

Pontiac Grand Prix LE
4 Door Sedan

 3.1L 140 hp Gas Fuel Inject.
 6 Cylinder "V"
 Automatic 3 Speed
 2 Wheel Front
 Automatic Seatbelts

Midsize

Purchase Price

Car Item	Dealer Cost	List
Base Price	**$13,327**	**$14,890**
Anti-Lock Brakes	$387	$450
4 Spd Auto	$172	$200
3.4L 210 hp Gas	$1,041	** $1,210
Air Conditioning	Std	Std
Power Steering	Std	Std
Cruise Control	$194	$225
All Wheel Drive	N/A	N/A
AM/FM Stereo Cassette	$120	$140
Steering Wheel, Tilt	$125	$145
Power Windows	$292	$340
*Options Price	$606	$705
*Total Price	$13,933	$15,595
Target Price		$15,032
Destination Charge		$505
Avg. Tax & Fees		$938
Total Target $		**$16,475**
Average Dealer Option Cost:	**86%**	

The 1993 Pontiac Grand Prix is available in five models - SE and GT Coupe, SE, LE, and STE Sedan. New for 1993, the LE Sedan offers the BYP Sport Appearance Package including lower aero ground effects, bucket seats with console and a rally gauge cluster. Standard features include bench seats, folding armrests and acoustical insulation. As options the LE offers the Generation II Head-Up Display and a Cold Climate Package that includes a rear window defogger.

Ownership Costs

Cost Area	5 Year Cost	Rate
Depreciation	$9,736	●
Financing ($331/month)	$2,713	
Insurance (Rating 4)	$6,658	○
State Fees	$645	
Fuel (Hwy 27 City 19)	$3,775	○
Maintenance	$4,954	○
Repairs	$709	○

Warranty/Maintenance Info

Major Tune-Up	$183	○
Minor Tune-Up	$134	○
Brake Service	$281	●
Overall Warranty	3 yr/36k	○
Drivetrain Warranty	3 yr/36k	○
Rust Warranty	6 yr/100k	○
Maintenance Warranty	N/A	
Roadside Assistance	3 yr/36k	

Ownership Cost By Year

$2,000 $4,000 $6,000 $8,000 $10,000

■ 1993
■ 1994
▨ 1995
▨ 1996
□ 1997

Resale Value

1993	1994	1995	1996	1997
$11,731	$10,208	$8,999	$7,833	$6,739

Cumulative Costs

	1993	1994	1995	1996	1997
Annual	$8,011	$4,794	$5,305	$4,248	$6,832
Total	$8,011	$12,805	$18,110	$22,358	$29,190

Ownership Costs (5yr)

Average	This Car
$27,214	$29,190
Cost/Mile 39¢	Cost/Mile 42¢

Ownership Cost Rating

● Poor

Pontiac Grand Prix SE
4 Door Sedan

 3.1L 140 hp Gas Fuel Inject.
 6 Cylinder "V"
 Automatic 3 Speed
 2 Wheel Front
 Automatic Seatbelts

Midsize

LE Model Shown

Purchase Price

Car Item	Dealer Cost	List
Base Price	**$14,166**	**$16,190**
Anti-Lock Brakes	$387	$450
4 Spd Auto	$172	$200
3.4L 210 hp Gas	$1,041	** $1,210
Air Conditioning	Std	Std
Power Steering	Std	Std
Cruise Control	$194	$225
All Wheel Drive	N/A	N/A
AM/FM Stereo Cassette	$120	$140
Steering Wheel, Tilt	$125	$145
Power Windows	$292	$340
*Options Price	$606	$705
*Total Price	$14,772	$16,895
Target Price		$15,966
Destination Charge		$505
Avg. Tax & Fees		$998
Total Target $		**$17,469**
Average Dealer Option Cost:	**86%**	

The 1993 Pontiac Grand Prix is available in five models - SE and GT Coupe, SE, LE, and STE Sedan. New for 1993, the SE Sedan offers the BYP Sport Appearance Package including lower aero ground effects, bucket seats with console and a rally gauge cluster. Standard features include flush-fitting tinted glass, front split bench seat, and a three-passenger rear bench seat with headrests. As options the SE offers a power glass sunroof, and a Cold Climate Package that includes a rear window defogger.

Ownership Costs

Cost Area	5 Year Cost	Rate
Depreciation	$10,132	●
Financing ($351/month)	$2,878	
Insurance (Rating 5)	$6,786	○
State Fees	$696	
Fuel (Hwy 27 City 19)	$3,775	○
Maintenance	$5,071	○
Repairs	$709	○

Warranty/Maintenance Info

Major Tune-Up	$183	○
Minor Tune-Up	$134	○
Brake Service	$281	●
Overall Warranty	3 yr/36k	○
Drivetrain Warranty	3 yr/36k	○
Rust Warranty	6 yr/100k	○
Maintenance Warranty	N/A	
Roadside Assistance	3 yr/36k	

Ownership Cost By Year

$2,000 $4,000 $6,000 $8,000 $10,000 $12,000

■ 1993
■ 1994
▨ 1995
▨ 1996
□ 1997

Resale Value

1993	1994	1995	1996	1997
$12,729	$11,112	$9,823	$8,555	$7,337

Cumulative Costs

	1993	1994	1995	1996	1997
Annual	$8,113	$4,976	$5,509	$4,397	$7,052
Total	$8,113	$13,089	$18,598	$22,995	$30,047

Ownership Costs (5yr)

Average	This Car
$28,544	$30,047
Cost/Mile 41¢	Cost/Mile 43¢

Ownership Cost Rating

◐ Worse Than Average

---

page 196

* Includes shaded options
** Other purchase requirements apply

● Poor   ◐ Worse Than Average   ○ Average   ◯ Better Than Average   ⬤ Excellent   ⊖ Insufficient Information

©1993 by IntelliChoice, Inc. (408) 554-8711   All Rights Reserved. Reproduction Prohibited.
Refer to *Section 3: Annotated Vehicle Charts* for an explanation of these charts.

Pontiac Grand Prix STE
4 Door Sedan

 3.1L 140 hp Gas Fuel Inject.
 6 Cylinder "V"
 Automatic 4 Speed
 2 Wheel Front
 Automatic Seatbelts

Midsize — LE Model Shown

Purchase Price

Car Item	Dealer Cost	List
Base Price	**$18,931**	**$21,635**
Anti-Lock Brakes	Std	Std
Manual 5 Speed	($194)	** ($225)
3.4L 210 hp Gas	$920	$1,070
Air Conditioning	Std	Std
Power Steering	Std	Std
Cruise Control	Std	Std
All Wheel Drive	N/A	N/A
AM/FM Stereo Cassette	Std	Std
Steering Wheel, Tilt	Std	Std
Power Windows	Std	Std
*Options Price	$0	$0
*Total Price	$18,931	$21,635
Target Price	$20,671	
Destination Charge	$505	
Avg. Tax & Fees	$1,280	
Total Target $	**$22,456**	
Average Dealer Option Cost:	86%	

Ownership Costs

Cost Area	5 Year Cost	Rate
Depreciation	$13,433	●
Financing ($451/month)	$3,699	
Insurance (Rating 8)	$7,223	○
State Fees	$885	
Fuel (Hwy 29 City 19)	$3,640	○
Maintenance	$5,119	○
Repairs	$709	○

Warranty/Maintenance Info

Major Tune-Up	$183	○
Minor Tune-Up	$134	○
Brake Service	$281	●
Overall Warranty	3 yr/36k	○
Drivetrain Warranty	3 yr/36k	○
Rust Warranty	6 yr/100k	○
Maintenance Warranty	N/A	
Roadside Assistance	3 yr/36k	

Ownership Cost By Year

Year	
1993	
1994	
1995	
1996	
1997	

Resale Value

1993	1994	1995	1996	1997
$15,601	$13,719	$11,985	$10,454	$9,023

Cumulative Costs

	1993	1994	1995	1996	1997
Annual	$10,673	$5,598	$6,232	$4,816	$7,389
Total	$10,673	$16,271	$22,503	$27,319	$34,708

Ownership Costs (5yr)

Average	This Car
$33,390	$34,708
Cost/Mile 48¢	Cost/Mile 50¢

Ownership Cost Rating: Worse Than Average

The 1993 Pontiac Grand Prix is available in five models - SE and GT Coupe, SE, LE, and STE Sedan. New for 1993, the STE Sedan upgrades the SE Sedan with power reclining front bucket seats, front floor storage console, electronic compass with trip computer, front overhead console with storage, illuminated entry and a rear window defogger. The exterior adds bodycolor lower aero extensions, bodycolor sport power mirrors and a power antenna finished in black.

Pontiac LeMans Value Leader Aerocoupe
2 Door Hatchback

 1.6L 74 hp Gas Fuel Inject.
 4 Cylinder In-Line
 Manual 4 Speed
 2 Wheel Front
Automatic Seatbelts

Compact — SE Model Shown

Purchase Price

Car Item	Dealer Cost	List
Base Price	**$7,624**	**$8,154**
Anti-Lock Brakes	N/A	N/A
Automatic Transmission	N/A	N/A
Optional Engine	N/A	N/A
Air Conditioning	N/A	N/A
Power Steering	N/A	N/A
Cruise Control	N/A	N/A
All Wheel Drive	N/A	N/A
AM/FM Stereo Cassette	$369	$429
Steering Wheel, Tilt	N/A	N/A
Power Windows	N/A	N/A
*Options Price	$369	$429
*Total Price	$7,993	$8,583
Target Price	$8,221	
Destination Charge	$345	
Avg. Tax & Fees	$517	
Total Target $	**$9,083**	
Average Dealer Option Cost:	86%	

Ownership Costs

Cost Area	5 Year Cost	Rate
Depreciation	$6,212	●
Financing ($183/month)	$1,497	
Insurance (Rating 13)	$8,061	●
State Fees	$356	
Fuel (Hwy 37 City 28)	$2,659	○
Maintenance	$4,366	○
Repairs	$659	○

Warranty/Maintenance Info

Major Tune-Up	$151	○
Minor Tune-Up	$81	○
Brake Service	$264	●
Overall Warranty	3 yr/36k	○
Drivetrain Warranty	3 yr/36k	○
Rust Warranty	6 yr/100k	○
Maintenance Warranty	N/A	
Roadside Assistance	3 yr/36k	

Resale Value

1993	1994	1995	1996	1997
$5,710	$4,858	$4,203	$3,558	$2,871

Cumulative Costs

	1993	1994	1995	1996	1997
Annual	$6,111	$3,811	$4,458	$4,073	$5,357
Total	$6,111	$9,922	$14,380	$18,453	$23,810

Ownership Costs (5yr)

Average	This Car
$21,864	$23,810
Cost/Mile 31¢	Cost/Mile 34¢

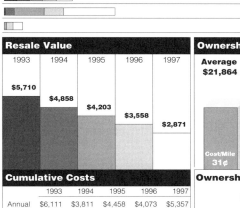
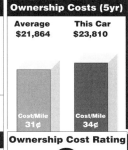

Ownership Cost Rating: Worse Than Average

The 1993 Pontiac LeMans is available in three models - Value Leader Aerocoupe, SE Aerocoupe, and SE Sedan. New for 1993, the LeMans Value Leader Aerocoupe has a new front and rear appearance along with high-gloss black bodyside moldings. Safety features include an energy-absorbing steering column, laminated safety glass windshield, electric rear window defogger and passenger-guard side door intrusion beams. The LeMans interiors have also been upgraded for a more sporty look.

Includes shaded options
** Other purchase requirements apply

● Poor ◐ Worse Than Average ○ Average ○ Better Than Average ○ Excellent ⊖ Insufficient Information

©1993 by IntelliChoice, Inc. (408) 554-8711 All Rights Reserved. Reproduction Prohibited.
Refer to *Section 3: Annotated Vehicle Charts* for an explanation of these charts.

Pontiac LeMans SE Aerocoupe
2 Door Hatchback
Compact

- 1.6L 74 hp Gas Fuel Inject.
- 4 Cylinder In-Line
- Manual 5 Speed
- 2 Wheel Front
- Automatic Seatbelts

Purchase Price

Car Item	Dealer Cost	List
Base Price	**$8,438**	**$9,054**
Anti-Lock Brakes	N/A	N/A
Automatic 3 Speed	$409	$475
Optional Engine	N/A	N/A
Air Conditioning	$606	$705
Power Steering	$194	** $225
Cruise Control	N/A	N/A
All Wheel Drive	N/A	N/A
AM/FM Stereo Cassette	$105	$122
Steering Wheel, Tilt	N/A	N/A
Power Windows	N/A	N/A
*Options Price	$1,120	$1,302
*Total Price	$9,558	$10,356
Target Price		$9,839
Destination Charge		$345
Avg. Tax & Fees		$616
Total Target $		**$10,800**
Average Dealer Option Cost:		*86%*

Ownership Costs

Cost Area	5 Year Cost	Rate
Depreciation	$7,185	●
Financing ($217/month)	$1,779	
Insurance (Rating 13)	$8,061	●
State Fees	$428	◐
Fuel (Hwy 32 City 26)	$2,977	○
Maintenance	$4,600	○
Repairs	$659	○

Warranty/Maintenance Info

Major Tune-Up	$151	○
Minor Tune-Up	$81	◐
Brake Service	$264	●
Overall Warranty	3 yr/36k	○
Drivetrain Warranty	3 yr/36k	○
Rust Warranty	6 yr/100k	○
Maintenance Warranty	N/A	
Roadside Assistance	3 yr/36k	

Ownership Cost By Year

Resale Value
1993: $6,942 | 1994: $5,939 | 1995: $5,163 | 1996: $4,410 | 1997: $3,615

Ownership Costs (5yr)
Average $23,386 | This Car $25,689
Cost/Mile 33¢ | Cost/Mile 37¢

Cumulative Costs

	1993	1994	1995	1996	1997
Annual	$6,791	$4,127	$4,825	$4,280	$5,666
Total	$6,791	$10,918	$15,743	$20,023	$25,689

Ownership Cost Rating
● Poor

The 1993 Pontiac LeMans is available in three models - Value Leader Aerocoupe, SE Aerocoupe, and SE Sedan. New for 1993, the LeMans SE Aerocoupe has a new front and rear appearance, a bodycolor rear hatch spoiler, bodycolor fascias and soft-ray tinted glass. Interior features include a 3-spoke sport steering wheel, door map pockets and fully reclining front seats with articulating headrests. For safety, an energy-absorbing steering column and laminated safety glass windshield are standard.

Pontiac LeMans SE
4 Door Sedan
Compact

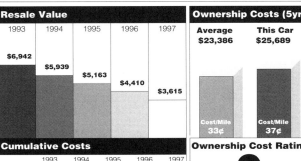

- 1.6L 74 hp Gas Fuel Inject.
- 4 Cylinder In-Line
- Manual 5 Speed
- 2 Wheel Front
- Automatic Seatbelts

Purchase Price

Car Item	Dealer Cost	List
Base Price	**$9,184**	**$9,854**
Anti-Lock Brakes	N/A	N/A
Automatic 3 Speed	$409	$475
Optional Engine	N/A	N/A
Air Conditioning	$606	$705
Power Steering	$194	** $225
Cruise Control	N/A	N/A
All Wheel Drive	N/A	N/A
AM/FM Stereo Cassette	$105	$122
Steering Wheel, Tilt	N/A	N/A
Power Windows	N/A	N/A
*Options Price	$1,120	$1,302
*Total Price	$10,304	$11,156
Target Price		$11,018
Destination Charge		$345
Avg. Tax & Fees		$683
Total Target $		**$12,046**
Average Dealer Option Cost:		*86%*

Ownership Costs

Cost Area	5 Year Cost	Rate
Depreciation	$8,190	●
Financing ($242/month)	$1,984	
Insurance (Rating 11)	$7,693	●
State Fees	$461	◐
Fuel (Hwy 32 City 26)	$2,977	○
Maintenance	$4,600	○
Repairs	$659	○

Warranty/Maintenance Info

Major Tune-Up	$151	○
Minor Tune-Up	$81	◐
Brake Service	$264	●
Overall Warranty	3 yr/36k	○
Drivetrain Warranty	3 yr/36k	○
Rust Warranty	6 yr/100k	○
Maintenance Warranty	N/A	
Roadside Assistance	3 yr/36k	

Resale Value
1993: $7,391 | 1994: $6,332 | 1995: $5,509 | 1996: $4,696 | 1997: $3,856

Ownership Costs (5yr)
Average $24,073 | This Car $26,564
Cost/Mile 34¢ | Cost/Mile 38¢

Cumulative Costs

	1993	1994	1995	1996	1997
Annual	$7,614	$4,184	$4,844	$4,285	$5,637
Total	$7,614	$11,798	$16,642	$20,927	$26,564

Ownership Cost Rating
● Poor

The 1993 Pontiac LeMans is available in three models - Value Leader Aerocoupe, SE Aerocoupe, and SE Sedan. New for 1993, the LeMans SE Sedan has a new front and rear appearance, rear combination lamps and decklid applique, rear quarter panels and soft-ray tinted glass. Safety features include an electric rear window defogger, child security door locks, laminated safety glass windshield and an energy-absorbing steering column. The SE also offers a removable sunroof and floor mats.

---

* Includes shaded options
** Other purchase requirements apply

● Poor ◐ Worse Than Average ○ Average ○ Better Than Average ○ Excellent ⊖ Insufficient Information

©1993 by IntelliChoice, Inc. (408) 554-8711 All Rights Reserved. Reproduction Prohibited.
Refer to *Section 3: Annotated Vehicle Charts* for an explanation of these charts.

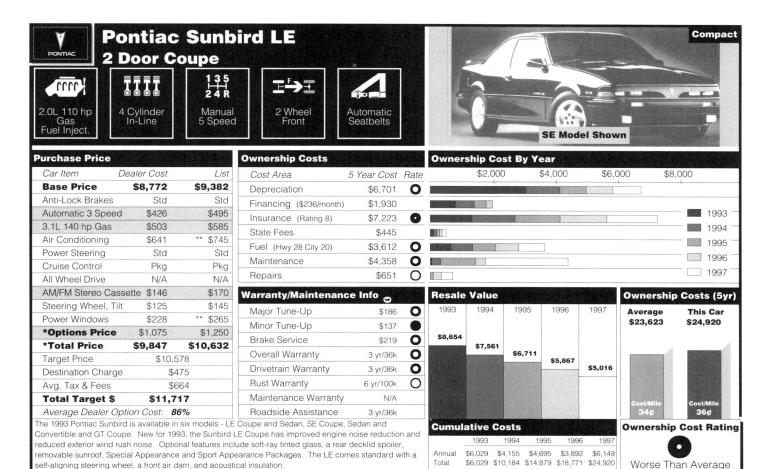

Pontiac Sunbird LE
2 Door Coupe

Compact

2.0L 110 hp Gas Fuel Inject. | 4 Cylinder In-Line | Manual 5 Speed | 2 Wheel Front | Automatic Seatbelts

SE Model Shown

Purchase Price

Car Item	Dealer Cost	List
Base Price	**$8,772**	**$9,382**
Anti-Lock Brakes	Std	Std
Automatic 3 Speed	$426	$495
3.1L 140 hp Gas	$503	$585
Air Conditioning	$641	** $745
Power Steering	Std	Std
Cruise Control	Pkg	Pkg
All Wheel Drive	N/A	N/A
AM/FM Stereo Cassette	$146	$170
Steering Wheel, Tilt	$125	$145
Power Windows	$228	** $265
***Options Price**	**$1,075**	**$1,250**
***Total Price**	**$9,847**	**$10,632**
Target Price		$10,578
Destination Charge		$475
Avg. Tax & Fees		$664
Total Target $		**$11,717**
Average Dealer Option Cost:		*86%*

Ownership Costs

Cost Area	5 Year Cost	Rate
Depreciation	$6,701	○
Financing ($236/month)	$1,930	
Insurance (Rating 8)	$7,223	◉
State Fees	$445	
Fuel (Hwy 28 City 20)	$3,612	○
Maintenance	$4,358	○
Repairs	$651	○

Warranty/Maintenance Info

Major Tune-Up	$186	○
Minor Tune-Up	$137	●
Brake Service	$219	○
Overall Warranty	3 yr/36k	○
Drivetrain Warranty	3 yr/36k	○
Rust Warranty	6 yr/100k	○
Maintenance Warranty	N/A	
Roadside Assistance	3 yr/36k	

Ownership Cost By Year

1993, 1994, 1995, 1996, 1997

Resale Value

1993	1994	1995	1996	1997
$8,654	$7,561	$6,711	$5,867	$5,016

Cumulative Costs

	1993	1994	1995	1996	1997
Annual	$6,029	$4,155	$4,695	$3,892	$6,149
Total	$6,029	$10,184	$14,879	$18,771	$24,920

Ownership Costs (5yr)

Average $23,623 | This Car $24,920
Cost/Mile 34¢ | Cost/Mile 36¢

Ownership Cost Rating

Worse Than Average

The 1993 Pontiac Sunbird is available in six models - LE Coupe and Sedan, SE Coupe, Sedan and Convertible and GT Coupe. New for 1993, the Sunbird LE Coupe has improved engine noise reduction and reduced exterior wind rush noise. Optional features include soft-ray tinted glass, a rear decklid spoiler, removable sunroof, Special Appearance and Sport Appearance Packages. The LE comes standard with a self-aligning steering wheel, a front air dam, and acoustical insulation.

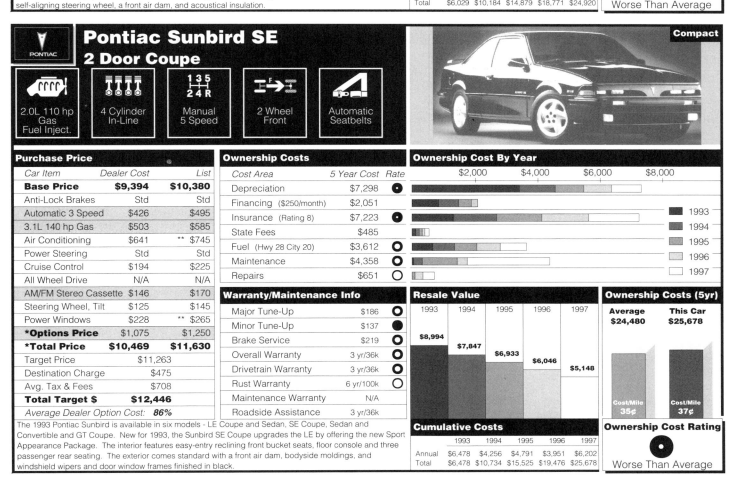

Pontiac Sunbird SE
2 Door Coupe

Compact

2.0L 110 hp Gas Fuel Inject. | 4 Cylinder In-Line | Manual 5 Speed | 2 Wheel Front | Automatic Seatbelts

Purchase Price

Car Item	Dealer Cost	List
Base Price	**$9,394**	**$10,380**
Anti-Lock Brakes	Std	Std
Automatic 3 Speed	$426	$495
3.1L 140 hp Gas	$503	$585
Air Conditioning	$641	** $745
Power Steering	Std	Std
Cruise Control	$194	$225
All Wheel Drive	N/A	N/A
AM/FM Stereo Cassette	$146	$170
Steering Wheel, Tilt	$125	$145
Power Windows	$228	** $265
***Options Price**	**$1,075**	**$1,250**
***Total Price**	**$10,469**	**$11,630**
Target Price		$11,263
Destination Charge		$475
Avg. Tax & Fees		$708
Total Target $		**$12,446**
Average Dealer Option Cost:		*86%*

Ownership Costs

Cost Area	5 Year Cost	Rate
Depreciation	$7,298	◉
Financing ($250/month)	$2,051	
Insurance (Rating 8)	$7,223	◉
State Fees	$485	
Fuel (Hwy 28 City 20)	$3,612	○
Maintenance	$4,358	○
Repairs	$651	○

Warranty/Maintenance Info

Major Tune-Up	$186	○
Minor Tune-Up	$137	●
Brake Service	$219	○
Overall Warranty	3 yr/36k	○
Drivetrain Warranty	3 yr/36k	○
Rust Warranty	6 yr/100k	○
Maintenance Warranty	N/A	
Roadside Assistance	3 yr/36k	

Resale Value

1993	1994	1995	1996	1997
$8,994	$7,847	$6,933	$6,046	$5,148

Cumulative Costs

	1993	1994	1995	1996	1997
Annual	$6,478	$4,256	$4,791	$3,951	$6,202
Total	$6,478	$10,734	$15,525	$19,476	$25,678

Ownership Costs (5yr)

Average $24,480 | This Car $25,678
Cost/Mile 35¢ | Cost/Mile 37¢

Ownership Cost Rating

Worse Than Average

The 1993 Pontiac Sunbird is available in six models - LE Coupe and Sedan, SE Coupe, Sedan and Convertible and GT Coupe. New for 1993, the Sunbird SE Coupe upgrades the LE by offering the new Sport Appearance Package. The interior features easy-entry reclining front bucket seats, floor console and three passenger rear seating. The exterior comes standard with a front air dam, bodyside moldings, and windshield wipers and door window frames finished in black.

---

* Includes shaded options
** Other purchase requirements apply

● Poor | ◉ Worse Than Average | ○ Average | ○ Better Than Average | ○ Excellent | ⊖ Insufficient Information

©1993 by IntelliChoice, Inc. (408) 554-8711 All Rights Reserved. Reproduction Prohibited.
Refer to *Section 3: Annotated Vehicle Charts* for an explanation of these charts.

page 199

Pontiac Sunbird GT
2 Door Coupe

3.1L 140 hp Gas Fuel Inject. | 6 Cylinder "V" | Manual 5 Speed | 2 Wheel Front | Automatic Seatbelts

Compact — SE Model Shown

Purchase Price

Car Item	Dealer Cost	List
Base Price	**$11,602**	**$12,820**
Anti-Lock Brakes	Std	Std
Automatic 3 Speed	$426	$495
Optional Engine	N/A	N/A
Air Conditioning	$641	$745
Power Steering	Std	Std
Cruise Control	$194	$225
All Wheel Drive	N/A	N/A
AM/FM Stereo Cassette	$146	$170
Steering Wheel, Tilt	$125	$145
Power Windows	$228	** $265
***Options Price**	**$1,213**	**$1,410**
***Total Price**	**$12,815**	**$14,230**
Target Price	$13,858	
Destination Charge	$475	
Avg. Tax & Fees	$864	
Total Target $	**$15,197**	
Average Dealer Option Cost:	**86%**	

The 1993 Pontiac Sunbird is available in six models - LE Coupe and Sedan, SE Coupe, Sedan and Convertible and GT Coupe. New for 1993, the Sunbird GT upgrades the SE with Soft-Ray tinted glass and rally gauges with tachometer. The top-of-the-line GT combines the best of the LE and SE into one standard package. Other standard features include Aero extensions, rear decklid spoiler, dual exhaust system, and fog lamps. The Acoustics package is optional, giving the Sunbird a more quiet ride.

Ownership Costs

Cost Area	5 Year Cost	Rate
Depreciation	$9,498	●
Financing ($305/month)	$2,503	
Insurance (Rating 11)	$7,693	◉
State Fees	$588	
Fuel (Hwy 28 City 20)	$3,612	○
Maintenance	$4,895	◉
Repairs	$651	○

Warranty/Maintenance Info

Major Tune-Up	$186	○
Minor Tune-Up	$137	●
Brake Service	$219	○
Overall Warranty	3 yr/36k	○
Drivetrain Warranty	3 yr/36k	○
Rust Warranty	6 yr/100k	○
Maintenance Warranty	N/A	
Roadside Assistance	3 yr/36k	

Ownership Cost By Year

1993, 1994, 1995, 1996, 1997

Resale Value

1993	1994	1995	1996	1997
$9,892	$8,651	$7,672	$6,704	$5,699

Cumulative Costs

	1993	1994	1995	1996	1997
Annual	$8,632	$4,604	$5,318	$4,181	$6,705
Total	$8,632	$13,236	$18,554	$22,735	$29,440

Ownership Costs (5yr)

Average	This Car
$26,713	$29,440
Cost/Mile 38¢	Cost/Mile 42¢

Ownership Cost Rating

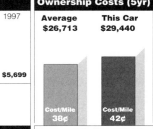

Poor

Pontiac Sunbird LE
4 Door Sedan

2.0L 110 hp Gas Fuel Inject. | 4 Cylinder In-Line | Manual 5 Speed | 2 Wheel Front | Automatic Seatbelts

Compact

Purchase Price

Car Item	Dealer Cost	List
Base Price	**$8,772**	**$9,382**
Anti-Lock Brakes	Std	Std
Automatic 3 Speed	$426	$495
3.1L 140 hp Gas	$503	$585
Air Conditioning	$641	** $745
Power Steering	Std	Std
Cruise Control	Pkg	Pkg
All Wheel Drive	N/A	N/A
AM/FM Stereo Cassette	$146	$170
Steering Wheel, Tilt	$125	$145
Power Windows	$284	** $330
***Options Price**	**$1,075**	**$1,250**
***Total Price**	**$9,847**	**$10,632**
Target Price	$10,578	
Destination Charge	$475	
Avg. Tax & Fees	$664	
Total Target $	**$11,717**	
Average Dealer Option Cost:	**86%**	

The 1993 Pontiac Sunbird is available in six models - LE Coupe and Sedan, SE Coupe, Sedan and Convertible and GT Coupe. New for 1993, the Sunbird LE Sedan offers a Sport Appearance Package, improved engine noise reduction, reduced exterior wind rush noise, and an Arctic White interior. Other standard features are a front air dam, acoustical insulation, and a side window defogger. For safety, the SE has a self-aligning steering wheel, and an energy absorbing steering column.

Ownership Costs

Cost Area	5 Year Cost	Rate
Depreciation	$6,857	○
Financing ($236/month)	$1,930	
Insurance (Rating 5)	$6,786	○
State Fees	$445	
Fuel (Hwy 28 City 20)	$3,612	○
Maintenance	$4,358	○
Repairs	$651	○

Warranty/Maintenance Info

Major Tune-Up	$186	○
Minor Tune-Up	$137	●
Brake Service	$219	○
Overall Warranty	3 yr/36k	○
Drivetrain Warranty	3 yr/36k	○
Rust Warranty	6 yr/100k	○
Maintenance Warranty	N/A	
Roadside Assistance	3 yr/36k	

Resale Value

1993	1994	1995	1996	1997
$8,643	$7,483	$6,600	$5,712	$4,860

Cumulative Costs

	1993	1994	1995	1996	1997
Annual	$5,959	$4,138	$4,641	$3,845	$6,056
Total	$5,959	$10,097	$14,738	$18,583	$24,639

Ownership Costs (5yr)

Average	This Car
$23,623	$24,639
Cost/Mile 34¢	Cost/Mile 35¢

Ownership Cost Rating

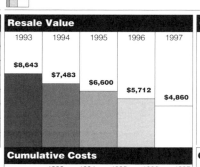

Worse Than Average

---

* Includes shaded options
** Other purchase requirements apply

● Poor ◉ Worse Than Average ○ Average ○ Better Than Average ○ Excellent ⊖ Insufficient Information

©1993 by IntelliChoice, Inc. (408) 554-8711 All Rights Reserved. Reproduction Prohibited.

Refer to *Section 3: Annotated Vehicle Charts* for an explanation of these charts.

Pontiac Sunbird SE
4 Door Sedan

Compact

LE Model Shown

- 2.0L 110 hp Gas Fuel Inject.
- 4 Cylinder In-Line
- Manual 5 Speed
- 2 Wheel Front
- Automatic Seatbelts

Purchase Price

Car Item	Dealer Cost	List
Base Price	**$9,394**	**$10,380**
Anti-Lock Brakes	Std	Std
Automatic 3 Speed	$426	$495
3.1L 140 hp Gas	$503	$585
Air Conditioning	$641	** $745
Power Steering	Std	Std
Cruise Control	$194	$225
All Wheel Drive	N/A	N/A
AM/FM Stereo Cassette	$146	$170
Steering Wheel, Tilt	$125	$145
Power Windows	$284	** $330
*Options Price	$1,075	$1,250
*Total Price	$10,469	$11,630
Target Price		$11,263
Destination Charge		$475
Avg. Tax & Fees		$708
Total Target $		**$12,446**
Average Dealer Option Cost:		86%

The 1993 Pontiac Sunbird is available in six models - LE Coupe and Sedan, SE Coupe, Sedan and Convertible and GT Coupe. New for 1993, the Sunbird SE Sedan offers a Sport Apperance Package, reduced exterior wind rush noise, Artic White interior and improved engine noise reduction. For safety the LE has an energy absorbing steering column, and a self-aligning steering wheel. Other standard features include acoustical insulation, side window defoggers and a front air dam.

Ownership Costs

Cost Area	5 Year Cost	Rate
Depreciation	$7,274	●
Financing ($250/month)	$2,051	
Insurance (Rating 5)	$6,786	◐
State Fees	$485	
Fuel (Hwy 28 City 20)	$3,612	○
Maintenance	$4,358	○
Repairs	$651	○

Warranty/Maintenance Info

Major Tune-Up	$186	○
Minor Tune-Up	$137	●
Brake Service	$219	○
Overall Warranty	3 yr/36k	○
Drivetrain Warranty	3 yr/36k	○
Rust Warranty	6 yr/100k	○
Maintenance Warranty	N/A	
Roadside Assistance	3 yr/36k	

Ownership Cost By Year

Resale Value

1993: $9,297
1994: $8,029
1995: $7,054
1996: $6,094
1997: $5,172

Cumulative Costs

	1993	1994	1995	1996	1997
Annual	$6,094	$4,293	$4,765	$3,933	$6,132
Total	$6,094	$10,387	$15,152	$19,085	$25,217

Ownership Costs (5yr)

Average	This Car
$24,480	$25,217
Cost/Mile 35¢	Cost/Mile 36¢

Ownership Cost Rating

○ Average

Pontiac Sunbird SE
2 Door Convertible

Compact

- 2.0L 110 hp Gas Fuel Inject.
- 4 Cylinder In-Line
- Manual 5 Speed
- 2 Wheel Front
- Automatic Seatbelts

Purchase Price

Car Item	Dealer Cost	List
Base Price	**$13,940**	**$15,403**
Anti-Lock Brakes	Std	Std
Automatic 3 Speed	$426	$495
3.1L 140 hp Gas	$503	$585
Air Conditioning	$641	$745
Power Steering	Std	Std
Cruise Control	$194	$225
All Wheel Drive	N/A	N/A
AM/FM Stereo Cassette	$146	$170
Steering Wheel, Tilt	$125	$145
Power Windows	Std	Std
*Options Price	$1,716	$1,995
*Total Price	$15,656	$17,398
Target Price		$17,023
Destination Charge		$475
Avg. Tax & Fees		$1,054
Total Target $		**$18,552**
Average Dealer Option Cost:		86%

The 1993 Pontiac Sunbird is available in six models - LE Coupe and Sedan, SE Coupe, Sedan and Convertible and GT Coupe. New for 1993, the Sunbird SE Convertible introduces a new glass rear window available with a rear window defogger replacing last year's vinyl rear window. Other standard features are Soft-Ray tinted glass, front seat armrest and storage bin, rear quarter courtesy lights, reclining front bucket seats, floor mats, and a rear compartment acoustics package.

Ownership Costs

Cost Area	5 Year Cost	Rate
Depreciation	$11,647	●
Financing ($373/month)	$3,056	
Insurance (Rating 13)	$8,061	◐
State Fees	$715	
Fuel (Hwy 28 City 20)	$3,612	○
Maintenance	$4,358	○
Repairs	$651	○

Warranty/Maintenance Info

Major Tune-Up	$186	○
Minor Tune-Up	$137	●
Brake Service	$219	○
Overall Warranty	3 yr/36k	○
Drivetrain Warranty	3 yr/36k	○
Rust Warranty	6 yr/100k	○
Maintenance Warranty	N/A	
Roadside Assistance	3 yr/36k	

Resale Value

1993: $11,381
1994: $10,090
1995: $9,059
1996: $7,976
1997: $6,905

Cumulative Costs

	1993	1994	1995	1996	1997
Annual	$10,828	$4,927	$5,321	$4,434	$6,590
Total	$10,828	$15,755	$21,076	$25,510	$32,100

Ownership Costs (5yr)

Average	This Car
$29,433	$32,100
Cost/Mile 42¢	Cost/Mile 46¢

Ownership Cost Rating

◐ Worse Than Average

---

* Includes shaded options
** Other purchase requirements apply

● Poor | ◐ Worse Than Average | ○ Average | ○ Better Than Average | ○ Excellent | ⊖ Insufficient Information

©1993 by IntelliChoice, Inc. (408) 554-8711 All Rights Reserved. Reproduction Prohibited.
Refer to *Section 3: Annotated Vehicle Charts* for an explanation of these charts.

Porsche 911 RS America
2 Door Coupe

3.6L 247 hp Gas Fuel Inject. | 6 Cylinder Opposing | Manual 5 Speed | 2 Wheel Rear | Driver/Psngr Airbags Std

Purchase Price

Car Item	Dealer Cost	List
Base Price	**$45,875**	**$54,800**
Anti-Lock Brakes	Std	Std
Automatic Transmission	N/A	N/A
Optional Engine	N/A	N/A
Auto Climate Control	$2,248	$2,805
Power Steering	N/A	N/A
Cruise Control	N/A	N/A
All Wheel Drive	N/A	N/A
AM/FM Stereo Cassette	$790	$986
Steering Wheel, Tilt	N/A	N/A
Power Windows	Std	Std
*Options Price	$3,038	$3,791
*Total Price	$48,913	$58,591
Target Price	$53,315	
Destination Charge	$725	
Avg. Tax & Fees	$3,295	
Luxury Tax	$2,404	
Total Target $	**$59,739**	

The 1993 911 America is available in two models: RS America and Roadster. New for 1993, the RS America features a large fixed-plane spoiler and carries special RS identification. Standard features include dual airbags, corduroy sport seats with electric height adjustment and one-key central locking and alarm system. Exterior features include windshield antenna with signal amplifier, rear window defogger and halogen headlights. Optional equipment includes slip differential, sunroof and air conditioning.

Ownership Costs

Cost Area	5 Year Cost	Rate
Depreciation	$16,782	○
Financing ($1,201/month)	$9,841	
Insurance (Rating 26 Sport+)	$18,179	●
State Fees	$2,373	
Fuel (Hwy 25 City 17 -Prem.)	$4,578	◐
Maintenance	$8,027	●
Repairs	$1,505	●

Warranty/Maintenance Info

Major Tune-Up	$378	●
Minor Tune-Up	$120	○
Brake Service	$380	●
Overall Warranty	2 yr/unlim. mi	◉
Drivetrain Warranty	2 yr/unlim. mi	◉
Rust Warranty	10 yr/unlim. mi	○
Maintenance Warranty	N/A	
Roadside Assistance	2 yr/unlim. mi	

Ownership Cost By Year
1993 / 1994 / 1995 / 1996 / 1997

Resale Value
1993	1994	1995	1996	1997
$49,799	$47,978	$46,562	$45,004	$42,957

Cumulative Costs
	1993	1994	1995	1996	1997
Annual	$18,888	$10,005	$10,997	$9,082	$12,313
Total	$18,888	$28,893	$39,890	$48,972	$61,285

Ownership Costs (5yr)
Average	This Car
$64,957	$61,285
Cost/Mile 93¢	Cost/Mile 88¢

Ownership Cost Rating
○ Excellent

Porsche 911 Carrera 2
2 Door Coupe

3.6L 247 hp Gas Fuel Inject. | 6 Cylinder Opposing | Manual 5 Speed | 2 Wheel Rear | Driver/Psngr Airbags Std

Purchase Price

Car Item	Dealer Cost	List
Base Price	**$54,405**	**$64,990**
Anti-Lock Brakes	Std	Std
Automatic 4 Speed	$2,635	$3,150
Optional Engine	N/A	N/A
Auto Climate Control	Std	Std
Power Steering	Std	Std
Cruise Control	Std	Std
All Wheel Drive	N/A	N/A
AM/FM Stereo Cassette	Std	Std
Steering Wheel, Tilt	N/A	N/A
Power Windows	Std	Std
*Options Price	$0	$0
*Total Price	$54,405	$64,990
Target Price	$59,301	
Destination Charge	$725	
Avg. Tax & Fees	$3,658	
Luxury Tax	$3,003	
Total Target $	**$66,687**	

The 1993 911 Carrera is available in six models: two- and four-wheel-drive coupes, Targas and Cabriolets. New for 1993, the 911 2WD Carrera Coupe offers standard equipment such as five-spoke light alloy wheels and Porsche 959-like exterior mirrors. Exterior features include a power sliding sun roof, tinted glass, and dual heated power mirrors. Interior features include a rear window defroster, power windows, automatic air conditioning and reclining bucket seats with electric height adjustment.

Ownership Costs

Cost Area	5 Year Cost	Rate
Depreciation	$20,285	○
Financing ($1,340/month)	$10,986	
Insurance (Rating 26 Sport+)	$18,179	●
State Fees	$2,628	
Fuel (Hwy 25 City 17 -Prem.)	$4,578	◐
Maintenance	$7,736	◉
Repairs	$1,505	●

Warranty/Maintenance Info

Major Tune-Up	$378	●
Minor Tune-Up	$120	○
Brake Service	$380	●
Overall Warranty	2 yr/unlim. mi	◉
Drivetrain Warranty	2 yr/unlim. mi	◉
Rust Warranty	10 yr/unlim. mi	○
Maintenance Warranty	N/A	
Roadside Assistance	2 yr/unlim. mi	

Resale Value
1993	1994	1995	1996	1997
$55,009	$52,739	$50,900	$48,912	$46,402

Cumulative Costs
	1993	1994	1995	1996	1997
Annual	$21,167	$10,869	$11,558	$9,640	$12,663
Total	$21,167	$32,036	$43,594	$53,234	$65,897

Ownership Costs (5yr)
Average	This Car
$69,277	$65,897
Cost/Mile 99¢	Cost/Mile 94¢

Ownership Cost Rating
○ Better Than Average

---

* Includes shaded options
** Other purchase requirements apply

● Poor · Worse Than Average ○ Average ○ Better Than Average ○ Excellent ⊖ Insufficient Information

©1993 by IntelliChoice, Inc. (408) 554-8711 All Rights Reserved. Reproduction Prohibited.
Refer to *Section 3: Annotated Vehicle Charts* for an explanation of these charts.

Porsche 911 Carrera 2 Targa
2 Door Coupe

 3.6L 247 hp Gas Fuel Inject. | 6 Cylinder Opposing | Manual 5 Speed | 2 Wheel Rear | Driver/Psngr Airbags Std

Sport — Base Model Shown

Purchase Price

Car Item	Dealer Cost	List
Base Price	**$55,765**	**$66,600**
Anti-Lock Brakes	Std	Std
Automatic 4 Speed	$2,635	$3,150
Optional Engine	N/A	N/A
Auto Climate Control	Std	Std
Power Steering	Std	Std
Cruise Control	Std	Std
All Wheel Drive	N/A	N/A
AM/FM Stereo Cassette	Std	Std
Steering Wheel, Tilt	N/A	N/A
Power Windows	Std	Std
*Options Price	$0	$0
*Total Price	$55,765	$66,600
Target Price		$60,784
Destination Charge		$725
Avg. Tax & Fees		$3,748
Luxury Tax		$3,151
Total Target $		**$68,408**

The 1993 911 Carrera is available in six models: two- and four-wheel-drive coupes, Targas and Cabriolets. New for 1993, the 911 2WD Carrera Targa offers such standard equipment such as five-spoke light alloy wheels and Porsche 959-like exterior mirrors. Standard features include dual airbags, power roof, reclining bucket seats with electric height adjustment and one-key central locking and alarm system. Other features include race-tuned suspension with self-stabilizing characteristics at the rear.

Ownership Costs

Cost Area	5 Year Cost	Rate
Depreciation	$21,456	○
Financing ($1,375/month)	$11,269	
Insurance (Rating 26 Sport+)	$18,179	●
State Fees	$2,693	
Fuel (Hwy 25 City 17 -Prem.)	$4,578	◐
Maintenance	$7,736	●
Repairs	$1,505	●

Warranty/Maintenance Info

Major Tune-Up	$378	●
Minor Tune-Up	$120	○
Brake Service	$380	●
Overall Warranty	2 yr/unlim. mi	◉
Drivetrain Warranty	2 yr/unlim. mi	◉
Rust Warranty	10 yr/unlim. mi	○
Maintenance Warranty	N/A	
Roadside Assistance	2 yr/unlim. mi	

Ownership Cost By Year

$5,000 / $10,000 / $15,000 / $20,000 / $25,000

Legend: 1993, 1994, 1995, 1996, 1997

Resale Value

1993	1994	1995	1996	1997
$54,451	$52,611	$51,002	$49,242	$46,952

Cumulative Costs

	1993	1994	1995	1996	1997
Annual	$23,580	$10,542	$11,397	$9,444	$12,453
Total	$23,580	$34,122	$45,519	$54,963	$67,416

Ownership Costs (5yr)

Average	This Car
$70,364	$67,416
Cost/Mile $1.01	Cost/Mile 96¢

Ownership Cost Rating

○ Average

Porsche 911 Carrera 2 Cabriolet
2 Door Convertible

 3.6L 247 hp Gas Fuel Inject. | 6 Cylinder Opposing | Manual 5 Speed | 2 Wheel Rear | Driver/Psngr Airbags Std

Sport

Purchase Price

Car Item	Dealer Cost	List
Base Price	**$62,070**	**$74,190**
Anti-Lock Brakes	Std	Std
Automatic 4 Speed	$2,635	$3,150
Optional Engine	N/A	N/A
Auto Climate Control	Std	Std
Power Steering	Std	Std
Cruise Control	Std	Std
All Wheel Drive	N/A	N/A
AM/FM Stereo Cassette	Std	Std
Steering Wheel, Tilt	N/A	N/A
Power Windows	Std	Std
*Options Price	$0	$0
*Total Price	$62,070	$74,190
Target Price		$67,656
Destination Charge		$725
Avg. Tax & Fees		$4,168
Luxury Tax		$3,838
Total Target $		**$76,387**

The 1993 911 Carrera is available in six models: two- and four-wheel-drive coupes, Targas and Cabriolets. New for 1993, the 911 2WD Carrera Cabriolet offers standard equipment such as five-spoke light alloy wheels and Porsche 959-like exterior mirrors. Exterior features include locking, cast alloy wheels and a power roof. Other features on all 1993 Carrera 2s include a suspension with self-stabilizing characteristics and high performance ZR-rated steel belted radial tires.

Ownership Costs

Cost Area	5 Year Cost	Rate
Depreciation	$20,458	○
Financing ($1,535/month)	$12,583	
Insurance (Rating 26 Sport+)	$18,179	●
State Fees	$2,995	
Fuel (Hwy 25 City 17 -Prem.)	$4,578	○
Maintenance	$7,736	●
Repairs	$1,505	●

Warranty/Maintenance Info

Major Tune-Up	$378	●
Minor Tune-Up	$120	○
Brake Service	$380	●
Overall Warranty	2 yr/unlim. mi	◉
Drivetrain Warranty	2 yr/unlim. mi	◉
Rust Warranty	10 yr/unlim. mi	○
Maintenance Warranty	N/A	
Roadside Assistance	2 yr/unlim. mi	

Resale Value

1993	1994	1995	1996	1997
$62,838	$61,115	$59,631	$58,065	$55,929

Cumulative Costs

	1993	1994	1995	1996	1997
Annual	$23,795	$10,904	$11,593	$9,398	$12,344
Total	$23,795	$34,699	$46,292	$55,690	$68,034

Ownership Costs (5yr)

Average	This Car
$75,488	$68,034
Cost/Mile $1.08	Cost/Mile 97¢

Ownership Cost Rating
○ Excellent

---

* Includes shaded options
** Other purchase requirements apply

Legend: ● Poor | ◐ Worse Than Average | ○ Average | ○ Better Than Average | ○ Excellent | ⊖ Insufficient Information

©1993 by IntelliChoice, Inc. (408) 554-8711 All Rights Reserved. Reproduction Prohibited.
Refer to *Section 3: Annotated Vehicle Charts* for an explanation of these charts.

page 203

Porsche 911 America
2 Door Convertible

 3.6L 247 hp Gas Fuel Inject.
 6 Cylinder Opposing
 Manual 5 Speed
 2 Wheel Rear
 Driver/Psngr Airbags Std

Sport

Purchase Price

Car Item	Dealer Cost	List
Base Price	**$74,805**	**$89,350**
Anti-Lock Brakes	Std	Std
Automatic 4 Speed	$2,635	$3,150
Optional Engine	N/A	N/A
Auto Climate Control	Std	Std
Power Steering	Std	Std
Cruise Control	Std	Std
All Wheel Drive	N/A	N/A
AM/FM Stereo Cassette	Std	Std
Steering Wheel, Tilt	N/A	N/A
Power Windows	Std	Std
*Options Price	$0	$0
*Total Price	$74,805	$89,350
Target Price		$81,537
Destination Charge		$725
Avg. Tax & Fees		$5,014
Luxury Tax		$5,226
Total Target $		**$92,502**

The 1993 911 America is available in two models: RS America and Roadster. New for 1993, the 911 America Roadster features wider fenders to cover its high performance 17-inch wheels and tires. The 911 Roadster also incorporates features found on the Carrera 2 models. These include a fully automatic handcrafted folding top and a speed-dependent extendable rear spoiler. Comfort and convenience features include an on-board six-function computerized information center and a one-key central-locking alarm system.

Ownership Costs

Cost Area	5 Year Cost	Rate
Depreciation	$28,861	○
Financing ($1,859/month)	$15,238	
Insurance (Rating 26 Sport+)	$18,179	●
State Fees	$3,604	
Fuel (Hwy 25 City 17 -Prem.)	$4,578	◐
Maintenance	$8,027	●
Repairs	$1,505	●

Warranty/Maintenance Info

Major Tune-Up	$378	●
Minor Tune-Up	$120	○
Brake Service	$380	●
Overall Warranty	2 yr/unlim. mi	⊙
Drivetrain Warranty	2 yr/unlim. mi	⊙
Rust Warranty	10 yr/unlim. mi	○
Maintenance Warranty	N/A	
Roadside Assistance	2 yr/unlim. mi	

Ownership Cost By Year

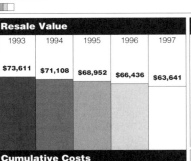

1993, 1994, 1995, 1996, 1997

Resale Value

1993	1994	1995	1996	1997
$73,611	$71,108	$68,952	$66,436	$63,641

Cumulative Costs

	1993	1994	1995	1996	1997
Annual	$30,395	$12,651	$13,053	$10,648	$13,245
Total	$30,395	$43,046	$56,099	$66,747	$79,992

Ownership Costs (5yr)

Average: $85,723 — This Car: $79,992
 Cost/Mile $1.22 — Cost/Mile $1.14

Ownership Cost Rating

○ Excellent

Porsche 911 Carrera 4
2 Door Coupe

 3.6L 247 hp Gas Fuel Inject.
 6 Cylinder Opposing
 Manual 5 Speed
 4 Wheel Full-Time
Driver/Psngr Airbags Std

Sport

Purchase Price

Car Item	Dealer Cost	List
Base Price	**$64,500**	**$77,050**
Anti-Lock Brakes	Std	Std
Automatic Transmission	N/A	N/A
Optional Engine	N/A	N/A
Auto Climate Control	Std	Std
Power Steering	Std	Std
Cruise Control	Std	Std
4 Wheel Full-Time Drive	Std	Std
AM/FM Stereo Cassette	Std	Std
Steering Wheel, Tilt	N/A	N/A
Power Windows	Std	Std
*Options Price	$0	$0
*Total Price	$64,500	$77,050
Target Price		$70,305
Destination Charge		$725
Avg. Tax & Fees		$4,330
Luxury/Gas Guzzler Tax		$5,403
Total Target $		**$80,763**

The 1993 911 Carrera is available in six models: two- and four-wheel-drive coupes, Targas and Cabriolets. New for 1993, the 911 4WD Carrera Coupe features newly styled, heated, power outside-mirrors and a rear parcel shelf. Interior equipment includes partial-leather front seats with electric height adjustment, power windows, electric sunroof, Porsche CR-1 cassette-radio and a one-key central-locking alarm system. Other features include five-spoke alloy wheels and 959-like exterior mirrors.

Ownership Costs

Cost Area	5 Year Cost	Rate
Depreciation	$27,194	○
Financing ($1,623/month)	$13,303	
Insurance (Rating 26 Sport+)	$18,179	●
State Fees	$3,111	
Fuel (Hwy 23 City 15 -Prem.)	$5,079	⊙
Maintenance	$7,736	⊙
Repairs	$1,505	●

Warranty/Maintenance Info

Major Tune-Up	$378	●
Minor Tune-Up	$120	○
Brake Service	$380	●
Overall Warranty	2 yr/unlim. mi	⊙
Drivetrain Warranty	2 yr/unlim. mi	⊙
Rust Warranty	10 yr/unlim. mi	○
Maintenance Warranty	N/A	
Roadside Assistance	2 yr/unlim. mi	

Ownership Cost By Year

Resale Value

1993	1994	1995	1996	1997
$61,762	$55,907	$54,127	$53,941	$53,569

Cumulative Costs

	1993	1994	1995	1996	1997
Annual	$29,669	$15,383	$12,155	$8,194	$10,706
Total	$29,669	$45,052	$57,207	$65,401	$76,107

Ownership Costs (5yr)

Average: $77,419 — This Car: $76,107
Cost/Mile $1.11 — Cost/Mile $1.09

Ownership Cost Rating

◐ Average

---

page 204

* Includes shaded options
** Other purchase requirements apply

● Poor   ◐ Worse Than Average   ○ Average   ○ Better Than Average   ○ Excellent   ⊖ Insufficient Information

©1993 by IntelliChoice, Inc. (408) 554-8711   All Rights Reserved. Reproduction Prohibited.
Refer to *Section 3: Annotated Vehicle Charts* for an explanation of these charts.

Porsche 911 Carrera 4 Targa
2 Door Coupe

 3.6L 247 hp Gas Fuel Inject.
 6 Cylinder Opposing
 Manual 5 Speed
 4 Wheel Full-Time
 Driver/Psngr Airbags Std

Sport — Base Model Shown

Purchase Price

Car Item	Dealer Cost	List
Base Price	**$65,855**	**$78,660**
Anti-Lock Brakes	Std	Std
Automatic Transmission	N/A	N/A
Optional Engine	N/A	N/A
Auto Climate Control	Std	Std
Power Steering	Std	Std
Cruise Control	Std	Std
4 Wheel Full-Time Drive	Std	Std
AM/FM Stereo Cassette	Std	Std
Steering Wheel, Tilt	N/A	N/A
Power Windows	Std	Std
*Options Price	$0	$0
***Total Price**	**$65,855**	**$78,660**
Target Price		$71,782
Destination Charge		$725
Avg. Tax & Fees		$4,419
Luxury/Gas Guzzler Tax		$5,551
Total Target $		**$82,477**

The 1993 911 Carrera is available in six models: two- and four-wheel-drive coupes, Targas and Cabriolets. New for 1993, the 911 4WD Carrera Targa features newly styled, heated, power outside-mirrors and a rear parcel shelf. Interior equipment includes partial-leather front seats with electric height adjustment, power windows, electric sunroof, Porsche CR-1 theft-coded digital-display stereo cassette-radio with eight-speaker system and a one-key central-locking alarm system.

Ownership Costs

Cost Area	5 Year Cost	Rate
Depreciation	$29,157	○
Financing ($1,658/month)	$13,587	
Insurance (Rating 26 Sport+)	$18,179	●
State Fees	$3,175	
Fuel (Hwy 23 City 15 -Prem.)	$5,079	●
Maintenance	$7,736	●
Repairs	$1,505	●

Warranty/Maintenance Info

Major Tune-Up	$378	●
Minor Tune-Up	$120	○
Brake Service	$380	●
Overall Warranty	2 yr/unlim. mi	●
Drivetrain Warranty	2 yr/unlim. mi	●
Rust Warranty	10 yr/unlim. mi	○
Maintenance Warranty	N/A	
Roadside Assistance	2 yr/unlim. mi	

Ownership Cost By Year

Legend: 1993, 1994, 1995, 1996, 1997

Resale Value

1993	1994	1995	1996	1997
$61,836	$56,269	$53,881	$53,694	$53,320

Cumulative Costs

	1993	1994	1995	1996	1997
Annual	$31,443	$15,198	$12,832	$8,228	$10,717
Total	$31,443	$46,641	$59,473	$67,701	$78,418

Ownership Costs (5yr)

Average	This Car
$78,506	$78,418
Cost/Mile $1.12	Cost/Mile $1.12

Ownership Cost Rating
○ Average

Porsche 911 Carrera 4 Cabriolet
2 Door Convertible

 3.6L 247 hp Gas Fuel Inject.
 6 Cylinder Opposing
 Manual 5 Speed
 4 Wheel Full-Time
 Driver/Psngr Airbags Std

Sport

Purchase Price

Car Item	Dealer Cost	List
Base Price	**$72,150**	**$86,250**
Anti-Lock Brakes	Std	Std
Automatic Transmission	N/A	N/A
Optional Engine	N/A	N/A
Auto Climate Control	Std	Std
Power Steering	Std	Std
Cruise Control	Std	Std
4 Wheel Full-Time Drive	Std	Std
AM/FM Stereo Cassette	Std	Std
Steering Wheel, Tilt	N/A	N/A
Power Windows	Std	Std
*Options Price	$0	$0
***Total Price**	**$72,150**	**$86,250**
Target Price		$78,644
Destination Charge		$725
Avg. Tax & Fees		$4,838
Luxury/Gas Guzzler Tax		$6,237
Total Target $		**$90,444**

The 1993 911 Carrera is available in six models: two- and four-wheel-drive coupes, Targas and Cabriolets. New for 1993, the 911 4WD Carrera Cabriolet features newly styled, heated, power outside-mirrors and a rear parcel shelf. The Cabriolet top is electrically raised and lowered including automatic latching and unlatching. Interior equipment includes partial-leather front seats, CR-1 theft-coded digital stereo cassette-radio with eight-speaker system and a one-key central-locking alarm system.

Ownership Costs

Cost Area	5 Year Cost	Rate
Depreciation	$29,042	○
Financing ($1,818/month)	$14,899	
Insurance (Rating 26 Sport+)	$18,179	●
State Fees	$3,480	
Fuel (Hwy 23 City 15 -Prem.)	$5,079	●
Maintenance	$7,736	●
Repairs	$1,505	●

Warranty/Maintenance Info

Major Tune-Up	$378	●
Minor Tune-Up	$120	○
Brake Service	$380	●
Overall Warranty	2 yr/unlim. mi	●
Drivetrain Warranty	2 yr/unlim. mi	●
Rust Warranty	10 yr/unlim. mi	○
Maintenance Warranty	N/A	
Roadside Assistance	2 yr/unlim. mi	

Resale Value

1993	1994	1995	1996	1997
$69,614	$63,882	$62,069	$61,784	$61,402

Cumulative Costs

	1993	1994	1995	1996	1997
Annual	$32,255	$15,842	$12,578	$8,474	$10,771
Total	$32,255	$48,097	$60,675	$69,149	$79,920

Ownership Costs (5yr)

Average	This Car
$83,630	$79,920
Cost/Mile $1.19	Cost/Mile $1.14

Ownership Cost Rating
◐ Better Than Average

---

* Includes shaded options
** Other purchase requirements apply

Legend: ● Poor, ◐ Worse Than Average, ◑ Average, ◒ Better Than Average, ○ Excellent, ⊖ Insufficient Information

©1993 by IntelliChoice, Inc. (408) 554-8711 All Rights Reserved. Reproduction Prohibited.
Refer to *Section 3: Annotated Vehicle Charts* for an explanation of these charts.

Porsche 928 GTS
2 Door Coupe

 5.4L 345 hp Gas Fuel Inject.
 8 Cylinder "V"
 Manual 5 Speed
 2 Wheel Rear
 Driver/Psngr Airbags Std

Sport

Purchase Price

Car Item	Dealer Cost	List
Base Price	**$66,925**	**$80,920**
Anti-Lock Brakes	Std	Std
Automatic 4 Speed	N/C	N/C
Optional Engine	N/A	N/A
Auto Climate Control	Std	Std
Power Steering	Std	Std
Cruise Control	Std	Std
All Wheel Drive	N/A	N/A
AM/FM Stereo Cassette	Std	Std
Steering Wheel, Tilt	Std	Std
Power Windows	Std	Std
*Options Price	$0	$0
*Total Price	$66,925	$80,920
Target Price		$73,618
Destination Charge		$700
Avg. Tax & Fees		$4,532
Luxury/Gas Guzzler Tax		$7,432
Total Target $		**$86,282**

The 1993 928 is available as one model edition. This year's GTS replaces the previous GT and S4 models. For 1993, the GTS coupe features vehicle information system, rear spoiler, Hi-fi sound system, locking light cast alloy wheels, tire pressure monitoring system, and dual airbags. Functional features include limited slip differential, vehicle diagnostic system, and sport suspension. Several Exclusive Packages will be available for the 1993 model year.

Ownership Costs

Cost Area	5 Year Cost	Rate
Depreciation	$47,612	●
Financing ($1,734/month)	$14,212	
Insurance (Rating 27++)	$17,028	◉
State Fees	$3,265	
Fuel (Hwy 19 City 12 -Prem.)	$6,250	●
Maintenance	$10,265	●
Repairs	$1,505	●

Warranty/Maintenance Info

Major Tune-Up	$456	●
Minor Tune-Up	$175	●
Brake Service	$430	●
Overall Warranty	2 yr/unlim. mi	◉
Drivetrain Warranty	2 yr/unlim. mi	◉
Rust Warranty	10 yr/unlim. mi	○
Maintenance Warranty	N/A	
Roadside Assistance	2 yr/unlim. mi	

Ownership Cost By Year

$5K $10K $15K $20K $25K $30K $35K $40K $45K $50K

■ 1993
■ 1994
■ 1995
□ 1996
□ 1997

Resale Value

1993	1994	1995	1996	1997
$56,130	$51,081	$46,529	$42,651	$38,670

Cumulative Costs

	1993	1994	1995	1996	1997
Annual	$41,242	$15,017	$15,793	$12,364	$15,721
Total	$41,242	$56,259	$72,052	$84,416	$100,137

Ownership Costs (5yr)

Average	This Car
$80,032	$100,137
Cost/Mile $1.14	Cost/Mile $1.43

Ownership Cost Rating
● Poor

Porsche 968
2 Door Coupe

 3.0L 236 hp Gas Fuel Inject.
 4 Cylinder In-Line
 Manual 6 Speed
 2 Wheel Rear
Driver/Psngr Airbags Std

Sport

Purchase Price

Car Item	Dealer Cost	List
Base Price	**$32,760**	**$39,950**
Anti-Lock Brakes	Std	Std
Automatic 4 Speed	$2,580	$3,150
Optional Engine	N/A	N/A
Auto Climate Control	Std	Std
Power Steering	Std	Std
Cruise Control	Std	Std
All Wheel Drive	N/A	N/A
AM/FM Stereo Cassette	Std	Std
Steering Wheel, Tilt	N/A	N/A
Power Windows	Std	Std
*Options Price	$0	$0
*Total Price	$32,760	$39,950
Target Price		$36,036
Destination Charge		$725
Avg. Tax & Fees		$2,245
Luxury Tax		$676
Total Target $		**$39,682**

The 1993 968 is available in two bodystyles - coupe and convertible. For 1993 the 968 features several functional upgrades including CFC-free air conditioning, new fade resistant brake fluid and new rear stereo speakers. The 968 coupe offers an optional sport suspension. Standard equipment includes electric rear wiper, power tilt/removable sunroof, roof-mounted antenna, graduated tinted glass and heated windshield washer nozzles. Many leather appointments are also available.

Ownership Costs

Cost Area	5 Year Cost	Rate
Depreciation		⊖
Financing ($798/month)	$6,537	
Insurance (Rating 24 Sport+)	$17,252	●
State Fees	$1,627	
Fuel (Hwy 26 City 17 -Prem.)	$4,488	○
Maintenance	$8,952	●
Repairs	$1,195	◉

Warranty/Maintenance Info

Major Tune-Up	$305	◉
Minor Tune-Up	$101	○
Brake Service	$425	●
Overall Warranty	2 yr/unlim. mi	◉
Drivetrain Warranty	2 yr/unlim. mi	◉
Rust Warranty	10 yr/unlim. mi	○
Maintenance Warranty	N/A	
Roadside Assistance	2 yr/unlim. mi	

Ownership Cost By Year

$5,000 $10,000 $15,000 $20,000

Insufficient Depreciation Information

■ 1993
■ 1994
■ 1995
□ 1996
□ 1997

Resale Value

Insufficient Information

Cumulative Costs

	1993	1994	1995	1996	1997
Annual	*Insufficient Information*				
Total	*Insufficient Information*				

Ownership Costs (5yr)

Insufficient Information

Ownership Cost Rating
⊖ Insufficient Information

---

page 206

* Includes shaded options
** Other purchase requirements apply

● Poor   ◉ Worse Than Average   ● Average   ○ Better Than Average   ○ Excellent   ⊖ Insufficient Information

©1993 by IntelliChoice, Inc. (408) 554-8711   All Rights Reserved. Reproduction Prohibited.
Refer to *Section 3: Annotated Vehicle Charts* for an explanation of these charts.

Porsche 968 Cabriolet
2 Door Convertible

Sport

 3.0L 236 hp Gas Fuel Inject.
 4 Cylinder In-Line
 Manual 6 Speed
 2 Wheel Rear
 Driver/Psngr Airbags Std

Purchase Price

Car Item	Dealer Cost	List
Base Price	**$42,555**	**$51,900**
Anti-Lock Brakes	Std	Std
Automatic 4 Speed	$2,580	$3,150
Optional Engine	N/A	N/A
Auto Climate Control	Std	Std
Power Steering	Std	Std
Cruise Control	Std	Std
All Wheel Drive	N/A	N/A
AM/FM Stereo Cassette	Std	Std
Steering Wheel, Tilt	N/A	N/A
Power Windows	Std	Std
*Options Price	$0	$0
*Total Price	$42,555	$51,900
Target Price		$46,810
Destination Charge		$725
Avg. Tax & Fees		$2,903
Luxury Tax		$1,754
Total Target $		**$52,192**

The 1993 968 is available in two bodystyles - coupe and convertible. For 1993 the 968 features several functional upgrades including CFC-free air conditioning, new fade resistant brake fluid and new rear stereo speakers. The 968 Cabriolet features standard windshield antenna, power top, graduated tinted glass and optional eight-speaker sound system. Many leather appointments are also available.

Ownership Costs

Cost Area	5 Year Cost	Rate
Depreciation		⊖
Financing ($1,049/month)	$8,598	
Insurance (Rating 26 Sport+)	$19,200	●
State Fees	$2,105	
Fuel (Hwy 26 City 17 -Prem.)	$4,488	○
Maintenance	$8,952	●
Repairs	$1,195	◉

Warranty/Maintenance Info

Major Tune-Up	$305	●
Minor Tune-Up	$101	○
Brake Service	$425	●
Overall Warranty	2 yr/unlim. mi	◉
Drivetrain Warranty	2 yr/unlim. mi	◉
Rust Warranty	10 yr/unlim. mi	○
Maintenance Warranty	N/A	
Roadside Assistance	2 yr/unlim. mi	

Ownership Cost By Year

Insufficient Depreciation Information

1993 / 1994 / 1995 / 1996 / 1997

Resale Value
Insufficient Information

Cumulative Costs
Annual: Insufficient Information
Total: Insufficient Information

Ownership Costs (5yr)
Insufficient Information

Ownership Cost Rating
⊖ Insufficient Information

Saab 9000 CS
4 Door Hatchback

Midsize

 2.3L 150 hp Gas Fuel Inject.
 4 Cylinder In-Line
 Manual 5 Speed
 2 Wheel Front
 Driver Airbag Psngr Belts

Purchase Price

Car Item	Dealer Cost	List
Base Price	**$22,123**	**$25,725**
Anti-Lock Brakes	Std	Std
Automatic 4 Speed	$736	$920
Optional Engine	N/A	N/A
Auto Climate Control	Std	Std
Power Steering	Std	Std
Cruise Control	Std	Std
All Wheel Drive	N/A	N/A
AM/FM Stereo Cassette	Std	Std
Steering Wheel, Tilt	N/A	N/A
Power Windows	Std	Std
*Options Price	$736	$920
*Total Price	$22,859	$26,645
Target Price		$24,734
Destination Charge		$440
Avg. Tax & Fees		$1,530
Total Target $		**$26,704**
Average Dealer Option Cost:		80%

The 1993 9000 is available in four models-CS and CSE hatchbacks; and CS and CDE sedans. New for 1993, the 9000 CS hatchback features a 2.3-liter 150 horsepower engine. The new CS is the only 5-door in Saab's 1993 9000 series. Standard features include a driver's-side airbag, anti-lock brakes, front and rear head restraints and an upgraded alarm system with central locking and child-proof rear door locks and windows. It is also covered by a 6 year/80,000 mile Major Systems Warranty.

Ownership Costs

Cost Area	5 Year Cost	Rate
Depreciation	$15,669	○
Financing ($537/month)	$4,399	
Insurance (Rating 13)	$7,797	○
State Fees	$1,085	
Fuel (Hwy 24 City 18)	$4,117	●
Maintenance	$5,504	◉
Repairs	$1,004	●

Warranty/Maintenance Info

Major Tune-Up	$134	○
Minor Tune-Up	$91	○
Brake Service	$280	◉
Overall Warranty	3 yr/40k	○
Drivetrain Warranty	6 yr/80k/$150/40k	○
Rust Warranty	6 yr/unlim. mi	○
Maintenance Warranty	N/A	
Roadside Assistance	3 yr/40k	

Resale Value

1993	1994	1995	1996	1997
$17,895	$15,763	$14,180	$12,473	$11,035

Cumulative Costs

	1993	1994	1995	1996	1997
Annual	$13,157	$6,264	$6,370	$5,359	$8,425
Total	$13,157	$19,421	$25,791	$31,150	$39,575

Ownership Costs (5yr)

Average	This Car
$38,513	$39,575
Cost/Mile 55¢	Cost/Mile 57¢

Ownership Cost Rating
○ Average

---

* Includes shaded options
** Other purchase requirements apply

● Poor ◉ Worse Than Average ○ Average ○ Better Than Average ○ Excellent ⊖ Insufficient Information

©1993 by IntelliChoice, Inc. (408) 554-8711 All Rights Reserved. Reproduction Prohibited.
Refer to *Section 3: Annotated Vehicle Charts* for an explanation of these charts.

Saab 9000 CS Turbo
4 Door Hatchback

Midsize

- 2.3L 200 hp Turbo Gas Fuel Inject.
- 4 Cylinder In-Line
- Manual 5 Speed
- 2 Wheel Front
- Driver Airbag Psngr Belts

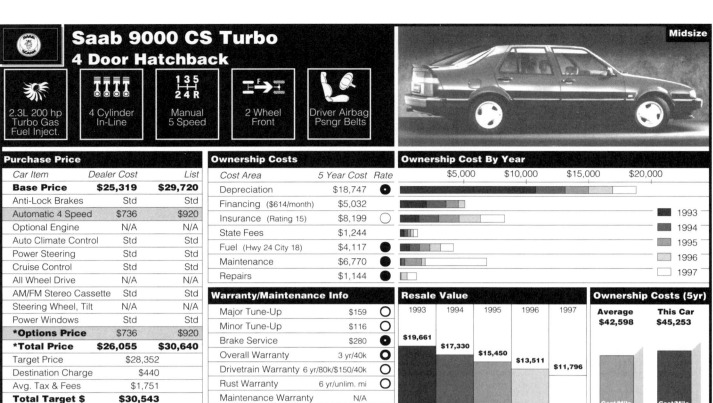

Purchase Price

Car Item	Dealer Cost	List
Base Price	**$25,319**	**$29,720**
Anti-Lock Brakes	Std	Std
Automatic 4 Speed	$736	$920
Optional Engine	N/A	N/A
Auto Climate Control	Std	Std
Power Steering	Std	Std
Cruise Control	Std	Std
All Wheel Drive	N/A	N/A
AM/FM Stereo Cassette	Std	Std
Steering Wheel, Tilt	N/A	N/A
Power Windows	Std	Std
***Options Price**	$736	$920
***Total Price**	**$26,055**	**$30,640**
Target Price	$28,352	
Destination Charge	$440	
Avg. Tax & Fees	$1,751	
Total Target $	**$30,543**	
Average Dealer Option Cost:	80%	

The 1993 9000 Turbo is available in four models-CS and CSE Turbo hatchbacks; and CD and CDE Turbo sedans. New for 1993, the 9000 CS Turbo hatchback features the Trionic engine management system. The Direct Ignition System is also incorporated into the Trionic logic. Other features include a system for improved side impact protection, Traction Control and a 150-watt audio system with Compact Disc Player/Equilizer. Other features include plush floormats and a leather wrapped steering wheel.

Ownership Costs

Cost Area	5 Year Cost	Rate
Depreciation	$18,747	●
Financing ($614/month)	$5,032	
Insurance (Rating 15)	$8,199	○
State Fees	$1,244	
Fuel (Hwy 24 City 18)	$4,117	●
Maintenance	$6,770	●
Repairs	$1,144	●

Warranty/Maintenance Info

Major Tune-Up	$159	○
Minor Tune-Up	$116	○
Brake Service	$280	●
Overall Warranty	3 yr/40k	○
Drivetrain Warranty	6 yr/80k/$150/40k	○
Rust Warranty	6 yr/unlim. mi	○
Maintenance Warranty	N/A	
Roadside Assistance	3 yr/40k	

Ownership Cost By Year

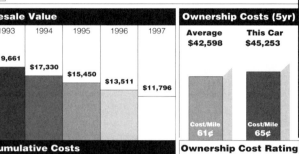

Resale Value

1993	1994	1995	1996	1997
$19,661	$17,330	$15,450	$13,511	$11,796

Cumulative Costs

	1993	1994	1995	1996	1997
Annual	$15,649	$6,842	$6,985	$5,858	$9,919
Total	$15,649	$22,491	$29,476	$35,334	$45,253

Ownership Costs (5yr)

Average	This Car
$42,598	$45,253
Cost/Mile 61¢	Cost/Mile 65¢

Ownership Cost Rating

Worse Than Average

Saab 9000 CSE
4 Door Hatchback

Luxury

- 2.3L 150 hp Gas Fuel Inject.
- 4 Cylinder In-Line
- Manual 5 Speed
- 2 Wheel Front
- Driver Airbag Psngr Belts

Purchase Price

Car Item	Dealer Cost	List
Base Price	**$25,687**	**$31,060**
Anti-Lock Brakes	Std	Std
Automatic 4 Speed	$736	$920
Optional Engine	N/A	N/A
Auto Climate Control	Std	Std
Power Steering	Std	Std
Cruise Control	Std	Std
All Wheel Drive	N/A	N/A
AM/FM Stereo CD	Std	Std
Steering Wheel, Tilt	N/A	N/A
Power Windows	Std	Std
***Options Price**	$736	$920
***Total Price**	**$26,423**	**$31,980**
Target Price	$28,771	
Destination Charge	$440	
Avg. Tax & Fees	$1,785	
Total Target $	**$30,996**	
Average Dealer Option Cost:	80%	

The 1993 9000 is available in four models-CS and CSE hatchbacks; and CS and CDE sedans. New for 1993, the 9000 CSE hatchback features a driver's-side airbag and anti-lock brakes. The CSE hatchback is powered by a 150 horsepower, 2.3-liter naturally-aspirated, balance-shaft engine. An upgraded alarm system is standard equipment in every CSE and also includes central locking with child-proof rear door locks and windows. The 9000 CSE hacthback is backed by a 6 year/80,000 mile Major Systems Warranty.

Ownership Costs

Cost Area	5 Year Cost	Rate
Depreciation	$19,591	●
Financing ($623/month)	$5,106	
Insurance (Rating 13)	$7,797	○
State Fees	$1,297	
Fuel (Hwy 24 City 18)	$4,117	○
Maintenance	$5,504	○
Repairs	$1,004	○

Warranty/Maintenance Info

Major Tune-Up	$134	○
Minor Tune-Up	$91	○
Brake Service	$280	●
Overall Warranty	3 yr/40k	○
Drivetrain Warranty	6 yr/80k/$150/40k	○
Rust Warranty	6 yr/unlim. mi	○
Maintenance Warranty	N/A	
Roadside Assistance	3 yr/40k	

Resale Value

1993	1994	1995	1996	1997
$20,201	$17,539	$15,376	$13,304	$11,405

Cumulative Costs

	1993	1994	1995	1996	1997
Annual	$15,491	$7,063	$7,132	$5,812	$8,918
Total	$15,491	$22,554	$29,686	$35,498	$44,416

Ownership Costs (5yr)

Average	This Car
$43,367	$44,416
Cost/Mile 62¢	Cost/Mile 63¢

Ownership Cost Rating

Average

Legend: ● Poor ◉ Worse Than Average ○ Average ◎ Better Than Average ○ Excellent ⊖ Insufficient Information

* Includes shaded options
** Other purchase requirements apply

©1993 by IntelliChoice, Inc. (408) 554-8711 All Rights Reserved. Reproduction Prohibited.
Refer to *Section 3: Annotated Vehicle Charts* for an explanation of these charts.

Saab 9000 CSE Turbo
4 Door Hatchback

 2.3L 200 hp Turbo Gas Fuel Inject.
 4 Cylinder In-Line
 Manual 5 Speed
 2 Wheel Front
 Driver Airbag Psngr Belts

Luxury — Base Model Shown

Purchase Price

Car Item	Dealer Cost	List
Base Price	$28,883	$35,055
Anti-Lock Brakes	Std	Std
Automatic 4 Speed	$736	$920
Optional Engine	N/A	N/A
Auto Climate Control	Std	Std
Power Steering	Std	Std
Cruise Control	Std	Std
All Wheel Drive	N/A	N/A
AM/FM Stereo CD	Std	Std
Steering Wheel, Tilt	N/A	N/A
Power Windows	Std	Std
*Options Price	$736	$920
*Total Price	$29,619	$35,975
Target Price	$32,432	
Destination Charge	$440	
Avg. Tax & Fees	$2,008	
Luxury Tax	$287	
Total Target $	**$35,167**	

The 1993 9000 is available in four models—CS and CSE Turbo hatchbacks; and CD and CDE Turbo sedans. New for 1993, the 9000 CSE Turbo hatchback features the Trionic engine management system. The Direct Ignition System is also incorporated into the Trionic logic. Standard features include driver's-side airbag, anti-lock brakes, three-point seat belts and head restraints. Other features include traction control, CFC-free air-conditioning and a system for improved side impact protection.

Ownership Costs

Cost Area	5 Year Cost	Rate
Depreciation	$21,318	●
Financing ($707/month)	$5,794	
Insurance (Rating 15)	$8,199	○
State Fees	$1,456	
Fuel (Hwy 24 City 18)	$4,117	◐
Maintenance	$6,770	●
Repairs	$1,144	◐

Warranty/Maintenance Info

Major Tune-Up	$159	○
Minor Tune-Up	$116	○
Brake Service	$280	●
Overall Warranty	3 yr/40k	○
Drivetrain Warranty	6 yr/80k/$150/40k	○
Rust Warranty	6 yr/unlim. mi	○
Maintenance Warranty	N/A	
Roadside Assistance	3 yr/40k	

Ownership Cost By Year
1993, 1994, 1995, 1996, 1997

Resale Value

1993	1994	1995	1996	1997
$23,209	$20,452	$18,198	$15,951	$13,849

Cumulative Costs

	1993	1994	1995	1996	1997
Annual	$17,095	$7,554	$7,552	$6,259	$10,338
Total	$17,095	$24,649	$32,201	$38,460	$48,798

Ownership Costs (5yr)
- Average: $46,357 — Cost/Mile 66¢
- This Car: $48,798 — Cost/Mile 70¢

Ownership Cost Rating
○ Average

Saab 9000 CD
4 Door Sedan

 2.3L 150 hp Gas Fuel Inject.
 4 Cylinder In-Line
Manual 5 Speed
 2 Wheel Front
 Driver Airbag Psngr Belts

Midsize

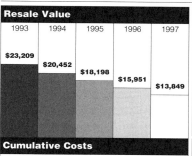
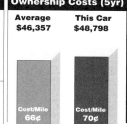

Purchase Price

Car Item	Dealer Cost	List
Base Price	$21,349	$24,825
Anti-Lock Brakes	Std	Std
Automatic 4 Speed	$736	$920
Optional Engine	N/A	N/A
Auto Climate Control	Std	Std
Power Steering	Std	Std
Cruise Control	Std	Std
All Wheel Drive	N/A	N/A
AM/FM Stereo Cassette	Std	Std
Steering Wheel, Tilt	N/A	N/A
Power Windows	Std	Std
*Options Price	$736	$920
*Total Price	$22,085	$25,745
Target Price	$23,864	
Destination Charge	$440	
Avg. Tax & Fees	$1,477	
Total Target $	**$25,781**	
Average Dealer Option Cost:	80%	

The 1993 9000 is available in four models—CS and CSE hatchbacks; and CS and CDE sedans. New for 1993, the 9000 CD sedan features two new colors: Imola Red and mica-based Ruby Red. Standard equipment for the CD sedan includes driver's-side airbag, anti-lock brakes, front seat pretensioners, front and rear head restraints, air conditioning, body-side moldings and a rear window defroster. An upgraded alarm system is standard equipment and also includes central locking with door locks and windows.

Ownership Costs

Cost Area	5 Year Cost	Rate
Depreciation	$16,586	●
Financing ($518/month)	$4,246	
Insurance (Rating 12)	$7,650	○
State Fees	$1,047	
Fuel (Hwy 24 City 18)	$4,117	●
Maintenance	$5,545	◐
Repairs	$1,004	●

Warranty/Maintenance Info

Major Tune-Up	$134	○
Minor Tune-Up	$91	○
Brake Service	$280	●
Overall Warranty	3 yr/40k	○
Drivetrain Warranty	6 yr/80k/$150/40k	○
Rust Warranty	6 yr/unlim. mi	○
Maintenance Warranty	N/A	
Roadside Assistance	3 yr/40k	

Resale Value

1993	1994	1995	1996	1997
$16,182	$14,011	$12,351	$10,643	$9,195

Cumulative Costs

	1993	1994	1995	1996	1997
Annual	$13,847	$6,219	$6,380	$5,311	$8,438
Total	$13,847	$20,066	$26,446	$31,757	$40,195

Ownership Costs (5yr)
- Average: $37,593 — Cost/Mile 54¢
- This Car: $40,195 — Cost/Mile 57¢

Ownership Cost Rating
● Poor

---

* Includes shaded options
** Other purchase requirements apply

Legend: ● Poor ◐ Worse Than Average ○ Average ◔ Better Than Average ○ Excellent ⊖ Insufficient Information

©1993 by IntelliChoice, Inc. (408) 554-8711 All Rights Reserved. Reproduction Prohibited.
Refer to *Section 3: Annotated Vehicle Charts* for an explanation of these charts.

page 209

Saab 9000 CD Turbo
4 Door Sedan

Midsize — Base Model Shown

- 2.3L 200 hp Turbo Gas Fuel Inject.
- 4 Cylinder In-Line
- Manual 5 Speed
- 2 Wheel Front
- Driver Airbag Psngr Belts

Purchase Price

Car Item	Dealer Cost	List
Base Price	**$24,545**	**$28,820**
Anti-Lock Brakes	Std	Std
Automatic 4 Speed	$736	$920
Optional Engine	N/A	N/A
Auto Climate Control	Std	Std
Power Steering	Std	Std
Cruise Control	Std	Std
All Wheel Drive	N/A	N/A
AM/FM Stereo Cassette	Std	Std
Steering Wheel, Tilt	N/A	N/A
Power Windows	Std	Std
*Options Price	$736	$920
*Total Price	$25,281	$29,740
Target Price		$27,472
Destination Charge		$440
Avg. Tax & Fees		$1,698
Total Target $		**$29,610**
Average Dealer Option Cost:		80%

The 1993 9000 Turbo is available in four models-CS and CSE Turbo hatchbacks; and CD and CDE Turbo sedans. New for 1993, the 9000 CD Turbo sedan features the Trionic engine management system. The Direct Ignition Sytem is also incorporated into the Trionic logic. The CD Turbo sedan offers power seats, power glass sunroof and a Luxury Package. Standard features include traction control, driver's-side airbag, anti-lock brakes and a 150-watt audio system with compact disc player/equilizer.

Ownership Costs

Cost Area	5 Year Cost	Rate
Depreciation	$18,132	●
Financing ($595/month)	$4,878	
Insurance (Rating 14+)	$9,547	○
State Fees	$1,207	
Fuel (Hwy 24 City 18)	$4,117	●
Maintenance	$6,811	●
Repairs	$1,144	●

Warranty/Maintenance Info

Major Tune-Up	$159	○
Minor Tune-Up	$116	○
Brake Service	$280	●
Overall Warranty	3 yr/40k	○
Drivetrain Warranty	6 yr/80k/$150/40k	○
Rust Warranty	6 yr/unlim. mi	○
Maintenance Warranty	N/A	
Roadside Assistance	3 yr/40k	

Ownership Cost By Year

1993, 1994, 1995, 1996, 1997

Resale Value

1993	1994	1995	1996	1997
$19,309	$17,012	$15,126	$13,179	$11,478

Cumulative Costs

	1993	1994	1995	1996	1997
Annual	$15,244	$7,012	$7,222	$6,127	$10,232
Total	$15,244	$22,256	$29,478	$35,605	$45,837

Ownership Costs (5yr)

Average: $41,678 — Cost/Mile 60¢
This Car: $45,837 — Cost/Mile 65¢

Ownership Cost Rating
● Poor

Saab 9000 CDE
4 Door Sedan

Luxury

- 2.3L 150 hp Gas Fuel Inject.
- 4 Cylinder In-Line
- Manual 5 Speed
- 2 Wheel Front
- Driver Airbag Psngr Belts

Purchase Price

Car Item	Dealer Cost	List
Base Price	**$25,496**	**$30,830**
Anti-Lock Brakes	Std	Std
Automatic 4 Speed	$736	$920
Optional Engine	N/A	N/A
Auto Climate Control	Std	Std
Power Steering	Std	Std
Cruise Control	Std	Std
All Wheel Drive	N/A	N/A
AM/FM Stereo CD	Std	Std
Steering Wheel, Tilt	N/A	N/A
Power Windows	Std	Std
*Options Price	$736	$920
*Total Price	$26,232	$31,750
Target Price		$28,553
Destination Charge		$440
Avg. Tax & Fees		$1,772
Total Target $		**$30,765**
Average Dealer Option Cost:		80%

The 1993 9000 is available in four models-CS and CSE hatchbacks; and CS and CDE sedans. New for 1993, the 9000 CDE sedan features a driver's-side airbag and anti-lock braking systems. Standard features include dual power leather seats, CD player with graphic equalizer, power glass sunroof, fog lights and body-side moldings. Other features include two new colors for 1993: Imola Red and mica-based Ruby Red. The 9000 CDE sedan is backed by a 6 year/80,000 mile Major Systems Warranty.

Ownership Costs

Cost Area	5 Year Cost	Rate
Depreciation	$19,475	●
Financing ($618/month)	$5,068	
Insurance (Rating 12)	$7,650	○
State Fees	$1,288	
Fuel (Hwy 24 City 18)	$4,117	○
Maintenance	$5,545	○
Repairs	$1,004	○

Warranty/Maintenance Info

Major Tune-Up	$134	○
Minor Tune-Up	$91	○
Brake Service	$280	●
Overall Warranty	3 yr/40k	○
Drivetrain Warranty	6 yr/80k/$150/40k	○
Rust Warranty	6 yr/unlim. mi	○
Maintenance Warranty	N/A	
Roadside Assistance	3 yr/40k	

Resale Value

1993	1994	1995	1996	1997
$19,963	$17,287	$15,203	$13,145	$11,290

Cumulative Costs

	1993	1994	1995	1996	1997
Annual	$15,452	$7,035	$7,016	$5,763	$8,881
Total	$15,452	$22,487	$29,503	$35,266	$44,147

Ownership Costs (5yr)

Average: $43,195 — Cost/Mile 62¢
This Car: $44,147 — Cost/Mile 63¢

Ownership Cost Rating
○ Average

---

page 210

* Includes shaded options
** Other purchase requirements apply

Legend: ● Poor | ◐ Worse Than Average | ○ Average | ○ Better Than Average | ○ Excellent | ⊖ Insufficient Information

©1993 by IntelliChoice, Inc. (408) 554-8711 All Rights Reserved. Reproduction Prohibited.
Refer to *Section 3: Annotated Vehicle Charts* for an explanation of these charts.

Saab 9000 CDE Turbo
4 Door Sedan

 2.3L 200 hp Turbo Gas Fuel Inject.
 4 Cylinder In-Line
 Manual 5 Speed
 2 Wheel Front
 Driver Airbag Psngr Belts

Luxury — Base Model Shown

Purchase Price

Car Item	Dealer Cost	List
Base Price	**$28,692**	**$34,825**
Anti-Lock Brakes	Std	Std
Automatic 4 Speed	$736	$920
Optional Engine	N/A	N/A
Auto Climate Control	Std	Std
Power Steering	Std	Std
Cruise Control	Std	Std
All Wheel Drive	N/A	N/A
AM/FM Stereo CD	Std	Std
Steering Wheel, Tilt	N/A	N/A
Power Windows	Std	Std
***Options Price**	$736	$920
***Total Price**	**$29,428**	**$35,745**
Target Price		$32,212
Destination Charge		$440
Avg. Tax & Fees		$1,995
Luxury Tax		$265
Total Target $		**$34,912**

The 1993 9000 Turbo is available in four models-CS and CSE Turbo hatchbacks; and CD and CDE Turbo sedans. New for 1993, the 9000 CDE Turbo sedan features the Trionic engine management system. The Direct Ignition System is also incorporated into the Trionic logic. Standard features include a driver's-side airbag, anti-lock brakes, front and rear head restraints along with three-point seat belt systems.

Ownership Costs

Cost Area	5 Year Cost	Rate
Depreciation	$21,120	●
Financing ($702/month)	$5,752	
Insurance (Rating 14+)	$9,547	○
State Fees	$1,447	
Fuel (Hwy 24 City 18)	$4,117	○
Maintenance	$6,811	●
Repairs	$1,144	○

Warranty/Maintenance Info

Major Tune-Up	$159	○
Minor Tune-Up	$116	○
Brake Service	$280	●
Overall Warranty	3 yr/40k	○
Drivetrain Warranty	6 yr/80k/$150/40k	○
Rust Warranty	6 yr/unlim. mi	○
Maintenance Warranty	N/A	
Roadside Assistance	3 yr/40k	

Ownership Cost By Year — 1993, 1994, 1995, 1996, 1997

Resale Value
1993: $23,312 | 1994: $20,536 | 1995: $18,224 | 1996: $15,919 | 1997: $13,792

Cumulative Costs
	1993	1994	1995	1996	1997
Annual	$16,967	$7,818	$7,869	$6,591	$10,694
Total	$16,967	$24,785	$32,654	$39,245	$49,939

Ownership Costs (5yr)
Average: $46,185 | This Car: $49,939
Cost/Mile: 66¢ | 71¢

Ownership Cost Rating: ◉ Worse Than Average

Saab 900 S
2 Door Hatchback

 2.1L 140 hp Gas Fuel Inject.
 4 Cylinder In-Line
 Manual 5 Speed
 2 Wheel Front
Driver Airbag Psngr Belts

Compact

Purchase Price

Car Item	Dealer Cost	List
Base Price	**$17,192**	**$20,345**
Anti-Lock Brakes	Std	Std
Automatic 3 Speed	$552	$690
Optional Engine	N/A	N/A
Air Conditioning	Std	Std
Power Steering	Std	Std
Cruise Control	Pkg	Pkg
All Wheel Drive	N/A	N/A
AM/FM Stereo Cassette	Std	Std
Steering Wheel, Tilt	N/A	N/A
Power Windows	Std	Std
***Options Price**	$552	$690
***Total Price**	**$17,744**	**$21,035**
Target Price		$19,032
Destination Charge		$440
Avg. Tax & Fees		$1,189
Total Target $		**$20,661**
Average Dealer Option Cost:		80%

The 1993 Saab 900 Series is available in three models - 900S Hatchback, 900S Convertible and 900S Sedan. New for 1993, the 900S hatchback is offered in two new exterior colors (Ruby Red and Imola Red). The 900S features pique upholstery, steel wheels with 16-spoke covers, air conditioning and 150 watts of sound from the standard audio system. The 900S also offers a Luxury Package that includes a power sunroof, cruise control, fog lamps, 15-spoke, light-alloy wheels and leather seating surfaces.

Ownership Costs

Cost Area	5 Year Cost	Rate
Depreciation	$11,890	●
Financing ($415/month)	$3,404	
Insurance (Rating 17)	$9,084	●
State Fees	$859	
Fuel (Hwy 21 City 18)	$4,430	●
Maintenance	$5,026	●
Repairs	$1,004	○

Warranty/Maintenance Info

Major Tune-Up	$151	○
Minor Tune-Up	$93	○
Brake Service	$261	●
Overall Warranty	3 yr/40k	○
Drivetrain Warranty	6 yr/80k/$150/40k	○
Rust Warranty	6 yr/unlim. mi	○
Maintenance Warranty	N/A	
Roadside Assistance	3 yr/40k	

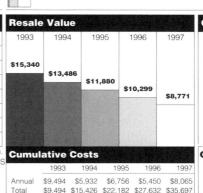

Ownership Cost By Year — 1993, 1994, 1995, 1996, 1997

Resale Value
1993: $15,340 | 1994: $13,486 | 1995: $11,880 | 1996: $10,299 | 1997: $8,771

Cumulative Costs
	1993	1994	1995	1996	1997
Annual	$9,494	$5,932	$6,756	$5,450	$8,065
Total	$9,494	$15,426	$22,182	$27,632	$35,697

Ownership Costs (5yr)
Average: $32,556 | This Car: $35,697
Cost/Mile: 47¢ | 51¢

Ownership Cost Rating: ● Poor

---

* Includes shaded options
** Other purchase requirements apply

● Poor | ◉ Worse Than Average | ○ Average | ◯ Better Than Average | Excellent | ⊖ Insufficient Information

©1993 by IntelliChoice, Inc. (408) 554-8711 All Rights Reserved. Reproduction Prohibited.
Refer to *Section 3: Annotated Vehicle Charts* for an explanation of these charts.

Saab 900 Turbo
2 Door Hatchback

 2.0L 160 hp Turbo Gas Fuel Inject.
 4 Cylinder In-Line
 Manual 5 Speed
 2 Wheel Front
Driver Airbag Psngr Belts

Compact

Purchase Price

Car Item	Dealer Cost	List
Base Price	**$25,575**	**$30,555**
Anti-Lock Brakes	Std	Std
Automatic 3 Speed	$552	$690
Optional Engine	N/A	N/A
Air Conditioning	Std	Std
Power Steering	Std	Std
Cruise Control	Std	Std
All Wheel Drive	N/A	N/A
AM/FM Stereo CD	Std	Std
Steering Wheel, Tilt	N/A	N/A
Power Windows	Std	Std
*Options Price	$552	$690
*Total Price	$26,127	$31,245
Target Price	$28,443	
Destination Charge	$440	
Avg. Tax & Fees	$1,761	
Total Target $	**$30,644**	
Average Dealer Option Cost:	80%	

Ownership Costs

Cost Area	5 Year Cost	Rate
Depreciation	$18,070	●
Financing ($616/month)	$5,049	
Insurance (Rating 22)	$11,167	◉
State Fees	$1,267	
Fuel (Hwy 23 City 19)	$4,111	●
Maintenance	$5,687	●
Repairs	$1,085	●

Warranty/Maintenance Info

Major Tune-Up	$168	○
Minor Tune-Up	$115	◉
Brake Service	$261	◉
Overall Warranty	3 yr/40k	○
Drivetrain Warranty	6 yr/80k/$150/40k	○
Rust Warranty	6 yr/unlim. mi	○
Maintenance Warranty	N/A	
Roadside Assistance	3 yr/40k	

Ownership Cost By Year
1993, 1994, 1995, 1996, 1997

Resale Value
1993	1994	1995	1996	1997
$22,242	$19,518	$17,116	$14,798	$12,574

Ownership Costs (5yr)
Average	This Car
$41,324	$46,436
Cost/Mile 59¢	Cost/Mile 66¢

Cumulative Costs
	1993	1994	1995	1996	1997
Annual	$13,729	$7,812	$8,564	$6,827	$9,504
Total	$13,729	$21,541	$30,105	$36,932	$46,436

Ownership Cost Rating
● Poor

The 1993 900 Turbo is available in two models - two door hatchback and convertible. New for 1993, the 900 Turbo Hatchback is powered by a 2.0-liter turbocharged 160 horsepower engine. It may be specified with either a 5-speed manual transmission or an optional automatic. Standard features include 15-inch wheels, new instrument graphics and standard power/heated sideview mirrors. Like every 1993 Saab, the 900 Turbo Hatchbacks are backed by a 6 year/80,000 mile Major Systems Warranty.

Saab 900 S
4 Door Sedan

 2.1L 140 hp Gas Fuel Inject.
 4 Cylinder In-Line
 Manual 5 Speed
 2 Wheel Front
Driver Airbag Psngr Belts

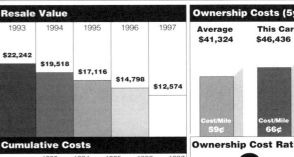

Compact

Purchase Price

Car Item	Dealer Cost	List
Base Price	**$17,711**	**$20,960**
Anti-Lock Brakes	Std	Std
Automatic 3 Speed	$552	$690
Optional Engine	N/A	N/A
Air Conditioning	Std	Std
Power Steering	Std	Std
Cruise Control	Pkg	Pkg
All Wheel Drive	N/A	N/A
AM/FM Stereo Cassette	Std	Std
Steering Wheel, Tilt	N/A	N/A
Power Windows	Std	Std
*Options Price	$552	$690
*Total Price	$18,263	$21,650
Target Price	$19,607	
Destination Charge	$440	
Avg. Tax & Fees	$1,223	
Total Target $	**$21,270**	
Average Dealer Option Cost:	80%	

Ownership Costs

Cost Area	5 Year Cost	Rate
Depreciation	$12,528	●
Financing ($427/month)	$3,504	
Insurance (Rating 18)	$9,453	●
State Fees	$885	
Fuel (Hwy 21 City 18)	$4,430	●
Maintenance	$5,026	●
Repairs	$1,004	○

Warranty/Maintenance Info

Major Tune-Up	$151	○
Minor Tune-Up	$93	○
Brake Service	$261	◉
Overall Warranty	3 yr/40k	○
Drivetrain Warranty	6 yr/80k/$150/40k	○
Rust Warranty	6 yr/unlim. mi	○
Maintenance Warranty	N/A	
Roadside Assistance	3 yr/40k	

Resale Value
1993	1994	1995	1996	1997
$15,267	$13,428	$11,821	$10,241	$8,742

Ownership Costs (5yr)
Average	This Car
$33,084	$36,830
Cost/Mile 47¢	Cost/Mile 53¢

Cumulative Costs
	1993	1994	1995	1996	1997
Annual	$10,291	$6,025	$6,856	$5,538	$8,120
Total	$10,291	$16,316	$23,172	$28,710	$36,830

Ownership Cost Rating
● Poor

The 1993 Saab 900 Series is available in three models - 900S Hatchback, 900S Convertible and 900S Sedan. New for 1993, the 900S sedan is offered in two new exterior colors (Ruby Red and Imola Red). The 900S features pique upholstery, steel wheels with 16-spoke covers, air conditioning and 150 watts of sound from the standard audio system. The 900S also offers a Luxury Package that includes a power sunroof, cruise control, fog lamps, 15-spoke, light-alloy wheels and leather seating surfaces.

---

* Includes shaded options
** Other purchase requirements apply

● Poor   ◉ Worse Than Average   ◐ Average   ○ Better Than Average   ○ Excellent   ⊖ Insufficient Information

©1993 by IntelliChoice, Inc. (408) 554-8711   All Rights Reserved. Reproduction Prohibited.
Refer to *Section 3: Annotated Vehicle Charts* for an explanation of these charts.

Saab 900 S
2 Door Convertible

 2.1L 140 hp Gas Fuel Inject.
 4 Cylinder In-Line
 Manual 5 Speed
 2 Wheel Front
 Driver Airbag Psngr Belts

Compact

Purchase Price

Car Item	Dealer Cost	List
Base Price	**$27,175**	**$32,160**
Anti-Lock Brakes	Std	Std
Automatic 3 Speed	$552	$690
Optional Engine	N/A	N/A
Air Conditioning	Std	Std
Power Steering	Std	Std
Cruise Control	Std	Std
All Wheel Drive	N/A	N/A
AM/FM Stereo Cassette	Std	Std
Steering Wheel, Tilt	N/A	N/A
Power Windows	Std	Std
*****Options Price**	**$552**	**$690**
*****Total Price**	**$27,727**	**$32,850**
Target Price		$30,270
Destination Charge		$440
Avg. Tax & Fees		$1,869
Luxury Tax		$71
Total Target $		**$32,650**

The 1993 Saab 900 Series is available in three models - 900S Hatchback, 900S Convertible and 900S Sedan. New for 1993, the 900S convertible is offered in two new exterior colors (a mica-based Ruby Red and Imola Red). The 900S convertible features power steering, alloy wheels, air conditioning and 150 watts of sound from the standard audio system. The 900S Convertible features an electrically-actuated fully-lined, snug-fitting, Cambria cloth top that includes an electrically-heated rear glass window.

Ownership Costs

Cost Area	5 Year Cost	Rate
Depreciation	$12,148	○
Financing ($656/month)	$5,378	
Insurance (Rating 22)	$10,640	◉
State Fees	$1,332	
Fuel (Hwy 21 City 18)	$4,430	●
Maintenance	$5,026	◉
Repairs	$1,004	○

Warranty/Maintenance Info

Major Tune-Up	$151	○
Minor Tune-Up	$93	○
Brake Service	$261	◉
Overall Warranty	3 yr/40k	○
Drivetrain Warranty	6 yr/80k/$150/40k	○
Rust Warranty	6 yr/unlim. mi	○
Maintenance Warranty	N/A	
Roadside Assistance	3 yr/40k	

Ownership Cost By Year

1993 / 1994 / 1995 / 1996 / 1997

Resale Value

1993	1994	1995	1996	1997
$28,317	$26,191	$24,397	$22,497	$20,502

Cumulative Costs

	1993	1994	1995	1996	1997
Annual	$9,733	$7,227	$7,740	$6,319	$8,939
Total	$9,733	$16,960	$24,700	$31,019	$39,958

Ownership Costs (5yr)

Average: $42,702 — This Car: $39,958
Cost/Mile: 61¢ — Cost/Mile: 57¢

Ownership Cost Rating
○ Better Than Average

Saab 900 Turbo
2 Door Convertible

 2.0L 160 hp Turbo Gas Fuel Inject.
 4 Cylinder In-Line
 Manual 5 Speed
 2 Wheel Front
 Driver Airbag Psngr Belts

Compact

Purchase Price

Car Item	Dealer Cost	List
Base Price	**$30,834**	**$37,060**
Anti-Lock Brakes	Std	Std
Automatic 3 Speed	$552	$690
Optional Engine	N/A	N/A
Air Conditioning	Std	Std
Power Steering	Std	Std
Cruise Control	Std	Std
All Wheel Drive	N/A	N/A
AM/FM Stereo CD	Std	Std
Steering Wheel, Tilt	N/A	N/A
Power Windows	Std	Std
*****Options Price**	**$552**	**$690**
*****Total Price**	**$31,386**	**$37,750**
Target Price		$34,368
Destination Charge		$440
Avg. Tax & Fees		$2,122
Luxury Tax		$481
Total Target $		**$37,411**

The 1993 900 Turbo is available in two models - two door hatchback and convertible. New for 1993, the 900 Turbo Convertible is powered by a 2.0-liter turbocharged 160 horsepower engine. Standard features include a theft alarm system with keyless entry, anti-lock brakes and driver's-side airbag. Other features include an electrically-actuated convertible top and a color-coordinated 3-piece, hard boot. The 900 Turbo Convertible is backed by a 6 year/80,000 mile Major Systems Warranty.

Ownership Costs

Cost Area	5 Year Cost	Rate
Depreciation	$14,541	○
Financing ($752/month)	$6,163	
Insurance (Rating 24)	$11,704	◉
State Fees	$1,528	
Fuel (Hwy 21 City 18)	$4,430	●
Maintenance	$5,687	●
Repairs	$1,085	●

Warranty/Maintenance Info

Major Tune-Up	$168	○
Minor Tune-Up	$115	◉
Brake Service	$261	◉
Overall Warranty	3 yr/40k	○
Drivetrain Warranty	6 yr/80k/$150/40k	○
Rust Warranty	6 yr/unlim. mi	○
Maintenance Warranty	N/A	
Roadside Assistance	3 yr/40k	

Ownership Cost By Year

1993 / 1994 / 1995 / 1996 / 1997

Resale Value

1993	1994	1995	1996	1997
$31,978	$29,501	$27,393	$25,141	$22,870

Cumulative Costs

	1993	1994	1995	1996	1997
Annual	$11,447	$8,136	$8,715	$7,065	$9,775
Total	$11,447	$19,583	$28,298	$35,363	$45,138

Ownership Costs (5yr)

Average: $46,909 — This Car: $45,138
Cost/Mile: 67¢ — Cost/Mile: 64¢

Ownership Cost Rating
○ Better Than Average

---

* Includes shaded options
** Other purchase requirements apply

● Poor   ◉ Worse Than Average   ○ Average   ○ Better Than Average   ○ Excellent   ⊖ Insufficient Information

©1993 by IntelliChoice, Inc. (408) 554-8711 All Rights Reserved. Reproduction Prohibited.
Refer to *Section 3: Annotated Vehicle Charts* for an explanation of these charts.

page 213

Saturn SC1
2 Door Coupe

Compact

- 1.9L 85 hp Gas Fuel Inject.
- 4 Cylinder In-Line
- Manual 5 Speed
- 2 Wheel Front
- Driver Airbag Psngr Belts

Purchase Price

Car Item	Dealer Cost	List
Base Price	**$9,786**	**$10,995**
Anti-Lock Brakes	$530	$595
Automatic 4 Speed	$667	$750
Optional Engine	N/A	N/A
Air Conditioning	$739	$830
Power Steering	Std	Std
Cruise Control	$196	$220
All Wheel Drive	N/A	N/A
AM/FM Stereo Cassette	$160	$180
Steering Wheel, Tilt	Std	Std
Power Windows	Pkg	Pkg
*Options Price	$1,566	$1,760
*Total Price	$11,352	$12,755
Target Price	$12,455	
Destination Charge	$300	
Avg. Tax & Fees	$769	
Total Target $	**$13,524**	
Average Dealer Option Cost:	89%	

Ownership Costs

Cost Area	5 Year Cost	Rate
Depreciation	$6,701	O
Financing ($272/month)	$2,228	
Insurance (Rating 8)	$7,223	●
State Fees	$521	
Fuel (Hwy 36 City 26)	$2,795	O
Maintenance	$3,668	O
Repairs	$610	O

Warranty/Maintenance Info

Major Tune-Up	$116	O
Minor Tune-Up	$86	O
Brake Service	$178	O
Overall Warranty	3 yr/36k	O
Drivetrain Warranty	3 yr/36k	O
Rust Warranty	6 yr/100k	O
Maintenance Warranty	N/A	
Roadside Assistance	3 yr/36k	

Ownership Cost By Year

1993, 1994, 1995, 1996, 1997

Resale Value

1993	1994	1995	1996	1997
$11,254	$9,985	$9,077	$7,802	$6,823

Cumulative Costs

	1993	1994	1995	1996	1997
Annual	$5,225	$4,214	$4,502	$4,859	$4,946
Total	$5,225	$9,439	$13,941	$18,800	$23,746

Ownership Costs (5yr)

Average: $25,446 | This Car: $23,746
Cost/Mile: 36¢ | Cost/Mile: 34¢

Ownership Cost Rating

Better Than Average

The Saturn SC coupe is available in SC1 and SC2 editions. The entry-level SC1 joins the line-up this year as Saturn's most affordable model. The SC1 comes standard with a custom cloth and vinyl interior, fully independent suspension system and all-season radials. Optional on the SC1 are cast aluminum "geartooth" wheels. Every Saturn comes with Saturn's Owner Protection Package which gives the owner the option of returning the vehicle for a refund within the first 30 days or 1,500 miles.

Saturn SC2
2 Door Coupe

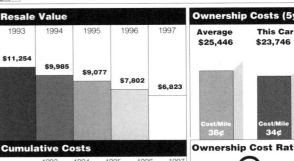

Compact

- 1.9L 124 hp Gas Fuel Inject.
- 4 Cylinder In-Line
- Manual 5 Speed
- 2 Wheel Front
- Driver Airbag Psngr Belts

Purchase Price

Car Item	Dealer Cost	List
Base Price	**$11,388**	**$12,795**
Anti-Lock Brakes	$530	$595
Automatic 4 Speed	$667	$750
Optional Engine	N/A	N/A
Air Conditioning	$739	$830
Power Steering	Std	Std
Cruise Control	$196	$220
All Wheel Drive	N/A	N/A
AM/FM Stereo Cassette	$160	$180
Steering Wheel, Tilt	Std	Std
Power Windows	Pkg	Pkg
*Options Price	$1,566	$1,760
*Total Price	$12,954	$14,555
Target Price	$14,255	
Destination Charge	$300	
Avg. Tax & Fees	$877	
Total Target $	**$15,432**	
Average Dealer Option Cost:	89%	

Ownership Costs

Cost Area	5 Year Cost	Rate
Depreciation	$7,934	O
Financing ($310/month)	$2,543	
Insurance (Rating 11)	$7,693	●
State Fees	$595	
Fuel (Hwy 33 City 24)	$3,039	O
Maintenance	$4,488	O
Repairs	$610	O

Warranty/Maintenance Info

Major Tune-Up	$196	●
Minor Tune-Up	$166	●
Brake Service	$178	O
Overall Warranty	3 yr/36k	O
Drivetrain Warranty	3 yr/36k	O
Rust Warranty	6 yr/100k	O
Maintenance Warranty	N/A	
Roadside Assistance	3 yr/36k	

Resale Value

1993	1994	1995	1996	1997
$12,487	$11,056	$10,022	$8,585	$7,498

Cumulative Costs

	1993	1994	1995	1996	1997
Annual	$6,180	$4,711	$5,153	$5,303	$5,555
Total	$6,180	$10,891	$16,044	$21,347	$26,902

Ownership Costs (5yr)

Average: $26,992 | This Car: $26,902
Cost/Mile: 39¢ | Cost/Mile: 38¢

Ownership Cost Rating

Average

The Saturn SC coupe is available in SC1 and SC2 editions. With the addition of the new entry-level SC1 to the coupe line up, the SC becomes the upscale SC2 in 1993. The SC2 upgrades the SC1 by adding a rear stabilizer bar and "teardrop" alloy wheels with performance tires as standard equipment. Every Saturn comes with Saturn's Owner Protection Package which includes a no-deductible, bumper-to-bumper, limited warranty and 24-hour roadside assistance.

---

page 214

* Includes shaded options
** Other purchase requirements apply

● Poor | ◐ Worse Than Average | ○ Average | ○ Better Than Average | ○ Excellent | ⊖ Insufficient Information

©1993 by IntelliChoice, Inc. (408) 554-8711 All Rights Reserved. Reproduction Prohibited.
Refer to *Section 3: Annotated Vehicle Charts* for an explanation of these charts.

Saturn SL
4 Door Sedan

Compact

 1.9L 85 hp Gas Fuel Inject.
 4 Cylinder In-Line
 Manual 5 Speed
 2 Wheel Front
Driver Airbag Psngr Belts

Purchase Price

Car Item	Dealer Cost	List
Base Price	**$8,184**	**$9,195**
Anti-Lock Brakes	$530	$595
Automatic Transmission	N/A	N/A
Optional Engine	N/A	N/A
Air Conditioning	$739	$830
Power Steering	N/A	N/A
Cruise Control	N/A	N/A
All Wheel Drive	N/A	N/A
AM/FM Stereo Cassette	$160	$180
Steering Wheel, Tilt	Std	Std
Power Windows	N/A	N/A
***Options Price**	$899	$1,010
***Total Price**	**$9,083**	**$10,205**
Target Price		$9,905
Destination Charge		$300
Avg. Tax & Fees		$615
Total Target $		**$10,820**
Average Dealer Option Cost:		89%

The Saturn SL sedan is available in SL, SL1 and SL2 editions. New this year each Saturn comes with standard driver's side airbag. The SL receives a redesigned instrument panel with reminder/warning lamps, two new exterior colors, and all-new fabric seating material. Also standard on the SL are black front and rear fascias and a specific interior trim. In an effort to attract targeted audiences, Saturn sponsors Black Excellence, Team Saturn, Saturn on Campus, and "3 on 3" Military Marketing.

Ownership Costs

Cost Area	5 Year Cost	Rate
Depreciation	$5,269	○
Financing ($217/month)	$1,781	
Insurance (Rating 3)	$6,520	●
State Fees	$421	
Fuel (Hwy 37 City 28)	$2,659	○
Maintenance	$3,471	○
Repairs	$610	○

Warranty/Maintenance Info

Major Tune-Up	$116	○
Minor Tune-Up	$86	○
Brake Service	$189	○
Overall Warranty	3 yr/36k	◐
Drivetrain Warranty	3 yr/36k	◐
Rust Warranty	6 yr/100k	○
Maintenance Warranty	N/A	
Roadside Assistance	3 yr/36k	

Ownership Cost By Year

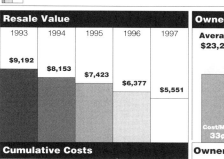

Resale Value

1993	1994	1995	1996	1997
$9,192	$8,153	$7,423	$6,377	$5,551

Cumulative Costs

	1993	1994	1995	1996	1997
Annual	$4,216	$3,662	$3,978	$4,406	$4,469
Total	$4,216	$7,878	$11,856	$16,262	$20,731

Ownership Costs (5yr)

Average: $23,257 — This Car: $20,731
Cost/Mile 33¢ — Cost/Mile 30¢

Ownership Cost Rating

Excellent

Saturn SL1
4 Door Sedan

Compact

 1.9L 85 hp Gas Fuel Inject.
 4 Cylinder In-Line
 Manual 5 Speed
 2 Wheel Front
Driver Airbag Psngr Belts

Purchase Price

Car Item	Dealer Cost	List
Base Price	**$8,896**	**$9,995**
Anti-Lock Brakes	$530	$595
Automatic 4 Speed	$667	$750
Optional Engine	N/A	N/A
Air Conditioning	$739	$830
Power Steering	Std	Std
Cruise Control	$196	$220
All Wheel Drive	N/A	N/A
AM/FM Stereo Cassette	$160	$180
Steering Wheel, Tilt	Std	Std
Power Windows	Pkg	Pkg
***Options Price**	$1,566	$1,760
***Total Price**	**$10,462**	**$11,755**
Target Price		$11,455
Destination Charge		$300
Avg. Tax & Fees		$709
Total Target $		**$12,464**
Average Dealer Option Cost:		89%

The Saturn SL sedan is available in SL, SL1 and SL2 editions. In addition to all the standard features of the base SL editon, the SL1 sedan features standard power steering and numerous optional items such as a power sunroof, power door locks, anti-lock braking system, cruise control and a new traction control system to aid in driving on slippery surfaces. Each Saturn continues to be built with rustproof, dent- and ding-resistant polymer exterior panels in areas such as the doors and fenders.

Ownership Costs

Cost Area	5 Year Cost	Rate
Depreciation	$6,060	○
Financing ($251/month)	$2,053	
Insurance (Rating 3)	$6,520	●
State Fees	$481	
Fuel (Hwy 36 City 26)	$2,795	○
Maintenance	$3,617	○
Repairs	$610	○

Warranty/Maintenance Info

Major Tune-Up	$116	○
Minor Tune-Up	$86	○
Brake Service	$178	○
Overall Warranty	3 yr/36k	◐
Drivetrain Warranty	3 yr/36k	◐
Rust Warranty	6 yr/100k	○
Maintenance Warranty	N/A	
Roadside Assistance	3 yr/36k	

Resale Value

1993	1994	1995	1996	1997
$10,525	$9,346	$8,508	$7,323	$6,404

Cumulative Costs

	1993	1994	1995	1996	1997
Annual	$4,681	$3,926	$4,249	$4,603	$4,677
Total	$4,681	$8,607	$12,856	$17,459	$22,136

Ownership Costs (5yr)

Average: $24,588 — This Car: $22,136
Cost/Mile 35¢ — Cost/Mile 32¢

Ownership Cost Rating

Excellent

---

* Includes shaded options
** Other purchase requirements apply

● Poor ◐ Worse Than Average ○ Average ○ Better Than Average ○ Excellent ⊖ Insufficient Information

©1993 by IntelliChoice, Inc. (408) 554-8711 All Rights Reserved. Reproduction Prohibited.
Refer to *Section 3: Annotated Vehicle Charts* for an explanation of these charts.

Saturn SL2
4 Door Sedan

 1.9L 124 hp Gas Fuel Inject.
 4 Cylinder In-Line
Manual 5 Speed
 2 Wheel Front
 Driver Airbag Psngr Belts

Compact

Purchase Price

Car Item	Dealer Cost	List
Base Price	**$10,231**	**$11,495**
Anti-Lock Brakes	$530	$595
Automatic 4 Speed	$667	$750
Optional Engine	N/A	N/A
Air Conditioning	$739	$830
Power Steering	Std	Std
Cruise Control	$196	$220
All Wheel Drive	N/A	N/A
AM/FM Stereo Cassette	$160	$180
Steering Wheel, Tilt	Std	Std
Power Windows	Pkg	Pkg
***Options Price**	**$1,566**	**$1,760**
***Total Price**	**$11,797**	**$13,255**
Target Price		$12,955
Destination Charge		$300
Avg. Tax & Fees		$799
Total Target $		**$14,054**
Average Dealer Option Cost:		*89%*

The Saturn SL sedan is available in SL, SL1 and SL2 editions. In addition to or replacing the standard features of the SL1 sedan, the SL2 offers a retuned fully independent suspension system, new T-series touring tires and 15 inch steel wheels with six-spoke trim covers. Refinements to the SL2 for 1993 include fresh body-colored fascias with sculptured flares along the lower edge of the rear bumper fascia, lowered front seats for additional headroom, and new optional fog lamps.

Ownership Costs

Cost Area	5 Year Cost	Rate
Depreciation	$7,097	○
Financing ($282/month)	$2,315	
Insurance (Rating 5)	$6,786	◉
State Fees	$542	
Fuel (Hwy 33 City 24)	$3,039	○
Maintenance	$4,436	◉
Repairs	$610	○

Warranty/Maintenance Info

Major Tune-Up	$196	●
Minor Tune-Up	$166	●
Brake Service	$178	○
Overall Warranty	3 yr/36k	○
Drivetrain Warranty	3 yr/36k	○
Rust Warranty	6 yr/100k	○
Maintenance Warranty	N/A	
Roadside Assistance	3 yr/36k	

Ownership Cost By Year

1993, 1994, 1995, 1996, 1997

Resale Value

1993	1994	1995	1996	1997
$11,543	$10,230	$9,286	$7,967	$6,957

Cumulative Costs

	1993	1994	1995	1996	1997
Annual	$5,471	$4,336	$4,826	$4,970	$5,222
Total	$5,471	$9,807	$14,633	$19,603	$24,825

Ownership Costs (5yr)

Average	This Car
$25,876	$24,825
Cost/Mile 37¢	Cost/Mile 35¢

Ownership Cost Rating

○ Better Than Average

Saturn SW1
4 Door Wagon

 1.9L 85 hp Gas Fuel Inject.
4 Cylinder In-Line
Manual 5 Speed
 2 Wheel Front
 Driver Airbag Psngr Belts

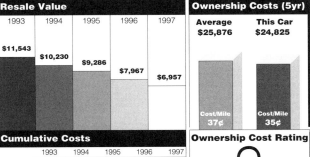
Compact Wagon

Purchase Price

Car Item	Dealer Cost	List
Base Price	**$9,697**	**$10,895**
Anti-Lock Brakes	$530	$595
Automatic 4 Speed	$667	$750
Optional Engine	N/A	N/A
Air Conditioning	$739	$830
Power Steering	Std	Std
Cruise Control	$196	$220
All Wheel Drive	N/A	N/A
AM/FM Stereo Cassette	$160	$180
Steering Wheel, Tilt	Std	Std
Power Windows	Pkg	Pkg
***Options Price**	**$1,566**	**$1,760**
***Total Price**	**$11,263**	**$12,655**
Target Price		$12,355
Destination Charge		$300
Avg. Tax & Fees		$763
Total Target $		**$13,418**
Average Dealer Option Cost:		*89%*

The all-new Saturn SW wagon is available in SW1 and SW2 editions. The Saturn wagon required a modification of the sedan's spaceframe to incorporate a cargo area which boasts a volume of 28.8 cubic feet. Standard features on the SW1 wagon include a 1.9L 4 Cylinder SOHC powertrain, power disc/drum brakes, rear seat heater ducts, intermittent windshield wipers, and remote tailgate release. Optional equipment includes an anti-lock braking system, air conditioning and cruise control.

Ownership Costs

Cost Area	5 Year Cost	Rate
Depreciation		⊖
Financing ($270/month)	$2,211	
Insurance (Rating 4)	$6,658	◉
State Fees	$520	
Fuel (Hwy 35 City 25)	$2,890	○
Maintenance	$3,617	○
Repairs	$610	○

Warranty/Maintenance Info

Major Tune-Up	$116	○
Minor Tune-Up	$86	○
Brake Service	$178	○
Overall Warranty	3 yr/36k	○
Drivetrain Warranty	3 yr/36k	○
Rust Warranty	6 yr/100k	○
Maintenance Warranty	N/A	
Roadside Assistance	3 yr/36k	

Ownership Cost By Year

Insufficient Depreciation Information

Resale Value

Insufficient Information

Ownership Costs (5yr)

Insufficient Information

Cumulative Costs

	1993	1994	1995	1996	1997
Annual	*Insufficient Information*				
Total	*Insufficient Information*				

Ownership Cost Rating

⊖ Insufficient Information

---

page 216

* Includes shaded options
** Other purchase requirements apply

● Poor ◉ Worse Than Average ◍ Average ○ Better Than Average ○ Excellent ⊖ Insufficient Information

©1993 by IntelliChoice, Inc. (408) 554-8711 All Rights Reserved. Reproduction Prohibited.
Refer to *Section 3: Annotated Vehicle Charts* for an explanation of these charts.

Saturn SW2
4 Door Wagon

Compact Wagon

 1.9L 124 hp Gas Fuel Inject.
 4 Cylinder In-Line
 Manual 5 Speed
 2 Wheel Front
 Driver Airbag Psngr Belts

Purchase Price

Car Item	Dealer Cost	List
Base Price	**$10,854**	**$12,195**
Anti-Lock Brakes	$530	$595
Automatic 4 Speed	$667	$750
Optional Engine	N/A	N/A
Air Conditioning	$739	$830
Power Steering	Std	Std
Cruise Control	$196	$220
All Wheel Drive	N/A	N/A
AM/FM Stereo Cassette	$160	$180
Steering Wheel, Tilt	Std	Std
Power Windows	Pkg	Pkg
*Options Price	$1,566	$1,760
*Total Price	$12,420	$13,955
Target Price	$13,655	
Destination Charge	$300	
Avg. Tax & Fees	$841	
Total Target $	**$14,796**	
Average Dealer Option Cost:	89%	

The all-new Saturn SW wagon is available in SW1 and SW2 editions. In addition to or replacing all the standard features of the SW1 wagon, the SW2 comes with a 1.9L 4 Cylinder DOHC powertrain which provides considerably more performance and larger tires. The SW2 has numerous optional items such as front fog lamps, 15 inch alloy "geartooth" wheels and a variety of AM/FM cassette stereos. The SW wagons feature corrostion-resistant sheet molded compound (SMC) exterior roofs and tailgates.

Ownership Costs

Cost Area	5 Year Cost	Rate
Depreciation		⊖
Financing ($297/month)	$2,437	
Insurance (Rating 5)	$6,786	●
State Fees	$571	
Fuel (Hwy 33 City 24)	$3,039	○
Maintenance	$4,436	○
Repairs	$610	○

Warranty/Maintenance Info

Major Tune-Up	$196	●
Minor Tune-Up	$166	●
Brake Service	$178	○
Overall Warranty	3 yr/36k	○
Drivetrain Warranty	3 yr/36k	○
Rust Warranty	6 yr/100k	○
Maintenance Warranty	N/A	
Roadside Assistance	3 yr/36k	

Ownership Cost By Year

Insufficient Depreciation Information

Legend: 1993, 1994, 1995, 1996, 1997

Resale Value

Insufficient Information

Cumulative Costs

	1993	1994	1995	1996	1997
Annual	Insufficient Information				
Total	Insufficient Information				

Ownership Costs (5yr)

Insufficient Information

Ownership Cost Rating

⊖ Insufficient Information

Subaru Justy
2 Door Hatchback

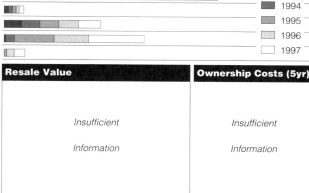

Subcompact

GL Model Shown

 1.2L 73 hp Gas Fuel Inject.
 3 Cylinder In-Line
 Manual 5 Speed
 2 Wheel Front
 Automatic Seatbelts

Purchase Price

Car Item	Dealer Cost	List
Base Price	**$6,726**	**$7,338**
Anti-Lock Brakes	N/A	N/A
Automatic Transmission	N/A	N/A
Optional Engine	N/A	N/A
Air Conditioning	$640	$799
Power Steering	N/A	N/A
Cruise Control	N/A	N/A
All Wheel Drive	N/A	N/A
AM/FM Stereo Cassette	$190	$290
Steering Wheel, Tilt	N/A	N/A
Power Windows	N/A	N/A
*Options Price	$0	$0
*Total Price	$6,726	$7,338
Target Price	$6,926	
Destination Charge	$445	
Avg. Tax & Fees	$447	
Total Target $	**$7,818**	
Average Dealer Option Cost:	75%	

The 1993 Subaru Justy is available in four models - (Base) Justy, GL, GL 4WD two-door hatchback and the GL 4WD four-door hatchback. New for 1993, the Justy (Base) receives standard cloth seats and more powerful engine. Other features include reclining front bucket seats and all-season radial tires. The Justy is available with an Electronic Continuously Variable Transmission (ECVT), which is designed for people who prefer automatic shifting and manual shift economy.

Ownership Costs

Cost Area	5 Year Cost	Rate
Depreciation	$4,169	○
Financing ($157/month)	$1,288	
Insurance (Rating 7)	$7,035	●
State Fees	$311	
Fuel (Hwy 37 City 33)	$2,470	○
Maintenance	$3,935	○
Repairs	$630	○

Warranty/Maintenance Info

Major Tune-Up	$189	○
Minor Tune-Up	$83	○
Brake Service	$219	○
Overall Warranty	3 yr/36k	○
Drivetrain Warranty	5 yr/60k	○
Rust Warranty	5 yr/unlim. mi	○
Maintenance Warranty	N/A	
Roadside Assistance	3 yr/36k	

Resale Value

1993	1994	1995	1996	1997
$6,438	$5,621	$5,004	$4,351	$3,649

Cumulative Costs

	1993	1994	1995	1996	1997
Annual	$3,792	$3,312	$3,719	$3,343	$5,672
Total	$3,792	$7,104	$10,823	$14,166	$19,838

Ownership Costs (5yr)

Average	This Car
$19,834	$19,838
Cost/Mile 28¢	Cost/Mile 28¢

Ownership Cost Rating

○ Average

---

* Includes shaded options
** Other purchase requirements apply

Legend: ● Poor | Worse Than Average | ○ Average | Better Than Average | Excellent | ⊖ Insufficient Information

©1993 by IntelliChoice, Inc. (408) 554-8711 All Rights Reserved. Reproduction Prohibited.
Refer to *Section 3: Annotated Vehicle Charts* for an explanation of these charts.

Subaru Justy GL
2 Door Hatchback

Subcompact

1.2L 73 hp Gas Fuel Inject. | 3 Cylinder In-Line | Continuously Variable Transmission | 2 Wheel Front | Automatic Seatbelts

Purchase Price

Car Item	Dealer Cost	List
Base Price	**$8,165**	**$9,113**
Anti-Lock Brakes	N/A	N/A
Manual Transmission	N/A	N/A
Optional Engine	N/A	N/A
Air Conditioning	$640	$799
Power Steering	N/A	N/A
Cruise Control	N/A	N/A
All Wheel Drive	N/A	N/A
AM/FM Stereo Cassette	N/A	N/A
Steering Wheel, Tilt	N/A	N/A
Power Windows	N/A	N/A
*Options Price	$0	$0
*Total Price	$8,165	$9,113
Target Price	$8,365	
Destination Charge	$445	
Avg. Tax & Fees	$537	
Total Target $	**$9,347**	
Average Dealer Option Cost:	*75%*	

Ownership Costs

Cost Area	5 Year Cost	Rate
Depreciation	$5,397	●
Financing ($188/month)	$1,539	
Insurance (Rating 10)	$7,479	●
State Fees	$382	
Fuel (Hwy 35 City 33)	$2,552	○
Maintenance	$4,371	○
Repairs	$630	○

Warranty/Maintenance Info

Major Tune-Up	$189	○
Minor Tune-Up	$83	○
Brake Service	$219	○
Overall Warranty	3 yr/36k	○
Drivetrain Warranty	5 yr/60k	◯
Rust Warranty	5 yr/unlim. mi	○
Maintenance Warranty	N/A	
Roadside Assistance	3 yr/36k	

Ownership Cost By Year
1993 / 1994 / 1995 / 1996 / 1997

Resale Value
1993	1994	1995	1996	1997
$6,790	$5,951	$5,316	$4,662	$3,950

Cumulative Costs
	1993	1994	1995	1996	1997
Annual	$5,189	$3,529	$4,116	$3,484	$6,032
Total	$5,189	$8,718	$12,834	$16,318	$22,350

Ownership Costs (5yr)
Average: $21,832 — Cost/Mile 31¢
This Car: $22,350 — Cost/Mile 32¢

Ownership Cost Rating
○ Average

The 1993 Subaru Justy is available in four models - (Base) Justy, GL, GL 4WD two-door hatchback and the GL 4WD four-door hatchback. New for 1993, the Justy GL features a power enhanced engine. Other features include reclining front bucket seats, front and rear anti-roll bars, rear window wiper/washer, full wheel covers and all-season radial tires. The Justy is available with an Electronic Continuously Variable Transmission (ECVT).

Subaru Justy GL 4WD
4 Door Hatchback

Subcompact

1.2L 73 hp Gas Fuel Inject. | 3 Cylinder In-Line | Manual 5 Speed | 4 Wheel On-Demand | Automatic Seatbelts

Purchase Price

Car Item	Dealer Cost	List
Base Price	**$8,495**	**$9,478**
Anti-Lock Brakes	N/A	N/A
Automatic Trans.	$477	$535
Optional Engine	N/A	N/A
Air Conditioning	$640	$799
Power Steering	N/A	N/A
Cruise Control	N/A	N/A
4 Whl On-Demand Dr.	Std	Std
AM/FM Stereo Cassette	N/A	N/A
Steering Wheel, Tilt	N/A	N/A
Power Windows	N/A	N/A
*Options Price	$0	$0
*Total Price	$8,495	$9,478
Target Price	$8,695	
Destination Charge	$445	
Avg. Tax & Fees	$556	
Total Target $	**$9,696**	
Average Dealer Option Cost:	*79%*	

Ownership Costs

Cost Area	5 Year Cost	Rate
Depreciation	$5,626	●
Financing ($195/month)	$1,597	
Insurance (Rating 12)	$7,898	●
State Fees	$396	
Fuel (Hwy 32 City 28)	$2,880	●
Maintenance	$4,175	○
Repairs	$740	●

Warranty/Maintenance Info

Major Tune-Up	$189	○
Minor Tune-Up	$83	○
Brake Service	$219	○
Overall Warranty	3 yr/36k	○
Drivetrain Warranty	5 yr/60k	◯
Rust Warranty	5 yr/unlim. mi	○
Maintenance Warranty	N/A	
Roadside Assistance	3 yr/36k	

Resale Value
1993	1994	1995	1996	1997
$6,786	$5,988	$5,378	$4,752	$4,070

Cumulative Costs
	1993	1994	1995	1996	1997
Annual	$5,707	$3,653	$4,176	$3,660	$6,116
Total	$5,707	$9,360	$13,536	$17,196	$23,312

Ownership Costs (5yr)
Average: $22,242 — Cost/Mile 32¢
This Car: $23,312 — Cost/Mile 33¢

Ownership Cost Rating
● Worse Than Average

The 1993 Subaru Justy is available in four models - (Base) Justy, GL, GL 4WD two-door hatchback and the GL 4WD four-door hatchback. New for 1993, the GL 4WD four-door receives a power enhanced engine. Other features include reclining front bucket seats and all-season radial tires. The Justy also offers On-Demand all-wheel drive, rear-window washer/wiper, body-side moldings and full wheel covers. The Justy is available with an Electronic Continuously Variable Transmission (ECVT).

---

page 218

* Includes shaded options
** Other purchase requirements apply

● Poor ◐ Worse Than Average ○ Average ◯ Better Than Average ○ Excellent ⊖ Insufficient Information

©1993 by *IntelliChoice, Inc.* (408) 554-8711 All Rights Reserved. Reproduction Prohibited.

Refer to *Section 3: Annotated Vehicle Charts* for an explanation of these charts.

Subaru Justy GL 4WD
2 Door Hatchback

Subcompact

4 Door Model Shown

- 1.2L 73 hp Gas Fuel Inject.
- 3 Cylinder In-Line
- Continuously Variable Transmission
- 4 Wheel On-Demand
- Automatic Seatbelts

Purchase Price

Car Item	Dealer Cost	List
Base Price	**$8,882**	**$9,913**
Anti-Lock Brakes	N/A	N/A
Manual Transmission	N/A	N/A
Optional Engine	N/A	N/A
Air Conditioning	$640	$799
Power Steering	N/A	N/A
Cruise Control	N/A	N/A
4 Whl On-Demand Dr.	Std	Std
AM/FM Stereo Cassette	N/A	N/A
Steering Wheel, Tilt	N/A	N/A
Power Windows	N/A	N/A
***Options Price**	**$0**	**$0**
***Total Price**	**$8,882**	**$9,913**
Target Price		$9,094
Destination Charge		$445
Avg. Tax & Fees		$581
Total Target $		**$10,120**
Average Dealer Option Cost:		75%

Ownership Costs

Cost Area	5 Year Cost	Rate
Depreciation	$5,934	●
Financing ($203/month)	$1,667	
Insurance (Rating 10)	$7,479	●
State Fees	$415	
Fuel (Hwy 31 City 31)	$2,813	○
Maintenance	$4,371	○
Repairs	$740	◉

Warranty/Maintenance Info

Major Tune-Up	$189	○
Minor Tune-Up	$83	○
Brake Service	$219	○
Overall Warranty	3 yr/36k	○
Drivetrain Warranty	5 yr/60k	◯
Rust Warranty	5 yr/unlim. mi	○
Maintenance Warranty	N/A	
Roadside Assistance	3 yr/36k	

Ownership Cost By Year

1993, 1994, 1995, 1996, 1997

Resale Value

1993	1994	1995	1996	1997
$6,994	$6,165	$5,538	$4,889	$4,186

Cumulative Costs

	1993	1994	1995	1996	1997
Annual	$5,867	$3,616	$4,209	$3,591	$6,136
Total	$5,867	$9,483	$13,692	$17,283	$23,419

Ownership Costs (5yr)

Average	This Car
$22,732	$23,419
Cost/Mile 32¢	Cost/Mile 33¢

Ownership Cost Rating

● Worse Than Average

The 1993 Subaru Justy is available in four models - (Base) Justy, GL, GL 4WD two-door hatchback and the GL 4WD four-door hatchback. New for 1993, the GL 4WD two-door receives a power enhanced engine. Other features include reclining front bucket seats and all-season radial tires. The Justy also offers On-Demand all-wheel drive, rear-window washer/wiper, body-side moldings and full wheel covers. The Justy is available with an Electronic Continuously Variable Transmission (ECVT).

Subaru Legacy L
4 Door Sedan

Compact

- 2.2L 130 hp Gas Fuel Inject.
- 4 Cylinder Opposing
- Manual 5 Speed
- 2 Wheel Front
- Driver Airbag Psngr Belts

Purchase Price

Car Item	Dealer Cost	List
Base Price	**$14,405**	**$16,250**
Anti-Lock Brakes	N/A	N/A
Automatic 4 Speed	$721	$800
Optional Engine	N/A	N/A
Air Conditioning	Std	Std
Power Steering	Std	Std
Cruise Control	Std	Std
All Wheel Drive	N/A	N/A
AM/FM Stereo Cassette	Std	Std
Steering, Tilt w/Memory	Std	Std
Power Windows	Std	Std
***Options Price**	**$721**	**$800**
***Total Price**	**$15,126**	**$17,050**
Target Price		$15,554
Destination Charge		$445
Avg. Tax & Fees		$975
Total Target $		**$16,974**
Average Dealer Option Cost:		90%

Ownership Costs

Cost Area	5 Year Cost	Rate
Depreciation	$7,421	○
Financing ($341/month)	$2,796	
Insurance (Rating 10)	$7,479	○
State Fees	$699	
Fuel (Hwy 29 City 22)	$3,390	○
Maintenance	$5,279	●
Repairs	$560	◯

Warranty/Maintenance Info

Major Tune-Up	$145	○
Minor Tune-Up	$89	○
Brake Service	$239	○
Overall Warranty	3 yr/36k	○
Drivetrain Warranty	5 yr/60k	◯
Rust Warranty	5 yr/unlim. mi	○
Maintenance Warranty	N/A	
Roadside Assistance	3 yr/36k	

Resale Value

1993	1994	1995	1996	1997
$13,558	$12,275	$12,036	$10,764	$9,553

Cumulative Costs

	1993	1994	1995	1996	1997
Annual	$6,807	$4,606	$4,208	$4,623	$7,380
Total	$6,807	$11,413	$15,621	$20,244	$27,624

Ownership Costs (5yr)

Average	This Car
$29,134	$27,624
Cost/Mile 42¢	Cost/Mile 39¢

Ownership Cost Rating

◯ Better Than Average

The 1993 Subaru Legacy is available in twelve models - L, LS, L 4WD, LS 4WD, LSi 4WD and Sport 4WD sedans; and L, LS, L 4WD, LS 4WD, LSi 4WD and Touring 4WD wagons. New for 1993, the L sedan receives a driver airbag, 60/40 rear split seats and one new exterior color (Jasper Green). Features include power windows, power door locks, rear-window defogger, four-wheel independent suspension, intermittent windshield wipers and front bucket seats.

---

* Includes shaded options
** Other purchase requirements apply

● Poor ◉ Worse Than Average ● Average ○ Better Than Average ◯ Excellent ⊖ Insufficient Information

©1993 by IntelliChoice, Inc. (408) 554-8711 All Rights Reserved. Reproduction Prohibited.
Refer to *Section 3: Annotated Vehicle Charts* for an explanation of these charts.

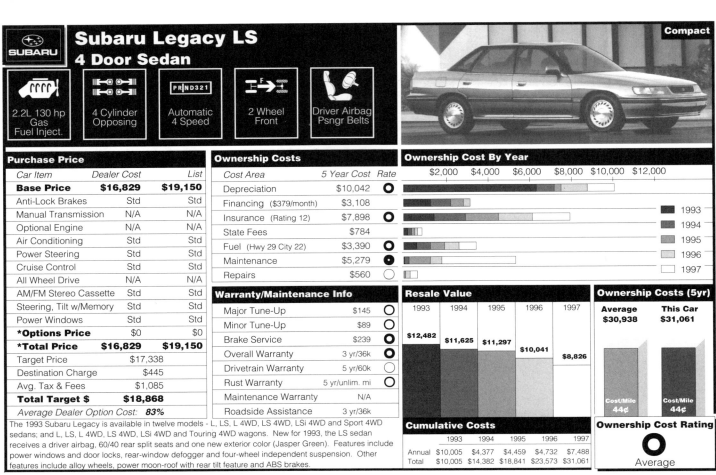

Subaru Legacy LS
4 Door Sedan

Compact

- 2.2L 130 hp Gas Fuel Inject.
- 4 Cylinder Opposing
- Automatic 4 Speed
- 2 Wheel Front
- Driver Airbag Psngr Belts

Purchase Price

Car Item	Dealer Cost	List
Base Price	**$16,829**	**$19,150**
Anti-Lock Brakes	Std	Std
Manual Transmission	N/A	N/A
Optional Engine	N/A	N/A
Air Conditioning	Std	Std
Power Steering	Std	Std
Cruise Control	Std	Std
All Wheel Drive	N/A	N/A
AM/FM Stereo Cassette	Std	Std
Steering, Tilt w/Memory	Std	Std
Power Windows	Std	Std
*Options Price	$0	$0
*Total Price	$16,829	$19,150
Target Price	$17,338	
Destination Charge	$445	
Avg. Tax & Fees	$1,085	
Total Target $	**$18,868**	
Average Dealer Option Cost:	83%	

The 1993 Subaru Legacy is available in twelve models - L, LS, L 4WD, LS 4WD, LSi 4WD and Sport 4WD sedans; and L, LS, L 4WD, LS 4WD, LSi 4WD and Touring 4WD wagons. New for 1993, the LS sedan receives a driver airbag, 60/40 rear split seats and one new exterior color (Jasper Green). Features include power windows and door locks, rear-window defogger and four-wheel independent suspension. Other features include alloy wheels, power moon-roof with rear tilt feature and ABS brakes.

Ownership Costs

Cost Area	5 Year Cost	Rate
Depreciation	$10,042	○
Financing ($379/month)	$3,108	
Insurance (Rating 12)	$7,898	○
State Fees	$784	
Fuel (Hwy 29 City 22)	$3,390	○
Maintenance	$5,279	◉
Repairs	$560	○

Warranty/Maintenance Info

Major Tune-Up	$145	○
Minor Tune-Up	$89	○
Brake Service	$239	○
Overall Warranty	3 yr/36k	○
Drivetrain Warranty	5 yr/60k	○
Rust Warranty	5 yr/unlim. mi	○
Maintenance Warranty	N/A	
Roadside Assistance	3 yr/36k	

Resale Value

1993	1994	1995	1996	1997
$12,482	$11,625	$11,297	$10,041	$8,826

Cumulative Costs

	1993	1994	1995	1996	1997
Annual	$10,005	$4,377	$4,459	$4,732	$7,488
Total	$10,005	$14,382	$18,841	$23,573	$31,061

Ownership Costs (5yr)

Average	This Car
$30,938	$31,061
Cost/Mile 44¢	Cost/Mile 44¢

Ownership Cost Rating: ○ Average

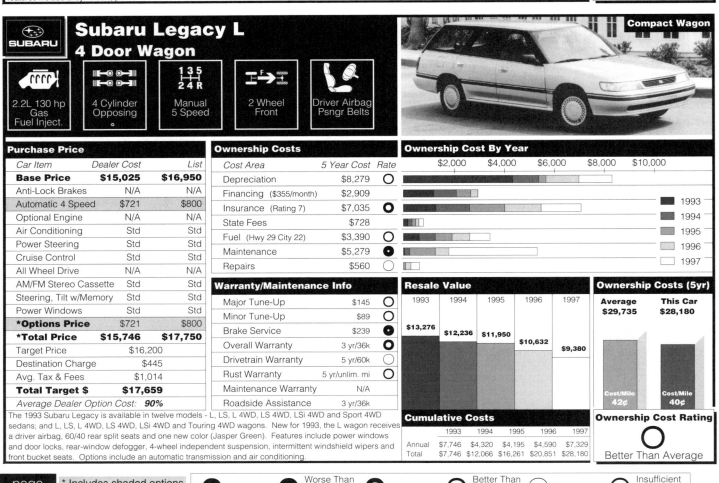

Subaru Legacy L
4 Door Wagon

Compact Wagon

- 2.2L 130 hp Gas Fuel Inject.
- 4 Cylinder Opposing
- Manual 5 Speed
- 2 Wheel Front
- Driver Airbag Psngr Belts

Purchase Price

Car Item	Dealer Cost	List
Base Price	**$15,025**	**$16,950**
Anti-Lock Brakes	N/A	N/A
Automatic 4 Speed	$721	$800
Optional Engine	N/A	N/A
Air Conditioning	Std	Std
Power Steering	Std	Std
Cruise Control	Std	Std
All Wheel Drive	N/A	N/A
AM/FM Stereo Cassette	Std	Std
Steering, Tilt w/Memory	Std	Std
Power Windows	Std	Std
*Options Price	$721	$800
*Total Price	$15,746	$17,750
Target Price	$16,200	
Destination Charge	$445	
Avg. Tax & Fees	$1,014	
Total Target $	**$17,659**	
Average Dealer Option Cost:	90%	

The 1993 Subaru Legacy is available in twelve models - L, LS, L 4WD, LS 4WD, LSi 4WD and Sport 4WD sedans; and L, LS, L 4WD, LS 4WD, LSi 4WD and Touring 4WD wagons. New for 1993, the L wagon receives a driver airbag, 60/40 rear split seats and one new color (Jasper Green). Features include power windows and door locks, rear-window defogger, 4-wheel independent suspension, intermittent windshield wipers and front bucket seats. Options include an automatic transmission and air conditioning.

Ownership Costs

Cost Area	5 Year Cost	Rate
Depreciation	$8,279	○
Financing ($355/month)	$2,909	
Insurance (Rating 7)	$7,035	◉
State Fees	$728	
Fuel (Hwy 29 City 22)	$3,390	○
Maintenance	$5,279	◉
Repairs	$560	○

Warranty/Maintenance Info

Major Tune-Up	$145	○
Minor Tune-Up	$89	○
Brake Service	$239	◉
Overall Warranty	3 yr/36k	○
Drivetrain Warranty	5 yr/60k	○
Rust Warranty	5 yr/unlim. mi	○
Maintenance Warranty	N/A	
Roadside Assistance	3 yr/36k	

Resale Value

1993	1994	1995	1996	1997
$13,276	$12,236	$11,950	$10,632	$9,380

Cumulative Costs

	1993	1994	1995	1996	1997
Annual	$7,746	$4,320	$4,195	$4,590	$7,329
Total	$7,746	$12,066	$16,261	$20,851	$28,180

Ownership Costs (5yr)

Average	This Car
$29,735	$28,180
Cost/Mile 42¢	Cost/Mile 40¢

Ownership Cost Rating: ○ Better Than Average

---

* Includes shaded options
** Other purchase requirements apply

● Poor ◉ Worse Than Average ○ Average ○ Better Than Average ○ Excellent ⊖ Insufficient Information

©1993 by IntelliChoice, Inc. (408) 554-8711 All Rights Reserved. Reproduction Prohibited.
Refer to *Section 3: Annotated Vehicle Charts* for an explanation of these charts.

Subaru Legacy LS
4 Door Wagon

Compact Wagon

- 2.2L 130 hp Gas Fuel Inject.
- 4 Cylinder Opposing
- Automatic 4 Speed (PRND321)
- 2 Wheel Front
- Driver Airbag Psngr Belts

Purchase Price

Car Item	Dealer Cost	List
Base Price	**$17,444**	**$19,850**
Anti-Lock Brakes	Std	Std
Manual Transmission	N/A	N/A
Optional Engine	N/A	N/A
Air Conditioning	Std	Std
Power Steering	Std	Std
Cruise Control	Std	Std
All Wheel Drive	N/A	N/A
AM/FM Stereo Cassette	Std	Std
Steering, Tilt w/Memory	Std	Std
Power Windows	Std	Std
*Options Price	$0	$0
***Total Price**	**$17,444**	**$19,850**
Target Price		$17,981
Destination Charge		$445
Avg. Tax & Fees		$1,124
Total Target $		**$19,550**
Average Dealer Option Cost:	**83%**	

The 1993 Subaru Legacy is available in twelve models - L, LS, L 4WD, LS 4WD, LSi 4WD and Sport 4WD sedans; and L, LS, L 4WD, LS 4WD, LSi 4WD and Touring 4WD wagons. New for 1993, the LS wagon receives a driver airbag, 60/40 rear split seats and one new exterior color (Jasper Green). Features include power windows and door locks, four-wheel independent suspension and front bucket seats. Other features include alloy wheels, power moon-roof with rear tilt feature and optional air conditioning.

Ownership Costs

Cost Area	5 Year Cost	Rate
Depreciation	$10,258	○
Financing ($393/month)	$3,221	
Insurance (Rating 13)	$8,061	○
State Fees	$812	
Fuel (Hwy 29 City 22)	$3,390	○
Maintenance	$5,279	◉
Repairs	$560	○

Warranty/Maintenance Info

Major Tune-Up	$145	○
Minor Tune-Up	$89	○
Brake Service	$239	◉
Overall Warranty	3 yr/36k	○
Drivetrain Warranty	5 yr/60k	○
Rust Warranty	5 yr/unlim. mi	○
Maintenance Warranty	N/A	
Roadside Assistance	3 yr/36k	

Ownership Cost By Year

1993, 1994, 1995, 1996, 1997

Resale Value

1993	1994	1995	1996	1997
$12,882	$12,038	$11,759	$10,502	$9,292

Cumulative Costs

	1993	1994	1995	1996	1997
Annual	$10,370	$4,438	$4,471	$4,780	$7,522
Total	$10,370	$14,808	$19,279	$24,059	$31,581

Ownership Costs (5yr)

Average	This Car
$31,539	$31,581
Cost/Mile 45¢	Cost/Mile 45¢

Ownership Cost Rating
○ Average

Subaru Legacy L AWD
4 Door Sedan

Compact

- 2.2L 130 hp Gas Fuel Inject.
- 4 Cylinder Opposing
- Manual 5 Speed
- 4 Wheel Full-Time
- Driver Airbag Psngr Belts

2WD Model Shown

Purchase Price

Car Item	Dealer Cost	List
Base Price	**$15,824**	**$17,850**
Anti-Lock Brakes	N/A	N/A
Automatic 4 Speed	$721	$800
Optional Engine	N/A	N/A
Air Conditioning	Std	Std
Power Steering	Std	Std
Cruise Control	Std	Std
4 Wheel Full-Time Drive	Std	Std
AM/FM Stereo Cassette	Std	Std
Steering, Tilt w/Memory	Std	Std
Power Windows	Std	Std
*Options Price	$721	$800
***Total Price**	**$16,545**	**$18,650**
Target Price		$17,032
Destination Charge		$445
Avg. Tax & Fees		$1,065
Total Target $		**$18,542**
Average Dealer Option Cost:	**90%**	

The 1993 Subaru Legacy is available in twelve models - L, LS, L 4WD, LS 4WD, LSi 4WD and Sport 4WD sedans; and L, LS, L 4WD, LS 4WD, LSi 4WD and Touring 4WD wagons. New for 1993, the L 4WD sedan receives a driver airbag, 60/40 rear split seats and one new exterior color (Jasper Green). Features include power windows, power door locks, rear-window defogger, four-wheel independent suspension, intermittent windshield wipers, cloth upholstery and front bucket seats.

Ownership Costs

Cost Area	5 Year Cost	Rate
Depreciation	$8,076	○
Financing ($373/month)	$3,055	
Insurance (Rating 10)	$7,479	○
State Fees	$765	
Fuel (Hwy 27 City 21)	$3,598	◉
Maintenance	$5,267	◉
Repairs	$630	○

Warranty/Maintenance Info

Major Tune-Up	$145	○
Minor Tune-Up	$89	○
Brake Service	$239	○
Overall Warranty	3 yr/36k	○
Drivetrain Warranty	5 yr/60k	○
Rust Warranty	5 yr/unlim. mi	○
Maintenance Warranty	N/A	
Roadside Assistance	3 yr/36k	

Resale Value

1993	1994	1995	1996	1997
$14,784	$13,392	$13,130	$11,770	$10,466

Cumulative Costs

	1993	1994	1995	1996	1997
Annual	$7,310	$4,851	$4,348	$4,801	$7,560
Total	$7,310	$12,161	$16,509	$21,310	$28,870

Ownership Costs (5yr)

Average	This Car
$30,508	$28,870
Cost/Mile 44¢	Cost/Mile 41¢

Ownership Cost Rating
○ Better Than Average

---

* Includes shaded options
** Other purchase requirements apply

● Poor ◉ Worse Than Average ○ Average ○ Better Than Average ○ Excellent ⊖ Insufficient Information

©1993 by IntelliChoice, Inc. (408) 554-8711 All Rights Reserved. Reproduction Prohibited.
Refer to *Section 3: Annotated Vehicle Charts* for an explanation of these charts.

Subaru Legacy LS AWD
4 Door Sedan

 2.2L 130 hp Gas Fuel Inject.
 4 Cylinder Opposing
 Automatic 4 Speed
 4 Wheel Full-Time
 Driver Airbag Psngr Belts

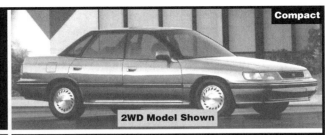
Compact — 2WD Model Shown

Purchase Price

Car Item	Dealer Cost	List
Base Price	**$18,234**	**$20,750**
Anti-Lock Brakes	Std	Std
Manual Transmission	N/A	N/A
Optional Engine	N/A	N/A
Air Conditioning	Std	Std
Power Steering	Std	Std
Cruise Control	Std	Std
4 Wheel Full-Time Drive	Std	Std
AM/FM Stereo Cassette	Std	Std
Steering, Tilt w/Memory	Std	Std
Power Windows	Std	Std
*****Options Price**	$0	$0
*****Total Price**	**$18,234**	**$20,750**
Target Price	$18,807	
Destination Charge	$445	
Avg. Tax & Fees	$1,175	
Total Target $	**$20,427**	
Average Dealer Option Cost:	83%	

The 1993 Subaru Legacy is available in twelve models - L, LS, L 4WD, LS 4WD, LSi 4WD and Sport 4WD sedans; and L, LS, L 4WD, LS 4WD, LSi 4WD and Touring 4WD wagons. New for 1993, the LS 4WD sedan receives a driver airbag, 60/40 rear split seats and one new exterior color (Jasper Green). Features include power windows and door locks, rear-window defogger, four-wheel independent suspension and front bucket seats. Other features include alloy wheels and power moon-roof with rear tilt feature.

Ownership Costs

Cost Area	5 Year Cost	Rate
Depreciation	$10,862	○
Financing ($411/month)	$3,365	
Insurance (Rating 13)	$8,061	○
State Fees	$848	
Fuel (Hwy 27 City 21)	$3,598	◉
Maintenance	$5,267	◉
Repairs	$630	○

Warranty/Maintenance Info

Major Tune-Up	$145	○
Minor Tune-Up	$89	○
Brake Service	$239	○
Overall Warranty	3 yr/36k	○
Drivetrain Warranty	5 yr/60k	○
Rust Warranty	5 yr/unlim. mi	○
Maintenance Warranty	N/A	
Roadside Assistance	3 yr/36k	

Ownership Cost By Year

1993 / 1994 / 1995 / 1996 / 1997

Resale Value

1993	1994	1995	1996	1997
$13,489	$12,569	$12,214	$10,864	$9,565

Cumulative Costs

	1993	1994	1995	1996	1997
Annual	$10,748	$4,607	$4,635	$4,948	$7,693
Total	$10,748	$15,355	$19,990	$24,938	$32,631

Ownership Costs (5yr)

Average: **$32,312** — This Car: **$32,631**
Cost/Mile: 46¢ — Cost/Mile: 47¢

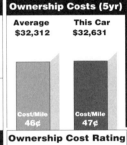

Ownership Cost Rating: Average

Subaru Legacy Sport AWD
4 Door Sedan

 2.2L 160 hp Gas Fuel Inject.
 4 Cylinder Opposing
 Manual 5 Speed
 4 Wheel Full-Time
 Driver Airbag Psngr Belts

Compact

Purchase Price

Car Item	Dealer Cost	List
Base Price	**$18,323**	**$20,850**
Anti-Lock Brakes	Std	Std
Automatic 4 Speed	$721	$800
Optional Engine	N/A	N/A
Air Conditioning	Std	Std
Power Steering	Std	Std
Cruise Control	Std	Std
4 Wheel Full-Time Drive	Std	Std
AM/FM Stereo Cassette	Std	Std
Steering, Tilt w/Memory	Std	Std
Power Windows	Std	Std
*****Options Price**	$721	$800
*****Total Price**	**$19,044**	**$21,650**
Target Price	$19,643	
Destination Charge	$445	
Avg. Tax & Fees	$1,225	
Total Target $	**$21,313**	
Average Dealer Option Cost:	90%	

The 1993 Subaru Legacy is available in twelve models - L, LS, L 4WD, LS 4WD, LSi 4WD and Sport 4WD sedans; and L, LS, L 4WD, LS 4WD, LSi 4WD and Touring 4WD wagons. New for 1993, the Sport 4WD sedan receives a driver airbag, 60/40 rear split seats and one new exterior color (Jasper Green). Features include power windows and door locks, alloy wheels, turbo-charged engine, power moonroof with tilt-feature, air conditioning, an 80-watt AM/FM cassette audio system with equalizer and sport seats.

Ownership Costs

Cost Area	5 Year Cost	Rate
Depreciation	$11,266	○
Financing ($428/month)	$3,510	
Insurance (Rating 14)	$8,238	○
State Fees	$885	
Fuel (Hwy 23 City 18)	$4,213	●
Maintenance	$5,686	●
Repairs	$630	○

Warranty/Maintenance Info

Major Tune-Up	$145	○
Minor Tune-Up	$89	○
Brake Service	$239	○
Overall Warranty	3 yr/36k	○
Drivetrain Warranty	5 yr/60k	○
Rust Warranty	5 yr/unlim. mi	○
Maintenance Warranty	N/A	
Roadside Assistance	3 yr/36k	

Resale Value

1993	1994	1995	1996	1997
$13,900	$13,939	$12,663	$11,348	$10,047

Cumulative Costs

	1993	1994	1995	1996	1997
Annual	$11,439	$3,853	$5,796	$5,253	$8,087
Total	$11,439	$15,292	$21,088	$26,341	$34,428

Ownership Costs (5yr)

Average: **$33,084** — This Car: **$34,428**
Cost/Mile: 47¢ — Cost/Mile: 49¢

Ownership Cost Rating: Average

---

* Includes shaded options
** Other purchase requirements apply

● Poor — ◐ Worse Than Average — ○ Average — ◯ Better Than Average — ◯ Excellent — ⊖ Insufficient Information

©1993 by IntelliChoice, Inc. (408) 554-8711 All Rights Reserved. Reproduction Prohibited.
Refer to *Section 3: Annotated Vehicle Charts* for an explanation of these charts.

Subaru Legacy LSi AWD
4 Door Sedan

 2.2L 130 hp Gas Fuel Inject.
 4 Cylinder Opposing
 Automatic 4 Speed
 4 Wheel Full-Time
 Driver Airbag Psngr Belts

Compact — LS 2WD Shown

Purchase Price

Car Item	Dealer Cost	List
Base Price	$19,026	$21,650
Anti-Lock Brakes	Std	Std
Manual Transmission	N/A	N/A
Optional Engine	N/A	N/A
Air Conditioning	Std	Std
Power Steering	Std	Std
Cruise Control	Std	Std
4 Wheel Full-Time Drive	Std	Std
AM/FM Stereo Cassette	Std	Std
Steering, Tilt w/Memory	Std	Std
Power Windows	Std	Std
*Options Price	$0	$0
*Total Price	$19,026	$21,650
Target Price	$19,636	
Destination Charge	$445	
Avg. Tax & Fees	$1,225	
Total Target $	$21,306	
Average Dealer Option Cost:	83%	

The 1993 Subaru Legacy is available in twelve models – L, LS, L 4WD, LS 4WD, LSi 4WD and Sport 4WD sedans; and L, LS, L 4WD, LS 4WD, LSi 4WD and Touring 4WD wagons. New for 1993, the LSi 4WD sedan receives a driver airbag, 60/40 rear split seats and one new exterior color (Jasper Green). Features include power windows and door locks, leather interior and four-wheel independent suspension. Other features include alloy wheels, power moon-roof with rear tilt feature and ABS brakes.

Ownership Costs

Cost Area	5 Year Cost	Rate
Depreciation	$10,706	O
Financing ($428/month)	$3,510	
Insurance (Rating 13)	$8,061	O
State Fees	$885	
Fuel (Hwy 27 City 21)	$3,598	●
Maintenance	$5,279	●
Repairs	$630	O

Warranty/Maintenance Info

Major Tune-Up	$145	O
Minor Tune-Up	$89	O
Brake Service	$239	O
Overall Warranty	3 yr/36k	O
Drivetrain Warranty	5 yr/60k	○
Rust Warranty	5 yr/unlim. mi	O
Maintenance Warranty	N/A	
Roadside Assistance	3 yr/36k	

Resale Value

1993	1994	1995	1996	1997
$14,895	$13,520	$13,260	$11,901	$10,600

Cumulative Costs

	1993	1994	1995	1996	1997
Annual	$10,290	$5,115	$4,576	$4,987	$7,701
Total	$10,290	$15,405	$19,981	$24,968	$32,669

Ownership Costs (5yr)

Average	This Car
$33,084	$32,669
Cost/Mile 47¢	Cost/Mile 47¢

Ownership Cost Rating: O Average

Subaru Legacy L AWD
4 Door Wagon

 2.2L 130 hp Gas Fuel Inject.
 4 Cylinder Opposing
 Manual 5 Speed
 4 Wheel Full-Time
Driver Airbag Psngr Belts

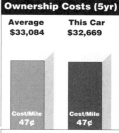
Compact Wagon — L 2WD Model Shown

Purchase Price

Car Item	Dealer Cost	List
Base Price	$16,444	$18,550
Anti-Lock Brakes	$873	$995
Automatic 4 Speed	$721	$800
Optional Engine	N/A	N/A
Air Conditioning	Std	Std
Power Steering	Std	Std
Cruise Control	Std	Std
4 Wheel Full-Time Drive	Std	Std
AM/FM Stereo Cassette	Std	Std
Steering, Tilt w/Memory	Std	Std
Power Windows	Std	Std
*Options Price	$721	$800
*Total Price	$17,165	$19,350
Target Price	$17,679	
Destination Charge	$445	
Avg. Tax & Fees	$1,104	
Total Target $	$19,228	
Average Dealer Option Cost:	89%	

The 1993 Subaru Legacy is available in twelve models – L, LS, L 4WD, LS 4WD, LSi 4WD and Sport 4WD sedans; and L, LS, L 4WD, LS 4WD, LSi 4WD and Touring 4WD wagons. New for 1993, the L 4WD wagon receives a driver airbag, 60/40 rear split seats and one new exterior color (Jasper Green). Features include power windows, power door locks, rear-window defogger, four-wheel independent suspension, intermittent windshield wipers, cloth upholstery and front bucket seats.

Ownership Costs

Cost Area	5 Year Cost	Rate
Depreciation	$8,983	O
Financing ($386/month)	$3,167	
Insurance (Rating 11)	$7,693	O
State Fees	$792	
Fuel (Hwy 27 City 21)	$3,598	O
Maintenance	$5,267	●
Repairs	$630	O

Warranty/Maintenance Info

Major Tune-Up	$145	O
Minor Tune-Up	$89	O
Brake Service	$239	●
Overall Warranty	3 yr/36k	O
Drivetrain Warranty	5 yr/60k	○
Rust Warranty	5 yr/unlim. mi	O
Maintenance Warranty	N/A	
Roadside Assistance	3 yr/36k	

Resale Value

1993	1994	1995	1996	1997
$14,429	$13,307	$12,995	$11,569	$10,245

Cumulative Costs

	1993	1994	1995	1996	1997
Annual	$8,444	$4,663	$4,467	$4,925	$7,631
Total	$8,444	$13,107	$17,574	$22,499	$30,130

Ownership Costs (5yr)

Average	This Car
$31,109	$30,130
Cost/Mile 44¢	Cost/Mile 43¢

Ownership Cost Rating: ○ Better Than Average

---

* Includes shaded options
** Other purchase requirements apply

● Poor ◐ Worse Than Average O Average ○ Better Than Average ○ Excellent ⊖ Insufficient Information

©1993 by IntelliChoice, Inc. (408) 554-8711 All Rights Reserved. Reproduction Prohibited.
Refer to *Section 3: Annotated Vehicle Charts* for an explanation of these charts.

page 223

Subaru Legacy LS AWD
4 Door Wagon

Compact Wagon — L 2WD Model Shown

- 2.2L 130 hp Gas Fuel Inject.
- 4 Cylinder Opposing
- Automatic 4 Speed
- 4 Wheel Full-Time
- Driver Airbag Psngr Belts

Purchase Price

Car Item	Dealer Cost	List
Base Price	**$18,850**	**$21,450**
Anti-Lock Brakes	Std	Std
Manual Transmission	N/A	N/A
Optional Engine	N/A	N/A
Air Conditioning	Std	Std
Power Steering	Std	Std
Cruise Control	Std	Std
4 Wheel Full-Time Drive	Std	Std
AM/FM Stereo Cassette	Std	Std
Steering, Tilt w/Memory	Std	Std
Power Windows	Std	Std
*Options Price	$0	$0
*Total Price	$18,850	$21,450
Target Price	$19,451	
Destination Charge	$445	
Avg. Tax & Fees	$1,214	
Total Target $	**$21,110**	
Average Dealer Option Cost:	83%	

The 1993 Subaru Legacy is available in twelve models - L, LS, L 4WD, LS 4WD, LSi 4WD and Sport 4WD sedans; and L, LS, L 4WD, LS 4WD, LSi 4WD and Touring 4WD wagons. New for 1993, the LS 4WD wagon receives a driver airbag, 60/40 rear split seats and one new exterior color (Jasper Green). Features include power windows and door locks, rear-window defogger, four-wheel independent suspension and front bucket seats. Other features include alloy wheels and power moon-roof with rear tilt feature.

Ownership Costs

Cost Area	5 Year Cost	Rate
Depreciation	$11,287	○
Financing ($424/month)	$3,477	
Insurance (Rating 14)	$8,238	○
State Fees	$875	
Fuel (Hwy 27 City 21)	$3,598	○
Maintenance	$5,267	◐
Repairs	$630	○

Warranty/Maintenance Info

Major Tune-Up	$145	○
Minor Tune-Up	$89	○
Brake Service	$239	◐
Overall Warranty	3 yr/36k	○
Drivetrain Warranty	5 yr/60k	○
Rust Warranty	5 yr/unlim. mi	○
Maintenance Warranty	N/A	
Roadside Assistance	3 yr/36k	

Ownership Cost By Year
1993, 1994, 1995, 1996, 1997

Resale Value
1993	1994	1995	1996	1997
$13,842	$12,894	$12,549	$11,158	$9,823

Cumulative Costs
	1993	1994	1995	1996	1997
Annual	$11,164	$4,710	$4,687	$5,040	$7,771
Total	$11,164	$15,874	$20,561	$25,601	$33,372

Ownership Costs (5yr)
- Average: $32,913
- This Car: $33,372
- Cost/Mile: 47¢ / 48¢

Ownership Cost Rating
○ Average

Subaru Legacy LSi AWD
4 Door Wagon

Compact Wagon — LS 2WD Shown

- 2.2L 130 hp Gas Fuel Inject.
- 4 Cylinder Opposing
- Automatic 4 Speed
- 4 Wheel Full-Time
- Driver Airbag Psngr Belts

Purchase Price

Car Item	Dealer Cost	List
Base Price	**$19,904**	**$22,650**
Anti-Lock Brakes	Std	Std
Manual Transmission	N/A	N/A
Optional Engine	N/A	N/A
Air Conditioning	Std	Std
Power Steering	Std	Std
Cruise Control	Std	Std
4 Wheel Full-Time Drive	Std	Std
AM/FM Stereo Cassette	Std	Std
Steering, Tilt w/Memory	Std	Std
Power Windows	Std	Std
*Options Price	$0	$0
*Total Price	$19,904	$22,650
Target Price	$20,556	
Destination Charge	$445	
Avg. Tax & Fees	$1,281	
Total Target $	**$22,282**	
Average Dealer Option Cost:	83%	

The 1993 Subaru Legacy is available in twelve models - L, LS, L 4WD, LS 4WD, LSi 4WD and Sport 4WD sedans; and L, LS, L 4WD, LS 4WD, LSi 4WD and Touring 4WD wagons. New for 1993, the LSi 4WD wagon receives a driver airbag, 60/40 rear split seats and one new exterior color (Jasper Green). Features include power windows and door locks, beige leather interior and four-wheel independent suspension. Other features include alloy wheels, power moon-roof with rear tilt feature and ABS brakes.

Ownership Costs

Cost Area	5 Year Cost	Rate
Depreciation	$11,192	○
Financing ($448/month)	$3,671	
Insurance (Rating 14)	$8,238	○
State Fees	$925	
Fuel (Hwy 27 City 21)	$3,598	○
Maintenance	$5,267	◐
Repairs	$630	○

Warranty/Maintenance Info

Major Tune-Up	$145	○
Minor Tune-Up	$89	○
Brake Service	$239	◐
Overall Warranty	3 yr/36k	○
Drivetrain Warranty	5 yr/60k	○
Rust Warranty	5 yr/unlim. mi	○
Maintenance Warranty	N/A	
Roadside Assistance	3 yr/36k	

Resale Value
1993	1994	1995	1996	1997
$15,562	$14,129	$13,880	$12,464	$11,090

Cumulative Costs
	1993	1994	1995	1996	1997
Annual	$10,709	$5,266	$4,640	$5,088	$7,818
Total	$10,709	$15,975	$20,615	$25,703	$33,521

Ownership Costs (5yr)
- Average: $33,943
- This Car: $33,521
- Cost/Mile: 48¢ / 48¢

Ownership Cost Rating
○ Average

---

*Includes shaded options
**Other purchase requirements apply

Legend: ● Poor · ◐ Worse Than Average · ○ Average · ◔ Better Than Average · ○ Excellent · ⊖ Insufficient Information

©1993 by IntelliChoice, Inc. (408) 554-8711 All Rights Reserved. Reproduction Prohibited.
Refer to *Section 3: Annotated Vehicle Charts* for an explanation of these charts.

Subaru Legacy Touring Wagon AWD
4 Door Wagon

Compact Wagon

 2.2L 160 hp Gas Fuel Inject.
 4 Cylinder Opposing
 Automatic 4 Speed
 4 Wheel Full-Time
 Driver Airbag Psngr Belts

Purchase Price

Car Item	Dealer Cost	List
Base Price	**$19,904**	**$22,650**
Anti-Lock Brakes	Std	Std
Manual Transmission	N/A	N/A
Optional Engine	N/A	N/A
Air Conditioning	Std	Std
Power Steering	Std	Std
Cruise Control	Std	Std
4 Wheel Full-Time Drive	Std	Std
AM/FM Stereo Cassette	Std	Std
Steering, Tilt w/Memory	Std	Std
Power Windows	Std	Std
*Options Price	$0	$0
*Total Price	$19,904	$22,650
Target Price	$20,556	
Destination Charge	$445	
Avg. Tax & Fees	$1,281	
Total Target $	**$22,282**	
Average Dealer Option Cost:	**83%**	

Ownership Costs

Cost Area	5 Year Cost	Rate
Depreciation		⊖
Financing ($448/month)	$3,671	
Insurance (Rating 14)	$8,238	◐
State Fees	$925	
Fuel (Hwy 23 City 18)	$4,213	●
Maintenance	$5,686	●
Repairs	$630	○

Warranty/Maintenance Info

Major Tune-Up	$145	○
Minor Tune-Up	$89	○
Brake Service	$239	●
Overall Warranty	3 yr/36k	○
Drivetrain Warranty	5 yr/60k	○
Rust Warranty	5 yr/unlim. mi	○
Maintenance Warranty	N/A	
Roadside Assistance	3 yr/36k	

Ownership Cost By Year

Insufficient Depreciation Information

1993, 1994, 1995, 1996, 1997

Resale Value

Insufficient Information

Ownership Costs (5yr)

Insufficient Information

Cumulative Costs

	1993	1994	1995	1996	1997
Annual	Insufficient Information				
Total	Insufficient Information				

Ownership Cost Rating

⊖ Insufficient Information

The 1993 Subaru Legacy is available in twelve models - L, LS, L 4WD, LS 4WD, LSi 4WD and Sport 4WD sedans; and L, LS, L 4WD, LS 4WD, LSi 4WD and Touring 4WD wagons. New for 1993, the Touring 4WD wagon receives a driver airbag, 60/40 rear split seats and one new exterior color (Jasper Green). Features include power moonroof with tilt feature, power windows and door locks, sport seats, an 80-watt stereo system with graphic equalizer, cruise control, alloy wheels and a functional hood scoop.

Subaru Loyale
4 Door Sedan

Subcompact

 1.8L 90 hp Gas Fuel Inject.
 4 Cylinder Opposing
 Manual 5 Speed
 2 Wheel Front
 Automatic Seatbelts

Purchase Price

Car Item	Dealer Cost	List
Base Price	**$9,313**	**$10,478**
Anti-Lock Brakes	N/A	N/A
Automatic 3 Speed	$488	$550
Optional Engine	N/A	N/A
Air Conditioning	Std	Std
Power Steering	Std	Std
Cruise Control	N/A	N/A
All Wheel Drive	N/A	N/A
AM/FM Stereo Cassette	N/A	N/A
Steering Wheel, Tilt	Std	Std
Power Windows	Std	Std
*Options Price	$0	$0
*Total Price	$9,313	$10,478
Target Price	$9,538	
Destination Charge	$445	
Avg. Tax & Fees	$608	
Total Target $	**$10,591**	
Average Dealer Option Cost:	**89%**	

Ownership Costs

Cost Area	5 Year Cost	Rate
Depreciation	$5,813	○
Financing ($213/month)	$1,745	
Insurance (Rating 7)	$7,035	●
State Fees	$436	
Fuel (Hwy 32 City 25)	$3,030	●
Maintenance	$4,043	○
Repairs	$560	○

Warranty/Maintenance Info

Major Tune-Up	$154	○
Minor Tune-Up	$88	○
Brake Service	$167	○
Overall Warranty	3 yr/36k	○
Drivetrain Warranty	5 yr/60k	○
Rust Warranty	5 yr/unlim. mi	○
Maintenance Warranty	N/A	
Roadside Assistance	3 yr/36k	

Resale Value

1993	1994	1995	1996	1997
$8,161	$7,132	$6,320	$5,544	$4,778

Ownership Costs (5yr)

Average	This Car
$23,367	$22,662
Cost/Mile 33¢	Cost/Mile 32¢

Cumulative Costs

	1993	1994	1995	1996	1997
Annual	$5,167	$3,808	$4,450	$3,640	$5,597
Total	$5,167	$8,975	$13,425	$17,065	$22,662

Ownership Cost Rating

○ Better Than Average

The 1993 Loyale is available in four models-(Base) Loyale and 4WD sedans; and (Base) Loyale and 4WD wagons. New for 1993, the Loyale features three new exterior colors: Teal, Misty Dawn and Freesia Red. The Base sedan includes power windows with driver's-side automatic up/down, power mirrors, power locks and tilt steering wheel. Additional equipment includes remote hood, and center console with covered storage bin. Options include fog lamps, security system and floor mats.

---

* Includes shaded options
** Other purchase requirements apply

● Poor | ◐ Worse Than Average | ○ Average | ○ Better Than Average | ○ Excellent | ⊖ Insufficient Information

©1993 by IntelliChoice, Inc. (408) 554-8711 All Rights Reserved. Reproduction Prohibited.
Refer to *Section 3: Annotated Vehicle Charts* for an explanation of these charts.

Subaru Loyale
4 Door Wagon

Subcompact Wag

 1.8L 90 hp Gas Fuel Inject. 4 Cylinder Opposing Manual 5 Speed 2 Wheel Front Automatic Seatbelts

Purchase Price

Car Item	Dealer Cost	List
Base Price	**$10,068**	**$11,328**
Anti-Lock Brakes	N/A	N/A
Automatic 3 Speed	$488	$550
Optional Engine	N/A	N/A
Air Conditioning	Std	Std
Power Steering	Std	Std
Cruise Control	N/A	N/A
All Wheel Drive	N/A	N/A
AM/FM Stereo Cassette	N/A	N/A
Steering Wheel, Tilt	Std	Std
Power Windows	Std	Std
***Options Price**	$0	$0
***Total Price**	**$10,068**	**$11,328**
Target Price	$10,318	
Destination Charge	$445	
Avg. Tax & Fees	$656	
Total Target $	**$11,419**	
Average Dealer Option Cost:	89%	

The Subaru Loyale is available in two models - (Base) Loyale and 4WD wagons. For 1993, the Base Loyale wagon features air conditioning, power steering, power windows with driver-side automatic up and down, power mirrors, power locks, AM/FM cassette audio system, tilt steering wheel, 60/40 split rear seat, rear-window defogger, all-season radial tires and a digital clock. Three new exterior colors are offered (Teal, Misty Dawn and Freesia Red).

Ownership Costs

Cost Area	5 Year Cost	Rate
Depreciation	$6,254	○
Financing ($229/month)	$1,881	
Insurance (Rating 8)	$7,223	◉
State Fees	$471	
Fuel (Hwy 30 City 25)	$3,140	◉
Maintenance	$4,043	○
Repairs	$560	◯

Warranty/Maintenance Info

Major Tune-Up	$154	○
Minor Tune-Up	$88	○
Brake Service	$167	◯
Overall Warranty	3 yr/36k	○
Drivetrain Warranty	5 yr/60k	◯
Rust Warranty	5 yr/unlim. mi	◯
Maintenance Warranty	N/A	
Roadside Assistance	3 yr/36k	

Ownership Cost By Year

1993, 1994, 1995, 1996, 1997

Resale Value

1993	1994	1995	1996	1997
$8,787	$7,681	$6,803	$5,974	$5,165

Cumulative Costs

	1993	1994	1995	1996	1997
Annual	$5,490	$3,993	$4,609	$3,771	$5,709
Total	$5,490	$9,483	$14,092	$17,863	$23,572

Ownership Costs (5yr)

Average	This Car
$24,324	$23,572
Cost/Mile 35¢	Cost/Mile 34¢

Ownership Cost Rating

○ Better Than Average

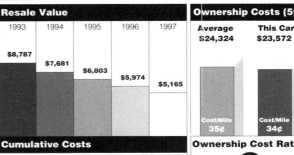

Subaru Loyale 4WD
4 Door Sedan

Subcompact

 1.8L 90 hp Gas Fuel Inject. 4 Cylinder Opposing Manual 5 Speed 4 Wheel On-Demand Automatic Seatbelts

2WD Model Shown

Purchase Price

Car Item	Dealer Cost	List
Base Price	**$10,512**	**$11,828**
Anti-Lock Brakes	N/A	N/A
Automatic 3 Speed	$488	$550
Optional Engine	N/A	N/A
Air Conditioning	Std	Std
Power Steering	Std	Std
Cruise Control	N/A	N/A
4 Whl On-Demand Dr.	Std	Std
AM/FM Stereo Cassette	N/A	N/A
Steering Wheel, Tilt	Std	Std
Power Windows	Std	Std
***Options Price**	$0	$0
***Total Price**	**$10,512**	**$11,828**
Target Price	$10,777	
Destination Charge	$445	
Avg. Tax & Fees	$684	
Total Target $	**$11,906**	
Average Dealer Option Cost:	89%	

The 1993 Loyale is available in four models-(Base) Loyale and 4WD sedans; and (Base) Loyale and 4WD wagons. New for 1993, the Loyale features three new exterior colors: Teal, Misty Dawn and Freesia Red. The Base 4WD sedan includes power windows with driver's-side automatic up/down, power mirrors, power locks and tilt steering wheel. Additional equipment includes remote hood, and center console with covered storage bin. Options include fog lamps, security system and floor mats.

Ownership Costs

Cost Area	5 Year Cost	Rate
Depreciation	$6,527	○
Financing ($239/month)	$1,961	
Insurance (Rating 8)	$7,223	◉
State Fees	$491	
Fuel (Hwy 29 City 24)	$3,258	◉
Maintenance	$4,028	○
Repairs	$560	◯

Warranty/Maintenance Info

Major Tune-Up	$154	◯
Minor Tune-Up	$88	○
Brake Service	$167	◯
Overall Warranty	3 yr/36k	○
Drivetrain Warranty	5 yr/60k	◯
Rust Warranty	5 yr/unlim. mi	◯
Maintenance Warranty	N/A	
Roadside Assistance	3 yr/36k	

Resale Value

1993	1994	1995	1996	1997
$9,140	$7,997	$7,084	$6,221	$5,379

Cumulative Costs

	1993	1994	1995	1996	1997
Annual	$5,684	$4,081	$4,594	$3,839	$5,850
Total	$5,684	$9,765	$14,359	$18,198	$24,048

Ownership Costs (5yr)

Average	This Car
$24,886	$24,048
Cost/Mile 36¢	Cost/Mile 34¢

Ownership Cost Rating

○ Better Than Average

---

page 226

* Includes shaded options
** Other purchase requirements apply

● Poor ◉ Worse Than Average ○ Average ◯ Better Than Average ○ Excellent ⊖ Insufficient Information

©1993 by IntelliChoice, Inc. (408) 554-8711 All Rights Reserved. Reproduction Prohibited.
Refer to *Section 3: Annotated Vehicle Charts* for an explanation of these charts.

Subaru Loyale 4WD
4 Door Wagon

Subcompact Wag

- 1.8L 90 hp Gas Fuel Inject.
- 4 Cylinder Opposing
- Manual 5 Speed
- 4 Wheel On-Demand
- Automatic Seatbelts

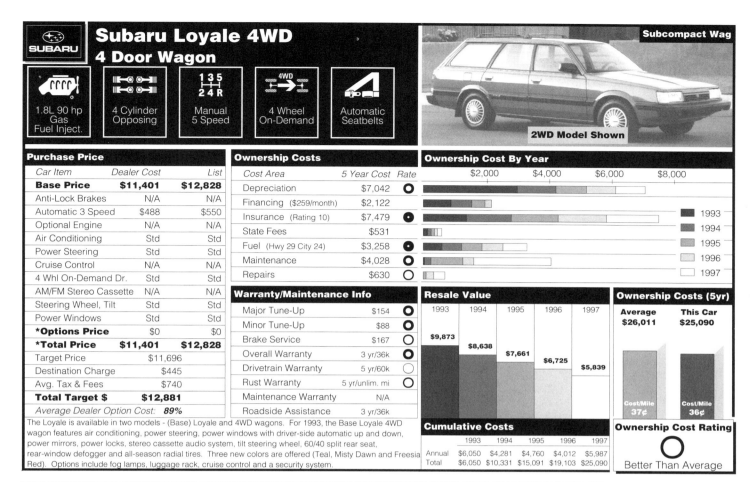

2WD Model Shown

Purchase Price

Car Item	Dealer Cost	List
Base Price	**$11,401**	**$12,828**
Anti-Lock Brakes	N/A	N/A
Automatic 3 Speed	$488	$550
Optional Engine	N/A	N/A
Air Conditioning	Std	Std
Power Steering	Std	Std
Cruise Control	N/A	N/A
4 Whl On-Demand Dr.	Std	Std
AM/FM Stereo Cassette	N/A	N/A
Steering Wheel, Tilt	Std	Std
Power Windows	Std	Std
*Options Price	$0	$0
*Total Price	$11,401	$12,828
Target Price		$11,696
Destination Charge		$445
Avg. Tax & Fees		$740
Total Target $		**$12,881**
Average Dealer Option Cost:		89%

The Loyale is available in two models - (Base) Loyale and 4WD wagons. For 1993, the Base Loyale 4WD wagon features air conditioning, power steering, power windows with driver-side automatic up and down, power mirrors, power locks, stereo cassette audio system, tilt steering wheel, 60/40 split rear seat, rear-window defogger and all-season radial tires. Three new colors are offered (Teal, Misty Dawn and Freesia Red). Options include fog lamps, luggage rack, cruise control and a security system.

Ownership Costs

Cost Area	5 Year Cost	Rate
Depreciation	$7,042	○
Financing ($259/month)	$2,122	
Insurance (Rating 10)	$7,479	◉
State Fees	$531	
Fuel (Hwy 29 City 24)	$3,258	◉
Maintenance	$4,028	○
Repairs	$630	○

Warranty/Maintenance Info

Major Tune-Up	$154	○
Minor Tune-Up	$88	○
Brake Service	$167	○
Overall Warranty	3 yr/36k	○
Drivetrain Warranty	5 yr/60k	○
Rust Warranty	5 yr/unlim. mi	○
Maintenance Warranty	N/A	
Roadside Assistance	3 yr/36k	

Ownership Cost By Year

$2,000 / $4,000 / $6,000 / $8,000
- 1993
- 1994
- 1995
- 1996
- 1997

Resale Value

1993	1994	1995	1996	1997
$9,873	$8,638	$7,661	$6,725	$5,839

Cumulative Costs

	1993	1994	1995	1996	1997
Annual	$6,050	$4,281	$4,760	$4,012	$5,987
Total	$6,050	$10,331	$15,091	$19,103	$25,090

Ownership Costs (5yr)

Average	This Car
$26,011	$25,090
Cost/Mile 37¢	Cost/Mile 36¢

Ownership Cost Rating
○ Better Than Average

Subaru SVX
2 Door Coupe

Sport

- 3.3L 230 hp Gas Fuel Inject.
- 6 Cylinder Opposing
- Automatic 4 Speed
- 4 Wheel Full-Time
- Driver Airbag Psngr Belts

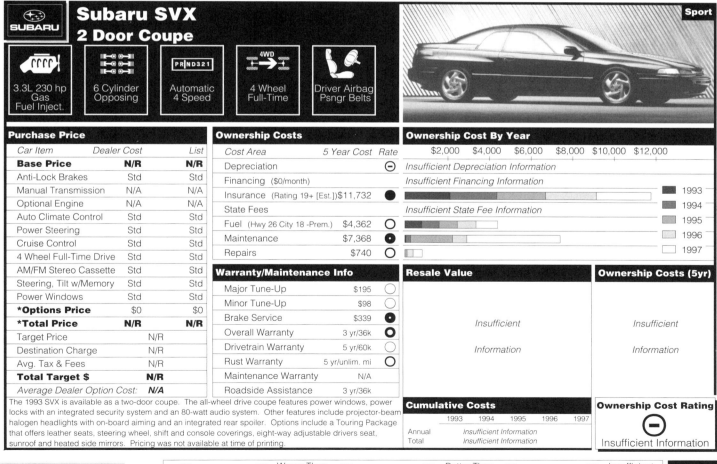

Purchase Price

Car Item	Dealer Cost	List
Base Price	**N/R**	**N/R**
Anti-Lock Brakes	Std	Std
Manual Transmission	N/A	N/A
Optional Engine	N/A	N/A
Auto Climate Control	Std	Std
Power Steering	Std	Std
Cruise Control	Std	Std
4 Wheel Full-Time Drive	Std	Std
AM/FM Stereo Cassette	Std	Std
Steering, Tilt w/Memory	Std	Std
Power Windows	Std	Std
*Options Price	$0	$0
*Total Price	N/R	N/R
Target Price		N/R
Destination Charge		N/R
Avg. Tax & Fees		N/R
Total Target $		**N/R**
Average Dealer Option Cost:		N/A

The 1993 SVX is available as a two-door coupe. The all-wheel drive coupe features power windows, power locks with an integrated security system and an 80-watt audio system. Other features include projector-beam halogen headlights with on-board aiming and an integrated rear spoiler. Options include a Touring Package that offers leather seats, steering wheel, shift and console coverings, eight-way adjustable drivers seat, sunroof and heated side mirrors. Pricing was not available at time of printing.

Ownership Costs

Cost Area	5 Year Cost	Rate
Depreciation		⊖
Financing ($0/month)		
Insurance (Rating 19+ [Est.])	$11,732	●
State Fees		
Fuel (Hwy 26 City 18 -Prem.)	$4,362	○
Maintenance	$7,368	◉
Repairs	$740	○

Warranty/Maintenance Info

Major Tune-Up	$195	○
Minor Tune-Up	$98	○
Brake Service	$339	◉
Overall Warranty	3 yr/36k	○
Drivetrain Warranty	5 yr/60k	○
Rust Warranty	5 yr/unlim. mi	○
Maintenance Warranty	N/A	
Roadside Assistance	3 yr/36k	

Ownership Cost By Year

$2,000 / $4,000 / $6,000 / $8,000 / $10,000 / $12,000

Insufficient Depreciation Information
Insufficient Financing Information
Insufficient State Fee Information

- 1993
- 1994
- 1995
- 1996
- 1997

Resale Value
Insufficient Information

Cumulative Costs

	1993	1994	1995	1996	1997
Annual	Insufficient Information				
Total	Insufficient Information				

Ownership Costs (5yr)
Insufficient Information

Ownership Cost Rating
⊖ Insufficient Information

---

* Includes shaded options
** Other purchase requirements apply

● Poor ◐ Worse Than Average ○ Average ◯ Better Than Average ◌ Excellent ⊖ Insufficient Information

©1993 by IntelliChoice, Inc. (408) 554-8711 All Rights Reserved. Reproduction Prohibited.
Refer to *Section 3: Annotated Vehicle Charts* for an explanation of these charts.

page 227

Suzuki Swift GA
2 Door Hatchback

 1.3L 70 hp Gas Fuel Inject.
 4 Cylinder In-Line
 Manual 5 Speed
 2 Wheel Front
 Automatic Seatbelts

Subcompact

Purchase Price

Car Item	Dealer Cost	List
Base Price	**$6,569**	**$7,299**
Anti-Lock Brakes	N/A	N/A
Automatic 3 Speed	$540	$600
Optional Engine	N/A	N/A
Air Conditioning	Dlr	Dlr
Power Steering	N/A	N/A
Cruise Control	N/A	N/A
All Wheel Drive	N/A	N/A
AM/FM Stereo Cassette	Dlr	Dlr
Steering Wheel, Tilt	N/A	N/A
Power Windows	N/A	N/A
*Options Price	$0	$0
*Total Price	$6,569	$7,299
Target Price	$6,912	
Destination Charge	$300	
Avg. Tax & Fees	$437	
Total Target $	**$7,649**	
Average Dealer Option Cost:	**90%**	

Swift is offered in two-door hatchback GA and GT editions and in four-door sedan GA and GS editions. New features for the GA two-door hatchback include sport wheel covers and an automatic front door locking system which automatically locks the front doors at eight miles per hour. Standard items include a rear window defogger, tinted window glass, and reclining front bucket seats. Available dealer installed accessories include front center armrests, theft deterrent system and manual sun roof.

Ownership Costs

Cost Area	5 Year Cost	Rate
Depreciation	$2,914	○
Financing ($154/month)	$1,261	
Insurance (Rating 10)	$7,479	●
State Fees	$304	
Fuel (Hwy 43 City 39)	$2,111	○
Maintenance	$4,325	⊙
Repairs	$640	○

Warranty/Maintenance Info

Major Tune-Up	$224	⊙
Minor Tune-Up	$77	○
Brake Service	$222	○
Overall Warranty	3 yr/36k	○
Drivetrain Warranty	3 yr/36k	○
Rust Warranty	3 yr/unlim. mi	⊙
Maintenance Warranty	N/A	
Roadside Assistance	3 yr/36k	

Ownership Cost By Year

1993, 1994, 1995, 1996, 1997

Resale Value

1993	1994	1995	1996	1997
$7,103	$6,546	$6,080	$5,425	$4,735

Cumulative Costs

	1993	1994	1995	1996	1997
Annual	$2,965	$3,048	$3,866	$3,511	$5,644
Total	$2,965	$6,013	$9,879	$13,390	$19,034

Ownership Costs (5yr)

Average: $19,791   This Car: $19,034
Cost/Mile: 28¢   Cost/Mile: 27¢

Ownership Cost Rating

○ Better Than Average

Suzuki Swift GT
2 Door Hatchback

 1.3L 100 hp Gas Fuel Inject.
4 Cylinder In-Line
 Manual 5 Speed
 2 Wheel Front
Automatic Seatbelts

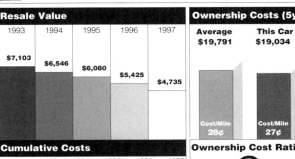

Subcompact

Purchase Price

Car Item	Dealer Cost	List
Base Price	**$8,699**	**$9,999**
Anti-Lock Brakes	N/A	N/A
Automatic Transmission	N/A	N/A
Optional Engine	N/A	N/A
Air Conditioning	Dlr	Dlr
Power Steering	Std	Std
Cruise Control	N/A	N/A
All Wheel Drive	N/A	N/A
AM/FM Stereo Cassette	Std	Std
Steering Wheel, Tilt	N/A	N/A
Power Windows	N/A	N/A
*Options Price	$0	$0
*Total Price	$8,699	$9,999
Target Price	$9,189	
Destination Charge	$300	
Avg. Tax & Fees	$577	
Total Target $	**$10,066**	
Average Dealer Option Cost:	**N/A**	

Swift is offered in two-door hatchback GA and GT editions and in four-door sedan GA and GS editions. In addition to or in place of the standard features of the GA, the GT hatchback comes with a more powerful engine, front air dam, mud guards, sport rocker panels, body striping, front fog lights, rear wiper, sport bucket seats and sport wheel covers. Dealer installed accessories include a 6-disc CD changer, door sill guard set, front mask/bra and roof rack.

Ownership Costs

Cost Area	5 Year Cost	Rate
Depreciation	$3,413	○
Financing ($202/month)	$1,658	
Insurance (Rating 10)	$7,479	●
State Fees	$412	
Fuel (Hwy 35 City 28)	$2,740	○
Maintenance	$4,830	⊙
Repairs	$640	○

Warranty/Maintenance Info

Major Tune-Up	$229	⊙
Minor Tune-Up	$81	○
Brake Service	$240	⊙
Overall Warranty	3 yr/36k	○
Drivetrain Warranty	3 yr/36k	○
Rust Warranty	3 yr/unlim. mi	⊙
Maintenance Warranty	N/A	
Roadside Assistance	3 yr/36k	

Resale Value

1993	1994	1995	1996	1997
$7,912	$7,870	$8,051	$7,477	$6,653

Cumulative Costs

	1993	1994	1995	1996	1997
Annual	$4,882	$2,808	$3,568	$3,658	$6,256
Total	$4,882	$7,690	$11,258	$14,916	$21,172

Ownership Costs (5yr)

Average: $22,828   This Car: $21,172
Cost/Mile: 33¢   Cost/Mile: 30¢

Ownership Cost Rating

○ Excellent

---

* Includes shaded options
** Other purchase requirements apply

● Poor   ◐ Worse Than Average   ○ Average   ⊙ Better Than Average   ○ Excellent   ⊖ Insufficient Information

©1993 by IntelliChoice, Inc. (408) 554-8711   All Rights Reserved. Reproduction Prohibited.
Refer to *Section 3: Annotated Vehicle Charts* for an explanation of these charts.

Suzuki Swift GA
4 Door Sedan

 1.3L 70 hp Gas Fuel Inject.
 4 Cylinder In-Line
Manual 5 Speed
2 Wheel Front
Automatic Seatbelts

Subcompact — GS Model Shown

Purchase Price

Car Item	Dealer Cost	List
Base Price	**$7,199**	**$7,999**
Anti-Lock Brakes	N/A	N/A
Automatic 3 Speed	$540	$600
Optional Engine	N/A	N/A
Air Conditioning	Dlr	Dlr
Power Steering	N/A	N/A
Cruise Control	N/A	N/A
All Wheel Drive	N/A	N/A
AM/FM Stereo Cassette	Dlr	Dlr
Steering Wheel, Tilt	N/A	N/A
Power Windows	N/A	N/A
*Options Price	$0	$0
*Total Price	$7,199	$7,999
Target Price		$7,584
Destination Charge		$300
Avg. Tax & Fees		$477
Total Target $		**$8,361**
Average Dealer Option Cost:		90%

Swift is offered in two-door hatchback GA and GT editions and in four-door sedan GA and GS editions. New features for the GA four-door sedan include sport wheel covers and an automatic front door locking system which automatically locks the front doors at eight miles per hour. A rear window defogger, power brakes, passenger mirror, child safety rear door locks and trip odometer are also standard equipment.

Ownership Costs

Cost Area	5 Year Cost	Rate
Depreciation	$3,575	○
Financing ($168/month)	$1,378	
Insurance (Rating 8)	$7,223	●
State Fees	$332	
Fuel (Hwy 43 City 39)	$2,111	○
Maintenance	$4,336	○
Repairs	$640	○

Warranty/Maintenance Info

Major Tune-Up	$224	●
Minor Tune-Up	$77	○
Brake Service	$225	●
Overall Warranty	3 yr/36k	○
Drivetrain Warranty	3 yr/36k	○
Rust Warranty	3 yr/unlim. mi	●
Maintenance Warranty	N/A	
Roadside Assistance	3 yr/36k	

Ownership Cost By Year

Resale Value

1993	1994	1995	1996	1997
$7,203	$6,636	$6,127	$5,472	$4,786

Cumulative Costs

	1993	1994	1995	1996	1997
Annual	$3,586	$3,053	$3,888	$3,458	$5,610
Total	$3,586	$6,639	$10,527	$13,985	$19,595

Ownership Costs (5yr)

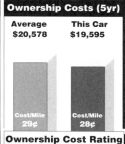

Average: $20,578 — This Car: $19,595
Cost/Mile: 29¢ — Cost/Mile: 28¢

Ownership Cost Rating: Excellent

Suzuki Swift GS
4 Door Sedan

 1.3L 70 hp Gas Fuel Inject.
 4 Cylinder In-Line — Manual 5 Speed
 2 Wheel Front
 Automatic Seatbelts

Subcompact

Purchase Price

Car Item	Dealer Cost	List
Base Price	**$8,271**	**$9,399**
Anti-Lock Brakes	N/A	N/A
Automatic 3 Speed	$528	$600
Optional Engine	N/A	N/A
Air Conditioning	Dlr	Dlr
Power Steering	Std	Std
Cruise Control	N/A	N/A
All Wheel Drive	N/A	N/A
AM/FM Stereo Cassette	Std	Std
Steering Wheel, Tilt	N/A	N/A
Power Windows	N/A	N/A
*Options Price	$0	$0
*Total Price	$8,271	$9,399
Target Price		$8,730
Destination Charge		$300
Avg. Tax & Fees		$549
Total Target $		**$9,579**
Average Dealer Option Cost:		88%

Swift is offered in two-door hatchback GA and GT editions and in four-door sedan GA and GS editions. In addition to or in place of the standard equipment of the GA sedan, the GS comes equipped with power steering, digital clock, remote fuel door and trunk releases, electric driver and passenger side mirrors, tachometer and AM/FM ETR cassette stereo. Available dealer installed accessories include a front mask/bra, roof rack and wheel locks.

Ownership Costs

Cost Area	5 Year Cost	Rate
Depreciation	$4,229	○
Financing ($193/month)	$1,578	
Insurance (Rating 8)	$7,223	●
State Fees	$388	
Fuel (Hwy 43 City 39)	$2,111	○
Maintenance	$4,336	○
Repairs	$640	○

Warranty/Maintenance Info

Major Tune-Up	$224	●
Minor Tune-Up	$77	○
Brake Service	$225	●
Overall Warranty	3 yr/36k	○
Drivetrain Warranty	3 yr/36k	○
Rust Warranty	3 yr/unlim. mi	●
Maintenance Warranty	N/A	
Roadside Assistance	3 yr/36k	

Resale Value

1993	1994	1995	1996	1997
$7,955	$7,319	$6,790	$6,080	$5,350

Cumulative Costs

	1993	1994	1995	1996	1997
Annual	$4,149	$3,196	$3,960	$3,538	$5,662
Total	$4,149	$7,345	$11,305	$14,843	$20,505

Ownership Costs (5yr)

Average: $22,153 — This Car: $20,505
Cost/Mile: 32¢ — Cost/Mile: 29¢

Ownership Cost Rating: Excellent

---

* Includes shaded options
** Other purchase requirements apply

Legend: ● Poor — ◐ Worse Than Average — ○ Average — ◑ Better Than Average — ○ Excellent — ⊖ Insufficient Information

©1993 by IntelliChoice, Inc. (408) 554-8711. All Rights Reserved. Reproduction Prohibited.
Refer to *Section 3: Annotated Vehicle Charts* for an explanation of these charts.

Toyota Camry DX
4 Door Sedan

 2.2L 130 hp Gas Fuel Inject. | 4 Cylinder In-Line | Manual 5 Speed | 2 Wheel Front | Driver Airbag Psngr Belts

V6 Model Shown — Compact

Purchase Price

Car Item	Dealer Cost	List
Base Price	**$12,809**	**$15,158**
Anti-Lock Brakes	$939	$1,145
Automatic 4 Speed	$676	$800
Optional Engine	N/A	N/A
Air Conditioning	$732	$915
Power Steering	Std	Std
Cruise Control	$184	$230
All Wheel Drive	N/A	N/A
AM/FM Stereo Cassette	$112	$150
Steering Wheel, Tilt	Std	Std
Power Windows	N/A	N/A
*Options Price	$1,520	$1,865
*Total Price	$14,329	$17,023
Target Price	$15,636	
Destination Charge	$295	
Avg. Tax & Fees	$970	
Total Target $	**$16,901**	
Average Dealer Option Cost:	82%	

The 1993 Camry is available in ten models - DX, LE, and XLE sedans; DX, LE, XLE, and SE V6 sedans; and DX, LE and LE V6 wagons. New for 1993, the DX offers a new multi-cone synchromesh gearing on four-cylinder manual transmission models for an improved shift feel. Two new colors have been added to the exterior, Red Pearl and Blue Pearl. Features include color-keyed bodyside molding, curved front seatbacks and recessed door panels.

Ownership Costs

Cost Area	5 Year Cost	Rate
Depreciation	$5,732	○
Financing ($340/month)	$2,784	
Insurance (Rating 5)	$6,786	○
State Fees	$693	
Fuel (Hwy 28 City 21)	$3,530	◐
Maintenance	$5,404	●
Repairs	$550	○

Warranty/Maintenance Info

Major Tune-Up	$185	○
Minor Tune-Up	$113	◐
Brake Service	$206	○
Overall Warranty	3 yr/36k	○
Drivetrain Warranty	5 yr/60k	○
Rust Warranty	5 yr/unlim. mi	○
Maintenance Warranty	N/A	
Roadside Assistance	N/A	

Ownership Cost By Year

Legend: 1993, 1994, 1995, 1996, 1997

Resale Value

1993	1994	1995	1996	1997
$14,600	$13,589	$12,867	$12,023	$11,169

Cumulative Costs

	1993	1994	1995	1996	1997
Annual	$5,586	$4,251	$5,060	$4,122	$6,460
Total	$5,586	$9,837	$14,897	$19,019	$25,479

Ownership Costs (5yr)

Average	This Car
$29,111	$25,479
Cost/Mile 42¢	Cost/Mile 36¢

Ownership Cost Rating
○ **Excellent**

Toyota Camry DX V6
4 Door Sedan

3.0L 185 hp Gas Fuel Inject. | 6 Cylinder "V" | Automatic 4 Speed | 2 Wheel Front | Driver Airbag Psngr Belts

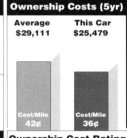
Compact

Purchase Price

Car Item	Dealer Cost	List
Base Price	**$15,141**	**$17,918**
Anti-Lock Brakes	$845	$1,030
Manual Transmission	N/A	N/A
Optional Engine	N/A	N/A
Air Conditioning	$732	$915
Power Steering	Std	Std
Cruise Control	$184	$230
All Wheel Drive	N/A	N/A
AM/FM Stereo Cassette	$112	$150
Steering Wheel, Tilt	Std	Std
Power Windows	N/A	N/A
*Options Price	$844	$1,065
*Total Price	$15,985	$18,983
Target Price	$17,544	
Destination Charge	$295	
Avg. Tax & Fees	$1,085	
Total Target $	**$18,924**	
Average Dealer Option Cost:	81%	

The 1993 Camry is available in ten models - DX, LE, and XLE sedans; DX, LE, XLE, and SE V6 sedans; and DX, LE and LE V6 wagons. New for 1993, the DX V6 is now available in two new colors, Red Pearl and Blue Pearl. Color-keyed bodyside molding is now standard. A oak colored interior is offered as well. Features include a fold-down 60/40 split rear seat with dual headrests, four-wheel disc brakes, auto-off headlamps, and intermittent wipers.

Ownership Costs

Cost Area	5 Year Cost	Rate
Depreciation	$6,774	○
Financing ($380/month)	$3,116	
Insurance (Rating 7)	$7,035	○
State Fees	$772	
Fuel (Hwy 24 City 18 -Prem.)	$4,547	●
Maintenance	$5,866	●
Repairs	$580	○

Warranty/Maintenance Info

Major Tune-Up	$211	●
Minor Tune-Up	$134	●
Brake Service	$191	○
Overall Warranty	3 yr/36k	○
Drivetrain Warranty	5 yr/60k	○
Rust Warranty	5 yr/unlim. mi	○
Maintenance Warranty	N/A	
Roadside Assistance	N/A	

Resale Value

1993	1994	1995	1996	1997
$16,042	$14,888	$14,046	$13,098	$12,150

Cumulative Costs

	1993	1994	1995	1996	1997
Annual	$6,562	$4,787	$5,654	$4,570	$7,117
Total	$6,562	$11,349	$17,003	$21,573	$28,690

Ownership Costs (5yr)

Average	This Car
$30,794	$28,690
Cost/Mile 44¢	Cost/Mile 41¢

Ownership Cost Rating
○ **Better Than Average**

---

page **230** — * Includes shaded options   ** Other purchase requirements apply

Legend: ● Poor | ◐ Worse Than Average | ○ Average | ○ Better Than Average | ○ Excellent | ⊖ Insufficient Information

©1993 by *IntelliChoice, Inc.* (408) 554-8711   All Rights Reserved. Reproduction Prohibited.
Refer to *Section 3: Annotated Vehicle Charts* for an explanation of these charts.

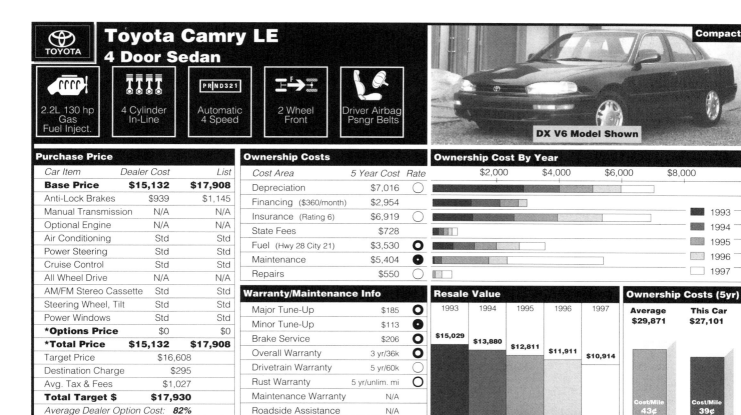

Toyota Camry LE
4 Door Sedan — Compact

- 2.2L 130 hp Gas Fuel Inject.
- 4 Cylinder In-Line
- Automatic 4 Speed
- 2 Wheel Front
- Driver Airbag Psngr Belts

DX V6 Model Shown

Purchase Price

Car Item	Dealer Cost	List
Base Price	**$15,132**	**$17,908**
Anti-Lock Brakes	$939	$1,145
Manual Transmission	N/A	N/A
Optional Engine	N/A	N/A
Air Conditioning	Std	Std
Power Steering	Std	Std
Cruise Control	Std	Std
All Wheel Drive	N/A	N/A
AM/FM Stereo Cassette	Std	Std
Steering Wheel, Tilt	Std	Std
Power Windows	Std	Std
***Options Price**	**$0**	**$0**
***Total Price**	**$15,132**	**$17,908**
Target Price		$16,608
Destination Charge		$295
Avg. Tax & Fees		$1,027
Total Target $		**$17,930**
Average Dealer Option Cost:		**82%**

The 1993 Camry is available in ten models - DX, LE, and XLE sedans; DX, LE, XLE, and SE V6 sedans; and DX, LE and LE V6 wagons. New for 1993, the LE is available in two fresh colors. A new oak colored interior has been added and includes standard bronze tinted safety glass. Features include color-keyed bumpers, child-protector door locks, and a fold-down split rear seat. Also included are power mirrors, a five-way adjustable driver's seat, and optional 14-inch aluminum alloy wheels.

Ownership Costs

Cost Area	5 Year Cost	Rate
Depreciation	$7,016	○
Financing ($360/month)	$2,954	
Insurance (Rating 6)	$6,919	○
State Fees	$728	
Fuel (Hwy 28 City 21)	$3,530	◐
Maintenance	$5,404	●
Repairs	$550	○

Warranty/Maintenance Info

Major Tune-Up	$185	○
Minor Tune-Up	$113	●
Brake Service	$206	○
Overall Warranty	3 yr/36k	○
Drivetrain Warranty	5 yr/60k	○
Rust Warranty	5 yr/unlim. mi	○
Maintenance Warranty	N/A	
Roadside Assistance	N/A	

Ownership Cost By Year

(Bars for 1993, 1994, 1995, 1996, 1997)

Resale Value

1993	1994	1995	1996	1997
$15,029	$13,880	$12,811	$11,911	$10,914

Cumulative Costs

	1993	1994	1995	1996	1997
Annual	$6,289	$4,476	$5,475	$4,225	$6,636
Total	$6,289	$10,765	$16,240	$20,465	$27,101

Ownership Costs (5yr)

- Average: **$29,871** — Cost/Mile 43¢
- This Car: **$27,101** — Cost/Mile 39¢

Ownership Cost Rating
Excellent

Toyota Camry SE V6
4 Door Sedan — Compact

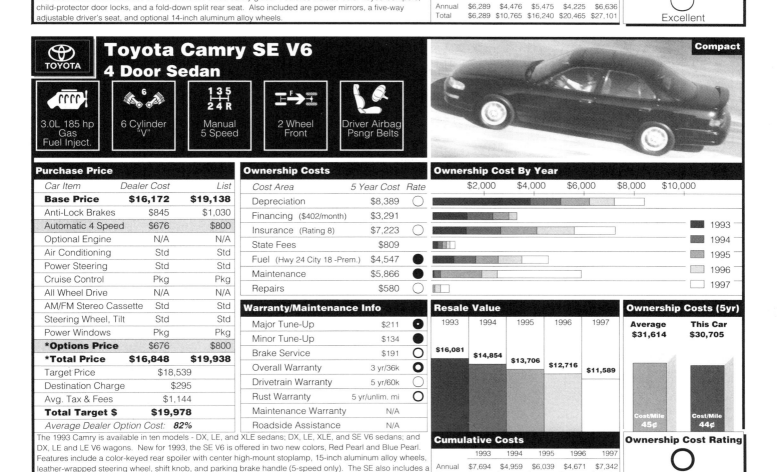

- 3.0L 185 hp Gas Fuel Inject.
- 6 Cylinder "V"
- Manual 5 Speed
- 2 Wheel Front
- Driver Airbag Psngr Belts

Purchase Price

Car Item	Dealer Cost	List
Base Price	**$16,172**	**$19,138**
Anti-Lock Brakes	$845	$1,030
Automatic 4 Speed	$676	$800
Optional Engine	N/A	N/A
Air Conditioning	Std	Std
Power Steering	Std	Std
Cruise Control	Pkg	Pkg
All Wheel Drive	N/A	N/A
AM/FM Stereo Cassette	Std	Std
Steering Wheel, Tilt	Std	Std
Power Windows	Pkg	Pkg
***Options Price**	**$676**	**$800**
***Total Price**	**$16,848**	**$19,938**
Target Price		$18,539
Destination Charge		$295
Avg. Tax & Fees		$1,144
Total Target $		**$19,978**
Average Dealer Option Cost:		**82%**

The 1993 Camry is available in ten models - DX, LE, and XLE sedans; DX, LE, XLE, and SE V6 sedans; and DX, LE and LE V6 wagons. New for 1993, the SE V6 is offered in two new colors, Red Pearl and Blue Pearl. Features include a color-keyed rear spoiler with center high-mount stoplamp, 15-inch aluminum alloy wheels, leather-wrapped steering wheel, shift knob, and parking brake handle (5-speed only). The SE also includes a tuned sport suspension.

Ownership Costs

Cost Area	5 Year Cost	Rate
Depreciation	$8,389	○
Financing ($402/month)	$3,291	
Insurance (Rating 8)	$7,223	○
State Fees	$809	
Fuel (Hwy 24 City 18 -Prem.)	$4,547	●
Maintenance	$5,866	●
Repairs	$580	○

Warranty/Maintenance Info

Major Tune-Up	$211	●
Minor Tune-Up	$134	●
Brake Service	$191	○
Overall Warranty	3 yr/36k	○
Drivetrain Warranty	5 yr/60k	○
Rust Warranty	5 yr/unlim. mi	○
Maintenance Warranty	N/A	
Roadside Assistance	N/A	

Resale Value

1993	1994	1995	1996	1997
$16,081	$14,854	$13,706	$12,716	$11,589

Cumulative Costs

	1993	1994	1995	1996	1997
Annual	$7,694	$4,959	$6,039	$4,671	$7,342
Total	$7,694	$12,653	$18,692	$23,363	$30,705

Ownership Costs (5yr)

- Average: **$31,614** — Cost/Mile 45¢
- This Car: **$30,705** — Cost/Mile 44¢

Ownership Cost Rating
Better Than Average

---

* Includes shaded options
** Other purchase requirements apply

Legend: ● Poor — ◐ Worse Than Average — ◑ Average — ○ Better Than Average — ○ Excellent — ⊖ Insufficient Information

©1993 by IntelliChoice, Inc. (408) 554-8711 All Rights Reserved. Reproduction Prohibited.
Refer to *Section 3: Annotated Vehicle Charts* for an explanation of these charts.

Toyota Camry LE V6
4 Door Sedan

3.0L 185 hp Gas Fuel Inject. | 6 Cylinder "V" | Automatic 4 Speed | 2 Wheel Front | Driver Airbag Psngr Belts

Compact

Purchase Price

Car Item	Dealer Cost	List
Base Price	**$16,788**	**$19,868**
Anti-Lock Brakes	$845	$1,030
Manual Transmission	N/A	N/A
Optional Engine	N/A	N/A
Air Conditioning	Std	Std
Power Steering	Std	Std
Cruise Control	Std	Std
All Wheel Drive	N/A	N/A
AM/FM Stereo Cassette	Std	Std
Steering Wheel, Tilt	Std	Std
Power Windows	Std	Std
***Options Price**	**$0**	**$0**
***Total Price**	**$16,788**	**$19,868**
Target Price	$18,501	
Destination Charge	$295	
Avg. Tax & Fees	$1,142	
Total Target $	**$19,938**	
Average Dealer Option Cost:	82%	

The 1993 Camry is available in ten models - DX, LE, and XLE sedans; DX, LE, XLE, and SE V6 sedans; and DX, LE and LE V6 wagons. New for 1993, the LE V6 now features two new exterior colors. An oak interior has been added to the line including a standard bronze tinted safety glass. Features include color-keyed bumpers, child-protector locks, and a 60/40 fold-down split rear seat. The LE also includes a five-way adjustable driver's seat, automatic illuminated entry/exit fade-out system and power mirrors.

Ownership Costs

Cost Area	5 Year Cost	Rate
Depreciation	$8,107	○
Financing ($401/month)	$3,285	
Insurance (Rating 7)	$7,035	○
State Fees	$806	
Fuel (Hwy 24 City 18 -Prem.)	$4,547	●
Maintenance	$5,828	●
Repairs	$580	○

Warranty/Maintenance Info

Major Tune-Up	$211	●
Minor Tune-Up	$134	●
Brake Service	$191	○
Overall Warranty	3 yr/36k	○
Drivetrain Warranty	5 yr/60k	○
Rust Warranty	5 yr/unlim. mi	○
Maintenance Warranty	N/A	
Roadside Assistance	N/A	

Ownership Cost By Year

(bars for 1993, 1994, 1995, 1996, 1997)

Resale Value

1993	1994	1995	1996	1997
$16,449	$15,166	$13,941	$12,934	$11,831

Cumulative Costs

	1993	1994	1995	1996	1997
Annual	$7,248	$4,976	$6,059	$4,648	$7,257
Total	$7,248	$12,224	$18,283	$22,931	$30,188

Ownership Costs (5yr)

Average	This Car
$31,554	$30,188
Cost/Mile 45¢	Cost/Mile 43¢

Ownership Cost Rating

○ Better Than Average

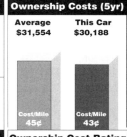

Toyota Camry XLE
4 Door Sedan

2.2L 130 hp Gas Fuel Inject. | 4 Cylinder In-Line | Automatic 4 Speed | 2 Wheel Front | Driver Airbag Psngr Belts

Compact — LE V6 Model Shown

Purchase Price

Car Item	Dealer Cost	List
Base Price	**$16,698**	**$19,878**
Anti-Lock Brakes	$939	$1,145
Manual Transmission	N/A	N/A
Optional Engine	N/A	N/A
Air Conditioning	Std	Std
Power Steering	Std	Std
Cruise Control	Std	Std
All Wheel Drive	N/A	N/A
AM/FM Stereo Cassette	Std	Std
Steering Wheel, Tilt	Std	Std
Power Windows	Std	Std
***Options Price**	**$0**	**$0**
***Total Price**	**$16,698**	**$19,878**
Target Price	$18,398	
Destination Charge	$295	
Avg. Tax & Fees	$1,137	
Total Target $	**$19,830**	
Average Dealer Option Cost:	81%	

The 1993 Camry is available in ten models - DX, LE, and XLE sedans; DX, LE, XLE, and SE V6 sedans; and DX, LE and LE V6 wagons. New for 1993, the XLE is available now in two new colors. An oak colored interior is offered (optional leather) and includes standard bronze tinted safety glass. Features include a seven-way power adjustable driver's seat, a fold-down 60/40 split rear seat, 14-inch aluminum alloy wheels, and an automatic illuminated entry/exit fade-out system.

Ownership Costs

Cost Area	5 Year Cost	Rate
Depreciation	$8,881	○
Financing ($399/month)	$3,267	
Insurance (Rating 7)	$7,035	○
State Fees	$807	
Fuel (Hwy 28 City 21)	$3,530	○
Maintenance	$5,404	●
Repairs	$550	○

Warranty/Maintenance Info

Major Tune-Up	$185	○
Minor Tune-Up	$113	●
Brake Service	$206	○
Overall Warranty	3 yr/36k	○
Drivetrain Warranty	5 yr/60k	○
Rust Warranty	5 yr/unlim. mi	○
Maintenance Warranty	N/A	
Roadside Assistance	N/A	

Resale Value

1993	1994	1995	1996	1997
$16,013	$14,588	$13,242	$12,134	$10,949

Cumulative Costs

	1993	1994	1995	1996	1997
Annual	$7,378	$4,890	$5,852	$4,493	$6,861
Total	$7,378	$12,268	$18,120	$22,613	$29,474

Ownership Costs (5yr)

Average	This Car
$31,563	$29,474
Cost/Mile 45¢	Cost/Mile 42¢

Ownership Cost Rating

○ Better Than Average

---

page 232

* Includes shaded options
** Other purchase requirements apply

● Poor | ◐ Worse Than Average | ○ Average | ○ Better Than Average | ○ Excellent | ⊖ Insufficient Information

©1993 by IntelliChoice, Inc. (408) 554-8711 All Rights Reserved. Reproduction Prohibited.
Refer to *Section 3: Annotated Vehicle Charts* for an explanation of these charts.

Toyota Camry XLE V6
4 Door Sedan

 3.0L 185 hp Gas Fuel Inject.
 6 Cylinder "V"
 Automatic 4 Speed
 2 Wheel Front
 Driver Airbag Psngr Belts

Compact — LE Model Shown

Purchase Price

Car Item	Dealer Cost	List
Base Price	**$18,378**	**$21,878**
Anti-Lock Brakes	$845	$1,030
Manual Transmission	N/A	N/A
Optional Engine	N/A	N/A
Air Conditioning	Std	Std
Power Steering	Std	Std
Cruise Control	Std	Std
All Wheel Drive	N/A	N/A
AM/FM Stereo Cassette	Std	Std
Steering Wheel, Tilt	Std	Std
Power Windows	Std	Std
*Options Price	$0	$0
*Total Price	$18,378	$21,878
Target Price	$20,333	
Destination Charge	$295	
Avg. Tax & Fees	$1,253	
Total Target $	**$21,881**	
Average Dealer Option Cost:	*81%*	

Ownership Costs

Cost Area	5 Year Cost	Rate
Depreciation	$9,824	○
Financing ($440/month)	$3,605	
Insurance (Rating 8)	$7,223	○
State Fees	$887	
Fuel (Hwy 24 City 18 -Prem.)	$4,547	●
Maintenance	$5,828	●
Repairs	$580	○

Warranty/Maintenance Info

Major Tune-Up	$211	●
Minor Tune-Up	$134	●
Brake Service	$191	○
Overall Warranty	3 yr/36k	○
Drivetrain Warranty	5 yr/60k	○
Rust Warranty	5 yr/unlim. mi	○
Maintenance Warranty	N/A	
Roadside Assistance	N/A	

Ownership Cost By Year

Resale Value

1993	1994	1995	1996	1997
$17,585	$16,028	$14,570	$13,362	$12,057

Cumulative Costs

	1993	1994	1995	1996	1997
Annual	$8,243	$5,405	$6,409	$4,925	$7,512
Total	$8,243	$13,648	$20,057	$24,982	$32,494

Ownership Costs (5yr)

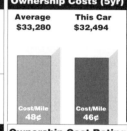

Average $33,280 — Cost/Mile 48¢
This Car $32,494 — Cost/Mile 46¢

Ownership Cost Rating
○ Better Than Average

The 1993 Camry is available in ten models - DX, LE, and XLE sedans; DX, LE, XLE, and SE V6 sedans; and DX, LE and LE V6 wagons. New for 1993, the XLE V6 is available in two fresh exterior colors, Red Pearl and Blue Pearl. An oak colored interior is offered which includes bronze tinted glass. Features include a seven-way adjustable power driver's seat, and a tilt-and-slide moon roof. Other features include alloy wheels, and an illuminated-entry system with maplights.

Toyota Camry DX
4 Door Wagon

 2.2L 130 hp Gas Fuel Inject.
 4 Cylinder In-Line
 Automatic 4 Speed
 2 Wheel Front
 Driver Airbag Psngr Belts

Compact Wagon — LE Model Shown

Purchase Price

Car Item	Dealer Cost	List
Base Price	**$14,608**	**$17,288**
Anti-Lock Brakes	$939	$1,145
Manual Transmission	N/A	N/A
Optional Engine	N/A	N/A
Air Conditioning	$732	$915
Power Steering	Std	Std
Cruise Control	$184	$230
All Wheel Drive	N/A	N/A
AM/FM Stereo Cassette	$112	$150
Steering Wheel, Tilt	Std	Std
Power Windows	N/A	N/A
*Options Price	$844	$1,065
*Total Price	$15,452	$18,353
Target Price	$16,937	
Destination Charge	$295	
Avg. Tax & Fees	$1,048	
Total Target $	**$18,280**	
Average Dealer Option Cost:	*81%*	

Ownership Costs

Cost Area	5 Year Cost	Rate
Depreciation	$6,740	○
Financing ($367/month)	$3,011	
Insurance (Rating 6)	$6,919	○
State Fees	$746	
Fuel (Hwy 28 City 21)	$3,530	○
Maintenance	$5,404	●
Repairs	$550	○

Warranty/Maintenance Info

Major Tune-Up	$185	●
Minor Tune-Up	$113	○
Brake Service	$206	○
Overall Warranty	3 yr/36k	○
Drivetrain Warranty	5 yr/60k	○
Rust Warranty	5 yr/unlim. mi	○
Maintenance Warranty	N/A	
Roadside Assistance	N/A	

Resale Value

1993	1994	1995	1996	1997
$15,728	$14,609	$13,552	$12,557	$11,540

Cumulative Costs

	1993	1994	1995	1996	1997
Annual	$5,968	$4,468	$5,477	$4,328	$6,659
Total	$5,968	$10,436	$15,913	$20,241	$26,900

Ownership Costs (5yr)

Average $30,253 — Cost/Mile 43¢
This Car $26,900 — Cost/Mile 38¢

Ownership Cost Rating
○ Excellent

The 1993 Camry is available in ten models - DX, LE, and XLE sedans; DX, LE, XLE, and SE V6 sedans; and DX, LE and LE V6 wagons. New for 1993, the DX has a multi-cone synchromesh gearing in manual transmission models which provides an improved shift feel. Color-keyed bodyside molding has been added. Features include color-keyed bumpers, a 60/40 split back seat with over 40 cu. ft. of cargo room (with the rear seat up), and an optional third rear seat that comes with 3-point seatbelts.

---

* Includes shaded options
** Other purchase requirements apply

● Poor   ◐ Worse Than Average   ◯ Average   ○ Better Than Average   ○ Excellent   ⊖ Insufficient Information

©1993 by IntelliChoice, Inc. (408) 554-8711 All Rights Reserved. Reproduction Prohibited.
Refer to *Section 3: Annotated Vehicle Charts* for an explanation of these charts.

Toyota Camry LE
4 Door Wagon

 2.2L 130 hp Gas Fuel Inject.
4 Cylinder In-Line
 Automatic 4 Speed
 2 Wheel Front
Driver Airbag Psngr Belts

Compact Wagon

Purchase Price

Car Item	Dealer Cost	List
Base Price	**$16,248**	**$19,228**
Anti-Lock Brakes	$939	$1,145
Manual Transmission	N/A	N/A
Optional Engine	N/A	N/A
Air Conditioning	Std	Std
Power Steering	Std	Std
Cruise Control	Std	Std
All Wheel Drive	N/A	N/A
AM/FM Stereo Cassette	Std	Std
Steering Wheel, Tilt	Std	Std
Power Windows	Std	Std
*Options Price	$0	$0
*Total Price	$16,248	$19,228
Target Price	$17,882	
Destination Charge	$295	
Avg. Tax & Fees	$1,104	
Total Target $	**$19,281**	
Average Dealer Option Cost:	81%	

The 1993 Camry is available in ten models - DX, LE, and XLE sedans; DX, LE, XLE, and SE V6 sedans; and DX, LE and LE V6 wagons. New for 1993, the LE has a new multi-cone synchromesh gearing added on manual transmission models for an improved shift feel. An oak colored interior has been introduced that includes standard bronze tinted glass. Features include a five-way adjustable driver's seat, color-keyed power outside mirrors, and an illuminated entry/exit fade-out system.

Ownership Costs

Cost Area	5 Year Cost	Rate
Depreciation	$7,716	○
Financing ($388/month)	$3,177	
Insurance (Rating 7)	$7,035	○
State Fees	$781	
Fuel (Hwy 28 City 21)	$3,530	◐
Maintenance	$5,404	●
Repairs	$550	○

Warranty/Maintenance Info

Major Tune-Up	$185	●
Minor Tune-Up	$113	○
Brake Service	$206	○
Overall Warranty	3 yr/36k	○
Drivetrain Warranty	5 yr/60k	
Rust Warranty	5 yr/unlim. mi	○
Maintenance Warranty	N/A	
Roadside Assistance	N/A	

Ownership Cost By Year

1993, 1994, 1995, 1996, 1997

Resale Value

1993: $15,619; 1994: $14,491; 1995: $13,462; 1996: $12,562; 1997: $11,565

Cumulative Costs

	1993	1994	1995	1996	1997
Annual	$7,178	$4,559	$5,512	$4,275	$6,669
Total	$7,178	$11,737	$17,249	$21,524	$28,193

Ownership Costs (5yr)

Average: $31,005 This Car: $28,193
Cost/Mile: 44¢ Cost/Mile: 40¢

Ownership Cost Rating
○ Excellent

Toyota Camry LE V6
4 Door Wagon

 3.0L 185 hp Gas Fuel Inject.
 6 Cylinder "V"
 Automatic 4 Speed
 2 Wheel Front
 Driver Airbag Psngr Belts

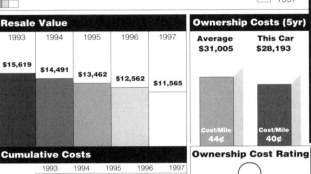

Compact Wagon — Base Model Shown

Purchase Price

Car Item	Dealer Cost	List
Base Price	**$17,921**	**$21,208**
Anti-Lock Brakes	$845	$1,030
Manual Transmission	N/A	N/A
Optional Engine	N/A	N/A
Air Conditioning	Std	Std
Power Steering	Std	Std
Cruise Control	Std	Std
All Wheel Drive	N/A	N/A
AM/FM Stereo Cassette	Std	Std
Steering Wheel, Tilt	Std	Std
Power Windows	Std	Std
*Options Price	$0	$0
*Total Price	$17,921	$21,208
Target Price	$19,805	
Destination Charge	$295	
Avg. Tax & Fees	$1,220	
Total Target $	**$21,320**	
Average Dealer Option Cost:	82%	

The 1993 Camry Wagon is available in three models - DX, LE, and LE V6. New for 1993, the LE V6 has two new exterior colors to choose from, a Red Pearl and a Blue Pearl. There is also a new oak interior color available that includes standard bronze tinted safety glass. Features include a five-way adjustable driver's seat, four-wheel disc brakes, color-keyed power outside mirrors, and an automatic illuminated entry/exit fade-out system, and an optional third rear seat with three-point seatbelts.

Ownership Costs

Cost Area	5 Year Cost	Rate
Depreciation	$8,774	○
Financing ($429/month)	$3,512	
Insurance (Rating 8)	$7,223	○
State Fees	$861	
Fuel (Hwy 24 City 18 -Prem.)	$4,547	●
Maintenance	$5,828	●
Repairs	$580	○

Warranty/Maintenance Info

Major Tune-Up	$211	●
Minor Tune-Up	$134	●
Brake Service	$191	○
Overall Warranty	3 yr/36k	○
Drivetrain Warranty	5 yr/60k	
Rust Warranty	5 yr/unlim. mi	○
Maintenance Warranty	N/A	
Roadside Assistance	N/A	

Resale Value

1993: $17,123; 1994: $15,868; 1995: $14,691; 1996: $13,688; 1997: $12,546

Cumulative Costs

	1993	1994	1995	1996	1997
Annual	$8,099	$5,068	$6,104	$4,709	$7,345
Total	$8,099	$13,167	$19,271	$23,980	$31,325

Ownership Costs (5yr)

Average: $32,705 This Car: $31,325
Cost/Mile: 47¢ Cost/Mile: 45¢

Ownership Cost Rating
○ Better Than Average

---

page 234

* Includes shaded options
** Other purchase requirements apply

● Poor ◐ Worse Than Average ◑ Average ○ Better Than Average ○ Excellent ⊖ Insufficient Information

©1993 by IntelliChoice, Inc. (408) 554-8711 All Rights Reserved. Reproduction Prohibited.
Refer to *Section 3: Annotated Vehicle Charts* for an explanation of these charts.

Toyota Celica ST
2 Door Coupe

 1.6L 103 hp Gas Fuel Inject. | 4 Cylinder In-Line | Manual 5 Speed | 2 Wheel Front | Driver Airbag Psngr Belts

Compact

Purchase Price

Car Item	Dealer Cost	List
Base Price	**$12,139**	**$14,198**
Anti-Lock Brakes	N/A	N/A
Automatic 4 Speed	$607	$710
Optional Engine	N/A	N/A
Air Conditioning	$732	$915
Power Steering	Std	Std
Cruise Control	$184	$230
All Wheel Drive	N/A	N/A
AM/FM Stereo Cassette	$112	$150
Steering Wheel, Tilt	Pkg	Pkg
Power Windows	N/A	N/A
*Options Price	$1,451	$1,775
*Total Price	$13,590	$15,973
Target Price	$14,805	
Destination Charge	$295	
Avg. Tax & Fees	$918	
Total Target $	**$16,018**	
Average Dealer Option Cost:	*82%*	

The 1993 Toyota Celica is available in six models - ST and GT Sport Coupes; GT, GT-S and Turbo All-Trac Liftbacks and GT Convertible. New for 1993, the Celica ST is offered in one new color (Turquoise Pearl). The ST Sport Coupe also offers cloth, reclining, sport bucket seats; a center console and dual mirrors. Exterior features include an All Weather Guard package, a remote control hatch release, a remote trunk release, bodyside molding and variable-assist power rack-and-pinion steering.

Ownership Costs

Cost Area	5 Year Cost	Rate
Depreciation	$6,191	○
Financing ($322/month)	$2,639	
Insurance (Rating 12)	$7,898	●
State Fees	$651	
Fuel (Hwy 31 City 24)	$3,140	○
Maintenance	$5,772	●
Repairs	$560	○

Warranty/Maintenance Info

Major Tune-Up	$201	●
Minor Tune-Up	$112	○
Brake Service	$222	○
Overall Warranty	3 yr/36k	○
Drivetrain Warranty	5 yr/60k	◐
Rust Warranty	5 yr/unlim. mi	○
Maintenance Warranty	N/A	
Roadside Assistance	N/A	

Ownership Cost By Year

Resale Value

1993	1994	1995	1996	1997
$12,976	$12,135	$11,375	$10,674	$9,827

Cumulative Costs

	1993	1994	1995	1996	1997
Annual	$6,389	$4,164	$5,356	$4,097	$6,845
Total	$6,389	$10,553	$15,909	$20,006	$26,851

Ownership Costs (5yr)

Average	This Car
$28,210	$26,851
Cost/Mile 40¢	Cost/Mile 38¢

Ownership Cost Rating
○ Better Than Average

Toyota Celica GT
2 Door Coupe

 2.2L 135 hp Gas Fuel Inject. | 4 Cylinder In-Line | Manual 5 Speed | 2 Wheel Front | Driver Airbag Psngr Belts

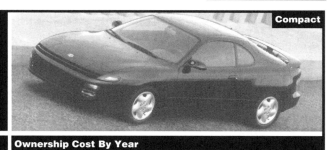

Compact

Purchase Price

Car Item	Dealer Cost	List
Base Price	**$14,202**	**$16,708**
Anti-Lock Brakes	$1,029	** $1,260
Automatic 4 Speed	$680	$800
Optional Engine	N/A	N/A
Air Conditioning	$732	$915
Power Steering	Std	Std
Cruise Control	$184	$230
All Wheel Drive	N/A	N/A
AM/FM Stereo Cassette	$135	$180
Steering Wheel, Tilt	Std	Std
Power Windows	Pkg	Pkg
*Options Price	$1,547	$1,895
*Total Price	$15,749	$18,603
Target Price	$17,245	
Destination Charge	$295	
Avg. Tax & Fees	$1,066	
Total Target $	**$18,606**	
Average Dealer Option Cost:	*83%*	

The 1993 Toyota Celica is available in six models - ST and GT Sport Coupes; GT, GT-S and Turbo All-Trac Liftbacks and GT Convertible. New for 1993, the Celica GT Coupe now features variable-assist power rack-and-pinion steering. Standard features include 15-inch wheels, 205/55VR15 speed-rated tires, dual color-keyed power side mirrors, rear spoiler, power antenna, a fold-down 50/50 split rear seat with security locks and tilt steering wheel. Options include a power sunroof and aluminum wheels.

Ownership Costs

Cost Area	5 Year Cost	Rate
Depreciation	$8,174	○
Financing ($374/month)	$3,066	
Insurance (Rating 14)	$8,238	●
State Fees	$755	
Fuel (Hwy 28 City 21)	$3,530	○
Maintenance	$6,674	●
Repairs	$600	○

Warranty/Maintenance Info

Major Tune-Up	$201	●
Minor Tune-Up	$112	○
Brake Service	$260	●
Overall Warranty	3 yr/36k	○
Drivetrain Warranty	5 yr/60k	◐
Rust Warranty	5 yr/unlim. mi	○
Maintenance Warranty	N/A	
Roadside Assistance	N/A	

Ownership Cost By Year

Resale Value

1993	1994	1995	1996	1997
$13,915	$12,990	$12,126	$11,365	$10,432

Cumulative Costs

	1993	1994	1995	1996	1997
Annual	$8,377	$4,545	$6,147	$4,374	$7,594
Total	$8,377	$12,922	$19,069	$23,443	$31,037

Ownership Costs (5yr)

Average	This Car
$30,468	$31,037
Cost/Mile 44¢	Cost/Mile 44¢

Ownership Cost Rating
○ Average

---

* Includes shaded options
** Other purchase requirements apply

● Poor ◐ Worse Than Average ○ Average ○ Better Than Average ○ Excellent — Insufficient Information

©1993 by IntelliChoice, Inc. (408) 554-8711 All Rights Reserved. Reproduction Prohibited.
Refer to *Section 3: Annotated Vehicle Charts* **for an explanation of these charts.**

Toyota Celica GT
2 Door Liftback

 2.2L 135 hp Gas Fuel Inject.
 4 Cylinder In-Line
 Manual 5 Speed
 2 Wheel Front
 Driver Airbag Psngr Belts

Compact

Purchase Price

Car Item	Dealer Cost	List
Base Price	**$14,321**	**$16,848**
Anti-Lock Brakes	$1,029	** $1,260
Automatic 4 Speed	$680	$800
Optional Engine	N/A	N/A
Air Conditioning	$732	$915
Power Steering	Std	Std
Cruise Control	$295	$365
All Wheel Drive	N/A	N/A
AM/FM Stereo Cassette	$135	$180
Steering Wheel, Tilt	Std	Std
Power Windows	Pkg	Pkg
*Options Price	$1,547	$1,895
*Total Price	$15,868	$18,743
Target Price	$17,380	
Destination Charge	$295	
Avg. Tax & Fees	$1,074	
Total Target $	**$18,749**	
Average Dealer Option Cost:	83%	

Ownership Costs

Cost Area	5 Year Cost	Rate
Depreciation	$8,124	○
Financing ($377/month)	$3,087	
Insurance (Rating 14)	$8,238	◉
State Fees	$760	
Fuel (Hwy 28 City 21)	$3,530	○
Maintenance	$6,674	●
Repairs	$600	○

Warranty/Maintenance Info

Major Tune-Up	$201	◉
Minor Tune-Up	$112	○
Brake Service	$260	◉
Overall Warranty	3 yr/36k	○
Drivetrain Warranty	5 yr/60k	○
Rust Warranty	5 yr/unlim. mi	○
Maintenance Warranty	N/A	
Roadside Assistance	N/A	

Ownership Cost By Year

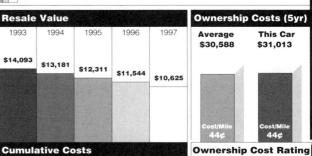

1993, 1994, 1995, 1996, 1997

Resale Value

1993	1994	1995	1996	1997
$14,093	$13,181	$12,311	$11,544	$10,625

Ownership Costs (5yr)

Average	This Car
$30,588	$31,013
Cost/Mile 44¢	Cost/Mile 44¢

Cumulative Costs

	1993	1994	1995	1996	1997
Annual	$8,352	$4,540	$6,158	$4,382	$7,581
Total	$8,352	$12,892	$19,050	$23,432	$31,013

Ownership Cost Rating

○ Average

The 1993 Toyota Celica is available in six models - ST and GT Sport Coupes; GT, GT-S and Turbo All-Trac Liftbacks and GT Convertible. New for 1993, the Celica GT Liftback now features variable-assist rack-and-pinion steering. Standard features include 15-inch wheels, 205/55VR15 speed-rated tires, dual color-keyed power side mirrors, rear spoiler, power antenna, a fold-down 50/50 split rear seat with security locks and tilt steering wheel. Options include a power sunroof and aluminum wheels.

Toyota Celica GT-S
2 Door Liftback

 2.2L 135 hp Gas Fuel Inject.
4 Cylinder In-Line
Manual 5 Speed
2 Wheel Front
 Driver Airbag Psngr Belts

Compact

Purchase Price

Car Item	Dealer Cost	List
Base Price	**$15,572**	**$18,428**
Anti-Lock Brakes	$1,029	** $1,260
Automatic 4 Speed	$676	$800
Optional Engine	N/A	N/A
Auto Climate Control	$880	$1,100
Power Steering	Std	Std
Cruise Control	$184	$230
All Wheel Drive	N/A	N/A
AM/FM Stereo Cassette	Std	Std
Steering Wheel, Tilt	Std	Std
Power Windows	Pkg	Pkg
*Options Price	$1,556	$1,900
*Total Price	$17,128	$20,328
Target Price	$18,819	
Destination Charge	$295	
Avg. Tax & Fees	$1,162	
Total Target $	**$20,276**	
Average Dealer Option Cost:	84%	

Ownership Costs

Cost Area	5 Year Cost	Rate
Depreciation	$9,737	○
Financing ($408/month)	$3,340	
Insurance (Rating 15)	$8,507	◉
State Fees	$825	
Fuel (Hwy 28 City 21)	$3,530	○
Maintenance	$6,633	●
Repairs	$550	○

Warranty/Maintenance Info

Major Tune-Up	$201	◉
Minor Tune-Up	$112	○
Brake Service	$260	◉
Overall Warranty	3 yr/36k	○
Drivetrain Warranty	5 yr/60k	○
Rust Warranty	5 yr/unlim. mi	○
Maintenance Warranty	N/A	
Roadside Assistance	N/A	

Ownership Cost By Year

1993, 1994, 1995, 1996, 1997

Resale Value

1993	1994	1995	1996	1997
$13,995	$13,097	$12,257	$11,456	$10,539

Ownership Costs (5yr)

Average	This Car
$31,949	$33,122
Cost/Mile 46¢	Cost/Mile 47¢

Cumulative Costs

	1993	1994	1995	1996	1997
Annual	$10,148	$4,671	$6,190	$4,482	$7,631
Total	$10,148	$14,819	$21,009	$25,491	$33,122

Ownership Cost Rating

○ Average

The 1993 Toyota Celica is available in six models - ST and GT Sport Coupes; GT, GT-S and Turbo All-Trac Liftbacks and GT Convertible. New for 1993, the Celica GT-S Liftback now features a full size spare tire. Features include aluminum alloy wheels, a five-way adjustable driver's seat, variable power-assist rack-and-pinion steering, tilt wheel, four-wheel independent strut suspension, dual color-keyed power side mirrors, six-speaker deluxe stereo system and rear-window intermittent wiper.

---

* Includes shaded options
** Other purchase requirements apply

● Poor   ◉ Worse Than Average   ○ Average   ○ Better Than Average   ○ Excellent   ⊖ Insufficient Information

©1993 by IntelliChoice, Inc. (408) 554-8711   All Rights Reserved. Reproduction Prohibited.
Refer to Section 3: Annotated Vehicle Charts for an explanation of these charts.

Toyota Celica GT
2 Door Convertible

Compact

 2.2L 135 hp Gas Fuel Inject.
 4 Cylinder In-Line
 Manual 5 Speed
 2 Wheel Front
 Driver Airbag Psngr Belts

Purchase Price

Car Item	Dealer Cost	List
Base Price	**$18,503**	**$21,768**
Anti-Lock Brakes	$1,029	** $1,260
Automatic 4 Speed	$680	$800
Optional Engine	N/A	N/A
Air Conditioning	$732	$915
Power Steering	Std	Std
Cruise Control	$184	$230
All Wheel Drive	N/A	N/A
AM/FM Stereo Cassette	$135	$180
Steering Wheel, Tilt	Std	Std
Power Windows	Pkg	Pkg
***Options Price**	**$1,547**	**$1,895**
***Total Price**	**$20,050**	**$23,663**
Target Price		$22,189
Destination Charge		$295
Avg. Tax & Fees		$1,364
Total Target $		**$23,848**
Average Dealer Option Cost:		81%

The 1993 Toyota Celica is available in six models - ST and GT Sport Coupes; GT, GT-S and Turbo All-Trac Liftbacks and GT Convertible. New for 1993, the GT Convertible features 4-Wheel disc brakes and variable power-assist steering. Other features include 14-inch wheels, dual color-keyed power side mirrors, a cloth, double-lined top, rear spoiler and a tilt wheel. Also included are a one piece folding seat, power rear-quarter windows and a pass-through passage between the trunk and interior.

Ownership Costs

Cost Area	5 Year Cost	Rate
Depreciation	$10,467	○
Financing ($479/month)	$3,928	
Insurance (Rating 17)	$9,084	◉
State Fees	$960	
Fuel (Hwy 27 City 21)	$3,598	◉
Maintenance	$5,926	●
Repairs	$600	○

Warranty/Maintenance Info

Major Tune-Up	$201	◉
Minor Tune-Up	$112	○
Brake Service	$236	○
Overall Warranty	3 yr/36k	○
Drivetrain Warranty	5 yr/60k	○
Rust Warranty	5 yr/unlim. mi	○
Maintenance Warranty	N/A	
Roadside Assistance	N/A	

Ownership Cost By Year

1993 / 1994 / 1995 / 1996 / 1997

Resale Value

1993	1994	1995	1996	1997
$16,928	$16,020	$15,211	$14,320	$13,381

Cumulative Costs

	1993	1994	1995	1996	1997
Annual	$11,184	$5,019	$6,128	$4,792	$7,440
Total	$11,184	$16,203	$22,331	$27,123	$34,563

Ownership Costs (5yr)

Average	This Car
$34,813	$34,563
Cost/Mile 50¢	Cost/Mile 49¢

Ownership Cost Rating

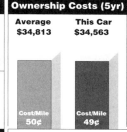
Average

Toyota Celica All-Trac Turbo
2 Door Liftback

Compact

 2.0L 200 hp Turbo Gas Fuel Inject.
 4 Cylinder In-Line
 Manual 5 Speed
 4 Wheel Full-Time
 Driver Airbag Psngr Belts

Purchase Price

Car Item	Dealer Cost	List
Base Price	**$23,912**	**$28,298**
Anti-Lock Brakes	Std	Std
Automatic Transmission	N/A	N/A
Optional Engine	N/A	N/A
Auto Climate Control	Std	Std
Power Steering	Std	Std
Cruise Control	Std	Std
4 Wheel Full-Time Drive	Std	Std
AM/FM Stereo CD	Std	Std
Steering Wheel, Tilt	N/A	N/A
Power Windows	Std	Std
***Options Price**	**$0**	**$0**
***Total Price**	**$23,912**	**$28,298**
Target Price		$26,816
Destination Charge		$295
Avg. Tax & Fees		$1,642
Total Target $		**$28,753**
Average Dealer Option Cost:		85%

The 1993 Toyota Celica is available in six models - ST and GT Sport Coupes; GT, GT-S and Turbo All-Trac Liftbacks and GT Convertible. New for 1993, the Celica Turbo All-Trac features a standard leather-trim package, including seven-way adjustable power seats, door trim, steering wheel, 5M shift knob and parking brake lever. Also standard is the "System 10" premium 3-in-1 AM/FM ETR/Cassette/CD sound system (including ten speakers), automatic air conditioner and power tilt/slide sunroof.

Ownership Costs

Cost Area	5 Year Cost	Rate
Depreciation	$15,684	○
Financing ($578/month)	$4,736	
Insurance (Rating 21)	$10,706	◉
State Fees	$1,144	
Fuel (Hwy 24 City 19)	$4,015	●
Maintenance	$6,713	●
Repairs	$640	○

Warranty/Maintenance Info

Major Tune-Up	$201	◉
Minor Tune-Up	$129	●
Brake Service	$260	◉
Overall Warranty	3 yr/36k	○
Drivetrain Warranty	5 yr/60k	○
Rust Warranty	5 yr/unlim. mi	○
Maintenance Warranty	N/A	
Roadside Assistance	N/A	

Resale Value

1993	1994	1995	1996	1997
$19,057	$17,477	$15,910	$14,495	$13,069

Cumulative Costs

	1993	1994	1995	1996	1997
Annual	$14,758	$6,433	$7,707	$5,886	$8,854
Total	$14,758	$21,191	$28,898	$34,784	$43,638

Ownership Costs (5yr)

Average	This Car
$38,793	$43,638
Cost/Mile 55¢	Cost/Mile 62¢

Ownership Cost Rating

● Poor

---

* Includes shaded options
** Other purchase requirements apply

● Poor   ◉ Worse Than Average   ○ Average   ○ Better Than Average   ○ Excellent   ⊖ Insufficient Information

©1993 by IntelliChoice, Inc. (408) 554-8711   All Rights Reserved. Reproduction Prohibited.
Refer to *Section 3: Annotated Vehicle Charts* for an explanation of these charts.

Toyota Corolla
4 Door Sedan

 1.6L 105 hp Gas Fuel Inject.
 4 Cylinder In-Line
 Manual 5 Speed
 2 Wheel Front
 Driver Airbag Psngr Belts

Subcompact

Purchase Price

Car Item	Dealer Cost	List
Base Price	**$9,966**	**$11,198**
Anti-Lock Brakes	$676	$825
Automatic 3 Speed	$445	$500
Optional Engine	N/A	N/A
Air Conditioning	$688	$860
Power Steering	$222	$260
Cruise Control	N/A	N/A
All Wheel Drive	N/A	N/A
AM/FM Stereo Cassette	$367	$490
Steering Wheel, Tilt	N/A	N/A
Power Windows	N/A	N/A
***Options Price**	**$222**	**$260**
***Total Price**	**$10,188**	**$11,458**
Target Price	$11,038	
Destination Charge	$295	
Avg. Tax & Fees	$685	
Total Target $	**$12,018**	
Average Dealer Option Cost:	**81%**	

The 1993 Corolla is available in four models - (Base) Corolla, Deluxe and LE sedans; and Deluxe wagon. New for 1993, the Corolla (Base) sedan has an all-new, larger, interior and exterior which moves it up one EPA size classification from subcompact to compact. The 1.6 liter engine is quieter, with increased 49-state horsepower. The Corolla also features wide protective bodyside moldings and a center console.

Ownership Costs

Cost Area	5 Year Cost	Rate
Depreciation	$5,722	○
Financing ($242/month)	$1,981	
Insurance (Rating 6 [Est.])	$6,919	○
State Fees	$471	
Fuel (Hwy 33 City 28)	$2,832	○
Maintenance	$5,519	●
Repairs	$540	○

Warranty/Maintenance Info

Major Tune-Up	$190	○
Minor Tune-Up	$118	●
Brake Service	$194	○
Overall Warranty	3 yr/36k	○
Drivetrain Warranty	5 yr/60k	○
Rust Warranty	5 yr/unlim. mi	○
Maintenance Warranty	N/A	
Roadside Assistance	N/A	

Ownership Cost By Year

(1993, 1994, 1995, 1996, 1997)

Resale Value

1993	1994	1995	1996	1997
$10,136	$9,055	$8,132	$7,186	$6,296

Cumulative Costs

	1993	1994	1995	1996	1997
Annual	$4,672	$3,917	$4,918	$4,012	$6,465
Total	$4,672	$8,589	$13,507	$17,519	$23,984

Ownership Costs (5yr)

Average $24,470 — This Car $23,984
Cost/Mile 35¢ — Cost/Mile 34¢

Ownership Cost Rating
○ Better Than Average

Toyota Corolla DX
4 Door Sedan

 1.8L 115 hp Gas Fuel Inject.
 4 Cylinder In-Line
 Manual 5 Speed
 2 Wheel Front
 Driver Airbag Psngr Belts

Subcompact

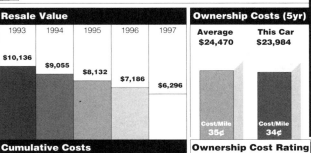

Purchase Price

Car Item	Dealer Cost	List
Base Price	**$10,575**	**$12,298**
Anti-Lock Brakes	$676	$825
Automatic 4 Speed	$688	$800
Optional Engine	N/A	N/A
Air Conditioning	$688	$860
Power Steering	Std	Std
Cruise Control	$184	$230
All Wheel Drive	N/A	N/A
AM/FM Stereo Cassette	$367	$490
Steering Wheel, Tilt	$124	$145
Power Windows	N/A	N/A
***Options Price**	**$0**	**$0**
***Total Price**	**$10,575**	**$12,298**
Target Price	$11,475	
Destination Charge	$295	
Avg. Tax & Fees	$715	
Total Target $	**$12,485**	
Average Dealer Option Cost:	**81%**	

The 1993 Corolla is available in four models - (Base) Corolla, Deluxe and LE sedans; and Deluxe wagon. New for 1993, the Corolla Deluxe sedan has an all-new, larger, interior and exterior which moves it up one EPA size classification from subcompact to compact. The 1.8 liter engine increases performance with more horsepower and strong low-end torque and a reduction in noise, vibration and harshness.

Ownership Costs

Cost Area	5 Year Cost	Rate
Depreciation	$6,050	○
Financing ($251/month)	$2,057	
Insurance (Rating 7 [Est.])	$7,035	○
State Fees	$504	
Fuel (Hwy 34 City 28)	$2,784	○
Maintenance	$5,463	●
Repairs	$540	○

Warranty/Maintenance Info

Major Tune-Up	$190	○
Minor Tune-Up	$118	●
Brake Service	$194	○
Overall Warranty	3 yr/36k	○
Drivetrain Warranty	5 yr/60k	○
Rust Warranty	5 yr/unlim. mi	○
Maintenance Warranty	N/A	
Roadside Assistance	N/A	

Resale Value

1993	1994	1995	1996	1997
$10,471	$9,345	$8,368	$7,374	$6,435

Cumulative Costs

	1993	1994	1995	1996	1997
Annual	$4,857	$4,006	$4,994	$4,085	$6,491
Total	$4,857	$8,863	$13,857	$17,942	$24,433

Ownership Costs (5yr)

Average $25,415 — This Car $24,433
Cost/Mile 36¢ — Cost/Mile 35¢

Ownership Cost Rating
○ Excellent

Legend: ● Poor · ◐ Worse Than Average · ○ Average · ○ Better Than Average · ○ Excellent · ⊖ Insufficient Information

* Includes shaded options
** Other purchase requirements apply

©1993 by IntelliChoice, Inc. (408) 554-8711 All Rights Reserved. Reproduction Prohibited.
Refer to *Section 3: Annotated Vehicle Charts* for an explanation of these charts.

Toyota Corolla LE
4 Door Sedan

 1.8L 115 hp Gas Fuel Inject.
 4 Cylinder In-Line
 Automatic 4 Speed
 2 Wheel Front
 Driver Airbag Psngr Belts

Subcompact

Purchase Price

Car Item	Dealer Cost	List
Base Price	**$13,042**	**$15,218**
Anti-Lock Brakes	$676	$825
Manual Transmission	N/A	N/A
Optional Engine	N/A	N/A
Air Conditioning	Std	Std
Power Steering	Std	Std
Cruise Control	Std	Std
All Wheel Drive	N/A	N/A
AM/FM Stereo Cassette	$112	$150
Steering Wheel, Tilt	Std	Std
Power Windows	Std	Std
*Options Price	$0	$0
*Total Price	$13,042	$15,218
Target Price	$14,240	
Destination Charge	$295	
Avg. Tax & Fees	$882	
Total Target $	**$15,417**	
Average Dealer Option Cost:	81%	

The 1993 Corolla is available in four models - (Base) Corolla, Deluxe and LE sedans; and Deluxe wagon. New for 1993, the Corolla LE sedan has an all-new, larger, interior and exterior which moves it up one EPA size classification from subcompact to compact. The 1.8 liter engine increases performance with more horsepower and strong low-end torque and a reduction in noise, vibration and harshness. Other features include a tachometer, air conditioning, tilt wheel and power windows and locks.

Ownership Costs

Cost Area	5 Year Cost	Rate
Depreciation	$8,011	◐
Financing ($310/month)	$2,540	
Insurance (Rating 11 [Est.])	$7,693	◐
State Fees	$621	
Fuel (Hwy 33 City 26)	$2,927	◐
Maintenance	$5,752	●
Repairs	$540	○

Warranty/Maintenance Info

Major Tune-Up	$190	◐
Minor Tune-Up	$118	●
Brake Service	$194	◐
Overall Warranty	3 yr/36k	◐
Drivetrain Warranty	5 yr/60k	○
Rust Warranty	5 yr/unlim. mi	◐
Maintenance Warranty	N/A	
Roadside Assistance	N/A	

Ownership Cost By Year

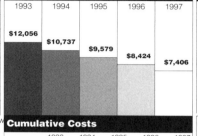

■ 1993
■ 1994
▨ 1995
▥ 1996
□ 1997

Resale Value

1993	1994	1995	1996	1997
$12,056	$10,737	$9,579	$8,424	$7,406

Cumulative Costs

	1993	1994	1995	1996	1997
Annual	$6,581	$4,530	$5,593	$4,469	$6,911
Total	$6,581	$11,111	$16,704	$21,173	$28,084

Ownership Costs (5yr)

Average	This Car
$28,700	$28,084
Cost/Mile 41¢	Cost/Mile 40¢

Ownership Cost Rating

○ Better Than Average

Toyota Corolla DX
4 Door Wagon

 1.8L 115 hp Gas Fuel Inject.
4 Cylinder In-Line
Manual 5 Speed
 2 Wheel Front
 Driver Airbag Psngr Belts

Subcompact Wag

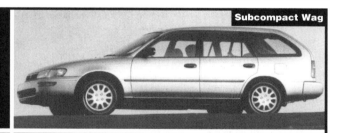

Purchase Price

Car Item	Dealer Cost	List
Base Price	**$11,160**	**$12,978**
Anti-Lock Brakes	$676	$825
Automatic 4 Speed	$689	$800
Optional Engine	N/A	N/A
Air Conditioning	$688	$860
Power Steering	Std	Std
Cruise Control	$184	$230
All Wheel Drive	N/A	N/A
AM/FM Stereo Cassette	$367	$490
Steering Wheel, Tilt	$124	$145
Power Windows	Pkg	Pkg
*Options Price	$0	$0
*Total Price	$11,160	$12,978
Target Price	$12,128	
Destination Charge	$295	
Avg. Tax & Fees	$754	
Total Target $	**$13,177**	
Average Dealer Option Cost:	81%	

The 1993 Corolla is available in four models - (Base) Corolla, Deluxe and LE sedans; and Deluxe wagon. New for 1993, the Corolla Deluxe wagon has an all-new, larger, interior and exterior which moves it up one EPA size classification from subcompact to compact. The 1.8 liter engine increases performance with more horsepower and strong low-end torque and a reduction in noise, vibration and harshness. Features include black bodyside molding, remote-control outside mirrors and rear defroster.

Ownership Costs

Cost Area	5 Year Cost	Rate
Depreciation	$6,342	○
Financing ($265/month)	$2,170	
Insurance (Rating 8 [Est.])	$7,223	◐
State Fees	$531	
Fuel (Hwy 34 City 28)	$2,784	◐
Maintenance	$5,463	●
Repairs	$540	○

Warranty/Maintenance Info

Major Tune-Up	$190	●
Minor Tune-Up	$118	●
Brake Service	$194	◐
Overall Warranty	3 yr/36k	◐
Drivetrain Warranty	5 yr/60k	○
Rust Warranty	5 yr/unlim. mi	◐
Maintenance Warranty	N/A	
Roadside Assistance	N/A	

Resale Value

1993	1994	1995	1996	1997
$10,776	$9,673	$8,723	$7,747	$6,835

Cumulative Costs

	1993	1994	1995	1996	1997
Annual	$5,333	$4,060	$5,031	$4,120	$6,509
Total	$5,333	$9,393	$14,424	$18,544	$25,053

Ownership Costs (5yr)

Average	This Car
$26,180	$25,053
Cost/Mile 37¢	Cost/Mile 36¢

Ownership Cost Rating

○ Excellent

---

* Includes shaded options
** Other purchase requirements apply

● Poor   ◐ Worse Than Average   ◑ Average   ○ Better Than Average   ○ Excellent   ⊖ Insufficient Information

page **239**

©1993 by IntelliChoice, Inc. (408) 554-8711   All Rights Reserved. Reproduction Prohibited.
Refer to *Section 3: Annotated Vehicle Charts* for an explanation of these charts.

Toyota MR2
2 Door Coupe

Sport

- 2.2L 135 hp Gas Fuel Inject.
- 4 Cylinder In-Line
- Manual 5 Speed
- 2 Wheel Rear
- Driver Airbag Psngr Belts

Purchase Price

Car Item	Dealer Cost	List
Base Price	**$15,521**	**$18,368**
Anti-Lock Brakes	$845	$1,130
Automatic 4 Speed	$676	$800
Optional Engine	N/A	N/A
Air Conditioning	$732	$915
Power Steering	$513	$600
Cruise Control	$212	$265
All Wheel Drive	N/A	N/A
AM/FM Stereo Cassette	Std	Std
Steering Wheel, Tilt	Std	Std
Power Windows	Pkg	Pkg
***Options Price**	**$1,245**	**$1,515**
***Total Price**	**$16,766**	**$19,883**
Target Price		$18,419
Destination Charge		$295
Avg. Tax & Fees		$1,138
Total Target $		**$19,852**
Average Dealer Option Cost:		81%

The 1993 MR-2 is available in two models - (Base) MR2 and MR2 Turbo. New for 1993, the Base MR2 features a new suspension including stiffer springs, a lowered ride height and larger 15-inch cast alloy wheels with low profile tires. New interior features include a shorter-throw gearshift, and an optional Black and Ivory Leather Trim Package. The 1993 MR2 also offers new exterior features such as an extended front air dam with a larger front air opening for additional engine cooling and breathing.

Ownership Costs

Cost Area	5 Year Cost	Rate
Depreciation	$8,361	○
Financing ($399/month)	$3,270	
Insurance (Rating 21 Sport+)	$14,988	●
State Fees	$807	
Fuel (Hwy 29 City 22)	$3,390	○
Maintenance	$6,171	◐
Repairs	$580	○

Warranty/Maintenance Info

Major Tune-Up	$212	○
Minor Tune-Up	$123	○
Brake Service	$215	○
Overall Warranty	3 yr/36k	◐
Drivetrain Warranty	5 yr/60k	○
Rust Warranty	5 yr/unlim. mi	○
Maintenance Warranty	N/A	
Roadside Assistance	N/A	

Ownership Cost By Year

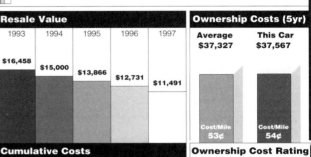

Resale Value

1993	1994	1995	1996	1997
$16,458	$15,000	$13,866	$12,731	$11,491

Cumulative Costs

	1993	1994	1995	1996	1997
Annual	$8,399	$6,505	$7,435	$6,167	$9,061
Total	$8,399	$14,904	$22,339	$28,506	$37,567

Ownership Costs (5yr)

Average	This Car
$37,327	$37,567
Cost/Mile 53¢	Cost/Mile 54¢

Ownership Cost Rating
○ Average

Toyota MR2 Turbo
2 Door Coupe

Sport

- 2.0L 200 hp Turbo Gas Fuel Inject.
- 4 Cylinder In-Line
- Manual 5 Speed
- 2 Wheel Rear
- Driver Airbag Psngr Belts

Purchase Price

Car Item	Dealer Cost	List
Base Price	**$20,278**	**$23,998**
Anti-Lock Brakes	$845	$1,130
Automatic Transmission	N/A	N/A
Optional Engine	N/A	N/A
Air Conditioning	Std	Std
Power Steering	$513	$600
Cruise Control	Std	Std
All Wheel Drive	N/A	N/A
AM/FM Stereo Cassette	Std	Std
Steering Wheel, Tilt	Std	Std
Power Windows	Std	Std
***Options Price**	**$513**	**$600**
***Total Price**	**$20,791**	**$24,598**
Target Price		$23,110
Destination Charge		$295
Avg. Tax & Fees		$1,419
Total Target $		**$24,824**
Average Dealer Option Cost:		79%

The 1993 MR-2 is available in two models - (Base) MR2 and MR2 Turbo. New for 1993, the Turbo MR2 features a new suspension including stiffer springs, a lowered ride height and larger 15-inch cast alloy wheels with low profile tires. New interior features include a shorter-throw gearshift, and an optional Black and Ivory Leather Trim Package. The 1993 MR2 also offers new exterior features such as an extended front air dam with a larger front air opening for additional engine cooling and breathing.

Ownership Costs

Cost Area	5 Year Cost	Rate
Depreciation	$11,467	○
Financing ($499/month)	$4,089	
Insurance (Rating 21 Sport+)	$14,988	●
State Fees	$995	
Fuel (Hwy 27 City 20)	$3,682	○
Maintenance	$6,319	◐
Repairs	$680	○

Warranty/Maintenance Info

Major Tune-Up	$212	○
Minor Tune-Up	$123	○
Brake Service	$215	○
Overall Warranty	3 yr/36k	◐
Drivetrain Warranty	5 yr/60k	○
Rust Warranty	5 yr/unlim. mi	○
Maintenance Warranty	N/A	
Roadside Assistance	N/A	

Resale Value

1993	1994	1995	1996	1997
$19,131	$17,311	$15,912	$14,583	$13,357

Cumulative Costs

	1993	1994	1995	1996	1997
Annual	$11,178	$7,260	$8,015	$6,575	$9,192
Total	$11,178	$18,438	$26,453	$33,028	$42,220

Ownership Costs (5yr)

Average	This Car
$41,851	$42,220
Cost/Mile 60¢	Cost/Mile 60¢

Ownership Cost Rating
○ Average

---

* Includes shaded options
** Other purchase requirements apply

● Poor | ◐ Worse Than Average | ○ Average | ○ Better Than Average | ○ Excellent | ⊖ Insufficient Information

©1993 by IntelliChoice, Inc. (408) 554-8711   All Rights Reserved. Reproduction Prohibited.
Refer to *Section 3: Annotated Vehicle Charts* for an explanation of these charts.

Toyota Paseo
2 Door Coupe

 1.5L 100 hp Gas Fuel Inject.
 4 Cylinder In-Line
 Manual 5 Speed
 2 Wheel Front
Driver Airbag Psngr Belts

Subcompact

Purchase Price

Car Item	Dealer Cost	List
Base Price	**$10,003**	**$11,498**
Anti-Lock Brakes	$676	$825
Automatic 4 Speed	$696	$800
Optional Engine	N/A	N/A
Air Conditioning	$672	$840
Power Steering	Std	Std
Cruise Control	$184	$230
All Wheel Drive	N/A	N/A
AM/FM Stereo Cassette	$210	$280
Steering Wheel, Tilt	N/A	N/A
Power Windows	N/A	N/A
*Options Price	$0	$0
*Total Price	$10,003	$11,498
Target Price		$10,839
Destination Charge		$325
Avg. Tax & Fees		$676
Total Target $		**$11,840**
Average Dealer Option Cost:	82%	

Ownership Costs

Cost Area	5 Year Cost	Rate
Depreciation		⊖
Financing ($238/month)	$1,951	
Insurance (Rating 11)	$7,693	●
State Fees	$473	
Fuel (Hwy 34 City 28)	$2,784	○
Maintenance	$5,074	●
Repairs	$560	○

Warranty/Maintenance Info

Major Tune-Up	$168	○
Minor Tune-Up	$90	○
Brake Service	$224	◉
Overall Warranty	3 yr/36k	○
Drivetrain Warranty	5 yr/60k	○
Rust Warranty	5 yr/unlim. mi	○
Maintenance Warranty	N/A	
Roadside Assistance	N/A	

Ownership Cost By Year

Insufficient Depreciation Information

1993, 1994, 1995, 1996, 1997

Resale Value

Insufficient Information

Cumulative Costs

	1993	1994	1995	1996	1997
Annual	Insufficient Information				
Total	Insufficient Information				

Ownership Costs (5yr)

Insufficient Information

Ownership Cost Rating

⊖ Insufficient Information

The 1993 Paseo two-door sport coupe is available in one model. New for 1993, the Paseo has black-out wheel wells, four-spoke steering wheel, new fabric and two new exterior colors (Sunfire Red Pearl and Stardust Blue Metallic). Standard features include power-assisted rack-and pinion steering, AM/FM ETR radio, tinted glass, digital quartz clock, remote fuel door and trunk releases, rear window defogger and a sport tuned suspension. Available options include air conditioning and a rear spoiler.

Toyota Tercel
2 Door Sedan

 1.5L 82 hp Gas Fuel Inject.
4 Cylinder In-Line
 Manual 4 Speed
 2 Wheel Front
Driver Airbag Psngr Belts

Subcompact

Purchase Price

Car Item	Dealer Cost	List
Base Price	**$7,180**	**$7,848**
Anti-Lock Brakes	$676	$825
Automatic Transmission	N/A	N/A
Optional Engine	N/A	N/A
Air Conditioning	$672	$840
Power Steering	N/A	N/A
Cruise Control	N/A	N/A
All Wheel Drive	N/A	N/A
AM/FM Stereo Cassette	Dlr	Dlr
Steering Wheel, Tilt	N/A	N/A
Power Windows	N/A	N/A
*Options Price	$0	$0
*Total Price	$7,180	$7,848
Target Price		$7,725
Destination Charge		$325
Avg. Tax & Fees		$485
Total Target $		**$8,535**
Average Dealer Option Cost:	81%	

Ownership Costs

Cost Area	5 Year Cost	Rate
Depreciation	$3,825	○
Financing ($172/month)	$1,406	
Insurance (Rating 7)	$7,035	●
State Fees	$327	
Fuel (Hwy 36 City 32)	$2,543	○
Maintenance	$4,411	○
Repairs	$540	○

Warranty/Maintenance Info

Major Tune-Up	$156	○
Minor Tune-Up	$85	○
Brake Service	$220	○
Overall Warranty	3 yr/36k	○
Drivetrain Warranty	5 yr/60k	○
Rust Warranty	5 yr/unlim. mi	○
Maintenance Warranty	N/A	
Roadside Assistance	N/A	

Resale Value

1993	1994	1995	1996	1997
$7,685	$6,798	$6,150	$5,421	$4,710

Ownership Costs (5yr)

Average	This Car
$20,408	$20,087
Cost/Mile 29¢	Cost/Mile 29¢

Cumulative Costs

	1993	1994	1995	1996	1997
Annual	$3,334	$3,448	$4,212	$4,138	$4,955
Total	$3,334	$6,782	$10,994	$15,132	$20,087

Ownership Cost Rating

○ Better Than Average

The 1993 Toyota Tercel is available in two bodystyles - two-door sedan with Base and Deluxe trim levels and four-door sedan with Deluxe and LE trim levels. New for 1993, the Tercel has a standard driver-side airbag. Other new features include a revised front grille and three added exterior colors are now offered (Frosted Mint Metallic, Teal Mist Metallic and Sunfire Red Pearl). Features include rear quarter storage pockets, analog gauges, a black outside driver's mirror and all-season tires.

---

* Includes shaded options
** Other purchase requirements apply

● Poor ◉ Worse Than Average ○ Average ○ Better Than Average ○ Excellent ⊖ Insufficient Information

©1993 by IntelliChoice, Inc. (408) 554-8711 All Rights Reserved. Reproduction Prohibited.
Refer to *Section 3: Annotated Vehicle Charts* for an explanation of these charts.

Toyota Tercel DX
2 Door Sedan

Subcompact

- 1.5L 82 hp Gas Fuel Inject.
- 4 Cylinder In-Line
- Manual 5 Speed
- 2 Wheel Front
- Driver Airbag Psngr Belts

Purchase Price

Car Item	Dealer Cost	List
Base Price	**$8,710**	**$9,678**
Anti-Lock Brakes	$676	$825
Automatic 3 Speed	$450	$500
Optional Engine	N/A	N/A
Air Conditioning	$672	$840
Power Steering	$222	$260
Cruise Control	N/A	N/A
All Wheel Drive	N/A	N/A
AM/FM Stereo Cassette	$367	$490
Steering Wheel, Tilt	N/A	N/A
Power Windows	N/A	N/A
*Options Price	$222	$260
*Total Price	$8,932	$9,938
Target Price		$9,647
Destination Charge		$325
Avg. Tax & Fees		$602
Total Target $		**$10,574**
Average Dealer Option Cost:		82%

Ownership Costs

Cost Area	5 Year Cost	Rate
Depreciation	$4,899	○
Financing ($213/month)	$1,742	
Insurance (Rating 10)	$7,479	●
State Fees	$411	
Fuel (Hwy 34 City 28)	$2,784	○
Maintenance	$4,496	○
Repairs	$540	○

Warranty/Maintenance Info

Major Tune-Up	$156	○
Minor Tune-Up	$85	○
Brake Service	$220	○
Overall Warranty	3 yr/36k	○
Drivetrain Warranty	5 yr/60k	○
Rust Warranty	5 yr/unlim. mi	○
Maintenance Warranty	N/A	
Roadside Assistance	N/A	

The 1993 Toyota Tercel is available in two bodystyles - two-door sedan with Base and Deluxe trim levels and four-door sedan with Deluxe and LE trim levels. New for 1993, the DX two-door sedan has a standard driver-side airbag. Other new features include a revised front grille, an upgraded interior fabric, tinted glass, bodyside moldings and optional ABS, which helps braking on wet roads. Three colors have been added as well (Frosted Mint Metallic, Teal Mist Metallic and Sunfire Red Pearl).

Ownership Cost By Year

1993, 1994, 1995, 1996, 1997

Resale Value

1993	1994	1995	1996	1997
$8,994	$7,987	$7,243	$6,445	$5,675

Cumulative Costs

	1993	1994	1995	1996	1997
Annual	$4,351	$3,823	$4,569	$4,389	$5,219
Total	$4,351	$8,174	$12,743	$17,132	$22,351

Ownership Costs (5yr)

Average	This Car
$22,760	$22,351
Cost/Mile 33¢	Cost/Mile 32¢

Ownership Cost Rating

○ Better Than Average

Toyota Tercel DX
4 Door Sedan

Subcompact

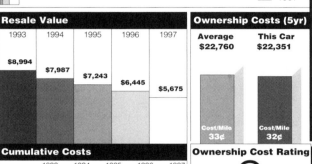

LE Model Shown

- 1.5L 82 hp Gas Fuel Inject.
- 4 Cylinder In-Line
- Manual 5 Speed
- 2 Wheel Front
- Driver Airbag Psngr Belts

Purchase Price

Car Item	Dealer Cost	List
Base Price	**$8,800**	**$9,778**
Anti-Lock Brakes	$676	$825
Automatic 3 Speed	$450	$500
Optional Engine	N/A	N/A
Air Conditioning	$672	$840
Power Steering	$222	$260
Cruise Control	N/A	N/A
All Wheel Drive	N/A	N/A
AM/FM Stereo Cassette	$367	$490
Steering Wheel, Tilt	N/A	N/A
Power Windows	N/A	N/A
*Options Price	$222	$260
*Total Price	$9,022	$10,038
Target Price		$9,746
Destination Charge		$325
Avg. Tax & Fees		$608
Total Target $		**$10,679**
Average Dealer Option Cost:		82%

Ownership Costs

Cost Area	5 Year Cost	Rate
Depreciation	$4,968	○
Financing ($215/month)	$1,760	
Insurance (Rating 8)	$7,223	●
State Fees	$415	
Fuel (Hwy 34 City 28)	$2,784	○
Maintenance	$4,496	○
Repairs	$540	○

Warranty/Maintenance Info

Major Tune-Up	$156	○
Minor Tune-Up	$85	○
Brake Service	$220	○
Overall Warranty	3 yr/36k	○
Drivetrain Warranty	5 yr/60k	○
Rust Warranty	5 yr/unlim. mi	○
Maintenance Warranty	N/A	
Roadside Assistance	N/A	

The 1993 Toyota Tercel is available in two bodystyles - two-door sedan with Base and Deluxe trim levels and four-door sedan with Deluxe and LE trim levels. New for 1993, the DX four-door sedan has a standard driver-side airbag. Other new features include a revised front grille, an upgraded interior fabric, color-keyed bumper, bodyside moldings and optional ABS, which helps braking on wet roads. Three colors have been added as well (Frosted Mint Metallic, Teal Mist Metallic and Sunfire Red Pearl).

Resale Value

1993	1994	1995	1996	1997
$9,049	$8,043	$7,292	$6,488	$5,711

Cumulative Costs

	1993	1994	1995	1996	1997
Annual	$4,362	$3,780	$4,529	$4,344	$5,171
Total	$4,362	$8,142	$12,671	$17,015	$22,186

Ownership Costs (5yr)

Average	This Car
$22,872	$22,186
Cost/Mile 33¢	Cost/Mile 32¢

Ownership Cost Rating

○ Better Than Average

---

* Includes shaded options
** Other purchase requirements apply

● Poor | ◐ Worse Than Average | ○ Average | ○ Better Than Average | ○ Excellent | ⊖ Insufficient Information

©1993 by IntelliChoice, Inc. (408) 554-8711 All Rights Reserved. Reproduction Prohibited.
Refer to Section 3: Annotated Vehicle Charts for an explanation of these charts.

Toyota Tercel LE
4 Door Sedan

Subcompact

 1.5L 82 hp Gas Fuel Inject.
 4 Cylinder In-Line
 Manual 5 Speed
 2 Wheel Front
 Driver Airbag Psngr Belts

Purchase Price

Car Item	Dealer Cost	List
Base Price	**$9,724**	**$11,308**
Anti-Lock Brakes	$676	$825
Automatic 3 Speed	$431	$500
Optional Engine	N/A	N/A
Air Conditioning	$672	$840
Power Steering	Std	Std
Cruise Control	N/A	N/A
All Wheel Drive	N/A	N/A
AM/FM Stereo Cassette	$210	$280
Steering Wheel, Tilt	N/A	N/A
Power Windows	N/A	N/A
*Options Price	$0	$0
*Total Price	$9,724	$11,308
Target Price	$10,529	
Destination Charge	$325	
Avg. Tax & Fees	$659	
Total Target $	**$11,513**	
Average Dealer Option Cost:	81%	

The 1993 Toyota Tercel is available in two bodystyles - two-door sedan with Base and Deluxe trim levels and four-door sedan with Deluxe and LE trim levels. New for 1993, the LE sedan has a standard driver-side airbag and power steering. Other new features include a revised front grille, color-keyed bumper, tinted glass, bodyside moldings and optional ABS, which helps braking on wet roads. Three colors have been added as well (Frosted Mint Metallic, Teal Mist Metallic and Sunfire Red Pearl).

Ownership Costs

Cost Area	5 Year Cost	Rate
Depreciation	$5,481	○
Financing ($231/month)	$1,897	
Insurance (Rating 11)	$7,693	●
State Fees	$465	
Fuel (Hwy 34 City 28)	$2,784	○
Maintenance	$4,496	○
Repairs	$540	○

Warranty/Maintenance Info

Major Tune-Up	$156	○
Minor Tune-Up	$85	○
Brake Service	$220	○
Overall Warranty	3 yr/36k	○
Drivetrain Warranty	5 yr/60k	◐
Rust Warranty	5 yr/unlim. mi	○
Maintenance Warranty	N/A	
Roadside Assistance	N/A	

Ownership Cost By Year
1993, 1994, 1995, 1996, 1997

Resale Value

1993	1994	1995	1996	1997
$9,754	$8,649	$7,782	$6,891	$6,032

Cumulative Costs

	1993	1994	1995	1996	1997
Annual	$4,646	$4,024	$4,776	$4,547	$5,363
Total	$4,646	$8,670	$13,446	$17,993	$23,356

Ownership Costs (5yr)

Average	This Car
$24,301	$23,356
Cost/Mile 35¢	Cost/Mile 33¢

Ownership Cost Rating: Excellent

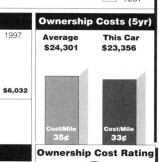

Volkswagen Cabriolet
2 Door Convertible

Subcompact

 1.8L 94 hp Gas Fuel Inject.
 4 Cylinder In-Line
 Manual 5 Speed
 2 Wheel Front
 Driver Airbag Psngr Belts

Purchase Price

Car Item	Dealer Cost	List
Base Price	**$16,307**	**$18,380**
Anti-Lock Brakes	N/A	N/A
Automatic 3 Speed	$557	$595
Optional Engine	N/A	N/A
Air Conditioning	$770	$895
Power Steering	Std	Std
Cruise Control	N/A	N/A
All Wheel Drive	N/A	N/A
AM/FM Stereo Cassette	Std	Std
Steering Wheel, Tilt	N/A	N/A
Power Windows	Std	Std
*Options Price	$0	$0
*Total Price	$16,307	$18,380
Target Price	$17,341	
Destination Charge	$390	
Avg. Tax & Fees	$1,075	
Total Target $	**$18,806**	
Average Dealer Option Cost:	89%	

The Cabriolet is available in two models - Base and Classic convertible coupe. New for 1993, the Base Cabriolet will feature front sport seats in white leatherette, sport-style light alloy wheels, leather steering wheel, power front windows and a six speaker sound system. Other standard features include dual remote mirrors, a reclining, height-adjustable drivers seat and a rear folding seat with trunk access. Safety features include safety cell construction and a driver side airbag.

Ownership Costs

Cost Area	5 Year Cost	Rate
Depreciation	$8,916	○
Financing ($378/month)	$3,097	
Insurance (Rating 21)	$10,706	●
State Fees	$751	
Fuel (Hwy 30 City 24)	$3,198	◉
Maintenance	$3,854	○
Repairs	$720	○

Warranty/Maintenance Info

Major Tune-Up	$186	○
Minor Tune-Up	$79	○
Brake Service	$207	○
Overall Warranty	2 yr/24k	●
Drivetrain Warranty	5 yr/50k	○
Rust Warranty	6 yr/unlim. mi	○
Maintenance Warranty	N/A	
Roadside Assistance	N/A	

Resale Value

1993	1994	1995	1996	1997
$14,254	$13,053	$12,026	$10,950	$9,890

Cumulative Costs

	1993	1994	1995	1996	1997
Annual	$8,643	$5,241	$5,777	$5,217	$6,364
Total	$8,643	$13,884	$19,661	$24,878	$31,242

Ownership Costs (5yr)

Average	This Car
$32,258	$31,242
Cost/Mile 46¢	Cost/Mile 45¢

Ownership Cost Rating: Better Than Average

---

* Includes shaded options
** Other purchase requirements apply

● Poor ◐ Worse Than Average ○ Average ○ Better Than Average ○ Excellent ⊖ Insufficient Information

©1993 by IntelliChoice, Inc. (408) 554-8711 All Rights Reserved. Reproduction Prohibited.
Refer to *Section 3: Annotated Vehicle Charts* for an explanation of these charts.

Volkswagen Cabriolet Classic
2 Door Convertible

Subcompact

1.8L 94 hp Gas Fuel Inject. | 4 Cylinder In-Line | Manual 5 Speed | 2 Wheel Front | Driver Airbag Psngr Belts

Base Model Shown

Purchase Price

Car Item	Dealer Cost	List
Base Price	**$17,678**	**$19,930**
Anti-Lock Brakes	N/A	N/A
Automatic 3 Speed	$557	$595
Optional Engine	N/A	N/A
Air Conditioning	Std	Std
Power Steering	Std	Std
Cruise Control	Std	Std
All Wheel Drive	N/A	N/A
AM/FM Stereo Cassette	Std	Std
Steering Wheel, Tilt	N/A	N/A
Power Windows	Std	Std
*Options Price	$0	$0
***Total Price**	**$17,678**	**$19,930**
Target Price	$18,841	
Destination Charge	$390	
Avg. Tax & Fees	$1,165	
Total Target $	**$20,396**	
Average Dealer Option Cost:	**94%**	

Ownership Costs

Cost Area	5 Year Cost	Rate
Depreciation	$10,117	○
Financing ($410/month)	$3,359	
Insurance (Rating 22)	$11,167	●
State Fees	$813	
Fuel (Hwy 30 City 24)	$3,198	◉
Maintenance	$3,854	○
Repairs	$720	○

Warranty/Maintenance Info

Major Tune-Up	$186	○
Minor Tune-Up	$79	○
Brake Service	$207	○
Overall Warranty	2 yr/24k	◉
Drivetrain Warranty	5 yr/50k	○
Rust Warranty	6 yr/unlim. mi	○
Maintenance Warranty	N/A	
Roadside Assistance	N/A	

The Cabriolet is available in two models - (Base) and Classic convertible coupe. New for 1993, the Classic model includes cruise control, fourteen-inch seven spoke forged alloy wheels, heatable front seats, leather-faced sport seats and a leather manual shift knob, boot and hand brake cover. Other standard features include dual remote mirrors, adjustable drivers seat and a rear folding seat with trunk access. Safety features include safety cell construction and a driver side airbag.

Ownership Cost By Year

$2,000 $4,000 $6,000 $8,000 $10,000 $12,000

1993, 1994, 1995, 1996, 1997

Resale Value

1993	1994	1995	1996	1997
$15,052	$13,739	$12,607	$11,430	$10,279

Cumulative Costs

	1993	1994	1995	1996	1997
Annual	$9,644	$5,537	$6,039	$5,444	$6,564
Total	$9,644	$15,181	$21,220	$26,664	$33,228

Ownership Costs (5yr)

Average	This Car
$34,002	$33,228
Cost/Mile 49¢	Cost/Mile 47¢

Ownership Cost Rating

○ Better Than Average

Volkswagen Corrado SLC
2 Door Coupe

Sport

2.8L 178 hp Gas Fuel Inject. | 6 Cylinder "V" | Manual 5 Speed | 2 Wheel Front | Automatic Seatbelts

Purchase Price

Car Item	Dealer Cost	List
Base Price	**$20,165**	**$22,870**
Anti-Lock Brakes	Std	Std
Automatic 4 Speed	$843	$875
Optional Engine	N/A	N/A
Air Conditioning	Std	Std
Power Steering	Std	Std
Cruise Control	Std	Std
All Wheel Drive	N/A	N/A
AM/FM Stereo Cassette	N/A	N/A
Steering, Tilt w/Memory	Std	Std
Power Windows	Std	Std
*Options Price	$0	$0
***Total Price**	**$20,165**	**$22,870**
Target Price	$21,577	
Destination Charge	$390	
Avg. Tax & Fees	$1,331	
Total Target $	**$23,298**	
Average Dealer Option Cost:	**94%**	

Ownership Costs

Cost Area	5 Year Cost	Rate
Depreciation	$11,493	○
Financing ($468/month)	$3,839	
Insurance (Rating 23)	$11,066	●
State Fees	$931	
Fuel (Hwy 25 City 18)	$4,031	○
Maintenance	$5,473	○
Repairs	$965	○

Warranty/Maintenance Info

Major Tune-Up	$172	○
Minor Tune-Up	$115	○
Brake Service	$239	○
Overall Warranty	2 yr/24k	●
Drivetrain Warranty	5 yr/50k	○
Rust Warranty	6 yr/unlim. mi	○
Maintenance Warranty	N/A	
Roadside Assistance	N/A	

The Corrado SLC is available as a two door sport coupe. New for 1993, the Corrado offers a new interior treatment, an improved stereo with optional CD changer control capability, an advanced alarm system, new alloy wheels and a longer list of exterior colors. Standard features include air conditioning, a speed-activated electric rear spoiler and a multi-function trip computer. Safety features include an energy-absorbing steering column and side impact protection.

Ownership Cost By Year

$2,000 $4,000 $6,000 $8,000 $10,000 $12,000

1993, 1994, 1995, 1996, 1997

Resale Value

1993	1994	1995	1996	1997
$16,763	$15,739	$14,180	$13,050	$11,805

Cumulative Costs

	1993	1994	1995	1996	1997
Annual	$11,203	$5,625	$7,353	$5,289	$8,328
Total	$11,203	$16,828	$24,181	$29,470	$37,798

Ownership Costs (5yr)

Average	This Car
$40,193	$37,798
Cost/Mile 57¢	Cost/Mile 54¢

Ownership Cost Rating

○ Excellent

---

*Includes shaded options
**Other purchase requirements apply

● Poor | ◉ Worse Than Average | ○ Average | ○ Better Than Average | ○ Excellent | ⊖ Insufficient Information

©1993 by IntelliChoice, Inc. (408) 554-8711 All Rights Reserved. Reproduction Prohibited.
Refer to *Section 3: Annotated Vehicle Charts* for an explanation of these charts.

Volkswagen Fox
2 Door Sedan

1.8L 81 hp Gas Fuel Inject. | 4 Cylinder In-Line | Manual 5 Speed | 2 Wheel Front | Automatic Seatbelts

Subcompact

Purchase Price

Car Item	Dealer Cost	List
Base Price	**$7,851**	**$8,490**
Anti-Lock Brakes	N/A	N/A
Automatic Transmission	N/A	N/A
Optional Engine	N/A	N/A
Air Conditioning	Std	Std
Power Steering	N/A	N/A
Cruise Control	N/A	N/A
All Wheel Drive	N/A	N/A
AM/FM Stereo Cassette	$279	$340
Steering Wheel, Tilt	N/A	N/A
Power Windows	N/A	N/A
*Options Price	$0	$0
*Total Price	$7,851	$8,490
Target Price		$8,235
Destination Charge		$375
Avg. Tax & Fees		$520
Total Target $		**$9,130**
Average Dealer Option Cost:		82%

The 1993 Fox is available in two models - two-door (Base) and four-door GL. New for 1993, the Base Fox comes to the U.S. in Wolfsburg Edition. New accessories include color-keyed integrated bumpers and full wheel covers. Features include air conditioning, radio prep, fully reclining front bucket seats, dual outside remote mirrors, electric rear-window defroster, a full-sized spare and side window defoggers. Metallic colors for 1993 are Huron Blue, Moon Dust Silver and Raspberry metallic.

Ownership Costs

Cost Area	5 Year Cost	Rate
Depreciation	$4,357	O
Financing ($183/month)	$1,503	
Insurance (Rating 13)	$8,061	●
State Fees	$355	
Fuel (Hwy 33 City 25)	$2,980	O
Maintenance	$3,894	O
Repairs	$720	O

Warranty/Maintenance Info

Major Tune-Up	$203	O
Minor Tune-Up	$89	O
Brake Service	$201	O
Overall Warranty	2 yr/24k	⊙
Drivetrain Warranty	5 yr/50k	O
Rust Warranty	6 yr/unlim. mi	O
Maintenance Warranty	N/A	
Roadside Assistance	N/A	

Ownership Cost By Year

1993, 1994, 1995, 1996, 1997

Resale Value

1993	1994	1995	1996	1997
$7,347	$6,615	$6,026	$5,430	$4,773

Cumulative Costs

	1993	1994	1995	1996	1997
Annual	$4,582	$3,646	$4,403	$4,008	$5,231
Total	$4,582	$8,228	$12,631	$16,639	$21,870

Ownership Costs (5yr)

Average	This Car
$21,131	$21,870
Cost/Mile 30¢	Cost/Mile 31¢

Ownership Cost Rating

Worse Than Average

Volkswagen Fox GL
4 Door Sedan

1.8L 81 hp Gas Fuel Inject. | 4 Cylinder In-Line | Manual 5 Speed | 2 Wheel Front | Automatic Seatbelts

Subcompact

Purchase Price

Car Item	Dealer Cost	List
Base Price	**$8,448**	**$9,290**
Anti-Lock Brakes	N/A	N/A
Automatic Transmission	N/A	N/A
Optional Engine	N/A	N/A
Air Conditioning	Std	Std
Power Steering	N/A	N/A
Cruise Control	N/A	N/A
All Wheel Drive	N/A	N/A
AM/FM Stereo Cassette	$279	$340
Steering Wheel, Tilt	N/A	N/A
Power Windows	N/A	N/A
*Options Price	$0	$0
*Total Price	$8,448	$9,290
Target Price		$8,870
Destination Charge		$375
Avg. Tax & Fees		$559
Total Target $		**$9,804**
Average Dealer Option Cost:		82%

The 1993 Fox is available in two models - two-door (Base) and four-door GL. New for 1993, the GL comes to the U.S. in Wolfsburg Edition. Standard features on the GL include air conditioning, reclining front bucket seats, dual outside remote mirrors, electric rear-window defroster, a full-sized spare and side window defoggers. The Fox GL adds velour upholstery, child-safety rear door locks and illuminated luggage compartment. The Fox model line comes in a choice six exterior colors.

Ownership Costs

Cost Area	5 Year Cost	Rate
Depreciation	$4,802	O
Financing ($197/month)	$1,615	
Insurance (Rating 11)	$7,693	●
State Fees	$387	
Fuel (Hwy 33 City 25)	$2,980	O
Maintenance	$3,894	O
Repairs	$720	O

Warranty/Maintenance Info

Major Tune-Up	$203	O
Minor Tune-Up	$89	O
Brake Service	$201	O
Overall Warranty	2 yr/24k	⊙
Drivetrain Warranty	5 yr/50k	O
Rust Warranty	6 yr/unlim. mi	O
Maintenance Warranty	N/A	
Roadside Assistance	N/A	

Resale Value

1993	1994	1995	1996	1997
$7,590	$6,820	$6,264	$5,662	$5,002

Cumulative Costs

	1993	1994	1995	1996	1997
Annual	$5,000	$3,655	$4,324	$3,952	$5,160
Total	$5,000	$8,655	$12,979	$16,931	$22,091

Ownership Costs (5yr)

Average	This Car
$22,031	$22,091
Cost/Mile 31¢	Cost/Mile 32¢

Ownership Cost Rating
O
Average

---

* Includes shaded options
** Other purchase requirements apply

● Poor   ⊙ Worse Than Average   O Average   ○ Better Than Average   ○ Excellent   ⊖ Insufficient Information

©1993 by IntelliChoice, Inc. (408) 554-8711 All Rights Reserved. Reproduction Prohibited.
Refer to *Section 3: Annotated Vehicle Charts* for an explanation of these charts.

Volkswagen Golf III GL
2 Door Hatchback

Subcompact

2.0L 115 hp Gas Fuel Inject. | 4 Cylinder In-Line | Manual 5 Speed | 2 Wheel Front | Automatic Seatbelts

Purchase Price

Car Item	Dealer Cost	List
Base Price	N/R	N/R
Anti-Lock Brakes	N/A	N/A
Automatic 4 Speed	N/R	N/R
Optional Engine	N/A	N/A
Air Conditioning	N/R	N/R
Power Steering	Std	Std
Cruise Control	N/A	N/A
All Wheel Drive	N/A	N/A
AM/FM Stereo Cassette	Std	Std
Steering Wheel, Tilt	N/A	N/A
Power Windows	N/A	N/A
*Options Price	$0	$0
*Total Price	N/R	N/R
Target Price		N/R
Destination Charge		N/R
Avg. Tax & Fees		N/R
Total Target $		**N/R**
Average Dealer Option Cost:		**N/A**

Ownership Costs

Cost Area	5 Year Cost	Rate
Depreciation		⊖
Financing ($0/month)		
Insurance (Rating 14 [Est.])	$8,238	●
State Fees		
Fuel (Hwy 32 City 24)	$3,088	◉
Maintenance	$3,899	○
Repairs	$720	○

Warranty/Maintenance Info

Major Tune-Up	$232	●
Minor Tune-Up	$91	○
Brake Service	$275	●
Overall Warranty	2 yr/24k	◉
Drivetrain Warranty	5 yr/50k	○
Rust Warranty	6 yr/unlim. mi	○
Maintenance Warranty	N/A	
Roadside Assistance	N/A	

Ownership Cost By Year

Insufficient Depreciation Information
Insufficient Financing Information
(1993)
Insufficient State Fee Information
(1994, 1995, 1996, 1997)

Resale Value

Insufficient Information

Ownership Costs (5yr)

Insufficient Information

Cumulative Costs

	1993	1994	1995	1996	1997
Annual	Insufficient Information				
Total	Insufficient Information				

Ownership Cost Rating

⊖ Insufficient Information

The 1993 Golf is available in two models - GL two-door and four-door. The 1993 GL two-door has been completely redesigned for this year. It is powered by a new 2.0 liter, 115 horsepower engine that uses a new engine management system. It rides on a wider stance, with larger 14-inch wheels, larger front-disc brakes and a re-tuned four-wheel independent suspension system. Standard features include CFC-free air conditioning and an anti-theft system. Pricing was not available at time of printing.

Volkswagen Golf III GL
4 Door Hatchback

Subcompact

2.0L 115 hp Gas Fuel Inject. | 4 Cylinder In-Line | Manual 5 Speed | 2 Wheel Front | Automatic Seatbelts

2 Door Model Shown

Purchase Price

Car Item	Dealer Cost	List
Base Price	N/R	N/R
Anti-Lock Brakes	N/A	N/A
Automatic 4 Speed	N/R	N/R
Optional Engine	N/A	N/A
Air Conditioning	N/R	N/R
Power Steering	Std	Std
Cruise Control	N/A	N/A
All Wheel Drive	N/A	N/A
AM/FM Stereo Cassette	Std	Std
Steering Wheel, Tilt	N/A	N/A
Power Windows	N/A	N/A
*Options Price	$0	$0
*Total Price	N/R	N/R
Target Price		N/R
Destination Charge		N/R
Avg. Tax & Fees		N/R
Total Target $		**N/R**
Average Dealer Option Cost:		**N/A**

Ownership Costs

Cost Area	5 Year Cost	Rate
Depreciation		⊖
Financing ($0/month)		
Insurance (Rating 10 [Est.])	$7,479	●
State Fees		
Fuel (Hwy 32 City 24)	$3,088	◉
Maintenance	$3,899	○
Repairs	$720	○

Warranty/Maintenance Info

Major Tune-Up	$232	●
Minor Tune-Up	$91	○
Brake Service	$275	●
Overall Warranty	2 yr/24k	◉
Drivetrain Warranty	5 yr/50k	○
Rust Warranty	6 yr/unlim. mi	○
Maintenance Warranty	N/A	
Roadside Assistance	N/A	

Ownership Cost By Year

Insufficient Depreciation Information
Insufficient Financing Information
(1993, 1994)
Insufficient State Fee Information
(1995, 1996, 1997)

Resale Value

Insufficient Information

Ownership Costs (5yr)

Insufficient Information

Cumulative Costs

	1993	1994	1995	1996	1997
Annual	Insufficient Information				
Total	Insufficient Information				

Ownership Cost Rating

⊖ Insufficient Information

The 1993 Golf is available in two models - GL two-door and four-door. The 1993 Golf has been completely redesigned for this year. It is powered by a new 2.0L 115 hp engine. Performance designed features include improved body rigidity, an available four-speed automatic transmission with closer gear ratios, Volkswagens patented track-correcting rear axle and power-assisted, variable ratio rack-and-pinion steering. Pricing was not available at time of printing.

---

* Includes shaded options
** Other purchase requirements apply

● Poor   ◉ Worse Than Average   ○ Average   ○ Better Than Average   ○ Excellent   ⊖ Insufficient Information

©1993 by IntelliChoice, Inc. (408) 554-8711   All Rights Reserved. Reproduction Prohibited.
Refer to *Section 3: Annotated Vehicle Charts* for an explanation of these charts.

Volkswagen GTI
2 Door Hatchback

 2.0L 115 hp Gas Fuel Inject. 4 Cylinder In-Line Manual 5 Speed 2 Wheel Front Automatic Seatbelts

Subcompact

Purchase Price

Car Item	Dealer Cost	List
Base Price	**N/R**	**N/R**
Anti-Lock Brakes	N/A	N/A
Automatic 3 Speed	N/R	N/R
Optional Engine	N/A	N/A
Air Conditioning	Std	Std
Power Steering	Std	Std
Cruise Control	N/A	N/A
All Wheel Drive	N/A	N/A
AM/FM Stereo Cassette	Std	Std
Steering Wheel, Tilt	Std	Std
Power Windows	N/A	N/A
*Options Price	$0	$0
*Total Price	N/R	N/R
Target Price		N/R
Destination Charge		N/R
Avg. Tax & Fees		N/R
Total Target $		**N/R**
Average Dealer Option Cost:		N/A

Ownership Costs

Cost Area	5 Year Cost	Rate
Depreciation		⊖
Financing ($0/month)		
Insurance (Rating 15 [Est.])	$8,507	●
State Fees		
Fuel (Hwy 32 City 24)	$3,088	◉
Maintenance	$3,948	○
Repairs	$629	○

Warranty/Maintenance Info

Major Tune-Up	$232	◉
Minor Tune-Up	$91	○
Brake Service	$255	●
Overall Warranty	2 yr/24k	◉
Drivetrain Warranty	5 yr/50k	○
Rust Warranty	6 yr/unlim. mi	○
Maintenance Warranty		N/A
Roadside Assistance		N/A

Ownership Cost By Year

Insufficient Depreciation Information
Insufficient Financing Information

■ 1993
■ 1994
■ 1995
■ 1996
□ 1997

Insufficient State Fee Information

Resale Value
Insufficient Information

Ownership Costs (5yr)
Insufficient Information

Cumulative Costs
	1993	1994	1995	1996	1997
Annual	Insufficient Information				
Total	Insufficient Information				

Ownership Cost Rating
⊖
Insufficient Information

The 1993 GTI is available as a two-door hatchback. New for 1993, the GTI is completely redesigned. The new 2.0L engine is matched to a close-ratio 5-speed manual transmission, power rack-and-pinion steering, larger four-wheel disc brakes and a sport tuned independent suspension. The exterior features a body-colored grille and mirror housings, integrated fog lamps and new design alloy wheels. An eight-speaker sound system is standard. Pricing was not available at time of printing.

Volkswagen Jetta GL III
4 Door Sedan

 2.0L 115 hp Gas Fuel Inject. 4 Cylinder In-Line Manual 5 Speed 2 Wheel Front Automatic Seatbelts

Subcompact

Purchase Price

Car Item	Dealer Cost	List
Base Price	**N/R**	**N/R**
Anti-Lock Brakes	N/A	N/A
Automatic 4 Speed	N/R	N/R
Optional Engine	N/A	N/A
Air Conditioning	N/R	N/R
Power Steering	Std	Std
Cruise Control	N/A	N/A
All Wheel Drive	N/A	N/A
AM/FM Stereo Cassette	Std	Std
Steering Wheel, Scope	Std	Std
Power Windows	N/A	N/A
*Options Price	$0	$0
*Total Price	N/R	N/R
Target Price		N/R
Destination Charge		N/R
Avg. Tax & Fees		N/R
Total Target $		**N/R**
Average Dealer Option Cost:		N/A

Ownership Costs

Cost Area	5 Year Cost	Rate
Depreciation		⊖
Financing ($0/month)		
Insurance (Rating 15 [Est.])	$8,507	●
State Fees		
Fuel (Hwy 30 City 23)	$3,260	◉
Maintenance	$3,972	○
Repairs	$720	○

Warranty/Maintenance Info

Major Tune-Up	$232	◉
Minor Tune-Up	$91	○
Brake Service	$275	●
Overall Warranty	2 yr/24k	◉
Drivetrain Warranty	5 yr/50k	○
Rust Warranty	6 yr/unlim. mi	○
Maintenance Warranty		N/A
Roadside Assistance		N/A

Ownership Cost By Year

Insufficient Depreciation Information
Insufficient Financing Information

■ 1993
■ 1994
■ 1995
■ 1996
□ 1997

Insufficient State Fee Information

Resale Value
Insufficient Information

Ownership Costs (5yr)
Insufficient Information

Cumulative Costs
	1993	1994	1995	1996	1997
Annual	Insufficient Information				
Total	Insufficient Information				

Ownership Cost Rating
⊖
Insufficient Information

The 1993 Jetta is available in three models - GL, GLS and GLX four-door sedans. New for 1993, the Jetta GL has been almost completely redesigned. The new GL has been designed for a roomier interior, more comfort features, better handling, advanced safety and more power. Handling and control are also designed for improvement with a 30% increase in body rigidity, a wider stance, larger front-disc brakes and a re-tuned four-wheel independent suspension system. Pricing was not available at time of printing.

---

* Includes shaded options
** Other purchase requirements apply

● Poor ◉ Worse Than Average ○ Average ○ Better Than Average ○ Excellent ⊖ Insufficient Information

©1993 by IntelliChoice, Inc. (408) 554-8711 All Rights Reserved. Reproduction Prohibited.
Refer to *Section 3: Annotated Vehicle Charts* for an explanation of these charts.

Volkswagen Jetta GLS III
4 Door Sedan

Subcompact

- 2.0L 115 hp Gas Fuel Inject.
- 4 Cylinder In-Line
- Manual 5 Speed
- 2 Wheel Front
- Automatic Seatbelts

GL Model Shown

Purchase Price

Car Item	Dealer Cost	List
Base Price	**N/R**	**N/R**
Anti-Lock Brakes	N/A	N/A
Automatic 4 Speed	N/R	N/R
Optional Engine	N/A	N/A
Air Conditioning	N/R	N/R
Power Steering	Std	Std
Cruise Control	N/R	N/R
All Wheel Drive	N/A	N/A
AM/FM Stereo Cassette	Std	Std
Steering Wheel, Scope	Std	Std
Power Windows	N/A	N/A
*Options Price	$0	$0
*Total Price	N/R	N/R
Target Price		N/R
Destination Charge		N/R
Avg. Tax & Fees		N/R
Total Target $		**N/R**
Average Dealer Option Cost:		N/A

Ownership Costs

Cost Area	5 Year Cost	Rate
Depreciation		⊖
Financing ($0/month)		
Insurance (Rating 16 [Est.])	$8,784	●
State Fees		
Fuel (Hwy 30 City 23)	$3,260	◉
Maintenance	$3,972	○
Repairs	$720	○

Warranty/Maintenance Info

Major Tune-Up	$232	●
Minor Tune-Up	$91	○
Brake Service	$275	●
Overall Warranty	2 yr/24k	◉
Drivetrain Warranty	5 yr/50k	○
Rust Warranty	6 yr/unlim. mi	○
Maintenance Warranty		N/A
Roadside Assistance		N/A

Ownership Cost By Year

Insufficient Depreciation Information
Insufficient Financing Information
Insufficient State Fee Information

- 1993
- 1994
- 1995
- 1996
- 1997

Resale Value
Insufficient Information

Ownership Costs (5yr)
Insufficient Information

Cumulative Costs
1993 1994 1995 1996 1997
Annual: Insufficient Information
Total: Insufficient Information

Ownership Cost Rating
⊖ Insufficient Information

The 1993 Jetta is available in three models - GL, GLS and GLX four-door sedans. New for 1993, the GLS has been almost completely redesigned. The new model has been designed for a roomier interior, more comfort features, better handling, advanced safety features and more power. Standard features include a remote trunk release, an alarm system, an eight-speaker sound system with optional CD changer, rear folding/split seats and a CFC-free air conditioning system. Pricing was not available at time of printing.

Volkswagen Jetta GLX III
4 Door Sedan

Subcompact

- 2.8L 172 hp Gas Fuel Inject.
- 6 Cylinder "V"
- Manual 5 Speed
- 2 Wheel Front
- Automatic Seatbelts

GL Model Shown

Purchase Price

Car Item	Dealer Cost	List
Base Price	**N/R**	**N/R**
Anti-Lock Brakes	N/A	N/A
Automatic 4 Speed	N/R	N/R
Optional Engine	N/A	N/A
Air Conditioning	N/R	N/R
Power Steering	Std	Std
Cruise Control	N/R	N/R
All Wheel Drive	N/A	N/A
AM/FM Stereo Cassette	Std	Std
Steering Wheel, Scope	Std	Std
Power Windows	N/A	N/A
*Options Price	$0	$0
*Total Price	N/R	N/R
Target Price		N/R
Destination Charge		N/R
Avg. Tax & Fees		N/R
Total Target $		**N/R**
Average Dealer Option Cost:		N/A

Ownership Costs

Cost Area	5 Year Cost	Rate
Depreciation		⊖
Financing ($0/month)		
Insurance (Rating 19 [Est.])	$9,777	●
State Fees		
Fuel (Hwy 25 City 18 [Est.])	$4,031	●
Maintenance	$4,590	◉
Repairs	$811	◉

Warranty/Maintenance Info

Major Tune-Up	$207	○
Minor Tune-Up	$122	●
Brake Service	$275	●
Overall Warranty	2 yr/24k	◉
Drivetrain Warranty	5 yr/50k	○
Rust Warranty	6 yr/unlim. mi	○
Maintenance Warranty		N/A
Roadside Assistance		N/A

Ownership Cost By Year

Insufficient Depreciation Information
Insufficient Financing Information
Insufficient State Fee Information

- 1993
- 1994
- 1995
- 1996
- 1997

Resale Value
Insufficient Information

Ownership Costs (5yr)
Insufficient Information

Cumulative Costs
1993 1994 1995 1996 1997
Annual: Insufficient Information
Total: Insufficient Information

Ownership Cost Rating
⊖ Insufficient Information

The 1993 Jetta is available in three models - GL, GLS and GLX four-door sedans. New for 1993, the GLX has been almost completely redesigned. The new model has been designed for a roomier interior, more comfort features, better handling, advanced safety and more power. Standard features include, depending on the model, an alarm system, an eight-speaker sound system with optional CD changer and CFC-free air conditioning system. Pricing was not available at time of printing.

---

* Includes shaded options
** Other purchase requirements apply

● Poor | ◉ Worse Than Average | ○ Average | ○ Better Than Average | ○ Excellent | ⊖ Insufficient Information

©1993 by IntelliChoice, Inc. (408) 554-8711 All Rights Reserved. Reproduction Prohibited.
Refer to *Section 3: Annotated Vehicle Charts* for an explanation of these charts.

Volkswagen Passat GLS
4 Door Sedan

Compact

- 2.0L 134 hp Gas Fuel Inject.
- 4 Cylinder In-Line
- Manual 5 Speed
- 2 Wheel Front
- Automatic Seatbelts

Purchase Price

Car Item	Dealer Cost	List
Base Price	**N/R**	**N/R**
Anti-Lock Brakes	Std	Std
Automatic 4 Speed	N/R	N/R
Optional Engine	N/A	N/A
Air Conditioning	Std	Std
Power Steering	Std	Std
Cruise Control	Std	Std
All Wheel Drive	N/A	N/A
AM/FM Stereo Cassette	Std	Std
Steering Wheel, Tilt	Std	Std
Power Windows	Std	Std
***Options Price**	$0	$0
***Total Price**	**N/R**	**N/R**
Target Price		N/R
Destination Charge		N/R
Avg. Tax & Fees		N/R
Total Target $		**N/R**
Average Dealer Option Cost:		N/A

Ownership Costs

Cost Area	5 Year Cost	Rate
Depreciation		⊖
Financing ($0/month)		
Insurance (Rating 11 [Est.])	$7,693	○
State Fees		
Fuel (Hwy 29 City 20)	$3,548	○
Maintenance	$4,968	◉
Repairs	$800	◉

Warranty/Maintenance Info

Major Tune-Up	$232	●
Minor Tune-Up	$97	○
Brake Service	$238	○
Overall Warranty	2 yr/24k	◉
Drivetrain Warranty	5 yr/50k	○
Rust Warranty	6 yr/unlim. mi	○
Maintenance Warranty	N/A	
Roadside Assistance	N/A	

Ownership Cost By Year

Insufficient Depreciation Information
Insufficient Financing Information
Insufficient State Fee Information

■ 1993
■ 1994
■ 1995
□ 1996
□ 1997

Resale Value

Insufficient Information

Ownership Costs (5yr)

Insufficient Information

Cumulative Costs

	1993	1994	1995	1996	1997
Annual	Insufficient Information				
Total	Insufficient Information				

Ownership Cost Rating
⊖ Insufficient Information

The 1993 Passat is available in five models - GL sedan and wagon, GLS sedan and GLX sedan and wagon. New for 1993, the GLS sedan offers four-wheel independent suspension, a front suspension system employing a wishbone lower control arm, stabilizer bar and MacPherson struts. Standard features include air conditioning, power windows, alloy wheels, power glass sunroof, cruise control, central locking with key-operated window closing and front and rear spoilers. Pricing was not available at time of printing.

Volkswagen Passat GL
4 Door Sedan

Compact

- 2.0L 134 hp Gas Fuel Inject.
- 4 Cylinder In-Line
- Manual 5 Speed
- 2 Wheel Front
- Automatic Seatbelts

Purchase Price

Car Item	Dealer Cost	List
Base Price	**$15,537**	**$17,610**
Anti-Lock Brakes	N/A	N/A
Automatic 4 Speed	$843	$875
Optional Engine	N/A	N/A
Air Conditioning	Std	Std
Power Steering	Std	Std
Cruise Control	Std	Std
All Wheel Drive	N/A	N/A
AM/FM Stereo Cassette	Std	Std
Steering Wheel, Tilt	Std	Std
Power Windows	Std	Std
***Options Price**	$843	$875
***Total Price**	**$16,380**	**$18,485**
Target Price		$17,397
Destination Charge		$390
Avg. Tax & Fees		$1,078
Total Target $		**$18,865**
Average Dealer Option Cost:		96%

Ownership Costs

Cost Area	5 Year Cost	Rate
Depreciation	$7,993	○
Financing ($379/month)	$3,107	
Insurance (Rating 13)	$8,061	○
State Fees	$755	
Fuel (Hwy 29 City 20)	$3,548	○
Maintenance	$4,968	◉
Repairs	$800	◉

Warranty/Maintenance Info

Major Tune-Up	$232	●
Minor Tune-Up	$97	○
Brake Service	$238	○
Overall Warranty	2 yr/24k	◉
Drivetrain Warranty	5 yr/50k	○
Rust Warranty	6 yr/unlim. mi	○
Maintenance Warranty	N/A	
Roadside Assistance	N/A	

Resale Value

1993	1994	1995	1996	1997
$15,389	$13,786	$13,300	$12,087	$10,872

Ownership Costs (5yr)

Average	This Car
$30,367	$29,232
Cost/Mile 43¢	Cost/Mile 42¢

Cumulative Costs

	1993	1994	1995	1996	1997
Annual	$7,155	$5,245	$5,193	$4,500	$7,139
Total	$7,155	$12,400	$17,593	$22,093	$29,232

Ownership Cost Rating
○ Better Than Average

The 1993 Passat is available in five models - GL sedan and wagon, GLS sedan and GLX sedan and wagon. New for 1993, the GL sedan includes an improved alarm system, deluxe cassette stereo with portable CD stereo input, a multi-function trip computer, lockable rear (split) folding seats, air conditioning, power windows, a front spoiler, power steering, cruise control, central locking (doors, trunk lid and fuel filler) with key-operated window closing and dual remote heated mirrors.

---

* Includes shaded options
** Other purchase requirements apply

● Poor ◉ Worse Than Average ○ Average ○ Better Than Average ○ Excellent ⊖ Insufficient Information

©1993 by IntelliChoice, Inc. (408) 554-8711 All Rights Reserved. Reproduction Prohibited.
Refer to *Section 3: Annotated Vehicle Charts* for an explanation of these charts.

Volkswagen Passat GLX
4 Door Sedan

Compact

2.8L 172 hp Gas Fuel Inject. | 6 Cylinder "V" | Manual 5 Speed | 2 Wheel Front | Automatic Seatbelts

Purchase Price

Car Item	Dealer Cost	List
Base Price	**$18,635**	**$21,130**
Anti-Lock Brakes	Std	Std
Automatic 4 Speed	$843	$875
Optional Engine	N/A	N/A
Air Conditioning	Std	Std
Power Steering	Std	Std
Cruise Control	Std	Std
All Wheel Drive	N/A	N/A
AM/FM Stereo Cassette	Std	Std
Steering Wheel, Tilt	Std	Std
Power Windows	Std	Std
*Options Price	$843	$875
*Total Price	$19,478	$22,005
Target Price	$20,791	
Destination Charge	$390	
Avg. Tax & Fees	$1,283	
Total Target $	**$22,464**	
Average Dealer Option Cost:	94%	

Ownership Costs

Cost Area	5 Year Cost	Rate
Depreciation	$10,418	O
Financing ($451/month)	$3,701	
Insurance (Rating 15)	$8,507	O
State Fees	$896	
Fuel (Hwy 24 City 17)	$4,232	●
Maintenance	$5,094	◉
Repairs	$965	●

Warranty/Maintenance Info

Major Tune-Up	$186	O
Minor Tune-Up	$115	O
Brake Service	$238	O
Overall Warranty	2 yr/24k	●
Drivetrain Warranty	5 yr/50k	O
Rust Warranty	6 yr/unlim. mi	O
Maintenance Warranty	N/A	
Roadside Assistance	N/A	

Ownership Cost By Year

1993, 1994, 1995, 1996, 1997

Resale Value

1993	1994	1995	1996	1997
$18,051	$15,987	$15,231	$13,630	$12,046

Cumulative Costs

	1993	1994	1995	1996	1997
Annual	$8,580	$6,164	$5,934	$5,257	$7,878
Total	$8,580	$14,744	$20,678	$25,935	$33,813

Ownership Costs (5yr)

Average	This Car
$33,389	$33,813
Cost/Mile 48¢	Cost/Mile 48¢

Ownership Cost Rating
O Average

The 1993 Passat is available in five models - GL sedan and wagon, GLS sedan and GLX sedan and wagon. New for 1993, the GLX sedan offers the VR6 engine. Standard features include air conditioning, alloy wheels, a power glass sunroof, power windows, a front and rear spoiler, premium sound system with CD changer control capability, power steering, cruise control, central locking (doors, trunk lid and fuel filler) with key-operated window closing and dual remote heated mirrors.

Volkswagen Passat GL
4 Door Wagon

   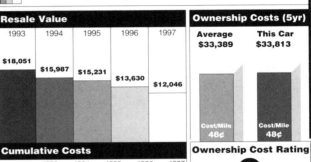

Compact Wagon

2.0L 134 hp Gas Fuel Inject. | 4 Cylinder In-Line | Manual 5 Speed | 2 Wheel Front | Automatic Seatbelts

Purchase Price

Car Item	Dealer Cost	List
Base Price	**N/R**	**N/R**
Anti-Lock Brakes	N/R	N/R
Automatic 4 Speed	N/R	N/R
Optional Engine	N/A	N/A
Air Conditioning	Std	Std
Power Steering	Std	Std
Cruise Control	Std	Std
All Wheel Drive	N/A	N/A
AM/FM Stereo Cassette	Std	Std
Steering Wheel, Tilt	Std	Std
Power Windows	Std	Std
*Options Price	$0	$0
*Total Price	N/R	N/R
Target Price	N/R	
Destination Charge	N/R	
Avg. Tax & Fees	N/R	
Total Target $	**N/R**	
Average Dealer Option Cost:	N/A	

Ownership Costs

Cost Area	5 Year Cost	Rate
Depreciation		⊖
Financing ($0/month)		
Insurance (Rating 14 [Est.])	$8,238	O
State Fees		
Fuel (Hwy 29 City 20)	$3,548	O
Maintenance	$4,968	O
Repairs	$800	●

Warranty/Maintenance Info

Major Tune-Up	$232	●
Minor Tune-Up	$97	O
Brake Service	$238	●
Overall Warranty	2 yr/24k	●
Drivetrain Warranty	5 yr/50k	O
Rust Warranty	6 yr/unlim. mi	O
Maintenance Warranty	N/A	
Roadside Assistance	N/A	

Ownership Cost By Year

Insufficient Depreciation Information
Insufficient Financing Information
Insufficient State Fee Information

Resale Value
Insufficient Information

Cumulative Costs

	1993	1994	1995	1996	1997
Annual	Insufficient Information				
Total	Insufficient Information				

Ownership Costs (5yr)
Insufficient Information

Ownership Cost Rating
⊖ Insufficient Information

The 1993 Passat is available in five models - GL sedan and wagon, GLS sedan and GLX sedan and wagon. New for 1993, the GL wagon includes an improved alarm system, deluxe cassette stereo with portable CD stereo input, a multi-function trip computer, lockable rear (split) folding seats, air conditioning, power windows, a front spoiler, power steering, cruise control, central locking with key-operated window closing and dual remote heated mirrors. Pricing was not available at time of printing.

---

page 250

* Includes shaded options
** Other purchase requirements apply

● Poor | ◉ Worse Than Average | O Average | ○ Better Than Average | ○ Excellent | ⊖ Insufficient Information

©1993 by IntelliChoice, Inc. (408) 554-8711 All Rights Reserved. Reproduction Prohibited.
Refer to *Section 3: Annotated Vehicle Charts* for an explanation of these charts.

Volkswagen Passat GLX
4 Door Wagon

Compact Wagon

 2.8L 172 hp Gas Fuel Inject.
 6 Cylinder "V"
 Manual 5 Speed
 2 Wheel Front
Automatic Seatbelts

Purchase Price

Car Item	Dealer Cost	List
Base Price	$19,013	$21,560
Anti-Lock Brakes	Std	Std
Automatic 4 Speed	$843	$875
Optional Engine	N/A	N/A
Air Conditioning	Std	Std
Power Steering	Std	Std
Cruise Control	Std	Std
All Wheel Drive	N/A	N/A
AM/FM Stereo Cassette	Std	Std
Steering Wheel, Tilt	Std	Std
Power Windows	Std	Std
*Options Price	$843	$875
*Total Price	$19,856	$22,435
Target Price	$21,207	
Destination Charge	$0	
Avg. Tax & Fees	$1,284	
Total Target $	**$22,491**	
Average Dealer Option Cost:	94%	

The 1993 Passat is available in five models – GL sedan and wagon, GLS sedan and GLX sedan and wagon. New for 1993, the GLX wagon offers the VR6 engine. Standard features include air conditioning, alloy wheels, a power glass sunroof, power windows, a front and rear spoiler, premium sound system with CD changer control capability, power steering, cruise control, central locking (doors, trunk lid and fuel filler) with key-operated window closing and dual remote heated mirrors.

Ownership Costs

Cost Area	5 Year Cost	Rate
Depreciation	$9,930	○
Financing ($452/month)	$3,705	
Insurance (Rating 15)	$8,507	◐
State Fees	$897	
Fuel (Hwy 24 City 17)	$4,232	◉
Maintenance	$5,094	◐
Repairs	$965	●

Warranty/Maintenance Info

Major Tune-Up	$186	◉
Minor Tune-Up	$115	◉
Brake Service	$238	◉
Overall Warranty	2 yr/24k	◉
Drivetrain Warranty	5 yr/50k	○
Rust Warranty	6 yr/unlim. mi	○
Maintenance Warranty	N/A	
Roadside Assistance	N/A	

Ownership Cost By Year

Resale Value

1993	1994	1995	1996	1997
$18,060	$15,925	$15,370	$13,946	$12,561

Cumulative Costs

	1993	1994	1995	1996	1997
Annual	$8,600	$6,236	$5,734	$5,080	$7,680
Total	$8,600	$14,836	$20,570	$25,650	$33,330

Ownership Costs (5yr)

Average	This Car
$33,758	$33,330
Cost/Mile 48¢	Cost/Mile 48¢

Ownership Cost Rating

○ Average

Volvo 240
4 Door Sedan

Compact

 2.3L 114 hp Gas Fuel Inject.
 4 Cylinder In-Line
 Manual 5 Speed
 2 Wheel Rear
Driver Airbag Psngr Belts

Purchase Price

Car Item	Dealer Cost	List
Base Price	$19,560	$21,820
Anti-Lock Brakes	Std	Std
Automatic 4 Speed	$585	$675
Optional Engine	N/A	N/A
Air Conditioning	Std	Std
Power Steering	Std	Std
Cruise Control	N/A	N/A
All Wheel Drive	N/A	N/A
AM/FM Stereo Cassette	Std	Std
Steering Wheel, Tilt	N/A	N/A
Power Windows	Std	Std
*Options Price	$585	$675
*Total Price	$20,145	$22,495
Target Price	$22,026	
Destination Charge	$395	
Avg. Tax & Fees	$1,350	
Total Target $	**$23,771**	
Average Dealer Option Cost:	87%	

The 1993 Volvo 240 is available in two bodystyles – sedan and wagon with base trimwear. New for 1993, the Volvo 240 sedan receives upgraded audio speakers, power operated and heated outside mirrors, metallic paint, a power antenna and a totally CFC-free air conditioning system which contains R134A, a refrigerant that has virtually no ozone depleting potential. The 240 sedan also features a matte black grille and trim package. Optional equipment includes leather seating and alloy wheels.

Ownership Costs

Cost Area	5 Year Cost	Rate
Depreciation	$10,515	○
Financing ($478/month)	$3,915	
Insurance (Rating 7)	$7,035	○
State Fees	$915	
Fuel (Hwy 25 City 20)	$3,837	◉
Maintenance	$3,346	○
Repairs	$760	◉

Warranty/Maintenance Info

Major Tune-Up	$237	●
Minor Tune-Up	$117	◉
Brake Service	$141	○
Overall Warranty	3 yr/50k	○
Drivetrain Warranty	3 yr/50k	○
Rust Warranty	8 yr/unlim. mi	○
Maintenance Warranty	N/A	
Roadside Assistance	3 yr/unlim. mi	

Resale Value

1993	1994	1995	1996	1997
$19,225	$17,575	$16,161	$14,710	$13,256

Cumulative Costs

	1993	1994	1995	1996	1997
Annual	$8,459	$5,274	$5,456	$4,942	$6,192
Total	$8,459	$13,733	$19,189	$24,131	$30,323

Ownership Costs (5yr)

Average	This Car
$33,810	$30,323
Cost/Mile 48¢	Cost/Mile 43¢

Ownership Cost Rating

○ Excellent

---

* Includes shaded options
** Other purchase requirements apply

● Poor ◉ Worse Than Average ○ Average ◐ Better Than Average ○ Excellent ⊖ Insufficient Information

©1993 by IntelliChoice, Inc. (408) 554-8711 All Rights Reserved. Reproduction Prohibited.
Refer to *Section 3: Annotated Vehicle Charts* for an explanation of these charts.

Volvo 240
4 Door Wagon

 2.3L 114 hp Gas Fuel Inject.
 4 Cylinder In-Line
 Manual 5 Speed
 2 Wheel Rear
Driver Airbag Psngr Belts

Compact Wagon

Purchase Price

Car Item	Dealer Cost	List
Base Price	$20,430	$22,820
Anti-Lock Brakes	Std	Std
Automatic 4 Speed	$585	$675
Optional Engine	N/A	N/A
Air Conditioning	Std	Std
Power Steering	Std	Std
Cruise Control	N/A	N/A
All Wheel Drive	N/A	N/A
AM/FM Stereo Cassette	Std	Std
Steering Wheel, Tilt	N/A	N/A
Power Windows	Std	Std
*Options Price	$585	$675
*Total Price	$21,015	$23,495
Target Price	$23,019	
Destination Charge	$395	
Avg. Tax & Fees	$1,410	
Total Target $	**$24,824**	
Average Dealer Option Cost:	87%	

Ownership Costs

Cost Area	5 Year Cost	Rate
Depreciation	$9,713	○
Financing ($499/month)	$4,089	
Insurance (Rating 7)	$7,035	○
State Fees	$955	
Fuel (Hwy 25 City 20)	$3,837	⊙
Maintenance	$3,394	○
Repairs	$760	⊙

Warranty/Maintenance Info

Major Tune-Up	$237	●
Minor Tune-Up	$117	⊙
Brake Service	$141	○
Overall Warranty	3 yr/50k	○
Drivetrain Warranty	3 yr/50k	○
Rust Warranty	8 yr/unlim. mi	○
Maintenance Warranty	N/A	
Roadside Assistance	3 yr/unlim. mi	

Ownership Cost By Year
1993, 1994, 1995, 1996, 1997

Resale Value
- 1993: $21,372
- 1994: $19,677
- 1995: $18,225
- 1996: $16,597
- 1997: $15,111

Cumulative Costs

	1993	1994	1995	1996	1997
Annual	$7,447	$5,382	$5,560	$5,139	$6,255
Total	$7,447	$12,829	$18,389	$23,528	$29,783

Ownership Costs (5yr)
- Average: $34,669
- This Car: $29,783
- Cost/Mile: 50¢ / 43¢

Ownership Cost Rating
 Excellent

The 1993 Volvo 240 is available in two bodystyles - sedan and wagon. New for 1993, the Volvo 240 wagon receives upgraded audio speakers, power operated and heated outside mirrors, metallic paint, a power antenna and a totally CFC-free air conditioning system which contains R134A, a refrigerant that has virtually no ozone depleting potential. The 240 wagon also features a matte black grille and trim package. Optional equipment includes leather seating and alloy wheels.

Volvo 850 GLT
4 Door Sedan

 2.4L 168 hp Gas Fuel Inject.
 5 Cylinder In-Line
 Manual 5 Speed
 2 Wheel Front
 Driver/Psngr Airbags Std

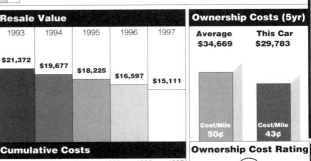
Midsize

Purchase Price

Car Item	Dealer Cost	List
Base Price	$21,115	$24,100
Anti-Lock Brakes	Std	Std
Automatic 4 Speed	$635	$750
Optional Engine	N/A	N/A
Air Conditioning	Std	Std
Power Steering	Std	Std
Cruise Control	Std	Std
All Wheel Drive	N/A	N/A
AM/FM Stereo Cassette	Std	Std
Steering Wheel, Scope	Std	Std
Power Windows	Std	Std
*Options Price	$635	$750
*Total Price	$21,750	$24,850
Target Price	$23,950	
Destination Charge	$395	
Avg. Tax & Fees	$1,469	
Total Target $	**$25,814**	
Average Dealer Option Cost:	84%	

Ownership Costs

Cost Area	5 Year Cost	Rate
Depreciation		⊖
Financing ($519/month)	$4,252	
Insurance (Rating 16)	$8,784	○
State Fees	$1,009	
Fuel (Hwy 28 City 20)	$3,612	○
Maintenance	$3,679	○
Repairs	$820	○

Warranty/Maintenance Info

Major Tune-Up	$166	○
Minor Tune-Up	$93	○
Brake Service	$129	○
Overall Warranty	3 yr/50k	○
Drivetrain Warranty	3 yr/50k	○
Rust Warranty	8 yr/unlim. mi	○
Maintenance Warranty	N/A	
Roadside Assistance	3 yr/unlim. mi	

Ownership Cost By Year
Insufficient Depreciation Information

Resale Value
Insufficient Information

Cumulative Costs

	1993	1994	1995	1996	1997
Annual	Insufficient Information				
Total	Insufficient Information				

Ownership Costs (5yr)
Insufficient Information

Ownership Cost Rating
⊖ Insufficient Information

The 1993 850 is an all-new luxury sport sedan available in only one model - GLT. The Volvo 850 GLT is front-wheel drive and its 168 horsepower, five-cylinder engine is mounted transversely. Other features include six-spoke aluminum alloy wheels, power operated, orthopedically-designed front bucket seats, analog gauges, tilt/telescopic steering wheel, power-assisted rack-and-pinion steering and an air conditioning system that uses a non-chlorofluorocarbon refrigerant (R134A).

---

page 252

* Includes shaded options
** Other purchase requirements apply

● Poor   ⊙ Worse Than Average   ○ Average   ◯ Better Than Average   ◯ Excellent   ⊖ Insufficient Information

©1993 by IntelliChoice, Inc. (408) 554-8711   All Rights Reserved. Reproduction Prohibited.
Refer to *Section 3: Annotated Vehicle Charts* for an explanation of these charts.

Volvo 940 — 4 Door Sedan — Midsize

2.3L 114 hp Gas Fuel Inject. | 4 Cylinder In-Line | Automatic 4 Speed | 2 Wheel Rear | Driver Airbag Psngr Belts

Purchase Price

Car Item	Dealer Cost	List
Base Price	**$22,070**	**$24,995**
Anti-Lock Brakes	Std	Std
Manual Transmission	N/A	N/A
Optional Engine	N/A	N/A
Auto Climate Control	Std	Std
Power Steering	Std	Std
Cruise Control	Std	Std
All Wheel Drive	N/A	N/A
AM/FM Stereo Cassette	Std	Std
Steering Wheel, Tilt	N/A	N/A
Power Windows	Std	Std
***Options Price**	$0	$0
***Total Price**	**$22,070**	**$24,995**
Target Price		$24,259
Destination Charge		$395
Avg. Tax & Fees		$1,487
Total Target $		**$26,141**
Average Dealer Option Cost:		85%

The 1993 Volvo 940 is available with GL and Turbo trim levels in both sedan and wagon. New for 1993, the Volvo 940 sedan will be available in two distinctive equipment levels. One level includes an automatic transmission, SRS, ABS, SIPS, all-season tires, power windows, central locking, and a totally CFC-free air conditioner. The other level includes all of the above equipment plus twenty-spoke alloy wheels, power operated sunroof, leather upolstery and an upgraded full-logic audio system.

Ownership Costs

Cost Area	5 Year Cost	Rate
Depreciation	$14,744	◐
Financing ($525/month)	$4,307	
Insurance (Rating 10)	$7,479	○
State Fees	$1,016	
Fuel (Hwy 27 City 19)	$3,775	◐
Maintenance	$4,100	◐
Repairs	$760	◐

Warranty/Maintenance Info

Major Tune-Up	$238	●
Minor Tune-Up	$122	○
Brake Service	$147	
Overall Warranty	3 yr/50k	◐
Drivetrain Warranty	3 yr/50k	◐
Rust Warranty	8 yr/unlim. mi	○
Maintenance Warranty	N/A	
Roadside Assistance	3 yr/unlim. mi	

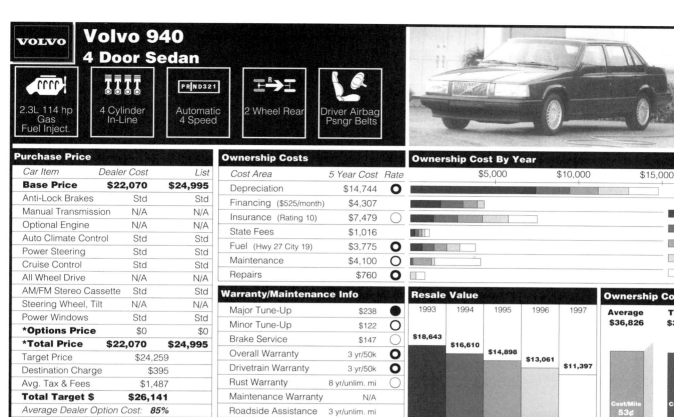

Resale Value: 1993 $18,643 | 1994 $16,610 | 1995 $14,898 | 1996 $13,061 | 1997 $11,397

Ownership Costs (5yr): Average $36,826 | This Car $36,181 — Cost/Mile 53¢ / 52¢

Cumulative Costs

	1993	1994	1995	1996	1997
Annual	$11,670	$5,875	$6,132	$5,187	$7,317
Total	$11,670	$17,545	$23,677	$28,864	$36,181

Ownership Cost Rating: ○ Average

Volvo 940 Turbo — 4 Door Sedan — Midsize

2.3L 162 hp Turbo Gas Fuel Inject. | 4 Cylinder In-Line | Automatic 4 Speed | 2 Wheel Rear | Driver Airbag Psngr Belts

Purchase Price

Car Item	Dealer Cost	List
Base Price	**$25,035**	**$28,495**
Anti-Lock Brakes	Std	Std
Manual Transmission	N/A	N/A
Optional Engine	N/A	N/A
Auto Climate Control	Std	Std
Power Steering	Std	Std
Cruise Control	Std	Std
All Wheel Drive	N/A	N/A
AM/FM Stereo Cassette	Std	Std
Steering Wheel, Tilt	N/A	N/A
Power Windows	Std	Std
***Options Price**	$0	$0
***Total Price**	**$25,035**	**$28,495**
Target Price		$27,690
Destination Charge		$395
Avg. Tax & Fees		$1,693
Total Target $		**$29,778**
Average Dealer Option Cost:		80%

The 1993 Volvo 940 is available with GL and Turbo trim levels in either sedan or wagon. New for 1993, the Volvo 940 Turbo wagon includes a four-cylinder turbocharged engine, automatic transmission, SRS, ABS, SIPS, all-season tires, power windows, central locking, totally CFC-free air conditioning system, turbo grille, power sunroof, full leather interior (optional), alloy wheels, an upgraded audio system with a full logic cassette deck, a turbo tailgate emblem and front foglights.

Ownership Costs

Cost Area	5 Year Cost	Rate
Depreciation	$16,605	○
Financing ($598/month)	$4,905	
Insurance (Rating 12)	$7,898	○
State Fees	$1,155	
Fuel (Hwy 23 City 19)	$4,111	●
Maintenance	$3,850	○
Repairs	$925	◐

Warranty/Maintenance Info

Major Tune-Up	$241	●
Minor Tune-Up	$125	○
Brake Service	$147	○
Overall Warranty	3 yr/50k	○
Drivetrain Warranty	3 yr/50k	○
Rust Warranty	8 yr/unlim. mi	○
Maintenance Warranty	N/A	
Roadside Assistance	3 yr/unlim. mi	

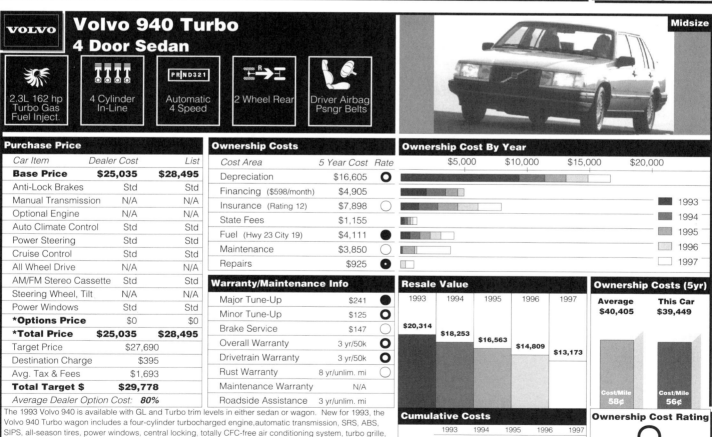

Resale Value: 1993 $20,314 | 1994 $18,253 | 1995 $16,563 | 1996 $14,809 | 1997 $13,173

Ownership Costs (5yr): Average $40,405 | This Car $39,449 — Cost/Mile 58¢ / 56¢

Cumulative Costs

	1993	1994	1995	1996	1997
Annual	$14,059	$6,266	$6,266	$5,401	$7,457
Total	$14,059	$20,325	$26,591	$31,992	$39,449

Ownership Cost Rating: ○ Better Than Average

---

* Includes shaded options
** Other purchase requirements apply

Legend: ● Poor | ◐ Worse Than Average | ○ Average | ○ Better Than Average | ○ Excellent | ⊖ Insufficient Information

©1993 by IntelliChoice, Inc. (408) 554-8711 All Rights Reserved. Reproduction Prohibited.
Refer to *Section 3: Annotated Vehicle Charts* for an explanation of these charts.

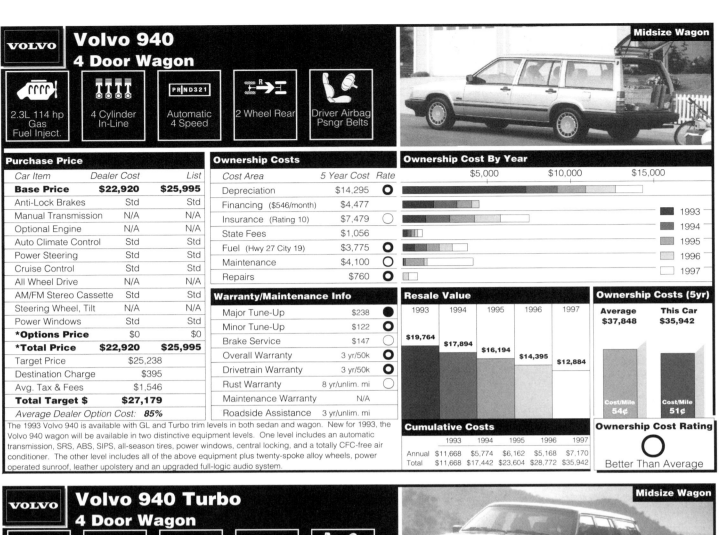

Volvo 940 — 4 Door Wagon
Midsize Wagon

2.3L 114 hp Gas Fuel Inject. | 4 Cylinder In-Line | Automatic 4 Speed | 2 Wheel Rear | Driver Airbag Psngr Belts

Purchase Price

Car Item	Dealer Cost	List
Base Price	**$22,920**	**$25,995**
Anti-Lock Brakes	Std	Std
Manual Transmission	N/A	N/A
Optional Engine	N/A	N/A
Auto Climate Control	Std	Std
Power Steering	Std	Std
Cruise Control	Std	Std
All Wheel Drive	N/A	N/A
AM/FM Stereo Cassette	Std	Std
Steering Wheel, Tilt	N/A	N/A
Power Windows	Std	Std
*Options Price	$0	$0
*Total Price	$22,920	$25,995
Target Price	$25,238	
Destination Charge	$395	
Avg. Tax & Fees	$1,546	
Total Target $	**$27,179**	

Average Dealer Option Cost: 85%

The 1993 Volvo 940 is available with GL and Turbo trim levels in both sedan and wagon. New for 1993, the Volvo 940 wagon will be available in two distinctive equipment levels. One level includes an automatic transmission, SRS, ABS, SIPS, all-season tires, power windows, central locking, and a totally CFC-free air conditioner. The other level includes all of the above equipment plus twenty-spoke alloy wheels, power operated sunroof, leather upolstery and an upgraded full-logic audio system.

Ownership Costs

Cost Area	5 Year Cost	Rate
Depreciation	$14,295	○
Financing ($546/month)	$4,477	
Insurance (Rating 10)	$7,479	◐
State Fees	$1,056	
Fuel (Hwy 27 City 19)	$3,775	○
Maintenance	$4,100	○
Repairs	$760	○

Warranty/Maintenance Info

Major Tune-Up	$238	●
Minor Tune-Up	$122	○
Brake Service	$147	◐
Overall Warranty	3 yr/50k	○
Drivetrain Warranty	3 yr/50k	○
Rust Warranty	8 yr/unlim. mi	◐
Maintenance Warranty	N/A	
Roadside Assistance	3 yr/unlim. mi	

Ownership Cost By Year
1993, 1994, 1995, 1996, 1997

Resale Value

1993	1994	1995	1996	1997
$19,764	$17,894	$16,194	$14,395	$12,884

Cumulative Costs

	1993	1994	1995	1996	1997
Annual	$11,668	$5,774	$6,162	$5,168	$7,170
Total	$11,668	$17,442	$23,604	$28,772	$35,942

Ownership Costs (5yr)
Average: $37,848 | This Car: $35,942
Cost/Mile: 54¢ | Cost/Mile: 51¢

Ownership Cost Rating
○ Better Than Average

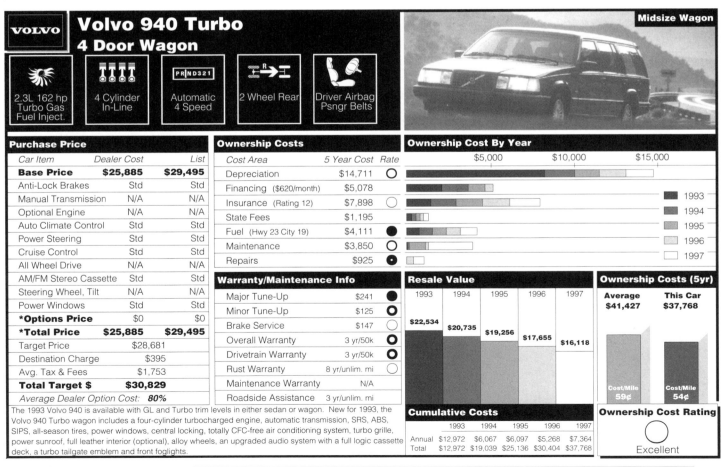

Volvo 940 Turbo — 4 Door Wagon
Midsize Wagon

2.3L 162 hp Turbo Gas Fuel Inject. | 4 Cylinder In-Line | Automatic 4 Speed | 2 Wheel Rear | Driver Airbag Psngr Belts

Purchase Price

Car Item	Dealer Cost	List
Base Price	**$25,885**	**$29,495**
Anti-Lock Brakes	Std	Std
Manual Transmission	N/A	N/A
Optional Engine	N/A	N/A
Auto Climate Control	Std	Std
Power Steering	Std	Std
Cruise Control	Std	Std
All Wheel Drive	N/A	N/A
AM/FM Stereo Cassette	Std	Std
Steering Wheel, Tilt	N/A	N/A
Power Windows	Std	Std
*Options Price	$0	$0
*Total Price	$25,885	$29,495
Target Price	$28,681	
Destination Charge	$395	
Avg. Tax & Fees	$1,753	
Total Target $	**$30,829**	

Average Dealer Option Cost: 80%

The 1993 Volvo 940 is available with GL and Turbo trim levels in either sedan or wagon. New for 1993, the Volvo 940 Turbo wagon includes a four-cylinder turbocharged engine, automatic transmission, SRS, ABS, SIPS, all-season tires, power windows, central locking, totally CFC-free air conditioning system, turbo grille, power sunroof, full leather interior (optional), alloy wheels, an upgraded audio system with a full logic cassette deck, a turbo tailgate emblem and front foglights.

Ownership Costs

Cost Area	5 Year Cost	Rate
Depreciation	$14,711	○
Financing ($620/month)	$5,078	
Insurance (Rating 12)	$7,898	◐
State Fees	$1,195	
Fuel (Hwy 23 City 19)	$4,111	●
Maintenance	$3,850	○
Repairs	$925	◉

Warranty/Maintenance Info

Major Tune-Up	$241	●
Minor Tune-Up	$125	○
Brake Service	$147	◐
Overall Warranty	3 yr/50k	○
Drivetrain Warranty	3 yr/50k	○
Rust Warranty	8 yr/unlim. mi	◐
Maintenance Warranty	N/A	
Roadside Assistance	3 yr/unlim. mi	

Resale Value

1993	1994	1995	1996	1997
$22,534	$20,735	$19,256	$17,655	$16,118

Cumulative Costs

	1993	1994	1995	1996	1997
Annual	$12,972	$6,067	$6,097	$5,268	$7,364
Total	$12,972	$19,039	$25,136	$30,404	$37,768

Ownership Costs (5yr)
Average: $41,427 | This Car: $37,768
Cost/Mile: 59¢ | Cost/Mile: 54¢

Ownership Cost Rating
○ Excellent

---

*Includes shaded options
**Other purchase requirements apply

● Poor | ◉ Worse Than Average | ◐ Average | ○ Better Than Average | ○ Excellent | ⊖ Insufficient Information

©1993 by IntelliChoice, Inc. (408) 554-8711 All Rights Reserved. Reproduction Prohibited.
Refer to *Section 3: Annotated Vehicle Charts* for an explanation of these charts.

Volvo 960 — 4 Door Sedan

Luxury

- 2.9L 201 hp Gas Fuel Inject.
- 6 Cylinder In-Line
- Automatic 4 Speed
- 2 Wheel Rear
- Driver/Psngr Airbags Std

Purchase Price

Car Item	Dealer Cost	List
Base Price	**$30,155**	**$35,675**
Anti-Lock Brakes	Std	Std
Manual Transmission	N/A	N/A
Optional Engine	N/A	N/A
Auto Climate Control	Std	Std
Power Steering	Std	Std
Cruise Control	Std	Std
All Wheel Drive	N/A	N/A
AM/FM Stereo CD	Std	Std
Steering Wheel, Tilt	Std	Std
Power Windows	Std	Std
*Options Price	$0	$0
*Total Price	$30,155	$35,675
Target Price	$33,623	
Destination Charge	$395	
Avg. Tax & Fees	$2,062	
Luxury Tax	$402	
Total Target $	**$36,482**	

Ownership Costs

Cost Area	5 Year Cost	Rate
Depreciation	$22,164	●
Financing ($733/month)	$6,010	
Insurance (Rating 14)	$8,238	○
State Fees	$1,443	
Fuel (Hwy 25 City 17)	$4,145	◐
Maintenance	$5,105	○
Repairs	$740	○

Warranty/Maintenance Info

Major Tune-Up	$161	○
Minor Tune-Up	$71	○
Brake Service	$173	○
Overall Warranty	3 yr/50k	◐
Drivetrain Warranty	3 yr/50k	◐
Rust Warranty	8 yr/unlim. mi	○
Maintenance Warranty	N/A	
Roadside Assistance	3 yr/unlim. mi	

Resale Value

1993	1994	1995	1996	1997
$21,977	$19,676	$17,911	$16,116	$14,318

Ownership Costs (5yr)

Average	This Car
$46,133	$47,845
Cost/Mile 66¢	Cost/Mile 68¢

Cumulative Costs

	1993	1994	1995	1996	1997
Annual	$19,704	$6,991	$6,981	$5,453	$8,716
Total	$19,704	$26,695	$33,676	$39,129	$47,845

Ownership Cost Rating: Average

The 1993 Volvo 960 is available in two bodystyles - sedan and wagon. New for 1993, the 960 sedan receives a passenger side supplemental restraint system (SRS), which consists of an airbag and a knee bolster. The interior also offers a new AM/FM stereo cassette with large easy to use controls and upgraded stereo speakers. Other safety features include pyrotechnique seatbelt pretensioners and integral padded front seat headrests. The 940 also features a side impact protection system.

Volvo 960 — 4 Door Wagon

Luxury

- 2.9L 201 hp Gas Fuel Inject.
- 6 Cylinder In-Line
- Automatic 4 Speed
- 2 Wheel Rear
- Driver/Psngr Airbags Std

Purchase Price

Car Item	Dealer Cost	List
Base Price	**$30,975**	**$36,675**
Anti-Lock Brakes	Std	Std
Manual Transmission	N/A	N/A
Optional Engine	N/A	N/A
Auto Climate Control	Std	Std
Power Steering	Std	Std
Cruise Control	Std	Std
All Wheel Drive	N/A	N/A
AM/FM Stereo Cassette	Std	Std
Steering Wheel, Tilt	Std	Std
Power Windows	Std	Std
*Options Price	$0	$0
*Total Price	$30,975	$36,675
Target Price	$34,537	
Destination Charge	$395	
Avg. Tax & Fees	$2,118	
Luxury Tax	$493	
Total Target $	**$37,543**	

Ownership Costs

Cost Area	5 Year Cost	Rate
Depreciation	$21,423	○
Financing ($755/month)	$6,184	
Insurance (Rating 15)	$8,507	○
State Fees	$1,483	
Fuel (Hwy 25 City 17)	$4,145	○
Maintenance	$5,105	○
Repairs	$740	○

Warranty/Maintenance Info

Major Tune-Up	$161	○
Minor Tune-Up	$71	○
Brake Service	$173	○
Overall Warranty	3 yr/50k	◐
Drivetrain Warranty	3 yr/50k	◐
Rust Warranty	8 yr/unlim. mi	○
Maintenance Warranty	N/A	
Roadside Assistance	3 yr/unlim. mi	

Resale Value

1993	1994	1995	1996	1997
$23,757	$21,507	$19,767	$17,924	$16,120

Ownership Costs (5yr)

Average	This Car
$46,881	$47,587
Cost/Mile 67¢	Cost/Mile 68¢

Cumulative Costs

	1993	1994	1995	1996	1997
Annual	$19,118	$7,055	$7,052	$5,576	$8,786
Total	$19,118	$26,173	$33,225	$38,801	$47,587

Ownership Cost Rating: Average

The 1993 960 is available in two bodystyles - sedan and wagon. New for 1993, the 960 wagon receives a passenger side supplemental restraint system (SRS), which consists of an airbag and a knee bolster. The rear seat of the wagon features a 3-point safety belt and a head restraint for the center passenger. The new rear seat incorporates a higher backrest, along with repositioned seat controls for easier folding of the seat. The side impact protection system is standard.

---

* Includes shaded options
** Other purchase requirements apply

Legend: ● Poor · ◐ Worse Than Average · ○ Average · ◯ Better Than Average · ◯ Excellent · ⊖ Insufficient Information

©1993 by IntelliChoice, Inc. (408) 554-8711 All Rights Reserved. Reproduction Prohibited.
Refer to Section 3: Annotated Vehicle Charts for an explanation of these charts.

Section Five
Appendices

Appendix A

The Complete Car Cost Guide

Best Overall Value

Subcompact

Under $12,000

Nissan Sentra E
4 Door 2WD Sedan
(2 Door Model Shown)

Suzuki Swift GS
4 Door 2WD Sedan

Mazda 323 SE
2 Door 2WD Hatchback
(Base Model Shown)

Over $12,000

Mitsubishi Eclipse GS
2 Door 2WD Coupe

Nissan 240 SX
2 Door 2WD Coupe

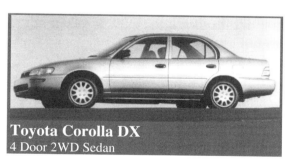
Toyota Corolla DX
4 Door 2WD Sedan

This appendix lists the best economic values in vehicles in the above class and price ranges. The best value does not mean the lowest cost to own, but rather the best relationship between the cost to own and the price to buy.
Note that we've only shown one member of a model family, even when multiple members are excellent.

Appendices

The Complete Car Cost Guide

Appendix A *Best Overall Value*

Compact

Under $16,500

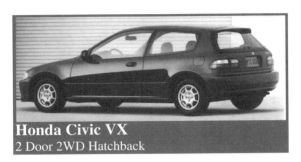
Honda Civic VX
2 Door 2WD Hatchback

Saturn SL
4 Door 2WD Sedan

Acura Integra RS
2 Door 2WD Hatchback

Over $16,500

Infiniti G20
4 Door 2WD Sedan

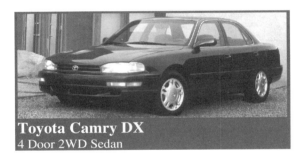
Toyota Camry DX
4 Door 2WD Sedan

Volvo 240
4 Door 2WD Sedan

Honda Prelude S
2 Door 2WD Coupe

This appendix lists the best economic values in vehicles in the above class and price ranges. The best value does not mean the lowest cost to own, but rather the best relationship between the cost to own and the price to buy.
Note that we've only shown one member of a model family, even when multiple members are excellent.

Appendices

Appendix A

The Complete Car Cost Guide

Best Overall Value

Midsize

Under $20,000

Honda Accord DX
2 Door 2WD Coupe

Mercury Sable GS
4 Door 2WD Sedan

Ford Taurus GL
4 Door 2WD Sedan

Over $20,000

BMW 318 i
4 Door 2WD Sedan

Volvo 940 Turbo
4 Door 2WD Sedan

This appendix lists the best economic values in vehicles in the above class and price ranges. The best value does not mean the lowest cost to own, but rather the best relationship between the cost to own and the price to buy.
Note that we've only shown one member of a model family, even when multiple members are excellent.

Appendices

Appendix A

The Complete Car Cost Guide

Best Overall Value

Large

Mercury Grand Marquis LS
4 Door 2WD Sedan

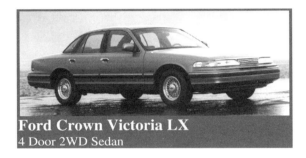

Ford Crown Victoria LX
4 Door 2WD Sedan

Mazda 929
4 Door 2WD Sedan

This appendix lists the best economic values in vehicles in the above class and price ranges. The best value does not mean the lowest cost to own, but rather the best relationship between the cost to own and the price to buy.
Note that we've only shown one member of a model family, even when multiple members are excellent.

Appendix A

The Complete Car Cost Guide

Best Overall Value

Luxury

Under $40,000

Mercedes Benz 190 E 2.3
4 Door 2WD Sedan

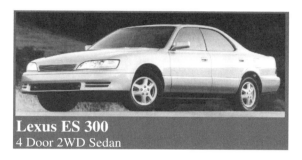
Lexus ES 300
4 Door 2WD Sedan

Acura Legend L
4 Door 2WD Sedan

BMW 525 i
4 Door 2WD Sedan

Over $40,000

Mercedes Benz 300 E
4 Door 2WD Sedan

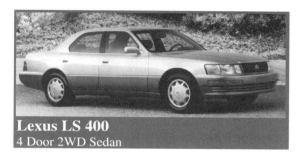
Lexus LS 400
4 Door 2WD Sedan

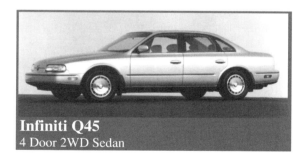
Infiniti Q45
4 Door 2WD Sedan

This appendix lists the best economic values in vehicles in the above class and price ranges. The best value does not mean the lowest cost to own, but rather the best relationship between the cost to own and the price to buy.
Note that we've only shown one member of a model family, even when multiple members are excellent.

Appendix A

The Complete Car Cost Guide

Best Overall Value

Sport

Under $25,000

Mazda MX-5 Miata
2 Door 2WD Convertible

Volkswagen Corrado SLC
2 Door 2WD Coupe

Over $25,000

Porsche 911 Carrera 2 Cabriolet
2 Door 2WD Convertible

Mitsubishi 3000GT SL
2 Door 2WD Coupe

This appendix lists the best economic values in vehicles in the above class and price ranges. The best value does not mean the lowest cost to own, but rather the best relationship between the cost to own and the price to buy.
Note that we've only shown one member of a model family, even when multiple members are excellent.

Appendices

The Complete Car Cost Guide

Appendix A

Best Overall Value

Subcompact/Compact Wagon

Under $15,000

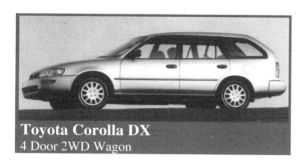

Toyota Corolla DX
4 Door 2WD Wagon

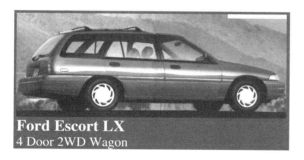

Ford Escort LX
4 Door 2WD Wagon

Over $15,000

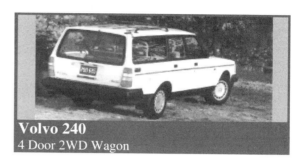

Volvo 240
4 Door 2WD Wagon

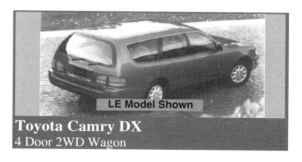

LE Model Shown

Toyota Camry DX
4 Door 2WD Wagon

This appendix lists the best economic values in vehicles in the above class and price ranges. The best value does not mean the lowest cost to own, but rather the best relationship between the cost to own and the price to buy.
Note that we've only shown one member of a model family, even when multiple members are excellent.

Appendices

The Complete Car Cost Guide

Appendix A
Best Overall Value

Midsize Wagon

Under $20,000

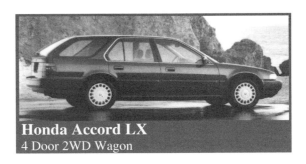
Honda Accord LX
4 Door 2WD Wagon

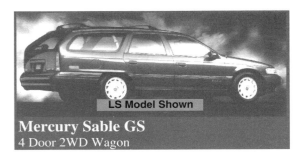
LS Model Shown
Mercury Sable GS
4 Door 2WD Wagon

Over $20,000

Volvo 940 Turbo
4 Door 2WD Wagon

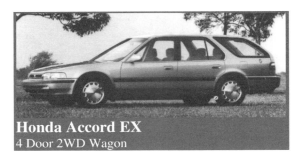
Honda Accord EX
4 Door 2WD Wagon

This appendix lists the best economic values in vehicles in the above class and price ranges. The best value does not mean the lowest cost to own, but rather the best relationship between the cost to own and the price to buy.
Note that we've only shown one member of a model family, even when multiple members are excellent.

Appendix A

The Complete Car Cost Guide

Best Overall Value

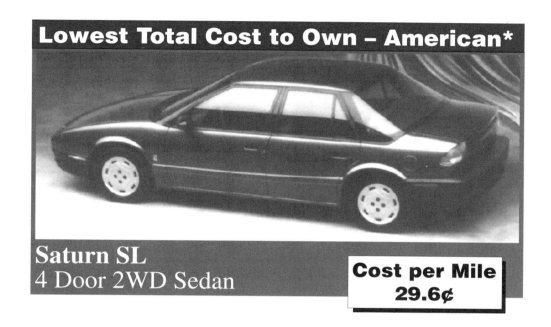

Lowest Total Cost to Own – American*

Saturn SL
4 Door 2WD Sedan

Cost per Mile 29.6¢

Lowest Total Cost to Own – Import**

Honda Civic CX
2 Door 2WD Hatchback

Cost per Mile 25.8¢

* An American vehicle is one determined to be domestically produced by the U.S. Environmental Protection Agency, and whose parent company is headquartered in the United States.
** An Import vehicle is one determined to be imported by the U.S. Environmental Protection Agency, or whose parent company is headquartered outside of the United States.

Appendices

Appendix A

The Complete Car Cost Guide

Best Overall Value

Best American* Car Value — under $13,000

Saturn SL
4 Door 2WD Sedan

* An American vehicle is one determined to be domestically produced by the U.S. Environmental Protection Agency, and whose parent company is headquartered in the United States.

The best value does not mean the lowest cost to own, but rather the best relationship between the cost to own and the price to buy.

Appendix A

Best Overall Value

Best American* Car Value — over $13,000

Buick Park Avenue
4 Door 2WD Sedan

* An American vehicle is one determined to be domestically produced by the U.S. Environmental Protection Agency, and whose parent company is headquartered in the United States.

The best value does not mean the lowest cost to own, but rather the best relationship between the cost to own and the price to buy.

Appendix A

The Complete Car Cost Guide

Best Overall Value

Best Import* Car under $13,000

Honda Civic VX
2 Door 2WD Hatchback

* An Import vehicle is one determined to be imported by the U.S. Environmental Protection Agency, or whose parent company is headquartered outside of the United States.

The best value does not mean the lowest cost to own, but rather the best relationship between the cost to own and the price to buy.

Appendices

Appendix A

The Complete Car Cost Guide

Best Overall Value

Best Import* Car   **over $13,000**

Infiniti G20
4 Door 2WD Sedan

* An Import vehicle is one determined to be imported by the U.S. Environmental Protection Agency, or whose parent company is headquartered outside of the United States.

The best value does not mean the lowest cost to own, but rather the best relationship between the cost to own and the price to buy.

Appendices

Appendix B

Lowest/Highest Maintenance Cost

The vehicles listed below have the lowest and highest costs in the maintenance area. The dollar figure to the right of each vehicle represents the total expected maintenance cost over a five-year period.

Lowest		Category	Highest	
Ford Escort LX	$3,417	Subcompact	Mazda MX-3 GS	$5,991
Dodge Daytona	$3,550		Toyota Corolla LE	$5,752
Ford Festiva L	$3,644		Mitsubishi Eclipse GSX 16V Tu. AWD	$5,642
Infiniti G20	$3,324	Compact	Toyota Celica All-Trac Turbo	$6,713
Volvo 240 sedan	$3,346		Honda Prelude Si	$6,329
Saturn SL	$3,471		Toyota Camry DX V6	$5,866
Volvo 850 GLT	$3,679	Midsize	Saab 9000 CD Turbo	$6,811
Volvo 940 Turbo sedan	$3,850		Mitsubishi Diamante LS	$6,296
Chrysler New Yorker Fifth Avenue	$4,086		Buick Regal Gran Sport series	$5,612
Chevrolet Caprice Classic	$3,792	Large	Acura Vigor series	$6,480
Ford Thunderbird LX	$4,001		Pontiac Bonneville SSEi	$6,474
Ford Crown Victoria	$4,003		Mazda 929	$6,298
Audi 100	$3,212	Luxury	Alfa Romeo 164 S	$8,872
Audi 90 S	$3,385		Cadillac Seville Touring Sedan	$7,737
Infiniti J30	$3,493		Jaguar XJ-S coupe	$7,169
BMW 525i	$3,721			
Mazda MX-5 Miata	$4,619	Sport	Porsche 928 GTS	$10,265
Dodge Daytona IROC	$4,674		Chevrolet Corvette ZR-1	$9,521
Alfa Romeo Spider	$4,983		Dodge Stealth R/T Turbo AWD	$8,228
Ford Escort LX	$3,417	Subcompact Wagon	Toyota Corolla DX	$5,463
Subaru Loyale 4WD	$4,028			
Volvo 240 Wagon	$3,394	Compact Wagon	Toyota Camry LE V6	$5,828
Saturn SW1	$3,617		Subaru Legacy Touring Wagon AWD	$5,686
Chevrolet Cavalier VL	$4,060		Mitsubishi Expo LRV AWD	$5,169
Volvo 940 Turbo	$3,850	Midsize Wagon	Mitsubishi Expo SP AWD	$5,196
Oldsmobile Cutlass Ciera Cruiser S	$4,171		Honda Accord EX	$5,061
Ford Taurus GL	$4,173		Buick Century series	$4,611

Appendix B

Lowest/Highest Fuel Cost

The vehicles listed below have the lowest and highest costs in the fuel area. The number to the right of each vehicle represents the expected fuel costs over a five year period, based on highway and city fuel estimates reported by the Environmental Protection Agency, and the assumptions described in this book.

Lowest

Vehicle	Cost	Category	Vehicle	Cost
Geo Metro XFi	$1,561	Subcompact	Volkswagen Jetta GLX III	$4,031
Suzuki Swift GA	$2,111		Plymouth Laser RS Turbo AWD	$3,837
Ford Festiva series	$2,243		Eagle Talon TSi Turbo AWD	$3,837
			Mitsubishi Eclipse GSX 16V Turbo AWD	$3,837
Honda Civic VX	$1,819	Compact	Toyota Camry DX V6	$4,547
Pontiac LeMans Value Leader Aero.	$2,659		Saab 900 S family	$4,430
Saturn SL	$2,659		Volkswagen Passat GLX	$4,232
Ford Taurus GL	$3,405	Midsize	Mitsubishi Diamante series	$4,547
Mercury Sable GS	$3,405		Saab 9000 CS	$4,117
Honda Accord series	$3,454		Volvo 940 Turbo	$4,111
Eagle Vision ESi	$3,612	Large	Ford Thunderbird Super Coupe	$4,777
Buick LeSabre Limited	$3,704		Buick Roadmaster	$4,272
Oldsmobile Eighty-Eight Royale LS	$3,704		Mercury Cougar XR7	$4,232
Pontiac Bonneville SE	$3,704			
Mercedes Benz 300 D 2.5 Turbo	$3,309	Luxury	Mercedes Benz 600 SEL	$6,821
Buick Park Avenue	$3,775		BMW 750 iL	$6,420
Oldsmobile Ninety-Eight series	$3,775		Mercedes Benz 500 SEL	$6,362
Mazda MX-5 Miata	$3,198	Sport	Porsche 928 GTS	$6,250
Toyota MR2	$3,390		BMW 850 Ci	$5,813
Alfa Romeo Spider	$3,677		BMW M5	$5,722
Ford Escort LX	$2,656	Subcompact Wagon	Subaru Loyale 4WD	$3,258
Mercury Tracer	$2,656			
Saturn SW1	$2,890	Compact Wagon	Toyota Camry LE V6	$4,547
Chevrolet Cavalier VL	$3,152		Volkswagen Passat GLX	$4,232
Mitsubishi Expo LRV	$3,385		Subaru Legacy Touring Wagon AWD	$4,213
Ford Taurus GL	$3,405	Midsize Wagon	Mitsubishi Expo SP AWD	$4,111
Mercury Sable GS	$3,405		Volvo 940 Turbo	$4,111
Honda Accord series	$3,598		Buick Century Special	$3,847

Appendix B

Lowest/Highest Insurance Cost

The vehicles listed below have the lowest and highest costs in the insurance area. The number to the right of each vehicle represents the expected insurance cost over a five-year period.

Lowest		Category	Highest	
Geo Prizm	$6,333	Subcompact	Mitsubishi Eclipse GSX 16V Turbo AWD	$11,732
Dodge Shadow/Plymouth Sundance	$6,919		Eagle Talon TSi Turbo AWD	$11,344
Toyota Corolla/Hyundai Elantra	$6,919		Nissan 240SX/VW Cabriolet Classic	$11,167
Dodge Spirit Highline	$6,333	Compact	Saab 900 Turbo	$11,704
Honda Civic CX	$6,520		Ford Mustang LX	$10,706
Saturn SL, SL1/Chevrolet Cavalier VL	$6,520		Toyota Celica All-Trac Turbo	$10,706
Plymouth Acclaim/Dodge Dynasty	$6,520			
Ford Taurus GL	$6,520	Midsize	BMW 318 iS	$10,706
Mercury Sable GS/Pont. Gr. Prix LE	$6,658		Saab 9000 CD Turbo	$9,547
Buick Cent. Spec./Oldsmobile Cut. Ciera S	$6,658		Chevrolet Lumina Z34	$9,227
Pontiac Bonneville SE	$7,035	Large	Eagle Vision TSi	$9,673
Chevrolet Caprice Classic	$7,035		Dodge Intrepid ES	$9,232
Buick LeSabre Custom	$7,040		Mazda 929	$8,784
Oldsmobile Ninety-Eight Regency	$7,035	Luxury	Mercedes Benz 500 SL	$20,170
Buick Park Avenue	$7,040		Cadillac Allante HT	$18,179
Saab 9000 CDE	$7,650		Jaguar XJS	$16,386
Dodge Daytona IROC	$8,784	Sport	Porsche 968 Cabriolet	$19,200
Ford Mustang LX 5.0L	$9,453		Porsche Carrera series	$18,179
Dodge Stealth	$9,777		Acura NSX	$18,179
Ford Escort LX	$7,223	Subcompact Wagon	Mercury Tracer	$7,693
Toyota Corolla DX	$7,223			
Subaru Loyale	$7,223			
Saturn SW1	$6,658	Compact Wagon	Volkswagen Passat GLX	$8,507
Chevrolet Cavalier VL	$6,658		Subaru Legacy LS AWD	$8,238
Mitsubishi Expo LRV/Toyota Camry DX	$6,919		Plymouth Colt Vista AWD	$7,479
Eagle Summit DL, LX	$6,919			
Ford Taurus GL	$6,658	Midsize Wagon	Volvo 940 Turbo	$7,898
Mercury Sable GS	$6,786		Honda Accord EX	$7,898
Oldsmobile Cutlass Ciera Cruiser S	$6,786		Mitsubishi Expo SP AWD	$7,693
Buick Century Special	$6,786			

Appendix B

Lowest/Highest Resale Value

The vehicles listed below have either the highest or lowest resale values after a five-year period. The percentage to the right of each vehicle represents how much of its original value the vehicle will retain at the end of five years. Resale value is based on the car's rate of depreciation. Vehicles with better resale values have slower rates of depreciation.

Highest		Category	Lowest	
Suzuki Swift GT	67%	Subcompact	Dodge Daytona ES	37%
Nissan Sentra E 2 door	64%		Hyundai Excel	41%
Plymouth Laser	63%		Mercury Tracer LTS	41%
Honda Civic CX	72%	Compact	Hyundai Sonata GLS	33%
Acura Integra RS Hatchback	68%		Pontiac LeMans Value Leader	33%
Toyota Camry DX	66%		Chrysler LeBaron Landau	37%
Honda Accord DX Coupe	69%	Midsize	Saab 9000 CD	36%
BMW 318 i	61%		Oldsmobile Cutlass Supreme Int'l.	37%
Chevrolet Lumina Euro	54%		Chrysler New Yorker Fifth Avenue	37%
Buick LeSabre Custom	51%	Large	Chevrolet Caprice Classic	36%
Mazda 929	50%		Ford Thunderbird LX	40%
Oldsmobile Eighty-Eight Royale	48%		Mercury Cougar XR7	41%
Mercedes Benz 300 SL	76%	Luxury	Chrysler Imperial	34%
Mercedes Benz 300 E	66%		Audi 90 S	34%
Lexus ES 300	66%			
Lexus LS 400	65%			
Porsche 911 Carrera 2 Cabriolet	75%	Sport	Alfa Romeo Spider	37%
Mazda MX-5 Miata	60%		Dodge Daytona IROC	38%
Toyota MR2	58%		Nissan 300ZX Turbo	42%
Chevrolet Corvette Convertible	56%			
Toyota Corolla DX	53%	Subcompact Wagon	Mercury Tracer	44%
Volvo 240	64%	Compact Wagon	Chevrolet Cavalier RS	42%
Toyota Camry DX	63%		Subaru Legacy LS AWD	46%
Volkswagen Passat GLX	56%		Mitsubishi Expo LRV AWD	48%
Honda Accord LX	64%	Midsize Wagon	Oldsmobile Cutlass Ciera SL	44%
Mercury Sable LS	57%		Buick Century Custom	46%
Ford Taurus GL	55%			
Volvo 940 Turbo	55%			

Appendix C

Financing Sources

When you consider a new car loan, it really pays to shop around and compare rates. (The same is also true for leasing!) Commercial banks, savings banks, credit unions, and even auto manufacturers offer a variety of loan terms for new cars. You probably know of a dozen or more local institutions willing to loan you money — but do you know which one has the lowest rate?

The chart below lists for each major metropolitan area the bank that has at one time offered the lowest rate among competitive banks in that area. Keep in mind, however, that rates change frequently, so there is no guarantee that the institution shown below currently offers the lowest rate in your area. While it is wise to consider the institutions listed here, make sure you also get price quotes from other financial institutions.

A Sample of Competitive New Auto Loans in Major Markets Around the Country
Based on $10,000 loan

Region	Institution	Recent rate	Term (months)
Atlanta	Trust Company Bank	7.90	48
Baltimore	Mercantile Safe Deposit	8.00	60
Boston	Bank of Boston	8.90	60
Chicago	Cole Taylor Bank	7.50	48
Cleveland	Society National Bank	7.90	60
Dallas	NationsBank	7.69	36
Denver	Bank Western	7.90	48
Detroit	First of America Bank	7.25	60
Houston	Texas Commerce Bank	7.79	36
Los Angeles	Community Bank	8.25	60
Miami	United National Bank	8.50	48
Minneapolis	National City Bank	8.25	36
New York City	Marine Midland Bank	8.25	48
N. New Jersey	Columbia Federal Savings	8.75	36
Philadelphia	Meridian Bank	8.75	48
Phoenix	Republic National Bank	8.25	60
Pittsburgh	Three Rivers Bank	8.00	60
St. Louis	Pulaski Bank	8.25	48
San Diego	Escondido National Bank	8.90	48
San Francisco	Sumitomo Bank	7.75	48
Seattle	Seafirst Bank	8.25	60
S. W. Connecticut	Union Savings Bank	7.75	60
Tampa	Sun Bank of Tampa Bay	7.50	48
Washington, D. C.	Chevy Chase Federal Savings	7.50	48

Rates subject to change without notice.
Source: HSH Associates, 1200 Route 23, Butler NJ 07457. © 1993, HSH Associates

Appendix D

Insurance Companies

Insurance is one of the largest of the seven ownership cost areas discussed throughout this book. If you live in a non-competitive insurance state (Massachusetts or Texas) then the rates are controlled by an insurance commission or rating bureau, and all insurers are required to charge the same premium. However, in all other states the rates are determined by each insurance company and may vary dramatically from one company to the next. Therefore, if you live in a competitive insurance state, it really pays to shop for the lowest rates.

The ten insurance companies listed below are the largest private passenger car insurers in the United States. To ensure that you are getting the most competitive rate, obtain quotes from several of the companies listed here, and compare them to smaller companies or independent agents. You can find a listing of insurance agents in your area by looking in the yellow pages under "Insurance." To help you make a side-by-side evaluation of different insurers, use the Insurance Comparison Worksheet on page 11A.

Although there are many differences between insurance companies, one thing you should seriously evaluate is the company's reputation when it comes to processing claims. Before you buy, it makes sense to check with a few local repair or body shops to find out which insurers have the best and the worst reputation. Also, you may want to check with the Council of Better Business Bureaus (Appendix on page 90A).

The following is a listing of the top 10 auto insurers:

Aetna Life & Casualty Insurance
151 Farmington Avenue
Hartford, CT 06156
(203) 273-0123

Allstate Insurance
Allstate Plaza
Northbrook, IL 60062
(708) 402-5000

Farmers Insurance
4680 Wilshire Boulevard
Los Angeles, CA 90010
(213) 930-4019

Geico Corporation Insurance
One GEICO Plaza
Washington, D.C. 20076
(301) 986-3000

ITT Hartford
Hartford Plaza
Hartford, CT 06115
(203) 547-5000

Liberty Mutual Insurance
175 Berkeley Street
Boston, MA 02117
(800) 225-2390

Nationwide Insurance
One Nationwide Plaza
Columbus, OH 43216
(800) 882-2822

State Farm Insurance
One State Farm Plaza
Bloomington, IL 61701
(309) 766-2311

Travelers Insurance
One Tower Square
Hartford, CT 06183
(203) 277-0111

USAA Insurance
USAA Building
San Antonio, TX 78288
(800) 531-8100

Appendix E

Service Contract Locator

If you decide to purchase an Extended Service Contract, your choices may be limited. In most cases, the dealer will offer you a contract that is backed by the manufacturer. He may also offer a contract from an independent provider.

You probably won't choose a dealer based on the Extended Service Contract he sells. However, if you are considering an Extended Service Contract that is offered by a provider other than the manufacturer, you may wish to learn more about that company's reputation. One source for this information is the Council of Better Business Bureaus (Appendix F). Also, you can directly contact the contract provider.

The following is a listing of the major independent Service Contract providers, and the insurance company that underwrites their policies:

GE Capital
Insurance Services Group
P.O. Box 14159
Denver, CO 80214-4159
(800) 445-4065

Backed by: Lexington Insurance
Other Info: In national service contract business approximately 5 years

General Insurance Administrators
16501 Ventura Blvd.
Suite 200
Encino, CA 91436
(818) 990-9590
(800) 242-9442

Backed by: Fireman's Fund/Western General Insurance Company
Other Info: In national service contract business approximately 4 years

JM&A Group
P.O. Box 1160
Deerfield Beach, FL 33443
(305) 429-2333

Backed by: Virginia Surety
Other Info: Available in 40 states

Ryan Warranty Services
123 N. Wacker Drive
Chicago, IL 60606
(312) 701-3700

Backed by: Virginia Surety
Other Info: In national service contract business approximately 14 years

Western National Warranty Corporation
4141 N. Scottsdale Road
Scottsdale, AZ 85252-2840
(800) 345-0191

Backed by: Continental Insurance Company
Other Info: In national service contract business approximately 11 years

Appendix F

Better Business Bureau Locations

Acquiring a new car usually involves many more business transactions than just the car deal. You will need insurance and financing, and you may deal with an extended service contract administrator. And after you've purchased your car, you'll require the services of many more businesses — service stations, automobile clubs, car washes, and repair shops.

A key element to a satisfying overall purchase and ownership experience is ensuring that the related businesses you require service from are credible, honest, and deliver good value.

The Council of Better Business Bureaus (BBB) can help. While it is still up to you to choose your service provider, the BBB does keep a log of complaints registered against businesses in your community. You can consult this log to help determine which businesses you'd rather not associate with.

If the company you want information about is not headquartered in your community, your local BBB office may be able to refer you to the BBB Office in the community of the business you're inquiring about.

The following is a listing of BBB offices and phone numbers:

UNITED STATES BUREAUS

ALABAMA

BIRMINGHAM, AL 35205
1210 South 20th Street
P.O. Box 55268 (35255-5268) 205/558-2222
DOTHAN, AL 36301
118 Woodburn Street 205/792-3804
HUNTSVILLE, AL 35801
501 Church Street, N.W.
P.O. Box 383 (35804) .... (24 hrs.)205/533-1604
MOBILE, AL 36602-3221
707 Van Antwerp Building 205/433-5494, 95
So. AL 800/554-4174
MONTGOMERY, AL 36104-3559
Union Bank Building, Commerce St., Suite 806 .... 205/262-5606

ALASKA

ANCHORAGE, AK 99503-5701
4011 Arctic Blvd.,#206 .... 907/562-0704

ARIZONA

PHOENIX, AZ 85014-4585
4428 North 12th Street 602/264-1721
TUCSON, AZ 85705-7353
50 W. Drachman St., Suite 103 .... Inq.602/622-7651
Comp. 622-7654
So. AZ only 800/696-2827

ARKANSAS

LITTLE ROCK, AR 72204-2605
1415 South University 501/664-7274

CALIFORNIA

BAKERSFIELD, CA 93301-4882
705 Eighteenth Street 805/322-2074
COLTON, CA 92324-0814
290 N. 10th St. Suite 206, P.O. Box 970 .... 714/825-7280
CYPRESS, CA 90630-3966
6101 Ball Road, Suite 309 .... Inq. & Comp. 714/527-0680
FRESNO, CA 93705-0341
1398 W. Indianapolis, #102 .... 209/222-8111
LOS ANGELES, CA 90020
3400 West 6th St., Suite 403 .... 213/251-9696
MONTEREY, CA 93940-2717
494 Alvarado St., Suite C 408/372-3149
OAKLAND, CA 94612-1564
510 16th St, Ste. 550 .... (24 hrs) 510/238-1000
SACRAMENTO, CA 95814-6997
400 S Street 916/443-6843
SAN DIEGO, CA 92108-1729
3111 Camino del Rio, N., Suite 600 .... (24 hrs) 619/521-5898
SAN FRANCISCO, CA 94105-4506
33 New Montgomery St. Tower, #290 .... 415/243-9999
SAN JOSE, CA 95125-5316
1505 Meridian Ave., Suite C 408/978-8700

CALIFORNIA Continued

SAN MATEO, CA 94402
400 S. El Camino Real, #450
P.O. Box 294 (94401-0294) 415/696-1240
SANTA BARBARA, CA 93101-0746
402 E. Carrillo St., Suite C, P.O. Box 746 .... 805/963-8657
SANTA ROSA, CA 95401-8541
300 B Street 707/577-0300
STOCKTON, CA 95202-1383
1111 North Center Street 209/948-4880, 81

COLORADO

COLORADO SPRINGS, CO 80907-5454
3022 North El Paseo, P.O. Box 7970 (80933-7970) . 719/636-1155
DENVER, CO 80222-4350
1780 S. Bellaire, Suite 700 .... (24 hrs.) Inq. 303/758-2100
.... Comp. 303/758-2212
FORT COLLINS, CO 80525-1073
1730 S. College Avenue, Suite 303 .... 303/484-1348
  S. WY only 800/878-3222
PUEBLO, CO 81003-3119
119 W. 6th St., Suite 203 .... 719/542-6464

CONNECTICUT

FAIRFIELD, CT 06430-3267
Fairfield Woods Plaza, 2345 Black Rock Turnpike
P.O. Box 1410 (06430-1410) 203/374-6161
WALLINGFORD, CT 06492-4395
100 S. Turnpike Rd .... Inq. 203/269-2700
  Comp. 269-4457

DELAWARE

WILMINGTON, DE 19808-5532
2055 Limestone Rd., Ste. 200 .... 302/996-9200

DISTRICT OF COLUMBIA

1012 14th St, N.W, 14th Floor (20005-3410) 202/393-8000

FLORIDA

CLEARWATER, FL 34620
13770 - 58th St, North, #309
P.O BOX 7950 (34618-7950) 813/535-5522
FORT MYERS, FL 33901-6003
2976-E Cleveland Ave 813/334-7331
  334-7152
JACKSONVILLE, FL 32216-2756
3100 University Blvd., South, Suite 239 .... 904/721-2288
MAITLAND; FL 32751-7147 (Orlando)
2605 Maitland Center Parkway 407/660-9500
MIAMI, FL 33014-6709
16291 N.W. - 57th Avenue .... Inq. 305/625-0307
  Comp. 625-1302
NEW PORT RICHEY, FL 34652
250 School Road Suite 11-W 813/842-5459
PENSACOLA, FL 32501
400 S. Alcaniz Street
P.O. Box 1511 (32597-1511) 904/433-6111
PORT ST. LUCIE, FL 34952
1950 Pt. St. Lucie Blvd., Suite 211 .... 407/878-2010
TAMPA, FL 33607
1111 N. Westshore Blvd, Suite 207 ..... Inq. & Comp. 813/854-1154
WEST PALM BEACH, FL 33409-3408
2247 Palm Beach Lakes Blvd., Suite 211 .... 407/686-2200

GEORGIA

ALBANY, GA 31707
611 N. Jefferson St, P.O. Box 3241 (31706-3241) 912/883-0744
ATLANTA, GA 30303-3075
100 Edgewood Avenue, Suite 1012 .... 404/688-4910
AUGUSTA, GA 30901-1463
624 Ellis St., Suite 106
P.O. Box 2085 (30903-2085) 404/722-1574
COLUMBUS, GA 31901-2151
8 13th Street, P.O. Box 2587 (31902-2587) 404/324-0712, 13
MACON, GA 31211-2499
1765 Shurling Drive 912/742-7999
SAVANNAH, GA 31405
6606 Abercorn Street Suite 108-C
P.O. Box 13956 (31416-0956) 912/354-7521

HAWAII

HONOLULU, HI 96814-3801
1600 Kapiolani Blvd., Suite 714 .... 808/942-2355

IDAHO

BOISE, ID 83702-5320
1333 West Jefferson 208/342-4649
  T-F ID only 800/339-8737
IDAHO FALLS, ID 83402-5026
1547 So. Blvd 208/523-9754

ILLINOIS

CHICAGO, IL 60606
211 West Wacker Drive .... Inq. 312/444-1188
  Comp. 346-3313
PEORIA, IL 61615-3770
3024 West Lake 309/688-3741
ROCKFORD, IL 61104
810 E. State St., 3rd Fl. 815/963-2222

The Complete Car Cost Guide

Appendix F *continued* — Better Business Bureau Locations

INDIANA

ELKHART, IN 46514-2988
722 W. Bristol St., Suite H-2
P.O. Box 405 (46515-0405) 219/262-8996
EVANSVILLE, IN 47715-2265
4004 Morgan Ave., Suite 201 .... 812/473-0202
FORT WAYNE, IN 46802-3493
1203 Webster Street 219/423-4433
  IN only 800-552-4631
GARY, IN 46408-2490
4231 Cleveland Street 219/980-1511
  No. IN only 800/637-2118
INDIANAPOLIS, IN 46204-3584
Victoria Centre, 22 E. Washington Street,
Suite 200 .... 317/488-2222
SOUTH BEND, IN 46637-4200
52303 Emmons Road, Suite 9 .... 219/277-9121
  No. IN only 800/439-5313

IOWA

BETTENDORF, IA 52722-4100
852 Middle Road., Suite 290 .... 319/355-6344
DES MOINES, IA 50309-2375
615 Insurance Exchange Building 515/243-8137
SIOUX CITY, IA 51101
318 Badgerow Building 712/252-4501

KANSAS

TOPEKA, KS 66607-1190
501 Jefferson, Suite 24 .... 913/232-0454
WICHITA, KS 67202-3857
212 S. Market St., #300 .... 316/263-3146

KENTUCKY

LEXINGTON, KY 40507-1203
311 W. Short Street 606/259-1008
LOUISVILLE, KY 40203-2186
844 S. Fourth Street 502/583-6546

LOUISIANA

ALEXANDRIA, LA 71301-6875
1605 Murray St., Suite 117 .... 318/473-4494
BATON ROUGE, LA 70806-1546
2055 Wooddale Blvd 504/926-3010
HOUMA, LA 70360-4455
501 E. Main St. 504/868-3456
LAFAYETTE, LA 70506
100 Huggins Rd., P.O. Box 30297 (70593-0297) 318/981-3497
LAKE CHARLES, LA 70605
3941-L Ryan St. P.O. Box 7314 (70606-7314) 318/478-6253
MONROE, LA 71201-7380
141 De Siard Street, Suite 808 .... 318/387-4600
NEW ORLEANS, LA 70130-5843
1539 Jackson Avenue, #400 .... (24 hrs) 504/581-6222
SHREVEPORT, LA 71105-2122
3612 Youree Dr. 318/861-6417

MAINE

PORTLAND, ME 04103-2648
812 Stevens Avenue 207/878-2715

MARYLAND

BALTIMORE, MD 21211-3215
2100 Huntingdon Avenue 301/347-3990

MASSACHUSETTS

BOSTON, MA 02116-4404
20 Park Plaza Suite 820 .... Inq. 617/426-9000
  802 Area only 800/4BBB-811
SPRINGFIELD, MA 01103-1402
293 Bridge Street, Suite 320 .... 413/734-3114
WORCESTER, MA 01608
32 Franklin Street, P.O. Box 379 (01601-0379) 508/755-2548

MICHIGAN

GRAND RAPIDS, MI 49503-3001
620 Trust Building 616/774-8236
SOUTHFIELD, MI 48076-7751 (Detroit)
30555 Southfield Road, Suite 200 .... Inq. 313/644-1012
  Comp. 644-9136

MINNESOTA

MINNEAPOLIS-ST. PAUL, MN 55116-2600
2706 Gannon Road 612/699-1111

MISSISSIPPI

JACKSON, MS 39206-3088
460 Briarwood Drive, Suite 340 .... 601/956-8282

MISSOURI

KANSAS CITY, MO 64106-2418
306 E. 12th Street, Suite 1024 .... 816/421-7800
ST. LOUIS, MO 63110-1400
5100 Oakland, Suite 200 .... Inq. 314/531-3300
SPRINGFIELD, MO 65806-1326
205 Park Central East, Suite 509 .... 417/862-9231

The Complete Car Cost Guide

NEBRASKA

LINCOLN, NE 68504-3491
719 North 48th Street 402/467-5261
OMAHA, NE 68102-2158
1613 Farnam Street, #417 .... 402/346-3033

NEVADA

LAS VEGAS, NV 89104-1515
1022 E. Sahara Avenue 702/735-6900
702/735-1969
RENO, NV 89502
991 Bible Way, P.O. Box 21269 (89515-1269) 702/322-0657

NEW HAMPSHIRE

CONCORD, NH 03301-3459
410 South Main Street 603/224-1991

NEW JERSEY

NEWARK, NJ 07102-3294
494 Broad Street 201/642-INFO
PARAMUS, NJ 07652-5291
2 Forest Avenue 201/845-4044
PARSIPPANY, NJ 07054
1300A Route #46, West, #215 .... 201/334-5990
TOMS RIVER, NJ 08753-8239
1721 Route 37 East 908/270-5577
TRENTON, NJ 08690-3596
1700 Whitehorse, Hamilton Square, Suite D-5 .... 201/588-0808
WESTMONT, NJ 08108-0303
16 Maple Avenue, P.O. Box 303 .... 609/854-8467

NEW MEXICO

ALBUQUERQUE, NM 87109-1292
4600-A Montgomery N.E., Suite 200 .... 505/884-0500
NM only 800/873-2224
FARMINGTON, NM 87401-5855
308 North Locke 505/326-6501
LAS CRUCES, NM 88005
2407 W. Picacho, Ste B-2 .... 505/524-3130

NEW YORK

BUFFALO, NY 14202-1899
346 Delaware Avenue 716/856-7180
FARMINGDALE, NY (Long Island) 11735-9998
266 Main Street 900/463-6222
NEW YORK, NY 10010-7384
257 Park Avenue, South 212/533-7500, 6200
SYRACUSE, NY 13202-2552
847 James St., #200 .... 315/479-6635
WHITE PLAINS, NY 10603-3213
30 Glenn Street 914/428-1230, 31

NORTH CAROLINA

ASHEVILLE, NC 28801-3418
801 BB&T Building 704/253-2392
CHARLOTTE, NC 28204-2626
1130 East 3rd St., Suite 400 .... (24 hrs.) 704/332-7151
CONOVER, NC 28613-9608
3305-10 16th Avenue, SE #303 .... 704/464-0372
GREENSBORO, NC 27410-4895
3608 West Friendly Avenue 919/852-4240, 41, 42
RALEIGH, NC 27604-1080
3125 Poplarwood Ct., Suite 308 .... 919/872-9240
East NC only 800/222-0950
Durham 919/688-6143
Chapel Hill 919/967-0296
WINSTON-SALEM, NC 27103-2516
2110 Cloverdale Ave., Suite 2-B 919/725-8348

OHIO

AKRON, OH 44303-2111
222 W Market Street 216/253-4590
CANTON, OH 44703-3135
1434 Cleveland Avenue, N.W 216/454-9401
OH only 800/362-0494
CINCINNATI, OH 45202-2097
898 Walnut Street 513/421-3015
CLEVELAND, OH 44115-1299
2217 East 9th Street #200 .... 216/241-7678
COLUMBUS, OH 43215-1000
1335 Dublin Street, #30A 614/486-6336
DAYTON, OH 45402-1828
40 West Fourth St, Suite 1250 .... 513/222-5825
LIMA, OH 45802-0269
121 W. High St., #370, P O. Box 269 .... 419/223-7010
TOLEDO, OH 43604-1055
425 Jefferson Avenue, Suite 909 .... 419/241-6276
YOUNGSTOWN, OH 44501-1495
1102 Mahoning Bank Building, P O. Box 1495 .... 216/744-3111

OKLAHOMA

OKLAHOMA CITY, OK 73102
17 S. Dewey .... Inq. 405/239-6081
Inq. 239-6860
Comp. 239-6083
TULSA, OK 74136-3327
6711 South Yale, Suite 230 .... 918/492-1266

OREGON

PORTLAND, OR 97205
610 SW. Alder Street, Suite 615 .... 503/226-3981
(OR/WA) 800/488-4155

Appendices 93A

Appendix F *continued*

Better Business Bureau Locations

PENNSYLVANIA

BETHLEHEM, PA 18018-5789
528 North New Street 215/866-8780
LANCASTER, PA 17602-5205
6 Marion Court 717/291-1151
Toll Free, York Co. Resident 846-2700
PHILADELPHIA, PA 19103-0297
1930 Chestnut St., P.O. Box 2297 .... 215/448-6100
PITTSBURGH, PA 15222-2578
610 Smithfield Street 412/456-2700
SCRANTON, PA 18503-2204
407 Connell Building
P.O. Box 993 (18501-0993) 717/342-9129

PUERTO RICO

SAN JUAN, PR 00936-3488
P. O. Box 363488 .... 809/756-5400

RHODE ISLAND

WARWICK, RI 02887-1300 (Providence)
Bureau Park, Box 1300 .... Inq 401/785-1212
.... Comp. 785-1213

SOUTH CAROLINA

COLUMBIA, SC 29201
1830 Bull Street, P.O. Box 8326 (29202-8326) 803/254-2525
GREENVILLE, SC 29605
113 Mills Avenue 803/242-5052
MYRTLE BEACH, SC 29577-1601
1601 Oak St., #403 .... 803/626-6881

TENNESSEE

BLOUNTVILLE, TN 37616
P.O. Box 1178 TCAS 615/323-6311
CHATTANOOGA, TN 37402-2614
1010 Market Street, #200 .... 615/266-6144
KNOXVILLE, TN 37919
2633 Kingston Pike, #2
P.O. Box 10327 (37939-0327) 615/522-2552
MEMPHIS, TN 38115
3792 South Mendenhall
P.O. Box 750704 (38175-0704) 901/795-8771
NASHVILLE, TN 37219-1778
Nations Bank Plaza, 414 Union St., #1830 .... 615/254-5872

TEXAS

ABILENE, TX 79605-5052
3300 S. 14th Street, Suite 307 .... 915/691-1533
AMARILLO, TX 79101-3408
1000 South Polk, P.O. Box 1905 (79105-1905) 806/379-6222
AUSTIN, TX 78701-3403
221 W. 6 Street, #450 .... 512/476-1616
BEAUMONT, TX 77704-2988
476 Oakland Ave., P.O Box 2988 (77701-2011) 409/835-5348
BRYAN, TX 77802-4413
4346 Carter Creek Pkwy. 409/260-2222
CORPUS CHRISTI, TX 78411-4418
4535 S. Padre Island Drive, #28 .... 512/854-2892
DALLAS, TX 75201-3093
2001 Bryan Street, Suite 850 .... 214/220-2000
EL PASO, TX 79903-4904
5160 Montano Ave., Lower Level 915/772-2727
FORT WORTH, TX 76102-3968
512 Main Street, #807 .... 817/332-7585
HOUSTON, TX 77008-1085
2707 North Loop West, Suite 900 .... 713/868-9500
LUBBOCK, TX 97401
1015 15th Street, P.O. Box 1178 (79408-1178) 806/763-0459
MIDLAND, TX 79711-0206
10100 County Rd, 118 West, P.O. Box 60206 .... 915/563-1880
.... 800/592-4433
SAN ANGELO, TX 76904
3121 Executive Dr., P.O. Box 3366 (76902-3366) 915/949-2989
SAN ANTONIO, TX 78217-5296
1800 Northeast Loop 410, Suite 400 .... 512/828-9441
TYLER, TX 75701
3600 Old Bullard Rd., #103-A
P.O. Box 6652 (75711-6652) 903/581-5704
WACO, TX 76710
6801 Sanger Avenue, Suite 125
P.O. Box 7203 (76714-7203) 817/772-7530
WESLACO, TX 78596-0069
609 Intl. Blvd., P.O Box 69 .... 512/968-3678
WICHITA FALLS, TX 76301-5079
1106 Brook Avenue 817/723-5526
.... TX only 800/388-1778

UTAH

SALT LAKE CITY, UT 84115-5382
1588 South Main Street 801/487-4656
.... UT only 800/388-1778

VIRGINIA

FREDERICKSBURG, VA 22407-4800
4022-B Plank Road 703/786-8397
NORFOLK, VA 23509-1499
3608 Tidewater Drive 804/627-5651
.... 804/851-9101 (Peninsula Area)
RICHMOND, VA 23219-2332
701 East Franklin, Suite 712 .... 804/648-0016
ROANOKE, VA 24011-1301
31 West Campbell Avenue 703/342-3455

WASHINGTON

KENNEWICK, WA 99336-3819
127 W. Canal St. 509/582-0222
SEATTLE, WA 98121-1857
2200 Sixth Avenue, #828 ....206/448-8888
(24 hrs.) 206-448-6222
SPOKANE, WA 99207-2356
E. 123 Indiana #106 ....509/328-2100
TACOMA, WA 98401-1274
1101 Fawcett Ave. #222 (98402), P.O. Box 1274 .....206/383-5561
YAKIMA, WA 98907-1584
424 Washington Mutual Bldg. (98901)
P.O. Box 1584 ....509/248-1326

WISCONSIN

MILWAUKEE, WI 53203-2478
740 North Plankinton Avenue 414/273-1600

INTERNATIONAL BUREAUS

NATIONAL HEADQUARTERS FOR CANADIAN BUREAUS

CONCORD, ONTARIO L4K 2Z5
2180 Steeles Avenue West, Suite 219 ....416/699-1248

ALBERTA

CALGARY, ALBERTA T2H 2H8
7330 Fisher Street, S.E., Suite 357 ....403/258-2920
EDMONTON, ALBERTA T5K 2L9
9707 - 110th Street 403/482-2341
Red Deer, Alberta 403/343-3200

BRITISH COLUMBIA

VANCOUVER, BC V6B 2M1
788 Beatty Street, Suite 404 ....604/682-2711
VICTORIA, BC V8W 1V7
201-1005 Langley Street 604/386-6348

MANITOBA

WINNIPEG, MANITOBA R3B 2K3
365 Hargrave Street, Room 204 ....204/943-1486

NEWFOUNDLAND

ST. JOHN'S, NEWFOUNDLAND A1E 2B6
360 Topsail Road, P.O. Box 516 (A1C 5K4) 709/364-2222

NOVA SCOTIA

HALIFAX, NOVA SCOTIA B3J 2A4
1731 Barrington Street
P.O. Box 2124 B35 3B7 ....Inq.902/422-6581
Comp. 902/422-6582

ONTARIO

HAMILTON, ONTARIO L8P 4V9
50 Bay Street, South 416/526-1111
KITCHENER, ONTARIO N2G 4L5
354 Charles Street, East 519/579-3080
LONDON, ONTARIO N6A 5C7
700 Richmond St., #402
P.O. Box 2153 (N6A 4E3) 519/673-3222
OTTAWA, ONTARIO K1P 5N2
71 Bank Street, 6th Floor 613/237-4856
ST. CATHERINES, ONTARIO L2R 3H6
11-101 King St 416/687-6686
TORONTO, ONTARIO M6P 4C7
One St. John's Rd., Suite 501 ....416/766-5744
WINDSOR, ONTARIO N9A 5K6
500 Riverside Drive West 519/258-7222

QUEBEC

MONTREAL, QUEBEC H3A 1V4
2055 Peel Street, Suite 460 ....514/286-9281
QUEBEC CITY, PQ G1R 1K2
475 rue Richelieu 418/523-2555

SASKATCHEWAN

REGINA, SASKATCHEWAN, S4N 6H4
1601 McAra Street 306/352-7601

Appendix G

Automobile Manufacturers

Acura
100 W. Alondra Blvd., Gardena, CA 90247 213-327-8280

Alfa Romeo Distributors of North America
8259 Exchange Drive, P.O. Box 598026, Orlando, FL 32859-8026 407-856-5000

Audi of America, Inc.
888 West Big Beaver Rd., Troy, MI 48007-3951 313-362-6000

BMW of North America, Inc.
300 Chestnut Ridge Rd., Livonia, MI 48150 201-307-4000

Buick Motor Division
General Motor Corp., 902 E. Hamilton Ave., Flint, MI 48550 313-236-5000

Cadillac Motor Car Division
General Motors Corp., 2860 Clark Street, Detroit, MI 48232 313-554-5067

Chevrolet Motor Division
General Motors Corp., 30007 Van Dyke Ave., Warren, MI 48090 313-492-8846

Chrysler Corp.
12000 Chrysler Drive, Highland Park, MI 48288-1919 313-956-5741

Dodge Division
12000 Chrysler Drive, Highland Park, MI 48288-1919 313-956-5741

Eagle Division
12000 Chrysler Drive, Highland Park, MI 48288-1919 313-956-5741

Ford Division
P.O. Box 43301, 300 Renaissance Center, Detroit, MI 48243 313-446-3800

Geo
General Motors Corp., 30007 Van Dyke Ave., Warren, MI 48090 313-492-8846

Honda - American Honda Motor Co., Inc.
100 W. Alondra Blvd., Gardena, CA 90247 213-327-8280

Hyundai Motor America
10550 Talbert Ave., Fountain Valley, CA 92728 714-965-3508

Infiniti Division
18501 Figueroa St., Carson, CA 90248 213-532-3111

Isuzu - American Isuzu Motors Corp.
2300 Pellissier Place, Whittier, CA 90601 213-949-0611

Jaguar Cars, Inc.
555 MacArthur Blvd., Mahwah, NJ 07430 201-818-8500

Jeep Division
12000 Chrysler Drive, Highland Park, MI 48288-1919 313-956-5741

Land Rover of North America Inc.
4390 Parliament Place, P.O. Box 1503, Lanham, MD 20706 301-731-9040

Lexus
19001 S. Western Ave., Torrence, CA 90509 213-618-4000

Lincoln Division - Ford Motor Co.
300 Renaissance Center, P.O. Box 43322, Detroit, MI 48243 313-446-4450

Mazda Motor of America, Inc.
7755 Irvine Center Drive, Irvine, CA 92718 714-727-1990

Mercedes Benz of North America, Inc.
One Mercedes Drive, Montvale, NJ 07645-0350 201-573-0600

Mercury Division - Ford Motor Co.
300 Renaissance Center, P.O. Box 43322, Detroit, MI 48243 313-446-4450

Mitsubishi Motor Sales of America, Inc.
6400 West Katella Ave., Cypress, CA 90248 714-372-6000

Nissan Motor Corp. in U.S.A.
18501 Figueroa St., Carson, CA 90248 213-532-3111

Oldsmobile Division
General Motors Corp., 920 Townsend St., Lansing, MI 48921 517-377-5000

Plymouth Division
12000 Chrysler Drive, Highland Park, MI 48288-1919 313-956-5741

Pontiac Motor Division
General Motors Corp., 1 Pontiac Plaza, Pontiac, MI 48058-3484 313-857-5000

Porsche Cars North America, Inc.
100 W. Liberty St., Reno, NV 89501 702-348-3000

Saab-Scania of America, Inc.
P.O. Box 697, Saab Drive, Orange, CT 06477 203-795-5671

Saturn Corp. - General Motors Corp.
1400 Stephenson Hwy., Troy, MI 48007-7025 313-524-5721

Subaru of America, Inc.
P.O. Box 6000, Cherry Hill, NJ 08034-6000 609-488-8500

Suzuki - American Suzuki Motor Corp.
3251 E. Imperial Hwy., Brea, CA 92621-6722 714-996-7040

Toyota Motor Sales, USA, Inc.
19001 S. Western Ave., Torrance, CA 90509 213-618-4000

Volkswagen of America, Inc.
888 Big Beaver Rd., P.O. Box 3951, Troy, MI 48099 313-362-6000

Volvo of North America Corp.
Rockleigh, NJ 07647 201-768-7300

Appendix H

Lease Interest Rates

While there are a number of differences between a lease and a loan, they are both similar transactions to a financial analyst. Both involve the exchange of a major asset (in this case, a car) between the "selling" party and the "using" party; both are for a fixed time period; both require periodic payments; and most importantly, both involve the use of an interest rate to set the periodic payment.

A loan's interest rate, or A.P.R. (Annual Percentage Rate) is the one and only true measure of the **cost** of a loan. With the APR, you can compare any number of loans on an apples-to-apples basis.

However, whereas lenders have strict disclosure requirements for automobile loans, lessors do not have the same stringent requirements to inform you of the interest rate being used to determine the monthly lease payments.

Same as with comparing APR between loans, the interest rate used to calculate the lease is the one true measure of the cost of a lease.

The following table will help you determine the interest rate in any lease transaction. You can use this table to help you compare different leases and/or compare a lease to a loan.

To use the tables, here are the things you need to know. The lessor should be able to provide you with this information.

- The acquisition price of the vehicle (Technically, since you're not purchasing the vehicle, the price is called the "capital cost" and is the price upon which the lease payments are based.)
- Any fees or "capital cost reduction" (down payment) required at the start of the lease
- The residual value percentage that is being used in the lease calculation
- The length of the lease
- The monthly lease payment

1. Find the table that corresponds to the length of your lease (24 months-pages 100-101, 36 months-pages 102-103, 48 months-pages 104-105, 60 months-pages 106-107).
2. There are eight tables for each lease term. Locate the table where the "Acquisition Price" is closest to the purchase price of the vehicle.
3. Read across the top row "Residual Value," and find the percent figure that is closest to the percent used to calculate your lease.
4. Trace down that column to the row corresponding to the monthly lease payment.
5. The figure at the intersection is an approximation of the interest rate of your lease.

You may find that the acquisition price of the vehicle you are considering is somewhere between the prices listed in these tables. (For example, the price of the vehicle is actually $22,500, which would be in between the $20,000 table and the $25,000 table.) If this is the case, ask the lessor to give you the lease payment as if the car cost $20,000 or $25,000, and then use the tables listed here.

Appendix H

The Complete Car Cost Guide

Lease Interest Rates

See instructions on page 99A.

24 Month Lease

$10,000 – Acquisition Price
Capital Cost Reduction: $500

Monthly Payment	Residual Value 64%	62%	60%	58%	56%	54%	52%	50%	48%	46%	44%	42%	40%	38%
$200	10.80%	9.65%	8.47%	7.26%	6.02%	4.75%	3.44%	2.10%	0.71%	N.A.	N.A.	N.A.	N.A.	N.A.
$210	12.33%	11.20%	10.04%	8.85%	7.63%	6.38%	5.09%	3.77%	2.41%	1.01%	N.A.	N.A.	N.A.	N.A.
$220	13.86%	12.74%	11.60%	10.43%	9.24%	8.01%	6.74%	5.44%	4.11%	2.73%	1.31%	N.A.	N.A.	N.A.
$230	15.39%	14.29%	13.17%	12.02%	10.84%	9.63%	8.39%	7.12%	5.80%	4.46%	3.06%	1.63%	0.15%	N.A.
$240	16.92%	15.84%	14.74%	13.61%	12.45%	11.26%	10.04%	8.79%	7.50%	6.18%	4.81%	3.41%	1.96%	0.46%
$250	18.45%	17.39%	16.30%	15.19%	14.05%	12.89%	11.69%	10.46%	9.19%	7.90%	6.56%	5.18%	3.76%	2.29%
$260	19.98%	18.94%	17.87%	16.78%	15.66%	14.51%	13.33%	12.13%	10.89%	9.61%	8.30%	6.95%	5.56%	4.12%
$270	21.52%	20.49%	19.44%	18.36%	17.26%	16.14%	14.98%	13.80%	12.58%	11.33%	10.04%	8.72%	7.36%	5.95%
$280	23.05%	22.04%	21.01%	19.95%	18.87%	17.76%	16.63%	15.46%	14.27%	13.04%	11.78%	10.48%	9.15%	7.77%
$290	24.59%	23.59%	22.58%	21.54%	20.47%	19.39%	18.27%	17.13%	15.96%	14.75%	13.52%	12.25%	10.94%	9.59%
$300	26.12%	25.15%	24.15%	23.12%	22.08%	21.01%	19.92%	18.79%	17.64%	16.46%	15.25%	14.01%	12.72%	11.40%

$12,500 – Acquisition Price
Capital Cost Reduction: $500

Monthly Payment	Residual Value 64%	62%	60%	58%	56%	54%	52%	50%	48%	46%	44%	42%	40%	38%
$250	10.10%	8.95%	7.77%	6.56%	5.32%	4.05%	2.74%	1.39%	N.A.	N.A.	N.A.	N.A.	N.A.	N.A.
$260	11.32%	10.18%	9.01%	7.82%	6.60%	5.34%	4.05%	2.72%	1.35%	N.A.	N.A.	N.A.	N.A.	N.A.
$270	12.53%	11.41%	10.26%	9.08%	7.87%	6.63%	5.36%	4.05%	2.70%	1.31%	N.A.	N.A.	N.A.	N.A.
$280	13.75%	12.64%	11.50%	10.34%	9.15%	7.93%	6.67%	5.38%	4.05%	2.68%	1.28%	N.A.	N.A.	N.A.
$290	14.96%	13.87%	12.75%	11.60%	10.43%	9.22%	7.98%	6.71%	5.40%	4.05%	2.67%	1.24%	N.A.	N.A.
$300	16.18%	15.10%	13.99%	12.86%	11.70%	10.51%	9.29%	8.04%	6.75%	5.42%	4.06%	2.65%	1.20%	N.A.
$310	17.39%	16.33%	15.24%	14.12%	12.98%	11.80%	10.60%	9.36%	8.09%	6.79%	5.44%	4.06%	2.63%	1.15%
$320	18.61%	17.56%	16.48%	15.38%	14.25%	13.10%	11.91%	10.69%	9.44%	8.15%	6.83%	5.47%	4.06%	2.61%
$330	19.83%	18.79%	17.73%	16.64%	15.53%	14.39%	13.22%	12.02%	10.78%	9.52%	8.21%	6.87%	5.49%	4.06%
$340	21.05%	20.02%	18.97%	17.90%	16.80%	15.68%	14.53%	13.34%	12.13%	10.88%	9.60%	8.28%	6.92%	5.51%
$350	22.27%	21.26%	20.22%	19.16%	18.08%	16.97%	15.83%	14.67%	13.47%	12.24%	10.98%	9.68%	8.34%	6.96%

$15,000 – Acquisition Price
Capital Cost Reduction: $500

Monthly Payment	Residual Value 64%	62%	60%	58%	56%	54%	52%	50%	48%	46%	44%	42%	40%	38%
$300	9.64%	8.49%	7.31%	6.10%	4.86%	3.58%	2.27%	0.92%	N.A.	N.A.	N.A.	N.A.	N.A.	N.A.
$320	11.65%	10.53%	9.37%	8.19%	6.97%	5.73%	4.45%	3.13%	1.78%	0.38%	N.A.	N.A.	N.A.	N.A.
$340	13.67%	12.57%	11.44%	10.28%	9.09%	7.87%	6.62%	5.34%	4.01%	2.65%	1.25%	N.A.	N.A.	N.A.
$360	15.69%	14.61%	13.50%	12.37%	11.21%	10.02%	8.80%	7.54%	6.25%	4.92%	3.56%	2.15%	0.69%	N.A.
$380	17.71%	16.65%	15.57%	14.46%	13.33%	12.16%	10.97%	9.74%	8.49%	7.19%	5.86%	4.49%	3.07%	1.61%
$400	19.73%	18.69%	17.64%	16.55%	15.44%	14.31%	13.14%	11.94%	10.72%	9.45%	8.16%	6.82%	5.44%	4.02%
$420	21.75%	20.74%	19.70%	18.64%	17.56%	16.45%	15.31%	14.14%	12.94%	11.71%	10.45%	9.15%	7.81%	6.43%
$440	23.77%	22.78%	21.77%	20.73%	19.67%	18.59%	17.48%	16.34%	15.17%	13.97%	12.74%	11.47%	10.17%	8.82%
$460	25.80%	24.83%	23.84%	22.83%	21.79%	20.73%	19.64%	18.53%	17.39%	16.22%	15.02%	13.79%	12.52%	11.21%
$480	27.83%	26.88%	25.91%	24.92%	23.91%	22.87%	21.81%	20.73%	19.61%	18.47%	17.30%	16.10%	14.87%	13.60%
$500	29.85%	28.93%	27.98%	27.01%	26.02%	25.01%	23.97%	22.92%	21.83%	20.72%	19.58%	18.41%	17.21%	15.98%

$17,500 – Acquisition Price
Capital Cost Reduction: $500

Monthly Payment	Residual Value 64%	62%	60%	58%	56%	54%	52%	50%	48%	46%	44%	42%	40%	38%
$350	9.31%	8.16%	6.98%	5.77%	4.53%	3.25%	1.94%	0.59%	N.A.	N.A.	N.A.	N.A.	N.A.	N.A.
$370	11.03%	9.90%	8.74%	7.56%	6.34%	5.09%	3.80%	2.48%	1.12%	N.A.	N.A.	N.A.	N.A.	N.A.
$390	12.76%	11.65%	10.51%	9.34%	8.15%	6.92%	5.66%	4.36%	3.03%	1.66%	0.25%	N.A.	N.A.	N.A.
$410	14.48%	13.39%	12.27%	11.13%	9.96%	8.75%	7.52%	6.25%	4.94%	3.60%	2.22%	0.79%	N.A.	N.A.
$430	16.21%	15.13%	14.04%	12.92%	11.76%	10.58%	9.37%	8.13%	6.85%	5.54%	4.19%	2.79%	1.36%	N.A.
$450	17.93%	16.88%	15.80%	14.70%	13.57%	12.42%	11.23%	10.01%	8.76%	7.48%	6.15%	4.79%	3.39%	1.94%
$470	19.66%	18.63%	17.57%	16.49%	15.38%	14.25%	13.09%	11.89%	10.67%	9.41%	8.12%	6.78%	5.41%	4.00%
$490	21.39%	20.37%	19.33%	18.27%	17.19%	16.08%	14.94%	13.77%	12.57%	11.34%	10.07%	8.77%	7.43%	6.05%
$510	23.11%	22.12%	21.10%	20.06%	19.00%	17.91%	16.79%	15.65%	14.47%	13.27%	12.03%	10.76%	9.45%	8.10%
$530	24.84%	23.87%	22.87%	21.85%	20.80%	19.74%	18.64%	17.52%	16.37%	15.19%	13.98%	12.74%	11.46%	10.14%
$550	26.57%	25.62%	24.64%	23.63%	22.61%	21.56%	20.49%	19.40%	18.27%	17.12%	15.93%	14.72%	13.47%	12.18%

The Complete Car Cost Guide

See instructions on page 99A.

24 Month Lease

$20,000 – Acquisition Price   Capital Cost Reduction: $500

Monthly Payment	Residual Value													
	64%	62%	60%	58%	56%	54%	52%	50%	48%	46%	44%	42%	40%	38%
$400	9.07%	7.92%	6.74%	5.52%	4.28%	3.01%	1.69%	0.34%	N.A.	N.A.	N.A.	N.A.	N.A.	N.A.
$420	10.57%	9.44%	8.28%	7.08%	5.86%	4.61%	3.32%	1.99%	0.63%	N.A.	N.A.	N.A.	N.A.	N.A.
$440	12.08%	10.96%	9.82%	8.64%	7.44%	6.21%	4.94%	3.64%	2.30%	0.92%	N.A.	N.A.	N.A.	N.A.
$460	13.58%	12.48%	11.36%	10.20%	9.02%	7.81%	6.56%	5.28%	3.97%	2.61%	1.22%	N.A.	N.A.	N.A.
$480	15.08%	14.00%	12.90%	11.76%	10.60%	9.41%	8.19%	6.93%	5.64%	4.31%	2.94%	1.53%	0.07%	N.A.
$500	16.59%	15.53%	14.44%	13.32%	12.18%	11.01%	9.81%	8.57%	7.30%	6.00%	4.66%	3.27%	1.85%	0.38%
$520	18.10%	17.05%	15.98%	14.88%	13.76%	12.61%	11.43%	10.21%	8.97%	7.69%	6.37%	5.02%	3.62%	2.18%
$540	19.60%	18.57%	17.52%	16.44%	15.34%	14.20%	13.04%	11.85%	10.63%	9.38%	8.09%	6.76%	5.39%	3.97%
$560	21.11%	20.10%	19.06%	18.00%	16.91%	15.80%	14.66%	13.49%	12.29%	11.06%	9.80%	8.49%	7.15%	5.77%
$580	22.62%	21.62%	20.60%	19.56%	18.49%	17.40%	16.28%	15.13%	13.95%	12.75%	11.50%	10.23%	8.91%	7.56%
$600	24.13%	23.15%	22.14%	21.12%	20.07%	19.00%	17.90%	16.77%	15.61%	14.43%	13.21%	11.96%	10.67%	9.34%

$25,000 – Acquisition Price   Capital Cost Reduction: $500

Monthly Payment	Residual Value													
	64%	62%	60%	58%	56%	54%	52%	50%	48%	46%	44%	42%	40%	38%
$500	8.73%	7.58%	6.39%	5.18%	3.94%	2.66%	1.35%	N.A.	N.A.	N.A.	N.A.	N.A.	N.A.	N.A.
$520	9.93%	8.79%	7.62%	6.43%	5.20%	3.94%	2.65%	1.31%	N.A.	N.A.	N.A.	N.A.	N.A.	N.A.
$540	11.13%	10.00%	8.85%	7.67%	6.46%	5.22%	3.94%	2.63%	1.28%	N.A.	N.A.	N.A.	N.A.	N.A.
$560	12.33%	11.22%	10.08%	8.91%	7.72%	6.49%	5.23%	3.94%	2.61%	1.24%	N.A.	N.A.	N.A.	N.A.
$580	13.52%	12.43%	11.31%	10.16%	8.98%	7.77%	6.53%	5.25%	3.94%	2.59%	1.20%	N.A.	N.A.	N.A.
$600	14.73%	13.64%	12.54%	11.40%	10.24%	9.05%	7.82%	6.56%	5.27%	3.94%	2.57%	1.16%	N.A.	N.A.
$620	15.93%	14.86%	13.76%	12.64%	11.50%	10.32%	9.11%	7.87%	6.60%	5.29%	3.94%	2.55%	1.12%	N.A.
$640	17.13%	16.07%	14.99%	13.89%	12.76%	11.60%	10.41%	9.18%	7.93%	6.64%	5.31%	3.94%	2.53%	1.08%
$660	18.33%	17.29%	16.22%	15.13%	14.01%	12.87%	11.70%	10.49%	9.26%	7.99%	6.68%	5.33%	3.94%	2.51%
$680	19.53%	18.50%	17.45%	16.37%	15.27%	14.14%	12.99%	11.80%	10.58%	9.33%	8.04%	6.72%	5.35%	3.94%
$700	20.73%	19.72%	18.68%	17.62%	16.53%	15.42%	14.28%	13.11%	11.91%	10.67%	9.41%	8.10%	6.76%	5.37%

$30,000 – Acquisition Price   Capital Cost Reduction: $500

Monthly Payment	Residual Value													
	64%	62%	60%	58%	56%	54%	52%	50%	48%	46%	44%	42%	40%	38%
$600	8.50%	7.35%	6.17%	4.96%	3.71%	2.44%	1.12%	N.A.	N.A.	N.A.	N.A.	N.A.	N.A.	N.A.
$625	9.75%	8.61%	7.44%	6.25%	5.02%	3.76%	2.47%	1.14%	N.A.	N.A.	N.A.	N.A.	N.A.	N.A.
$650	10.99%	9.87%	8.72%	7.54%	6.33%	5.09%	3.82%	2.50%	1.16%	N.A.	N.A.	N.A.	N.A.	N.A.
$675	12.24%	11.13%	10.00%	8.83%	7.64%	6.42%	5.16%	3.87%	2.54%	1.17%	N.A.	N.A.	N.A.	N.A.
$700	13.49%	12.39%	11.27%	10.13%	8.95%	7.74%	6.51%	5.23%	3.92%	2.58%	1.19%	N.A.	N.A.	N.A.
$725	14.74%	13.66%	12.55%	11.42%	10.26%	9.07%	7.85%	6.60%	5.31%	3.98%	2.61%	1.21%	N.A.	N.A.
$750	15.98%	14.92%	13.83%	12.71%	11.57%	10.40%	9.19%	7.96%	6.69%	5.38%	4.04%	2.65%	1.23%	N.A.
$775	17.23%	16.18%	15.11%	14.00%	12.88%	11.72%	10.53%	9.32%	8.07%	6.78%	5.46%	4.10%	2.69%	1.24%
$800	18.48%	17.44%	16.38%	15.30%	14.18%	13.05%	11.88%	10.68%	9.45%	8.18%	6.88%	5.54%	4.16%	2.74%
$825	19.73%	18.71%	17.66%	16.59%	15.49%	14.37%	13.22%	12.04%	10.82%	9.58%	8.30%	6.98%	5.62%	4.22%
$850	20.98%	19.97%	18.94%	17.88%	16.80%	15.69%	14.56%	13.40%	12.20%	10.98%	9.71%	8.42%	7.08%	5.71%

$40,000 – Acquisition Price   Capital Cost Reduction: $1,000

Monthly Payment	Residual Value													
	64%	62%	60%	58%	56%	54%	52%	50%	48%	46%	44%	42%	40%	38%
$800	9.07%	7.92%	6.74%	5.52%	4.28%	3.01%	1.69%	0.34%	N.A.	N.A.	N.A.	N.A.	N.A.	N.A.
$825	10.01%	8.87%	7.70%	6.50%	5.27%	4.01%	2.71%	1.37%	0.00%	N.A.	N.A.	N.A.	N.A.	N.A.
$850	10.95%	9.82%	8.66%	7.47%	6.26%	5.01%	3.72%	2.40%	1.05%	N.A.	N.A.	N.A.	N.A.	N.A.
$875	11.89%	10.77%	9.62%	8.45%	7.24%	6.01%	4.74%	3.43%	2.09%	0.71%	N.A.	N.A.	N.A.	N.A.
$900	12.83%	11.72%	10.59%	9.42%	8.23%	7.01%	5.75%	4.46%	3.13%	1.77%	0.36%	N.A.	N.A.	N.A.
$925	13.77%	12.67%	11.55%	10.40%	9.22%	8.01%	6.77%	5.49%	4.18%	2.83%	1.44%	N.A.	N.A.	N.A.
$950	14.71%	13.62%	12.51%	11.37%	10.21%	9.01%	7.78%	6.52%	5.22%	3.89%	2.51%	1.09%	N.A.	N.A.
$975	15.65%	14.57%	13.47%	12.35%	11.19%	10.01%	8.79%	7.54%	6.26%	4.94%	3.58%	2.18%	0.74%	N.A.
$1,000	16.59%	15.53%	14.44%	13.32%	12.18%	11.01%	9.81%	8.57%	7.30%	6.00%	4.66%	3.27%	1.85%	0.38%
$1,025	17.53%	16.48%	15.40%	14.30%	13.17%	12.01%	10.82%	9.60%	8.34%	7.06%	5.73%	4.36%	2.96%	1.50%
$1,050	18.47%	17.43%	16.36%	15.27%	14.15%	13.01%	11.83%	10.62%	9.38%	8.11%	6.80%	5.45%	4.06%	2.63%

Appendices

Appendix H *continued*

Lease Interest Rates

See instructions on page 99A.

36 Month Lease

$10,000 – Acquisition Price — *Capital Cost Reduction: $500*

Monthly Payment	Residual Value													
	58%	56%	54%	52%	50%	48%	46%	44%	42%	40%	38%	36%	34%	32%
$150	7.44%	6.65%	5.84%	5.01%	4.16%	3.28%	2.38%	1.45%	0.49%	N.A.	N.A.	N.A.	N.A.	N.A.
$160	9.01%	8.24%	7.45%	6.65%	5.82%	4.97%	4.09%	3.19%	2.26%	1.30%	0.30%	N.A.	N.A.	N.A.
$170	10.57%	9.83%	9.06%	8.28%	7.47%	6.65%	5.79%	4.92%	4.02%	3.09%	2.12%	1.13%	0.10%	N.A.
$180	12.14%	11.41%	10.67%	9.91%	9.12%	8.32%	7.49%	6.64%	5.77%	4.87%	3.94%	2.98%	1.98%	0.96%
$190	13.71%	13.00%	12.27%	11.53%	10.77%	9.99%	9.19%	8.36%	7.51%	6.64%	5.74%	4.81%	3.85%	2.86%
$200	15.27%	14.58%	13.87%	13.15%	12.41%	11.65%	10.87%	10.07%	9.25%	8.41%	7.53%	6.64%	5.71%	4.75%
$210	16.83%	16.16%	15.47%	14.77%	14.05%	13.31%	12.56%	11.78%	10.98%	10.16%	9.32%	8.45%	7.56%	6.63%
$220	18.39%	17.74%	17.07%	16.39%	15.69%	14.97%	14.24%	13.48%	12.71%	11.92%	11.10%	10.26%	9.40%	8.50%
$230	19.95%	19.31%	18.66%	18.00%	17.32%	16.62%	15.91%	15.18%	14.43%	13.66%	12.87%	12.06%	11.22%	10.36%
$240	21.51%	20.89%	20.25%	19.61%	18.95%	18.27%	17.58%	16.87%	16.15%	15.40%	14.64%	13.85%	13.04%	12.21%
$250	23.06%	22.46%	21.84%	21.22%	20.57%	19.92%	19.25%	18.56%	17.86%	17.13%	16.39%	15.63%	14.85%	14.05%

$12,500 – Acquisition Price — *Capital Cost Reduction: $500*

Monthly Payment	Residual Value													
	58%	56%	54%	52%	50%	48%	46%	44%	42%	40%	38%	36%	34%	32%
$200	8.51%	7.75%	6.96%	6.15%	5.32%	4.47%	3.59%	2.69%	1.75%	0.79%	N.A.	N.A.	N.A.	N.A.
$220	11.01%	10.27%	9.52%	8.74%	7.95%	7.13%	6.30%	5.44%	4.55%	3.63%	2.69%	1.71%	0.71%	N.A.
$240	13.49%	12.79%	12.07%	11.33%	10.57%	9.79%	8.99%	8.17%	7.32%	6.45%	5.56%	4.63%	3.68%	2.69%
$260	15.98%	15.30%	14.61%	13.90%	13.18%	12.43%	11.67%	10.89%	10.08%	9.26%	8.41%	7.53%	6.63%	5.69%
$280	18.45%	17.81%	17.15%	16.47%	15.78%	15.07%	14.34%	13.59%	12.83%	12.04%	11.23%	10.40%	9.55%	8.66%
$300	20.93%	20.31%	19.68%	19.03%	18.37%	17.69%	17.00%	16.28%	15.56%	14.81%	14.04%	13.25%	12.44%	11.61%
$320	23.40%	22.81%	22.20%	21.58%	20.95%	20.30%	19.64%	18.96%	18.27%	17.56%	16.83%	16.08%	15.31%	14.52%
$340	25.87%	25.30%	24.72%	24.13%	23.52%	22.91%	22.27%	21.63%	20.97%	20.29%	19.60%	18.89%	18.16%	17.42%
$360	28.33%	27.79%	27.23%	26.67%	26.09%	25.50%	24.90%	24.28%	23.66%	23.01%	22.36%	21.68%	20.99%	20.28%
$380	30.79%	30.27%	29.74%	29.20%	28.65%	28.09%	27.51%	26.93%	26.33%	25.72%	25.09%	24.45%	23.80%	23.13%
$400	33.25%	32.75%	32.24%	31.73%	31.20%	30.66%	30.12%	29.56%	28.99%	28.41%	27.82%	27.21%	26.59%	25.96%

$15,000 – Acquisition Price — *Capital Cost Reduction: $500*

Monthly Payment	Residual Value													
	58%	56%	54%	52%	50%	48%	46%	44%	42%	40%	38%	36%	34%	32%
$240	8.19%	7.42%	6.63%	5.83%	4.99%	4.14%	3.26%	2.36%	1.42%	0.46%	N.A.	N.A.	N.A.	N.A.
$260	10.26%	9.52%	8.76%	7.98%	7.18%	6.35%	5.51%	4.64%	3.74%	2.82%	1.87%	0.88%	N.A.	N.A.
$280	12.32%	11.61%	10.87%	10.12%	9.35%	8.56%	7.75%	6.91%	6.05%	5.16%	4.25%	3.31%	2.34%	1.33%
$300	14.38%	13.69%	12.99%	12.26%	11.52%	10.76%	9.97%	9.17%	8.35%	7.50%	6.62%	5.72%	4.79%	3.83%
$320	16.44%	15.78%	15.09%	14.39%	13.68%	12.95%	12.19%	11.42%	10.63%	9.82%	8.98%	8.12%	7.23%	6.31%
$340	18.50%	17.85%	17.20%	16.52%	15.83%	15.13%	14.41%	13.66%	12.90%	12.12%	11.32%	10.49%	9.64%	8.77%
$360	20.55%	19.93%	19.29%	18.65%	17.98%	17.30%	16.61%	15.90%	15.17%	14.42%	13.65%	12.86%	12.05%	11.21%
$380	22.60%	22.00%	21.39%	20.76%	20.13%	19.47%	18.80%	18.12%	17.42%	16.70%	15.96%	15.21%	14.43%	13.63%
$400	24.65%	24.07%	23.48%	22.88%	22.26%	21.63%	20.99%	20.33%	19.66%	18.97%	18.27%	17.54%	16.80%	16.04%
$420	26.69%	26.13%	25.57%	24.99%	24.39%	23.79%	23.17%	22.54%	21.89%	21.23%	20.56%	19.86%	19.15%	18.42%
$440	28.73%	28.20%	27.65%	27.09%	26.52%	25.94%	25.34%	24.74%	24.12%	23.49%	22.84%	22.17%	21.49%	20.80%

$17,500 – Acquisition Price — *Capital Cost Reduction: $500*

Monthly Payment	Residual Value													
	58%	56%	54%	52%	50%	48%	46%	44%	42%	40%	38%	36%	34%	32%
$280	7.96%	7.19%	6.40%	5.59%	4.76%	3.91%	3.03%	2.12%	1.19%	0.22%	N.A.	N.A.	N.A.	N.A.
$300	9.73%	8.98%	8.22%	7.43%	6.63%	5.80%	4.95%	4.07%	3.17%	2.24%	1.28%	0.29%	N.A.	N.A.
$320	11.49%	10.77%	10.03%	9.27%	8.49%	7.69%	6.86%	6.02%	5.15%	4.25%	3.32%	2.37%	1.38%	0.36%
$340	13.25%	12.55%	11.83%	11.10%	10.34%	9.57%	8.77%	7.95%	7.11%	6.25%	5.35%	4.43%	3.48%	2.50%
$360	15.01%	14.33%	13.64%	12.92%	12.19%	11.44%	10.67%	9.88%	9.07%	8.23%	7.37%	6.49%	5.57%	4.63%
$380	16.77%	16.11%	15.44%	14.74%	14.04%	13.31%	12.56%	11.80%	11.02%	10.21%	9.38%	8.53%	7.65%	6.74%
$400	18.53%	17.89%	17.23%	16.56%	15.87%	15.17%	14.45%	13.72%	12.96%	12.18%	11.38%	10.56%	9.71%	8.84%
$420	20.28%	19.66%	19.02%	18.37%	17.71%	17.03%	16.33%	15.62%	14.89%	14.14%	13.37%	12.58%	11.77%	10.93%
$440	22.03%	21.43%	20.81%	20.18%	19.54%	18.88%	18.21%	17.52%	16.82%	16.09%	15.35%	14.59%	13.80%	13.00%
$460	23.78%	23.20%	22.60%	21.99%	21.37%	20.73%	20.08%	19.42%	18.73%	18.04%	17.32%	16.59%	15.83%	15.06%
$480	25.53%	24.96%	24.38%	23.79%	23.19%	22.58%	21.95%	21.30%	20.64%	19.97%	19.28%	18.57%	17.85%	17.10%

The Complete Car Cost Guide

See instructions on page 99A.

36 Month Lease

$20,000 – Acquisition Price
Capital Cost Reduction: $500

Monthly Payment	_____Residual Value_____													
	58%	56%	54%	52%	50%	48%	46%	44%	42%	40%	38%	36%	34%	32%
$320	7.79%	7.02%	6.23%	5.42%	4.59%	3.73%	2.85%	1.95%	1.01%	0.05%	N.A.	N.A.	N.A.	N.A.
$340	9.33%	8.58%	7.81%	7.03%	6.22%	5.39%	4.53%	3.65%	2.75%	1.81%	0.85%	N.A.	N.A.	N.A.
$360	10.87%	10.14%	9.40%	8.63%	7.84%	7.03%	6.20%	5.35%	4.47%	3.56%	2.63%	1.67%	0.67%	N.A.
$380	12.41%	11.70%	10.97%	10.23%	9.46%	8.68%	7.87%	7.04%	6.19%	5.31%	4.41%	3.47%	2.51%	1.51%
$400	13.95%	13.26%	12.55%	11.82%	11.08%	10.32%	9.53%	8.73%	7.90%	7.05%	6.17%	5.27%	4.34%	3.37%
$420	15.48%	14.81%	14.12%	13.41%	12.69%	11.95%	11.19%	10.41%	9.61%	8.78%	7.93%	7.06%	6.16%	5.23%
$440	17.02%	16.36%	15.69%	15.00%	14.30%	13.58%	12.84%	12.08%	11.31%	10.51%	9.68%	8.84%	7.97%	7.07%
$460	18.55%	17.91%	17.26%	16.59%	15.91%	15.21%	14.49%	13.75%	13.00%	12.22%	11.43%	10.61%	9.77%	8.90%
$480	20.08%	19.46%	18.82%	18.17%	17.51%	16.83%	16.13%	15.42%	14.69%	13.93%	13.16%	12.37%	11.56%	10.72%
$500	21.61%	21.00%	20.38%	19.75%	19.11%	18.45%	17.77%	17.08%	16.37%	15.64%	14.89%	14.13%	13.34%	12.53%
$520	23.14%	22.55%	21.94%	21.33%	20.70%	20.06%	19.40%	18.73%	18.04%	17.34%	16.61%	15.87%	15.11%	14.33%

$25,000 – Acquisition Price
Capital Cost Reduction: $500

Monthly Payment	_____Residual Value_____													
	58%	56%	54%	52%	50%	48%	46%	44%	42%	40%	38%	36%	34%	32%
$420	8.78%	8.03%	7.25%	6.46%	5.65%	4.81%	3.95%	3.06%	2.15%	1.21%	0.24%	N.A.	N.A.	N.A.
$440	10.01%	9.27%	8.52%	7.74%	6.94%	6.13%	5.29%	4.42%	3.53%	2.61%	1.66%	0.68%	N.A.	N.A.
$460	11.24%	10.52%	9.78%	9.02%	8.24%	7.44%	6.62%	5.77%	4.90%	4.01%	3.08%	2.13%	1.14%	0.12%
$480	12.46%	11.76%	11.03%	10.29%	9.53%	8.75%	7.94%	7.12%	6.27%	5.40%	4.50%	3.57%	2.61%	1.62%
$500	13.69%	13.00%	12.29%	11.56%	10.82%	10.05%	9.27%	8.46%	7.64%	6.78%	5.91%	5.00%	4.07%	3.10%
$520	14.92%	14.24%	13.54%	12.83%	12.10%	11.36%	10.59%	9.80%	9.00%	8.17%	7.31%	6.43%	5.52%	4.58%
$540	16.14%	15.47%	14.79%	14.10%	13.39%	12.66%	11.91%	11.14%	10.35%	9.54%	8.71%	7.85%	6.97%	6.05%
$560	17.36%	16.71%	16.05%	15.37%	14.67%	13.96%	13.23%	12.48%	11.71%	10.92%	10.10%	9.27%	8.41%	7.52%
$580	18.58%	17.95%	17.29%	16.63%	15.95%	15.25%	14.54%	13.81%	13.05%	12.28%	11.49%	10.68%	9.84%	8.98%
$600	19.80%	19.18%	18.54%	17.89%	17.23%	16.54%	15.85%	15.13%	14.40%	13.65%	12.88%	12.08%	11.27%	10.43%
$620	21.02%	20.41%	19.79%	19.15%	18.50%	17.83%	17.15%	16.46%	15.74%	15.01%	14.25%	13.48%	12.69%	11.87%

$30,000 – Acquisition Price
Capital Cost Reduction: $500

Monthly Payment	_____Residual Value_____													
	58%	56%	54%	52%	50%	48%	46%	44%	42%	40%	38%	36%	34%	32%
$500	8.41%	7.66%	6.88%	6.09%	5.27%	4.43%	3.56%	2.67%	1.76%	0.81%	N.A.	N.A.	N.A.	N.A.
$525	9.69%	8.95%	8.19%	7.42%	6.62%	5.80%	4.95%	4.08%	3.19%	2.27%	1.32%	0.34%	N.A.	N.A.
$550	10.97%	10.25%	9.50%	8.74%	7.96%	7.16%	6.34%	5.49%	4.62%	3.72%	2.79%	1.84%	0.85%	N.A.
$575	12.25%	11.54%	10.81%	10.07%	9.30%	8.52%	7.72%	6.89%	6.04%	5.17%	4.27%	3.33%	2.37%	1.38%
$600	13.52%	12.83%	12.12%	11.39%	10.64%	9.88%	9.10%	8.29%	7.46%	6.61%	5.73%	4.82%	3.89%	2.93%
$625	14.79%	14.11%	13.42%	12.71%	11.98%	11.23%	10.47%	9.68%	8.88%	8.04%	7.19%	6.31%	5.40%	4.46%
$650	16.06%	15.40%	14.72%	14.03%	13.32%	12.59%	11.84%	11.07%	10.28%	9.48%	8.64%	7.79%	6.90%	5.99%
$675	17.33%	16.68%	16.02%	15.34%	14.65%	13.94%	13.21%	12.46%	11.69%	10.90%	10.09%	9.26%	8.40%	7.51%
$700	18.60%	17.97%	17.32%	16.66%	15.98%	15.28%	14.57%	13.84%	13.09%	12.32%	11.53%	10.72%	9.89%	9.03%
$725	19.87%	19.25%	18.62%	17.97%	17.30%	16.63%	15.93%	15.22%	14.49%	13.74%	12.97%	12.18%	11.37%	10.53%
$750	21.14%	20.53%	19.91%	19.28%	18.63%	17.97%	17.29%	16.59%	15.88%	15.15%	14.40%	13.63%	12.84%	12.03%

$40,000 – Acquisition Price
Capital Cost Reduction: $1,000

Monthly Payment	_____Residual Value_____													
	58%	56%	54%	52%	50%	48%	46%	44%	42%	40%	38%	36%	34%	32%
$650	8.17%	7.41%	6.63%	5.82%	5.00%	4.15%	3.27%	2.37%	1.45%	0.49%	N.A.	N.A.	N.A.	N.A.
$675	9.14%	8.39%	7.62%	6.83%	6.01%	5.18%	4.32%	3.44%	2.53%	1.59%	0.62%	N.A.	N.A.	N.A.
$700	10.10%	9.36%	8.61%	7.83%	7.03%	6.21%	5.37%	4.50%	3.61%	2.69%	1.74%	0.76%	N.A.	N.A.
$725	11.07%	10.34%	9.59%	8.83%	8.04%	7.24%	6.41%	5.56%	4.69%	3.78%	2.85%	1.89%	0.90%	N.A.
$750	12.03%	11.31%	10.58%	9.83%	9.06%	8.27%	7.45%	6.62%	5.76%	4.88%	3.96%	3.02%	2.05%	1.04%
$775	12.99%	12.28%	11.56%	10.83%	10.07%	9.29%	8.49%	7.67%	6.83%	5.96%	5.07%	4.15%	3.20%	2.21%
$800	13.95%	13.26%	12.55%	11.82%	11.08%	10.32%	9.53%	8.73%	7.90%	7.05%	6.17%	5.27%	4.34%	3.37%
$825	14.91%	14.23%	13.53%	12.82%	12.09%	11.34%	10.57%	9.78%	8.97%	8.13%	7.27%	6.39%	5.48%	4.53%
$850	15.87%	15.20%	14.51%	13.81%	13.09%	12.36%	11.60%	10.83%	10.03%	9.21%	8.37%	7.50%	6.61%	5.69%
$875	16.83%	16.17%	15.49%	14.81%	14.10%	13.38%	12.63%	11.87%	11.09%	10.29%	9.47%	8.62%	7.74%	6.84%
$900	17.78%	17.14%	16.47%	15.80%	15.10%	14.39%	13.67%	12.92%	12.15%	11.37%	10.56%	9.72%	8.87%	7.98%

Appendices

The Complete Car Cost Guide

Appendix H *continued* — Lease Interest Rates

See instructions on page 99A.

48 Month Lease

$10,000 – Acquisition Price *Capital Cost Reduction: $500*

Monthly Payment	Residual Value													
	52%	50%	48%	46%	44%	42%	40%	38%	36%	34%	32%	30%	28%	26%
$140	8.21%	7.63%	7.04%	6.44%	5.82%	5.18%	4.52%	3.84%	3.14%	2.41%	1.66%	0.89%	0.08%	N.A.
$150	9.82%	9.27%	8.70%	8.12%	7.52%	6.91%	6.28%	5.63%	4.96%	4.26%	3.55%	2.81%	2.04%	1.25%
$160	11.42%	10.89%	10.35%	9.79%	9.22%	8.63%	8.02%	7.40%	6.76%	6.10%	5.42%	4.71%	3.98%	3.23%
$170	13.02%	12.51%	11.99%	11.45%	10.90%	10.34%	9.76%	9.16%	8.55%	7.92%	7.27%	6.60%	5.91%	5.19%
$180	14.62%	14.13%	13.62%	13.11%	12.58%	12.04%	11.48%	10.91%	10.33%	9.72%	9.10%	8.47%	7.81%	7.13%
$190	16.21%	15.74%	15.25%	14.76%	14.25%	13.73%	13.20%	12.65%	12.09%	11.52%	10.92%	10.32%	9.69%	9.04%
$200	17.80%	17.34%	16.87%	16.40%	15.91%	15.41%	14.90%	14.38%	13.85%	13.30%	12.73%	12.15%	11.55%	10.94%
$210	19.38%	18.94%	18.49%	18.03%	17.57%	17.09%	16.60%	16.10%	15.59%	15.06%	14.52%	13.97%	13.40%	12.82%
$220	20.95%	20.53%	20.10%	19.66%	19.21%	18.76%	18.29%	17.81%	17.32%	16.82%	16.30%	15.77%	15.23%	14.67%
$230	22.53%	22.12%	21.71%	21.29%	20.86%	20.42%	19.97%	19.51%	19.04%	18.56%	18.07%	17.56%	17.05%	16.52%
$240	24.09%	23.70%	23.31%	22.90%	22.49%	22.07%	21.64%	21.20%	20.75%	20.29%	19.82%	19.34%	18.85%	18.34%

$12,500 – Acquisition Price *Capital Cost Reduction: $500*

Monthly Payment	Residual Value													
	52%	50%	48%	46%	44%	42%	40%	38%	36%	34%	32%	30%	28%	26%
$180	8.46%	7.89%	7.31%	6.71%	6.10%	5.47%	4.82%	4.15%	3.46%	2.75%	2.01%	1.25%	0.46%	N.A.
$200	11.01%	10.48%	9.93%	9.37%	8.80%	8.21%	7.60%	6.98%	6.34%	5.67%	4.99%	4.28%	3.55%	2.80%
$220	13.55%	13.05%	12.54%	12.01%	11.48%	10.92%	10.36%	9.78%	9.18%	8.56%	7.93%	7.27%	6.60%	5.90%
$240	16.08%	15.61%	15.13%	14.64%	14.13%	13.62%	13.09%	12.54%	11.99%	11.41%	10.82%	10.22%	9.60%	8.95%
$260	18.60%	18.16%	17.70%	17.24%	16.77%	16.28%	15.79%	15.28%	14.76%	14.23%	13.68%	13.12%	12.55%	11.95%
$280	21.10%	20.69%	20.26%	19.83%	19.39%	18.93%	18.47%	18.00%	17.51%	17.02%	16.51%	15.99%	15.46%	14.91%
$300	23.60%	23.20%	22.81%	22.40%	21.98%	21.56%	21.13%	20.69%	20.24%	19.78%	19.30%	18.82%	18.32%	17.82%
$320	26.08%	25.71%	25.34%	24.96%	24.57%	24.17%	23.77%	23.35%	22.93%	22.50%	22.07%	21.62%	21.16%	20.69%
$340	28.55%	28.21%	27.86%	27.50%	27.13%	26.76%	26.38%	26.00%	25.61%	25.21%	24.80%	24.38%	23.96%	23.52%
$360	31.02%	30.69%	30.36%	30.03%	29.68%	29.34%	28.98%	28.63%	28.26%	27.89%	27.51%	27.12%	26.72%	26.32%
$380	33.47%	33.17%	32.86%	32.54%	32.22%	31.90%	31.57%	31.23%	30.89%	30.54%	30.19%	29.83%	29.46%	29.09%

$15,000 – Acquisition Price *Capital Cost Reduction: $500*

Monthly Payment	Residual Value													
	52%	50%	48%	46%	44%	42%	40%	38%	36%	34%	32%	30%	28%	26%
$220	8.62%	8.06%	7.49%	6.89%	6.29%	5.66%	5.02%	4.36%	3.67%	2.97%	2.24%	1.48%	0.70%	N.A.
$230	9.68%	9.14%	8.58%	8.00%	7.41%	6.80%	6.18%	5.53%	4.87%	4.18%	3.48%	2.75%	1.99%	1.21%
$240	10.74%	10.21%	9.66%	9.10%	8.53%	7.93%	7.33%	6.70%	6.06%	5.39%	4.71%	4.00%	3.27%	2.51%
$250	11.80%	11.28%	10.74%	10.20%	9.64%	9.06%	8.47%	7.87%	7.24%	6.60%	5.93%	5.25%	4.54%	3.81%
$260	12.85%	12.34%	11.83%	11.29%	10.75%	10.19%	9.61%	9.02%	8.42%	7.79%	7.15%	6.49%	5.80%	5.09%
$270	13.90%	13.41%	12.90%	12.38%	11.85%	11.31%	10.75%	10.18%	9.59%	8.98%	8.36%	7.72%	7.05%	6.37%
$280	14.95%	14.47%	13.98%	13.47%	12.96%	12.43%	11.88%	11.33%	10.75%	10.17%	9.56%	8.94%	8.30%	7.64%
$290	16.00%	15.53%	15.05%	14.56%	14.05%	13.54%	13.01%	12.47%	11.92%	11.34%	10.76%	10.16%	9.53%	8.89%
$300	17.04%	16.59%	16.12%	15.64%	15.15%	14.65%	14.14%	13.61%	13.07%	12.52%	11.95%	11.36%	10.76%	10.14%
$310	18.09%	17.64%	17.18%	16.72%	16.24%	15.75%	15.26%	14.74%	14.22%	13.68%	13.13%	12.57%	11.98%	11.38%
$320	19.13%	18.69%	18.25%	17.79%	17.33%	16.86%	16.37%	15.87%	15.37%	14.84%	14.31%	13.76%	13.20%	12.62%

$17,500 – Acquisition Price *Capital Cost Reduction: $500*

Monthly Payment	Residual Value													
	52%	50%	48%	46%	44%	42%	40%	38%	36%	34%	32%	30%	28%	26%
$260	8.74%	8.18%	7.61%	7.02%	6.42%	5.80%	5.16%	4.50%	3.82%	3.12%	2.40%	1.65%	0.88%	0.07%
$280	10.55%	10.02%	9.47%	8.91%	8.33%	7.74%	7.13%	6.50%	5.86%	5.20%	4.51%	3.80%	3.07%	2.31%
$300	12.35%	11.84%	11.32%	10.78%	10.23%	9.67%	9.09%	8.49%	7.88%	7.25%	6.60%	5.93%	5.23%	4.52%
$320	14.15%	13.66%	13.16%	12.65%	12.12%	11.58%	11.03%	10.46%	9.88%	9.28%	8.66%	8.03%	7.37%	6.70%
$340	15.94%	15.47%	14.99%	14.50%	14.00%	13.49%	12.96%	12.42%	11.87%	11.30%	10.71%	10.11%	9.49%	8.85%
$360	17.72%	17.27%	16.82%	16.35%	15.87%	15.38%	14.88%	14.36%	13.84%	13.30%	12.74%	12.17%	11.58%	10.98%
$380	19.50%	19.07%	18.63%	18.18%	17.73%	17.26%	16.78%	16.29%	15.79%	15.28%	14.75%	14.21%	13.66%	13.09%
$400	21.27%	20.86%	20.44%	20.01%	19.58%	19.13%	18.67%	18.21%	17.73%	17.24%	16.74%	16.23%	15.71%	15.17%
$420	23.04%	22.64%	22.24%	21.83%	21.42%	20.99%	20.56%	20.11%	19.66%	19.20%	18.72%	18.24%	17.74%	17.23%
$440	24.80%	24.42%	24.04%	23.65%	23.25%	22.84%	22.43%	22.01%	21.57%	21.13%	20.68%	20.22%	19.75%	19.27%
$460	26.55%	26.19%	25.83%	25.45%	25.07%	24.68%	24.29%	23.89%	23.48%	23.06%	22.63%	22.19%	21.74%	21.29%

Appendices

The Complete Car Cost Guide

See instructions on page 99A.

48 Month Lease

$20,000 – Acquisition Price   *Capital Cost Reduction: $500*

Monthly Payment	Residual Value													
	52%	50%	48%	46%	44%	42%	40%	38%	36%	34%	32%	30%	28%	26%
$300	8.83%	8.27%	7.70%	7.12%	6.52%	5.90%	5.26%	4.61%	3.94%	3.24%	2.52%	1.78%	1.00%	0.20%
$320	10.41%	9.87%	9.32%	8.76%	8.19%	7.59%	6.98%	6.36%	5.71%	5.05%	4.36%	3.65%	2.92%	2.16%
$340	11.98%	11.47%	10.94%	10.40%	9.85%	9.28%	8.69%	8.09%	7.48%	6.84%	6.18%	5.51%	4.81%	4.09%
$360	13.55%	13.06%	12.55%	12.03%	11.50%	10.95%	10.39%	9.82%	9.23%	8.62%	7.99%	7.35%	6.68%	6.00%
$380	15.11%	14.64%	14.15%	13.65%	13.14%	12.62%	12.08%	11.53%	10.96%	10.38%	9.79%	9.17%	8.54%	7.88%
$400	16.67%	16.22%	15.75%	15.27%	14.78%	14.27%	13.76%	13.23%	12.69%	12.13%	11.56%	10.98%	10.37%	9.75%
$420	18.23%	17.79%	17.34%	16.87%	16.40%	15.92%	15.43%	14.92%	14.40%	13.87%	13.33%	12.77%	12.19%	11.60%
$440	19.78%	19.35%	18.92%	18.48%	18.02%	17.56%	17.09%	16.60%	16.11%	15.60%	15.08%	14.54%	14.00%	13.43%
$460	21.32%	20.91%	20.50%	20.07%	19.64%	19.19%	18.74%	18.27%	17.80%	17.31%	16.82%	16.31%	15.78%	15.25%
$480	22.86%	22.47%	22.07%	21.66%	21.24%	20.82%	20.38%	19.94%	19.48%	19.02%	18.54%	18.05%	17.56%	17.04%
$500	24.40%	24.02%	23.64%	23.24%	22.84%	22.43%	22.01%	21.59%	21.15%	20.71%	20.25%	19.79%	19.31%	18.83%

$25,000 – Acquisition Price   *Capital Cost Reduction: $500*

Monthly Payment	Residual Value													
	52%	50%	48%	46%	44%	42%	40%	38%	36%	34%	32%	30%	28%	26%
$375	8.63%	8.08%	7.51%	6.92%	6.32%	5.70%	5.07%	4.41%	3.74%	3.04%	2.32%	1.57%	0.80%	N.A.
$400	10.21%	9.67%	9.12%	8.56%	7.98%	7.39%	6.78%	6.15%	5.51%	4.84%	4.15%	3.44%	2.71%	1.95%
$425	11.78%	11.26%	10.73%	10.19%	9.64%	9.07%	8.48%	7.88%	7.27%	6.63%	5.97%	5.30%	4.60%	3.87%
$450	13.34%	12.85%	12.34%	11.82%	11.29%	10.74%	10.18%	9.60%	9.01%	8.40%	7.78%	7.13%	6.46%	5.78%
$475	14.90%	14.42%	13.93%	13.43%	12.92%	12.40%	11.86%	11.31%	10.74%	10.16%	9.56%	8.95%	8.31%	7.66%
$500	16.45%	15.99%	15.52%	15.04%	14.55%	14.05%	13.53%	13.01%	12.46%	11.91%	11.34%	10.75%	10.14%	9.52%
$525	18.00%	17.56%	17.11%	16.65%	16.18%	15.69%	15.20%	14.69%	14.17%	13.64%	13.09%	12.53%	11.96%	11.36%
$550	19.55%	19.12%	18.69%	18.24%	17.79%	17.33%	16.85%	16.37%	15.87%	15.36%	14.84%	14.30%	13.76%	13.19%
$575	21.09%	20.68%	20.26%	19.83%	19.40%	18.95%	18.50%	18.03%	17.56%	17.07%	16.57%	16.06%	15.54%	15.00%
$600	22.62%	22.23%	21.83%	21.42%	21.00%	20.57%	20.13%	19.69%	19.23%	18.77%	18.29%	17.80%	17.30%	16.79%
$625	24.15%	23.77%	23.39%	22.99%	22.59%	22.18%	21.76%	21.34%	20.90%	20.45%	20.00%	19.53%	19.06%	18.57%

$30,000 – Acquisition Price   *Capital Cost Reduction: $500*

Monthly Payment	Residual Value													
	52%	50%	48%	46%	44%	42%	40%	38%	36%	34%	32%	30%	28%	26%
$450	8.50%	7.95%	7.38%	6.79%	6.19%	5.57%	4.94%	4.28%	3.60%	2.91%	2.18%	1.44%	0.67%	N.A.
$475	9.81%	9.28%	8.72%	8.16%	7.57%	6.98%	6.36%	5.73%	5.08%	4.41%	3.71%	3.00%	2.25%	1.49%
$500	11.12%	10.60%	10.06%	9.51%	8.95%	8.37%	7.78%	7.17%	6.54%	5.90%	5.23%	4.54%	3.83%	3.09%
$525	12.42%	11.92%	11.40%	10.87%	10.32%	9.77%	9.19%	8.60%	8.00%	7.38%	6.73%	6.07%	5.39%	4.68%
$550	13.72%	13.23%	12.73%	12.22%	11.69%	11.15%	10.60%	10.03%	9.45%	8.85%	8.23%	7.59%	6.94%	6.26%
$575	15.02%	14.54%	14.06%	13.56%	13.05%	12.53%	11.99%	11.45%	10.88%	10.31%	9.71%	9.10%	8.47%	7.82%
$600	16.31%	15.85%	15.38%	14.90%	14.41%	13.90%	13.39%	12.86%	12.31%	11.76%	11.18%	10.60%	9.99%	9.37%
$625	17.60%	17.15%	16.70%	16.23%	15.75%	15.27%	14.77%	14.26%	13.74%	13.20%	12.65%	12.08%	11.50%	10.90%
$650	18.88%	18.45%	18.01%	17.56%	17.10%	16.63%	16.15%	15.65%	15.15%	14.63%	14.10%	13.56%	13.00%	12.42%
$675	20.16%	19.75%	19.32%	18.88%	18.44%	17.98%	17.52%	17.04%	16.56%	16.06%	15.55%	15.02%	14.49%	13.93%
$700	21.44%	21.04%	20.62%	20.20%	19.77%	19.33%	18.88%	18.42%	17.96%	17.47%	16.98%	16.48%	15.96%	15.43%

$40,000 – Acquisition Price   *Capital Cost Reduction: $1,000*

Monthly Payment	Residual Value													
	52%	50%	48%	46%	44%	42%	40%	38%	36%	34%	32%	30%	28%	26%
$600	8.83%	8.27%	7.70%	7.12%	6.52%	5.90%	5.26%	4.61%	3.94%	3.24%	2.52%	1.78%	1.00%	0.20%
$625	9.82%	9.27%	8.72%	8.15%	7.56%	6.96%	6.34%	5.70%	5.05%	4.37%	3.67%	2.95%	2.20%	1.43%
$650	10.80%	10.27%	9.73%	9.17%	8.60%	8.02%	7.41%	6.79%	6.15%	5.50%	4.82%	4.12%	3.39%	2.64%
$675	11.78%	11.27%	10.74%	10.20%	9.64%	9.07%	8.48%	7.88%	7.26%	6.62%	5.96%	5.28%	4.57%	3.85%
$700	12.77%	12.26%	11.74%	11.22%	10.67%	10.12%	9.54%	8.96%	8.35%	7.73%	7.09%	6.43%	5.75%	5.04%
$725	13.75%	13.25%	12.75%	12.23%	11.70%	11.16%	10.60%	10.03%	9.44%	8.84%	8.22%	7.58%	6.92%	6.23%
$750	14.72%	14.24%	13.75%	13.25%	12.73%	12.20%	11.66%	11.10%	10.53%	9.94%	9.34%	8.72%	8.08%	7.41%
$775	15.70%	15.23%	14.75%	14.26%	13.75%	13.24%	12.71%	12.17%	11.61%	11.04%	10.45%	9.85%	9.23%	8.59%
$800	16.67%	16.22%	15.75%	15.27%	14.78%	14.27%	13.76%	13.23%	12.69%	12.13%	11.56%	10.98%	10.37%	9.75%
$825	17.65%	17.20%	16.74%	16.27%	15.79%	15.30%	14.80%	14.29%	13.76%	13.22%	12.67%	12.10%	11.51%	10.91%
$850	18.62%	18.18%	17.73%	17.28%	16.81%	16.33%	15.84%	15.34%	14.83%	14.31%	13.77%	13.21%	12.65%	12.06%

Appendices

Appendix H *continued*

Lease Interest Rates

The Complete Car Cost Guide

See instructions on page 99A.

60 Month Lease

$10,000 – Acquisition Price
Capital Cost Reduction: $500

Monthly Payment	Residual Value													
	48%	46%	44%	42%	40%	38%	36%	34%	32%	30%	28%	26%	24%	22%
$120	6.94%	6.48%	6.00%	5.51%	5.01%	4.49%	3.95%	3.40%	2.83%	2.24%	1.63%	1.00%	0.34%	N.A.
$130	8.58%	8.14%	7.69%	7.22%	6.75%	6.25%	5.75%	5.23%	4.69%	4.14%	3.56%	2.97%	2.35%	1.71%
$140	10.21%	9.79%	9.36%	8.92%	8.47%	8.00%	7.53%	7.04%	6.53%	6.01%	5.47%	4.91%	4.34%	3.74%
$150	11.83%	11.43%	11.03%	10.61%	10.18%	9.74%	9.29%	8.82%	8.35%	7.85%	7.35%	6.83%	6.29%	5.73%
$160	13.44%	13.07%	12.68%	12.28%	11.88%	11.46%	11.03%	10.59%	10.14%	9.68%	9.21%	8.71%	8.21%	7.69%
$170	15.05%	14.69%	14.32%	13.94%	13.56%	13.17%	12.76%	12.35%	11.92%	11.49%	11.04%	10.58%	10.11%	9.62%
$180	16.65%	16.30%	15.95%	15.60%	15.23%	14.86%	14.48%	14.08%	13.68%	13.27%	12.85%	12.42%	11.98%	11.52%
$190	18.23%	17.91%	17.58%	17.24%	16.89%	16.54%	16.18%	15.81%	15.43%	15.04%	14.65%	14.24%	13.82%	13.39%
$200	19.82%	19.51%	19.19%	18.87%	18.54%	18.21%	17.86%	17.51%	17.16%	16.79%	16.42%	16.04%	15.64%	15.24%
$210	21.39%	21.10%	20.80%	20.49%	20.18%	19.86%	19.54%	19.21%	18.87%	18.53%	18.17%	17.81%	17.45%	17.07%
$220	22.96%	22.68%	22.39%	22.10%	21.81%	21.51%	21.20%	20.89%	20.57%	20.25%	19.91%	19.57%	19.23%	18.87%

$12,500 – Acquisition Price
Capital Cost Reduction: $500

Monthly Payment	Residual Value													
	48%	46%	44%	42%	40%	38%	36%	34%	32%	30%	28%	26%	24%	22%
$160	7.92%	7.48%	7.02%	6.55%	6.06%	5.57%	5.05%	4.53%	3.98%	3.42%	2.83%	2.23%	1.61%	0.96%
$170	9.22%	8.79%	8.35%	7.90%	7.44%	6.96%	6.47%	5.97%	5.45%	4.91%	4.36%	3.78%	3.19%	2.57%
$180	10.51%	10.10%	9.68%	9.25%	8.80%	8.35%	7.88%	7.40%	6.90%	6.39%	5.86%	5.32%	4.75%	4.17%
$190	11.80%	11.41%	11.00%	10.58%	10.16%	9.72%	9.27%	8.81%	8.34%	7.85%	7.35%	6.83%	6.30%	5.75%
$200	13.08%	12.70%	12.31%	11.91%	11.51%	11.09%	10.66%	10.22%	9.77%	9.30%	8.83%	8.33%	7.83%	7.30%
$210	14.36%	13.99%	13.62%	13.24%	12.85%	12.45%	12.04%	11.62%	11.18%	10.74%	10.29%	9.82%	9.34%	8.84%
$220	15.63%	15.28%	14.92%	14.55%	14.18%	13.79%	13.40%	13.00%	12.59%	12.17%	11.73%	11.29%	10.83%	10.36%
$230	16.89%	16.56%	16.21%	15.86%	15.50%	15.14%	14.76%	14.38%	13.98%	13.58%	13.17%	12.74%	12.31%	11.86%
$240	18.15%	17.83%	17.50%	17.16%	16.82%	16.47%	16.11%	15.74%	15.37%	14.98%	14.59%	14.18%	13.77%	13.34%
$250	19.41%	19.10%	18.78%	18.46%	18.13%	17.79%	17.45%	17.10%	16.74%	16.37%	15.99%	15.61%	15.22%	14.81%
$260	20.66%	20.36%	20.06%	19.75%	19.43%	19.11%	18.78%	18.44%	18.10%	17.75%	17.39%	17.03%	16.65%	16.27%

$15,000 – Acquisition Price
Capital Cost Reduction: $500

Monthly Payment	Residual Value													
	48%	46%	44%	42%	40%	38%	36%	34%	32%	30%	28%	26%	24%	22%
$200	8.57%	8.13%	7.69%	7.23%	6.76%	6.27%	5.77%	5.26%	4.73%	4.19%	3.62%	3.04%	2.43%	1.80%
$210	9.64%	9.22%	8.79%	8.35%	7.89%	7.43%	6.95%	6.45%	5.94%	5.42%	4.88%	4.32%	3.74%	3.14%
$220	10.71%	10.31%	9.89%	9.46%	9.02%	8.57%	8.11%	7.63%	7.14%	6.64%	6.12%	5.58%	5.03%	4.45%
$230	11.78%	11.39%	10.98%	10.57%	10.15%	9.71%	9.27%	8.81%	8.34%	7.85%	7.35%	6.84%	6.31%	5.76%
$240	12.84%	12.46%	12.07%	11.67%	11.26%	10.85%	10.42%	9.97%	9.52%	9.06%	8.58%	8.08%	7.57%	7.05%
$250	13.90%	13.53%	13.16%	12.77%	12.38%	11.97%	11.56%	11.13%	10.70%	10.25%	9.79%	9.32%	8.83%	8.32%
$260	14.96%	14.60%	14.24%	13.86%	13.48%	13.09%	12.69%	12.28%	11.87%	11.43%	10.99%	10.54%	10.07%	9.59%
$270	16.01%	15.66%	15.31%	14.95%	14.58%	14.21%	13.82%	13.43%	13.03%	12.61%	12.19%	11.75%	11.30%	10.84%
$280	17.06%	16.72%	16.38%	16.03%	15.68%	15.32%	14.95%	14.57%	14.18%	13.78%	13.37%	12.95%	12.52%	12.08%
$290	18.10%	17.78%	17.45%	17.11%	16.77%	16.42%	16.06%	15.70%	15.32%	14.94%	14.55%	14.15%	13.73%	13.31%
$300	19.14%	18.83%	18.51%	18.19%	17.86%	17.52%	17.17%	16.82%	16.46%	16.09%	15.72%	15.33%	14.93%	14.53%

$17,500 – Acquisition Price
Capital Cost Reduction: $500

Monthly Payment	Residual Value													
	48%	46%	44%	42%	40%	38%	36%	34%	32%	30%	28%	26%	24%	22%
$220	7.18%	6.73%	6.27%	5.79%	5.30%	4.79%	4.27%	3.73%	3.18%	2.61%	2.01%	1.40%	0.76%	0.10%
$240	9.03%	8.60%	8.16%	7.71%	7.25%	6.77%	6.28%	5.78%	5.26%	4.73%	4.17%	3.60%	3.01%	2.40%
$260	10.86%	10.45%	10.04%	9.61%	9.18%	8.73%	8.27%	7.80%	7.32%	6.82%	6.30%	5.77%	5.22%	4.65%
$280	12.67%	12.29%	11.90%	11.50%	11.09%	10.67%	10.24%	9.80%	9.35%	8.88%	8.40%	7.90%	7.39%	6.87%
$300	14.48%	14.12%	13.75%	13.37%	12.99%	12.59%	12.19%	11.78%	11.35%	10.91%	10.47%	10.01%	9.53%	9.04%
$320	16.28%	15.94%	15.59%	15.23%	14.87%	14.50%	14.12%	13.73%	13.33%	12.93%	12.51%	12.08%	11.64%	11.18%
$340	18.06%	17.74%	17.41%	17.08%	16.74%	16.39%	16.03%	15.67%	15.29%	14.91%	14.52%	14.12%	13.71%	13.29%
$360	19.84%	19.54%	19.23%	18.91%	18.59%	18.26%	17.92%	17.58%	17.23%	16.88%	16.51%	16.14%	15.75%	15.36%
$380	21.61%	21.32%	21.03%	20.73%	20.43%	20.12%	19.80%	19.48%	19.15%	18.82%	18.48%	18.13%	17.77%	17.40%
$400	23.36%	23.09%	22.82%	22.54%	22.25%	21.96%	21.66%	21.36%	21.06%	20.74%	20.42%	20.09%	19.76%	19.42%
$420	25.11%	24.86%	24.60%	24.33%	24.06%	23.79%	23.51%	23.23%	22.94%	22.65%	22.35%	22.04%	21.73%	21.41%

The Complete Car Cost Guide

See instructions on page 99A.

60 Month Lease

$20,000 – Acquisition Price — Capital Cost Reduction: $500

Monthly Payment	\-\- Residual Value \-\-													
	48%	46%	44%	42%	40%	38%	36%	34%	32%	30%	28%	26%	24%	22%
$260	7.76%	7.32%	6.86%	6.39%	5.91%	5.42%	4.91%	4.38%	3.84%	3.28%	2.71%	2.11%	1.49%	0.84%
$280	9.36%	8.94%	8.51%	8.07%	7.61%	7.14%	6.66%	6.17%	5.65%	5.13%	4.58%	4.02%	3.44%	2.84%
$300	10.96%	10.56%	10.15%	9.73%	9.29%	8.85%	8.39%	7.93%	7.44%	6.95%	6.44%	5.91%	5.37%	4.80%
$320	12.55%	12.17%	11.77%	11.37%	10.96%	10.54%	10.11%	9.67%	9.22%	8.75%	8.27%	7.77%	7.26%	6.73%
$340	14.13%	13.76%	13.39%	13.01%	12.62%	12.22%	11.81%	11.40%	10.97%	10.53%	10.07%	9.61%	9.13%	8.64%
$360	15.70%	15.35%	15.00%	14.63%	14.27%	13.89%	13.50%	13.11%	12.70%	12.29%	11.86%	11.42%	10.97%	10.51%
$380	17.26%	16.93%	16.59%	16.25%	15.90%	15.54%	15.18%	14.80%	14.42%	14.03%	13.62%	13.21%	12.79%	12.36%
$400	18.81%	18.50%	18.18%	17.85%	17.52%	17.18%	16.84%	16.48%	16.12%	15.75%	15.37%	14.98%	14.59%	14.18%
$420	20.36%	20.06%	19.76%	19.45%	19.13%	18.81%	18.48%	18.15%	17.81%	17.46%	17.10%	16.73%	16.36%	15.98%
$440	21.90%	21.61%	21.33%	21.03%	20.73%	20.43%	20.12%	19.80%	19.48%	19.15%	18.81%	18.47%	18.11%	17.75%
$460	23.43%	23.16%	22.89%	22.61%	22.32%	22.03%	21.74%	21.44%	21.13%	20.82%	20.50%	20.18%	19.85%	19.51%

$25,000 – Acquisition Price — Capital Cost Reduction: $500

Monthly Payment	Residual Value													
	48%	46%	44%	42%	40%	38%	36%	34%	32%	30%	28%	26%	24%	22%
$325	7.60%	7.16%	6.70%	6.23%	5.75%	5.25%	4.75%	4.22%	3.68%	3.12%	2.54%	1.94%	1.32%	0.67%
$350	9.20%	8.78%	8.34%	7.90%	7.44%	6.97%	6.49%	5.99%	5.48%	4.96%	4.41%	3.85%	3.27%	2.66%
$375	10.79%	10.39%	9.98%	9.55%	9.12%	8.68%	8.22%	7.75%	7.27%	6.77%	6.26%	5.73%	5.19%	4.62%
$400	12.37%	11.99%	11.60%	11.20%	10.79%	10.36%	9.93%	9.49%	9.03%	8.57%	8.08%	7.59%	7.08%	6.55%
$425	13.94%	13.58%	13.21%	12.83%	12.44%	12.04%	11.63%	11.21%	10.78%	10.34%	9.89%	9.42%	8.94%	8.44%
$450	15.51%	15.16%	14.81%	14.45%	14.08%	13.70%	13.31%	12.92%	12.51%	12.09%	11.67%	11.23%	10.78%	10.31%
$475	17.07%	16.74%	16.40%	16.06%	15.70%	15.35%	14.98%	14.60%	14.22%	13.83%	13.43%	13.01%	12.59%	12.15%
$500	18.61%	18.30%	17.98%	17.65%	17.32%	16.98%	16.63%	16.28%	15.92%	15.55%	15.17%	14.78%	14.38%	13.97%
$525	20.16%	19.86%	19.55%	19.24%	18.93%	18.60%	18.28%	17.94%	17.60%	17.25%	16.89%	16.52%	16.15%	15.76%
$550	21.69%	21.41%	21.12%	20.82%	20.52%	20.22%	19.90%	19.59%	19.26%	18.93%	18.59%	18.25%	17.90%	17.53%
$575	23.22%	22.95%	22.67%	22.39%	22.11%	21.82%	21.52%	21.22%	20.91%	20.60%	20.28%	19.96%	19.62%	19.28%

$30,000 – Acquisition Price — Capital Cost Reduction: $500

Monthly Payment	Residual Value													
	48%	46%	44%	42%	40%	38%	36%	34%	32%	30%	28%	26%	24%	22%
$400	8.03%	7.59%	7.14%	6.68%	6.21%	5.72%	5.22%	4.70%	4.17%	3.62%	3.05%	2.47%	1.86%	1.23%
$425	9.35%	8.94%	8.50%	8.06%	7.61%	7.15%	6.67%	6.18%	5.67%	5.15%	4.61%	4.05%	3.47%	2.87%
$450	10.68%	10.27%	9.86%	9.44%	9.01%	8.56%	8.10%	7.63%	7.15%	6.65%	6.14%	5.61%	5.07%	4.50%
$475	11.99%	11.61%	11.21%	10.81%	10.39%	9.97%	9.53%	9.08%	8.62%	8.15%	7.66%	7.16%	6.64%	6.10%
$500	13.30%	12.93%	12.55%	12.16%	11.77%	11.36%	10.94%	10.52%	10.08%	9.63%	9.16%	8.69%	8.19%	7.69%
$525	14.61%	14.25%	13.89%	13.52%	13.14%	12.75%	12.35%	11.94%	11.52%	11.09%	10.65%	10.20%	9.73%	9.25%
$550	15.90%	15.56%	15.21%	14.86%	14.49%	14.12%	13.74%	13.35%	12.95%	12.54%	12.12%	11.69%	11.25%	10.80%
$575	17.20%	16.87%	16.54%	16.19%	15.85%	15.49%	15.13%	14.75%	14.37%	13.98%	13.58%	13.17%	12.75%	12.32%
$600	18.48%	18.17%	17.85%	17.52%	17.19%	16.85%	16.50%	16.15%	15.78%	15.41%	15.03%	14.64%	14.24%	13.83%
$625	19.77%	19.46%	19.16%	18.84%	18.53%	18.20%	17.87%	17.53%	17.18%	16.83%	16.46%	16.09%	15.71%	15.32%
$650	21.04%	20.75%	20.46%	20.16%	19.85%	19.54%	19.22%	18.90%	18.57%	18.23%	17.89%	17.53%	17.17%	16.80%

$40,000 – Acquisition Price — Capital Cost Reduction: $1,000

Monthly Payment	Residual Value													
	48%	46%	44%	42%	40%	38%	36%	34%	32%	30%	28%	26%	24%	22%
$550	8.96%	8.54%	8.10%	7.65%	7.19%	6.71%	6.22%	5.72%	5.20%	4.67%	4.12%	3.55%	2.96%	2.34%
$575	9.96%	9.55%	9.13%	8.69%	8.24%	7.78%	7.31%	6.83%	6.33%	5.81%	5.28%	4.73%	4.17%	3.58%
$600	10.96%	10.56%	10.15%	9.73%	9.29%	8.85%	8.39%	7.93%	7.44%	6.95%	6.44%	5.91%	5.37%	4.80%
$625	11.95%	11.56%	11.17%	10.76%	10.34%	9.91%	9.47%	9.02%	8.55%	8.08%	7.58%	7.08%	6.55%	6.01%
$650	12.94%	12.57%	12.18%	11.78%	11.38%	10.96%	10.54%	10.10%	9.66%	9.19%	8.72%	8.23%	7.73%	7.21%
$675	13.93%	13.56%	13.19%	12.81%	12.41%	12.01%	11.60%	11.18%	10.75%	10.31%	9.85%	9.38%	8.90%	8.40%
$700	14.91%	14.56%	14.19%	13.82%	13.45%	13.06%	12.66%	12.25%	11.84%	11.41%	10.97%	10.52%	10.05%	9.58%
$725	15.89%	15.55%	15.20%	14.84%	14.47%	14.10%	13.71%	13.32%	12.92%	12.50%	12.08%	11.65%	11.20%	10.74%
$750	16.87%	16.53%	16.19%	15.85%	15.49%	15.13%	14.76%	14.38%	13.99%	13.59%	13.19%	12.77%	12.34%	11.90%
$775	17.84%	17.52%	17.19%	16.85%	16.51%	16.16%	15.80%	15.43%	15.06%	14.67%	14.28%	13.88%	13.47%	13.04%
$800	18.81%	18.50%	18.18%	17.85%	17.52%	17.18%	16.84%	16.48%	16.12%	15.75%	15.37%	14.98%	14.59%	14.18%

Appendices

Appendix I

Monthly Payments

The Complete Car Cost Guide

If you know the amount you want to borrow, this appendix will allow you to calculate your monthly payment. It will also help you decide how much you can afford to spend on a new car.

To use this table to determine monthly payments:
1. Choose one of the following payment terms (months).
2. Read across the top row, "Loan Amount," and find the dollar figure that is closest to the actual amount of your loan.
3. Read down the column to the row corresponding to the interest rate closest to your actual rate.
4. The figure at the intersection is your monthly payment.

To use this table to determine a car price:
1. Choose one of the following payment terms (months).
2. Read down the column to find the interest rate closest to your expected rate.
3. Read across the row to the dollar figure which you can afford to pay each month.
4. Read up the column to the top row to find the dollar amount you can afford to pay for your new car.

24 Months Loan Amount

Interest Rate	$5,000	$7,000	$9,000	$10,000	$11,000	$12,000	$13,000	$14,000	$15,000	$20,000	$25,000
5.00%	$219	$307	$395	$439	$483	$526	$570	$614	$658	$877	$1,097
7.00%	$224	$313	$403	$448	$492	$537	$582	$627	$672	$895	$1,119
9.00%	$228	$320	$411	$457	$503	$548	$594	$640	$685	$914	$1,142
9.50%	$230	$321	$413	$459	$505	$551	$597	$643	$689	$918	$1,148
9.75%	$230	$322	$414	$460	$506	$552	$598	$644	$690	$921	$1,151
10.00%	$231	$323	$415	$461	$508	$554	$600	$646	$692	$923	$1,154
10.25%	$231	$324	$416	$463	$509	$555	$601	$648	$694	$925	$1,157
10.50%	$232	$325	$417	$464	$510	$557	$603	$649	$696	$928	$1,159
10.75%	$232	$325	$418	$465	$511	$558	$604	$651	$697	$930	$1,162
11.00%	$233	$326	$419	$466	$513	$559	$606	$653	$699	$932	$1,165
12.00%	$235	$330	$424	$471	$518	$565	$612	$659	$706	$941	$1,177

36 Months Loan Amount

Interest Rate	$5,000	$7,000	$9,000	$10,000	$11,000	$12,000	$13,000	$14,000	$15,000	$20,000	$25,000
5.00%	$150	$210	$270	$300	$330	$360	$390	$420	$450	$599	$749
7.00%	$154	$216	$278	$309	$340	$371	$401	$432	$463	$618	$772
9.00%	$159	$223	$286	$318	$350	$382	$413	$445	$477	$636	$795
9.50%	$160	$224	$288	$320	$352	$384	$416	$448	$480	$641	$801
9.75%	$161	$225	$289	$321	$354	$386	$418	$450	$482	$643	$804
10.00%	$161	$226	$290	$323	$355	$387	$419	$452	$484	$645	$807
10.25%	$162	$227	$291	$324	$356	$389	$421	$453	$486	$648	$810
10.50%	$163	$228	$293	$325	$358	$390	$423	$455	$488	$650	$813
10.75%	$163	$228	$294	$326	$359	$391	$424	$457	$489	$652	$816
11.00%	$164	$229	$295	$327	$360	$393	$426	$458	$491	$655	$818
12.00%	$166	$233	$299	$332	$365	$399	$432	$465	$498	$664	$830

The Complete Car Cost Guide

42 Months Loan Amount

Interest Rate	$5,000	$7,000	$9,000	$10,000	$11,000	$12,000	$13,000	$14,000	$15,000	$20,000	$25,000
5.00%	$130	$182	$234	$260	$286	$312	$338	$364	$390	$520	$650
7.00%	$135	$188	$242	$269	$296	$323	$350	$377	$404	$538	$673
9.00%	$139	$195	$251	$278	$306	$334	$362	$390	$418	$557	$696
9.50%	$140	$197	$253	$281	$309	$337	$365	$393	$421	$562	$702
9.75%	$141	$197	$254	$282	$310	$338	$367	$395	$423	$564	$705
10.00%	$142	$198	$255	$283	$311	$340	$368	$396	$425	$566	$708
10.25%	$142	$199	$256	$284	$313	$341	$370	$398	$427	$569	$711
10.50%	$143	$200	$257	$286	$314	$343	$371	$400	$428	$571	$714
10.75%	$143	$201	$258	$287	$315	$344	$373	$401	$430	$573	$717
11.00%	$144	$202	$259	$288	$317	$346	$374	$403	$432	$576	$720
12.00%	$146	$205	$263	$293	$322	$351	$381	$410	$439	$586	$732

48 Months Loan Amount

Interest Rate	$5,000	$7,000	$9,000	$10,000	$11,000	$12,000	$13,000	$14,000	$15,000	$20,000	$25,000
5.00%	$115	$161	$207	$230	$253	$276	$299	$322	$345	$461	$576
7.00%	$120	$168	$216	$239	$263	$287	$311	$335	$359	$479	$599
9.00%	$124	$174	$224	$249	$274	$299	$324	$348	$373	$498	$622
9.50%	$126	$176	$226	$251	$276	$301	$327	$352	$377	$502	$628
9.75%	$126	$177	$227	$252	$278	$303	$328	$353	$379	$505	$631
10.00%	$127	$178	$228	$254	$279	$304	$330	$355	$380	$507	$634
10.25%	$127	$178	$229	$255	$280	$306	$331	$357	$382	$510	$637
10.50%	$128	$179	$230	$256	$282	$307	$333	$358	$384	$512	$640
10.75%	$129	$180	$232	$257	$283	$309	$334	$360	$386	$514	$643
11.00%	$129	$181	$233	$258	$284	$310	$336	$362	$388	$517	$646
12.00%	$132	$184	$237	$263	$290	$316	$342	$369	$395	$527	$658

60 Months Loan Amount

Interest Rate	$5,000	$7,000	$9,000	$10,000	$11,000	$12,000	$13,000	$14,000	$15,000	$20,000	$25,000
5.00%	$94	$132	$170	$189	$208	$226	$245	$264	$283	$377	$472
7.00%	$99	$139	$178	$198	$218	$238	$257	$277	$297	$396	$495
9.00%	$104	$145	$187	$208	$228	$249	$270	$291	$311	$415	$519
9.50%	$105	$147	$189	$210	$231	$252	$273	$294	$315	$420	$525
9.75%	$106	$148	$190	$211	$232	$253	$275	$296	$317	$422	$528
10.00%	$106	$149	$191	$212	$234	$255	$276	$297	$319	$425	$531
10.25%	$107	$150	$192	$214	$235	$256	$278	$299	$321	$427	$534
10.50%	$107	$150	$193	$215	$236	$258	$279	$301	$322	$430	$537
10.75%	$108	$151	$195	$216	$238	$259	$281	$303	$324	$432	$540
11.00%	$109	$152	$196	$217	$239	$261	$283	$304	$326	$435	$544
12.00%	$111	$156	$200	$222	$245	$267	$289	$311	$334	$445	$556

Appendix J

Discount Financing

This appendix will show you the dollar value of a discount financing "deal." To use this table, you need to know the "discounted" interest rate, the length and amount of the loan, and the interest rate you would normally pay. If need be, you can estimate any of the above.

To use the tables:
1. Choose the chart on this page or on the following pages to compare discount financing rates from 2.9% to 7.9%.
2. Choose one of the following payment terms.

2.90% "Discount" Interest Rate

24 Months Loan Amount

Normal rate	$5,000	$7,000	$9,000	$10,000	$11,000	$12,000	$13,000	$14,000	$15,000	$20,000	$25,000
8.00%	$253	$354	$456	$506	$557	$608	$658	$709	$760	$1,013	$1,266
9.00%	$301	$421	$541	$601	$662	$722	$782	$842	$902	$1,203	$1,504
9.25%	$313	$438	$563	$625	$688	$750	$813	$875	$938	$1,250	$1,563
9.50%	$324	$454	$584	$648	$713	$778	$843	$908	$973	$1,297	$1,621
9.75%	$336	$470	$605	$672	$739	$806	$873	$941	$1,008	$1,344	$1,680
10.00%	$348	$487	$626	$695	$765	$834	$904	$973	$1,043	$1,390	$1,738
10.25%	$359	$503	$647	$718	$790	$862	$934	$1,006	$1,078	$1,437	$1,796
10.50%	$371	$519	$667	$742	$816	$890	$964	$1,038	$1,112	$1,483	$1,854
10.75%	$382	$535	$688	$765	$841	$918	$994	$1,070	$1,147	$1,529	$1,912
11.00%	$394	$551	$709	$788	$866	$945	$1,024	$1,103	$1,181	$1,575	$1,969
11.50%	$417	$583	$750	$833	$917	$1,000	$1,083	$1,167	$1,250	$1,667	$2,083

36 Months Loan Amount

Normal rate	$5,000	$7,000	$9,000	$10,000	$11,000	$12,000	$13,000	$14,000	$15,000	$20,000	$25,000
8.00%	$367	$514	$660	$734	$807	$880	$954	$1,027	$1,101	$1,467	$1,834
9.00%	$434	$608	$782	$869	$956	$1,042	$1,129	$1,216	$1,303	$1,737	$2,172
9.25%	$451	$631	$812	$902	$992	$1,082	$1,173	$1,263	$1,353	$1,804	$2,255
9.50%	$468	$655	$842	$935	$1,029	$1,122	$1,216	$1,309	$1,403	$1,870	$2,338
9.75%	$484	$678	$871	$968	$1,065	$1,162	$1,259	$1,355	$1,452	$1,936	$2,421
10.00%	$501	$701	$901	$1,001	$1,101	$1,201	$1,301	$1,401	$1,502	$2,002	$2,503
10.25%	$517	$724	$930	$1,034	$1,137	$1,240	$1,344	$1,447	$1,551	$2,067	$2,584
10.50%	$533	$746	$960	$1,066	$1,173	$1,279	$1,386	$1,493	$1,599	$2,132	$2,665
10.75%	$549	$769	$989	$1,098	$1,208	$1,318	$1,428	$1,538	$1,648	$2,197	$2,746
11.00%	$565	$791	$1,018	$1,131	$1,244	$1,357	$1,470	$1,583	$1,696	$2,261	$2,827
11.50%	$597	$836	$1,075	$1,194	$1,314	$1,433	$1,553	$1,672	$1,792	$2,389	$2,986

48 Months Loan Amount

Normal rate	$5,000	$7,000	$9,000	$10,000	$11,000	$12,000	$13,000	$14,000	$15,000	$20,000	$25,000
8.00%	$476	$666	$856	$951	$1,047	$1,142	$1,237	$1,332	$1,427	$1,903	$2,379
9.00%	$562	$786	$1,011	$1,123	$1,235	$1,348	$1,460	$1,572	$1,685	$2,246	$2,808
9.25%	$583	$816	$1,049	$1,165	$1,282	$1,398	$1,515	$1,631	$1,748	$2,331	$2,913
9.50%	$604	$845	$1,087	$1,207	$1,328	$1,449	$1,569	$1,690	$1,811	$2,414	$3,018
9.75%	$624	$874	$1,124	$1,249	$1,374	$1,499	$1,624	$1,748	$1,873	$2,498	$3,122
10.00%	$645	$903	$1,161	$1,290	$1,419	$1,548	$1,677	$1,806	$1,935	$2,581	$3,226
10.25%	$666	$932	$1,198	$1,331	$1,464	$1,598	$1,731	$1,864	$1,997	$2,663	$3,328
10.50%	$686	$961	$1,235	$1,372	$1,509	$1,647	$1,784	$1,921	$2,058	$2,744	$3,430
10.75%	$706	$989	$1,271	$1,413	$1,554	$1,695	$1,837	$1,978	$2,119	$2,825	$3,532
11.00%	$727	$1,017	$1,308	$1,453	$1,598	$1,744	$1,889	$2,034	$2,180	$2,906	$3,633
11.50%	$766	$1,073	$1,379	$1,533	$1,686	$1,839	$1,993	$2,146	$2,299	$3,066	$3,832

Appendices

The Complete Car Cost Guide

3. Read across the top row, "Loan Amount," and find the dollar figure that is closest to the actual amount that you are financing.
4. Read down the column to the row corresponding to your normal, non-discounted interest rate.
5. The figure in the intersection is the dollar value of your discount financing arrangement. Compare this figure to any "cash rebate" offered. If the rebate amount is higher, choose the rebate. If the rebate amount is lower, choose the discount financing.

3.90% "Discount" Interest Rate

24 Months Loan Amount

Normal rate	$5,000	$7,000	$9,000	$10,000	$11,000	$12,000	$13,000	$14,000	$15,000	$20,000	$25,000
8.00%	$204	$286	$368	$408	$449	$490	$531	$572	$613	$817	$1,021
9.00%	$252	$353	$454	$504	$555	$605	$656	$706	$757	$1,009	$1,261
9.25%	$264	$370	$475	$528	$581	$634	$687	$739	$792	$1,056	$1,320
9.50%	$276	$386	$497	$552	$607	$662	$717	$773	$828	$1,104	$1,380
9.75%	$288	$403	$518	$576	$633	$691	$748	$806	$863	$1,151	$1,439
10.00%	$300	$419	$539	$599	$659	$719	$779	$839	$899	$1,198	$1,498
10.25%	$311	$436	$560	$623	$685	$747	$809	$872	$934	$1,245	$1,556
10.50%	$323	$452	$581	$646	$711	$775	$840	$904	$969	$1,292	$1,615
10.75%	$335	$468	$602	$669	$736	$803	$870	$937	$1,004	$1,338	$1,673
11.00%	$346	$485	$623	$692	$762	$831	$900	$969	$1,039	$1,385	$1,731
11.50%	$369	$517	$665	$739	$813	$886	$960	$1,034	$1,108	$1,477	$1,847

36 Months Loan Amount

Normal rate	$5,000	$7,000	$9,000	$10,000	$11,000	$12,000	$13,000	$14,000	$15,000	$20,000	$25,000
8.00%	$296	$415	$533	$593	$652	$711	$770	$830	$889	$1,185	$1,481
9.00%	$365	$511	$657	$730	$803	$876	$949	$1,021	$1,094	$1,459	$1,824
9.25%	$382	$534	$687	$763	$840	$916	$993	$1,069	$1,145	$1,527	$1,909
9.50%	$399	$558	$717	$797	$877	$957	$1,036	$1,116	$1,196	$1,594	$1,993
9.75%	$415	$581	$748	$831	$914	$997	$1,080	$1,163	$1,246	$1,661	$2,077
10.00%	$432	$605	$778	$864	$950	$1,037	$1,123	$1,210	$1,296	$1,728	$2,160
10.25%	$449	$628	$807	$897	$987	$1,076	$1,166	$1,256	$1,346	$1,794	$2,243
10.50%	$465	$651	$837	$930	$1,023	$1,116	$1,209	$1,302	$1,395	$1,860	$2,325
10.75%	$481	$674	$867	$963	$1,059	$1,155	$1,252	$1,348	$1,444	$1,926	$2,407
11.00%	$498	$697	$896	$996	$1,095	$1,195	$1,294	$1,394	$1,493	$1,991	$2,489
11.50%	$530	$742	$954	$1,060	$1,166	$1,272	$1,378	$1,484	$1,590	$2,121	$2,651

48 Months Loan Amount

Normal rate	$5,000	$7,000	$9,000	$10,000	$11,000	$12,000	$13,000	$14,000	$15,000	$20,000	$25,000
8.00%	$385	$539	$693	$770	$846	$923	$1,000	$1,077	$1,154	$1,539	$1,924
9.00%	$472	$661	$850	$945	$1,039	$1,134	$1,228	$1,322	$1,417	$1,889	$2,362
9.25%	$494	$691	$889	$988	$1,086	$1,185	$1,284	$1,383	$1,482	$1,975	$2,469
9.50%	$515	$721	$927	$1,030	$1,133	$1,237	$1,340	$1,443	$1,546	$2,061	$2,576
9.75%	$536	$751	$966	$1,073	$1,180	$1,288	$1,395	$1,502	$1,609	$2,146	$2,682
10.00%	$558	$781	$1,004	$1,115	$1,227	$1,338	$1,450	$1,561	$1,673	$2,230	$2,788
10.25%	$579	$810	$1,041	$1,157	$1,273	$1,388	$1,504	$1,620	$1,736	$2,314	$2,893
10.50%	$599	$839	$1,079	$1,199	$1,319	$1,438	$1,558	$1,678	$1,798	$2,397	$2,997
10.75%	$620	$868	$1,116	$1,240	$1,364	$1,488	$1,612	$1,736	$1,860	$2,480	$3,100
11.00%	$641	$897	$1,153	$1,281	$1,409	$1,537	$1,665	$1,794	$1,922	$2,562	$3,203
11.50%	$681	$954	$1,226	$1,363	$1,499	$1,635	$1,771	$1,908	$2,044	$2,725	$3,406

The Complete Car Cost Guide

Appendix J *continued*

Discount Financing

This appendix will show you the dollar value of a discount financing "deal." To use this table, you need to know the "discounted" interest rate, the length and amount of the loan, and the interest rate you would normally pay. If need be, you can estimate any of the above.

To use the tables:
1. Choose the chart on this page or on the pages preceding or following to compare discount financing rates from 2.9% to 7.9%.
2. Choose one of the following payment terms.

4.90% "Discount" Interest Rate

24 Months Loan Amount

Normal rate	$5,000	$7,000	$9,000	$10,000	$11,000	$12,000	$13,000	$14,000	$15,000	$20,000	$25,000
8.00%	$155	$217	$279	$310	$341	$372	$403	$434	$465	$619	$774
9.00%	$203	$285	$366	$407	$447	$488	$529	$569	$610	$813	$1,017
9.25%	$215	$302	$388	$431	$474	$517	$560	$603	$646	$862	$1,077
9.50%	$227	$318	$409	$455	$500	$546	$591	$637	$682	$909	$1,137
9.75%	$239	$335	$431	$479	$526	$574	$622	$670	$718	$957	$1,197
10.00%	$251	$352	$452	$502	$553	$603	$653	$703	$754	$1,005	$1,256
10.25%	$263	$368	$473	$526	$579	$631	$684	$737	$789	$1,052	$1,315
10.50%	$275	$385	$495	$550	$605	$660	$715	$770	$825	$1,099	$1,374
10.75%	$287	$401	$516	$573	$631	$688	$745	$803	$860	$1,147	$1,433
11.00%	$298	$418	$537	$597	$656	$716	$776	$835	$895	$1,193	$1,492
11.50%	$322	$450	$579	$643	$708	$772	$836	$901	$965	$1,287	$1,608

36 Months Loan Amount

Normal rate	$5,000	$7,000	$9,000	$10,000	$11,000	$12,000	$13,000	$14,000	$15,000	$20,000	$25,000
8.00%	$225	$315	$405	$450	$495	$540	$585	$630	$675	$900	$1,125
9.00%	$295	$412	$530	$589	$648	$707	$766	$825	$884	$1,178	$1,473
9.25%	$312	$436	$561	$624	$686	$748	$811	$873	$935	$1,247	$1,559
9.50%	$329	$460	$592	$658	$724	$789	$855	$921	$987	$1,315	$1,644
9.75%	$346	$484	$623	$692	$761	$830	$899	$968	$1,038	$1,383	$1,729
10.00%	$363	$508	$653	$726	$798	$871	$943	$1,016	$1,088	$1,451	$1,814
10.25%	$380	$531	$683	$759	$835	$911	$987	$1,063	$1,139	$1,518	$1,898
10.50%	$396	$555	$713	$793	$872	$951	$1,030	$1,110	$1,189	$1,585	$1,982
10.75%	$413	$578	$743	$826	$909	$991	$1,074	$1,156	$1,239	$1,652	$2,065
11.00%	$430	$601	$773	$859	$945	$1,031	$1,117	$1,203	$1,289	$1,718	$2,148
11.50%	$462	$647	$832	$925	$1,017	$1,110	$1,202	$1,295	$1,387	$1,850	$2,312

48 Months Loan Amount

Normal rate	$5,000	$7,000	$9,000	$10,000	$11,000	$12,000	$13,000	$14,000	$15,000	$20,000	$25,000
8.00%	$293	$410	$527	$585	$644	$702	$761	$819	$878	$1,171	$1,463
9.00%	$382	$535	$688	$764	$840	$917	$993	$1,069	$1,146	$1,528	$1,910
9.25%	$404	$565	$727	$808	$889	$969	$1,050	$1,131	$1,212	$1,616	$2,020
9.50%	$426	$596	$766	$851	$937	$1,022	$1,107	$1,192	$1,277	$1,703	$2,129
9.75%	$447	$626	$805	$895	$984	$1,074	$1,163	$1,253	$1,342	$1,790	$2,237
10.00%	$469	$656	$844	$938	$1,032	$1,125	$1,219	$1,313	$1,407	$1,876	$2,345
10.25%	$490	$686	$883	$981	$1,079	$1,177	$1,275	$1,373	$1,471	$1,961	$2,451
10.50%	$512	$716	$921	$1,023	$1,125	$1,228	$1,330	$1,432	$1,535	$2,046	$2,558
10.75%	$533	$746	$959	$1,065	$1,172	$1,278	$1,385	$1,491	$1,598	$2,130	$2,663
11.00%	$554	$775	$996	$1,107	$1,218	$1,329	$1,439	$1,550	$1,661	$2,214	$2,768
11.50%	$595	$833	$1,071	$1,190	$1,309	$1,428	$1,547	$1,666	$1,785	$2,380	$2,975

Appendices

The Complete Car Cost Guide

3. Read across the top row, "Loan Amount," and find the dollar figure that is closest to the actual amount that you are financing.
4. Read down the column to the row corresponding to your normal, non-discounted interest rate.
5. The figure in the intersection is the dollar value of your discount financing arrangement. Compare this figure to any "cash rebate" offered. If the rebate amount is higher, choose the rebate. If the rebate amount is lower, choose the discount financing.

5.90% "Discount" Interest Rate

24 Months Loan Amount

Normal rate	$5,000	$7,000	$9,000	$10,000	$11,000	$12,000	$13,000	$14,000	$15,000	$20,000	$25,000
8.00%	$105	$147	$189	$210	$231	$253	$274	$295	$316	$421	$526
9.00%	$154	$216	$278	$308	$339	$370	$401	$432	$463	$617	$771
9.25%	$166	$233	$299	$333	$366	$399	$433	$466	$499	$665	$832
9.50%	$178	$250	$321	$357	$393	$428	$464	$500	$535	$714	$892
9.75%	$191	$267	$343	$381	$419	$457	$495	$534	$572	$762	$953
10.00%	$203	$284	$365	$405	$446	$486	$527	$567	$608	$810	$1,013
10.25%	$215	$300	$386	$429	$472	$515	$558	$601	$644	$858	$1,073
10.50%	$226	$317	$408	$453	$498	$544	$589	$634	$679	$906	$1,132
10.75%	$238	$334	$429	$477	$524	$572	$620	$667	$715	$953	$1,192
11.00%	$250	$350	$450	$500	$550	$600	$651	$701	$751	$1,001	$1,251
11.50%	$274	$383	$493	$548	$602	$657	$712	$767	$821	$1,095	$1,369

36 Months Loan Amount

Normal rate	$5,000	$7,000	$9,000	$10,000	$11,000	$12,000	$13,000	$14,000	$15,000	$20,000	$25,000
8.00%	$153	$214	$276	$306	$337	$368	$398	$429	$459	$613	$766
9.00%	$224	$313	$403	$448	$492	$537	$582	$627	$671	$895	$1,119
9.25%	$241	$338	$434	$482	$531	$579	$627	$675	$724	$965	$1,206
9.50%	$259	$362	$465	$517	$569	$620	$672	$724	$776	$1,034	$1,293
9.75%	$276	$386	$496	$552	$607	$662	$717	$772	$827	$1,103	$1,379
10.00%	$293	$410	$527	$586	$644	$703	$762	$820	$879	$1,172	$1,465
10.25%	$310	$434	$558	$620	$682	$744	$806	$868	$930	$1,240	$1,550
10.50%	$327	$458	$589	$654	$719	$785	$850	$916	$981	$1,308	$1,635
10.75%	$344	$481	$619	$688	$757	$825	$894	$963	$1,032	$1,376	$1,720
11.00%	$361	$505	$649	$721	$794	$866	$938	$1,010	$1,082	$1,443	$1,804
11.50%	$394	$552	$709	$788	$867	$946	$1,025	$1,104	$1,182	$1,577	$1,971

48 Months Loan Amount

Normal rate	$5,000	$7,000	$9,000	$10,000	$11,000	$12,000	$13,000	$14,000	$15,000	$20,000	$25,000
8.00%	$199	$279	$359	$399	$439	$479	$519	$558	$598	$798	$997
9.00%	$291	$407	$523	$581	$639	$697	$755	$813	$872	$1,162	$1,453
9.25%	$313	$438	$563	$626	$688	$751	$814	$876	$939	$1,252	$1,564
9.50%	$335	$469	$603	$670	$737	$804	$871	$938	$1,005	$1,341	$1,676
9.75%	$357	$500	$643	$714	$786	$857	$929	$1,000	$1,072	$1,429	$1,786
10.00%	$379	$531	$683	$758	$834	$910	$986	$1,062	$1,138	$1,517	$1,896
10.25%	$401	$561	$722	$802	$882	$962	$1,043	$1,123	$1,203	$1,604	$2,005
10.50%	$423	$592	$761	$845	$930	$1,014	$1,099	$1,183	$1,268	$1,691	$2,113
10.75%	$444	$622	$799	$888	$977	$1,066	$1,155	$1,244	$1,332	$1,777	$2,221
11.00%	$466	$652	$838	$931	$1,024	$1,117	$1,210	$1,303	$1,397	$1,862	$2,328
11.50%	$508	$711	$914	$1,016	$1,117	$1,219	$1,320	$1,422	$1,524	$2,031	$2,539

The Complete Car Cost Guide

Appendix J *continued*

This appendix will show you the dollar value of a discount financing "deal." To use this table, you need to know the "discounted" interest rate, the length and amount of the loan, and the interest rate you would normally pay. If need be, you can estimate any of the above.

Discount Financing

To use the tables:
1. Choose the chart on this page or on the previous pages to compare discount financing rates from 4.9% to 7.9%.
2. Choose one of the following payment terms.

6.90% "Discount" Interest Rate

24 Months Loan Amount

Normal rate	$5,000	$7,000	$9,000	$10,000	$11,000	$12,000	$13,000	$14,000	$15,000	$20,000	$25,000
8.00%	$55	$77	$100	$111	$122	$133	$144	$155	$166	$221	$276
9.00%	$105	$147	$189	$210	$231	$252	$272	$293	$314	$419	$524
9.25%	$117	$164	$211	$234	$258	$281	$304	$328	$351	$468	$585
9.50%	$129	$181	$233	$259	$284	$310	$336	$362	$388	$517	$646
9.75%	$141	$198	$255	$283	$311	$340	$368	$396	$424	$566	$707
10.00%	$154	$215	$276	$307	$338	$369	$399	$430	$461	$614	$768
10.25%	$166	$232	$298	$331	$365	$398	$431	$464	$497	$663	$829
10.50%	$178	$249	$320	$356	$391	$427	$462	$498	$533	$711	$889
10.75%	$190	$266	$342	$380	$418	$455	$493	$531	$569	$759	$949
11.00%	$202	$282	$363	$403	$444	$484	$525	$565	$605	$807	$1,009
11.50%	$226	$316	$406	$451	$496	$541	$586	$632	$677	$902	$1,128

36 Months Loan Amount

Normal rate	$5,000	$7,000	$9,000	$10,000	$11,000	$12,000	$13,000	$14,000	$15,000	$20,000	$25,000
8.00%	$81	$113	$145	$161	$177	$193	$209	$226	$242	$322	$403
9.00%	$152	$213	$274	$305	$335	$365	$396	$426	$457	$609	$761
9.25%	$170	$238	$306	$340	$374	$408	$442	$476	$510	$680	$850
9.50%	$188	$263	$338	$375	$413	$450	$488	$525	$563	$750	$938
9.75%	$205	$287	$369	$410	$451	$492	$533	$574	$615	$820	$1,025
10.00%	$222	$311	$400	$445	$489	$534	$578	$623	$667	$890	$1,112
10.25%	$240	$336	$432	$480	$528	$576	$624	$671	$719	$959	$1,199
10.50%	$257	$360	$463	$514	$566	$617	$668	$720	$771	$1,028	$1,285
10.75%	$274	$384	$494	$548	$603	$658	$713	$768	$823	$1,097	$1,371
11.00%	$291	$408	$524	$583	$641	$699	$757	$816	$874	$1,165	$1,456
11.50%	$325	$455	$585	$650	$715	$780	$845	$910	$976	$1,301	$1,626

48 Months Loan Amount

Normal rate	$5,000	$7,000	$9,000	$10,000	$11,000	$12,000	$13,000	$14,000	$15,000	$20,000	$25,000
8.00%	$105	$147	$189	$210	$231	$252	$273	$294	$315	$420	$525
9.00%	$198	$277	$356	$396	$435	$475	$515	$554	$594	$792	$990
9.25%	$221	$309	$397	$442	$486	$530	$574	$618	$662	$883	$1,104
9.50%	$243	$341	$438	$487	$536	$584	$633	$682	$730	$974	$1,217
9.75%	$266	$372	$479	$532	$585	$638	$692	$745	$798	$1,064	$1,330
10.00%	$288	$404	$519	$577	$634	$692	$750	$807	$865	$1,153	$1,442
10.25%	$311	$435	$559	$621	$683	$745	$808	$870	$932	$1,242	$1,553
10.50%	$333	$466	$599	$665	$732	$798	$865	$931	$998	$1,331	$1,663
10.75%	$355	$496	$638	$709	$780	$851	$922	$993	$1,064	$1,418	$1,773
11.00%	$376	$527	$678	$753	$828	$903	$979	$1,054	$1,129	$1,506	$1,882
11.50%	$420	$587	$755	$839	$923	$1,007	$1,091	$1,175	$1,259	$1,678	$2,098

Appendices

The Complete Car Cost Guide

3. Read across the top row, "Loan Amount," and find the dollar figure that is closest to the actual amount that you are financing.
4. Read down the column to the row corresponding to your normal, non-discounted interest rate.
5. The figure in the intersection is the dollar value of your discount financing arrangement. Compare this figure to any "cash rebate" offered. If the rebate amount is higher, choose the rebate. If the rebate amount is lower, choose the discount financing.

7.90% "Discount" Interest Rate

24 Months Loan Amount

Normal rate	$5,000	$7,000	$9,000	$10,000	$11,000	$12,000	$13,000	$14,000	$15,000	$20,000	$25,000
8.00%	$5	$7	$9	$10	$11	$12	$13	$14	$15	$20	$25
9.00%	$55	$77	$99	$110	$121	$132	$143	$154	$165	$220	$275
9.25%	$67	$94	$121	$135	$148	$162	$175	$189	$202	$270	$337
9.50%	$80	$112	$144	$160	$176	$192	$207	$223	$239	$319	$399
9.75%	$92	$129	$166	$184	$203	$221	$239	$258	$276	$368	$461
10.00%	$104	$146	$188	$209	$230	$250	$271	$292	$313	$417	$522
10.25%	$117	$163	$210	$233	$256	$280	$303	$326	$350	$466	$583
10.50%	$129	$180	$232	$258	$283	$309	$335	$361	$386	$515	$644
10.75%	$141	$197	$254	$282	$310	$338	$366	$395	$423	$564	$705
11.00%	$153	$214	$275	$306	$337	$367	$398	$428	$459	$612	$765
11.50%	$177	$248	$319	$354	$390	$425	$460	$496	$531	$708	$885

36 Months Loan Amount

Normal rate	$5,000	$7,000	$9,000	$10,000	$11,000	$12,000	$13,000	$14,000	$15,000	$20,000	$25,000
8.00%	$7	$10	$13	$15	$16	$18	$19	$21	$22	$29	$37
9.00%	$80	$112	$144	$160	$176	$192	$208	$224	$240	$320	$401
9.25%	$98	$137	$177	$196	$216	$235	$255	$275	$294	$392	$490
9.50%	$116	$162	$209	$232	$255	$278	$301	$325	$348	$464	$580
9.75%	$134	$187	$241	$267	$294	$321	$348	$374	$401	$535	$668
10.00%	$151	$212	$272	$303	$333	$363	$394	$424	$454	$606	$757
10.25%	$169	$237	$304	$338	$372	$406	$439	$473	$507	$676	$845
10.50%	$186	$261	$336	$373	$410	$448	$485	$522	$559	$746	$932
10.75%	$204	$285	$367	$408	$449	$489	$530	$571	$612	$816	$1,019
11.00%	$221	$310	$398	$442	$487	$531	$575	$619	$664	$885	$1,106
11.50%	$256	$358	$460	$511	$562	$613	$665	$716	$767	$1,022	$1,278

48 Months Loan Amount

Normal rate	$5,000	$7,000	$9,000	$10,000	$11,000	$12,000	$13,000	$14,000	$15,000	$20,000	$25,000
8.00%	$10	$13	$17	$19	$21	$23	$25	$27	$29	$38	$48
9.00%	$104	$146	$188	$209	$229	$250	$271	$292	$313	$417	$521
9.25%	$128	$179	$230	$255	$281	$306	$332	$357	$383	$510	$638
9.50%	$151	$211	$271	$301	$332	$362	$392	$422	$452	$603	$753
9.75%	$174	$243	$313	$347	$382	$417	$451	$486	$521	$695	$868
10.00%	$196	$275	$354	$393	$432	$472	$511	$550	$589	$786	$982
10.25%	$219	$307	$394	$438	$482	$526	$570	$614	$657	$877	$1,096
10.50%	$242	$338	$435	$483	$532	$580	$628	$677	$725	$967	$1,208
10.75%	$264	$370	$475	$528	$581	$634	$686	$739	$792	$1,056	$1,320
11.00%	$286	$401	$515	$572	$630	$687	$744	$801	$859	$1,145	$1,431
11.50%	$330	$462	$594	$660	$726	$793	$859	$925	$991	$1,321	$1,651

Index

Index by Vehicle

Page	Vehicle	Manufacturer	Page	Vehicle	Manufacturer
12	100 (S, 4 door)	Audi	125	626 DX (S, 4 door)	Mazda
13	100 CS (S, 4 door)	Audi	126	626 ES (S, 4 door)	Mazda
14	100 CS Quattro (S, 4WD, 4 door)	Audi	126	626 LX (S, 4 door)	Mazda
14	100 CS Quattro (W, 4WD, 4 door)	Audi	20	740 i (S, 4 door)	BMW
13	100 S (S, 4 door)	Audi	21	740 iL (S, 4 door)	BMW
9	164 L (S, 4 door)	Alfa Romeo	21	750 iL (S, 4 door)	BMW
9	164 S (S, 4 door)	Alfa Romeo	22	850 Ci (C, 2 door)	BMW
131	190E 2.3 (S, 4 door)	Mercedes Benz	252	850 GLT (S, 4 door)	Volvo
132	190E 2.6 (S, 4 door)	Mercedes Benz	11	90 CS (S, 4 door)	Audi
251	240 (S, 4 door)	Volvo	12	90 CS Quattro Sport (S, 4WD, 4 door)	Audi
252	240 (W, 4 door)	Volvo	11	90 S (S, 4 door)	Audi
161	240SX (C, 2 door)	Nissan	213	900 S (CN, 2 door)	Saab
163	240SX (CN, 2 door)	Nissan	211	900 S (H, 2 door)	Saab
162	240SX (H, 2 door)	Nissan	212	900 S (S, 4 door)	Saab
161	240SX SE (C, 2 door)	Nissan	213	900 Turbo (CN, 2 door)	Saab
162	240SX SE (H, 2 door)	Nissan	212	900 Turbo (H, 2 door)	Saab
132	300 CE (C, 2 door)	Mercedes Benz	209	9000 CD (S, 4 door)	Saab
135	300 CE (CN, 2 door)	Mercedes Benz	210	9000 CD Turbo (S, 4 door)	Saab
133	300 D 2.5 Turbo (S, 4 door)	Mercedes Benz	210	9000 CDE (S, 4 door)	Saab
134	300 E (S, 4 door)	Mercedes Benz	211	9000 CDE Turbo (S, 4 door)	Saab
133	300 E 2.8 (S, 4 door)	Mercedes Benz	207	9000 CS (H, 4 door)	Saab
137	300 E 4Matic (S, 4WD, 4 door)	Mercedes Benz	208	9000 CS Turbo (H, 4 door)	Saab
134	300 SD (S, 4 door)	Mercedes Benz	208	9000 CSE (H, 4 door)	Saab
135	300 SE (S, 4 door)	Mercedes Benz	209	9000 CSE Turbo (H, 4 door)	Saab
136	300 SL (CN, 2 door)	Mercedes Benz	204	911 America (CN, 2 door)	Porsche
136	300 TE (W, 4 door)	Mercedes Benz	202	911 Carrera 2 (C, 2 door)	Porsche
137	300 TE 4Matic (W, 4WD, 4 door)	Mercedes Benz	203	911 Carrera 2 Cabriolet (CN, 2 door)	Porsche
148	3000GT (C, 2 door)	Mitsubishi	203	911 Carrera 2 Targa (C, 2 door)	Porsche
148	3000GT SL (C, 2 door)	Mitsubishi	204	911 Carrera 4 (C, 4WD, 2 door)	Porsche
149	3000GT VR-4 (C, 4WD, 2 door)	Mitsubishi	205	911 Carrera 4 Cabriolet (CN, 4WD, 2 door)	Porsche
163	300ZX (C, 2 door)	Nissan	205	911 Carrera 4 Targa (C, 4WD, 2 door)	Porsche
165	300ZX (CN, 2 door)	Nissan	202	911 RS America (C, 2 door)	Porsche
164	300ZX 2+2 (C, 2 door)	Nissan	206	928 GTS (C, 2 door)	Porsche
164	300ZX Turbo (C, 2 door)	Nissan	127	929 (S, 4 door)	Mazda
16	318 i (S, 4 door)	BMW	253	940 (S, 4 door)	Volvo
16	318 iS (C, 2 door)	BMW	254	940 (W, 4 door)	Volvo
124	323 (H, 2 door)	Mazda	253	940 Turbo (S, 4 door)	Volvo
125	323 SE (H, 2 door)	Mazda	254	940 Turbo (W, 4 door)	Volvo
17	325 i (S, 4 door)	BMW	255	960 (S, 4 door)	Volvo
18	325 iC (CN, 2 door)	BMW	255	960 (W, 4 door)	Volvo
17	325 iS (C, 2 door)	BMW	206	968 (C, 2 door)	Porsche
138	400 E (S, 4 door)	Mercedes Benz	207	968 Cabriolet (CN, 2 door)	Porsche
138	400 SEL (S, 4 door)	Mercedes Benz	182	Acclaim (S, 4 door)	Plymouth
139	500 E (S, 4 door)	Mercedes Benz	99	Accord DX (C, 2 door)	Honda
139	500 SEL (S, 4 door)	Mercedes Benz	101	Accord DX (S, 4 door)	Honda
140	500 SL (CN, 2 door)	Mercedes Benz	100	Accord EX (C, 2 door)	Honda
18	525 i (S, 4 door)	BMW	102	Accord EX (S, 4 door)	Honda
20	525 i Touring (W, 4 door)	BMW	103	Accord EX (W, 4 door)	Honda
19	535 i (S, 4 door)	BMW	99	Accord LX (C, 2 door)	Honda
140	600 SEL (S, 4 door)	Mercedes Benz	101	Accord LX (S, 4 door)	Honda

| H-Hatchback | C-Coupe | S-Sedan | CN-Convertible | 4WD-4-Wheel Drive | W-Wagons |

Index by Vehicle

Page	Vehicle	Manufacturer
103	Accord LX (W, 4 door)	Honda
100	Accord SE (C, 2 door)	Honda
102	Accord SE (S, 4 door)	Honda
173	Achieva S (C, 2 door)	Oldsmobile
174	Achieva S (S, 4 door)	Oldsmobile
173	Achieva SL (C, 2 door)	Oldsmobile
174	Achieva SL (S, 4 door)	Oldsmobile
35	Allante (CN, 2 door)	Cadillac
36	Allante HT (CN, 2 door)	Cadillac
167	Altima GLE (S, 4 door)	Nissan
166	Altima GXE (S, 4 door)	Nissan
166	Altima SE (S, 4 door)	Nissan
165	Altima XE (S, 4 door)	Nissan
40	Beretta (C, 2 door)	Chevrolet
41	Beretta GT (C, 2 door)	Chevrolet
41	Beretta GTZ (C, 2 door)	Chevrolet
190	Bonneville SE (S, 4 door)	Pontiac
190	Bonneville SSE (S, 4 door)	Pontiac
191	Bonneville SSEi (S, 4 door)	Pontiac
243	Cabriolet (CN, 2 door)	Volkswagen
244	Cabriolet Classic (CN, 2 door)	Volkswagen
42	Camaro (C, 2 door)	Chevrolet
42	Camaro Z28 (C, 2 door)	Chevrolet
230	Camry DX (S, 4 door)	Toyota
233	Camry DX (W, 4 door)	Toyota
231	Camry DX V6 (S, 4 door)	Toyota
230	Camry LE (S, 4 door)	Toyota
234	Camry LE (W, 4 door)	Toyota
232	Camry LE V6 (S, 4 door)	Toyota
234	Camry LE V6 (W, 4 door)	Toyota
231	Camry SE V6 (S, 4 door)	Toyota
232	Camry XLE (S, 4 door)	Toyota
233	Camry XLE V6 (S, 4 door)	Toyota
141	Capri (CN, 2 door)	Mercury
141	Capri XR2 (CN, 2 door)	Mercury
43	Caprice Classic (S, 4 door)	Chevrolet
44	Caprice Classic (W, 4 door)	Chevrolet
43	Caprice Classic LS (S, 4 door)	Chevrolet
45	Cavalier RS (C, 2 door)	Chevrolet
47	Cavalier RS (CN, 2 door)	Chevrolet
46	Cavalier RS (S, 4 door)	Chevrolet
48	Cavalier RS (W, 4 door)	Chevrolet
44	Cavalier VL (C, 2 door)	Chevrolet
46	Cavalier VL (S, 4 door)	Chevrolet
48	Cavalier VL (W, 4 door)	Chevrolet
45	Cavalier Z24 (C, 2 door)	Chevrolet
47	Cavalier Z24 (CN, 2 door)	Chevrolet
237	Celica All-Trac Turbo (H, 4WD, 2 door)	Toyota
235	Celica GT (C, 2 door)	Toyota
237	Celica GT (CN, 2 door)	Toyota
236	Celica GT (H, 2 door)	Toyota
236	Celica GT-S (H, 2 door)	Toyota
235	Celica ST (C, 2 door)	Toyota
22	Century Custom (C, 2 door)	Buick
23	Century Custom (S, 4 door)	Buick
25	Century Custom (W, 4 door)	Buick
24	Century Limited (S, 4 door)	Buick
23	Century Special (S, 4 door)	Buick
24	Century Special (W, 4 door)	Buick
106	Civic CX (H, 2 door)	Honda
105	Civic del Sol S (C, 2 door)	Honda
105	Civic del Sol Si (C, 2 door)	Honda
104	Civic DX (C, 2 door)	Honda
106	Civic DX (H, 2 door)	Honda
108	Civic DX (S, 4 door)	Honda
104	Civic EX (C, 2 door)	Honda
109	Civic EX (S, 4 door)	Honda
108	Civic LX (S, 4 door)	Honda
107	Civic Si (H, 2 door)	Honda
107	Civic VX (H, 2 door)	Honda
59	Colt (C, 2 door)	Dodge
182	Colt (C, 2 door)	Plymouth
60	Colt (S, 4 door)	Dodge
183	Colt (S, 4 door)	Plymouth
60	Colt GL (C, 2 door)	Dodge
183	Colt GL (C, 2 door)	Plymouth
61	Colt GL (S, 4 door)	Dodge
184	Colt GL (S, 4 door)	Plymouth
184	Colt Vista (W, 3 door)	Plymouth
185	Colt Vista AWD (W, 4WD, 3 door)	Plymouth
185	Colt Vista SE (W, 3 door)	Plymouth
53	Concorde (S, 4 door)	Chrysler
121	Continental Executive (S, 4 door)	Lincoln
122	Continental Signature (S, 4 door)	Lincoln
238	Corolla (S, 4 door)	Toyota
238	Corolla DX (S, 4 door)	Toyota
239	Corolla DX (W, 4 door)	Toyota
239	Corolla LE (S, 4 door)	Toyota
244	Corrado SLC (C, 2 door)	Volkswagen
49	Corsica LT (S, 4 door)	Chevrolet
49	Corvette (C, 2 door)	Chevrolet
50	Corvette (CN, 2 door)	Chevrolet
50	Corvette ZR-1 (C, 2 door)	Chevrolet
142	Cougar XR7 (C, 2 door)	Mercury
78	Crown Victoria (S, 4 door)	Ford
79	Crown Victoria LX (S, 4 door)	Ford
176	Cutlass Ciera Cruiser S (W, 4 door)	Oldsmobile
176	Cutlass Ciera Cruiser SL (W, 4 door)	Oldsmobile
175	Cutlass Ciera S (S, 4 door)	Oldsmobile
175	Cutlass Ciera SL (S, 4 door)	Oldsmobile

H-Hatchback   C-Coupe   S-Sedan   CN-Convertible   4WD-4-Wheel Drive   W-Wagons

Index

Index by Vehicle

Page	Vehicle	Manufacturer
179	Cutlass Supreme Convertible (CN, 2 door)	Oldsmobile
177	Cutlass Supreme International (C, 2 door)	Oldsmobile
178	Cutlass Supreme International (S, 4 door)	Oldsmobile
177	Cutlass Supreme S (C, 2 door)	Oldsmobile
178	Cutlass Supreme S (S, 4 door)	Oldsmobile
61	Daytona (H, 2 door)	Dodge
62	Daytona ES (H, 2 door)	Dodge
62	Daytona IROC (H, 2 door)	Dodge
63	Daytona IROC R/T (H, 2 door)	Dodge
36	DeVille Coupe (C, 2 door)	Cadillac
37	DeVille Sedan (S, 4 door)	Cadillac
37	DeVille Touring Sedan (S, 4 door)	Cadillac
149	Diamante ES (S, 4 door)	Mitsubishi
150	Diamante LS (S, 4 door)	Mitsubishi
63	Dynasty (S, 4 door)	Dodge
64	Dynasty LE (S, 4 door)	Dodge
150	Eclipse (C, 2 door)	Mitsubishi
151	Eclipse GS (C, 2 door)	Mitsubishi
151	Eclipse GS 16V (C, 2 door)	Mitsubishi
152	Eclipse GS 16V Turbo (C, 2 door)	Mitsubishi
152	Eclipse GSX 16V Turbo AWD (C, 4WD, 2 dr)	Mitsubishi
179	Eighty-Eight Royale (S, 4 door)	Oldsmobile
180	Eighty-Eight Royale LS (S, 4 door)	Oldsmobile
110	Elantra (S, 4 door)	Hyundai
111	Elantra GLS (S, 4 door)	Hyundai
38	Eldorado (C, 2 door)	Cadillac
119	ES 300 (S, 4 door)	Lexus
81	Escort GT (H, 2 door)	Ford
80	Escort LX (H, 2 door)	Ford
80	Escort LX (H, 4 door)	Ford
81	Escort LX (S, 4 door)	Ford
82	Escort LX (W, 4 door)	Ford
82	Escort LX-E (S, 4 door)	Ford
79	Escort Pony (H, 2 door)	Ford
111	Excel (H, 2 door)	Hyundai
112	Excel (S, 4 door)	Hyundai
113	Excel GL (S, 4 door)	Hyundai
112	Excel GS (H, 2 door)	Hyundai
154	Expo (W, 4 door)	Mitsubishi
155	Expo AWD (W, 4WD, 4 door)	Mitsubishi
153	Expo LRV (W, 3 door)	Mitsubishi
154	Expo LRV AWD (W, 4WD, 3 door)	Mitsubishi
153	Expo LRV Sport (W, 3 door)	Mitsubishi
155	Expo SP (W, 4 door)	Mitsubishi
156	Expo SP AWD (W, 4WD, 4 door)	Mitsubishi
83	Festiva GL (H, 2 door)	Ford
83	Festiva L (H, 2 door)	Ford
191	Firebird (C, 2 door)	Pontiac
192	Firebird Formula (C, 2 door)	Pontiac
192	Firebird Trans Am (C, 2 door)	Pontiac

Page	Vehicle	Manufacturer
38	Fleetwood (S, 4 door)	Cadillac
245	Fox (S, 2 door)	Volkswagen
245	Fox GL (S, 4 door)	Volkswagen
116	G20 (S, 4 door)	Infiniti
157	Galant ES (S, 4 door)	Mitsubishi
157	Galant LS (S, 4 door)	Mitsubishi
156	Galant S (S, 4 door)	Mitsubishi
246	Golf III GL (H, 2 door)	Volkswagen
246	Golf III GL (H, 4 door)	Volkswagen
193	Grand Am GT (C, 2 door)	Pontiac
194	Grand Am GT (S, 4 door)	Pontiac
193	Grand Am SE (C, 2 door)	Pontiac
194	Grand Am SE (S, 4 door)	Pontiac
142	Grand Marquis GS (S, 4 door)	Mercury
143	Grand Marquis LS (S, 4 door)	Mercury
195	Grand Prix GT (C, 2 door)	Pontiac
196	Grand Prix LE (S, 4 door)	Pontiac
195	Grand Prix SE (C, 2 door)	Pontiac
196	Grand Prix SE (S, 4 door)	Pontiac
197	Grand Prix STE (S, 4 door)	Pontiac
247	GTI (H, 2 door)	Volkswagen
54	Imperial (S, 4 door)	Chrysler
2	Integra GS (H, 2 door)	Acura
4	Integra GS (S, 4 door)	Acura
3	Integra GS-R (H, 2 door)	Acura
1	Integra LS (H, 2 door)	Acura
4	Integra LS (S, 4 door)	Acura
2	Integra LS Special (H, 2 door)	Acura
1	Integra RS (H, 2 door)	Acura
3	Integra RS (S, 4 door)	Acura
64	Intrepid (S, 4 door)	Dodge
65	Intrepid ES (S, 4 door)	Dodge
116	J30 (S, 4 door)	Infiniti
247	Jetta GL III (S, 4 door)	Volkswagen
248	Jetta GLS III (S, 4 door)	Volkswagen
248	Jetta GLX III (S, 4 door)	Volkswagen
217	Justy (H, 2 door)	Subaru
218	Justy GL (H, 2 door)	Subaru
219	Justy GL 4WD (H, 4WD, 2 door)	Subaru
218	Justy GL 4WD (H, 4WD, 4 door)	Subaru
186	Laser (H, 2 door)	Plymouth
186	Laser RS (H, 2 door)	Plymouth
187	Laser RS Turbo (H, 2 door)	Plymouth
187	Laser RS Turbo AWD (H, 2 door)	Plymouth
54	LeBaron (C, 2 door)	Chrysler
57	LeBaron (CN, 2 door)	Chrysler
55	LeBaron GTC (C, 2 door)	Chrysler
57	LeBaron GTC (CN, 2 door)	Chrysler
56	LeBaron Landau (S, 4 door)	Chrysler
56	LeBaron LE (S, 4 door)	Chrysler

H-Hatchback C-Coupe S-Sedan CN-Convertible 4WD-4-Wheel Drive W-Wagons

Index by Vehicle

Page	Vehicle	Manufacturer
55	LeBaron LX (C, 2 door)	Chrysler
58	LeBaron LX (CN, 2 door)	Chrysler
219	Legacy L (S, 4 door)	Subaru
220	Legacy L (W, 4 door)	Subaru
221	Legacy L AWD (S, 4WD, 4 door)	Subaru
223	Legacy L AWD (W, 4WD, 4 door)	Subaru
220	Legacy LS (S, 4 door)	Subaru
221	Legacy LS (W, 4 door)	Subaru
222	Legacy LS AWD (S, 4WD, 4 door)	Subaru
224	Legacy LS AWD (W, 4WD, 4 door)	Subaru
223	Legacy LSi AWD (S, 4WD, 4 door)	Subaru
224	Legacy LSi AWD (W, 4WD, 4 door)	Subaru
222	Legacy Sport AWD (S, 4WD, 4 door)	Subaru
225	Legacy Touring Wagon AWD (W, 4WD, 4 dr)	Subaru
6	Legend (S, 4 door)	Acura
5	Legend L (C, 2 door)	Acura
6	Legend L (S, 4 door)	Acura
5	Legend LS (C, 2 door)	Acura
7	Legend LS (S, 4 door)	Acura
198	LeMans SE (S, 4 door)	Pontiac
198	LeMans SE Aerocoupe (H, 2 door)	Pontiac
197	LeMans Value Leader Aerocoupe (H, 2 dr)	Pontiac
25	LeSabre Custom (S, 4 door)	Buick
26	LeSabre Limited (S, 4 door)	Buick
225	Loyale (S, 4 door)	Subaru
226	Loyale (W, 4 door)	Subaru
226	Loyale 4WD (S, 4WD, 4 door)	Subaru
227	Loyale 4WD (W, 4WD, 4 door)	Subaru
120	LS 400 (S, 4 door)	Lexus
51	Lumina (C, 2 door)	Chevrolet
52	Lumina (S, 4 door)	Chevrolet
51	Lumina Euro (C, 2 door)	Chevrolet
53	Lumina Euro (S, 4 door)	Chevrolet
52	Lumina Z34 (C, 2 door)	Chevrolet
19	M5 (S, 4 door)	BMW
122	Mark VIII (C, 2 door)	Lincoln
167	Maxima GXE (S, 4 door)	Nissan
168	Maxima SE (S, 4 door)	Nissan
94	Metro (H, 2 door)	Geo
95	Metro (H, 4 door)	Geo
96	Metro LSi (CN, 2 door)	Geo
95	Metro LSi (H, 2 door)	Geo
96	Metro LSi (H, 4 door)	Geo
94	Metro XFi (H, 2 door)	Geo
158	Mirage ES (C, 2 door)	Mitsubishi
160	Mirage ES (S, 4 door)	Mitsubishi
159	Mirage LS (C, 2 door)	Mitsubishi
160	Mirage LS (S, 4 door)	Mitsubishi
158	Mirage S (C, 2 door)	Mitsubishi
159	Mirage S (S, 4 door)	Mitsubishi
240	MR2 (C, 2 door)	Toyota
240	MR2 Turbo (C, 2 door)	Toyota
87	Mustang GT (CN, 2 door)	Ford
86	Mustang GT (H, 2 door)	Ford
84	Mustang LX (C, 2 door)	Ford
86	Mustang LX (CN, 2 door)	Ford
85	Mustang LX (H, 2 door)	Ford
84	Mustang LX 5.0L (C, 2 door)	Ford
87	Mustang LX 5.0L (CN, 2 door)	Ford
85	Mustang LX 5.0L (H, 2 door)	Ford
127	MX-3 (C, 2 door)	Mazda
128	MX-3 GS (C, 2 door)	Mazda
128	MX-5 Miata (CN, 2 door)	Mazda
129	MX-6 (C, 2 door)	Mazda
129	MX-6 LS (C, 2 door)	Mazda
59	New Yorker Fifth Avenue (S, 4 door)	Chrysler
58	New Yorker Salon (S, 4 door)	Chrysler
180	Ninety-Eight Regency (S, 4 door)	Oldsmobile
181	Ninety-Eight Regency Elite (S, 4 door)	Oldsmobile
181	Ninety-Eight Touring Sedan (S, 4 door)	Oldsmobile
7	NSX (C, 2 door)	Acura
168	NX 1600 (C, 2 door)	Nissan
169	NX 2000 (C, 2 door)	Nissan
26	Park Avenue (S, 4 door)	Buick
27	Park Avenue Ultra (S, 4 door)	Buick
241	Paseo (C, 2 door)	Toyota
249	Passat GL (S, 4 door)	Volkswagen
250	Passat GL (W, 4 door)	Volkswagen
249	Passat GLS (S, 4 door)	Volkswagen
250	Passat GLX (S, 4 door)	Volkswagen
251	Passat GLX (W, 4 door)	Volkswagen
109	Prelude S (C, 2 door)	Honda
110	Prelude Si (C, 2 door)	Honda
97	Prizm (S, 4 door)	Geo
97	Prizm LSi (S, 4 door)	Geo
88	Probe (H, 2 door)	Ford
88	Probe GT (H, 2 door)	Ford
130	Protege DX (S, 4 door)	Mazda
130	Protege LX (S, 4 door)	Mazda
117	Q45 (S, 4 door)	Infiniti
27	Regal Custom (C, 2 door)	Buick
29	Regal Custom (S, 4 door)	Buick
28	Regal Gran Sport (C, 2 door)	Buick
30	Regal Gran Sport (S, 4 door)	Buick
28	Regal Limited (C, 2 door)	Buick
29	Regal Limited (S, 4 door)	Buick
30	Riviera (C, 2 door)	Buick
31	Roadmaster (S, 4 door)	Buick
32	Roadmaster Estate Wagon (W, 4 door)	Buick
31	Roadmaster Limited (S, 4 door)	Buick

H-Hatchback	C-Coupe	S-Sedan	CN-Convertible	4WD-4-Wheel Drive	W-Wagons

Index by Vehicle

Page	Vehicle	Manufacturer	Page	Vehicle	Manufacturer
131	RX-7 (C, 2 door)	Mazda	75	Summit AWD (W, 4WD, 3 door)	Eagle
15	S4 (S, 4WD, 4 door)	Audi	72	Summit DL (C, 2 door)	Eagle
143	Sable GS (S, 4 door)	Mercury	73	Summit DL (S, 4 door)	Eagle
144	Sable GS (W, 4 door)	Mercury	74	Summit DL (W, 3 door)	Eagle
144	Sable LS (S, 4 door)	Mercury	72	Summit ES (C, 2 door)	Eagle
145	Sable LS (W, 4 door)	Mercury	73	Summit ES (S, 4 door)	Eagle
120	SC 300 (C, 2 door)	Lexus	74	Summit LX (W, 3 door)	Eagle
121	SC 400 (C, 2 door)	Lexus	200	Sunbird GT (C, 2 door)	Pontiac
214	SC1 (C, 2 door)	Saturn	199	Sunbird LE (C, 2 door)	Pontiac
214	SC2 (C, 2 door)	Saturn	200	Sunbird LE (S, 4 door)	Pontiac
113	Scoupe (C, 2 door)	Hyundai	199	Sunbird SE (C, 2 door)	Pontiac
114	Scoupe LS (C, 2 door)	Hyundai	201	Sunbird SE (CN, 2 door)	Pontiac
114	Scoupe Turbo (C, 2 door)	Hyundai	201	Sunbird SE (S, 4 door)	Pontiac
169	Sentra E (S, 2 door)	Nissan	188	Sundance (H, 2 door)	Plymouth
170	Sentra E (S, 4 door)	Nissan	188	Sundance (H, 4 door)	Plymouth
172	Sentra GXE (S, 4 door)	Nissan	189	Sundance Duster (H, 2 door)	Plymouth
171	Sentra SE (S, 2 door)	Nissan	189	Sundance Duster (H, 4 door)	Plymouth
172	Sentra SE-R (S, 2 door)	Nissan	227	SVX (C, 4WD, 2 door)	Subaru
170	Sentra XE (S, 2 door)	Nissan	216	SW1 (W, 4 door)	Saturn
171	Sentra XE (S, 4 door)	Nissan	217	SW2 (W, 4 door)	Saturn
39	Seville (S, 4 door)	Cadillac	228	Swift GA (H, 2 door)	Suzuki
39	Seville Touring Sedan (S, 4 door)	Cadillac	229	Swift GA (S, 4 door)	Suzuki
65	Shadow (H, 2 door)	Dodge	229	Swift GS (S, 4 door)	Suzuki
66	Shadow (H, 4 door)	Dodge	228	Swift GT (H, 2 door)	Suzuki
68	Shadow ES (CN, 2 door)	Dodge	75	Talon DL (C, 2 door)	Eagle
66	Shadow ES (H, 2 door)	Dodge	76	Talon ES (C, 2 door)	Eagle
67	Shadow ES (H, 4 door)	Dodge	76	Talon TSi Turbo (C, 2 door)	Eagle
67	Shadow Highline (CN, 2 door)	Dodge	77	Talon TSi Turbo AWD (C, 4WD, 2 door)	Eagle
40	Sixty Special Sedan (S, 4 door)	Cadillac	89	Taurus GL (S, 4 door)	Ford
32	Skylark Custom (C, 2 door)	Buick	90	Taurus GL (W, 4 door)	Ford
34	Skylark Custom (S, 4 door)	Buick	89	Taurus LX (S, 4 door)	Ford
33	Skylark Gran Sport (C, 2 door)	Buick	91	Taurus LX (W, 4 door)	Ford
35	Skylark Gran Sport (S, 4 door)	Buick	90	Taurus SHO (S, 4 door)	Ford
33	Skylark Limited (C, 2 door)	Buick	91	Tempo GL (S, 2 door)	Ford
34	Skylark Limited (S, 4 door)	Buick	92	Tempo GL (S, 4 door)	Ford
215	SL (S, 4 door)	Saturn	92	Tempo LX (S, 4 door)	Ford
215	SL1 (S, 4 door)	Saturn	241	Tercel (S, 2 door)	Toyota
216	SL2 (S, 4 door)	Saturn	242	Tercel DX (S, 2 door)	Toyota
115	Sonata (S, 4 door)	Hyundai	242	Tercel DX (S, 4 door)	Toyota
115	Sonata GLS (S, 4 door)	Hyundai	243	Tercel LE (S, 4 door)	Toyota
10	Spider (CN, 2 door)	Alfa Romeo	93	Thunderbird LX (C, 2 door)	Ford
10	Spider Veloce (CN, 2 door)	Alfa Romeo	93	Thunderbird Super Coupe (C, 2 door)	Ford
69	Spirit ES (S, 4 door)	Dodge	145	Topaz GS (S, 2 door)	Mercury
68	Spirit Highline (S, 4 door)	Dodge	146	Topaz GS (S, 4 door)	Mercury
69	Stealth (C, 2 door)	Dodge	124	Town Car Cartier Designer (S, 4 door)	Lincoln
70	Stealth ES (C, 2 door)	Dodge	123	Town Car Executive (S, 4 door)	Lincoln
70	Stealth R/T (C, 2 door)	Dodge	123	Town Car Signature (S, 4 door)	Lincoln
71	Stealth R/T Turbo AWD (C, 4WD, 2 door)	Dodge	146	Tracer (S, 4 door)	Mercury
98	Storm 2+2 (C, 2 door)	Geo	147	Tracer (W, 4 door)	Mercury
98	Storm 2+2 GSi (C, 2 door)	Geo	147	Tracer LTS (S, 4 door)	Mercury

H-Hatchback C-Coupe S-Sedan CN-Convertible 4WD-4-Wheel Drive W-Wagons

Index by Vehicle

Page	Vehicle	Manufacturer
15	V8 Quattro (S, 4WD, 4 door)	Audi
8	Vigor GS (S, 4 door)	Acura
8	Vigor LS (S, 4 door)	Acura
71	Viper RT/10 (CN, 2 door)	Dodge
77	Vision ESi (S, 4 door)	Eagle
78	Vision TSi (S, 4 door)	Eagle
117	XJ6 (S, 4 door)	Jaguar
118	XJ6 Vanden Plas (S, 4 door)	Jaguar
118	XJS (C, 2 door)	Jaguar
119	XJS (CN, 2 door)	Jaguar

H-Hatchback C-Coupe S-Sedan CN-Convertible 4WD-4-Wheel Drive W-Wagons

For ArmChair Compare® orders – *Please complete information on other side of this form*

CUSTOMER INFORMATION

Name _____

Address _____

City _____ State _____ Zip _____

Home Phone ( ) _____ – _____

Charge Information:

☐ VISA ☐ Mastercard

Name on card _____

Account # _____

Expiration date _____

Signature _____

Include order form with personal check or charge information for appropriate amount and mail to:

IntelliChoice, Inc.
**1135 S. Saratoga-Sunnyvale Rd.
San Jose, CA 95129-3660**

Note: CA residents must add sales tax. **(408) 554-8711**

JUST THE FACTS™

$14.⁹⁵

Send following information for each model line report requested:

Year _____ Make _____

Model _____

Number of Doors _____

<u>Drive</u>   ☐ 2 WD   ☐ 4 WD

<u>BodyStyle</u>   ☐ Sedan   ☐ Hatch   ☐ Coupe

                  ☐ Wagon   ☐ Convertible

For additional report:

Year _____ Make _____

Model _____

Number of Doors _____

<u>Drive</u>   ☐ 2 WD   ☐ 4 WD

<u>BodyStyle</u>   ☐ Sedan   ☐ Hatch   ☐ Coupe

                  ☐ Wagon   ☐ Convertible

IntelliChoice Report Order Form

For ArmChair Compare® orders – *Please complete information on other side of this form*

CUSTOMER INFORMATION

Name _____

Address _____

City _____ State _____ Zip _____

Home Phone ( ) _____ – _____

Charge Information:

☐ VISA ☐ Mastercard

Name on card _____

Account # _____

Expiration date _____

Signature _____

Include order form with personal check or charge information for appropriate amount and mail to:

IntelliChoice, Inc.
**1135 S. Saratoga-Sunnyvale Rd.
San Jose, CA 95129-3660**

Note: CA residents must add sales tax. **(408) 554-8711**

JUST THE FACTS™

$14.⁹⁵

Send following information for each model line report requested:

Year _____ Make _____

Model _____

Number of Doors _____

<u>Drive</u>   ☐ 2 WD   ☐ 4 WD

<u>BodyStyle</u>   ☐ Sedan   ☐ Hatch   ☐ Coupe

                  ☐ Wagon   ☐ Convertible

For additional report:

Year _____ Make _____

Model _____

Number of Doors _____

<u>Drive</u>   ☐ 2 WD   ☐ 4 WD

<u>BodyStyle</u>   ☐ Sedan   ☐ Hatch   ☐ Coupe

                  ☐ Wagon   ☐ Convertible

IntelliChoice Report Order Form

Please complete information on both sides of this form

THE ARMCHAIR COMPARE® REPORT $19.⁹⁵

Your Annual Mileage _ _ , _ _ _

Desired Down Payment: ____%, OR $_____ , OR ☐ Trade-in value or minimum

Vehicle One
Describe the vehicle below:

Year _____ Make _____

Model _____

Number of Doors _____

<u>Drive</u> ☐ 2 WD ☐ 4 WD

<u>BodyStyle</u> ☐ Sedan ☐ Hatch ☐ Coupe
☐ Wagon ☐ Convertible ☐ Other_____

Vehicle Two
Describe the vehicle below:

Year _____ Make _____

Model _____

Number of Doors _____

<u>Drive</u> ☐ 2 WD ☐ 4 WD

<u>BodyStyle</u> ☐ Sedan ☐ Hatch ☐ Coupe
☐ Wagon ☐ Convertible

Complete this area if you have a 1981-1992 vehicle to trade-in.

Year _____ Make _____

Model _____

Number of Doors _____

<u>Condition</u> ☐ Excellent ☐ Good ☐ Poor

<u>BodyStyle</u> ☐ Sedan ☐ Hatch ☐ Coupe
☐ Wagon ☐ Convertible ☐ Other_____

Loan Balance (if known) $_____

If balance not known, complete the following:

Purchase Date: Month _____ Year _____

Months of loan _____ Monthly Pmt. $_____

Interest Rate (if known) _____%

IntelliChoice Report Order Form

Please complete information on both sides of this form

THE ARMCHAIR COMPARE® REPORT $19.⁹⁵

Your Annual Mileage _ _ , _ _ _

Desired Down Payment: ____%, OR $_____ , OR ☐ Trade-in value or minimum

Vehicle One
Describe the vehicle below:

Year _____ Make _____

Model _____

Number of Doors _____

<u>Drive</u> ☐ 2 WD ☐ 4 WD

<u>BodyStyle</u> ☐ Sedan ☐ Hatch ☐ Coupe
☐ Wagon ☐ Convertible ☐ Other_____

Vehicle Two
Describe the vehicle below:

Year _____ Make _____

Model _____

Number of Doors _____

<u>Drive</u> ☐ 2 WD ☐ 4 WD

<u>BodyStyle</u> ☐ Sedan ☐ Hatch ☐ Coupe
☐ Wagon ☐ Convertible

Complete this area if you have a 1981-1992 vehicle to trade-in.

Year _____ Make _____

Model _____

Number of Doors _____

<u>Condition</u> ☐ Excellent ☐ Good ☐ Poor

<u>BodyStyle</u> ☐ Sedan ☐ Hatch ☐ Coupe
☐ Wagon ☐ Convertible ☐ Other_____

Loan Balance (if known) $_____

If balance not known, complete the following:

Purchase Date: Month _____ Year _____

Months of loan _____ Monthly Pmt. $_____

Interest Rate (if known) _____%

IntelliChoice Report Order Form